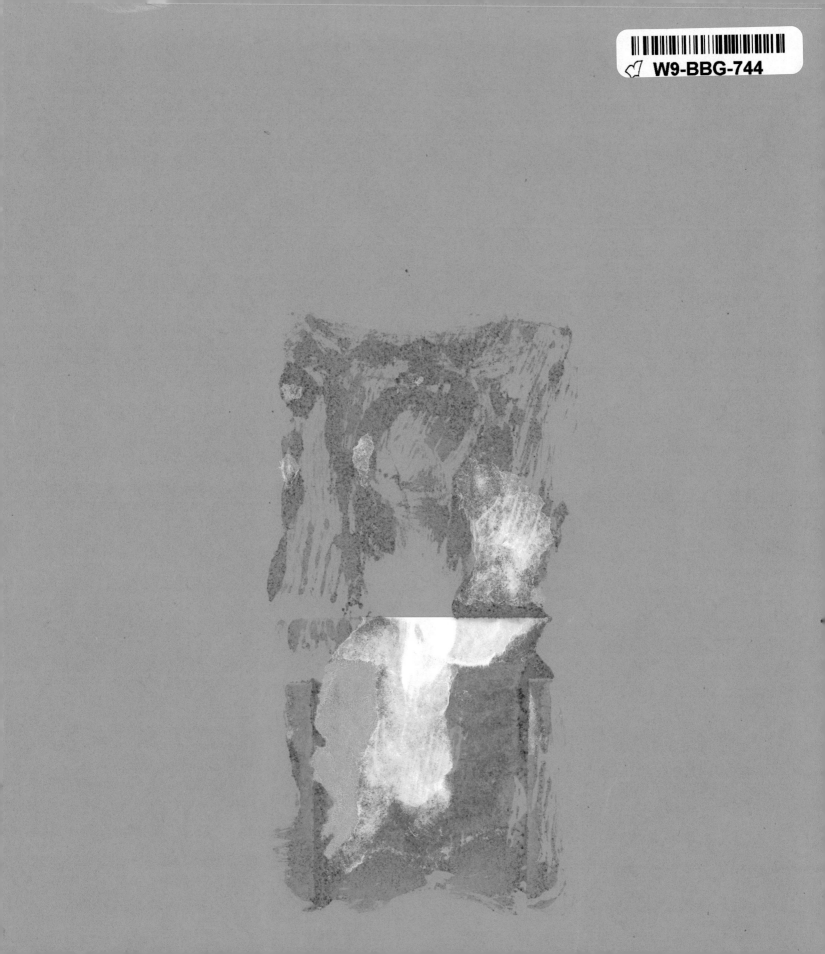

Plant Physiology
Second Edition

Frank B. Salisbury
Utah State University

Cleon W. Ross
Colorado State University

Wadsworth Publishing Company, Inc. 1978
Belmont, California

Biology books under the editorship of William A. Jensen, University of California, Berkeley

Biology: Kormandy *et al.*

Biology: The Foundations, Wolfe

Botany: An Ecological Approach, Jensen and Salisbury

Biology of the Cell, Wolfe

Plant Physiology, Second Edition, Salisbury and Ross

Plant Physiology Laboratory Manual, Ross

An Evolutionary Survey of the Plant Kingdom, Scagel *et al.*

Plant Diversity: An Evolutionary Approach, Scagel *et al.*

Plants and Civilization, Third Edition, Baker

Plants and the Ecosystem, Third Edition, Billings

Other books in the Wadsworth biology series

Biology: The Unity and Diversity of Life, Kirk, Taggart, and Starr

Living in the Environment: Concepts, Problems, and Alternatives, Miller

Replenish the Earth: A Primer in Human Ecology, Miller

Energy and Environment: Four Energy Crises, Miller

Oceanology: An Introduction, Ingmanson and Wallace

Biology Editor: Jack C. Carey

Editing and Production: Greg Hubit Bookworks

Cover photo by Ernest Braun

Printed in the United States of America

2 3 4 5 6 7 8 9 10—82 81 80 79 78

Library of Congress Cataloging in Publication Data

Salisbury, Frank B.
  Plant physiology.

  Bibliography: p.
  Includes index.
  1. Plant physiology.   I. Ross, Cleon,
1934—   joint author.
QK711.2.S23–1978   581.1   78-2581

Preface

Our science has grown at a phenomenal rate since May of 1969, when we put the finishing touches on the first edition of this book. Much has been learned, some important problems have been solved or brought closer to a solution, and interest has diversified into several exciting new areas. Our subject has become increasingly broad and challenging.

This situation has required us to write a new book rather than just revise the first edition. Although we have used many figures from the first edition, most have been redrawn. Probably only 10 to 20 percent of the written material was used in preparing this edition, and that has been extensively revised.

A Text for Beginning Students: Organization We have written a prologue that reviews some necessary background material including a discussion of the cell, and we have written 24 core chapters that we think are essential to a reasonably broad survey of our science. We believe the organization of these chapters is simple and logical. The first seven chapters (Section One) deal with physical processes in plants, the middle seven chapters (Section Two) with metabolism and biochemistry, and the eight chapters of Section Three with growth and development. The last two chapters (Section Four) draw upon the other 22 in relating plant function to the environment that so strongly influences it.

The broad organization is the same as in the first edition, but we have relegated chapters on water and reactive surfaces to the Special Topics section; moved the mineral nutrition chapter to Section One; and greatly reduced the discussion of nucleic acids and protein synthesis (often called molecular biology), since most students now study molecular biology before plant physiology. We have expanded the topic of stress physiology, originally discussed in the physiological ecology chapter, to an entire chapter (Chapter 24). As in the first edition, our emphasis is on higher plants, but lower plants, microorganisms, and even animals are considered when appropriate. (Discussing the biological clock without reference to animals, for example, would be highly artificial.)

We have made a valiant attempt to keep all chapters to a reasonable length. Nearly all chapters are shorter than they were in the first edition, and, indeed, the core chapters of our book form a unit that is much shorter than our first edition. Our goal has been to present basic principles in a modern context, sometimes with a little history, and often illustrated by examples from the recent literature.

Guest Essays We have asked several of our friends to write personal essays telling of their involvement in plant physiology and what it has meant to them in their lives. These are placed at the ends of appropriate chapters; we hope that you will enjoy them as much as we have.

Special Topics Several special topics are placed at the end of the book. Most of these discuss matters that are central to plant physiology but may be drawn from a related field or present reference material in a summary fashion. We have numbered them according to the chapters they relate to.

Applications We have included two types: the use of a principle in an applied field such as agronomy or range science, and the application of a principle in helping us understand the nature of life better. There are many ways that plants solve the problems they face, and recognition of these adaptive features often provides much of the wonder of plant physiology.

Our Obligation to the Science In writing a text on plant physiology, we have obligations to present the most modern thinking on each topic, to represent accurately the work we are reporting, and to reflect the "balance" of the field. Though it is not easy to reach perfection in any of these goals, we have taken the following steps in our attempts to do so.

The Reviewers Thirty-two reviewers helped insure that the text is accurate and up to date. The reviewers were extremely helpful. Indeed, it took us almost a year to revise our early manuscript in response to their comments. We are most grateful, but of course we remain responsible for the present status of the text.

References to the Literature of Plant Physiology We have tried, more than in the first edition, to provide the reader with references to the scientific literature, documenting most topics. A student or instructor can find a given paper in the bibliography for each chapter by looking for the author's name (mentioned in the chapter) or by scanning titles in the chapter references. All author names actually mentioned in the text are indicated in the references, which follow the Special Topics.

Writing Style We have tried to write simply and directly, define our terms when first introduced (using boldface type), and not take anything for granted. Although we assume that the student has some background in botany or biology and in organic chemistry, we realize that each of us forgets much necessary background material after the first or second encounter. Hence, opportunity for review is often provided. Plant physiology consists of a complex multidimensional network of knowledge that can be approached from many vantage points and pursued along many pathways. Often the trails overlap, so redundancy and cross-referencing become an unavoidable part of such a text. We hope that the redundancy will help the learning process.

We have spent considerable time on the chapter introductions to provide a suitable perspective. We have tried to provide clear and relevant drawings, graphs, and other illustrations, and we have included brief ''box essays'' to stimulate your imagination.

So this is plant physiology as we see it today. We hope we can help you learn plant physiology while transferring to you some of the enthusiasm and joy we experience in learning about functioning plants.

We have provided portrait photos of all our guest authors, all but three having been taken by F.B.S. So it seemed appropriate that you should also see what the authors look like.

Frank B. Salisbury

Cleon W. Ross

Reviewers

Luke S. Albert University of Rhode Island
J. Clair Batty Utah State University
Caroline Bledsoe University of Washington
John S. Boyer University of Illinois
C. H. Brueske Mount Union College
R. Dean Decker University of Richmond
Donald B. Fisher University of Georgia
D. J. Friend University of Hawaii
Donald R. Geiger University of Dayton
Shirley Gould Lewis and Clark College
Richard H. Hageman University of Illinois
John Hendrix Colorado State University
Theodore Hsiao University of California, Davis
William T. Jackson Dartmouth College
Andre Jagendorf Cornell University
Edwin H. Liu University of South Carolina

James A. Lockhart University of Massachusetts
Jerry McClure Miami University
Linda McMahan Utica College
Robert Mellor University of Arizona
Joseph E. Miller Argonne National Laboratory
Murray Nabors Colorado State University
Dave Newman Miami University, Ohio
James O'Leary University of Arizona
Herbert D. Papenfuss Boise State University
Wayne S. Pierce California State College
Donald W. Rains University of California, Davis
George Spomer University of Idaho
Larry L. Tieszen Augustana College
Herman Weibe Utah State University
Conrad J. Weiser Oregon State University
Jan A. D. Zeevaart Michigan State University

Contents in Brief

Prologue   1

Section One
Water, Solutions, and Surfaces

1   Water Potential   8

2   Osmosis   18

3   The Photosynthesis-Transpiration
     Compromise   32

4   The Ascent of Sap   49

5   Absorption of Mineral Salts   64

6   Mineral Nutrition   79

7   Phloem Transport   93

Section Two
Plant Biochemistry

8   Enzymes, Proteins, and Amino acids   110

9   Photosynthesis   123

10   Carbon Dioxide Fixation and
     Carbohydrate Synthesis   136

11   Photosynthesis: Environmental and
     Agricultural Aspects   160

12   Respiration   174

13   Assimilation of Nitrogen and Sulfur   192

14   Lipids and Aromatic Compounds   206

Section Three
Plant Development

15   Growth and Development   224

16   Hormones and Growth Regulators:
     Auxins and Gibberellins   240

17   Hormones and Growth Regulators:
     Cytokinins, Ethylene, and Abscisic Acid   258

18   Differentiation and Differential Growth   272

19   Photomorphogenesis   290

20   The Biological Clock   304

21   Plant Responses to Temperature and
     Related Phenomena   317

22   Photoperiodism   332

Section Four
Environmental Physiology

23   Principles of Environmental Physiology   350

24   Stress Physiology   361

Section Five
Special Topics   375

Bibliography   405

Author Index   423

Index of Species and Subjects   425

Contents in Detail

Prologue		**1**

P. 1	The Physical Sciences	1
P. 2	Vitalism versus Mechanism	1
P. 3	The Cell	2
P. 4	The Cellular Organelles	2
P. 5	Macromolecules	4
P. 6	Structure and Function	4
P. 7	Tissues and Organs	5
P. 8	Development or Morphogenesis	5
P. 9	Response of Organisms to Environment	5
P.10	A Definition of Life	5

Section One
Water, Solutions, and Surfaces

1	**Water Potential**	**8**
1.1	Water and Minerals in the Plant: How to Understand Them	8
1.2	Energy	9
1.3	Kinetic Theory	9
1.4	Thermodynamics	10
1.5	Bulk Flow and Diffusion	13
1.6	Chemical and Water Potential Gradients	14
1.7	The Rate of Diffusion	15
1.8	Vapor Pressure	15

2	**Osmosis**	**18**

2.1	An Osmotic System	18
2.2	The Components of Water Potential	20
2.3	Energy and Pressure Units for Water Potential	20
2.4	Dilution	22
2.5	The Membrane	22
2.6	Measuring the Components of Water Potential	23

3	**The Photosynthesis-Transpiration Compromise**	**32**
3.1	Measurement of Transpiration	32
3.2	The Paradox of Pores	34
3.3	Stomatal Mechanics	35
3.4	Stomatal Mechanisms	37
3.5	The Role of Transpiration: Water Stress	40
3.6	The Role of Transpiration: Energy Exchange	42
3.7	Plant Energy Exchange in Ecosystems	46

4	**The Ascent of Sap**	**49**
4.1	The Problem	49
4.2	The Cohesion Hypothesis of the Ascent of Sap	50
4.3	The Driving Force: A Water Potential Gradient	51
4.4	The Pathway	53
4.5	Cohesion	58
4.6	Some Conclusions	61

5   Absorption of Mineral Salts   64

5.1   Roots as Absorbing Surfaces   64

5.2   Mycorrhizae   65

5.3   Soils and Their Mineral Elements   67

5.4   Ion Traffic into the Root   68

5.5   The Nature of Membranes   70

5.6   Some Principles of Solute Absorption   72

5.7   Characteristics of Ion Absorption   72

5.8 pH Changes and Maintenance of Electrical Neutrality   75

5.9   Electropotentials Across Roots Modify Ion Absorption   76

5.10   Ion Accumulation by Roots Requires Respiration and ATP Production   76

6   Mineral Nutrition   79

6.1   The Elements in Dry Matter   79

6.2   Methods of Studying Plant Nutrition: Hydroponics   80

6.3   The Essential Elements   81

6.4   Quantitative Requirements and Tissue Analysis   84

6.5   Chelating Agents   85

6.6   Functions of Essential Elements: Some Principles   87

6.7   Nutrient Deficiency Symptoms   88

7   Phloem Transport   93

7.1   Surgical Experiments   93

7.2   The Use of Tracers   94

7.3   In What Form?   95

7.4   Where and When?   97

7.5   How Fast?   97

7.6   Phloem Loading   98

7.7   An Introduction to Phloem Transport Theories   99

7.8   Phloem Anatomy   101

7.9   Bidirectional Phloem Transport   105

7.10   Pressure in the Phloem   106

7.11   Different Transport Velocities for Different Substances   107

7.12   The Role of Metabolism   107

7.13   Phloem Transport Theories: A Final Consideration   108

7.14   Some Final Problems   108

**Section Two
Plant Biochemistry**

8   Enzymes, Proteins, and Amino Acids   110

8.1   Enzymes in Cells   110

8.2   Properties of Enzymes   110

8.3   Mechanisms of Enzymes Action   117

8.4   Denaturation   118

8.5   Factors Influencing Rates of Enzymatic Reactions   119

8.6   Allosteric Proteins and Feedback Control   121

9   Photosynthesis   123

9.1   Historical Summary of Photosynthesis   123

9.2   Chloroplasts: Structures, Pigments, and Development   125

9.3   The Role of Light in Photosynthesis   129

9.4   The Emerson Enhancement Effect: Cooperating Photosystems   131

9.5   Transport of Electrons From H_2O 131

10   Carbon Dioxide Fixation and Carbohydrate Synthesis   136

10.1   Products of Carbon Dioxide Fixation   136

10.2   The Calvin Cycle   137

10.3   The C-4 Carboxylic Acid Pathway: Some Species Fix CO_2 Differently   140

10.4	Control of Photosynthetic Enzymes by Light	145
10.5	CO₂ Fixation in Succulent Species	145
10.6	Formation of Sucrose and Starch	146

11 Photosynthesis: Environmental and Agricultural Aspects 160

11.1	The Daily Course of Photosynthesis	161
11.2	Photosynthetic Rates in Various Species	161
11.3	Environmental Factors Affecting Photosynthesis	162
11.4	Photosynthetic Rates and Crop Production	172

12 Respiration 174

12.1	The Respiratory Quotient	174
12.2	Formation of Hexrose Sugars from Reserve Polysaccharides	174
12.3	Glycolysis and Fermentation	177
12.4	Mitochondria	179
12.5	The Krebs Cycle	180
12.6	The Electron Transport System and Oxidative Phosphorylation	182
12.7	The Pentose Phosphate Pathway	184

13 Assimilation of Nitrogen and Sulfur 192

13.1	The Nitrogen Cycle	192
13.2	Nitrogen Fixation	194
13.3	Metabolism of Nitrate Ions	197
13.4	Assimilation of NH₄ into Organic Compounds	199
13.5	Transamination Reactions	201
13.6	Nitrogen Transformations During Plant Development	201
13.7	Metabolism of Sulfur	204

14 Lipids and Aromatic Compounds 206

14.1	Fats and Oils	206
14.2	Phospholipids and Glycolipids	212
14.3	Waxes, Cutin, and Suberin: Plant Protective Coats	213
14.4	The Isoprenoid Compounds	213
14.5	Phenolic Compounds and Their Relatives	216
14.6	Lignin	218
14.7	The Flavonoids	220
14.8	Betalains	221

**Section Three
Plant Development**

15 Growth and Development 224

15.1	What Is Meant by Growth?	224
15.2	Growth in Relation to Time (Growth Kinetics)	225
15.3	Some Features of Plant Growth	227
15.4	Growth of Roots	228
15.5	Growth of Stems	230
15.6	Leaf Growth	232
15.7	Growth and Development of Flowers	233
15.8	Seed and Fruit Development	234
15.9	Relations Between Vegetative and Reproductive Growth	235
15.10	Growth at the Cellular Level	236

16 Hormones and Growth Regulators: Auxins and Gibberellins 240

| 16.1 | The Auxins | 240 |
| 16.2 | The Gibberellins | 247 |

17 Hormones and Growth Regulators: Cytokinins, Ethylene, and Abscisic Acid 258

| 17.1 | The Cytokinins | 258 |
| 17.2 | Ethylene, a Volatile Hormone | 264 |

17.3 Growth Inhibitors and Miscellaneous Regulating Compounds   267

17.4 Abscisic Acid   267

17.5 Hormones in Senescence and Abscission   269

17.6 Hormone Interactions   271

18 Differentiation and Differential Growth   272

18.1 Some Principles of Differentiation   272

18.2 Totipotency and Tissue Cultures   276

18.3 Juvenility and Phase Changes   276

18.4 Differential Growth and Turgor Movements   277

18.5 Phototropism   278

18.6 Geotropism   280

18.7 Reaction Wood   284

18.8 Nyctinasty   284

18.9 Thigmonasty   285

18.10 Thigmomorphogenesis   287

18.11 Leaf-Folding and Leaf-Rolling Movements in Grasses   289

19 Photomorphogenesis   290

19.1 Discovery of Phytochrome   290

19.2 Physical and Chemical Properties of Phytochrome   292

19.3 Distribution of Phytochrome Among Species, Tissues, and Cells   293

19.4 Phototransformations of Phytochrome and Their Relation to Photomorphogenesis   293

19.5 The Role of Light in Seed Germination   294

19.6 The Role of Light in Seedling Establishment and Vegetative Growth   297

19.7 Photoperiodic Effects of Light   302

19.8 Light-Enhanced Flavonoid Synthesis   302

19.9 Effects of Light on Chloroplast Arrangements   303

20 The Biological Clock   304

20.1 Endogenous or Exogenous?   304

20.2 Circadian and Other Rhythms   307

20.3 Basic Concepts and Terminology   309

20.4 Rhythm Characteristics: Light   309

20.5 Rhythm Characteristics: Temperature   311

20.6 Rhythm Characteristics: Applied Chemicals   313

20.7 Photoperiodism   313

20.8 Time Memory   315

20.9 Celestial Navigation   315

20.10 The Biological Clock in Nature   315

21 Plant Responses to Temperature and Related Phenomena   317

21.1 The Temperature-Enzyme Dilemma   317

21.2 Vernalization   318

21.3 Dormancy   321

21.4 Seed Longevity and Germination   322

21.5 Seed Dormancy   323

21.6 Bud Dormancy   325

21.7 Underground Storage Organs   328

21.8 Thermoperiodism   329

21.9 Mechanisms of the Low Temperature Response   330

22 Photoperiodism   332

22.1 Detecting Seasonal Time by Measuring Daylength   332

22.2 Some General Principles of Flowering Physiology   333

22.3 The Response Types   333

22.4 Ripeness to Respond   338

22.5 Phytochrome and the Role of the Dark Period   338

22.6 Time Measurement in Photoperiodism   340

22.7 The Florigen Concept: Flowering Hormones and Inhibitors   343

22.8 Responses to Applied Plant Hormones and Growth Regulators   345

22.9 The Induced State and Floral Development   346

22.10 Other Responses to Photoperiod   346

Section Four
Environmental Physiology

23 Principles of Environmental Physiology   350

23.1 The Problems of Environmental Physiology   351

23.2 Some Principles of Plant Response to Environment   351

23.3 How It's Done: Field and Laboratory   357

23.4 The Environment and Plant Response   359

24 Stress Physiology   361

24.1 Drought   361

24.2 High Temperatures   366

24.3 Low Temperatures   366

24.4 The Alpine Tundra   368

24.5 Salt   371

24.6 Soil pH 373

24.7 A Final Note   374

Section Five
Special Topics

1.1 The Water Milieu   376

2.1 Colloids   378

3.1 Radiant Energy: Some Definitions   383

4.1 Soil Water—**Jerome K. Jurinak**   388

7.1 What is a Plant? The Significance of Vacuoles and Cells Walls—**Herman H. Wiebe**   389

9.1 A Theory of ATP Synthesis   392

16.1 Herbicides—**John O. Evans**   393

19.1 Possible Mechanisms of Light Action in Photomorphogenesis   395

23.1 Ultraviolet Radiation and Plants—**Martyn M. Caldwell**   397

23.2 Physiological Plant Pathology—**Penelope J. Hanchey-Bauer**   399

1.2 An Introduction to the Laws of Thermodynamics—**J. Clair Batty**   402

Bibliography   405

Author Index   423

Index of Species and Subjects   425

Guest Essays

Do Fluids Penetrate Through Stomates?
Herman H. Wiebe   41

The Biophysical Ecologist **David M. Gates**   47

Pursuing the Questions of Soil–Plant–Atmosphere
Water Relations **Ralph O. Slatyer**   62

Studying Water, Minerals, and Roots
Paul J. Kramer   77

Exploring the Path of Carbon in Photosynthesis (I)
James A. Bassham   148

Exploring the Path of Carbon in Photosynthesis (II)
Melvin Calvin   152

Try a Little Diversity in Your Science
Peter Albersheim   156

The Hazards of Hubris
Bastiaan Jacob Dirk Meeuse   188

Why a Biologist? Some Reflections **Frits W. Went**   250

Discovering the Wall-Loosening Factor
David L. Rayle   252

A Role for Tissue Culture in Plant Breeding
Murray W. Nabors   282

The Sense of Touch in Plants **Mark J. Jaffe**   286

Potato Cellars, Trains, and Dreams: Discovering
the Biological Clock **Erwin Bünning**   312

Women in Science **Beatrice M. Sweeney**   314

The Annual Cycle in Fruit Trees
Schuyler D. Seeley   327

Cabbages and Humidity **T. W. Tibbitts**   360

(See also authors of Special Topics)

Prologue

Life and Plant Function

Imagine a plant, one of your favorites, growing in the sunlight where it was planted. Perhaps you see a stately elm on the lawn of a Virginia plantation, or one corn plant among many in an Iowa field, or an avalanche lily below a melting Rocky Mountain snow bank. Probably the plant is swaying slightly, or its leaves are rustling in the breeze, but the overwhelming impression is one of inactivity and peace: a commonplace object passively soaking up its environment.

Actually, you know better. The promise of plant physiology, which you are about to study, is that plants are complex and fascinating functioning systems. Understanding them is challenging and can even be exciting, both for you as you learn about what is known and for the scientist as he or she seeks new knowledge.

Just how challenging? How much is known about plant function, and how much could you learn if you were willing to spend the time and effort? What portion of this human endeavor called plant physiology is available to you?

How are living organisms unique? You've had biology courses before this one. How would you answer this question? What is the essential nature of life? Science today understands much about life and how it works. (We must hasten to add that much remains a mystery, and we can be grateful for the remaining challenges.) But could you summarize the things you have learned about biology—from grade school to college—by formulating a few general ideas and principles, important discoveries along the road to our present understanding? And what remains to be discovered by science? What do *you* expect to discover by studying plant physiology?

In writing a prologue for this text, we have an excellent opportunity to put plant physiology in perspective by attempting to formulate such a summary of past discoveries, their relation to plant function, and future challenges. (It would be valuable for you if you would try such a summary yourself before reading our version.) We don't have much space, so we will have to limit ourselves to the most important and most general ideas, those vital key insights that help us comprehend the "big picture." Virtually all of them will be discussed in more detail in later chapters.

P.1 The Physical Sciences

Before physical science had begun to develop, much could be done in biology by way of describing and organizing the dif-ferent kinds of organisms, but it was virtually impossible to understand anything about plant and animal function without first knowing something about the principles of physics and chemistry. Thus, our kind of biology (plant physiology) could not develop much until physical science began to consist of correct and verifiable principles. Of course, physical science is still developing, but it had come far enough by about 1800 to allow the development of considerable understanding about plant and animal function.

There are two extremely important aspects of all this that will form important themes throughout this text: energy and its transfer in living plants, and the molecular structures that constitute organisms. In the early chapters, we shall be concerned with how water and dissolved substances move within plants. Modern understanding of these matters is based upon thermodynamics: the science devoted to an understanding of energy and how it is transferred. The biochemistry that fills the second section of this text may seem to be more a discussion of molecules, their structures, and their interactions, than anything else, It is also strongly dependent upon thermodynamics, however, since energy is the driving force for all the chemical reactions that we shall discuss. Thus, our first principle might be stated as follows: **The physical sciences occupy a fundamental position in modern biology.**

P.2 Vitalism versus Mechanism

Sometimes the problem of understanding life as a functioning machine has proved so difficult that philosophers have proposed an alternative: that life's functions depend upon some vital force (a spirit or entelechy) beyond the limits of physics, chemistry, and engineering. Furthermore, it can be logically argued that life is so complex that it could only have been created by an intelligent designer: a god or gods. Such viewpoints are called **vitalism** as contrasted to the concept of **mechanism,** which states that life can be understood in terms of physical principles.

In the history of science, vitalism has been unproductive, while mechanism provides the only approach that allows experimentation, verification of results, and hence application of the scientific method. Those are the rules of the game, and we shall abide by them in this text. Thus we shall state as our second principle: **Only mechanism is productive in science.**

This is not to say that certain vitalistic principles may not be true, only that there is no satisfying, empirical way to study them in the laboratory. You may have your own convictions (as do we), based upon other kinds of evidence, relating to the existence of a Creator.

P.3 The Cell

Is there a key to understanding life function? Scientists at the time of the American Revolution must have asked that question. The cell proved to be such a key. Cells had been observed as early as 1665, but microscopes did not develop enough for us to recognize that virtually all cells have nuclei and other features in common until about the 1830s. At that time it was possible for the botanist Matthias Schleiden (1838) and zoologist Theodor Schwann (1839) in Germany to state the **cell theory,** which is our third principle: **The cell is the fundamental unit of life. Or, all organisms consist of cells.**

As happens with so many great generalizations of this type, some real and some apparent exceptions have since come to light. Coenocytic organisms such as certain algae, slime molds, and other fungi have the basic parts of cells (nuclei, mitochondria, chloroplasts, and so on), but these units are not separated by cell membranes and cell walls into individual cells. Thus, in a technical sense, the coenocytes are exceptions to the cell theory, but this is not disturbing to anyone because they consist of the fundamental cell structures anyway. The important point is to realize that the discovery of the cell was the real key to understanding of plant and animal structure and function.

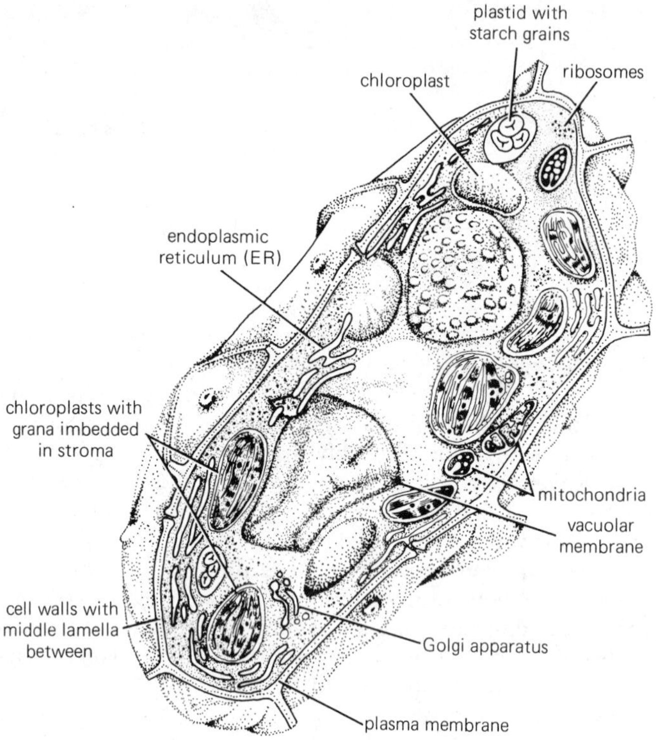

P.4 The Cellular Organelles

In the 1950s, biologists began to use several "new" tools to study cells: for example, the ultracentrifuge, electrophoresis, chromatography, but especially the electron microscope. These tools were available before then, but everything seemed to fall into place in the late 1950s, during the 1960s, and continuing until the present. By now our understanding of cells has progressed to the point where we can describe the structures and functions of many cell parts. Indeed, much of the discussion in this text will be related to the events associated with specific cellular **organelles.** Detailed discussion of these subcellular structures will have to wait for the appropriate context, but for the sake of review and preview, let's list the important cellular organelles, indicating briefly some of their known functions.

Figure P-1 illustrates a "typical" plant cell based on electron micrographs, and Table P-1 summarizes some cellular and other dimensions. Such tables and figures are valuable because they allow you to put a few things in perspective, but such generalizations about the cell can also be misleading, because there is no such thing as a "typical" cell. Cells and cellular organelles vary in many ways.

You should already be familiar with the cellular organelles.

Figure P-1 A "typical" plant cell. The interpretive drawing is based on electron micrographs such as many shown in later figures.

Nevertheless, to help you review, we will use **boldface type** for the name of each organelle or cell part (a practice we shall follow throughout the text when terms are introduced and defined, usually by their context).

Most of the cellular organelles are contained in **protoplasm,** which consists of **cytoplasm** and typically one **nucleus.** Within the nucleus is at least one **nucleolus,** the chief function of which is to synthesize RNA (a nucleic acid). When the nucleus divides, **chromosomes** become visible, especially when preparations are stained with certain dyes. (The term chromosome is derived from the Greek word *khroma* for color.) When the cell is not dividing, the chromosomes are dispersed in the nucleus as **chromatin,** which is mostly protein and DNA, another nucleic acid. DNA is the stuff of the **genes,** which are the units of heredity that control all inherited features of organisms through their ability to help form different kinds of RNA molecules.

Within the cytoplasm, there are **ribosomes,** consisting partially of RNA made in the nucleolus. These are the sites of protein synthesis, responding both to messenger RNA arriving from the nucleus and to transfer RNA. The messenger RNA carries the genetic information from the genes, and the transfer RNA carries the amino acids that will form protein. Also imbedded in the cytoplasm are **mitochondria,** where cellular respiration takes place. A principal function of mitochondria is to produce **ATP,** which is the energy source for many of the processes going on in cells. Hence, mitochondria are important "powerhouses" in cells. Sometimes 15 to 20 percent of cytoplasm volume consists of mitochondria.

There is an intricate network of membranes within the cytoplasm called the **endoplasmic reticulum (ER).** When ribosomes are attached to the ER, it is called "rough ER"; without ribosomes, it is called "smooth ER." Some of the cellular metabolic activities are associated with enzymes located on the ER membranes, and these membranes are also involved in transport processes within the cell.

Cells have **dictyosomes,** which appear in electron micrographs as flattened sacs stacked in a characteristic way. The dictyosomes, which may be referred to collectively as the **Golgi apparatus,** are concerned with cell wall synthesis and other important cellular functions. There are **microtubules** in

the cytoplasm of virtually all cells, and these participate in separation of chromosomes during mitosis as well as in cell wall formation and probably other processes. Plant cells also include **microbodies,** the two major types being **peroxisomes** and **glyoxysomes.** Animal cells contain **lysosomes,** which contain numerous hydrolytic or digestive enzymes, and there is inconclusive evidence that lysosomes also occur in plant cells. Other organelles include **plastids: leucoplasts,** which are colorless, and **chromoplasts,** which contain pigments. An example of a leucoplast is an **amyloplast,** which contains one or more starch grains. The most important example of a chromoplast is the **chloroplast,** which contains chlorophyll and is the site of photosynthesis. Some anatomists reserve the term "chromoplast" for all colored plastids *except* chloroplasts.

The cell and all the organelles we have mentioned except for ribosomes are surrounded by **membranes.** In mature plant cells, there is almost always a large membrane-bound volume

Table P-1 Metric Units of Length

Unit and Abbreviation	Metric Equivalent	English Equivalent	Objects in Size Range of Unit
Meter (m)	0.001 kilometer 10 decimeters 100 centimeters 1,000 millimeters 1,000,000 micrometers 1,000,000,000 nanometers 10,000,000,000 angstroms	1.094 yards 39.37 inches	Large mammals, adult humans (usually less than 2m tall), shrubs, large herbs, etc.
Centimeter (cm)	.01 meter 10 millimeters	0.3937 inches (1 inch = 2.54 cm)	Width of typical human male little fingernail, female thumbnail, iris of the eye, several flower petals, etc.
Millimeter (mm)	0.001 meter 0.1 centimeter 1,000 micrometers	0.03937 inch	Very large cells (heavy cardboard used for posters is about 1 mm thick, thumbnail is about 0.3 mm thick).
Micrometer (micron[a]) (μm or μ[a])	0.000001 meter 0.001 millimeter 1,000 nanometers	(Meaningless)	Mitochondrion is 0.5 to 1.0 μm wide, 5 to 10 μm long; chloroplasts 2 to 4 μm wide, 5 to 10 μm long; primary cell walls 1 to 3 μm thick.
Nanometer (millimicron[a]) (nm or mμ[a])	0.000000001 meter 0.000001 millimeter 0.001 micrometer 10 angstroms		(Nanometer is now being used more often, angstrom less often.) Ribosomes are 15 to 25 nm in diameter; microtubules, 20 to 30 nm; and DNA molecules are 2 nm thick
Angstrom[b] (A or Å)	0.0000000001 meter 0.1 nanometer		DNA is 20 A thick; protein molecules 25 to 50 A or more in diameter; a carbon to hydrogen bond is 1.09 A; a carbon to carbon single bond is 1.54 A

[a]Found in the literature, but no longer to be used; not an official International System (SI) unit.

[b]Also not an official SI unit, but still used by many workers in reference to wavelengths and atomic or molecular dimensions.

of dilute water solution called the **vacuole.** It lies in the center of the cell, interior to the cytoplasm, and usually occupies some 90 percent of the volume of a mature cell. It, too, is surrounded by a membrane, the **tonoplast.** Membranes are fundamental to life, since they control the passage of materials in and out of cells and cellular organelles. Most membranes freely transmit water and dissolved gases but pass other dissolved substances only at slower and controlled rates. The nucleus has a rather special double membrane perforated with what appear to be holes, but the membranes surrounding the cytoplasm of the cell itself (**the plasmalemma**) and many organelles, including mitochondria, plastids, ER, dictyosomes, and microbodies seem to be quite similar to each other. Some of these organelles have double membrane systems. In electron micrographs, a membrane may appear as two dark lines with a lighter area between: the **unit membrane.** Plant cells have a **cell wall** surrounding the plasmalemma and in intimate contact with it. Between the cell walls of adjacent cells there is a layer rich in pectins called the **middle lamella.** The cell wall is perforated with minute pores called **plasmodesmata.** Cytoplasm passes through the plasmodesmata, connecting adjacent cells.

The cells of bacteria, blue-green algae, and mycoplasms are without organized nuclei or other large organelles such as mitochondria or chloroplasts and are called **procaryotic** cells. Most cells of all other plants and animals do have organized nuclei and other organelles and are said to be **eucaryotic.** The procaryotic cells do have organized subcellular systems such as ribosomes and photosynthetic membranes that make their functions possible.

Plant cells are unique in certain ways. Many contain chloroplasts capable of photosynthesis, whereas animal and fungal cells do not. Even more characteristically, they have cell walls, and most plant cells have large vacuoles. The presence of cell walls and vacuoles allows a plant to have a large surface area for a relatively small amount of cytoplasm. This makes possible the efficient absorption of sunlight and transfer of gases in photosynthesis. (Herman Wiebe has written Special Topic 7.1, discussing this characteristic and highly important feature of plant cells.) Our fourth principle, then, is: **Cellular organelles and other substructures are the fundamental units of cellular function.**

P.5 Macromolecules

Large molecules are characteristic of living things. Some of these, such as **starch** and **cellulose,** are polymers not really different in principle from the large crystal molecules found in the nonliving world. They consist of *repeating* identical building blocks or units. Yet we should recognize that starch and cellulose are *not* found in the inorganic world and are of an order of complexity considerably higher than that encountered in most crystals. This is especially true because most of the carbon atoms in the sugars that constitute these compounds have four different groups attached and thus are

stereoisomers that can occur in a "right"- or "left"-handed form. These polymers are food storage products (starch) or structural materials (cellulose).

Certain of the macromolecules of life, however, the **proteins** and the **nucleic acids,** have a structural feature that is not, as far as we know, encountered in any system other than one produced by life. These molecules are also made of smaller building blocks (**amino acids** in proteins and **nucleotides** in nucleic acids), but the building blocks are not all identical, nor are they arranged in a repeating fashion as are the units in a crystal. The order of these units may appear random, yet this order is highly specific, is characteristic of a given molecule, and is preserved from generation to generation as these molecules are duplicated. As the letters in this sentence are not arranged in any repeating fashion but are, nevertheless, anything but random and contain specific information, so the sequences of amino acids and nucleotides in proteins and nucleic acids contain "information" that makes life function possible.

The information is carried from generation to generation by the nucleic acid DNA, each molecule of which consists of four kinds of building blocks (actually five nucleotides, two of which act the same) arranged in sequences characteristic of that molecule (which is a gene). There are 20 different amino acids in most protein molecules, and it is the genetic information (the nucleotide sequence) that determines the amino acid sequences in proteins. Many proteins are **enzymes,** the functioning machinery of cells, which are actually the key to understanding life. One could argue convincingly that enzymes provide the most fundamental central concept of biology, since it is their special structures that catalyze biochemical reactions, which in turn are the actual basis of life function. Thus, our fifth principle is: **Life owes much of its nature to the special kinds of macromolecules upon which it is based, especially the enzymes, genes, and other important proteins and nucleic acids.**

P.6 Structure and Function

It should now be apparent that the functions of living organisms depend upon the special structures that make them possible. Divorce is not allowed in the marriage between structure and function. It is the structures that function, but the functions that create the structures.

Physiology is the study of organism function, so **plant physiology** is the study of plant function. Though physiology is concerned with the functions of cells and whole organisms, it must continually consider the structures that are functioning, and it is almost the only biological science that considers the molecular structures. Other basic sciences would include **morphology, anatomy, genetics, taxonomy, ecology,** and even **sociology** and **political science.** Some are more concerned with structure, others with function; some (e.g., ecology) have branches devoted to each. The parent sciences are **botany** and **zoology,** but the division is not always clear; it is difficult to decide whether some cells are plant or animal in their characteristics. The basic sciences are complemented by such applied sciences as **agriculture** (**horticulture** and **agronomy**), **medicine, forestry,** and **range management.** In each case, the structure and function relationship can be seen, so our sixth principle is: **In living organisms, as in other machines, structure and function are intimately wedded.**

P.7 Tissues and Organs

In a zoology class, it is valid to say that similar cells are organized into **tissues,** and that tissues, in turn, are organized into **organs.** The same approach can be used with plants, but the distinctions are far less clear. The basic organs of plants are **roots, stems, leaves, flowers,** and **fruits,** but similar tissues occur in all the organs, and a single tissue (e.g., xylem or phloem) often includes several cell types. It is sometimes tempting to subdivide these in ways related to the organ systems of animals that are based upon specific functions. For example, in animals the skeletal system provides support. In plants, support is partially the function of the thick cell walls of fiber and other xylem cells in stems, but also of the turgid cells in leaves and even the roots that are anchored in the soil. Long-distance transport takes place in plants through xylem and phloem tissues but also occurs over shorter distances in cells of other tissues in all plant organs. Animals have gills or lungs for gas exchange, but plants have only the porous nature of their leaf tissue, with stomates allowing passage of gases in and out of a leaf by diffusion. Absorption of nutrients occurs not through a special gastrointestinal tract but through the roots and, to a lesser extent, through other plant organs.

Plants do have specialized reproductive systems with flowers, spore tissues, and other features, but these seldom appear at the beginning of plant development as germ cells do at the beginning of animal development. They typically appear only after the plant has reached a certain vegetative maturity. Since plants do not move as animals do, they must have special devices for dispersal of seeds and spores. The seed itself has virtually no counterpart in the animal kingdom. Plants also have special systems for protection, such as thorns and poisonous or distasteful substances. In spite of the difficulties in making the tissue-organ distinction in plants, it is probably valid to state our seventh principle as: **In multicellular organisms, cells are organized into functioning tissues and organs.**

P.8 Development or Morphogenesis

Consider a redwood tree. It started out as a single cell, a zygote, developed to an embryo in a seed, germinated, thrust its roots into the soil and its stem and leaves toward the sky until it became a massive giant stretching upward nearly 100 meters. It is an intricate assemblage of live and dead cells organized into tissues and organs capable of continually gener-

ating more of itself and of producing seeds that can grow into new redwood trees. Or consider yourself: You too started out as a single cell, a zygote.

How can all this happen? This is the problem of **development** or **morphogenesis,** the origin of form. We can talk about the problem in a fairly intelligent way, but so far we have no solution. It is probably the most important unsolved problem in modern biology. We know that development happens by cell divisions and cell specializations, but we don't know why a given cell should divide at a given time and place and specialize into a given differentiated cell. We know that hormones are sometimes involved, but we simply do not know most of the controls or the ultimate developmental program. Nevertheless, though our understanding is limited, we can state our eighth principle: **A living organism is a self-generating structure.**

P.9 Response of Organisms to Environment

Plant roots generally grow toward gravity and away from light; plant stems usually grow away from gravity and toward the light—although, of course, there are complications and exceptions, such as horizontally growing branches and roots. Plants respond to temperature and to the length of day. They also measure the time during the day, anticipating changes. In numerous complex and fascinating ways, plants and animals respond to and interact with their environments. Much has been learned about the pigments that absorb light, mediating various photochemical responses. And much will be said about response of organisms to temperature, including activities of their enzymes, balances of saturated and unsaturated fatty acids in their membranes, and other responses.

In responding to environment, organisms often have special mechanisms that control internal conditions at levels appropriate for life function. Warm-blooded animals have systems to control their body temperatures within narrow limits, providing what is probably the best example of **homeostasis.** Though less exact, plant leaves also have temperature control mechanisms, as we shall see. In such systems, the organism detects a change in some environmental factor and responds internally to adjust to the new conditions. Often these systems include **negative feedback loops,** as when the product of an enzymatic reaction or reaction chain feeds back to the enzyme of another in the chain to slow the reaction, thereby maintaining a desirable level of product. We can state our ninth principle: **Organisms grow and develop within environments and interact with these environments and with each other in many special ways, including mechanisms of homeostasis and feedback.**

P.10 A Definition of Life

Based upon the things we have been discussing, we can make a first attempt at defining life. Of course, there are difficulties:

The special macromolecular structures might exist on the surface of a dead world somewhere in the universe, and surely they exist after an organism has died, so in that sense they are not characteristic of life. Viruses display many of the properties of life, but only when they are associated with other living organisms. Nevertheless, we believe that a definition of life can be formulated at this time and that it can serve as a final summary of our summary: **Life is a peculiar series of functions associated with a peculiar series of organized structures in which certain macromolecules, having building blocks arranged in nonrepeating but reproducing sequences, are capable of reproduction, transfer and utilization of information, and catalysis of metabolic reactions. All this is organized at least to the level of a cell, which level allows the functions of growth, metabolism, irritability or response to environment, and reproduction.**

Section One

Water, Solutions, and Surfaces

Young oat plants with drops of guttation water at their leaf tips.

1

Water Potential

Science is a process of devising analogies to help us understand nature. These analogies have been called *hypotheses, theories,* or *laws,* but we usually call them *models.* Sometimes they can be constructed as real physical models, but much more frequently they are strictly conceptual. Best of all, they are reduced to mathematical expressions. The word **model** is a good term. It accurately describes the product of a scientist's creative thought. Plant physiology is no exception to the analogizing and model-building bent of science. And the water relations of plants provide an excellent topic to begin our study of plant physiology; attempts to understand these plant functions have resulted in much scientific model building. How does the water "run uphill" in moving to the top of a tall tree? A physical analogue and a conceptual model, continually being modified by new information and new ideas, exist to help us understand. How is the sugar made in a leaf transported to the roots? Again, a model is at hand, and we can build a piece of laboratory apparatus that provides insight. How is the energy in sunlight absorbed by a leaf and dissipated again to the environment? Mathematical formulations help us to understand. How do water and dissolved substances move within the plant from cell to cell without a heart to pump them? The several models devised to give us insight into such physiological phenomena are the topics of discussion in these first chapters.

1.1 Water and Minerals in the Plant: How to Understand Them

Wherever a plant grows, it faces water and mineral problems. Even a submerged aquatic, such as a single-celled green alga, must maintain a water and mineral balance with its surroundings. Its metabolic activities require different mineral concentrations inside the cell from those outside, and dissolved substances are synthesized within plant cells. These dissolved minerals and metabolites lead to *osmosis,* as we shall see, which causes cells to take up water and build up pressures that could lead to bursting. Plant cells have two important solutions to these problems: cell membranes that regulate mineral uptake and cell walls that resist bursting. (Single-celled aquatic animals have membranes but not walls; they may also have contrac-

tile vacuoles to expel the excess water, preventing pressure buildup.)

Land plants face more serious mineral and water problems. Minerals are all in the soil, whereas much of the metabolizing plant is in the atmosphere, and concentrations of minerals in the soil water are seldom adequate or appropriate for the plant's metabolic activities. Walls and membranes are essential but not enough. There must be an absorption and a distribution system. Furthermore, if the leaves are to exchange gases and absorb sunlight in photosynthesis, they must have large surface areas compared to their volumes, and in the process of exchanging carbon dioxide and oxygen with the atmosphere, they will usually lose water by evaporation (*transpiration*). Solutions to these problems are complex, but they include the large vacuoles in plant cells that allow for thin layers of cytoplasm (a large surface for a small volume of cytoplasm) and the stomatal mechanisms that regulate gas diffusion in and out of leaves.

The problems faced by land plants are greatly intensified in some locations: deserts, where water is scarce during much of the year; salt marshes or salty soils, where it is osmotically difficult for plants to absorb water; tundras, where low temperatures reduce metabolic activities and consequently mineral and water absorption; and other special locations. Problems also change with the seasons and even with the time of day.

These problems exist because of the physical properties of water (see Special Topic 1.1) and of minerals in solution. Furthermore, it is the *energy relations* of water and solutions in plants that are often of most immediate concern. To understand these things, then, we must understand the physics, especially the *energetics* (thermodynamics), of water and solutions. We must understand the models that have been formulated.

But what kinds of models? Many plant scientists have been busily engaged in the art of scientific model building at least since the time of Stephen Hales in 1727, when he published his "Vegetable Staticks" (see the boxed essay on p. 16). Understanding some of these models requires advanced mathematics and physical chemistry. On the other hand, some of the models used over the years are so simple that they do not accurately portray nature. Models not only give scientists insight into how nature functions, but also are teaching devices. Thus, the models we will describe that relate to plant water relations represent neither the simple models presented in basic biology courses nor the most sophisticated models presented in the literature of plant physiology. (See Kramer, Nobel, Slatyer, and Sutcliffe for these more involved formulations.) We hope that the approach taken here will challenge you without overwhelming you with the detail more appropriate to graduate courses. We'll begin by presenting the necessary background from physical science—again at an appropriate level, we hope.

1.2 Energy

Matter has mass and occupies space, but what is energy? **Energy** is something that can transform or act on matter, although it occupies no space and has no mass. We observe energy only by observing its effects on matter. So energy is not easy to deal with, and a marvelous insight occurred during the last century when it was recognized that energy can be changed in many ways but cannot be destroyed. (Now we know that it can be converted to matter.) Perhaps this recognition of the conservation of energy was the first really significant step leading to the formulation of three related and even overlapping systems of conceptual thought (models) upon which we are presently so dependent in our understanding of energy.

One of these models concerns the discrete levels of energy within individual atoms and molecules and includes radiant energy transfers; in its most modern form it is called **quantum mechanics.** It is discussed in Special Topic 3.1 but has little direct bearing upon internal plant water relations and thus is not discussed in this chapter. Quantum mechanics is, however, important in other areas of plant physiology. The second system or model, the **kinetic theory** of heat (that atomic-sized particles are always in motion) formulated by James C. Maxwell and his contemporaries, is extremely important and will be discussed. The third system of conceptual thought, **thermodynamics,** we will also discuss briefly in this chapter. In a sense, thermodynamics is the mother discipline, encompassing kinetic theory and even quantum mechanics.

1.3 Kinetic Theory

The concept that the elementary particles (atoms, ions, or molecules) are in constant motion above the absolute zero of temperature so permeates our science that aspects of it are taught in elementary school. In one way or another, we apply various components of this theory in almost any consideration of matter and the changes that it undergoes.

The theory states that the average energy of a particle in a homogenous substance rises as temperature increases but is constant for various substances at a given temperature. It is instructive when using this model to consider some of the

actual velocities and masses of the moving particles. These can be readily calculated for gases, although it is much more difficult to obtain values for liquids and solids. The average velocity of particles in a gas (V_{ave}) is calculated by the following formula (see modern texts on statistical thermodynamics):

$$V_{ave} = \sqrt{\frac{8RT}{\pi M}} \qquad (1)$$

where
V_{ave} = average velocity in centimeters/second (cm/s)
R = ideal gas constant (8.31×10^7 ergs/mole degree)
T = absolute temperature in degrees kelvin (K)
M = molecular weight in grams/mole (g/mol)
π = 3.1416

As this equation shows, the averge velocity is proportional to the square root of the absolute temperature; that is, the higher the temperature, the faster the motion of the atomic-sized particle. At the same time, the average velocity is inversely proportional to the square root of the mass; the smaller the atomic-sized particle, the faster it moves at a given temperature.

Applying this and other equations produces some rather amazing numbers, as illustrated in Table 1-1. Average velocities are surprisingly high. Hydrogen molecules near room temperature are moving on the average close to 2 km/s—or 6,433 km/hr (3,997 mph)! Even the much heavier CO_2 molecule has an average speed of 1,372 km/hr. At the same time, however, the atomic-sized particles do not move very far between collisions—only 150 to 400 times their own diameters! (The **mean-free-path,** which is an average distance traversed between collisions, is much longer at low gas pressures; about 1 cm for oxygen at 1.3×10^{-6} atm or 0.01 mm of Hg, for example.) With such high velocities and such short pathways between collisions, the number of collisions for each molecule is enormous; on the order of billions/second. In liquids, for which no one has yet written satisfactory equations (i.e.,

Table 1-1 Some Molecular Values for Three Gases

	H₂	O₂	CO₂
Molecular weight of gas	2.01	32.0	44.0
Average velocity at 0 C, meters/second (m/s)	1,695	425	362
Average velocity at 30 C, m/s	1,787	448	381
Mean-free-path between collisions with other molecules, 0 C, 1 atmosphere (atm) pressure, nanometers	112	63	39
Number of collisions of 1 molecule/sec, billions, which is the number shown × 10⁹	15.1	6.8	9.4
Diameter of each molecule, nanometers	0.272	0.364	0.462
Number of molecules (× 10¹⁹), 1 atm pressure, in 1 cm³	2.70	2.71	2.72

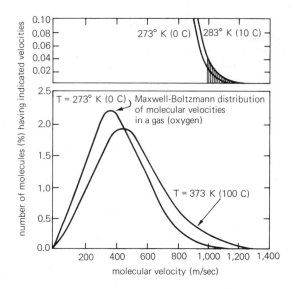

Figure 1-1 The Maxwell-Boltzmann distribution of molecular velocities in a gas at two temperatures 100 C apart. The curves at the top show the high-velocity portion for a gas at two temperatures 10 C apart. The area under the curves indicated by the vertical lines represents the number of highly energetic particles and approximately doubles in going from the lower to the higher temperature.

models), the velocities are of the same order of magnitude at room temperatures, but the mean-free-paths are much shorter, and thus the number of collisions is much greater. In solids, the particles are more or less held in place but vibrate against each other. Table 1-1 also shows that changing the temperature from 0 to 30 C,* which is much of the range of life functions, increases the average particle velocities only about 5 percent.

It is important to realize that the *actual* velocities of individual particles in a homogenous substance vary widely from the *average* velocity. Indeed, the actual velocities are distributed according to the Maxwell-Boltzmann equation, which produces curves such as those for oxygen at two temperatures shown in Fig. 1-1. We suspect that this distribution would be similar for liquids and solutes dissolved in them. It is easy to understand why particle speeds would vary so widely if we think of the totally random manner in which particle collisions occur. In such collisions, one of the partners may gain energy while the other loses it; hence, the actual speed of a given particle will probably change on the order of billions of times per second (i.e., at virtually every collision). Yet statistically the particle velocities will be distributed at any instant as in Fig. 1-1.

High-speed (high-energy) particles are the ones most likely to cause melting, evaporation, and chemical reactions,

*Several modern publications, including the journal *Plant Physiology,* now omit the degree symbol when temperatures are given. Thus: 30°C = 30 C.

while particles with lower energies cause freezing or condensation from the vapor to the liquid state. These are important processes in living plants.

You'll note from Fig. 1-1 (see also Table 1-1) that average particle velocity does not change much even with an increase of 100 K (equivalent to a change of 100 C), yet the number of high-velocity particles (under the right-hand tails of the curves) increases considerably even with a 10 K rise in temperature. This is a feature of the shape of the curves. If the number of these high-energy particles doubles with such an increase, and it is these particles that take part in chemical reactions, then we can understand why many chemical reactions double in reaction rate with an increase of 10 C. The factor by which a reaction increases with a 10 C increase in temperature is called the **Q_{10}.***

1.4 Thermodynamics

Thermodynamics, the science of energy transformation, is a system of thinking devised during the last century to help us understand heat and machines, especially steam engines. Thanks to J. Willard Gibbs and others, the principles of thermodynamics now apply to energy in general (not just heat) and are widely used in virtually every field of science. These principles are best developed step by step, with each new concept introduced along with its accepted symbols. Such a procedure is not difficult but requires many pages of text and several hours of a student's time. There simply isn't room for such an approach in this text. Yet three thermodynamic concepts are especially helpful in a modern discussion of plant water relations. You can gain some intuitive feeling for these concepts by seeing them defined in elementary ways too brief to be logically complete and then by applying the concepts in discussions of plant water relations. (See also Special Topic, 1.2, p. 403, by J. Clair Batty.)

Entropy (S) and the Laws of Thermodynamics The **First Law of Thermodynamics** is already familiar. It can be stated in several ways, such as: **In all chemical and physical changes,**

*Q_{10} values may be calculated when reaction rates are known at any two temperatures:

$$Q_{10} = \left(\frac{k_2}{k_1}\right)^{\frac{10}{T_2 - T_1}}$$

or

$$\log Q_{10} = \left(\frac{10}{T_2 - T_1}\right) \log \frac{k_2}{k_1}$$

where T_1 = lower temperature
  T_2 = higher temperature
  k_1 = rate at lower temperature
  k_2 = rate at higher temperature

energy is neither created nor destroyed but only transformed from one form to another. Or, In any process, the total energy of the system (i.e., the collection of matter under consideration) plus its surroundings (i.e., the rest of the universe) remains constant. Or, You can't get something for nothing.

This law puts some important limitations on what can and what cannot be done. The energy trapped in organic molecules by photosynthesis, for example, can never exceed the light energy absorbed. The First Law is basic.*

The **Second Law of Thermodynamics** can also be stated in several ways: **Heat cannot be completely converted into work without leaving a change in some part of the system. Or, Every time energy is converted from one form to another or transferred from one compound to another, there is a reduction in the amount of useful energy. Or, Any system plus its surroundings tends spontaneously toward increasing disorder (entropy—see following). Or, In any energy conversion, some energy is transferred as heat to the environment. Or, No real process can be 100 percent efficient. Or, There can never be a perpetual motion machine. Or, You can't even break even.**

The consequences of the Second Law are extremely important. For example, photosynthesis will never be 100 percent efficient, because some of the light energy driving the process will be converted to heat. Because some of the energy driving *any* process will be converted to or will remain as heat, there will never be a perpetual motion machine. The statement that randomness or disorder must always increase for the universe as a whole is especially important. If randomness decreases in one system (that is, if that system becomes more ordered), then that is only because energy is being put into the system—and randomness is increasing somewhere else.

The measure of this randomness is called **entropy** (S). There is no way to calculate an absolute value for entropy for any given system or part of a system,† but we can calculate the entropy changes (ΔS) that take place in a system and its surroundings during various physical and chemical processes.

---

*It is also in reality somewhat more complex than we have indicated. As it is stated, this law is logically meaningless, since it is impossible to define energy without reference to the "First Law." Yet it can have meaning for you if you accept an intuitive definition of energy. Modern thermodynamics suggests that the law we have stated is only a useful derivation from the more complex First Law, which is beyond our discussion here. Furthermore, we know that energy actually can be destroyed—or created; turned into matter or created from the destruction of matter, according to Einstein's famous equation: $E = mc^2$, where E = Energy, m = mass, and c = the velocity of light. But this equation doesn't really apply in plant physiology—except as it applies in the nuclear-reacting sun, where the light for photosynthesis is produced, or in the disintegration of some atomic nuclei used as tracers in plant physiological research. Perhaps we should state as a "First Law of Learning about Nature": **All things are more complex than they first seem to be!**

†The so-called Third Law of Thermodynamics *assumes* that entropy of any substance is zero at the absolute zero. We are, nevertheless, more concerned with entropy *changes* (ΔS) than with absolute values.

If the entropy of a system plus its surroundings increases during such a process, then it is a **spontaneous process** (which is one way of defining such a process). As a rule, if entropy for the system alone decreases, it is because the process is being driven by an input of external energy; it is not spontaneous. A simple spontaneous cooling process is an exception, however.

Spontaneous processes are **irreversible** unless there is an input of energy that increases entropy in the surroundings. Perfume molecules do not spontaneously condense back into an open bottle, for example. And as Sir James Jeans said: "You can't unscramble eggs." When the entropy reaches its maximum level for a system and its surroundings, then the system is said to be at **equilibrium.** At equilibrium there is no net change in entropy or any other property of the system. We shall see that there are other ways of defining equilibrium, but one way is to define it as the state of maximum entropy.

Is life an exception to the Second Law of Thermodynamics? As a zygote grows into a mature organism, order and complexity increase greatly. Large amounts of energy and raw materials must be consumed to make this possible, so overall entropy is increasing as organisms grow. In that sense, life is not a spontaneous process and is not an exception to the Second Law. Yet things certainly *seem* to grow spontaneously. What are the control mechanisms in living organisms that account for this gathering of raw materials and energy to decrease entropy? How did they come about? These are the unsolved problems of organism development.

Entropy is a valuable concept, because it forces us to think about the degree of orderliness in the universe and how the changes that we observe around us are always spontaneously driven by the overall increase in disorder somewhere. The nuclear energy concentrated in the stars, including our sun, for example, is being randomly distributed throughout the universe—but it drives life and the geologic processes along its way to maximum entropy. Our musings about plant function can often be aided by the entropy concept.

The Gibbs Free Energy (G) Entropy is a valuable concept, but we need to know more about a system than just its entropy. Gibbs in the 1870s worked out another concept that allows us to think of the energy passing across the boundary between a system and its surroundings that is available to do work. His formulation of what is now called the **Gibbs free energy** proves to be exceptionally valuable in understanding plant water relations and metabolic reactions. Indeed, we can begin to appreciate this by realizing that it is the input of free energy that sustains all plant activities: **The Gibbs free energy is a measure of the maximum energy available for conversion to work (at constant temperature and pressure).**

To begin with, free energy includes all components of the internal energy (E) in a substance. These include the velocity or translational kinetic motions of the particles discussed previously as well as their rotations and vibrations. Internal energy also includes the electronic energies discussed in

Special Topic 3.1 that involve levels of electrons in molecules as these are influenced by absorbing radiant energy, as well as the molecular electronic configurations that we refer to collectively as chemical bonds. Part of the internal energy of gasoline, for example, is the energy locked in the bonds holding its atoms of hydrogen and carbon together. Clearly, internal energy, like entropy, cannot be known exactly. But this limitation is not serious, as we shall see.

We can think of energy as existing in two conditions: the chemical bond energy in gasoline is **potential energy;** when the gasoline burns, converting the bond energy to the translation, rotation, and vibration of the atoms and molecules, which we observe as an increase in temperature, this molecular activity is called **kinetic energy.** In the engine, some of this heat is converted to the mechanical energy of the moving cylinders, which is another form of kinetic energy. Many of the processes in plants involve conversions from potential to kinetic energy, or vice versa. Photosynthesis converts kinetic light energy to potential bond energy, and respiration releases it again as kinetic heat or mechanical or other energies. In a sense, we can think of potential energy as being due to position or condition and kinetic energy as being due to motion (of objects, molecules, photons, electrons, and so on). So internal energy is an important part of free energy.

Molecules may also have potential energy due to their position relative to the Earth's center of gravity. This gravitational energy is of considerable importance when we are discussing how the water gets to the top of tall trees, but it is not properly a part of internal energy. Internal energy does include nuclear energy, but this has no direct influence on plant water relations because it does not change.

In the basic equation for the Gibbs free energy (G), the unavailable energy (the entropy factor) is subtracted from the total energy (at constant temperature and pressure). The entropy factor is the entropy (S) multiplied by the absolute temperature (T). The total energy includes the important internal energy (E) just discussed and the pressure (P) multiplied by the volume (V):

$$G = E + PV - TS \qquad (2)$$

where G = Gibbs free energy
   E = internal energy
   PV = pressure-volume product
   TS = entropy or disorder factor

The pressure-volume product is of special interest. It has units of energy, and it describes one of the ways that energy can flow across the boundary of a system or otherwise interact with the surroundings. We'll see that "pressure" can have a special meaning for solutes in solutions.

Since S and E cannot be known exactly, we cannot calculate G either, but the *change* in G between two states of a system ($\Delta G = G_2 - G_1$) can be calculated. For example, a standard free energy change ($\Delta G°$) of a chemical reaction can be determined by the following equation:

$$\Delta G° = -RT \ln K_{eq} \qquad (3)$$

where $\Delta G°$ = standard free energy change* in calories (cal)
   R = ideal gas constant (1.987 cal/mol degree)
   T = absolute temperature
   ln = natural logarithm
   K_{eq} = equilibrium constant
   = $\dfrac{\text{activities of products multiplied together}}{\text{activities of reactants multiplied together}}$

The results of the calculation are the same as would be obtained if it were possible to subtract the absolute free energies of the products from those of the reactants. *Thousands of experiments have shown that free energy decreases for spontaneous reactions and increases for nonspontaneous reactions.* For spontaneous processes, then, ΔG will be *negative*, and this is another way to define such processes. In equation 3, when a reaction is spontaneous, at equilibrium the activities of products are higher than those of reactants; the equilibrium constant is greater than 1, so ln K_{eq} is positive and ΔG is negative. If there are more reactants than products at equilibrium, the equilibrium constant is a fraction, and ΔG is positive.

For our purposes, it is important to get some feeling for the components of G, especially those incorporated in E. Further, it is important to know that, while the entropy *increases* in spontaneous processes, the free energy that is *lost* from the system ($-\Delta G$) is an indication of the *maximum* work that could be done by the process if it were 100 percent efficient (which it never can be). If the process is not spontaneous (if ΔG is positive), then ΔG indicates the *minimum* energy that must be *added* to the system to make the process go. When equilibrium is reached, all macroscopic properties cease changing, so there is no longer any change in free energy; that is, $\Delta G = 0$.

The Chemical Potential (Water Potential) We can speak of the free energy change of the total system being considered, or of the free energy change of any one of its components. We note, however, that a large volume of water has more free energy than a smaller one under otherwise identical conditions. Hence, it is convenient to consider the free energy of a substance in relation to some unit quantity of the substance. The free energy per gram molecular weight (i.e., the free energy/mol) is called the **chemical potential.** As such, it is a term analogous to solute concentration or temperature. It becomes independent of the quantity of substance being considered.

Chemical potential is not as simple as concentration, however, because the term PV is itself a measure of concentration. So chemical potential is a *function* of concentration and cannot be exactly analogous to concentration. In a solution,

*The standard free energy change is the maximum useful work that can be obtained when one mole of each reactant is converted to one mole of each product.

P in the PV term is the "pressure" (actually, the equivalent partial pressure) that the solute particles would have if they existed as a gas at the same "concentration"—i.e., quantity/volume. Furthermore, even the solution concentration must be corrected, since the solute particles may interact with each other, usually to decrease effective concentration. After the correction has been made, we speak of **effective concentration** or **activity** of the solution.

The chemical potential of water is an extremely valuable concept in plant physiology; it is referred to as the **water potential*** and is symbolized by the Greek letter psi (ψ). The reason that water potential is so important is that water moves in response to gradients in water potential. When water potential is higher in one part of a system than in another and nothing (e.g., some impermeable membrane) prevents water movement, water moves from the high point of water potential to the low point. In the process, the system's free energy decreases; that is, free energy is released to the surroundings, so the process is spontaneous. This released energy has the potential to do work, as when water is osmotically lifted up stems against gravity in the phenomenon known as root pressure. The maximum possible work is equivalent to the free energy released, as noted. Sometimes no work is done, but the free energy simply appears in the system and its surroundings as heat or increased entropy. In any case, it is important to remember that equilibrium is reached when the change in free energy (ΔG) or the water potential difference ($\Delta\psi$) is equal to zero. At this point, entropy for the system and its surroundings will be at a maximum, but entropy change (ΔS) will equal zero.

Based on kinetic theory and thermodynamics, we will now consider water movement in various processes that occur in plants and other situations. Studying water movement will help us to gain a better feeling for the conceptual models. After all, that's the way it worked in the first place: Scientists developed the models by observing and considering nature.

1.5 Bulk Flow and Diffusion

When the particles of gases or liquids (both often referred to as *fluids*) all move together in response to a pressure gradient, this is **bulk flow.** It is the simplest form of fluid movement. Sometimes the pressure gradient is established by gravity (the weight of fluid), in which case we speak of **hydrostatic pres-**

---

*Actually, as you might have predicted from the "First Law of Learning about Nature" or the "Footnote Law" (p. 10), chemical potential of water and water potential are not *exactly* equivalent: The water potential of water in a system assumes that the chemical potential of pure water at atmospheric pressure *and the same temperature* has been subtracted. Furthermore, as we shall see in Chapter 2, chemical potential is usually expressed in *energy* terms (e.g., calories), whereas a simple conversion is made so water potential can be expressed as *pressure* (see p. 21).

sures. In other cases, the pressure is produced by some mechanical compression applied to all or part of the system. In plants, pressures develop inside cells, but bulk flow usually cannot occur from the inside to the outside, because it is prevented by the membrane. If the membrane is punctured, then the cell contents flow in bulk out through the puncture.

Water molecules can move through a membrane before it is punctured by the process of **diffusion,** which is a nonbulk, net movement of individual molecules from one point in space to another point in space due to the random kinetic activities of the molecules. But thinking of the membrane is an unnecessary complication in understanding the rather formidable definition presented in the last sentence. Let's begin by imagining a model: two rooms connected through an opening, one containing white balls in free motion, the other containing black balls also in motion. Our imaginary balls lose no energy as heat when they bounce off the walls or each other; like molecules, they are "perfect bouncers." Obviously, the chances of a black ball going through the opening into the other room in a given interval of time will depend upon the speed and concentration (number per unit volume) of black balls. At the beginning, the concentration of black balls is higher in one room than the other, but as some black balls go through the opening, the concentration builds up in the other room. Gradually a condition of **dynamic equilibrium** is approached in which the concentration of black balls is the same in both rooms. Now balls are still passing through the opening, but the chances that a black ball will go in one direction through the opening are the same as its chances for going in the opposite direction. Of course, the same is true for the white balls, and indeed the direction of diffusion before equilibrium of each will be independent of the species of the other, provided the two do not stick to each other (i.e., interact chemically). The equilibrium is *dynamic,* because balls are still passing through the window, but there is no *net* movement.

When the balls were separated in the two different rooms, there was a relatively high degree of order; entropy was relatively low. At equilibrium, there is much less order; a maximum state of randomness has developed as the black and white balls became equally mixed. Entropy is at a maximum.

To separate the white and the black balls at the beginning of the experiment required work; free energy had to be put into the system. As mixing occurred, the free energy of the black balls in the one room decreased as the balls diffused to the other room. The energy of this dispersion could have been used to do work. That is, the free energy of the black balls at the beginning was high, because the balls were more concentrated; as they become dispersed and intermingled with the white balls (as their "partial pressure" decreases), their free energy decreases. In the Gibbs equation ($G = E + PV - TS$), the pressure-volume decreases for the system more than the entropy term increases.

It is easy to apply our model to nature in terms of the kinetic theory of matter. We visualize, for example, the molecules of a crystal of dye placed in a beaker of water. Their collisions are perfectly elastic (they are "perfect bouncers"), and they are in motion. The entropy is relatively low because the molecules are all ordered at one point in the system. Free energy is high because of the orderliness (the TS term, which is subtracted, is small) and because of the concentration of molecules in the crystal (the PV term, concentration, is high). After the crystal has dissolved and diffused throughout the system, entropy is much higher and free energy is much lower. Of course, we don't need to start with a crystal; we could consider the higher concentration of a mineral nutrient in solution outside of a cell and think of the entropy increase and the free energy decrease as the nutrient diffused through the membrane into the cell.

1.6 Chemical and Water Potential Gradients

Observation of nature, especially plants, has shown us that other factors besides concentration can be responsible for diffusion. *Indeed, in the case of water diffusion, differences in water concentration can readily be ignored.* Thermodynamic theory has helped us to realize that diffusion occurs not just in response to concentration differences but in response to gradients in free energy or chemical potential. And concentration is only one of the components of free energy or chemical potential. Let us consider the five most important factors in diffusion of **solutes** (dissolved substances) or of water (**solvent**) in the soil-plant-air continuum.

Concentration or Activity When we are discussing movement of solute particles (mineral ions, sugars, and so on), the activity (effective concentration) is by far the most common and important factor in establishing the free energy gradients that drive diffusion. In this text, when we are discussing the movement of solute particles in and out of cells and otherwise throughout the plant, we will think almost exclusively of activity differences. Particles diffuse from points of high to points of low activity.

Water is the common solvent in plants. It is for practical purposes incompressible, and thus its concentration remains nearly constant at about 55.2 to 55.5 moles/liter (mol/l), depending on temperature. Concentration does change with addition of solutes, the extent of the change depending upon the kind and amount of solute. Usually the addition of solute causes a slight to considerable expansion of the system so that the water becomes less concentrated. In a few cases, there is virtually no effect; in a few others, the system actually contracts slightly so that water becomes more concentrated. So a model based strictly upon water concentration will not explain diffusion of water in plants. Yet such an oversimplified and essentially incorrect model appears in many basic textbooks.

Temperature In the Gibbs equation ($G = E + PV - TS$), increasing the temperature increases the internal energy (E)

and, consequently, the free energy (chemical potential or water potential). Thus, diffusion should occur from points of high to low temperature, all other factors being equal. We all know that water vapor diffuses from frozen food to the colder freezer coils in a refrigerator. By the same process, liquid water or water vapor may diffuse from deep in the soil to the surface when it is cooled at night—and back into the soil during the day.

Usually this phenomenon is ignored in discussions of plant water relations, but it is easy to imagine situations in which striking temperature gradients might exist in plants. Consider, for example, the plants of the Arctic or alpine tundras, sometimes only a few centimeters high. Their roots may be in soil close to freezing while their leaves are warmed by the sun to over 20 C. Or think of temperate zone plants in springtime, when their buds are beginning to become active although their roots are still in soil covered by snow.

Pressure Increasing pressure increases the free energy and hence the chemical potential in a system. Imagine a closed container separated into two parts by a rigid membrane that will permit only passage of solvent (water, we'll assume) molecules (Fig. 1-2). If pressure is applied to the solution on one side of the membrane but not on the other, the water potential will be increased, and water molecules will diffuse more rapidly toward the compartment of lower pressure. This is an extremely important consideration with plants, because the contents of most plant cells are under pressure compared to their surroundings, and fluids in the xylem can be under tension (negative pressure).

Effects of Solute on Solvent Chemical Potential It has been observed that solute particles decrease the chemical potential of the solvent molecules. This decrease is independent of any effect on solvent (water) concentration, which, as we have seen, may be negative, innocuous, or positive, according to the species of solute. Rather, it depends primarily on the number of solute particles (ions or molecules) compared to the number of solvent particles: the **mole fraction.** We still don't know why the *number* of particles, almost regardless of their size or charge, should be so important, but observation shows that it is.

Consider again the closed container divided into two compartments by a membrane permeable to water (Fig. 1-2). If there is pure water on one side and a solution on the other side, a water potential gradient will exist, with water potential being lower on the solution side of the membrane. Water will diffuse through the membrane from the pure water side into the solution. This special case of diffusion is **osmosis.** It is often responsible for water movement from the soil into the plant and movement from one living plant cell to another.

Of course, a sharp chemical potential gradient exists for the solute particles, and if they can penetrate the membrane, they will move from the solution side (where their chemical

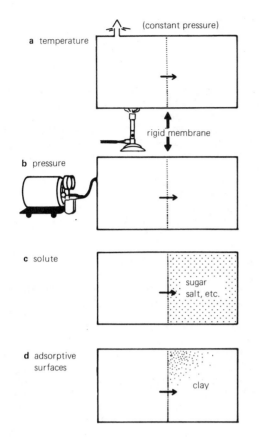

a temperature (constant pressure)

rigid membrane

b pressure

c solute
sugar
salt, etc.

d adsorptive surfaces
clay

Figure 1-2 Models of diffusional systems.

potential is high) to the other side, where initially their chemical potential is infinitely low. When they are equally concentrated on each side, there will be no more net movement. Differences in chemical potential of solutes across cell membranes is a major factor in movement of ions from the soil into the plant and for transport of ions and nonionized solutes into and out of cells within the plant. We shall see that solutes also move across membranes *against* chemical potential gradients, using metabolic energy to do so.

Matric Effects Many charged surfaces, such as those of clay particles, proteins, cellulose, or starch, have a great affinity for water molecules. These surfaces usually have a net negative charge that attracts the slightly positive sides of the polar water molecules. Because of hydrogen bonding, even surfaces such as starch that have no net charge also bind water. A material with surfaces that bind water is called a **matrix.** As binding occurs, free energy is released (ΔG is negative), so it is a spontaneous process. In our imaginary double compartment, we might have pure water on one side of the membrane

and dry protein or starch particles on the other. This condition establishes a steep gradient in water potential, with the water potential being high in the pure water and extremely low in the dry clay. Thus, water molecules will diffuse through the membrane down the free energy or water potential gradient into the protein or starch. This process of hydration is primarily responsible for the first phase of water uptake by a seed prior to germination. (See Special Topic 2.1 on colloids.)

1.7 The Rate of Diffusion

The tendency for diffusion to occur and the direction it takes depend upon the chemical potential gradient; if there is no gradient, there will be no diffusion. But the *rate* is controlled by other factors as well.

The steeper the chemical potential gradient (i.e., the greater the difference in chemical potential per unit distance), the more rapid will be the rate of diffusion, all other factors being equal. Diffusion will also be more rapid if the medium through which it occurs is more permeable to the diffusing substances. Permeability can be expressed in terms of the quantity of substance that diffuses through a unit cross section of medium in a unit of time and with a given gradient in chemical potential. In general, the more permeable the medium to a given substance, the more substance will diffuse through it. This applies to solutes diffusing through a solvent or through a membrane, such as minerals diffusing through the soil water in contact with a root and then into root cells through their membranes. It also applies to water diffusing through membranes or cell walls, some of which are more permeable to the water than others.

Increasing the temperature increases the average velocity of all the molecular-sized particles and thus increases the rate of diffusion. As we saw in Table 1-1, this effect is not great over the relatively narrow range of temperatures at which organisms are normally active. The Q_{10} for diffusion of many gases is about 1.03 (meaning that a gas diffuses only 1.03 times faster when temperature is raised 10 C; see page 10). Solutes in water, however, have Q_{10} values of 1.2 to 1.4 for diffusion. This is because increasing temperature breaks hydrogen bonds in the water so that solutes can diffuse more rapidly; permeability of water to solutes is increased. Since smaller particles have higher average velocities at a given temperature, they will diffuse more rapidly than larger particles, all other factors being equal.

1.8 Vapor Pressure

Effects of temperature, pressure, and solutes on water potential are well illustrated by their effects on vapor pressure. This is because at a given temperature and pressure, the number of solvent molecules that can exist in the vapor state and exert pressure in a closed volume above a free surface of solvent or solution will be a function of the solvent's chemical potential.

How We Arrived: Some History of Plant Water Relations

One of the earliest investigators to apply modern scientific methods to an understanding of such plant processes as transpiration and root pressure was Stephen Hales in Middlesex, England. His book *Vegetable Staticks,* published two and a half centuries ago in 1727, established the field of plant physiology. Hales was also the first to measure animal blood pressure. Yet real understanding of plant water relations awaited the formulation 125 years later (1857–1859) of the kinetic theory of heat (applied to gases) by James C. Maxwell, who was then in Aberdeen, Scotland. Ludwig Eduard Boltzmann, an Austrian physicist, helped develop the theory. In 1877, Wilhelm F. P. Pfeffer in Germany measured the osmotic pressure of several solutions, thereby laying the foundation for a real understanding of plant water relations. Pfeffer wrote a plant physiology text in two volumes published in 1881. Jacobus Hendricus van't Hoff, a Dutch chemist and the recipient of the first Nobel prize in chemistry, applied Pfeffer's data in 1887 to develop an empirical equation that would estimate the osmotic pressure that could be developed by a solution of known concentration (equation 6, p. 21).

An extremely important advance in our understanding of nature was provided at about the same time (1876–1878) by the spectacular mathematical insights of J. Willard Gibbs of Yale University, perhaps, as some contend, the most brilliant scientist that America has yet produced. He showed us how to apply the developing science of thermodynamics to chemistry—and thus to plants as well. His penetrating mind ranged from vector analysis to crystallography, planets and comets, optics, statistical mechanics, and even a railroad brake for which he held a patent!

Real understanding of water in the soil-plant-atmosphere continuum depends upon the concept of water potential. Otto Renner in Germany provided this concept in 1915. In 1938, it was formulated by Bernard S. Meyer of Ohio State University into a scheme much like the one discussed in Chapter 2. Conceptually, the model of water potential and its components had been born. It differed, however, from our present models in the names applied to its various units and in the mathematical sign of those units. What we now refer to as water potential was called by Renner the **Saugkraft** (suction force) and by Meyer the **diffusion pressure deficit.** Numerous other terms have been applied to this concept. Solutions not under pressure have a positive *Saugkraft* or diffusion pressure deficit but a negative water potential.

The modern water potential terminology was introduced in the 1960s and gained wide acceptance by the early 1970s. Most plant physiologists had recognized that the terminology needed to be revised and brought in line with thermodynamic theory. Several people were involved in this effort, including Paul J. Kramer, Ralph O. Slatyer, Sterling Taylor (a soil physicist), and others.

Many scientists are still studying the models that help us to understand plant water relations. Some (Kramer, Slatyer, and Herman H. Wiebe) have written essays for this volume. We will mention others when it seems particularly appropriate, but of course it is impossible to do justice to all the good workers in the field in a limited space. Modern books on the subject have been written by Kramer, Slatyer, and Park S. Nobel.

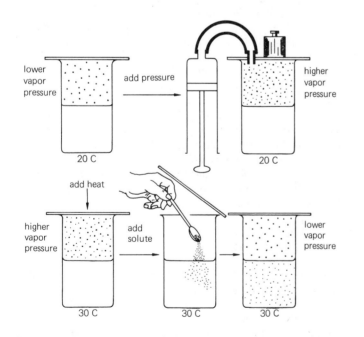

Figure 1-3 Illustrating some principles of vapor pressure.

The higher the chemical potential of the solvent, the more molecules will go into the gaseous state, so the higher the pressure produced by the solvent molecules in the vapor state (the **vapor pressure**). Increasing temperature and pressure increases the chemical potential of the solvent and hence the vapor pressure, but the presence of dissolved solutes decreases the solvent chemical potential and hence the vapor pressure (Fig. 1-3). **Raoult's law,** formulated in 1887, states that the vapor pressure over perfect solutions is proportional to the mole fraction of solvent:

$$p = X_1 p^\circ \tag{4}$$

where p = the vapor pressure of the solution
X_1 = the mole fraction of solvent
$$= \frac{\text{moles of solvent}}{\text{moles of solvent} + \text{moles of solute}}$$
p° = vapor pressure of pure solvent

Perfect solutions (those that exactly follow such laws) are seldom, if ever, encountered in nature, but Raoult's law provides a close approximation for real solutions. Furthermore, it is beautifully simple.

Since the solvent is usually present in quantities much greater than the solute, the actual lowering of the vapor pressure is usually not very great. Consider 1 gram molecular weight (1 mol) of sucrose (342.3 grams) dissolved in 1,000 g (55.508 mol) of water. The mole fraction of water in this solution is calculated as follows:

$$X \, H_2O = \frac{55.508}{55.508 + 1.000} = 0.9823 \, (98.23\%)$$

At equilibrium, the vapor pressure above pure water in a closed container is by definition equal to 100 percent relative humidity. At 20 C, this vapor pressure equals 17.535 mm of Hg, so by Raoult's law the vapor pressure above a 1.00 molal solution of sucrose equals $17.535 \times 0.9823 = 17.225$ mm of Hg. Reversing the calculation, you see that 17.225 is 98.23 percent of 17.535, so the relative humidity above a 1.00 molal sucrose solution is 98.23 percent. Thus, relative humidity above a solution (at equilibrium) is equal to the mole fraction of solvent expressed as percent.

With modern methods discussed in Chapter 2, it is possible to measure relative humidities to a small fraction of a percent, so that vapor pressure above a solution in a closed container can be accurately determined. From this, as we shall see, the water potential can be calculated. Water potentials of leaves or other plant materials placed in closed containers can be determined this way.

2

Osmosis

There is nothing at all spectacular about turning on a water faucet or flushing a toilet. We are perfectly familiar with water movement as a bulk flow phenomenon—our plumbing systems see to that. But in the world around us, vast quantities of water are moving by diffusion or in bulk due to pressure gradients established by diffusion.

It takes some mental effort to visualize this more unfamiliar aspect of the real world. With our mind's eye (there is no other way) we must see those water molecules, flying and bouncing billions of times each second in the vapor state, holding each other in the liquid state with their hydrogen bonding—positive side of one to negative of another—even while their kinetic motions tend to make them fly apart. We must sense the entropy, free energies, and chemical potentials and how these properties can drive the molecules to diffuse down a gradient. We must feel how pressure increases these quantities even while solute particles and matric surfaces decrease them. With these visions of the ultramicroscopic world playing on the wide screen that we see when our eyes are shut—that is, with the models in mind as they have so far been developed—we are ready to extend our concepts to the cells of plants. We are ready to discuss osmosis and related matters.

2.1 An Osmotic System

A device that measures osmosis is an **osmometer.** This is usually a laboratory device, but a living cell may be thought of as an osmotic system (Fig. 2-1). In both cases, two things are present: Solutions or pure water are isolated by a membrane that restricts the movement of solute particles more than it restricts the movement of solvent molecules, and there must be some means of allowing pressure to build up. In the laboratory osmometer, pressures usually build up hydrostatically by raising the solution in the tube against gravity, but other means are also used, such as a piston for automatically increasing the pressure in the system as soon as the volume of liquid begins to expand by the first small increment. In the cell, the rigidity of the plant cell wall is responsible for increase in pressure.

It is important to differentiate the cell wall from the mem-

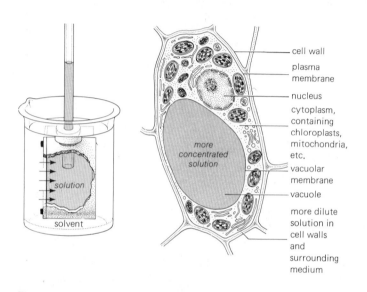

Figure 2-1 A mechanical osmometer (left), and the cell as an osmotic system. (From Salisbury and Park, 1965.)

brane. The membrane allows water molecules to pass more readily than solute particles; the cell wall normally allows both to pass readily. It is the plant cell membrane that makes osmosis possible but the cell wall that provides the rigidity to allow a buildup in pressure. Animal cells do not have walls, so if pressures built up in them nearly as much as in plant cells, they would burst. Turgid cells provide much of the rigidity of herbaceous plants, while animals are supported by a skeleton.

Consider at first a **perfect osmometer.** In such a device, the membrane is **semipermeable,*** allowing ready passage of solvent (water), but *no* passage of solute, and the solution is so confined that movement of water into the osmometer

---

*Many teachers have been trained to react violently and irrationally at the sight of this term! Please note how it is defined—as an ideal and theoretical concept—and note that we will soon introduce the concept of a *differentially permeable membrane.*

causes no significant increase in solution volume. A nearly perfect osmometer can be constructed in the laboratory, but a cell is never a perfect osmotic system.

As we saw in the previous chapter, restricting the diffusion of solute particles compared to solvent molecules can result in the establishment of a water-potential gradient. If there is pure water on one side of the membrane and a solution on the other side (typically inside the laboratory osmometer or the cell), then the water potential of the solution will be lower than that of the pure water. By convention, **water potential of pure water at atmospheric pressure and the temperature of the solution being considered is set equal to zero, so the water potential of an aqueous solution at atmospheric pressure will be some negative number (less than zero).** Hence, water molecules will diffuse from a higher water potential on the outside to a lower water potential in the cell solution; that is, water will diffuse "down" a water-potential gradient into the solution. The result will be a buildup of pressure within the system, either a raising of liquid in the tube of the laboratory osmometer or a pressure upon the cell wall. *Increasing pressure will raise the water potential,* so the water potential within the osmotic system will begin to increase toward zero. This is illustrated by the first part of Fig. 2-2.

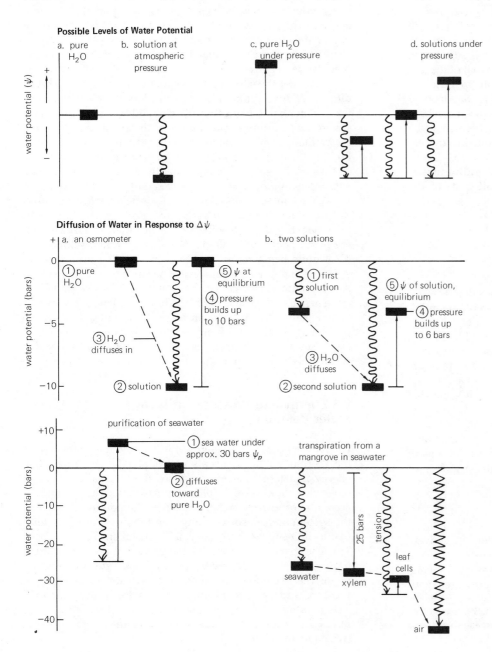

Figure 2-2 Effects of various conditions on water potentials, and diffusion of water in response to gradients in water potential.

The situation is analogous to the scale of a thermometer, but in this case we are dealing almost exclusively with values below zero. Adding solute decreases the water potential to some level below zero, and adding pressure raises this toward zero. If pure water is on one side of the membrane, pressure on the other side will increase until the water potential of the solution is equal to zero; that is, equal to the water potential of the pure water on the other side. **When water potentials (ψ) are equal on both sides, the water potential difference ($\Delta\psi$) between the two sides of the membrane is zero, and equilibrium has been achieved ($\Delta\psi = \psi_1 - \psi_2 = 0$).**

If on one side of the membrane there is a solution and on the other side another solution of different concentration, osmosis will still occur. The more concentrated solution will have the lower (more negative) water potential, so water will diffuse into it from the other solution until its pressure builds up to the point where its water potential equals that of the less concentrated solution, upon which there is no pressure. At this time, both solutions will have a water potential of some negative *but equal* value. Equilibrium will have been reached. Actually, the process is completely general. There could be pressure on both solutions, or the solution outside the osmometer might be more concentrated (water would move out, as in plasmolysis; see later), but **when equilibrium is achieved, water potential will be equal in all parts of the system. ($\psi_1 = \psi_2 = \psi$; hence, $\Delta\psi = 0$.)**

2.2 The Components of Water Potential

In the preceding paragraphs, we have considered water potential and two of its components: **pressure potential,** which is due to the addition of pressure and is equal to the real pressure in the part of the system being considered, and **osmotic potential,** which is due to the presence of solute particles. Most authors use the symbols ψ_P for pressure potential and ψ_π for osmotic potential. This emphasizes that these values are *components* of water potential (ψ). A somewhat simpler approach, appropriate for chalkboard discussions, is to use P for pressure potential and π for osmotic potential.

In simple systems at constant temperatures, the water potential results from the combined but opposite actions of the pressure potential and the osmotic potential (Fig. 2-2):

$$\psi = \psi_P + \psi_\pi \qquad (5)$$
$$(\psi = P + \pi)$$

The pressure potential may have any value. By convention, $\psi_P = 0$ at *atmospheric* pressure. Addition of pressure results in a positive pressure potential, and **tension*** (pulling; the opposite of pressure) results in a negative pressure poten-

---

*An archaic use of the term *tension* is as *pressure,* the opposite of its correct meaning. This survives in *hypertension,* which means high blood pressure.

tial. Pressure potential is usually positive in living cells but is often negative in dead xylem elements. The osmotic potential is always negative. That is, in our experience, adding solute particles always decreases water potential. Since pressure potential can be positive and very high, and osmotic potential can be either zero or negative, water potential can be either negative, zero, or positive. In plants it is almost always negative. Water potential of pure water at atmospheric pressure is defined as zero. In a solution at atmospheric pressure, water potential will be negative. In pure water under some external pressure above atmospheric, water potential will be positive. In a solution under some pressure other than atmospheric, water potential may be negative (osmotic potential is more negative than pressure potential is positive), zero (pressure potential equals osmotic potential but is opposite in sign), or positive (pressure potential is more positive than osmotic potential is negative).

Consider the situation in a higher plant. Under most conditions (relative humidity somewhat less than 100 percent), water potential is highest in the soil and lowest in the atmosphere, with intermediate values in the plant; that is, there is a gradient from the soil through the plant to the atmosphere. But the *components* of water potential vary considerably. In the soil water, $\psi_P = 0$, ψ_π is only slightly negative since the soil solution is dilute, so ψ is also only slightly negative. Water in the xylem contains few solutes, so ψ_π is only slightly negative, but the water may be under *tension* (ψ_P is negative), so ψ is more negative than for soil water, which therefore moves into the plant. Leaf cells contain a more concentrated solution, so ψ_π is quite negative; water moves in, building up a positive ψ_P, but ψ in the cells remains more negative than in the xylem. Atmospheric ψ (which we have not yet discussed) is still more negative. During a rainstorm or heavy dew, a few species may actually absorb water through the leaves and build up positive ψ_P values and hence positive ψ values in the xylem (see discussion of *Prosopis tamarugo* in Section 24.1), but this is probably rare. So ψ in the plant is usually negative.

2.3 Energy and Pressure Units for Water Potential

Let's review these components of water potential in relation to the thermodynamic concepts of Chapter 1. The Gibbs free energy is a term that encompasses the internal energies of molecules in a system, plus a pressure-volume term that accounts for energy exchanges across the boundary between the system and its surroundings, and minus a disorder term that includes the entropy (as you'll remember: $G = E + PV - TS$). The free energy indicates the maximum energy available to do work. The free energy per mole of water in a system minus the free energy per mole of pure water at atmospheric pressure and the same temperature is the water potential. This is influenced by components due to solutes and matric forces (which decrease the water potential) and pressure (which increases it).

Water potential expresses the ability of the part of the system under consideration to do work compared with that ability in a comparable quantity of pure water at atmospheric pressure and the same temperature. Water potential and its components may be thought of in units either of energy or of pressure. The osmotic potential of a solution is negative because the solvent water in the solution can do *less* work than pure water. As pressure on the solution increases, the solvent's ability to do work—and thus the water potential of the solution—also increases.

In practice, the work is performed by movement of pure water into the solution. In a laboratory osmometer, for example, a 1.0 molal* sugar solution at 28 C has an osmotic potential of -10.77 cal/mol (-2503 joules/kg, -2.503×10^7 ergs/g), and this indicates the maximum work that can be done as pure water comes into equilibrium with the solution in the osmometer (see equation 6, following). It is the energy required to raise the level of the liquid to the point in the laboratory osmometer where the pressure exerted increases the water potential of the solution to zero. This pressure in a perfect osmometer is equal to 25.18 bars, which is the metric unit of preferred use (25 bars equals 25×10^6 dynes/cm², 24.67 atm, 18.75 mm of Hg, 738.25 in. of Hg, or 25.49 kg/cm²). In the case of the cell, the work is done by stretching the cell wall. It is important to remember that the work is actually done by the pure water, which has the higher water potential.

Water potential or its components expressed in terms of energy might more accurately be referred to as **molal** or **specific water potentials** (-10.77 cal/mol for the 1.0 *m* sugar solution earlier). Dividing this value by the molar volume (the volume of 1 mol of H_2O, or 18 cm³/mol) or specific volume (1 cm³/g) of water changes the expression to units of pressure.† The result may be thought of as the **volumetric water potential** (-25 bars for the same solution). For many applications in plant physiology, expressing water potential and its components in terms of pressure offers important advantages. We are often dealing with real pressures or tensions in plants.

---

*Molality = m = moles of solute/1000 g of H_2O. This somewhat more accurately expresses osmotic relationships than molarity = M = moles of solute/liter of final solution.

†For example, energy units are: ergs/g; divided by the molar volume, cm³/g:

$$\frac{\text{ergs/g}}{\text{cm}^3/\text{g}} = \text{ergs/cm}^3 = \text{dyn cm/cm}^3 = \text{dyn/cm}^2$$

the dimensions of pressure. Specifically, since 1.0 erg = 2.390×10^{-8} cal, 1.0 mol H_2O = 18 g, and 1 bar = 10^6 dyn/cm², then:

$$-10.77 \frac{\text{cal}}{\text{mol}} \times \frac{1.0 \text{ erg}}{2.390 \times 10^{-8}\text{cal}} \times \frac{1.0 \text{ mol}}{18 \text{ g}} \times \frac{1.0 \text{ g}}{1.0 \text{ cm}^3}$$

$$= -25.03 \times 10^6 \text{ ergs/cm}^3 \text{ or dyn/cm}^2$$

and

$$-25.03 \times 10^6 \text{ dyn/cm}^2 \times \frac{1.0 \text{ bar}}{10^6 \text{ dyn/cm}^2} = -25.03 \text{ bar}$$

Hence, we will use pressure values for water potential in the rest of this text. **Bars** are the most convenient metric unit, but some workers feel that atmospheres are appropriate if atmospheric pressure is a part of the discussion (e.g., as in discussions of the ascent of sap). One bar is approximately equal to 1 atm anyway (1 bar = 0.987 atm).

In 1887, J. H. van't Hoff discovered an empirical relationship that allows the calculation of an approximate osmotic potential from the molal concentration of a solution. He plotted osmotic potentials from direct osmometer readings as a function of molal concentrations, obtaining the following relationship, which resembles exactly the law for perfect gases:

$$\psi_\pi = -miRT \qquad (6)$$

where
ψ_π = osmotic potential
m = molality of the solution (moles of solute/1000 g H_2O)
i = a constant that accounts for ionization of the solute and/or other deviations from perfect solutions (i.e., *activity*, as mentioned in Chapter 1)
R = the gas constant (0.0831 liter bars/mol deg, or 0.08205 liter atm/mol degree, or 0.0357 liter cal/mol degree)
T = absolute temperature (deg K)

If i, m, and T are known for a given solution, then osmotic potential may be readily calculated (in the absence of highly active colloids). For nonionized molecules such as sucrose or mannitol, i may be equal or close to one, but in other cases i may vary with concentration (activity), partially because the extent to which a salt or acid ionizes may depend upon its concentration. Total ψ_π for a complex solution such as cell sap is the sum of all osmotic potentials due to all solutes. (It may be expressed as **osmolality**.)

Consider some examples. A 1.0 *m* sugar solution at 30 C:

$$\psi_\pi = -(1.0 \ \frac{\text{mol}}{\text{liter } H_2O})(1.0)(0.0831 \ \frac{\text{liter bar}}{\text{mol deg}})(303 \text{ deg K})$$
$$\psi_\pi = -25.18 \text{ bars (at 30 C = 303 K)}$$

Under standard conditions of 0 C:

$$\psi_\pi = -(1.0 \ m)(1.0)(0.0831 \ \frac{\text{l bar}}{\text{mol deg}})(273 \text{ K})$$
$$\psi_\pi = -22.68 \text{ bars (at 0 C = 273 K)}$$

Note that the osmotic potential is less negative than at 30 C, hence water will diffuse from the warm to the cold solution in a suitable apparatus.

Or a 1.0 *m* NaCl solution at 20 C ($i = 1.8$):

$$\psi_\pi = -(1.0 \ m)(1.8) \ 0.0831 \ \frac{\text{l bar}}{\text{mol deg}} \ (293 \text{ K})$$

$$\psi_\pi = -44.0 \text{ bars}$$

Since the van't Hoff equation has the same form as the equation for perfect gases, the **osmotic pressure** (the real pressure developed in an osmometer rather than the osmotic *potential*) is the same as would be exerted on the walls of the container if the solute particles existed as a gas in an equivalent volume. Under standard conditions (0 C), the pressure of 1 mol of a perfect gas in a volume of 1 liter is 22.7 bars (25.2 bars at 30 C), and 22.7 bars is the osmotic pressure developed by a 1-molal solution of nonionizing solute under standard conditions. This interesting relationship, which once caused considerable confusion, is now accepted as essentially coincidental; it is certainly incorrect to think of solute particles as exerting pressure on the walls of their container as though they were a gas. Pressures on the walls of either an open beaker of solution or an ordinary osmometer at equilibrium are only hydrostatic pressures due to the weight of the solution. Incidentally, as the gas law applies only to "perfect" gases, the van't Hoff equation is at best only an approximation for certain solutions.

The van't Hoff equation is not just coincidence, however, since it can be derived from Raoult's law (equation 4), plus other thermodynamic considerations. Thus, it is not simply empirical. Its derivation was an early confirmation of developing thermodynamic principles.

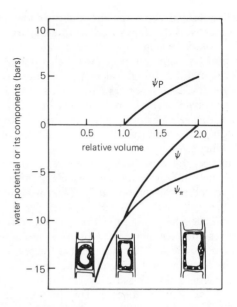

Figure 2-3 The Höfler diagram, using the concept of water potential. The components of water potential are shown as they change with changing volume.

2.4 Dilution

We have neglected one factor that is usually important in a real osmotic system as contrasted to a perfect osmometer. As water diffuses across the membrane, it not only causes an increase in pressure but also dilutes the solution. This will increase the osmotic potential in the solution (make it less negative), so the pressure required to reach equilibrium will be less than would have been predicted from the original osmotic potential.

The relationship between water potential and its two primary components during dilution is well illustrated by the so-called Höfler diagram (Fig. 2-3). The concept of this diagram was devised by K. Höfler in Germany in 1920. It describes the changing magnitudes of water potential, pressure potential, and osmotic potential as volume changes, assuming that the system expands only by taking up water and that no solutes move out or in during expansion. The curve for osmotic potential is derived from the simple dilution relationship that holds to a close approximation for dilute molal solutions ($\psi_{\pi 1} V_1 = \psi_{\pi 2} V_2$). The curve for pressure potential, on the other hand, is purely hypothetical. It depends upon the tube diameter of the osmometer or the stretching properties of the cell wall, being steep if the tube is narrow or the wall is rigid, and less steep if the tube is wide or the wall is less rigid. The water potential curve is the algebraic summation of the pressure potential and the osmotic potential curves as given by equation 5.

The Höfler diagram of Fig. 2-3 provides a good way to visualize the principles of equation 5 plus the complication of dilution. It describes nicely what happens when mature cells are placed in solutions of different osmotic potentials so that water is either gained or lost from the cells but the solutes within remain relatively constant. We shall discuss some of these approaches later in this chapter. The Höfler diagram may also describe what happens in some plant cells under normal conditions, at least over short time intervals, but the model upon which Fig. 2-3 is based is too simple to represent a *growing* plant cell, because it assumes that solutes do not pass in or out of the cell and that the cell wall is elastic like the wall of a rubber balloon, exerting more pressure the more it is stretched. Neither assumption holds for growing plant cells. As they expand in volume, the osmotic potential seldom becomes less negative and may even become more negative as solutes are absorbed and/or produced in the cell. Furthermore, when cells grow, their walls become softened (Chapter 15) so that they stretch irreversibly (plastically), and pressures may not increase at all.

2.5 The Membrane

Membranes exist in a wide variety, but osmosis will occur regardless of how the membrane functions, as long as solute movement is restricted compared to water movement (Fig. 2-4). The "membrane" could consist of a layer of liquid immiscible in the solvent of the osmotic system but allowing more solvent molecules than solute particles to pass through. A layer of air between two water solutions provides a perfect osmotic barrier. Water will pass between different parts of the system in the vapor state, but nonvolatile solutes will not. We can imagine a membrane filled with small pores containing air, or a membrane in which the solvent molecules can dissolve, as in the layer-of-liquid analogy. The third membrane model we can visualize is a sieve-type membrane with holes of such a size that water molecules could pass but the larger solute particles could not. We shall see that cell membranes have properties of both the "layer-of-liquid" and the sieve models (Chapter 5), and in dry soils water may sometimes pass in the vapor state from soil particles to the root.

In 1960, Peter Ray brought an interesting problem to the attention of plant physiologists. Calculations of the thickness of certain membranes and the rates of osmotic water movement across them showed that this movement could not occur strictly by diffusion. Rates were too high. Ray suggested that the actual zone of diffusion was very thin: an interface, say, between water in the pores of the membrane and the solution inside the osmometer. At this interface, the water-potential gradient would be extremely steep, resulting in a rapid diffusion. This rapid movement of water across the interface into the solution would create a tension in the water in the pore, pulling it along in a bulk flow (Fig. 2-4). This fourth membrane mechanism again illustrates the complexities of nature. Note, however, that the thermodynamic relations still hold.

The vapor "membrane" is a good example of a truly semipermeable membrane. It is found, however, that most and probably all true membranes occurring in plants allow some solute to pass as well as water. In such cases, we are dealing with **differentially permeable membranes** rather than truly semipermeable ones. The membrane is permeable to both solvent and solute but usually much *more* permeable to solvent. This introduces an important complication into our consideration of osmosis. The permeability of the membrane to solute particles will determine the rate at which the equilibrium established by solute concentration and pressure will gradually shift as osmotic potentials on either side of the membrane change in response to the passage of solute particles.

2.6 Measuring the Components of Water Potential

Soon after the water-potential concept was formulated in 1915, methods were developed for measuring water potential and its components. Recently, newer methods have been introduced. The older methods can help us understand plant water relations, but the newer methods are more useful in both agriculture and for current research. They are also instructive. We will summarize both older and newer methods, not as a "cookbook" of techniques that you might use but as an illustration and application of the principles we have been discussing.

Water Potential Probably the most meaningful property we can measure in the soil-plant-air system is the water potential. Not only is this the final determinant of diffusional water movement, but bulk water movement can also occur in response to pressure gradients set up by such diffusional movement. Furthermore, both in principle and in practice, water potential is probably the simplest component of an osmotic system to measure. You'll remember that at equilibrium, $\Delta\psi = 0$; that is, ψ is equal throughout all parts of the system. Thus, a plant part can be introduced into a closed system, and after equilibrium has been achieved, ψ may be known or determined for any other part of the system, indicating its value for the plant part. There are several possibilities for applying this principle. Three general approaches are most widely used (Fig. 2-5).

In the **tissue volume method,** sample pieces of the tissue in question are placed in a series of solutions usually of sucrose, mannitol or, better yet, polyethylene glycol (PEG) of varying but known concentrations. *The object is to find that solution in which the tissue volume does not change,* indicating neither a gain nor a loss in water. Such a situation implies that the tissue and the solution were in equilibrium to begin with, so the water potential of the tissue must be equal to that of the solution. At atmospheric pressure, when $\psi_p = 0$, $\psi = \psi_\pi$, and ψ_π can be calculated from equation 6.

In practice, there are several ways to determine this volume change. One way is to measure the volume of the tissue before placing it in each solution (usually standard

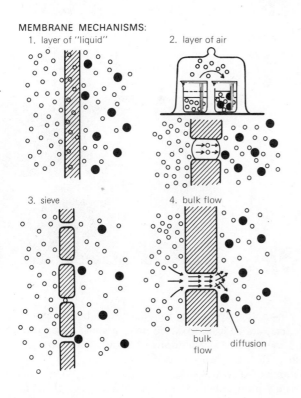

MEMBRANE MECHANISMS:
1. layer of "liquid"
2. layer of air
3. sieve
4. bulk flow

bulk flow   diffusion

Figure 2-4 Schematic diagram of the four conceivable membrane mechanisms.

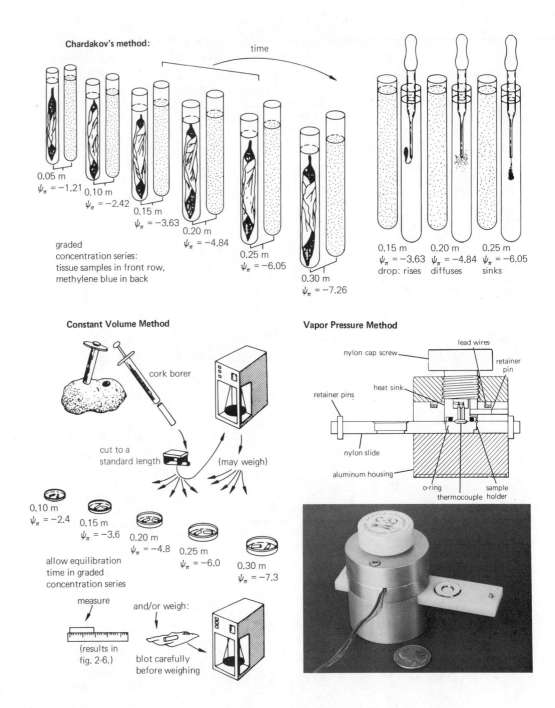

Chardakov's method:

time

0.05 m
$\psi_\pi = -1.21$
0.10 m
$\psi_\pi = -2.42$
0.15 m
$\psi_\pi = -3.63$
0.20 m
$\psi_\pi = -4.84$
0.25 m
$\psi_\pi = -6.05$
0.30 m
$\psi_\pi = -7.26$

graded
concentration series:
tissue samples in front row,
methylene blue in back

0.15 m
$\psi_\pi = -3.63$
drop: rises
0.20 m
$\psi_\pi = -4.84$
diffuses
0.25 m
$\psi_\pi = -6.05$
sinks

Constant Volume Method

cork borer

cut to a
standard length

(may weigh)

0.10 m
$\psi_\pi = -2.4$
0.15 m
$\psi_\pi = -3.6$
0.20 m
$\psi_\pi = -4.8$
0.25 m
$\psi_\pi = -6.0$
0.30 m
$\psi_\pi = -7.3$

allow equilibration
time in graded
concentration series

measure

(results in
fig. 2-6.)

and/or weigh:

blot carefully
before weighing

Vapor Pressure Method

lead wires
nylon cap screw
retainer pin
heat sink
retainer pins
nylon slide
aluminum housing
o-ring
thermocouple
sample holder

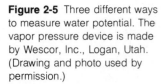

Figure 2-5 Three different ways to measure water potential. The vapor pressure device is made by Wescor, Inc., Logan, Utah. (Drawing and photo used by permission.)

volumes are cut), and then to measure the volume (or only the length) again after time has been allowed for exchange of water (Fig. 2-5, lower left). Volume change may be plotted as a function of solution concentration, indicating a gain of volume in relatively dilute solutions and a loss of volume in relatively concentrated ones. On such a plot (Fig. 2-6), the point at which the volume curve crosses the zero line indicates the solution that had the same water potential as that of the tissue at the start of the experiment.

In another approach (not illustrated), tissue samples (e.g., of a leaf) are allowed to equilibrate in small closed containers with the vapor over solutions of known concentration rather than with the solutions themselves. Thus, the liquid is not contaminated with solutes from the tissue. Weights are usually measured rather than volumes when using this method.

Rather than measuring changes in the tissue, one might measure the **concentration of the test solution.** If it becomes less concentrated, the tissue will have lost water. This is a better approach than measuring tissue volumes, which often gives ψ values that are too negative, since some solutes may be absorbed from the test solution.

In 1953, a Russian scientist, V. S. Chardakov, devised a simple yet efficient method of determining the test solution in which no change in concentration occurs (Fig. 2-5, top). **Chardakov's method** can often be used in the field. Test tubes or vials containing the graded solutions of known concentrations are colored slightly by dissolving a small crystal of a dye such as methylene blue. (Addition of the crystal does not change osmotic potential significantly.) Tissue samples are placed in test tubes containing equivalent solutions but with no dye. Some time is allowed for exchange of water. It is not essential that the tissue reach equilibrium with the solution, only that a certain amount of water exchange occurs. This may happen in as little as 5 to 15 minutes. After this, tissue is removed, and a small drop of the comparable colored solution is added to the test tube. If the colored drop rises, the solution in which the tissue was incubated has become more dense, indicating that the tissue has taken up water. Thus, the tissue must have had a lower (more negative) water potential than the original solution. If the drop sinks, the solution is now less dense than originally, having absorbed some water from the tissue. The solution, then, had a lower water potential than the original tissue. If the drop diffuses evenly out into the solution without sinking, then no change in concentration has occurred, and the water potential of the solution equals that of the tissue. When several solutions having different concentrations are used, one can usually be found in which the drop neither sinks nor floats. With equation 6 we can calculate its ψ_π, which at $\psi_p = 0$ is equal to its ψ and thus the average ψ of the tissue.

In the **vapor pressure method,** tissue is placed in a small closed volume of air, and the water potential of the air comes into equilibrium with the water potential of the tissue, which will change only insignificantly in the process. The water potential of the air is measured by measuring vapor pressure (humidity) at a known temperature.

Knowing the absolute temperature (T), the vapor pressure of pure water at that temperature ($p°$), the vapor pressure in the test chamber (p), and the molar volume of water (V_1 in liters/mol), water potential is calculated from the following formula derived from Raoult's law:

$$\psi = -\frac{RT}{V_1} \ln \frac{p°}{p} \qquad (7)$$

The ratio $100p/p°$ is the relative humidity (RH); converting to common logarithms, and using numerical values for R and V_1, equation 14 simplifies to:

$$\psi \text{ bars} = -10.6T \log_{10}\left(\frac{100}{\text{RH}}\right) \qquad (8)$$

The principle of water potential measurement by this method is simple, but in practice a number of difficulties are involved. These were solved only rather recently, but now this approach is the predominant method. To begin with, temperature must be uniform within at least a hundredth of a degree centigrade, if the method is to be sufficiently accurate. This is because slight changes in relative humidity indicate large changes in water potential, and slight changes in temperature result in significant humidity changes. The second problem involves measurement of the humidity inside the test chamber. An ingenious method was first developed in 1951 by D. C. Spanner in England and since has been improved by several workers. Two thermocouple junctions are built into the test chamber. One of these has a relatively large mass and thus remains at the temperature of the air in the chamber. The second is very minute, and when a weak current is passed in the right direction through the two junctions, the small one cools rapidly by the Peltier effect.* As this thermocouple cools, a minute drop of moisture condenses on it from the air inside the chamber. This point of moisture then acts as the "wet bulb," and the difference between its temperature and that of the dry thermocouple indicates the water potential of the air in the chamber. Uniform air temperatures between the two

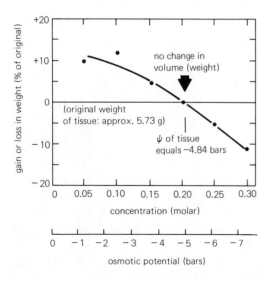

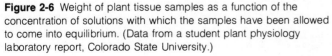

Figure 2-6 Weight of plant tissue samples as a function of the concentration of solutions with which the samples have been allowed to come into equilibrium. (Data from a student plant physiology laboratory report, Colorado State University.)

*A thermocouple operates on the principle of the thermoelectric effect: When a circuit consists of two wires made of different metals (e.g., copper and the alloy constantan), and the two junctions between the two wires are at different temperatures, a current flows around the circuit. When one junction is at a known temperature and the current is measured, the temperature of the other junction (the thermocouple) can be calculated. The Peltier effect is the opposite of the thermoelectric effect: When a current is passed through a circuit of two different metals, the two junctions will have different temperatures. One will heat and one will cool, depending on the direction of the current.

thermocouples are maintained by immersing the chamber in a water bath or by placing it in an aluminum block or smaller silver block (since silver is the best conductor of heat). In practice, the drop evaporates so rapidly that actual temperatures cannot be measured. Rather, the system is arbitrarily calibrated using solutions of known concentrations. Typically, measurements (which require less than a minute) are made at regular intervals until they stabilize after an hour or two, indicating that the tissue has reached equilibrium with the air in the chamber (Fig. 2-5).

Still another method for measurement of water potential in plant stems involves use of a pressure bomb. This is discussed later (pages 27–28) and in Section 4.5.

Osmotic Potential Since the absolute value of osmotic potential (which is negative) is equivalent to real pressure (positive) developed in a "perfect" osmometer, this property of a solution may be measured directly. Initially, many measurements were made, particularly in 1877 by Wilhelm F. P. Pfeffer, who made nearly perfect, rigid, semipermeable membranes by soaking a porous clay cup in potassium ferrocyanide and then in cupric sulfate, precipitating cupric ferrocyanide in the pores. Columns of mercury were used to determine pressure. Data obtained in this way were the basis for equation 6, the van't Hoff equation. With our developing understanding of the properties of solutions, it became apparent that other, simpler measurements could be made, and the data could then be converted to osmotic potential. Excellent ways of doing this with free liquids have been developed, but no completely satisfactory method is yet available for measuring the osmotic potential of the liquid in plant cells. Attempts to measure it almost invariably result in changing it.

The **vapor pressure method** described for the measurement of water potential by equilibration of the tissue sample with air in a closed container applies equally well to the measurement of osmotic potential of a free liquid. This is now performed routinely on human body fluids using a device similar to that shown in Fig. 2-5. But how can the pressure in cells be reduced to zero so that only the osmotic component influences the vapor pressure? A tissue may be frozen and thawed and its water potential measured by the vapor pressure method. Rapid freezing in the laboratory produces ice crystals that rupture all the membranes, so pressure within the cells becomes zero, and $\psi = \psi_\pi$. The treatment probably results in some mixture of cytoplasm and vacuolar sap with a consequent change in osmotic potential, but the method is in frequent current use.

The freezing point, as well as the vapor pressure, is a function of the mole fraction and hence the osmotic potential. Properties of solutions that are a function of the mole fraction are called the **colligative properties.** They include freezing point, boiling point, vapor pressure, and osmotic potential. Hence, when freezing point is determined, osmotic potential can be calculated. This is called the **cryoscopic** or **freezing point method.** As mentioned, the osmotic potential of a molal

nonionized solution at 0 C is ideally −22.7 bars. Its freezing point (Δ_f) proves to be −1.86 C. Thus, the osmotic potential of an unknown dilute solution may be estimated from its freezing point by the relationship:

$$\frac{\psi_\pi}{\Delta_f} = \frac{-22.7\,\text{bars}}{-1.86\,\text{C}} \tag{9}$$

or

$$\psi_\pi \text{ bars} = 12.2\Delta_f \qquad (\Delta_f \text{ in degrees C}) \tag{10}$$

Determination of the freezing point proves to be relatively simple. Highly accurate mercury thermometers (to 0.01 C) are commercially available for this purpose, and thermocouples can be even more sensitive and easier to operate. A time-temperature curve is determined (Fig. 2-7) to account for supercooling and the subsequent warming upon freezing due to release of the heat of fusion. Results are compared with those using pure water.

Obtaining a pure plant sap is far more difficult. One may squeeze out the sap with a press, freeze the tissue to rupture the cells and then squeeze out the sap, or homogenize the tissue in a blender and filter the sap. All these methods applied to the same tissue typically give different results, and the difference may be as high as 50 percent. Blender values are usually the most negative (most concentrated); hand squeezed through cheesecloth, the least. A principal problem is that the various methods involve different degrees of mixing of cyto-

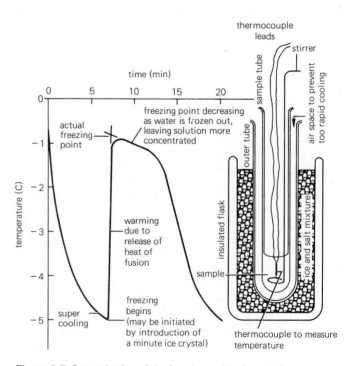

Figure 2-7 Determination of the freezing point of a solution.

plasmic contents and other substances with the vacuolar sap, which is primarily responsible for the osmotic behavior of plant tissues. On the other hand, pressing living cells gives almost pure water by osmotic filtration. In spite of the limitations, the cryoscopic method has been widely used for many years.

If the sap within plant tissue were in osmotic equilibrium with some outside surrounding solution ($\Delta\psi = 0$) and *no pressure or tension existed within the tissue*, then the osmotic potential of the sap would be equal to the osmotic potential of the solution. The problem with such a measurement is to obtain zero pressure within the tissue without changing the other osmotic properties any more than necessary (which may be too much!). This is the method of measuring osmotic potential by observing **incipient plasmolysis**. Samples of tissue are placed in a graded solution series of known osmotic potentials. Sucrose or mannitol solutions may be used, but substances such as PEG have the advantage of not penetrating or being changed metabolically as readily by the tissue. After an equilibration period (usually 30 minutes to an hour), the tissue is examined under a microscope. It has been arbitrarily assumed by plant physiologists that incipient plasmolysis occurs in tissue in which about half the cells are just beginning to **plasmolyze** (protoplasts are just beginning to pull away from the cell wall), and that this represents an internal pressure of zero. If this assumption is true, then the osmotic potential of the solution producing incipient plasmolysis is equivalent to the osmotic potential of the cells within the tissue, *after they have come to equilibrium with the solution*.

If this is true for the tissue at equilibrium (and its being true depends only upon the assumption that incipient plasmolysis truly represents zero pressure), then we must question how much the tissue changed as incipient plasmolysis developed. Pulling away of the protoplasts from the wall is a shrinkage or decrease in volume, so the sap solution inside, by becoming more concentrated, developed a more negative osmotic potential. (Remember the Höfler diagram, Fig. 2-3.) If careful volume measurements of original tissue and tissue at incipient plasmolysis are made (either the overall volume of the tissue or, better still, the dimensions of a fairly large sample of protoplasts), then the change in osmotic potential due to change in volume can be calculated. When this is not done, values of osmotic potential obtained by the plasmolytic method are too negative, often by a value of 1 to several bars (5 to 10 percent or more).

This traditional method continues to be used when it is appropriate, although it is severely limited by the ability of the experimenter to detect incipient plasmolysis in many tissues. Is there an alternative way to determine when turgor pressure (and hence ψ_p) equals zero? James A. Lockhart suggested an interesting approach in 1959, although little use seems to have been made of his method. He found that stems or other tissue could not be easily deformed until the cellular turgor pressure reached zero. Stem sections are placed in a graded osmotic series for an appropriate time (depending on the tissue), held horizontally at one end, and a small weight

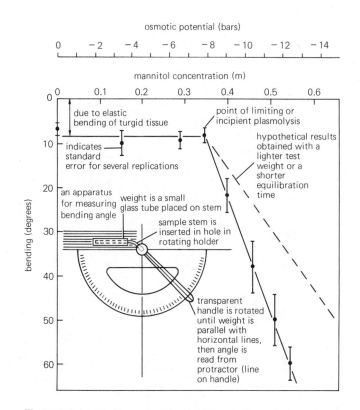

Figure 2-8 Lockhart's method for detecting limiting plasmolysis.

is placed on the other end. The angle of deformation is measured and plotted as a function of the osmotic potential of the solution in which the tissue had been placed. The point at which the deformation curve extrapolates to zero indicates the point of incipient plasmolysis, where osmotic potential of the tissue equals that of the solution with which it has come to equilibrium (Fig. 2-8). At that time, tissue osmotic potential may be more negative than that of the original tissue, as indicated earlier. This simple little technique dramatically illustrates the idea that a plant's form, its rigidity, is often determined by the turgor within its cells.

During recent years, another somewhat sophisticated approach has been coming into wide use. This method uses a pressure bomb and is capable of providing several data points relating to the water status of a plant. To measure osmotic potential, a leaf or branch is removed from a plant and allowed to hydrate for a matter of hours or even overnight; that is, the cut end is placed in pure water, and the specimen is covered with a plastic bag to ensure 100 percent RH. Studies indicate that osmotic potential changes only slightly during this time. Or pure water may be forced into the specimen under pressure to effect rapid hydration. The hydrated branch is then placed in the pressure bomb (Fig. 2-9) with the cut end protruding. Pressure is applied, and sap begins to exude from the cut end. This proves to be almost pure water. Pressure is usually released slightly to stop the flow (which may otherwise take

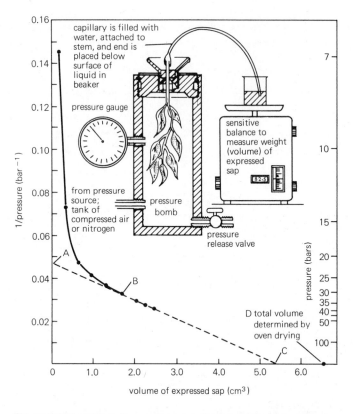

Figure 2-9 Illustrating the pressure bomb as a method for measuring several parameters concerning the water relations of plants. Point A: the water potential of hydrated tissue, about 21 bars. Point B: incipient plasmolysis, indicating an osmotic potential of about 31 bars. Point C: volume of free water in the tissue, about 5.35 cm³. Point D: total volume of tissue water, about 6.72 cm³. Bound water equals D minus C, or 1.37 cm³.

awhile to be complete due to resistance in the tissue), and then the volume exuded is measured and noted along with the so-called **balancing pressure,** the gauge pressure at that point.

The reciprocal of the volume of exuded sap ($1/V$) is plotted as a function of the balancing pressure (P) to obtain a characteristic curve as shown in Fig. 2-9. With increasing pressure, the result is at first a curved line, but then (at point B) it becomes linear. For reasons that we shall not discuss (see papers of Hellkvist et al., 1973; Tyree and Dainty, 1973; and Tyree and Hammel, 1972), the cells at this point are at incipient plasmolysis, and the negative value of the pressure in the bomb is equal to the osmotic potential in the cells. Extrapolating the line back to an expressed volume of zero (point A) gives osmotic potential of the cells of the initial hydrated tissue, and a Höfler diagram can be constructed to calculate volume of water in the tissue at that point, or at any other pressure. Extrapolating the line ahead to infinite pressure (point C) gives the volume of liquid originally present in the cells. Water remaining after the highest pressure has been applied can be measured by weighing before and after drying in an oven.

The excess over that calculated at point C is assumed to be bound water held with great force to hydrophilic surfaces.

Measurements of osmotic potentials by these methods (especially freezing points) have yielded results varying from values of about −1 bar in aquatic plants to extremely negative values, −200 or more bar, in salt-containing halophytes. The values are subject to all the problems just described, plus others; the extreme halophyte values, for example, are far too negative due to salt crystals on the leaf surfaces. Osmotic

Table 2-1 Some Examples of Empirically Determined Osmotic Potentials of Leaves

Species	Osmotic Potential ψ_π (bars)
Shadscale (*Atriplex confertifolia*)[a]	−24 to −205
(The most negative values are highly questionable.)	
Pickleweed (*Allenrolfea occidentalis*)[b]	−89
Sagebrush (*Artemisia tridentata*)[a]	−14 to −74
Salicornia (*Salicornia rubra*)[a]	−32 to −73
Blue spruce (*Picea pungens*)[b]	−52
Mandarin orange (*Citrus reticulata*)[b]	−48
Willow (*Salix babylonica*)[b]	−36
Cottonwood (*Populus deltoides*)[c]	−21
White oak (*Quercus alba*)[c]	−20
Sunflower (*Helianthus annuus*)[c]	−19
Red Maple (*Acer rubrum*)[c]	−17
Waterlily (*Nymphaea odorata*)[c]	−15
Bluegrass (*Poa pratensis*)[c]	−14
Dandelion (*Taraxacum officinale*)[c]	−14
Cocklebur (*Xanthium spp.*)[c]	−12
Chickweed (*Stellaria media*)[c]	−7.4
Wandering Jew (*Zebrina pendula*)[c]	−4.9
White pine (*Pinus monticola*)[d], dry site, August, exposed to sun	−25
White pine (*Pinus monticola*)[d], moist site, April, more shaded	−20
Big sage (*Artemisia tridentata*)[d], March to June, 1973	−14 to −23
Big sage (*Artemisia tridentata*)[d], July to August, 1973	−38 to −59
Herbs of moist forests[e]	−6 to −14
Herbs of dry forests[e]	−11 to −30
Deciduous trees and shrubs[e]	−14 to −25
Evergreen conifers and Ericaceous plants[e]	−16 to −31
Herbs of the alpine zone[e]	−7 to −17

[a]Harris, 1934, pp. 65, 70, 110. Harris made thousands of measurements by observing freezing point depressions of expressed sap. The ranges shown indicate the variability he encountered within a single species, in a single state (e.g., Utah), and in a single year. Obviously, little certainty can be attached to the specific values shown for individual species. Presence of salt crystals on leaf surfaces must account for the extremely negative values in some cases.

[b]Student reports, Plant Physiology class, Colorado State University, Fort Collins.

[c]Meyer and Anderson, 1939.

[d]Cline and Campbell, 1976; Campbell and Harris, 1975.

[e]Pisek, 1956.

potentials of halophyte cells are probably -50 to -80 bars, and osmotic potentials of most plant saps lie between -4 and -20 bars (Table 2-1). As might be expected, values vary within a growing plant, those of the youngest leaves being most negative. Osmotic potentials of root cells also must change in response to drying soils.

Pressure Potential In a laboratory osmometer, pressure is measured directly, but direct measurement of pressure in plant cells is much more difficult. Usually the pressure is calculated after water potential and osmotic potential have been determined:

$$\psi_P = \psi - \psi_\pi \tag{11}$$

Paul B. Green and Frederick W. Stanton (1967), then at the University of Pennsylvania, described a method for the direct measurement of turgor pressure in large cells, such as those of the alga *Nitella axillaris*. The method has since been applied with some success to higher plant cells, especially phloem cells (discussed in Section 7.10).

A minute manometer is made by fusing closed one end of a capillary tube (diameter 40 μm) and fashioning the other end into a tip like that of a syringe needle. As the tube is observed with the microscope, some of the water surrounding the cell will be seen to enter the open end due to capillarity, somewhat compressing the air inside the tube. The position of the meniscus inside the tube is noted and the volume of air calculated. As the open end of the tube punctures the cell, pressure in the cell is transferred to the air in the tube, compressing it further as indicated by movement of the meniscus (Fig. 2-10). The final pressure in the tube is equal to the pressure before penetration of the cell multiplied by the ratio of the original to the final volume (according to Boyle's law). The pressure before penetration can be determined by multiplying atmospheric pressure by the ratio of the original volume to the volume after entrance of water by capillarity. There will be a slight change in pressure within the cell upon penetration of the tube, but even this can be ascertained by penetrating the cell with a second tube while observing the change in pressure in the first. The method measures actual pressure in the cell, but according to convention this will be 1 atm higher than the **turgor pressure** (the term used to refer to pressure in a cell—equal to pressure potential of the water in the cell).

Green and Stanton found a turgor pressure of 5.1 atm in a *Nitella* cell in equilibrium with pure water, implying that the osmotic potential was -5.1 atm ($\psi = 0$, so $-\psi_\pi = \psi_P$). As the cell was allowed to equilibrate with sugar solutions of increasing concentrations (more negative osmotic potentials), turgor pressure decreased exactly as predicted by the Höfler diagram. It reached a value of zero with the cell in a solution of $\psi_\pi = -5.7$, confirming that the cell must shrink before incipient plasmolysis becomes apparent. The osmotic potential in the cells (which at $\psi_P = 0$ equals the osmotic potential of the outside solution) had thus become more negative (changed from -5.1 to -5.7) during plasmolysis.

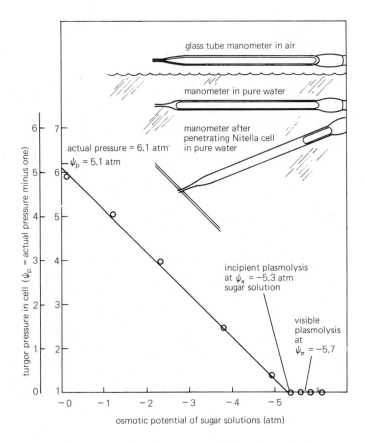

Figure 2-10 Green and Stanton's experiment. (From Green and Stanton, 1967.)

Matric Potential Hydrophilic surfaces (e.g., of such colloids as protein, starch, and clay—see Special Topic 2.1) may adsorb water, and the tenacity with which the molecules of water are adsorbed is not only a function of the nature of the surface but also of the distance between the surface and the molecules: Those located directly on the absorbing surface will be held extremely tightly; those some distance from the surface will be held much less tightly. The adsorption of water by hydrophilic surfaces is called **hydration,** or (in older texts) **imbibition.**

The **matric potential** (ψ_m) is a measure at atmospheric pressure of the tendency for the matrix to adsorb additional water molecules. This is equal to the average tenacity with which the least tightly held (most distant) layer of water molecules is adsorbed. It is expressed in the units of water potential. A dry colloid or hydrophilic surface, such as filter paper, wood, soil, gelatin, or the stipe of a brown alga, may have an extremely low matric potential (as low as $-3,000$ bars), while the same colloid in a large volume of pure water at atmospheric pressure will have a matric potential of zero (it is completely saturated). When the colloid at atmospheric pressure has come into equilibrium with its surroundings, the least tightly held water molecules will have the same free energy as water molecules in the surroundings, so the matric potential of the colloid will be equal to the water potential of the surroundings.

A common modern means of measuring matric potential is instructive. A hydrated colloid is enclosed in a pressure chamber with a membrane filter. Pores of the membrane are about 2 μm in diameter, large enough to allow passage of both solutes and water but not the colloid. Surface tension also prevents passage of air through the wet membrane, which must be supported by a special screen that can withstand high pressure. Assume in the simplest case that the colloid is wet with pure water only (no solutes). Compressed gas is introduced into the pressure filter. Increasing the pressure raises the potential of the water adsorbed on the colloid toward and finally above zero, so that water molecules begin to diffuse out through the membrane. Further increases in pressure on the colloid result in additional but smaller increments of water movement from the colloid out through the membrane.

When water movement stops, water on the colloid under pressure will be in equilibrium through the membrane with water at atmospheric pressure on the outside of the membrane. That is, the water potential of the least tightly adsorbed water molecules on the colloid under pressure is equal to zero when they are in equilibrium with pure water at atmospheric pressure on the outside of the membrane. If pressure and the matric potential are the only components influencing water potential inside the membrane, as indicated by the following formula:

$$\psi = \psi_p + \psi_m \qquad (12)$$

then, when water potential is zero, the absolute numerical value of the negative matric potential will be equal to the positive pressure potential that is produced by the compressed air ($-\psi_m = \psi_p$). The pressure is known and positive, so the matric potential is also known but negative.

Usually, after equilibrium has been reached at a given pressure, water content of the colloid is determined by weighing before and after drying in an oven. Matric potential plotted as a function of water content provides the **moisture release curve** (Fig. 2-11).

A test of the assumption that the pressure in the pressure membrane apparatus is a measurement of matric potential is to measure the water potential of the colloid at atmospheric pressure by the vapor method just described. The two measurements agree closely, indicating the validity of the approach.

It is evident from this definition of matric potential that, at equilibrium, the matric potential of any hydrophilic surfaces in a system must be equal to the water potential of that system. It is a primary postulate that water potentials at equilibrium must be equal throughout a system. Thus the water potential of the layer of least tightly adsorbed molecules on any surface at equilibrium will have to be equal to the water potential of the system as a whole. If it is more negative, water will be adsorbed; if less negative, water will be lost.

The term "matric potential" has been used in a different sense, however. In a complex system such as a cell containing

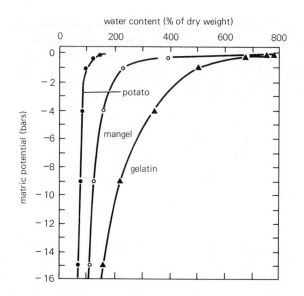

Figure 2-11 Moisture release curves for two plant materials and gelatin. (From Wiebe, 1966.)

molecules of colloidal protein, starch grains, and other hydrophilic surfaces as well as simple solute molecules, the final water potential will be determined not only by the solute particles and the pressure, but also by the matric effects of the proteins and other surfaces.

We can imagine a number of complex interactions in such a system. Some of the polar solutes, especially ions, will be adsorbed on some of the hydrophilic surfaces, influencing their water-adsorbing properties. Water molecules will also be adsorbed on these surfaces, decreasing in a sense the total quantity of water involved in the rest of the system not influenced by the matrix. That is, the rest of the solution is more concentrated than it otherwise would be. Indeed, it is possible that no water molecules in the system will be far enough away from a hydrating surface (e.g., of a protein molecule) that they are completely unaffected by such surfaces. This must be especially true of soil water with its dissolved solutes. Virtually no soil water would be beyond the influence of the soil's clay particles.

Clearly, we have a long way to go before we can completely describe such highly complex systems. In the meantime, to help us visualize these systems, plant physiologists, soil scientists, and others have used the term "matric potential" in a sense different from the definition just given. They have characterized it as one of the components of water potential. Indeed, in the simplest sense, we can think of the water potential of cells and other systems containing hydrophilic colloids as being influenced by three factors: pressure, the decrease in the free energy of water molecules due to true solute particles, and the decrease in free energy due to the presence of matric surfaces. The pressure effect is the pressure potential (ψ_p), the solute effect is the osmotic potential (ψ_π)

and, after equilibrium has been achieved, the combined effects of all hydrating surfaces in the system may be called the **matric component** (ψ_c). The term "component" distinguishes this effect from the matric potential (ψ_m), which (as defined earlier) has to be equal to the water potential of the system under consideration rather than being a component of it. For complex systems, then, we might expand equation 5 as follows:

$$\psi = \psi_P + \psi_\pi + \psi_c \qquad (13)$$

In most growing cells, the matric component is thought to be rather small, but in dry seeds, water uptake at first is largely due to matric effects. The surfaces of proteins, cellulose, starch, and so on must become hydrated before germination can begin. Growing fungi and perhaps roots and other systems may also depend upon hydration for the powerful expansion forces they often develop. These forces may split rocks, for example. Understanding water potential with its components is thus essential to understanding plants.

3

The Photosynthesis-Transpiration Compromise

During the summer of 1974, John Hanks, a soil scientist at Utah State University, kept careful track of the amount of water required to grow a crop of maize on the college Greenville farm. To mature the crop, water equivalent to 60 cm (23.6 in.) of rain was added to the field. A fourth of this evaporated from the soil, but most of the remaining 45 cm passed through the plants into the atmosphere. This evaporation of water from plants (and animals, according to most dictionaries) is called **transpiration.** Continuing the calculations, Hanks showed that 600 kg of water were transpired by the corn plants to produce 1 kg of dry corn (grain). As a matter of fact, 225 kg of water had to be transpired to produce 1 kg of dried plant material, including leaves, stems, roots, cobs, and seeds. This figure is typical, although there are substantial differences among species.

Why is so much water lost by transpiration to mature a crop? Because an essential part of those dry corn seeds and all other plant parts are the carbon atoms that form the skeletons of the organic molecules of which those plant parts consist, and virtually all this carbon must come from the atmosphere. It enters the plant as carbon dioxide (CO_2) through the stomatal pores, mostly on leaf surfaces, and water exits by diffusion through these same pores as long as they are open. This is the dilemma faced by the plant: How to get as much CO_2 as possible from an atmosphere in which it is extremely dilute (about 0.03 percent by volume) and at the same time retain as much water as possible, water that must fill and keep turgid all the cells and provide the medium in which the CO_2 can be photosynthetically changed into the molecules of life. It is also the challenge faced by the agriculturalist: How to achieve a maximum crop yield with a minimum of applied irrigation water, an important natural resource.

Understanding the environmental factors and how they influence transpiration from and CO_2 absorption into a given leaf in the field at any given moment turns out to be a difficult assignment. This is because the factors interact with each other in so many ways. Not only do environmental factors influence the physical processes of evaporation and diffusion, but most (over 90 percent) of the transpired water and the CO_2 pass through the stomates on the leaf's surface, and their apertures are also strongly influenced by environment. Increasing leaf temperature, for example, promotes evaporation considerably and diffusion slightly—but may eventually cause

stomates to close. At dawn, stomates open in response to the increasing light intensity, and the light increases the temperature of the leaf. The increasing temperature means the air can hold more moisture, so evaporation is promoted, and perhaps stomatal aperture is affected. Wind brings more CO_2 and blows away the vapor, causing an increase in evaporation and CO_2 uptake. But if the leaf is warmed above air temperature by sunlight, wind will lower its temperature, causing a decrease in transpiration. When soil moisture becomes limiting, transpiration and CO_2 uptake are inhibited because stomates close. So if one were to visit the Greenville farm at any odd moment, hoping to make a statement about the extent of transpiration or CO_2 uptake, he would clearly have to be armed with a rather formidable arsenal of environmental measuring devices, not to mention an electronic pocket calculator.

3.1 Measurement of Transpiration

How could we measure transpiration at the Greenville farm to see if our developing models and calculations were valid? For the moment, we shall be more concerned with transpiration than with CO_2 uptake. It is perhaps more challenging, since CO_2 concentration in the atmosphere is fairly constant, but transpiration studies must consider evaporation as well as diffusion. Anyway, understanding transpiration will give us a good foundation for understanding CO_2 uptake.

In an elaborate method for measuring transpiration (Fig. 3-1), a tent of transparent plastic is placed over a number of plants, and humidities, temperatures, and carbon dioxide levels are measured for air entering and leaving the tent. From these data, the amount of transpiration and photosynthesis can be calculated. But the problem is obvious: How can we be sure that the environment surrounding the corn plants is not influenced by the tent? With elaborate instrumentation, it is possible to control temperatures, humidities, and gas concentrations within the tent, but this is no simple assignment for a basic plant physiology student laboratory. Thousands of dollars must be invested in such equipment.

A much simpler approach is to weigh at intervals on a sensitive balance a potted plant with its soil sealed against water loss. Stephen Hales did this 200 years ago, and it is

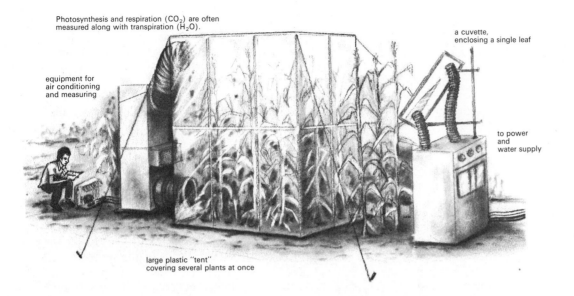

Photosynthesis and respiration (CO_2) are often measured along with transpiration (H_2O).

equipment for air conditioning and measuring

a cuvette, enclosing a single leaf

to power and water supply

large plastic "tent" covering several plants at once

Figure 3-1 Tent and cuvette methods for measuring transpiration in the field.

known as the **lysimeter method.** Since the amount of water used in plant growth is less than 1 percent of the final dry weight of the plant (note Hanks' figures), virtually all the change in weight can be ascribed to transpiration. The problem here is being sure that the potted plant is really representative of other plants in the field, after being moved from its location in the field to the balance and back, and with its roots confined. Hanks has considerably expanded this simple approach to arrive at figures that are fairly trustworthy. His lysimeters on the Greenville farm are large containers (several cubic meters) full of soil and buried so their top surface is level with the field's surface. The container is placed on a large plastic bag buried beneath it and filled with fluid (water-antifreeze) that extends into a standpipe above the surface (Fig. 3-2). The level of liquid in the pipe is a measure of the weight of the lysimeter, so it changes with the water content of the soil in the lysimeter, which depends only upon evaporation from the soil surface and transpiration from plants growing in the soil. Evaporation from soil and transpiration combined are called **evapotranspiration.** Lysimeters provide the most reliable field method for studying evapotranspiration, although they are obviously expensive, involved, and cannot be moved around on the spur of the moment. Though not universally available, lysimeters are widely used.

A convenient and thus widely used method is to remove a leaf from a plant in the field and put it immediately on a sensitive balance. The loss of weight during the first minute or two has often been used as an indication of transpiration. Since the very act of detaching the leaf from the plant significantly changes the rate of transpiration, this method should be avoided.

The potted plant method works nicely in the laboratory. Another laboratory approach is to detach a plant and insert the cut stem into a device that allows easy measurement of

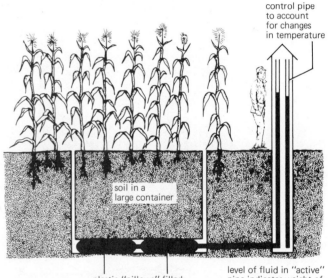

control pipe to account for changes in temperature

soil in a large container

plastic "pillows" filled with fluid (antifreeze)

level of fluid in "active" pipe indicates weight of the lysimeter

Figure 3-2 Diagram of a large field lysimeter, operating on a hydraulic principle.

the water absorbed by the plant (which just might differ from that transpired!). Sometimes the stem is attached to a burette, or water absorption might be measured as an air bubble moves through a capillary tube connected to the closed water reservoir from which the plant is absorbing water (a **potometer**). This can be useful when relative transpiration rates must be studied for brief time intervals, but of course no one is comfortable with the idea that a detached plant acts like a plant with its roots intact.

The **cuvette** (illustrated in field use in Fig. 3-1) is widely used in the laboratory for study of single leaves. As in the tent method, moisture or CO_2 entering and exiting is measured. In the laboratory, where purposes are special, cuvettes and potometers may be highly appropriate. Instead of trying to follow natural transpiration, laboratory studies may be aimed at understanding the effects of individual environmental factors upon transpiration. Factors can be standardized and measured under steady-state conditions.

3.2 The Paradox of Pores

Nature often proves to be more complex than our frequently simplistic approach to things might lead us to expect. Suppose we measure the evaporation rate from the surface of a bottle of water. Then suppose we cover half that surface with metal strips or a similar material and measure evaporation again. We would expect the second rate to be about half the first. Now let's cover all but about 1 percent of the surface. We might use a thin piece of foil with holes equal to about 1 percent

of the total area. Will we measure about 1 percent as much evaporation from the surface? Not if the holes have about the same size and spacing as the stomates found in the epidermis of a leaf. We will in fact measure about half as much evaporation (50 percent) as from the open surface.

How can this be? Why isn't evaporation directly proportional to area? It certainly seems paradoxical that stomatal openings on the leaf make up only about 1 percent of the surface area, whereas the leaf sometimes transpires half as much water as would evaporate from an equivalent area of wet filter paper. We resolve this apparent paradox by realizing that evaporation is a diffusion process from water surface to atmosphere and that the rate of diffusion, all other factors being equal, is strongly influenced by the steepness of the water potential or vapor pressure gradient. Water molecules evaporating from a free water surface will be part of a relatively dense column of molecules extending above the surface and not having a very steep concentration gradient from surface to atmosphere. Water molecules diffusing through a pore, however, can go in any direction within an imaginary hemisphere centered above the pore. In such a situation, the

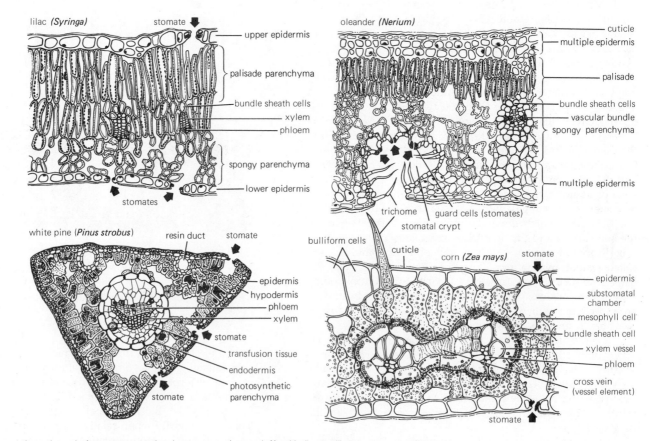

Figure 3-3 Cross sections through four representative leaves, one (upper left) with "normal" stomates, one (lower left) with slightly sunken stomates, one (lower right) a grass, and one (upper right) with stomates deeply sunken in a substomatal cavity. Note other details of differing leaf anatomy. Pine leaves do not have a palisade layer.

concentration gradient from pore to atmosphere will be relatively much steeper than above the free water surface. Diffusion through the pore will thus be much more rapid than from a free water surface, but then there is much less pore area.

Many studies have been made to determine the effects of pore size, shape, and distribution on diffusion rates. The stomates of typical plants prove to be nearly optimal for maximum gas or vapor diffusion. Thus, plants are ideally adapted for CO_2 absorption from the atmosphere—but also for loss of water by transpiration. The stomates can *close,* however, and they are adapted to close when CO_2 absorption is unnecessary because photosynthesis is not possible.

3.3 Stomatal Mechanics

Stomates come in a considerable variety. Figure 3-3 shows drawings of cross sections through four kinds of leaves and labels some of the important anatomical features. The waxy **cuticle** on the leaf's surface restricts diffusion so that most

water vapor and other gases must pass through the openings between the **guard cells.** Technically, the term **"stomate"*** refers only to this opening, but often the term is applied to the entire **stomatal apparatus,** including the guard cells. Adjacent to each guard cell are usually one or two other modified epidermal cells called **accessory** or **subsidiary cells.** Water evaporates inside the leaf from the **palisade parenchyma** and **spongy parenchyma** cell walls into the **intercellular spaces,** which are continuous with the outside air when the stomates are open. Carbon dioxide follows the reverse diffusional path into the leaf. Many of the cell walls of both palisade and spongy parenchyma cells (collectively called the **mesophyll** cells) are exposed to the internal leaf atmosphere, although this is seldom evident for palisade cells in drawings and photomicrographs of leaf cross sections. It becomes much more apparent in sections through the palisade and parallel to the leaf surface and is also strikingly apparent in scanning electron micrographs such as those of Fig. 3-4.

Stomates may occur only on the bottom of leaves but often

---

*Some anatomists prefer **stoma** (singular) and **stomata** (plural).

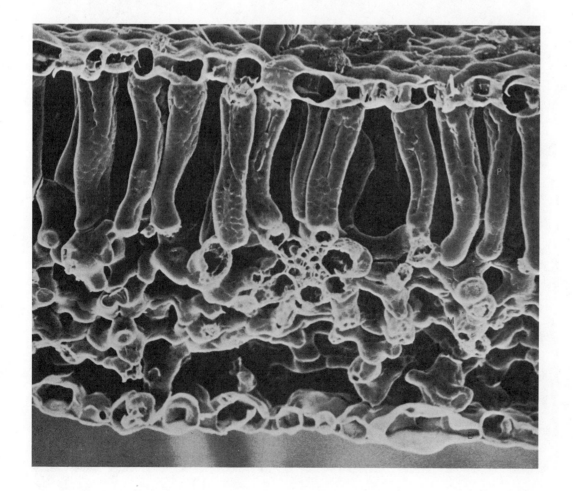

Figure 3-4 Transverse view of a mature broadbean leaf. The internal organization of cells in this leaf is characteristic of many plants, consisting of a layer of palisade cells (P) in the upper half of the leaf, spongy mesophyll cells (M) in the lower half, and bounded on both sides by the epidermis (E). Note the large air gaps between the palisade cells as well as the spongy mesophyll cells. Most of the surface of the cells is exposed to the air; the area of cell wall exposed is about ten times the surface area of the leaf. The proportion of air volume to cell volume in a leaf can vary from 10 percent to 80 percent between different types of plants. × 420. (From Troughton and Donaldson, 1972.)

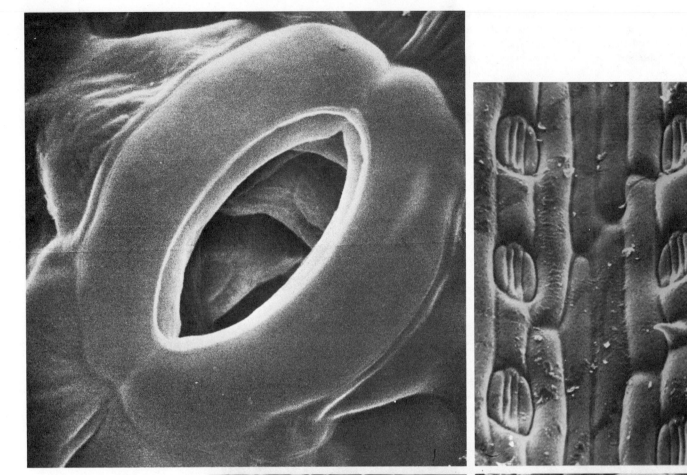
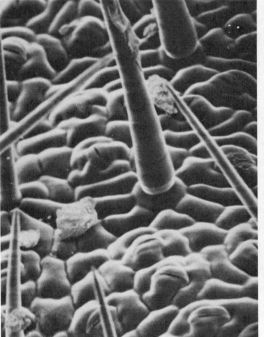
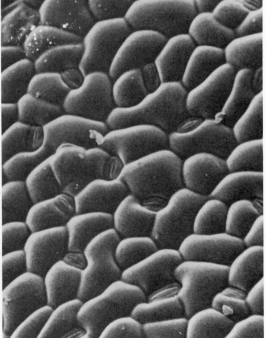

Figure 3-5 Upper left: Spongy mesophyll cells, as seen through a stomate on the lower leaf surface of cucumber. × 7900. (From Troughton and Donaldson, 1972.) Upper right: upper surface of wheat (*Triticum* sp.); note the characteristic monocot stomates. Lower right: upper surface of lambsquarter (*Chenopodium album*). Lower left: upper surface of velvet leaf. (*Limnocharis flava*); note the many hairs. (Scanning electron micrographs courtesy Dan Hess.)

are found on both top and bottom (perhaps more on the bottom). Lily pads have stomates only on top, but submerged plants have no stomates. Grasses usually have about equal numbers on both sides. Sometimes, as in the oleander or pine, the stomates occur in a substomatal crypt. Such sunken stomates may be an adaptation to reduce transpiration.

Figure 3-5 shows scanning electron micrographs of stomates from four species. Typical stomates of dicots consist of two kidney-shaped guard cells, but grass guard cells tend to be somewhat more elongate ("dumbbell" or "telephone receiver" shaped). Guard cells have a few chloroplasts, while their neighboring epidermal cells seldom do. In some species, at least, there seem to be fewer or no plasmodesmata connecting the cytoplasms of guard cells and accessory cells, although there may be plasmodesmata between guard cells and the mesophyll cells below.

Stomates open as the guard cells take up water and swell. At first, this seems paradoxical; one might imagine that the swelling cells would force the stomate to close rather than open. Stomates function the way they do because of special features in the submicroscopic anatomy of their cell walls. The cellulose microfibrils or micelles that make up the plant cell wall are arranged around the circumference of the slightly elongated guard cells. It is as though they radiated out from a point or region at the center of the stomate (Fig. 3-6). This arrangement of microfibrils is called **radial micellation.** The result is that when a guard cell expands by taking up water, it cannot increase much in diameter, since the microfibrils do not stretch much along their length, but the guard cell can increase somewhat in length. The guard cell thus curves during opening, partially because each guard cell is attached at each end to its partner.

It was noticed as long ago as 1856 that the guard cells of some species were slightly more thickened along the concave wall adjacent to the stomatal opening. Since then, most authors have suggested that the thickening was itself responsible for the opening when the guard cell takes up water. Donald

Figure 3-7 Two balloons representing a guard cell pair. Top photograph shows the balloons in their "relaxed" state with masking tape applied to represent both the "radial micellation" and the thickening along part of the ventral walls. Bottom photograph shows the balloon pair in an inflated state. Balloons were glued together at the ends with rubber cement before inflating (which weakened the rubber and caused eight pairs to burst when inflated before achieving success with the pair shown!).

E. Aylor, Jen-Yves Parlange, and A. D. Krikorian (1973) at the Connecticut Agricultural Experiment Station reinvestigated a 1938 discovery of H. Ziegenspeck. They showed both with rubber balloon models (Fig. 3-7) and through mathematical modeling that the radial micellation is much more important in stomate opening than the thickening of one wall, and that radial micellation was as important in the grass-type stomate as in the familiar dicot guard cells.

3.4 Stomatal Mechanisms

What is it that causes the guard cells to take up water, so that the stomates open? This classical problem of plant physiology has been discussed and studied for many decades. A prime suspect for the cause of stomatal opening would be some osmotic relationship that resulted in swelling of the guard cells. There are several possibilities: If the osmotic potential of the guard cell protoplast became more negative in relation to surrounding cells, water might be expected to move in by osmosis, causing an increase in pressure and a swelling of the guard cell. Another possibility would be for the guard cell wall to become less rigid (more elastic) so that it might stretch more easily, decreasing ψ_P and allowing uptake of

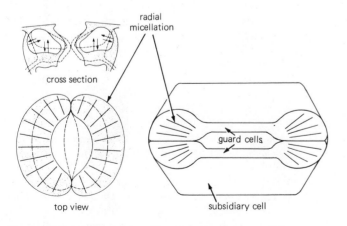

Figure 3-6 Schematic drawing of two stomates, showing radial micellation. Left, a dicot stomate; right, a monocot (grass) stomate.

water and hence swelling. Or the surrounding accessory cells might shrink, allowing guard cells to swell. Numerous measurements indicate that the osmotic potential of guard cells does become much more negative when stomates open, but no changes in elasticity or surrounding pressure have been found. For example, Humble and Raschke (1971) measured values of -19 bars for the osmotic potential of broadbean (*Vicia faba*) guard cells with closed stomates and −35 bars when the stomates were open. Since the guard cells about double in volume during opening, this increase in solutes is in spite of dilution. The increased solutes result in a transfer of water from the accessory cells to the guard cells. We can now restate the question more precisely: *What causes the change in osmotic potential in guard cells that results in stomatal opening?*

Environmental Effects on Stomates It is well known that many changes in environmental factors will influence stomatal aperture. Any theory purporting to explain guard cell action must account for the known environmental effects, so it is appropriate to review these effects before considering how osmotic potentials change in the guard cells.

Stomates of most plants open at sunrise and close in darkness, allowing for entry of the CO_2 needed for photosynthesis during the daytime. Opening generally requires about an hour, while closing is gradual throughout the afternoon (Fig. 3-8). Stomates close faster if plants are placed directly in darkness. Certain succulents native to hot, dry conditions (e.g., cacti, *Kalanchoe*, *Bryophyllum*) act in an opposite manner: They open their stomates at night, fix carbon dioxide into organic acids in the dark, and close their stomates during the day. This is a handy way to absorb CO_2 through open stomates but conserve water during the heat of the day. For opening in nonsucculent plants, there appears to be a minimum light intensity of about 1/1000 to 1/30 of full sunlight, just enough to cause some net photosynthesis and reduction in the CO_2 concentration in the leaf. Light intensity influences not only the rate of opening but also final aperture size, bright light causing a wider aperture.

Low concentrations of CO_2 cause stomates to open, and removal of CO_2 during photosynthesis by parenchyma and mesophyll cells is the main reason stomates of most species open in light. On the other hand, succulents fix CO_2 into organic acids at night, and this also causes stomatal opening in them. If CO_2-free air is swept across nonsucculent leaves even in darkness, their slightly open stomates open wider. Conversely, high CO_2 causes the stomates to close, and this occurs in the light as well as the dark. The CO_2 *inside* the leaf controls this response. When the stomates are *completely* closed, which is unusual, *external* CO_2-free air has no effect. There is good reason to believe that other environmental factors that influence photosynthesis or respiration have their effects upon stomatal opening and closing by acting indirectly through the internal CO_2 concentration. Such coupling of stomatal action to photosynthesis has obvious survival value.

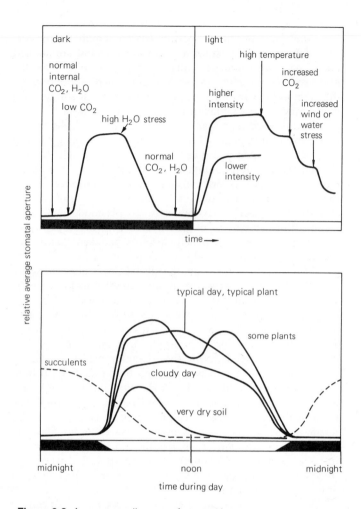

Figure 3-8 A summary diagram of stomatal response to several environmental conditions.

The water potential within a leaf also has a powerful control over stomatal opening and closing. As water potential decreases (water stress increases), the stomates close. This effect can predominate over low CO_2 levels and bright light. Its protective value during drought is obvious.

High temperatures (30 to 35 C) usually cause stomatal closing. This might be due to water stress, or a rise in respiration rate might cause an increase in CO_2 within the leaf. High CO_2 is probably the correct explanation for some species, since closing in response to high temperature can be prevented by flushing the leaf continuously with CO_2-free air. But in some plants, high temperatures cause stomatal opening instead of closing.

Sometimes stomates partially close when the leaf is exposed to gentle breezes, probably because additional CO_2 is brought close to stomates where it can diffuse in faster. Wind may also increase transpiration under some circumstances, leading to water stress and stomatal closing.

Control of Guard Cell Osmotic Potential Since stomates open because the guard cells take up water, and water uptake is due to more solute and hence a more negative osmotic potential, what is the solute and where does it come from? There have been many theories. It was soon apparent that the obvious ones wouldn't work. For example, it seemed reasonable that sugars produced in photosynthesis by the guard cell chloroplasts might account for the osmotic effect, but they are not produced nearly fast enough nor in large enough quantities. Indeed, guard cells may be incapable of significant photosynthesis. Furthermore, such a theory would not account for dark opening of nearly closed stomates in CO_2-free air.

A theory was developed in the 1920s and 1930s that remained the predominant one until recently. It suggested that photosynthesis in the guard cells increased their pH, which in turn resulted in a breakdown of starch to sugar and thus a more negative osmotic potential. The theory seemed to account for most of the observed facts but certainly not all of them.

Guard Cell Uptake of Potassium Ions Beginning in about 1968, we have experienced a revolution in our understanding of stomatal physiology. Actually, the initial discoveries were made earlier than that, but few plant physiologists paid much attention. Japanese scientists were the first to observe that when the stomates open, relatively large quantities of potassium ions (K^+) moved from the surrounding cells into the guard cells. An early paper by S. Imamura was published in 1943; M. Fujino published a more definitive paper in 1959 and expanded on his observations in 1968, the year when R. A.

Fischer and Theodore C. Hsiao in Davis, California, and other scientists around the world began to substantiate the observations that K^+ moved into guard cells when stomates opened. Representative data are shown in Fig. 3-9.

The quantities of K^+ building up in vacuoles of guard cells during stomatal opening are easily sufficient to account for the opening, assuming that the K^+ was associated with a suitable anion. Increases of up to 0.5 m in the K^+ concentration are observed, enough to decrease the osmotic potential by about 20 bars. Stomatal opening and K^+ movement into the guard cells are closely correlated in every case investigated. Light causes a buildup of K^+ in guard cells, as does CO_2-free air. When leaves are transferred to the dark, K^+ moves out of guard cells into the surrounding cells, and stomates close. These observations are true for at least 50 species sampled from all levels of the plant kingdom (e.g., mosses, ferns, conifers, monocots, dicots), and for the stomates occurring on various plant organs (e.g., moss sporophytes, leaves, stems, sepals). When strips of epidermal tissue are removed from the broadbean *(Vicia faba)*, most epidermal cells are broken, but the guard cells remain intact. When these strips are floated on solutions, stomates will not open unless the solutions contain K^+. So guard cells must obtain K^+ ions from accessory cells. Stomates will also close in response to the application of abscisic acid, a plant hormone (see later); application of this hormone causes loss of K^+ from the guard cells. So all evidences agree that the K^+ transport from accessory to guard cells is the cause of more negative osmotic potentials and hence stomatal opening, and that reverse transport is the cause of closing.

The situation is certainly not simple, however. Having learned about the K^+ flux, we are now in a position to ask the next question: *What is the mechanism of K^+ movement?* Consideration of this question forces us to realize that the science of plant physiology has become increasingly complex in recent years. It is no longer possible to consider such an apparently simple phenomenon as stomatal opening and closing only in terms of the thermodynamic concepts of water relations that have thus far been the topic of this text.

To begin with, stomatal opening is not as simple as an active metabolically controlled transport (pumping) of K^+ into the guard cells by energy provided by light. Such a mechanism would not account for dark opening in CO_2-free air, which is also accompanied by K^+ influx. It has also been observed that increasing the pH of guard cells (e.g., by exposing them to ammonium vapor) causes opening, and further, that stomates will sometimes open in response to low oxygen levels. Neither of these observations is accounted for by a simple light-driven uptake of K^+ in the guard cells.

In some species, Cl^- ions or other anions accompany K^+ into and out of guard cells, but Klaus Raschke and G. D. Humble (1973) observed that *no* anion accompanied the movement of K^+ into guard cells of *Vicia faba* leaves. Instead, as potassium ions went into the guard cells, an equivalent number of hydrogen ions came out. That is, the pH was increasing in the guard cells, because H^+ was being secreted (Fig. 3-9). Where does

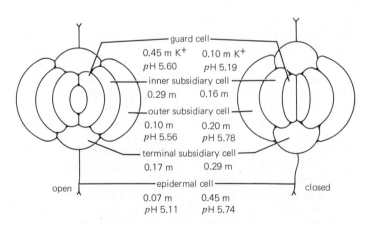

Figure 3-9 Quantitative changes in K^+ concentrations and pH values of the vacuoles in several cells making up the stomatal complex of *Commelina communis*. Values are given for the open (left) and closed conditions of the stomatal pore. (Data of Penny and Bowling, 1974 and 1975.)

the H⁺ come from? Apparently organic acids are synthesized in the guard cells in response to factors that cause stomatal opening. Probably malic acid is most responsible. The hydrogen ions come from the organic acids, and the K^+ that moves into the guard cells neutralizes the negative charges on the organic acid anions, thereby making the osmotic potential more negative. What is the origin of the organic acids? They are made from starch and perhaps sugars stored in the cell. It has long been known that starch disappears in the guard cells as stomates open, although careful measurements could never demonstrate that sugars appeared in the guard cells. The increasingly negative osmotic potential is due then to an uptake of KCl in some species and a build-up of potassium salts of organic acids, mostly dipotassium malate. But how? No satisfactory answer has been found, but theories are developing (see, for example, Levitt, 1974).

The Abscisic Acid Effect on Stomates Another observation of recent years has been almost as revolutionary as that of K^+ uptake. Abscisic acid (ABA) exercises a powerful control over guard cell action. Application of ABA at extremely low concentrations (e.g., $10^{-6} M$) causes stomates to close. It was observed several years ago that stomates do not close immediately after an applied water stress; rather, there is a lag of 10 to 15 minutes. Thus, the stomates do not lose water directly in response to water stress. They respond to a messenger instead, and that messenger is probably ABA. Furthermore, another naturally occurring substance similar to ABA, called *farnesol*, increases under drought conditions in a *Sorghum* species and could be more important than ABA.

Based upon the available evidence, it appears that there are at least two **feedback loops** that control stomatal opening and closing. When CO_2 decreases in the intercellular spaces and thus in the guard cells, K^+ moves into them and the stomates open, allowing CO_2 to diffuse in and completing the first loop. This meets the needs of photosynthesis. In non-succulents, it also leads to transpiration. If water stress de-

velops, ABA begins to appear in the water moving to the guard cells, so the stomates close, completing the second loop. The two loops interact: The degree of stomatal response to ABA depends upon CO_2 concentration in the guard cells, and response to CO_2 depends upon ABA. One feedback loop provides CO_2 for photosynthesis; the other protects against excessive water loss (Fig. 3-10). Stomates, as Raschke (1976) has said, have been ''delegated the task of providing food while preventing thirst.''

Light Quality and Stomatal Response Since stomates will open in the dark in response to CO_2-free air, low oxygen levels, and other factors, light is not essential for their opening. When light is effective, it is because photosynthesis in the leaf mesophyll cells lowers the CO_2 in the intercellular spaces, and the guard cells respond by taking up more K^+ ions. We would expect, then, that the same wavelengths of light (colors) that are effective in photosynthesis would be effective in opening stomates. At fairly high light intensities, this is exactly what is observed: Blue and red light are effective in both photosynthesis and stomatal opening. When the experiment is done quantitatively, however, it is seen that blue light is more effective relative to red light in causing stomatal opening than it is in photosynthesis. Furthermore, at low light intensities blue light may cause stomatal opening when red light has no effect at all. Thus, there seems to be an enhancement of stomatal opening by blue light acting directly on guard cells and in addition to the effects on photosynthesis in mesophyll cells and perhaps in guard cells. The blue light causes movement of K^+ but seems to be independent of effects of CO_2 or ABA. Further research will be required to understand this phenomenon.

3.5 The Role of Transpiration: Water Stress

What good is transpiration? Many philosophers of science would immediately object to such a question, labeling it as **teleological**. Such a question assumes that all things in the universe must have their purpose—an assumption that cannot be scientifically demonstrated. Yet the biologist avoids the question of teleology simply by rewording the question: *What is the selective advantage of transpiration?* Evolutionary theory implies that features of living organisms must either confer some advantage upon their possessors or be innocuous. Otherwise, if there is an alternative, a harmful feature will be eliminated by the process of natural selection. So what is the advantage of transpiration to a plant?

It is conceivable that the ''advantage'' might be a sort of ''win by default'': It is not only advantageous but completely essential to the life of a land plant to be able to absorb carbon dioxide from the atmosphere; it seems that the stomatal mechanism has evolved to permit this, and the consequence is transpiration. What about oxygen? Since stomates are typically

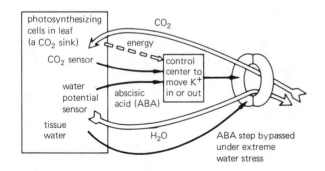

Figure 3-10 Two important feedback loops (one for CO_2, one for H_2O) that control stomatal action.

If the intercellular spaces of leaves are filled with air and the stomates open, why don't these spaces fill with rain water like a sponge? There are two reasons. First, the mesophyll cells are themselves also covered with a thin layer of cuticle that repels water but that is readily wetted by fat solvents such as gasoline or ether. These readily infiltrate the leaf when stomates are open and are even used as a crude test of stomatal aperture. Infiltrated areas of the leaf have a different, translucent, often darker appearance and can be readily recognized. Leaves can be **vacuum infiltrated** with water by placing them under water in a vacuum chamber, removing the air from the leaf by applying vacuum (reducing pressure), and then returning the leaf to atmospheric-pressure. Water penetrates the air spaces due to the pressure difference, and the leaf that floated before infiltration now sinks because of the added weight. Sometimes portions of a leaf may be naturally infiltrated during a particularly driving rain; these portions usually die unless the water is absorbed by the cells within an hour.

The second reason why water infiltration is prevented is that stomates almost without exception have a very thin sharp ridge or collar of cuticle lining the outer edge of the aperture. This is also hydrophobic and, because of the large wetting angle of water, prevents it from entering the aperture. Instead, the water forms a film over the aperture, which runs off or evaporates after the rain. Fat solvents, in contrast, have a small wetting angle and therefore readily penetrate the stomates. (Actually, water does penetrate the leaves of a few species. See the discussion of a special desert mesquite in Section 24.1.)

Herbicide specialists are aware of the difficulty in getting herbicides (weed killers) to penetrate the leaf. Herbicides remaining on the external leaf surface are obviously harmless. Better penetration is achieved in several ways: The herbicide compound itself may be modified to make it more soluble in the cuticle so that it can penetrate better. Ester formulations of 2,4-D are examples. Or surfactants, carriers, or wetting agents may be added to reduce the wetting angle and help the solution penetrate the stomates or cuticle.

Let's try another question: What are the chances that plant water loss may be reduced by spraying crops with **antitranspirants,** compounds such as ABA that close stomates and reduce transpiration, perhaps without reducing photosynthesis significantly? It might just be possible to achieve this, since transpiration is somewhat more sensitive to partial stomatal closure than is CO_2 diffusion (for reasons we shall not discuss here; see Raschke, 1976). Indeed, field trials with ABA have produced some promising results, but as a rule, photosynthesis and transpiration are affected about equally. Even this could be valuable to reduce the water requirements of ornamentals along freeways or during drought.

Another approach would be to cover plants with a substance that is much more permeable to CO_2 than water. So far, such a substance that is nontoxic to plants is unknown, but some materials (e.g., hexadecanol) have properties that are nearly suitable. Actually, the cuticle has similar properties, especially in some plants that continue to absorb considerable CO_2 even when stomates are closed. We'll keep searching.

closed at night, they obviously don't need to be open to absorb the oxygen used in respiration. This is probably because of the vast amounts of oxygen, compared to CO_2, in the atmosphere. During the day, when leaves are photosynthesizing and stomates are open, oxygen diffuses out through them. But their real reason for being is CO_2 absorption.

It can be argued that transpiration is neither essential nor of any advantage (other than as a consequence of the need for CO_2 uptake), since many plants can be grown through their life cycles in atmospheres of 100 percent relative humidity where transpiration is greatly reduced. Think of plants growing in a closed terrarium. Indeed, it is a common observation that many plants grow better in atmospheres of high relative humidity (see boxed essay on p. 360). Certain alpine land plants (e.g., *Caltha leptosepala*) will even grow for days to weeks when completely submerged in water.

Nevertheless, careful investigation and thought have revealed several situations in which transpiration *per se* does seem to be of some benefit to the plant. A necessity may be turned into an advantage. In most of these cases, it is possible for the plant to grow without transpiration, but when transpiration occurs, it seems to confer some benefit.

Mineral Transport Minerals absorbed into the roots typically move up through the plant in the transpiration stream. But is the transpiration stream *essential* for this movement? As it turns out, there is a sort of circulation in the plant (see Chapter 7). Water moves from assimilating organs to utilizing organs in the phloem tissue, and even when there is no transpiration, this water will return through the xylem tissue. Such a circulation has been demonstrated with radioactive tracers. Thus, transpiration is not essential for movement of minerals within the plant. Actually, the rate at which minerals arrive in the leaves will be a function only of the rate of movement into the xylem tissue, providing there is any xylem movement at all. The rate at which goods are delivered by an endless belt is a function only of the rate of loading.

Nevertheless, when transpiration occurs, it may aid mineral absorption from the soil and transport in the plant. In one study, for example, tomato plants were grown in a greenhouse with high humidity. Transpiration was reduced considerably, although not completely, since leaves in sunlight remain somewhat warmer than the air. Plants exhibited some calcium deficiency in the young leaves, as though calcium transport required an active transpiration stream.

Optimum Turgidity: The Physiology of Water Deficits If plants do not grow as well at 100 percent relative humidity as they do when some transpiration is allowed to occur, as is sometimes reported, it could be because cells function best when there is some water deficit. That is, there may be an "optimum turgidity" for cells, so that the various plant functions are slower or less efficient both above and below this level. If transpiration is not allowed to occur, plants will become overly turgid and thus will not grow as well as when there is some water stress.

The problem of too much water, more than enough for optimum turgidity, is not the problem that concerns plant physiologists and agriculturists. The problem is not enough water: **water stress** or a negative plant-water potential. Cell expansion requires the turgor pressure established in cells by osmosis (Chapters 15 and 24), so cell expansion—growth—is the first plant response to be inhibited by increasing water stress. Tree ring studies provide an apt example: In years when there is ample moisture, cells are large and tree rings wide; in drier years, both are narrow. Furthermore, if one considers overall productivity of the world's ecosystems, it is soon apparent that the overriding environmental factor is the availability of water: the more water, the greater the productivity.

As water stress increases, many plant processes are affected. Synthesis of protein and of cell walls is quite sensitive, and at higher stress levels nitrogen metabolism and other processes are reduced. We have seen that stomates close with increasing water stress, most likely in response to increasing ABA, so photosynthesis is soon inhibited. Finally, at fairly high stress levels, the amino acid proline begins to accumulate, and in some cases sugar accumulates. These compounds decrease osmotic potential and may provide storage for reduced carbon and nitrogen during stress (see Chapter 24).

3.6 The Role of Transpiration: Energy Exchange

For years, plant physiologists argued about whether transpiration was necessary to cool a leaf being warmed by the sun. On one hand, transpiration is certainly a cooling process; on the other, it was argued, if transpiration does not cool the leaf, other physical processes will—although in the absence of transpiration, it was conceded, leaves might be a few degrees warmer than otherwise. Growth of plants in atmospheres of 100 percent relative humidity where transpiration is considerably reduced were cited to support this view. As we have come to understand more about how the leaf exchanges energy with its environment, the argument has lost much of its impact. Truly, if transpiration doesn't remove the heat from a leaf, other processes will, but emphasizing this idea tends to obscure the important fact that in nature transpiration often plays an extremely important role in cooling the leaf.

Evaporation of water is a powerful cooling process. Remember the Maxwell distribution of molecular velocities (Fig. 1-1). It is the molecules with high velocities that evaporate, and as they leave the liquid, the average velocity of the remaining molecules is lower, which is the same as saying that the liquid is cooler. Greenhouses in dry climates are often cooled by evaporative cooling: Air is drawn through a wet fibrous pad. Some 540 calories are required to convert 1 g of water at the boiling point into vapor. When 1 g of water at 20 C evaporates, it absorbs 586 cal from its environment; at 30 C, the **latent heat of vaporization** is 580 cal/g. Plants evaporate tremendous amounts of water into their environments, and each gram of water so transpired absorbs about 580 cal from the leaf and its environment.

Sometimes transpiration is the only important means of dissipating heat in the environment. Consider the large fan-shaped leaf of a native palm tree (*Washingtonia fillifera*) growing at a southern California oasis. Such a leaf may be cooler than the surrounding air, even though it is in full sunlight. Since it is cooler than the surrounding air, it is absorbing some heat from the air, and it is absorbing more radiant energy from sunlight than it is radiating into its environment. It is cooler than air only because it is evaporating large quantities of water.

Investigations of the energy exchange between a plant and its environment provide an interesting example of plant biophysics. In the remainder of this chapter we shall consider some of the principles involved, without applying mathematics. Yet the mathematics are not difficult (see, for example, Gates, 1968 and 1971).

Leaf Temperature Consider the various factors that influence the temperature of a leaf: Transpiration cools the leaf, as we have seen. Condensation of moisture or ice on the leaf (dew or frost) releases the **latent heat of condensation** to the environment, and some of this heat is absorbed by the leaf. Incoming radiation absorbed by the leaf warms the leaf. The leaf is itself radiating energy to its environment, however, and this is a cooling process. If the leaf is at a different temperature from the surrounding air, it will exchange heat directly, first by **conduction** (in which the energies of molecules on the leaf's surface are exchanged directly with the contacting molecules of air) and then by **convection** (in which a quantity of air that has been warmed expands, becomes lighter, and thereby rises). In our discussion we shall arbitrarily refer to the combination of conduction and convection simply as convection.

If leaf temperature is changing, as is usually the case, the leaf is storing or losing heat. If a thin leaf stores a given amount of heat, its temperature rises rapidly; the same amount stored in a cactus would raise the temperature much less. But we shall consider only a leaf in a steady state of equilibrium with its environment; that is, at constant temperature. Some of the light energy absorbed by the leaf is converted to chemical energy in the process of photosynthesis. This is usually only a small percent of total light energy, however, so we shall ignore it also. Some energy is being produced in respiration and other metabolic processes; this is also small enough to

Figure 3-11 Plants showing high reflectivity in the infrared portion of the spectrum. (**a**) Photograph taken with ordinary panchromatic film, which gives gray tones similar to the color intensities experienced by the human eye. (**b**) Photograph taken on infrared-sensitive film through a dark red filter that excludes most of the visible spectrum. Note the brilliant white appearance of the vegetation and compare in particular the clumps of grass on the right.

ignore. Under steady-state conditions, three factors primarily influence leaf temperature: radiation, convection, and transpiration. Each of these is worthy of our consideration.

Radiation Special Topic 3.1 discusses the principles of radiant energy, including those that apply in heat transfer studies. The thing to remember from the standpoint of leaf temperature is that it is the **net radiation** that is important. The leaf absorbs light and invisible infrared radiation from its surroundings but also radiates infrared energy. If the leaf is absorbing more radiant energy than it radiates, then the excess will have to be dissipated by convection and/or transpiration. At night, the leaf may be radiating more energy than is absorbed. In this case, it will probably absorb heat from the air and possibly also from water condensing as dew on its surface. There are three important things to keep in mind when discussing the net radiation of a plant leaf:

First is the absorption spectrum of the leaf. Of the energy incident on a leaf, some will be transmitted, some reflected, and some absorbed. The energy absorbed will depend upon its wavelength. Leaves irradiated with white light absorb most of the blue and red wavelengths and a considerable

portion of the green. Much of the green is reflected and transmitted, however, which is why leaves appear green. Leaves absorb very little of the near-infrared part of the spectrum; most is either transmitted or reflected. Thus, if our eyes could see in this part of the spectrum, vegetation would appear very bright to us. We can get this impression by photographing vegetation with infrared sensitive film (Fig. 3-11). Virtually all the far-infrared part of the spectrum is absorbed. If our eyes were sensitive only to that part of the spectrum, vegetation would appear as black as black velvet. Figure 3-12 illustrates an absorption spectrum of a leaf, presenting these ideas quantitatively.

Second, the quality of light falling on a leaf may vary considerably. Figure 3.1-3, page 385, illustrates the emission spectra for several light sources. The sun and the filament of an incandescent lamp emit light because of their high temperature. The higher the temperature, the more the peak of the emission spectrum is shifted toward the blue (see *Wien's law* in Special Topic 3.1). The temperature of the sun's surface is considerably above that of an incandescent filament in a light bulb, and thus sunlight is richer in blue and green wavelengths than is light from an incandescent lamp.

The sun's light is further modified by passing through

the atmosphere. Much of the ultraviolet is removed, and radiant energy is also absorbed by the atmosphere at several discrete wavelengths in the infrared part of the spectrum. Most of the ultraviolet is absorbed by ozone in the upper atmosphere, and the infrared absorption bands are caused primarily by water and carbon dioxide.

Fluorescent light, often used in plant growth chambers, is produced by fluorescence rather than incandescence. It is usually rich in blue; by using suitable phosphors, it can be enriched in the red part of the spectrum as well. Most fluorescent tubes give off virtually no infrared or even far-red radiation [wavelengths beyond 700 nanometers (nm)—still visible]. Special fluorescent tubes have been developed for plant growth. These are especially rich in blue and red with smaller quantities of green light in their spectra; that is, they are purple. Plants illuminated with these tubes appear dark in color, since the green wavelengths normally reflected by the plants are at relatively low intensity. The idea of these lamps is to match their spectra with the absorption spectrum of a typical leaf. This makes good sense, yet several studies comparing the growth of plants under these tubes with growth under ordinary fluorescent tubes (or, more commonly, a mixture of ordinary fluorescent and incandescent bulbs) have shown little if any advantage of the special tubes.

Sodium and mercury vapor lamps now used widely in street lighting operate on still another principle. A current passing through their hot vapor causes them to emit light at highly specific wavelength: orange for the sodium vapor and blue and green for the mercury vapor lamps. They have sometimes been used in plant growth studies, but results are not impressive. Their spectra simply don't match the leaf's absorption spectrum.

All objects at temperatures above the absolute zero emit radiation. For objects at ordinary temperatures, most of this is in the far-infrared part of the spectrum, so plants are receiving this radiation from all their surroundings, including air molecules in the atmosphere. The quantity may be a sizable portion (e.g., 50 percent) of the total radiant environment, including sunlight.

The light absorbed by a plant will be a function of the leaf's absorption spectrum and the emission spectrum of the light source illuminating the plant. The actual percentage of light absorbed will, therefore, vary considerably, since both absorption and emission spectra vary. On the order of 44 to 88 percent may be absorbed in common situations. Absorption is high when plants are illuminated with fluorescent light, since the leaf absorbs strongly in most of the wavelengths emitted by the fluorescent tube. Absorption is somewhat less when plants are illuminated with incandescent light, which proves to be rich in the near-infrared part of the spectrum that is most poorly absorbed by plants.

Third, the radiant energy emitted by the plant is emitted in the far-infrared part of the spectrum. The quantity of energy emitted can be calculated by application of the **Stefan-Boltzmann law** (Special Topic 3.1). This law states that the energy emitted is a function of the fourth power of the absolute temperature. Thus, as a leaf's temperature increases in sunlight, radiant energy emitted by it increases sharply, so it heats up less than would otherwise be the case. Nevertheless, on the absolute scale, the normal temperature range of plants (from about 273 K to 310 K) is not very large. Energy emitted varies by only about 50 percent, but this can be quite significant. Even when a plant is being illuminated by sunlight and is also receiving far-infrared radiation from all the surrounding objects in its environment (e.g., the sky, clouds, trees, rocks, and soil), the radiant energy emitted from the leaf is usually more than 50 percent of the radiant energy absorbed.

Convection Heat is conducted-convected from the leaf to the atmosphere in response to a temperature difference between leaf and atmosphere. If the leaf is warmer, heat will move from the leaf to the atmosphere. *The temperature difference is the driving force*; the greater the difference, the greater the driving force for convection.

With a given temperature difference, the rate of convective heat transfer will depend upon the resistance to convection. The situation is exactly analogous to Ohm's law: **The rate is proportional to the driving force and inversely proportional to the resistance.**

Understanding the resistance to convective heat transfer is greatly aided by the concept of the **boundary layer** (sometimes called the **unstirred layer**). This is the transfer zone of fluid (gas or liquid) in contact with an object (in this case, the leaf) in which the temperature, vapor pressure, or velocity of the fluid is influenced by the object (Fig. 3-13). Beyond the

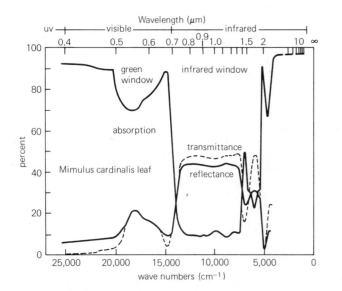

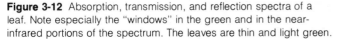

Figure 3-12 Absorption, transmission, and reflection spectra of a leaf. Note especially the "windows" in the green and in the near-infrared portions of the spectrum. The leaves are thin and light green.

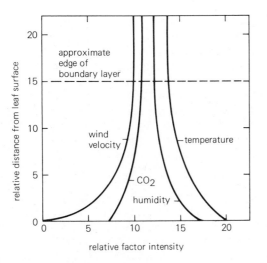

Figure 3-13 Relative factor intensities as a function of distance from a surface, illustrating the principle of the boundary layer.

boundary layer, there is no influence of the object upon the medium. For a given temperature difference between the leaf and the air beyond the boundary layer, the convective transfer of heat will be more rapid when the boundary layer is thin and slower when it is thicker. That is, the thickness of the boundary layer is inversely proportional to the steepness of the temperature gradient, so a thicker boundary layer indicates greater resistance to convective heat transfer.

Usually there is at least some air movement around a leaf, from a slight breeze to a high-velocity wind. The more rapid the air movement, the thinner the boundary layer (the steeper the temperature gradient). The boundary layer is also thinnest next to a leaf's leading edge (the edge that faces into the wind). If the leaf's surface is parallel to the direction of wind movement, then the boundary layer becomes thicker from the leading edge toward the center of the leaf. Small leaves, especially conifer needles, will have the thinnest boundary layers, so that convection will be most effective with such leaves. Large leaves, such as those of the desert fan palms (*Washingtonia Filifera*) will have the thickest boundary layers. To summarize: The boundary layer will be thinnest (least resistance to convective heat transfer) for small leaves and high wind velocities. Convective heat transfer will be most efficient under such conditions, so smaller leaves will have temperatures closer to air temperature than larger leaves, especially if there is a wind.

Transpiration In some ways, transpiration is closely analogous to convective heat transfer; in other ways, it is different. *The driving force for transpiration is the difference in vapor pressure of water within the leaf and in the atmosphere beyond the boundary layer.* The resistance to transpiration is partially the resistance

of the boundary layer. To this extent, convection and transpiration are closely analogous. There is an additional resistance to transpiration: the stomates. If the stomates are closed or nearly closed, resistance to transpiration can be very high; if they are open, resistance is relatively low. There may be other resistances within the leaves besides that of the stomates, but others usually remain fairly constant. It is conceivable that the resistance of the cuticle to passage of water depends upon atmospheric humidity, temperature, and perhaps light or other factors, but it is always relatively high, so this is seldom considered. The important thing to realize is that *there is always some leaf resistance;* that is, the leaf is never simply like a piece of wet blotter paper. It is also important to know that the leaf resistance can vary over a wide range as environmental factors influence stomatal apertures.

The vapor pressure gradient is influenced primarily by two factors: humidity and temperature. Usually as a first approximation, we imagine that the relative humidity in the internal spaces of the leaf approaches 100 percent. Actually, it will be somewhat less, because at equilibrium the water potential of the internal leaf atmosphere must be equal to the water potential of the surfaces from which the water evaporates, and this will probably be −5 to −30 bars. (If equilibrium is not achieved, water potential of the leaf atmosphere will be even lower.) Nevertheless, this will be equivalent to a relative humidity of about 98 percent or higher (see calculations following equation 4, p. 16). Such high relative humidities will seldom be present in the atmosphere beyond the boundary layer, so *even if the leaf is at exactly the same temperature as the atmosphere beyond the boundary layer, there will, under most conditions, be a vapor pressure gradient with higher values inside the leaf compared to the atmosphere.*

Temperature can greatly accentuate the vapor pressure gradient, since vapor pressure is strongly a function of temperature (Fig. 3-14). The vapor pressure approximately doubles for every 10 C (or 20 F) increase in temperature. Thus, air at 100 percent relative humidity at 21 C will have 50 percent relative humidity if heated to about 33 C. Warm air can hold more water than cold air. An examination of Fig. 3-14 shows, for example, that air at 20 C and an atmospheric humidity of 10 percent will establish a vapor pressure difference of about 21 millibars (mbars) between the leaf and the air, provided both are at the same temperature, and the atmosphere within the leaf approaches 100 percent relative humidity. (At 20 C, saturated vapor pressure is about 23 mbars; 10 percent of this about 2 mbars.) If the leaf is at 30 C, however, and the atmospheric humidity is 90 percent (at 20 C), there will still be a vapor pressure difference of about 22 mbars (At 30 C, vapor pressure is 43 mbars, 90 percent of 23 mbars is 21 mbars, which subtracted from 43 mbars leaves a gradient of 22 mbars.) Thus, if the leaf is considerably warmer than the air (a common phenomenon in sunlight), transpiration may occur into an atmosphere having a relative humidity of 100 percent. As the vapor goes beyond the boundary layer, it may condense—as when a forest steams under the sun after a rainstorm.

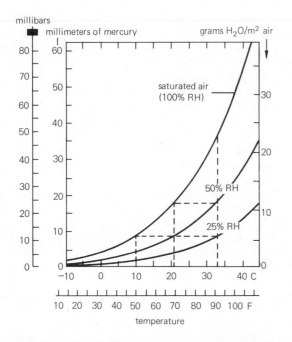

Figure 3-14 Relationship between moisture in the atmosphere (partial pressure or water vapor) and temperature for saturated air and for air at 50 percent and 25 percent relative humidity. Dashed lines indicate, for example, that air at 10 C, 100 percent RH has the same amount of moisture as air at 21 C, 50 percent RH, and air at 33 C, 25 percent RH.

3.7 Plant Energy Exchange in Ecosystems

Application of heat transfer principles in field situations has provided considerable insight into the functioning of plant communities. To illustrate this, we shall consider a few of the principles discussed in the previous sections as they apply to plants in the desert, the alpine tundra, and some other situations.

The Desert In the desert, high temperatures are combined with high radiation, low humidity, and little available water. It is to the plant's advantage to conserve moisture and to maintain a relatively low leaf temperature. Many desert plants, perhaps most, have small leaves, providing a thin boundary layer with consequent efficient convective heat transfer. Their temperature is closely coupled to air temperature, so at least they are not heated well above air temperature by sunlight. But such a situation might also produce a high rate of evaporation unless leaf resistance is high. Some desert plants have sunken stomates and other mechanisms that result in high leaf resistance and retard transpiration. Some are also light gray in color, reflecting much of the sun's radiation. We saw that desert succulents have the ability to close their stomates

during the day and to fix CO_2 into organic acids at night, thereby conserving water.

Desert air temperatures are often extreme, so the ideal solution might be air conditioning by evaporative cooling—a technique applied widely by man living in desert cities. A few desert species with access to unlimited water (their roots reach the water table) achieve such evaporative cooling. Typically, they have large leaves, as in the palm growing in an oasis, and their high rates of transpiration result in leaf temperatures several degrees below air temperature. This is possible because of the low ambient humidity, which provides an ample vapor pressure difference even when leaf temperature is relatively low, and because the large leaf has a thick boundary layer, resulting in a low rate of convective heat uptake by the leaf from the surrounding, warmer air. Leaf temperatures of large cocklebur (*Xanthium strumarium*) leaves (not native to the desert) growing along a ditch bank in southern Arizona were as much as 9 to 11 C cooler than the surrounding air, which was 36 C under a clear, midsummer sky. On the other hand, leaf temperatures of this same species measured in Oregon on a cool summer morning proved to be several degrees *warmer* than the surrounding air. Wind tunnel studies showed that this plant does exercise considerable control over its leaf temperature by controlling transpiration, being warmer than cold air and cooler than hot air (Fig. 3-15).

The Alpine Tundra Alpine plants are faced with the opposite situation. Cool, fairly humid, gusty air combines with solar radiation intensities that may be extremely high. Water is seldom limiting, but evaporative cooling is no advantage in a situation where ambient temperatures are often well below the range considered optimum for metabolic processes. Several hundred measurements of leaf temperatures in the alpine

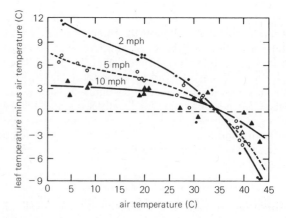

Figure 3-15 The difference between leaf and air temperatures as a function of air temperature for three wind velocities. Light intensity 1.3 cal/cm²/min. Curves are third-order polynomials drawn by computer to match the data. (Data for both figures from Drake, 1967.)

The Biophysical Ecologist

David M. Gates

F.B.S. 1975

What might drive a man to give up one successful career and begin another one? Or to remold one career into another? David M. Gates, Director of the Biological Station (Pellston), University of Michigan, Ann Arbor, recalls such an experience in the following essay.

I had the remarkable good fortune to have as a father the distinguished ecologist and botanical systematist, Frank C. Gates. Equally important were the summers spent in one of the world's most interesting scientific communities, the Biological Station of the University of Michigan on Douglas Lake. It was here that I lived for eight weeks every summer among some of the world's great field biologists, including Henry Allan Gleason, Paul S. Welch, Herbert B. Hungerford, William W. Cort, and many others.

I had a serious interest in biology, particularly entomology, as a youngster and listened intently to lectures and discussions by these experts. However, I often wondered what I could do on the frontier of biological knowledge that would be significant beyond their accomplishments. Despite my interest in the biosphere, I began to wonder more about how the physical world worked—gravity, electrostatics, radioactivity, atomic structure, quantum theory—and the physics books piled up about me. I majored in physics at the University of Michigan and had the good fortune to have superb, inspiring teachers. My Ph.D. thesis in 1948 was on the molecular spectroscopy of hydrocarbons in the far infrared.

When I went to the University of Denver to teach physics in 1947, I decided to turn my attention to the use of spectroscopy for the identification of atmospheric molecular constituents. The choice to study the atmosphere came from a conscious recognition on my part that it was a significant part of the biosphere. The molecular composition of the troposphere and stratosphere strongly affects the heat balance of our planet. Sunlight is reflected, transmitted, and absorbed by atmosphere, clouds, and ground, and radiant heat emitted toward space is intercepted by these. On the surface is an abundance of life in sea and soil, but a life made possible by the flow of radiant energy. The notion that the spectral distribution of infrared radiation absorbed by plant leaves was important and had not been adequately measured led my physics student, W. Tantraporn, and me to apply our newly acquired infrared spectrophotometer to this problem in 1950. The paper published was a landmark and an early signal that my appetite for biological phenomena had been whetted. But it was nearly a decade before I was to return to biological science with a vengeance. These were years in which the biological ideas and facts from earlier times merged in my mind with the research experiences from atmospheric physics. They fermented, crystallized, and grew into a vision of analytical ecology.

Suddenly I could see clearly that what the science of ecology needed was a sound theoretical foundation based on physical principles. How were plants and animals interacting with their environments? How did temperature, radiation, wind, and water affect a plant or animal? It was the flow of energy and the energy status of a plant or animal that was an important determinant in how it functioned and survived. These ideas were launched in a lecture at the University of Michigan Biological Station semicentennial celebration in 1959 and a series at the University of Minnesota in 1960 initiated by Professor Donald Lawrence. These lectures were published as a small book entitled "Energy Exchange in the Biosphere," which became popular. During these years, while working in the Boulder, Colorado, laboratories of the National Bureau of Standards, I devoted every spare moment during nights, weekends, and holidays to experiments concerning energy exchange and transpiration between plant leaves and their environment.

About this time, I traveled on a lecture tour on the west coast for National Sigma Xi and RESA, where I lectured on this subject at thirty universities and corporations. Midway in the tour, I decided that this was the time to secure a professorship in botany somewhere in the country. Being a physicist working in atmospheric physics, however, was not the easiest route to a professorship of biology. Dr. John Marr, Professor of Biology at the University of Colorado, had expressed confidence in me as early as 1958 and negotiated an adjunct appointment for me in the Institute of Arctic and Alpine Research, where I then began a close collaboration with Professor Frank Kreith of the engineering faculty. We built a wind tunnel and began a series of experiments to support our theoretical work. This collaboration resulted in a series of papers concerning the energy budgets of plants and animals. I was fortunate to have at that time some excellent students, many of whom have gone on to distinction in ecology. In 1964, I became Professor of Natural History at the University of Colorado, the first full-time academic appointment I ever held with an opportunity to do biology.

In 1965, I became Director of the Missouri Botanical Garden in St. Louis and Professor of Botany at Washington University. With generous support from the Ford Foundation, we built up our program there in biophysical ecology. Through a combination of theoretical work, laboratory measurements, and field observations, we were able to advance greatly our understanding of productivity by plant leaves and to initiate our major work with the energetics of animals. The model for photosynthesis by a whole leaf involving energy exchange, gas exchange, and chemical kinetics, which was accomplished there, is a cornerstone for much of our future research.

Finally, in 1971, my life's cycle became complete when I returned to Douglas Lake and the University of Michigan as Director of the Biological Station, where I had lived as a boy. The work on energy exchange continues.

My travels have taken me to most continents of the world and to many extreme environments, including Antarctica, the arctic, high mountains, and many deserts. Everywhere I go, I observe plants and animals and interpret their behavior in terms of the fundamental principles that we have established in biophysical ecology.

tundra by Salisbury and George Spomer showed that when the sun is irradiating the leaves, their temperatures may often be as high as 30 C, which may be as much as 20 C above the temperature of the surrounding air. Alpine plants also typically have small or finely divided leaves, but the plants grow in a layer only about 10 cm above the ground, where wind velocities are greatly reduced. Many have a rosette form that results in a thick boundary layer determined by the entire plant instead of by individual leaves. High leaf temperatures clearly indicate that transpiration does not provide much cooling power under these conditions. The reasons why are not always apparent, so integrated field and laboratory research would be in order.

Other Situations: Effects of Wind The ecological problems of transpiration and heat transfer are somewhat less challenging in environments less extreme than those of the deserts and the alpine tundra. Nevertheless, careful study of plant heat transfer properties should provide some interesting insights. The extreme variety of leaf shapes even in moderate environments is certainly suggestive.

Data obtained during the first half of this century often seemed quite contradictory. Some indicated that wind increased transpiration (as it always increases evaporation from a free surface); others, that wind decreased transpiration. When radiation loads are relatively low and leaf resistance is also low, transpiration will certainly be increased by wind. If leaf temperature is below air temperature, increasing wind velocity always tends to increase transpiration. Yet it is now clear that transpiration may indeed be decreased by wind when the radiation heat load is high, particularly if leaf resistance is also high. Under such conditions, the leaf temperature may be far above the air temperature, accounting for a high transpiration rate. The wind cools the leaf, and this cooling effect is more important in reducing transpiration than the wind is in increasing evaporation.

It is now time to reinvestigate ecological problems of transpiration, taking into account the principles of heat transfer.

4

The Ascent of Sap

According to the *Guinness Book of World Records* (1976 edition), the tallest tree in the world is the Howard Libbey *Sequoia* in Redwood Creek Grove, Humboldt County, California. The height in 1970 was estimated to be 111.6 meters (366.2 feet). The book goes on to say that, in 1872, a *Eucalyptus regnans,* found in Victoria, Australia, measured 133 m from its roots to the point where the trunk had been broken off by its fall—at which point the trunk's diameter was nearly 1 m, so the overall height may have been over 150 m. Another specimen was reported to be 141 m in 1868, and a Douglas fir felled in British Columbia in 1940 was claimed to be 127 m. All these claims for trees taller than the Howard Libbey tree remain unsubstantiated. In any case, water has probably moved in some of the tallest trees from the roots to the uppermost leaves a vertical distance of well over 120 m. What, asks the plant physiologist, is the mechanism of this movement?

4.1 The Problem

Although we tend to take tall trees for granted, the more we cogitate on how water moves uphill at rapid rates to their tops, the more we appreciate the challenge of the problem. A suction pump can lift water only to the **barometric height,** which is that height supported by atmospheric pressure from below (about 10.3 m or 34 ft at one atmosphere, the normal air pressure at sea level). If a long horizontal pipe is filled with water, sealed at one end, and then placed in an upright position with the open end down and in water, atmospheric pressure will support the water column to 10.3 m. At this height the pressure equals zero, and above this height the water will turn to vapor; that is, at zero pressure water will "normally" **vacuum boil** even at 0 C when the pressure is reduced in a column of water so that vapor forms or air bubbles appear (the air coming out of solution). The column is said to **cavitate** (**cavitation** occurs). A laboratory barometer (Fig. 4-1) contains mercury instead of water. One atmosphere of pressure supports a column of mercury 760 mm high; 1 bar supports 10.2 m of water and 750 mm of mercury.

To raise water from the ground level to the top of the Howard Libbey tree, a pressure at the base of 10.83 atm would be required, plus additional pressure to overcome resistance in the water's pathway and to maintain a flow. If overcoming the resistance required a pressure about equal to that required to raise the water, then a total of about 22 atm (11 and 11) would be necessary. To raise water to the top of the tallest tree that ever lived (say, 150 m), a total of about 30 atm or bars (15 and 15) might be required. Clearly, water is not pushed to the top of tall trees by the normal atmospheric pressure.

Root pressures have been observed in several species. If the stem is cut from a grapevine or other plant, for example, and a tube with a mercury-manometer is attached, it can be

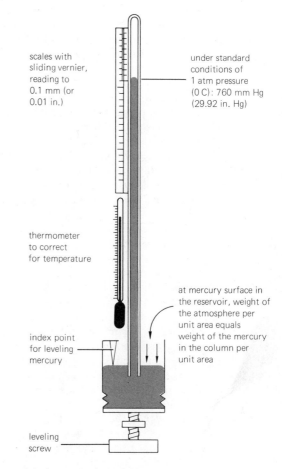

scales with sliding vernier, reading to 0.1 mm (or 0.01 in.)

under standard conditions of 1 atm pressure (0 C): 760 mm Hg (29.92 in. Hg)

thermometer to correct for temperature

at mercury surface in the reservoir, weight of the atmosphere per unit area equals weight of the mercury in the column per unit area

index point for leveling mercury

leveling screw

Figure 4-1 A mercury barometer.

seen that water may indeed be forced from roots under considerable pressure. Pressures of 5 or 6 bars have been recorded, although in most species values do not exceed 1 bar. Root pressures appear in most plants but only when ample moisture is present in the soil and when humidities are high; that is, when transpiration is exceptionally low. It is possible to see droplets of water exuded by the tips of grass leaves, a phenomenon called **guttation.** When plants are exposed to relatively dry atmospheres and low soil moisture conditions, root pressures don't occur because water in their stems is under *tension* rather than *pressure.* Root pressures are not seen in conifer trees (including the sequoia and the Douglas fir) under any conditions, although slight pressures have been observed in excised conifer roots. Furthermore, rates of movement by root pressure are too slow to account for water movement in trees. So we must reject root pressure as the means of moving water to the top of tall trees, although it does operate in some plants sometimes.

How about capillarity? That is the mechanism visualized by most people who are unacquainted with the problem. **Capillarity** is the rise of liquids in small tubes. It occurs because the liquid wets the side of the tube (by adhesion) and is pulled up. This is evident from the curved **meniscus** at the top of the liquid column. As Fig. 4-2 shows, it is simple to calculate that liquids rise higher in tubes of smaller diameter. By the same token, it is easy to calculate that water will rise less than half a meter by capillarity in the xylem elements of plant stems, falling short by a factor of over 300 of accounting for the rise of sap in tall trees. Furthermore, a little reflection shows that capillarity does not function in plants at all, at least not in the usual sense. Water rises in a small capillary tube because of the open meniscus at the top of the water's surface, but the xylem tubes in plants are mostly filled with water; they do not have open menisci. We can think of submicroscopic menisci existing between the microfibrils of cellulose in the cell walls of plant leaves and other tissues. Such submicroscopic menisci would be the *holding points* for water, not the source of movement. But we shall discuss the condition of water in cell walls in terms of hydration rather than capillarity.

Many years ago, it was suggested that water was moved up the trunk in response to some living function or pumping action of stem cells. We can lift water to any height by pumping it over successive intervals, each one less than the barometric height. But careful anatomical study has failed to reveal any pumping cells. Indeed, most water moves in the *dead* xylem elements. This has been clearly demonstrated by the use of radioactive (tritiated) water. Furthermore, in 1891 Eduard A. Strasburger (a pioneer investigator of mitosis and meiosis in plants) sawed off trees 20 m tall but left them suspended upright in buckets of copper sulfate, picric acid, or other poisons. The fluids ascended all the way to the leaves, killing the bark as well as the scattered living cells in the wood. Water continued to move up through the trunk, although eventually, after the leaves had been killed, water flow ceased. He also scalded long sections of a wisteria vine, but sap continued to rise above 10 m.

Capillarity

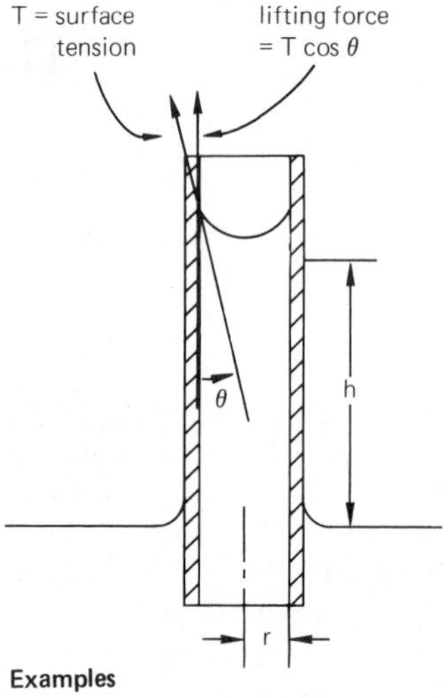

T = surface tension

lifting force = $T \cos \theta$

total lifting force = $T \cos \theta\, 2r\, \pi$

weight of liquid = $\pi r^2 hdg$

(d = density of liquid
g = acceleration due to gravity)
$\pi = 3.1416$

hence:
$T \cos \theta\, 2r\pi = \pi r^2\, hdg$

or:

$$h = \frac{2T \cos \theta}{dgr}$$

for water in glass:

$$h = \frac{0.153 \text{ (cm)}}{r}$$

(cellulose is similar to glass)

Examples

tube radius (r)		height	
cm	microns	cm	
0.0001	1	1,530.0	
0.001	10	153.0	
0.01	100	15.3	
0.1	1,000	1.53	
0.004	40	38.3	typical tracheid
5.0×10^{-9}	0.005	$3.0 \times 10^5 = 3$ km	cell wall microcapillaries

Figure 4-2 The principle of capillarity and the mathematics to predict the height a liquid may be expected to reach in a tube. Examples of calculated heights are also shown.

The importance of living cells for sap flow in the wood cannot be overemphasized, however. The dead xylem in a plant was built by living cells, and new wood is laid down each year by the living, water-filled cambial cells. When the leaves die, water movement stops. Yet the idea that water is pumped up through the trunk by cells along the way must be rejected. So how *does* water "flow uphill" to the tops of tall trees?

4.2 The Cohesion Hypothesis of the Ascent of Sap

Around the turn of the century, a model was formulated to account for the rise of sap in tall trees. One of its elements, the cohesion property of water, was not familiar from everyday experience, so the model proved to be somewhat controversial. As a good hypothesis should, however, it suggested several consequences by which it could be tested. Now, after three quarters of a century, the numerous data that have accumulated support the model. Most difficulties and criticisms have been laid to rest, although we must still accept a visualization of reality that is not a familiar part of everyday experience.

There are three basic elements of the **cohesion theory** for the ascent of sap: the *driving force, hydration* in the pathway, and the *cohesion of water*. The driving force is the gradient in decreasing (more negative) water potentials from the soil through the plant to the atmosphere. Water moves in the pathway from the soil through the epidermis, cortex, and endodermis into the vascular tissues of the root, up through the xylem elements in the wood, into the leaves and finally, by transpiration, through the stomates into the atmosphere. It is the special structure of this pathway (the relatively small diameter and thick walls of the tubes) and the hydration properties of leaf parenchyma cell walls that make the system functional. The hydration force between water molecules and the cell walls is due to hydrogen bonding and is called **adhesion**, which is an attractive force between *unlike* molecules. **Cohesion** is the third key. This is the mutual force of attraction (also due to hydrogen bonding—see Special Topic 1.1) between the water molecules in the pathway. In that special environment, these cohesive forces are so great (water has such a high tensile strength) that water can be pulled from the top of a tall tree by evaporation into the atmosphere, and the pull can be extended all the way down through the trunk and the roots into the soil. Whereas a water column in a vertical pipe of macrodimensions would cavitate as just discussed, cohesion works in the plant and cavitation does not occur because of the plant's highly specialized anatomy. With this brief preview in mind, let us examine each of the three key points.

4.3 The Driving Force: A Water Potential Gradient

What is required to generate a pressure of 30 bars (430 lb/in²!), which could be the maximum pressure difference required to raise water to the tops of the tallest trees (either as a push from below or as a pull from above)? The key to understanding is the great capacity of dry air to hold water vapor. As the relative humidity of air drops below 100 percent, its affinity for water increases dramatically. This is shown by the rapid drop in the water potential (ψ) of increasingly dry air (Fig. 4-3 and equations 7 and 8, p. 25). At 100 percent RH, ψ of air equals zero (at 20 C); at 98 percent RH, it has already dropped to −27.5 bars (enough to hold a column of water 281 m high!); at 90 percent RH, $\psi = -143$ bars; at 50 percent RH, $\psi = -944$ bars; and at 10 percent RH, $\psi = -3135$ bars. Thus, air doesn't have to be very dry to establish a steep water potential gradient from the soil, through the plant, and into the atmosphere.

When there is ample water in the soil, containing minimal dissolved salts, the soil water has a potential near zero. If this water is to be pulled through the trunk to the uppermost leaves on the tallest possible trees, the leaves will have to have a water potential as negative as about −30 bars. If the cells in these leaves are to remain turgid, their osmotic potentials will

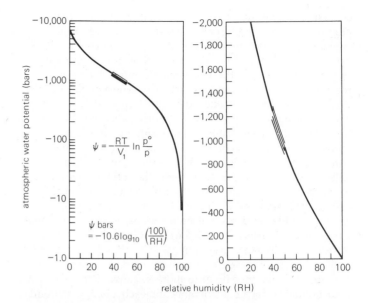

Figure 4-3 The relationship of atmospheric water potential (20 C) to relative humidity plotted on a logarithmic scale (left) and on a linear scale (right). The four thin lines are for different temperatures: 0 (bottom line), 10, 20, and 30 C (upper line). The curves were calculated using the equations shown, which are equations 7 and 8 on page 25.

have to be somewhat more negative than this or water will diffuse out, causing them to collapse. Measured osmotic potentials of leaves are on the order of −20 to −40 bars, as we saw in Table 2-1. Cells with high salt concentrations may have osmotic potentials more negative than this, but then osmotic potentials of soil water may also be somewhat more negative in the salty soils where halophytes usually grow. Clearly, actual water potentials in soil water, roots, stem or trunk, leaves, and in the atmosphere will depend upon conditions: the soil, the plant species, the climate, and so on. But if water is to move via the cohesion mechanism, the driving force—the gradient in water potential from soil to atmosphere through the plant—must exceed the pressure difference required to lift and move water from soil to atmosphere. If this restriction is not met, water will not move at all. If cohesion is operating, tension must exist in the stem water.

But how can we observe the water potential gradient by measuring *tensions* or negative pressures in a plant stem? In 1965, Per Scholander and his colleagues at the Scripps Institute of Oceanography in California published the elegantly simple and satisfactory **pressure bomb method** (see Fig. 2-9), which provided a way to measure tension in stems.* Scholander reasoned that water contained within a stem is under tension,

---

*It is an interesting historical footnote that Henry Dixon, who developed the cohesion theory, understood the principle of the pressure bomb and constructed glass models. After two rather serious explosions, he abandoned this approach!

because the pressure outside the stem is greater than the pressure inside. Therefore, when a stem is cut, allowing pressure inside to equal pressure outside, the water column in the xylem should recede from the cut surface. If the same difference is reestablished, the water should move back exactly to the cut. The method, then, consists of cutting a branch or twig from a tree or shrub, placing the branch in a pressure bomb, and increasing the gas pressure on the branch until water in the xylem tubes can be observed through a binocular microscope or hand lens to come back to the cut surface. The pressure in the bomb should then be equivalent to the absolute value of the tension in the stem before the cut. We have already encountered the pressure bomb in modern measurements of osmotic potential and of matric potential (Section 2.6).

Scholander and his colleagues made a number of measurements of tensions in stems under a variety of conditions. Results for different environments are shown in Fig. 4-4. Tensions were observed under all conditions, varying from a few atmospheres (negative) to more negative than −80 atm. Measurements vary considerably for each plant, but trends

are clear: Forest and freshwater species have the least negative tensions, with much more negative tensions in desert and seashore plants. As might be expected, tensions at night are somewhat less negative. Positive pressures that could result from root pressures were never observed under these conditions, even at night.

In another interesting study (Fig. 4-5), twigs were shot with a high-powered rifle from a tall Douglas fir at two heights above the ground (30 and 74 m) and quickly placed in the pressure bomb. Just as might be expected, tensions varied with time of day, being most negative around noon when light intensities were highest and humidities were low. The difference in heights between the two samples was about 45 m. The hydrostatic pressure difference for 45 m would be 4.37 atm, which is close to the observed value of a little over 5 atm. Resistance within the plant could account for the small discrepancy. In any case, the pressure difference for the two samples remained constant throughout the period of sampling.

Other methods for measuring tensions in trunks have also been developed in recent years. One is to measure vapor

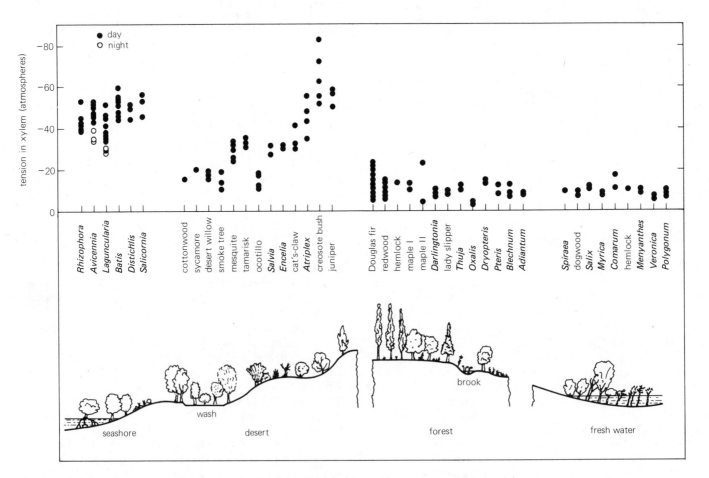

Figure 4-4 Sap pressure in a variety of flowering plants, conifers, and ferns. Most measurements were taken during the daytime in strong sunlight. Night values in all cases are apt to be several atmospheres higher (less negative). (From Scholander et al., 1965.)

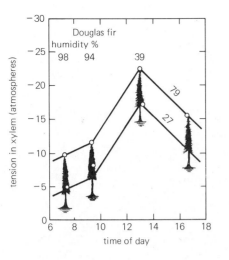

Figure 4-5 Hydrostatic gradients in a Douglas fir as a function of time of day, measured by shooting down twigs and placing them in the pressure bomb. Lines indicate true elevations and circles denote pressures at a given clock time, taking the top measure as a standard. (From Scholander et al., 1965.)

pressure in the trunk of a tree (e.g., by the cooled junction thermocouple method discussed in Section 2.6); another is to insert electrical resistance blocks (e.g., of gypsum) such as those used to measure negative water potentials of soils. Such studies also support the concept of water tensions within plants. The gradient from soil to air via the plant constitutes the driving force for water movement.

4.4 The Pathway

Water movement via the cohesion mechanism is possible in plants only because of their highly specialized anatomy. Thus, to understand movement by this mechanism, it is essential to understand the anatomy of plants. The following discussion and figures provide a review; more details are in your elementary botany or biology books and in books devoted exclusively to the topic of plant anatomy.

Anatomy of the Pathway In the preceding chapter on transpiration, we considered the anatomy of the leaf (see Fig. 3-3, p. 34). Important features include the cuticle, the stomatal apparatus, and the considerable intercellular space with exposed moist cell surfaces (the mesophyll cells). Figure 4-6 shows the relationship between leaf cells and nearby vascular elements. It is obvious that no cell in a leaf is far from a vascular element.

Water with a few dissolved substances, particularly mineral salts, ascends in a plant mainly through the **xylem tissue.** This tissue exists in the **vascular bundles** of herbaceous stems and constitutes the **wood** of woody stems. Details vary considerably in different species of vascular plants, but gymnosperms have only **tracheids,** while most angiosperms have both **vessel elements** and tracheids (Fig. 4-7). Both are somewhat elongate cells that function as dead elements. That is, after they have been produced by growth and differentiation of meristematic cells, they die, and their protoplasts are absorbed by other cells. Before death, however, some important changes in the walls occur that are important for water flow through them. One change is the formation of a **secondary**

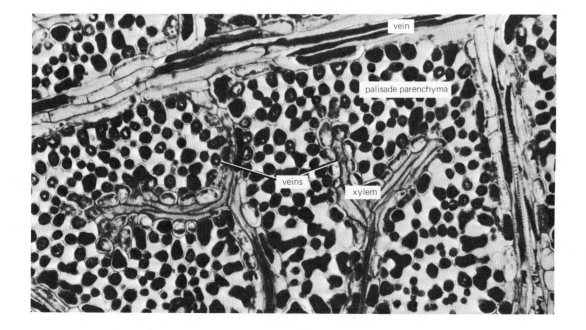

Figure 4-6 A section cut parallel to the surface of a dicot leaf, showing the intimate relationship of the vascular tissue to the mesophyll cells.

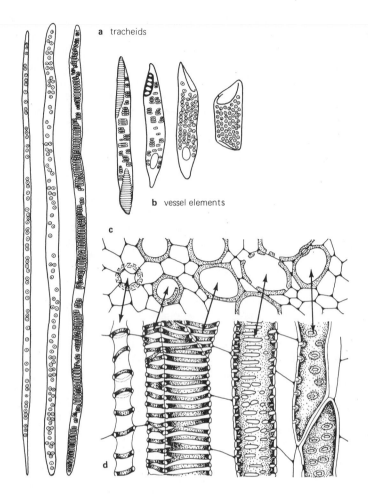

a tracheids

b vessel elements

c

d

Figure 4-7 (**a** and **b**) Tracheids and vessel elements shown isolated from the tissue, (**c**) in cross section, and (**d**) in longitudinal section. (**e**) A scanning electron microgaph of xylem vessels in a cucumber stem, showing extensive pitting. (From Troughton and Donaldson, 1972.)

e

wall, consisting largely of cellulose and lignin and covering most of the **primary wall.** This gives considerable compression strength to the cells and prevents them from collapsing under the extreme tensions that often exist in them. The lignified walls are also not as permeable to water as the primary walls, but secondary walls fail to form over small, round areas called **pits,** which are round thin places between cells where the cells are separated only by the primary walls, which are quite permeable to water. Sometimes the pits are **simple** (just a round opening), but often in both vessel elements and tracheids they are somewhat more complex structures called **bordered pits** with the secondary walls extending over the center, and the primary walls are also swollen in the center of the pit to form a **torus** (Fig. 4-8). The structure is difficult to describe in words, but you can see from the figure that the torus could act as a valve, closing when pressure on one side is greater than pressure on the other.

Bordered pits allow passage of water through the slanted ends of the tracheids and through lateral walls of both tracheids and vessel elements. Vessel elements typically are somewhat

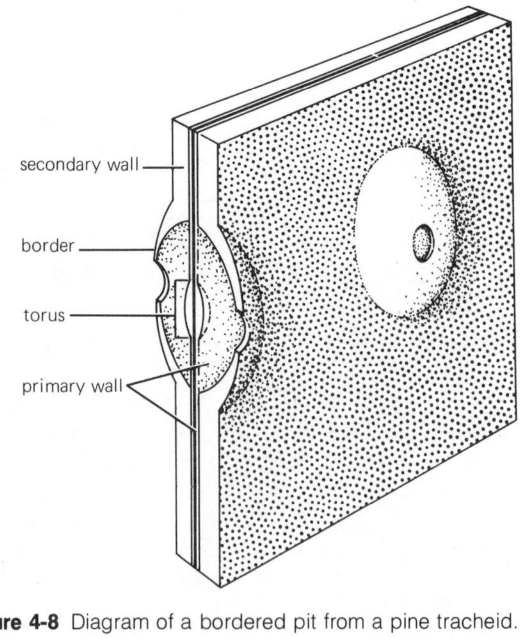

secondary wall

border

torus

primary wall

Figure 4-8 Diagram of a bordered pit from a pine tracheid. If pressure on one side exceeds pressure on the other, the torus will plug the hole, shutting off the flow.

wider and shorter than tracheids, and in addition to pits, they may be strengthened by spiral or other thickenings like the tubing used in many vacuum cleaners (Fig. 4-7). They also have **perforation plates** on their ends, plates with openings where not only secondary wall fails to form but primary wall and middle lamella dissolve away. These openings allow rapid movement of water. Vessel elements are aligned to form long tubes called **vessels,** which sometimes extend for several meters in tall trees. Because of the perforation plates, resistance to water flow is usually considerably less in angiosperms than in the less porous tracheids of gymnosperms, and flow velocities are correspondingly greater in angiosperms.

Some spiral vessels exhibit another important feature, the ability to elongate and grow while continuously conducting water under tension. Such growth occurs regularly in stems of most grasses, which support full-grown photosynthesizing and transpiring leaves.

Xylem tissue usually also includes **fiber cells** that have thickened walls and appear to function primarily in support rather than transport of water. They also contain no protoplasts. In xylem tissue there are also living cells called **xylem parenchyma.** These occur most abundantly in **rays** that run radially through the wood of a tree, but some are scattered more randomly. Figure 4-9 shows cross sections of young stems and a root. A discussion of xylem anatomy should also emphasize the **cambium,** consisting of living cells that divide to produce the **spring** and **summer wood** (both **secondary xylem tissues**) and the cells in the **apical meristems** that produce the **primary xylem tissues.** These are important features, because the cambium and meristems produce tracheids and vessels filled with sap, sometimes under tension.

Water moving through the stem and leaves originally enters the plant through the roots. Figure 4-10 shows a longitudinal section of a young root tip. The xylem tissue in the root center is continuous with the xylem tissue in the stem. It is also closely associated with **phloem tissue,** the conducting tissue through which dissolved sugars and other assimilates move. As the root grows in diameter, cells between the xylem and phloem form a vascular cambium that produces mostly xylem tissue to the inside and phloem tissue to the outside.

The xylem and phloem elements are surrounded by a layer of living cells called the **pericycle.** The vascular tissue and the pericycle form a tube of conducting cells called the **stele.** Just outside the stele is a layer of cells called the **endodermis.** The endodermal cells are especially interesting and important from the standpoint of water movement in the plant, because their radial and transverse cell walls are impregnated with **suberin,** which, like lignin and the cutin in the cuticle, is quite impermeable to water. The endodermal tangential walls (the inside and outside walls parallel to the surface of the root) are not so impregnated. Thus, water, with its dissolved substances, cannot pass around the endodermal cells via their walls but must pass directly through the cell itself (the **protoplast** = cell contents excluding the wall). Fig. 4-9 also illustrates the radial water-impermeable thickenings around endodermal cells, which are called the **Casparian strips.**

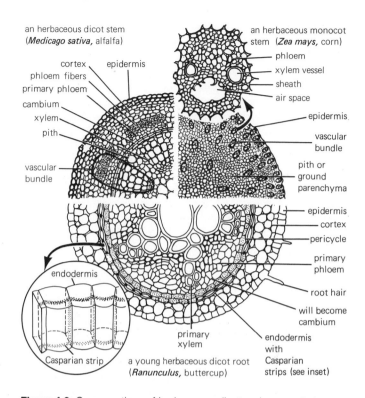

Figure 4-9 Cross sections of herbaceous dicot and monocot stems and of a young root. The three-dimensional drawing of the endodermal cells (inset) shows the position of the Casparian strips in the walls.

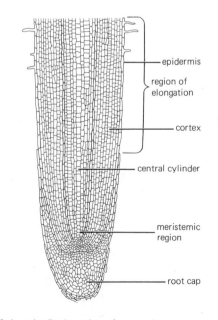

Figure 4-10 Longitudinal section of a root tip.

Outside the endodermis are several layers of relatively large, thin-walled living cells, often with considerable intercellular space at their corners, called the **cortex.** Their cell walls are highly permeable to water and its dissolved solutes, and thus it is quite likely that water can enter a root, moving via the cell wall pathway in the cortex until it encounters the endodermal layer, at which point it must pass directly through the cell (the protoplast) rather than circumventing the membranes of living cells by moving through the primary cell walls.

The layer of somewhat flattened cells on the outside of the cortex is called the **epidermis.** This layer of cells seldom has a layer of waxy cuticle comparable to that found on the epidermis of stems and leaves. Instead, water can readily penetrate through the epidermal layer. Indeed, some of the cells of the epidermis develop long projections called **root hairs** (see Fig. 5-2, p. 65), which extend out among the soil particles around the root, greatly increasing the root surface capable of absorbing water and the volume of soil penetrated. Most of the water and minerals absorbed by roots is absorbed through root hairs.

The root tips are continualy growing through the soil, encountering new regions of moisture. A **root cap** protects the delicate meristematic cells and is continually sloughed off at the forefront and replenished by divisions of these meristematic cells. Since the tissues of the stele and the endodermis are formed as the cells in the meristem divide, enlarge, elongate, and differentiate, the stele will be open at the end where it is being formed. Could water enter through that end, bypassing the endodermal layer? Studies using dyes and radioactive (tritiated) water indicate that this does not happen. Perhaps the cells in the meristematic region are so small and dense that resistance to water movement is too high. Most of the water enters through the root hairs and their associated epidermal cells in the region of a young root where xylem vessels are mature.

Root Pressure and the Apoplast-Symplast Concept As it turns out, water with its dissolved solutes can freely diffuse into a certain portion of the plant, probably the cell walls and those few intercellular spaces not filled with gas. That portion of the volume of the plant that can be in free diffusion equilibrium with the plant's surroundings is called the **apparent free space.** In the basic experiments, plant tissues (e.g., a root system or slices of tuber tissue) are allowed to come into diffusion equilibrium in a solution of known composition and concentration. By knowing the quantity and concentration of a solute in the solution at the beginning and after equilibrium has been achieved, it is possible to calculate how much solute has penetrated the tissue. Of course, some solutes will cross membranes into protoplasts, and one must correct for this. But knowing the quantity of solute that has penetrated the tissue by pure diffusion (i.e., did not cross any membranes) and that can diffuse out readily if the tissue is placed in pure water, it is possible to calculate that portion of the volume of

The Experiment

10 cm^3 of plant tissue is added to 100 ml of 0.2014 M $SO_4^=$:

time for equilibrium

The Results

Solution is 0.2000 M $SO_4^=$; so tissue contains 0.00014 mol of $SO_4^=$ (0.0134 g)

The Calculation of AFS (Apparent Free Space)

1. Assume that $SO_4^=$ in the free space is in diffusion equilibrium with outside solution—hence it is at the same concentration (0.200 M $SO_4^=$)

2. How much volume does it occupy?

   $$\text{Concentration} = \frac{\text{quantity}}{\text{volume}} \text{ or } C = \frac{A}{V}$$

   Hence: $V = \dfrac{A}{C} = \dfrac{0.00014 \text{ mol}}{0.2 \text{ mol/liter}} = 0.0007$ liter or $\boxed{0.7 \text{ cm}^3}$

3. So 0.7 cm^3 of 10 cm^3 is occupied by 0.200 M $SO_4^=$, or

   $$\text{AFS} = \frac{0.7 \times 100}{10} = \boxed{7.0\%}$$
   But
   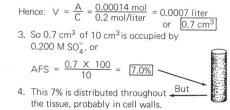

4. This 7% is distributed throughout the tissue, probably in cell walls.

Figure 4-11 The measurement and calculation of apparent free space.

the tissue occupied by the solute. The principle of measurement and calculation is shown in Fig. 4-11.

The free space consists of about 7 to 10 percent of the volume of many tissues. Careful measurements of micrographs of plant tissue indicate that cell walls and those intercellular spaces that are filled with water (relatively few) could account for this. Furthermore, based upon what we know of the ultramicroscopic structure of the cell wall, it is probable that water with its dissolved solutes can freely penetrate by diffusion.

In 1932, before the concept of free space had been formulated, E. Münch in Germany introduced a closely related concept and terminology that prove valuable in our discussion of the pathway of water and solute movement through the plant. He suggested that the interconnecting walls and the water-filled xylem elements should be considered as a single system and called the **apoplast.** This is, in a sense, the "dead" part of the plant. It would consist of all cell walls in the root cortex and, technically, the cell walls of the endodermis, although for our purposes the Casparian strip portion of the endodermal cells would not be included. All the tracheids and vessels in the xylem tissue would be part of the apoplast, as would the cell walls in the rest of the plant, including those in such tissues as the phloem, other cells in the bark, and the leaves. The apparent free space of a root would not be equivalent to the apoplast, because diffusion inward would stop at the Casparian strips. Except for the thickenings in the endodermal cells, the ascent of sap in a plant would take place in the apoplast, particularly the xylem portion, but including the cell walls of the cortex and even the living cells in the leaves.

The rest of the plant—the "living" part of the plant—Münch called the **symplast.** This would include the cytoplasm of all the cells in the plant, although it would probably not include the large vacuoles surrounded by cytoplasm in most of the cells. As it turns out, the cytoplasm of adjoining cells is connected through minute pores in the cell walls. The cytoplasmic connections are called **plasmodesmata** (Fig. 5-7, p. 69).

Based upon the apoplast-symplast concept, Alden S. Crafts and Theodore C. Broyer in 1938 proposed a mechanism to account for root pressure. A slightly modified version of their model still seems reasonable. Assume that the root is in contact with a soil solution. Ions will diffuse into the root via the free space or apoplast (i.e., cell walls) across the epidermis, through the cortex, and up to the endodermal layer. Along the way, ions will be passing across the cell membranes from the apoplast into the symplast in an active, respiration-requiring process. The result is a buildup in concentration of ions inside the cells within the symplast to levels higher than those outside in the apoplast. The symplast is continuous across the endodermal layer, so ions can move freely into the pericycle and other living cells within the stele (Fig. 4-9). This might occur by diffusion through the plasmodesmata, and the velocity of movement inward might be increased by the **cytoplasmic streaming** or circular flowing of the cytoplasm that is often observed within such cells.

Within the stele, there is less oxygen than in more external cells. Crafts and Broyer suggested that this might result in a less efficient retention, so ions might leak into the apoplast (xylem) inside the stele. Now it appears that ions are actively pumped into the stele. In either case, the result would be a buildup in concentration of solutes within the apoplast in the stele to a level higher than that of the soil solution, so ψ_π in the stele is more negative than ψ_π in the soil. Because water must pass through the protoplasts of the endodermal layer, this layer would act as a differentially permeable membrane, and the root becomes an osmotic system. The buildup in pressure in this osmotic system would be the cause of root pressure.

It is important to realize that root pressure is a relatively rare phenomenon in plants. It occurs when transpiration is greatly reduced by cool humid conditions and when there is ample water with some dissolved minerals in the soil. During much of the life of most plants, transpiration is creating negative water potentials in the leaves so that water is being pulled into the plant and moved through the xylem system so rapidly that ionic concentrations cannot build up in the stele within the roots, and the roots never get a chance to act like an osmometer. Rather than pressures being built up due to osmosis, tensions are established by transpiration and cohesion.

From Soil to Air In summary, consider the movement of water along the pathway from the soil through the plant to the atmosphere. If the roots are in contact with liquid water and its dissolved solutes, this solution will move by diffusion into the root via the apoplast. If the plant is rapidly transpiring so that tensions are created in the leaves and transmitted through the stem to the roots, these tensions will lower the water potential within the xylem of the roots considerably below that of the soil water. This gradient in water potential (more negative inside the roots) will increase the rate of diffusion.

It has been suggested that sometimes the root may not be in contact with liquid water. The soil may be so dry that all liquid water in the near vicinity of the root has already been absorbed by the plant (see Special Topic 4.1). Even under these conditions, however, water might conceivably continue to enter the root. If the water potential within the root is more negative than water in the soil (even though it may be at some distance from the root), diffusion of water into the root will occur, but it might move to the root from its location in the soil in the vapor state rather than as a liquid. It will distill across the vapor gap. This could help account for apparent discrepancies that have been noted in the relationship between water and solute uptake. If there is ample water in the soil, then it is likely that water molecules will diffuse more readily into the root, carrying some solutes with them, where they can be actively absorbed and carried in the transpiration stream to the rest of the plant. If the soil is dry, however, then water molecules can cross the vapor gap, but solute particles cannot. Under these conditions, the uptake of water may be completely uncoupled from the uptake of solutes. Actually, solute and water uptake are never closely coupled, since solute uptake is an **active,** energy-requiring process (Chapter 5), while water uptake is **passive.**

Once the water has moved across the epidermis and cortex into the stele, its movement up through the xylem conduits of the stem will be subject to the laws that apply to water under tension. If the cohesion mechanism is at work, one important question immediately comes to mind: What happens if a bubble forms? That is, what happens if cavitation occurs in this water under tension? In our normal experience, in a system with negative pressure, a bubble of air or water vapor would instantly expand until it filled the entire cross section of the transport system, forming a "vapor lock," and water movement would stop. In a plant, this does not happen because of the highly specialized anatomy of tracheids and vessel elements. As we saw, the secondary walls of these tubes are typically perforated by minute holes (the pits), but these may be closed by primary walls with such a fine porosity that bubbles of air or water vapor cannot pass through them.

The torus in bordered pits provides an especially efficient valve. Liquid water can move through these pits, but the system forms a **check valve** against the passage of air or vapor.

We can understand this somewhat better by thinking of the surface-tension properties of small droplets of liquid in air. As a drop of liquid gets smaller, its surface/volume ratio increases, and the forces of surface tension on it become stronger, compressing the droplet more and more. This increase of pressure in the droplet increases the free energy, and this in turn results in a higher vapor pressure. As one watches such a small droplet under a microscope, it tends to shrink (dry) faster and faster, until, as the vapor pressure increases to the point where the drop boils at room temperature, it suddenly disappears.

A bubble of air in liquid also exhibits the powerful force of a curved meniscus. When the bubble is very small, the interfacial tensions acting on it are proportionately large, giving it a high stability against deformation. The result is that the surface of a bubble cannot be bent enough to pass through the minute pores in the pits. Hence, if the water in one column should cavitate, as sometimes happens, producing a bubble of air or vapor, this break is not transferred laterally to another xylem element or tube. As long as some tubes remain filled, water will continue to move upward, passing laterally from tube to tube when necessary.

It is important to remember the role of hydration as the water moves through the xylem and into the leaves. Water is held to the cell walls against considerable tensions, approaching probably −1,000 to −3,000 bars. This is the case in the xylem elements and cortex cells as well as in the leaf mesophyll cells from which evaporation to the atmosphere occurs. Water potential of the atmosphere may reach negative values from tens to hundreds or thousands of bars, creating a gradient along which water can move from the leaf cells through the stomates to the atmosphere. Yet hydration forces are great enough to retain the water in the cell walls of leaves and the xylem elements, and to retain some of the water in the roots. Regardless of the tensions that might build up in the system, it cannot collapse as water is pulled away from the cell walls. Water is held too tightly.

4.5 Cohesion

An Irish botanist, Henry H. Dixon (1914), carefully formulated the hypothesis that the tensions created by evaporation and the subsequent hydration of water by the cell walls, drawing water from the pathway within the plant, were relieved by an upward movement of water from below, the columns being held together by cohesion (Fig. 4-12). Actually, Dixon and John Joley, a physicist, had independently begun to develop the idea as early as 1894, as had E. Askenasy in 1895 and Otto R. Renner in 1911. But it was published by Dixon in book form with a vast body of supporting data in 1914.

Since the original formulation of the cohesion theory, plant physiologists all over the world have attempted to

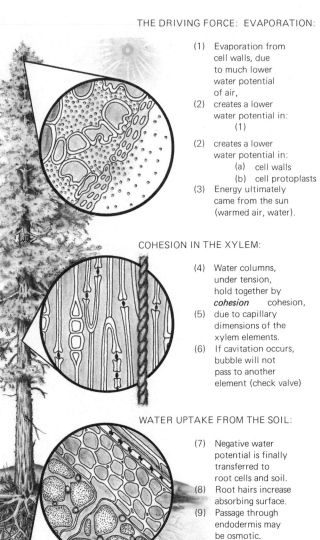

(1) Evaporation from cell walls, due to much lower water potential of air,

(2) creates a lower water potential in:
  (1)

(2) creates a lower water potential in:
  (a) cell walls
  (b) cell protoplasts

(3) Energy ultimately came from the sun (warmed air, water).

COHESION IN THE XYLEM:

(4) Water columns, under tension, hold together by *cohesion* cohesion,

(5) due to capillary dimensions of the xylem elements.

(6) If cavitation occurs, bubble will not pass to another element (check valve)

WATER UPTAKE FROM THE SOIL:

(7) Negative water potential is finally transferred to root cells and soil.

(8) Root hairs increase absorbing surface.

(9) Passage through endodermis may be osmotic.

Figure 4-12 The cohesion theory of the ascent of sap summarized.

devise experiments capable of testing it critically. Some of these experiments were suggested by the nature of the theory, others arose because of objections to it, and still others were formulated to test possible alternatives. The cohesion hypothesis succeeds well at suggesting experimental approaches to its investigation. We shall summarize some of the questions that have been asked, sometimes reviewing points already mentioned.

Does Water Have a High Enough Tensile Strength? The problem is whether water can sustain tensions of up to −30 bars without cavitating; whether the cohesive forces of water are as negative as −30 bars. Determination of the tensile strength of water has proved to be an extremely difficult task, and there are many conflicting data. Three approaches are worth discussing, and the third may have provided us with a final solution.

First, information relating to hydrogen bonding in water suggests that potential cohesive strength under ideal conditions is extremely high, enough to resist a tension of several thousand bars. But since our theories of liquids remain so imperfect, many assumptions must be made in such an approach, and the results are hardly convincing.

Second, several experimental measurements have been made. The force required to separate steel plates held together by a water film has been measured. Fern annuli that are pulled apart by water under tension have been studied. Glass tubes have been sealed while full of hot expanded water, and then cavitation has been observed as the water cools and contracts. These and still other methods have produced values for the tensile strength of water on the order of −100 to −300 bars, although a few workers could measure only lower values, on the order of only −1 to −30 bars. These methods measure both cohesion of water molecules to each other and adhesion of water to the container. Results represent minimal tensile strengths for either adhesion or cohesion.

Third, the most clear-cut experimental approach is one introduced by Lyman Briggs in 1950, using capillary glass tubes bent in the form of a "Z" (Fig. 4-13). These tubes are centrifuged, causing tension on the water at the center of the tube. The tension present when the water column breaks can easily be calculated. Such observations have produced an important conclusion: *The smaller the capillary, the higher the tensile strength of water.* With rather fine capillaries, values as negative as −264 bars were measured. Using a capillary tube with a diameter of 0.5 mm (considerably thicker than most xylem elements), air-saturated tap water did not cavitate under tensions of −20 bars, although cavitation did occur when the center of the Z-tube was frozen with dry ice. (Air is virtually insoluble in ice, so freezing forces the dissolved air out of water, forming bubbles.) The cohesive forces in water seem to be quite sufficient for the cohesion mechanism of the ascent of sap, providing that the water is held in tubes of small enough diameter.

It is important to realize that the tensile strength of water is inversely proportional to the thickness of a water column, although it is not yet clear why this should be the case. Thus, it is the special anatomy of the xylem tissue that makes the cohesion mechanism work. Scholander, et al. (1965) have said that the water transport system of a plant "combines capillary dimensions with check-valved compartmentalization." He has further suggested that no other solution of the movement of water to the top of tall trees may be possible.

Are the Water Columns Really Continuous? Again, it is easy to find contradictory evidence in the literature. Some direct observations indicate continuity, while others do not. But virtually any method of observation is itself likely to introduce discontinuity. The observed flow velocities in stems clearly imply that the xylem elements are indeed filled and continuous. Modern velocity measurements apply a pulse of heat at one point on the stem, measuring the time it takes to detect the warmed water at some level above. Velocities vary greatly depending upon environmental conditions, but values ranging from 50 to 4,360 cm/hr have been measured. Furthermore, tensions developing in one part of a tree are rapidly transferred to other parts, again implying continuous columns.

Are the Columns Really under Tension? The pressure bomb measurements described earlier (Section 4.3) clearly demonstrate that water in the plant is under tension. Some older observations are also instructive. For example, the stem of a transpiring plant may be immersed in a dye solution and cut. The dye instantly moves a considerable distance both up and down the stem inside the xylem elements and then stops rather suddenly. It is difficult to interpret this in any way other than as a sudden relief in the tension on the walls of the xylem tubes. The sudden movement is due to the elasticity of the cell walls.

Renner in 1911 performed an elegantly simple experiment. He attached a leafy branch to a burette to measure water uptake. He constricted the stem with a clamp to produce a high resistance. After measuring uptake under these conditions, he cut off the leafy end and applied suction with a pump that produced a tension of about −1 bar. In response to the pump, only about one tenth as much water moved as in response to the leafy branch. We seem compelled to conclude that the leaves must exert a pull of some −10 bars.

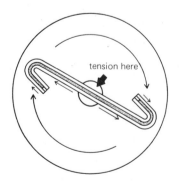

Figure 4-13 Method of measuring the cohesive properties of water utilizing a centrifuged Z-tube. Small arrows indicate direction of centrifugal force and principle of balancing due to the Z-tube.

What if the Columns Cavitate? Do field observations agree with the cohesion theory? Perhaps the most important question raised by plant physiologists about the cohesion hypothesis concerns what would happen if the continuous columns of water should in some way be broken. Say, for example, that a strong wind disturbance causes the columns of water in the stem to cavitate, or that the water in the tree trunk freezes, forming bubbles of gas. Or what would happen if the columns were broken by sawing part way through the trunk? Many investigators have taken many approaches to these and other problems. Figure 4-14 illustrates some further elegant experiments of Scholander and his coworkers. The experiments are described in the figure caption. They nicely support the cohesion theory.

In another set of experiments, Scholander and his coworkers (1961) studied grapevines and tropical lianas. He made crosscuts in the vines at different heights and from opposite sides so that *all* large vessels were disrupted. Yet water uptake continued at a rapid rate! It could be shown that this occurred only at the expense of a much higher tension gradient across the treated section of stem. Apparently the

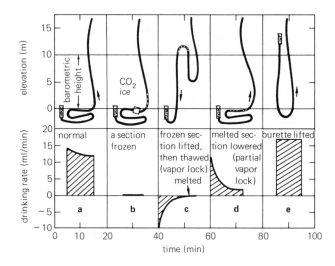

Figure 4-14 Scholander's experiments with tropical rattan vines (*Calamus* sp.). (**a**) The vine is cut off under water and a burette is attached, allowing measurement of the rate of water uptake. If the burette is stoppered, water continues to be taken up anyway until a vacuum is created in the burette, and the water boils. (**b**) To freeze the water in the vine, the burette first had to be taken off so that air entered all the xylem elements, vapor-locking the system. Then after freezing, the vapor-locked portion (about 2 m) was cut off under water and the burette attached again. There was still no water uptake, indicating that freezing had indeed blocked the system. (**c**) If the vapor-locked portion was hoisted above the barometric height and allowed to thaw, some water ran out, but there was no uptake, indicating that the system was now vapor-locked. (**d**) If the vapor-locked portion was lowered to the ground, there was a rapid initial uptake as vapor condensed to water, breaking the vapor lock, but then uptake was slower than originally because some air had been excluded from freezing. (**e**) When the burette was elevated 11 m, rate of water uptake returned to the original level, indicating that the vapor lock had now been completely eliminated. (From Scholander et al., 1961.)

stem consists of a system made up of macropores (the vessels and tracheids) surrounded by micropores (probably the cell walls). Resistance is considerably less through the macropores, but movement can occur through the micropores, providing the force gradient is high enough. In experiments in which the stem is cut off and allowed to take up air, such as those described in Fig. 4-14, even the micropores become vapor locked and blocked.

An interesting approach to the problem of cavitation in the field has been to apply sensitive microphones to growing corn plants in the field. Popping noises are heard that are probably due to cavitation in certain vessels! As the air dries and transpiration stress increases, the number of cavitation sounds also increases.

What about Air Excluded from Solution in the Stem by Freezing in Northern Trees? Microscopic observation has shown that air blockage occurs when some trees in cold climates are frozen, just as when water was frozen in the spinning Z-tube by dry ice. Inability to restore the water columns in the spring may well be the factor that excludes certain trees and especially vines with large vessels from these regions. How do trees that grow in such regions manage? Observations have shown that the blocked vessels are indeed restored in such trees. But how?

Several explanations have been proposed. In one of the most convincing of these, E. Sucoff (1968) studied the mathematics of bubbles in a liquid. He showed that large bubbles can expand easier than small bubbles, especially if the liquid is placed under tension. Imagine a northern tree thawing in the spring. As the ice melts, the tracheids become filled with liquid containing the many bubbles of air that had been forced out by freezing. As melting continues and transpiration begins, tension begins to develop in the xylem. Sucoff showed that a critical point would be reached for a large bubble so that it would expand suddenly under tension as water turned to vapor; indeed, the expansion would be *explosive*, occurring in a fraction of a second. This would be confined to the tracheid in which it occurred, due to its anatomy, but it would send a shock wave to surrounding tracheids, effectively driving their small air bubbles back into solution. What about the tracheid in which bubble expansion occurred? It would be vapor locked and lost to sap movement. Study of wood in the spring indicated that about 10 percent of the tracheids were indeed filled with vapor, but the remaining 90 percent should be ample to handle sap movement. So, according to this explanation, spring recovery of the xylem transport system in northern trees is enabled by miniature explosions in about 10 percent of the tracheids!

It is also helpful to remember that cambial cells divide in the spring, producing new water-filled xylem elements. In some ring-porous trees, virtually all the water moves in these newly formed tubes. Clearly, gymnosperms with their tracheids are especially well adapted to cold climates. Trees, and especially vines with large, long vessels, are practically absent from cold climates but are abundant in the tropics.

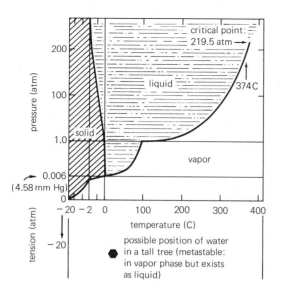

Figure 4-15 Phase diagram for water. To show the shape of the curves close to 0 C and also at much higher temperatures and pressures, it was necessary to change the scales sharply. Note that negative pressures must always result in water in the vapor state, in which state negative pressures cannot be maintained. Yet water in plants exists under tension.

4.6 Some Conclusions

Life in the Metastable State When matter is in a state contrary to our expectations, it is said to be **metastable.** Thus, a super-cooled or a supersaturated solution are in metastable states. Water under tension is metastable water. Based upon our common laboratory observations, when pressure in water reaches zero, cavitation occurs and the water vacuum boils. On a phase diagram for water, which shows the expected states for given temperatures and pressures, only vapor can exist at temperatures above freezing and pressures below zero (Fig. 4-15). So what about water in the xylem that is known to be under tension?

Obviously, our concept of a metastable state is a rather arbitrary one. It refers only to our common experiences in the macroworld. In the xylem tissue, the normal state of liquid water is tension. This seems to be due to the microdimensions of the tracheids and vessels in which sap movement occurs. Still, it is interesting to consider the extent of "metastable" water on the earth's surface. All the tall trees must contain "metastable" water. Even short herbaceous plants contain "metastable" water due to the high resistances in their stems. Low soil moisture may also produce "metastable" water in any plant. Pressure bomb studies demonstrated the universal occurrence of water under tension in plants.

A Final Problem We may wonder how a *turgid* plant with positive pressures in its leaf mesophyll cells can contain water under *tension* in the xylem. Obviously, as we have seen, it does. But how? Water in the transpiration pathway will have a negative water potential due almost entirely to its negative pressure. Why doesn't water move into the pathway from surrounding cells, so that they lose turgor and wilt?

Since turgor is maintained in the surrounding cells, their water potentials must be at least equal to that of the water in the xylem. In fact, since their growth is driven by water absorbed from the pathway, their water potentials must be slightly more negative than that of the xylem. Photosynthetic production of sugars and absorption of soil minerals are the primary causes of such negative water potentials. The osmotic potential of typical leaf cells has been measured and found to be on the order of −20 to −30 bars. It would appear, then, that the water potential of the water in the pathway will of necessity be less negative than −20 to −30 bars.

Scholander (1968) studied a situation in which the validity of this principle was clearly demonstrated (Fig. 4-16). Tropical

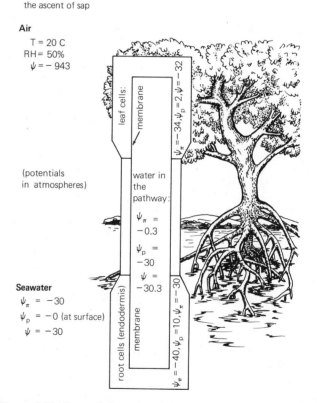

Figure 4-16 Water relations of a mangrove tree growing with its roots immersed in sea water. The diagram indicates the "essential" parts of the mangrove tree from the standpoint of water relations, particularly the membranes of the endodermis and of the leaf cells. It is important to note that if leaf membranes should suddenly cease to be differentially permeable, salt and subsequently water would move from leaf cells into the pathway due to the high tension of water there, and thus the leaf cells would collapse. (Data from Scholander et al., 1965.)

mangrove trees grow with their roots bathed in sea water, which has an osmotic potential of about −30 bars. As water enters the tree through the roots, salts are excluded, probably by the endodermal layer, so that the xylem sap is almost pure water with an osmotic potential of nearly zero. (Several other halophytes have the ability to exclude salt.) The reason that the xylem sap remains in the tree and does not move out through the roots into the sea water is that it has a water potential of −30 bars or lower. This is achieved by constant xylem *tensions* of −30 bars or more negative. Scholander was able with his pressure bomb to measure the predicted tensions. By this mechanism, the mangroves avoid a lethal buildup of salt in their leaves.

Obviously, if the xylem sap in mangroves has a water potential of −30 bars, the protoplasts of leaf cells must also have water potentials of −30 bars if they are to remain turgid. If they have pressure potentials of about 5 bars, then their osmotic potentials will have to be about −35 bars, values that can be readily measured. Ordinary trees that don't have their roots in salt water have similar, if somewhat less spectacular, water relations. Tensions often exist in their xylem tissues, but their turgid cells have osmotic potentials even less negative than the water potentials of the xylem sap with its negative pressures. In any case, the importance of osmosis in the plant is clearly demonstrated. Without the differentially permeable membranes around the living cells and the highly negative osmotic potentials of the sap inside, the high tensions in the xylem system would lead to collapse of the living tissue. Without osmosis, plants would collapse!

Pursuing the Questions of Soil-Plant-Atmosphere Water Relations **Ralph O. Slatyer**

F.B.S. 1971

Ralph O. Slatyer was educated in Australia in the 1950s. Since 1967, he has been a Professor of Biology at the Australian National University in Canberra City. His essay well illustrates the international nature of plant physiology; it also reinforces and expands several topics that we have been considering.

To me, the most exciting and stimulating part of scientific research is when you make an observation or generate an hypothesis that you think is original and may, in addition, be at variance with accepted phenomena or attitudes.

Of course, in most cases, more careful observation, more thorough reading of the literature, or a critical discussion with colleagues leads to an awareness either that your ideas or observations won't hold up, or that someone else has anticipated you. But occasionally there is the breakthrough, and scientific knowledge and understanding moves forward.

In my own case, the first such heady moments came in the mid-1950s when I was investigating the effects of progressive and prolonged water stress on physiological plant responses. At that time, it was widely accepted that the permanent wilting percentage was a soil constant, the soil water content below which no plant growth would occur and further transpiration would cease. This concept had strong empirical support, mainly through painstaking work by Veihmeyer and Hendrickson at the University of California, Davis, on irrigated fruit crops, but dated back to previous works of Briggs and Shantz in the early part of the century. Associated with the concept was the notion that soil water was equally freely available to the plant at water contents down to the permanent wilting percentage, although this had come under challenge, particularly from scientists at the United States Salinity Laboratory at Riverside, California.

My initial research work, associated with my master's and doctoral studies, was concerned with crop and native species responses in arid and semiarid regions, and I repeatedly found evidence that consistently contradicted the established dogma. Accordingly, I set off to spend what was essentially a brief postdoctoral period with Professor Paul Kramer at Duke University. I found him receptive to my ideas and full of encouragement.

Basically, all I had done was to propose that as water stress was progressively imposed by soil water depletion, the turgor pressure of the leaf cells decreased until it reached zero, when the leaf water potential equaled the osmotic potential. I argued that at this point the leaves would be permanently wilted and that growth might be reasonably expected to have ceased. Even with stomatal closure, however, continued transpiration would be expected to continue to deplete soil water until desiccation of the plant itself reached lethal levels. From this argument it followed that the permanent wilting percentage should not be a soil constant, but merely the soil water content at which the soil water potential and the plant water potential were balanced, at a level equal to the leaf cell osmotic potential, so that zero turgor pressure existed.

Somewhat surprisingly, these views, published in both experimental and review papers, were rapidly accepted by the scientific community and have since been reconfirmed in general aspects by numerous investigators. In the process, of course, some of the more specific assertions have required qualification, but overall the dynamic nature of the soil-plant-water interaction and the permanent wilting percentage was clearly established.

This approach also seemed to provide a better basis for interaction between plant and soil scientists interested in plant-environment interactions, and it led to a requirement for a more integrated term to describe the state of water in plants and soils. Through the 1950s, "diffusion pressure deficit," "total soil moisture stress," and related terms were being used by these two groups of scientists who were basically talking about the same thing. This matter finally came to a head, informally, over dinner in a restaurant in Madrid during a UNESCO Plant Water Relations conference, attended by, among others, Sterling Taylor, Wilford Gardner, Robert Hagan, Fred Milthorpe, and myself. We proposed the term "water potential" (already suggested many years earlier by soil scientists), based thermodynamically on the chemical potential of water, as a single term for both soil and plant scientists, to be divided into component potentials, as appropriate. The meeting asked Sterling Taylor and me to draft a letter to *Nature* and a more definitive paper on the subject, and from this rather informal and personal beginning, the new terminology was launched. As far as I know, it is now used almost universally, although it too has been improved by modification and qualification.

The next challenge was to look at the plant-atmosphere interface. Although the plant-water system is intimately and tightly coupled to the soil-water system, it is loosely coupled to the atmosphere, since the phase change of water accompanying transpiration and the interposition of the stomata in the gas phase provide land plants with a powerful regulatory device to insulate themselves against rapid desiccation by evaporation. Several workers, notably Maskell, and Penman and Schofield, in Britain, had pioneered work on the degree to which control over water transport was exercised by the stomata rather than at the soil-root interface. Furthermore, van den Honert in the Netherlands had emphasized the essential nature of stomatal control in a paper that, while oversimplifying the processes involved, clearly demonstrated that effective control over water transport must ultimately be exercised in the gas phase by the stomata.

I have been only one of many people concerned with the evaluation of the gas exchanges between leaf and atmosphere. My colleagues and I have concentrated on evaluating the relative significance on photosynthesis of impeded CO_2 supply, through progressive stomatal closure, compared with damage to the photosynthetic apparatus as stress is imposed on a plant. We have also paid special attention to factors affecting relative impedance to CO_2 and water vapor exchange, changes in water use efficiency with stress imposition, and the different relationships in C_3, C_4, and CAM plants. Although our primary concern has been with water stress, we have also been concerned with high and low temperature stress, high and low light environments, and aging.

To a surprising extent, all this work has underlined the considerable degree to which initial inhibition of photosynthesis, associated with stress imposition, can be attributed to increased stomatal impedance to CO_2 supply. Naturally, as stress is increased, progressive damage is caused to the biochemical apparatus, but in many species, particularly more xeromorphic ones, biochemical lesions do not become detectable until a stress level corresponding approximately to permanent wilting.

I began this brief essay by referring to the excitement of scientific discovery. I conclude by referring to the spirit of cooperation that has existed in that part of the scientific community with which I have been associated. While it has always been a challenge to be first with a new piece of work, my life has been enriched by the warm personal relationships that have been developed between both my immediate colleagues and those further afield, whom one first meets by correspondence, or at conferences, and then by sharing space and facilities in a common laboratory.

5

Absorption of Mineral Salts

In previous chapters we have been primarily concerned with movement of water in plants, and it has been advantageous to ignore the movement of solutes. Indeed, osmosis occurs because transport of solutes across membranes is much slower than transport of water. But solutes do move from cell to cell and from one organelle to another in plants, and this movement is essential to life. Except for atmospheric CO_2 and O_2 sources, the elements that plants (and animals) contain are absorbed by roots from the soil by what has aptly been called a "solution mining" process. Just as leaves have to absorb CO_2 from a low atmospheric carbon dioxide concentration, roots must accumulate the mineral elements needed for the entire plant from very low available concentrations in the soil. In this chapter we shall examine plant roots and then discuss the remarkable process by which mineral absorption occurs. We shall also discuss transport of both ions and organic compounds across intercellular and intracellular membranes.

5.1 Roots as Absorbing Surfaces

The problem of absorbing sufficient water and mineral elements from the soil is countered in nature by the ability of plants to form root systems with extremely large surface areas. The overall shapes of root systems in various species is quite variable and is both genetically and environmentally controlled. Grasses such as corn and other monocots generally have fibrous and highly branched root systems near the soil surface, while most dicots usually form a tap root that can extend several feet down into the soil. Many trees have intermediate systems.

The surprisingly large surface area of roots is illustrated by measurements of a rye plant (*Secale cereale*). One such four-month-old plant exhibited, upon painstaking measurements, roots estimated to total 626 km (387 miles) in length with 233 m² (2,554 sq ft) in surface area. Estimating the root hairs brought the totals to nearly 11,300 km and 638 m²! The extensive surface area sometimes added by root hairs is illustrated for a radish seedling in Fig. 5-1, although this plant was not grown in soil where the presence of microorganisms and other soil conditions usually inhibits such extensive root hair development (Mosse, 1975; Reynolds, 1975). Each root hair on all species is a single modified epidermal cell.

Figure 5-1 Scanning electron micrograph of root hairs on a radish seedling. The seed was germinated on a moist filter paper at a high relative humidity. (From Troughton and Donaldson, 1972.)

Figure 5-2 A stand of root sprouts from a single 40-year-old sweetgum tree. The sprouts range in age from 1 to 15 years; some are over 5 inches in diameter. (From Kormanik and Brown, 1967.)

Certain tree species such as quaking aspen (*Poplulus tremuloides*) and sweetgum (*Liquidamber styraciflua*) form groups whose individual shoot systems develop from adventitious buds on the roots. Therefore, numerous individual trees are connected to a common root system, as shown in Fig. 5-2. Considering branch roots and root hairs, a vast volume of soil is explored by such systems.

An important factor facilitating exploration of the soil by roots is the constantly changing location of branch (feeder) root development. When roots growing in soil next to the wall of a glass container are observed or their growth habits followed by time-lapse photography, an individual root is found to elongate only several days or a few weeks. Branch roots subsequently develop from this root, each of which itself elongates only a few weeks. Some of these become part of the permanent root system of a perennial plant, but most small branch roots seem to decay within a year or two and are replaced with others. Some desert shrubs replace up to a fourth of their root systems each year, in the process absorbing both water and minerals from new locations (Caldwell and Camp, 1974).

The region of soil occupied by roots of a given species is controlled by the sites of available moisture and mineral elements and by soil depth. Roots have been shown to proliferate in zones fertilized with phosphate, nitrate, or ammonium, for example (Drew, 1975). This type of control and the similar proliferation where moisture is adequate are due simply to enhanced root growth in the specific "desirable" zones, and almost surely not to any tropic growth of the roots toward elements or water. (That is, the roots probably don't detect the elements or water from a distance and then grow toward them.) Because growth of roots is stimulated in fertile, moist soils, so is growth of the entire plant.

5.2 Mycorrhizae

Both students and teachers usually learn about root structures from plants that were grown in a greenhouse, sometimes even in the absence of soil (hydroponically; see Section 6.2). But in nature the appearance of young roots of most species (perhaps 97 percent) is somewhat different, not just because they aren't confined in a pot, but because they become infected with fungi to form **mycorrhizae.** A mycorrhiza (fungus-root) is a symbiotic association between a nonpathogenic or only weakly pathogenic fungus and living root cells, primarily cortical and epidermal cells. The fungi receive food from the plant and, in turn, alter the appearance and improve nutrient and water-absorbing properties of such roots.

Generally only the tender young roots become infected, perhaps because on the older parts a protective layer of suberin develops (Section 14.3). Root hair production either slows or ceases upon infection, so mycorrhizae often have few such hairs. This would greatly decrease the absorbing surface, ex-

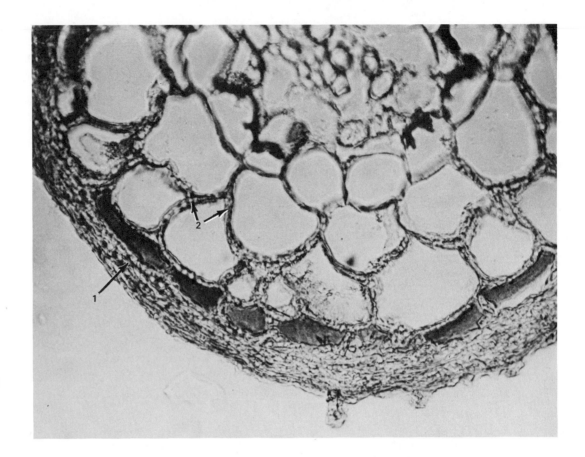

Figure 5-3 Ectomycorrhiza formed between *Pinus taeda* and the fungus *Thelephora terrestris*. Note the external mantle (1) and the Hartig net (2) between the cells. (Courtesy C. P. P. Reid.)

cept that the soil volume penetrated is greatly increased by the slender fungal hyphae extending from the mycorrhizae.

Two main groups of mycorrhizae are recognized: the **ectomycorrhizae** (Marks and Kozlowski, 1973) and the **endomycorrhizae** (Gerdemann, 1975; Sanders et al., 1976) although a more rare group with intermediate properties, the **ectendotrophic,** is sometimes encountered. In the ectomycorrhizae, common to gymnosperms and to many woody angiosperms, the fungal hyphae form a mantle outside the root and also inside the root in the intercellular spaces of the epidermis and cortex. No intracellular penetration into epidermal or cortical cells occurs in the ectomycorrhizae, but an extensive network called the **Hartig net** is formed between these cells. Figure 5-3 shows a cross section of a *Pinus taeda* mycorrhiza with the external hyphal mantle and the Hartig net between cells of the cortex. As usual, no fungal penetration through the endodermis has occurred. Figure 5-4 shows a scanning electron micrograph of *Pinus contorta* roots infected with an ectomycorrhiza fungal species.

The endomycorrhizae are present on thousands of herbaceous angiosperms, including prairie grasses, and also occur on the gymnosperm genera *Cupressus, Thuja, Taxodium, Juniperus,* and *Sequoia*. The fungus, usually a member of the Endogonaceae, forms no dense external or internal network, but instead exists primarily within the cortical cells, although hyphae extend out into the soil where nutrients are absorbed.

Apparently the fungal partner of both kinds of mycorrhizae receives sugars from the host plant. Plants grown in shade and deficient in sugars often have poor mycorrhizae development. The most well-documented advantage to the host plant is a considerable increase in the rate of phosphate absorption by fungi in both mycorrhizal types, although absorption of other nutrients is also facilitated. Mycorrhizae usually develop best in relatively unfertile soils, yet nurserymen and horticulturists often grow tree seedlings in containers with well-fertilized soils that suppress mycorrhizal development. If these seedlings were first transplanted into somewhat unfertile container soils, mycorrhizal formation could be encouraged and better survival and faster growth of subsequently transplanted seedlings could be attained.

Ectomycorrhizae seem to offer a large advantage to trees growing on unfertile forest soils. In fact, without the nutrient-absorbing properties of mycorrhizae, many communities of trees could not exist. For example, some European pines introduced to the United States grew poorly until they were inoculated with the mycorrhizal fungi from soils of their native habitats. Apparently, a considerable potential exists for population of certain mine waste areas, landfills, roadsides, and other infertile soils through introduction of plants inoculated with fungi capable of forming mycorrhizal associations.

5.3 Soils and Their Mineral Elements

Soils are a heterogenous and variable mixture of clay mineral (weathered rock) particles, decaying organic matter, and various living microorganisms, along with air, water, and various ions and organic molecules dissolved in the water. The clay particles and parts of the organic matter are colloids and thus exhibit the surface properties of particles having high surface-to-volume ratios discussed in Special Topic 2.1. The clay mineral fraction is the most abundant in the soil, consisting of kaolinite (most common in older, well-weathered soils) and of montmorillonite and illite (more common in younger soils). All three types are composed of crystal lattices of silicon,

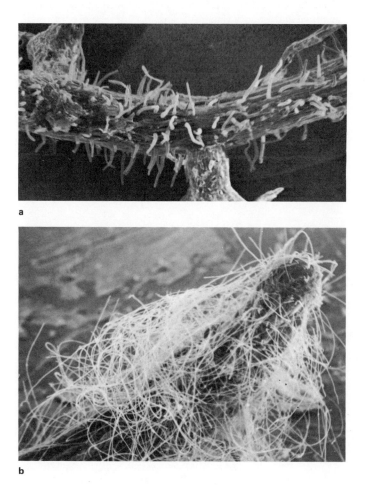

a

b

Figure 5-4 (**a**) Scanning electron micrograph of dichotomous roots and root hairs of *Pinus contorta*. Note the absence of a fungal mantle. (**b**) Scanning electron micrograph of ectomycorrhiza of *Pinus contorta* inoculated with *Cenococcum graniforme*. (Courtesy John G. Mexal, Edwin L. Burke, and C. P. P. Reid.)

oxygen, and aluminum, and all become negatively charged to varying degrees by partial replacement of silicon or aluminum ions with a cation of lesser positive charge. For example, Al^{+3} can replace Si^{+4}, and Mg^{+2} or Fe^{+2} can replace Al^{+3}. The negative charges on clay minerals are important, for these attract and hold (adsorb) cations essential to plant growth, thus minimizing their loss by solution and leaching when water moves down through the soil. The ions are adsorbed with different tenacities (Special Topic 2.1) as follows:

$$Al^{+3} > H^+ > Ca^{+2} > Mg^{+2} > K^+ = NH_4^+ > Na^+$$

However, Al^{+3} is a minor component of the adsorbed cations, and aluminum usually exists in solution as the $AlOH^{+2}$ form.

When rainfall is high, H^+ and $AlOH^{+2}$ become more abundant on the mineral surfaces as the other ions are leached out. Such soils become acidic, because rainwater contains dissolved CO_2, which combines with water to form carbonic acid: $CO_2 + H_2O \rightleftharpoons H_2CO_3$. The carbonic acid in turn ionizes to form H^+ plus bicarbonate and a small amount of carbonate:

$$H_2CO_3 \rightleftharpoons H^+ + HCO_3^- \rightleftharpoons H^+ + CO_3^=.$$

As ions except for $AlOH^{+2}$ are replaced by H^+ and leached to the water table, the increased acidity promotes solution of hydroxides such as those of iron and aluminum. Thus, soils of high rainfall areas are acidic and poor in mineral ions. In cold climates, especially under conifers, silica builds up, producing the grey podzolic soils. In tropical laterite soils, silica is broken down and leached away, and the reddish oxides of iron and aluminum (the **sesquoxides**) accumulate. In dry climates (deserts), soluble salts produced by gradual weathering of rock are not leached away, so Ca^{+2}, Na^+, Mg^{+2}, and/or K^+ become abundant, sometimes so much so that they are toxic to plants. Irrigation often makes these soils highly fertile by leaching away excess salts.

The organic matter fraction of soils also has an excess of negative charges resulting mainly from ionized carboxyl groups ($-COOH \rightarrow -COO^- + H^+$), but also from ionized OH groups ($-OH \rightarrow -O^- + H^+$) from phenolic compounds present in decaying lignin, an important compound in wood. Cations are adsorbed onto organic matter surfaces as well as to clay surfaces. In either case, the cations can exchange reversibly with identical or different cations dissolved in the soil solution, as illustrated in Fig. 5-5. Potassium replacement by Ca^{+2} is shown as an example of exchange between different cations. This phenomenon is called **cation exchange.** With **equivalent solutions** (the same number of ionic charges per unit volume), cations will replace each other according to the series given earlier, with Al^{+3} replacing all the others. The absorption of cations from the **soil solution** (soil water with its solutes) by roots, microorganisms, or mycorrhizae removes ions from solution, so some of those ions remaining adsorbed to the soil particles will be released into solution to maintain equilibrium. In general, soils having relatively high amounts of exchangeable cations (having a high cation exchange

Figure 5-5 Cation exchange on a clay particle. The lower right reaction illustrates replacement of K^+ by Ca^{++}.

capacity) are more fertile than those having a low cation exchange capacity.

Roots and microorganisms also influence binding of cations to soil minerals and organic matter through the CO_2 they release in respiration. The H^+ ions from the resulting carbonic acid are strongly attracted to negative sites of the soil particles and constantly replace some of the adsorbed cations. The replaced cations are then more available to the roots and microorganisms.

Thus far, we have spoken only about adsorbed and dissolved cations. Phosphate ions ($H_2PO_4^-$ or $HPO_4^=$, depending upon pH) also occur in solution in low concentrations, although most of the soil phosphate seems to be reversibly precipitated in solid phase aluminum, calcium, or iron salts. Anions such as nitrate (NO_3^-), sulfate ($SO_4^=$), and chloride (Cl^-) are not effectively attracted to organic matter or clay minerals, and their salts are highly soluble, so they are primarily dissolved in the soil solution where their charges are balanced by one or more cations. As a result of its extreme solubility and lack of adsorption, NO_3^- is readily leached from soils. Another form of nitrogen, NH_4^+, is adsorbed on colloids, but most soils contain little NH_4^+, because it is so rapidly oxidized to NO_3^- by bacteria and fungi, which obtain energy from this oxidation process. Because of the high requirement of plants for nitrogen and the loss of NO_3^- by leaching, most crops (except legumes) require relatively high amounts of nitrogen fertilizers. If nitrogen can be added in organic matter such as animal manure or dead vegetation (compost), it will be released only by decay and thus will be available to plants for a longer time. The organic matter will also improve the physical characteristics of the soil, including its cation exchange capacity, and will provide many nutrients other than nitrogen.

5.4 Ion Traffic into the Root

The most readily available anions and cations are those in the soil solution, even though their concentrations are usually very low. For example, in one survey of more than 100 agricultural soils at or near field capacity, more than half had dissolved NO_3^- concentrations of less than 2 mM, phosphate of less than 0.001 mM, K^+ of less than 1.2 mM, and $SO_4^=$ of

less than 0.5 mM (Reisenauer, 1966). Concentrations in the plant may reach 10 to 1000 times these values. Such nutrients reach the roots in three ways: (1) by diffusing through the soil solution, (2) by being passively carried along as water moves into the roots, and (3) by roots growing toward them.

Early experimenters working with nonmycorrhizal roots traced ion uptake processes with radioactive elements and found that the meristematic region of plant roots just back of the root cap accumulated ions more effectively than did older regions where cell elongation and differentiation were occurring or were complete. The accumulation of salts near the root tips is probably due to a high respiratory activity in this region and to the absence of conducting xylem to transport the salts away. In 1954, Herman Wiebe and Paul Kramer of Duke University found that although the meristem region effectively absorbs salts, it retains them. Older regions, in which root hairs are present and in which the xylem is well differentiated, are far more effective in releasing ions to the xylem where they can move to the shoot (Fig. 5-6). This work has been confirmed, and the evidence collectively indicates that salts are absorbed through and transported from regions of the root containing root hairs and also through much older regions several cm from the root tip (Clarkson, 1974).

Roots of trees and other perennials having secondary growth (thickening due to division of vascular cambium) eventually lose their epidermal and most of their cortical cells, and the endodermis is also destroyed. These roots become covered with a thin layer of corky tissue, frequently containing numerous specialized holes called **lenticels** through which oxygen can penetrate. Much of the nutrients and water

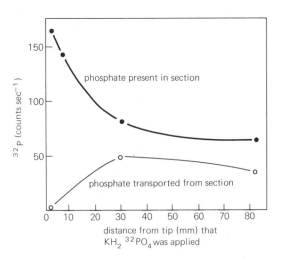

Figure 5-6 Ability of various regions of barley roots to accumulate phosphate and to translocate it to other parts of the plant. Phosphate containing ^{32}P was provided to 3 mm sections of the root at various distances from the apex for uptake periods of 6 hours. The plant was harvested, and the treated section and other organs were analyzed for ^{32}P. (Drawn from data of Wiebe and Kramer, 1954.)

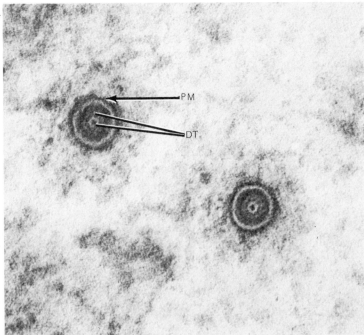

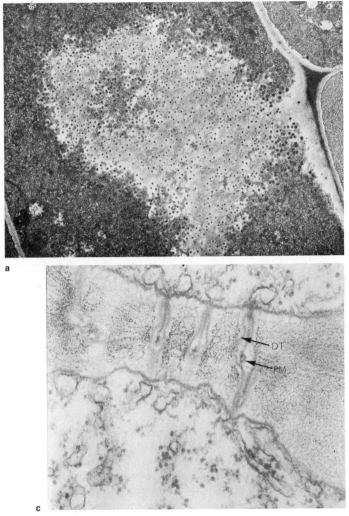

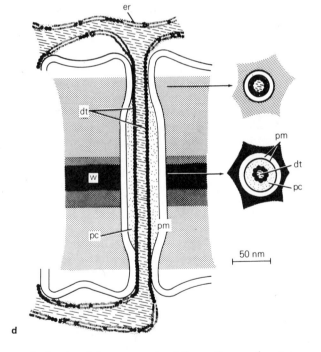

Figure 5-7 The structure of plasmodesmata. (**a**) Hundreds of plasmodesmata (small black dots) in a primary pit field in the end wall of a young barley endodermal cell. (**b**) High magnification of a few such plasmodesmata, showing their tubular nature. (**c**) Longitudinal view of plasmodesmata across two adjacent young endodermal cells. PM, plasma membranes; DT, desmotubule. (Courtesy A. W. Robards, 1976.) (**d**) An interpretation of plasmodesmatal structure; endoplasmic reticulum (ER), center of cell walls (W), plasmodesmatal cavity (PC). (From Robards, 1976.)

must penetrate from the soil through such roots, especially during periods of reduced feeder root growth, even though a layer of water-repellent suberin material nearly covers the outer surface of these corky cells. This layer is discontinuous in certain places, allowing some water and salts to enter, while other entry paths include the highly permeable lenticels and holes left where some of the branch roots have died (Chung and Kramer, 1975).

In Section 4.4, we examined the pathway of water movement into a root (see Figs. 4-9 and 4-10). There are three apparent paths through which water and ions might move into the xylem cells of nonmycorrhizal roots: (1) through the cell walls (apoplast) of epidermal and cortical cells, (2) through the cytoplasmic (symplast) system, moving from cell to cell, and (3) from vacuole to vacuole of the living root cells (where the cytoplasm of each cell would necessarily form part of the pathway). Ions adsorbed into root cell vacuoles usually move out slowly, so the third pathway is less important than the other two. So far, too little attention has been given to the mycorrhizal fungi in solute transport. Since they are so widespread in nature, pathways of transport from the fungal hyphae in the soil to the hyphae in the cortex and then into the root cells must also be evaluated.

As we saw in Section 4.4, movement of water and ions occurs readily through the walls of cortex cells until restricted by the impermeable Casparian strips of endodermal cells or the suberin layer on the outer tangential wall of the endodermis next to the cortex (see Fig. 4-9). At this point, further progress of solutes toward the xylem is controlled by the plasma membranes of the endodermal cells. These membranes help control the rates of ion absorption and the kinds of solutes absorbed. Soil colloids that might otherwise get into the xylem are thought to be restricted at this site (Clarkson and Robards, 1975). As dissolved ions move along in the walls of epidermal and cortical cells, many are absorbed by these cells into the cytoplasmic pathway. Some of these ions are further absorbed into the vacuole, where they contribute greatly to the negative osmotic potentials of roots, facilitating water uptake, turgor pressure, and growth of the roots through the soil.

An ion in the cytoplasmic pathway moving toward the xylem must often pass through the epidermis, several cortical cells, the endodermis, and the pericycle, depending upon which cell first absorbed it. This movement could involve transport directly through each of the two plasma membranes, primary walls, and middle lamella between the cytoplasm of adjacent cells. Alternatively, the ion could move across through **plasmodesmata,** which are tubular structures formed across adjacent cell walls of nearly all living plant cells (Fig. 5-7). Their frequencies are commonly greater than 1 million per mm² (Robards, 1975; 1976; Robards and Clarkson, 1976). They are lined by plasma membranes continuous between the adjacent cells. In their central region lies a tubular structure called a **desmotubule** that is apparently part of the endoplasmic reticulum, compressed in the wall region. Flow of solutes from one cell to the next through plasmodesmata has been demonstrated in various ways: for example, by precipitation of Cl⁻ with Ag⁺ salts. However, we are not sure whether solutes pass *only* through the desmotubule, in which case they must have first penetrated the endoplasmic reticulum, or whether they can also move through the plasmodesmatal cavity (PC in Fig. 5-7). Regardless of the overall contribution of plasmodesmata to solute movement across cells, direct movement through membranes is also involved in ion uptake by roots. To understand this process, we need to understand membranes.

5.5 The Nature of Membranes

Electron micrographs show that most biological membranes, regardless of their location, are similar. They are generally 7.5 nm to more than 10 nm thick and usually appear in cross-sectional high resolution electron micrographs as two

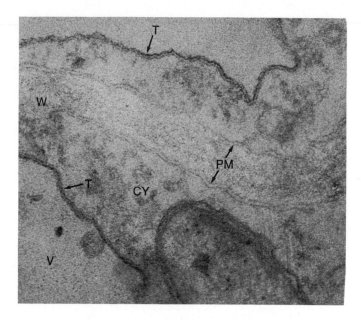

Figure 5-8 The three-layered appearance of the tonoplast (T) and plasmalemma (PM) in two root tip cells of potato. The cell wall (W), cytoplasm (CY), and part of the vacuole (V) are also shown. (Courtesy Paul Grun, 1963.)

dark (electron-dense) lines separated by a lighter (electron-transparent) layer (Fig. 5-8). The complete structure of three apparent layers is often referred to as a **unit membrane.**

Figure 5-9 shows an electron micrograph of a pea root membrane in surface view. During preparation of the tissue, the knife tore away part of the membrane and exposed the cellulose microfibrils in the wall below. The tiny bumps on the membrane represent protein molecules.

Membranes consist largely of proteins and lipids, the proteins usually representing about one half to two thirds of the membrane dry weight. The kinds and proportions of both proteins and lipids vary with the membrane and the physiological state of the cell in which it occurs, so some differences exist between plasmalemma, tonoplast, endoplasmic reticulum, and membranes of dictyosomes, chloroplasts, nuclei, mitochondria, and microbodies (peroxisomes and glyoxysomes). Important composition differences in a given membrane such as the plasmalemma also exist among various species. Nevertheless, the most abundant lipids of most plant membranes are **phospholipids, glycolipids,** and **sterols** (Chapter 14). These membrane lipids have a polar hydrophilic portion that can hydrogen-bond with water (at either surface of the membrane) and a nonpolar hydrophobic portion that extends from that surface toward the interior of the membrane.

The nature of membrane proteins has been less well investigated, partly because it was not previously known how to solubilize them. Membranes contain catalytic proteins (enzymes), and apparently some structural proteins. Another essential component of membranes is an ionic fraction, the most notable component of which is calcium. Although we do not yet know just how this ion is bound in the membrane, we do know that it is essential for normal membrane function.

The arrangement of proteins, lipids, and ions in membranes provides a continuing problem receiving much attention from biologists and chemists. Numerous structural models have been proposed, but none is universally accepted. A fairly recent model called the **fluid mosaic model** has received considerable support (Fig. 5-10). Protein molecules are imbedded like a mosaic in a fluid bilayer of lipids (Singer and Nicolson, 1972). Some of the hydrophobic proteins or their hydrophobic parts (those parts containing hydrophobic amino acids) penetrate well into the lipid-rich interior. Furthermore, some of the proteins extend all the way through the bilayer as shown. Most of the lipids present in the plasma membrane are phospholipids such as phosphatidyl choline, illustrated in Section 14.2, but glycolipids are more abundant in chloroplast membranes. Both phospholipids and glycolipids have a hydrophilic portion at either surface of the membrane and two long-chain fatty acids that provide the hydrophobic interior of the membrane.

One of the important things learned recently about membranes is that most of the lipids and many of the proteins in a membrane can move laterally over several molecular distances (hence the fluidity term in the fluid-mosaic model).

Figure 5-9 Scanning electron micrograph showing part of plasma-membrane of a pea root cell in surface view. The tissue was rapidly frozen before sectioning by a freeze fracture technique that splits the membrane in half between the lipid bilayers. As a result, only the lower half of the membrane adjacent to the cell wall is shown. Tiny bumps in the membrane represent protein molecules. The large hole in the membrane at the center was caused by the sectioning knife; this fortuitously allows visualization of strands of cellulose (microfibrils) in the wall below. (Courtesy Dan Hess.)

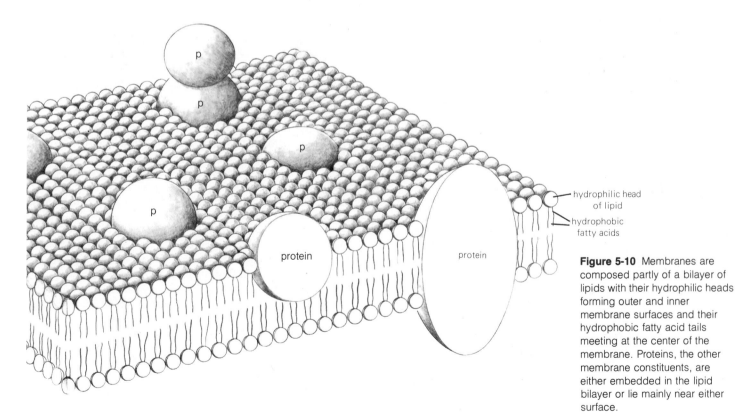

hydrophilic head of lipid

hydrophobic fatty acids

protein

protein

Figure 5-10 Membranes are composed partly of a bilayer of lipids with their hydrophilic heads forming outer and inner membrane surfaces and their hydrophobic fatty acid tails meeting at the center of the membrane. Proteins, the other membrane constituents, are either embedded in the lipid bilayer or lie mainly near either surface.

When different membranes fuse, as frequently occurs, the differences present at the moment of fusion gradually disappear, presumably due to the formation of a mosaic between formerly distinct components. One example of this occurs when vesicles of dictyosomes (Golgi vesicles) fuse with the plasma membrane, thus increasing its surface area during growth (see Fig. 10-16).

5.6 Some Principles of Solute Absorption

It is one thing to study membranes by studying their chemistry and their appearance in electron micrographs, but any membrane model must account for past observations of what passes through membranes. For many decades, before membranes could be isolated or seen with the electron microscope, this was the only approach. Numerous investigators studied the uptake of countless substances by myriad cell types and tried to imagine membrane models that would account for the observations (Collander, 1959). Though the picture that emerged was highly complex and marred by many exceptions to virtually every generalization, a few principles are of broad application. We shall consider four of them:

1. *Cells must be alive and metabolizing, or their membranes become much more permeable to virtually any solute.* If a cell is killed by high temperatures or poisons or if its metabolism is inhibited by low temperatures, nonlethal high temperatures, or specific inhibitors, many of the ions and relatively small molecules in the cell begin to leak out through the membranes.

2. *Water molecules and dissolved gases, such as O_2, N_2, and CO_2, penetrate all cell membranes very rapidly.*

3. *Nonionized, water soluble, hydrophilic molecules penetrate cells at rates inversely related to their molecular size (i.e., small molecules penetrate fastest).* As early as 1867, Moritz Traube in Germany made this observation and suggested the **molecular sieve hypothesis** or model, which says that cell membranes must contain numerous minute holes in the size range of the molecules being studied. Clearly, the smaller the molecule, the greater its expected chance of encountering a pore to penetrate. Nevertheless, even the smallest molecules penetrate much slower than water or dissolved gases, a central fact in our discussion of osmosis.

4. *The penetration of hydrophobic materials occurs at rates positively related to their lipid solubility.* In 1895, Charles E. Overton in Switzerland observed that regardless of the size of hydrophobic molecules, penetration is much faster if they are highly lipid soluble. Overton suggested the **lipid model** for membranes, which proposed that the entire membrane was essentially a lipid and that those hydrophobic substances that could dissolve in the membrane would penetrate best.

It occurred to subsequent investigators that the models of Traube and Overton could be nicely reconciled by a membrane that was a **lipid matrix with watery holes.** Lipid-soluble materials would go through the matrix, and hydrophilic materials through the holes. This was a forerunner of the fluid mosaic model discussed earlier. In the more modern model, the protein molecules with their associated water molecules might provide the hydrophilic "holes" through which water-soluble molecules penetrate. Yet the model is still too simple, because it does not explain the active control of solute passage into and out of cells (Singer, 1974). In particular, many ionized substances and some molecules do not follow the rules of Traube or Overton.

5.7 Characteristics of Ion Absorption

Ion Absorption at Various Ion Concentrations What are the facts of ion uptake by cells? Considerable attention has been given to the relation between the rate of absorption of mineral salts and the concentration of such solutes in the external medium. Many early experiments showed that if a dissolved solute moves across a cell membrane by simple diffusion, the rate of transport is directly proportional to the external concentration (see lower dashed line in Fig. 5-11a). However, when plant cells of many types are exposed to low concentrations of ions such as those encountered in nature, the uptake rate is much faster than predicted by simple diffusion and varies as shown by the upper solid line of Fig. 5-11a. The absorption rate does not rise linearly but appears to approach a maximum value as solute concentration in the external medium increases. Detailed mathematical analysis of such curves suggests that solute movement is being facilitated by some transport system that becomes saturated by the solute when the solute is present at the higher concentrations. This is one of the evidences for the existence of **carriers** (also called transporters, translocases, or permeases) that speed the movement of solutes across membranes (Epstein, 1972; 1973).

If the tissues are exposed to still higher solute concentrations (greater than about 1 mM), the expected saturation does not occur. Instead, the absorption rate first shows a rather sharp rise followed by another leveling off, then still other jumps and plateaus (Fig. 5-11b). Such results have been obtained for many ions and many different plant parts; even sugar absorption is multiphasic (Nissen, 1974).

We do not yet understand what such complex kinetics mean. They have been interpreted by some as suggesting that a proteinaceous ion carrier in the membrane becomes nearly saturated, but that further increases in the ion concentration cause the protein to change its shape so as to expose a new ion-binding site or to make the original site more accessible, perhaps by allosteric effects (see Section 8.6).

Numerous results such as those of Fig. 5-11 imply that the concentration of an ion outside plant cells is of primary importance in determining its rate of absorption, yet thermodynamic laws tell us that the internal effective concentration of a solute, its activity, is also important. Therefore, ion uptake rates should be more closely related to the difference in ex-

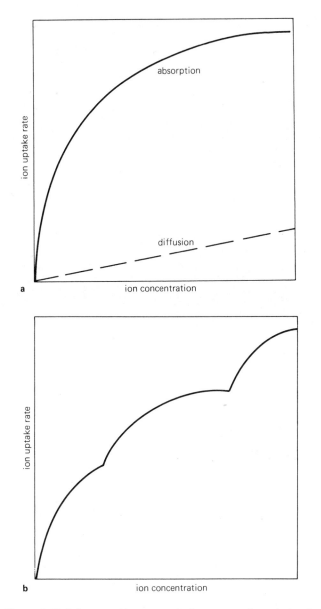

Figure 5-11 Influence of ion concentration surrounding plant cells on the rate of ion uptake. (**a**) If free diffusion were responsible for uptake, the rate would be low and essentially proportional to concentration, but actual rates are considerably higher and show saturation kinetics. (**b**) If the ion concentration shown in (**a**) is increased even further, additional rapid rises in uptake rates followed by evidence of saturation appear.

ternal and internal activities; that is, to the activity gradient. Research with numerous species supports this conclusion by showing that uptake by roots already containing relatively high concentrations of an ion is considerably slower than that of roots grown in more dilute solutions. Furthermore, it is probably the activity of an ion in the cytoplasm, not the entire cell, that influences absorption; cytoplasmic concentrations of many ions are suspected to be much lower than those of vacuoles, even when the average values for the entire cell are high. The slower absorption of ions by cells already containing relatively large amounts of these ions also applies to algae and other plant cells. In lakes, the rate of NH_4^+ absorption by algae often inversely reflects the amount of soluble nitrogen that has previously arrived in the lake.

Accumulation One of the most remarkable facts about absorption of ions by root or other cells is that, although slow compared to water absorption, such cells can ultimately absorb ions and other solutes to much higher concentrations than exist in the soil solution or other medium surrounding them. The process is called **accumulation.** For example, plant roots or sections from storage tissues such as slices of potato tubers placed in nutrient solutions often deplete the concentration of external ions to nearly zero within a day or two. During this time some of the ions attain concentrations inside the cell vacuoles over a thousandfold higher than those that are finally present in the surrounding solution. Some ions, including nitrate, ammonium, phosphate, and sulfate, are incorporated into organic compounds such as proteins and nucleic acids. This reduces the internal concentration or activity of the free ions, yet such concentrations often remain higher than outside. Others, such as chloride and potassium, are not incorporated into compounds and hence remain relatively free in solution.

Consider potassium. Plant tissues usually contain at least 1 percent of K^+ on a dry weight basis. When alive, such tissues typically contain about 90 percent water, so the living plant must contain about 0.1 percent or 25 mM K^+. This value is often attained by plants having roots growing in soils in which the dissolved K^+ is at no more than 0.1 mM, indicating accumulation ratios of about 250 to 1. The thermodynamic laws discussed in Chapter 1 clearly indicate that free diffusion not involving expenditure of the cell's energy could not be responsible for such accumulation ratios.

The accumulation of certain ions is not restricted to higher plants but is probably a universal phenomenon of living cells. Algae living in salty water must accumulate solutes to prevent plasmolysis, and one of the adaptive mechanisms of algae to a more saline environment is to increase their internal salt concentrations. Table 5-1 shows accumulation data for two species of algae living in different sea waters. Cells of both species have large central vacuoles from which solution was obtained for analysis. We may conclude from the table that whether or not an ion is accumulated depends upon both the ion and the species. Note that both algae accumulated K^+, *Nitella* also accumulated Cl^- and to a lesser extent Na^+, while *Halicystis* appeared to be in equilibrium with respect to Cl^- and actually restricted entry of Na^+.

This restriction or exclusion of Na^+ is of widespread occurrence in both plant and animal cells and depends (at least in animals) on an ATP-dependent "pump" (**sodium pump**) in the membrane to remove the Na^+ that slowly but continuously diffuses in (Skou, 1975). Most plant cells apparently have similar pumping mechanisms to remove Na^+ from their cytoplasm, although such pumps have only occasionally been clearly demonstrated. A pump in the plasmalemma causes efflux of the Na^+ that diffuses in, as in animal cells, and a tonoplast pump apparently transports some of the cytoplasmic Na^+ into the vacuole. In roots of most species, much of the Na^+ in the vacuoles is never transported to the shoots. In some salt-tolerant species, however, appreciable transport of Na^+ to the shoot does occur; here it is likely that the translocated Na^+ ions are those in the cytoplasm, not in the vacuoles, since transport of most ions out of the vacuoles is relatively slow.

Selectivity of Ion Absorption That solutes are absorbed and accumulated by selective processes is further indicated by other studies, many of which were initially performed by Emanuel Epstein at the University of California in Davis with roots excised from barley seedlings. The seedlings were grown in a solution that was allowed to become largely depleted of nutrients as growth occurred (Fig. 5-12a). Such "low-salt" roots have a high capacity for subsequent absorption of several ions, and this capacity is maintained for several hours even if the roots are cut off from the shoots (Epstein, 1972). Figure 5-12b shows a method by which such excised roots can be used in ion uptake studies. Aeration of both the excised roots and those attached to the seedlings during the prior growth period is essential for normal ion accumulation in most species (note tubes for forcing air through the solutions).

If such roots are provided with a solution containing dilute (about 0.2 mM) KCl and with about 0.5 mM Ca^{+2} to maintain normal membrane functions, the rate of absorption of K^+ is unaffected by the presence of similar concentrations of Na^+ salts. This is true even though Na^+ is chemically similar to K^+. The process of K^+ accumulation is therefore selective and uninfluenced by a related ion under these conditions. As ex-

Figure 5-12 (**a**) Barley seedlings grown as a source of low-salt roots. Seeds are germinated in water, then grown 3 days in darkness in 0.2 mM $CaSO_4$ to maintain membrane permeability but partially to deplete the roots of nutrient salts. (**b**) A 1-g sample of excised barley roots enclosed in an open-weave cheesecloth "teabag" and immersed in aerated experimental solution. In actual experiments, arrays of flasks are kept in temperature-controlled water baths. (From Epstein, 1972.)

pected, several other monovalent and less related divalent ions also have no influence on K^+ uptake. In similar experiments, the absorption of chloride is indifferent to the presence of the related halides, fluoride and iodide, as well as to NO_3^-, $SO_4^=$, or $H_2PO_4^-$. Calcium ions are essential for this selectivity, because without them potassium absorption, for example, becomes inhibited by similar low concentrations of sodium.

In spite of this apparent high selectivity, the uptake mechanisms can sometimes be "fooled." Potassium absorption is inhibited in the presence of Rb^+, and indeed the penetration of membranes by these two ions appears to follow the same mechanism. Similar competitive results are often obtained with Cl^- and Br^-, with Ca^{+2} and Sr^{+2}, and with $SO_4^=$ and selenate ($SeO_4^=$).

Failure of Absorbed Ions to Leak Out Once ions have been absorbed into the cytoplasm or vacuoles of root cells originally deficient in such ions, they do not readily leak out (efflux is almost undetectable). Leakage can be induced by damaging the membrane with heat, poisons, lack of O_2, and to some extent by removing Ca^{+2}, but these are abnormal situations. This indicates that absorption in low-salt roots is primarily a

Table 5-1 Concentrations of Major Ions in Sea Water Compared with Their Concentrations in Vacuoles of Algae Living There

	Nitella obtusa[a]—Baltic Sea		*Halicystis ovalis*[b]	
Ion	Vacuole Concentration	Sea Water Concentration	Vacuole Concentration	Sea Water Concentration
Na^+	54 mM	30 mM	257 mM	488 mM
K^+	113 mM	0.65 mM	337 mM	12 mM
Cl^-	206 mM	35 mM	543 mM	523 mM

[a]Data for *Nitella* are from J. Dainty, 1962, Annual Review of Plant Physiology 13:379.

[b]Data for *Halicystis* are from R. W. Blount and B. H. Levedahl, 1960, Acta Physiologia Scandinavia 49:1.

unidirectional influx. Efflux is much more extensive in roots that contain relatively high ion concentrations.

If the concentration of an ion in a low-salt tissue is measured as a function of time to which the cells are exposed to that ion, graphs like that of Fig. 5-13 are usually obtained. Under normal conditions of temperature and aeration (upper curve), there is an initial rapid inward movement of ions, although most of this simply represents filling of the free space in the walls (see Fig. 4-11), not actual penetration of the selective plasma membranes. Subsequently, the rate of absorption becomes essentially constant, often for as long as several hours. Now suppose the tissues are removed from the solutions at the time indicated by the arrows, placed in water, and the amount of the ion retained by the tissues followed with time. Only a small fraction of the ions present leak out (dashed line of Fig. 5-13). The ions that are lost represent mainly those that were present in the free space, not those actually inside the cells. Those present in the cytoplasm and vacuoles remain there for many hours and, as the figure shows, they represent by far the major fraction in such experiments.

If only cations are considered, there is less efflux when tissues are placed into water than with anions, because negatively charged cell wall constituents, apparently mostly ionized carboxyl groups of galacturonic acid molecules, strongly attract and hold the cations. These cations are adsorbed as they are on clay particles and can take part in cation exchange with other cations.

Figure 5-13 also shows the time course of accumulation when metabolic processes (especially respiration) are inhibited by lack of O_2, cold temperature, or numerous respiratory poisons. The initial rapid "absorption" phase is not greatly altered, but the absorption rate then quickly falls to nearly zero (i.e., the curve becomes horizontal). When these tissues are transferred to water, nearly all the ion is lost into the washing medium. This loss is of the same magnitude as that from the healthy tissues, and both occur mostly from the cell wall network. Such results illustrate both the failure of truly absorbed ions to leak out of such tissues and the inhibition of such absorption and accumulation by conditions not conducive to normal respiration. More will be said of the relation of respiration and ion uptake in Section 5.10.

Ion Uptake by Growing Cells We mentioned earlier that much of the ion uptake occurs in the root hair region and some even in older parts of the root. Most of these cells are nearly or entirely full grown, and we may ask whether ion absorption rates differ for growing and nongrowing cells. To answer this, first consider that water uptake is the driving force for growth (Chapters 2, 3, and 15). As the cell grows, the internal concentration of ions might be expected to decrease proportionately to the volume increase. According to the Höfler diagram (see Fig. 2-3), this dilution would make the osmotic potential and water potential of the cell less negative, and would decrease the pressure potential or turgor in the cell. Eventually the walls would yield no more, and growth would stop. We would predict, therefore, that ion uptake by growing cells would be faster than by nongrowing cells, simply because of the dilution effect by absorbed water upon the ion concentration gradient across the membranes of growing cells. Although our prediction is generally correct, there may be additional reasons for the more rapid ion uptake rate in growing cells, because a buildup of turgor inside the cells by injection of water from a microsyringe can rapidly reduce ion absorption rates, and a decrease in turgor can increase such rates. How turgor changes control ion uptake is unknown, but since water movement accompanies ion transport, the cell is provided with an important feedback mechanism allowing maintenance of a more constant turgor and therefore a more constant growth rate (Hsiao et al., 1976).

5.8 pH Changes and Maintenance of Electrical Neutrality

It is usually found that the uptake of nutrient anions such as NO_3^-, $H_2PO_4^-$, Cl^-, and $SO_4^=$ from soils exceeds the uptake of cations such as K^+, Na^+, NH_4^+, Ca^{+2}, and Mg^{+2}, although occasionally the cations are absorbed more rapidly (Noggle, 1966). A critical factor determining whether uptake of anions is more rapid than cations or vice versa is the relative concentration of *monovalent* ions, especially K^+ and NO_3^- (or NH_4^+ in soils recently fertilized with an ammonium salt). All three of these ions are absorbed much more rapidly than are divalent ions such as Ca^{+2}, Mg^{+2}, or $SO_4^=$.

Whenever cation and anion absorption are not equal, problems of electrical neutrality in the soil solution and inside

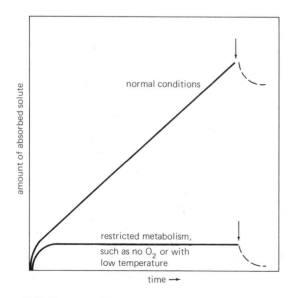

Figure 5-13 Progress of ion uptake with time under various conditions. For explanation, see text.

the plant arise, because the total number of positive and negative charges in the soil must almost exactly balance, and the same must be true within plant cells (Davies, 1973; Higinbotham, 1973; Raven and Smith, 1974). Electrical neutrality could be obtained in either of two ways. If uptake of nutrient anions exceeds nutrient cation uptake, HCO_3^- or OH^- could be transported outwardly from the cells to satisfy the charge difference. Alternatively, H^+ could be absorbed along with the excess anions. We cannot readily distinguish between these possibilities, since both lead to an increased pH in the soil or nutrient solution in which the roots exist.

When cation uptake exceeds anion uptake, H^+ could be released from the roots or HCO_3^- or OH^- might be absorbed as charge balancers. Either effect causes a *decrease* in soil or nutrient pH. Evidence now indicates that an H^+ and OH^- pump dependent on ATP causes H^+ to accumulate outside the roots, allowing cations to enter and neutralize the remaining cytoplasmic anions. It must be emphasized, however, that H^+ would accumulate outside the plasma membrane if the ATP-dependent pump transfers OH^- inwardly, just as well as if H^+ were pumped outwardly. Thus, when we speak of an H^+ pump, we cannot distinguish between movement of OH^- inwardly and H^+ outwardly; in fact, both might occur simultaneously.

Now what goes on inside the root? When cation uptake is in excess and H^+ is lost from the cells, there is an increased production of organic acids, especially malic acid (as in the stomatal mechanism discussed in Section 3.4). These acids ionize, the resulting H^+ partly replaces those that were exchanged for incoming excess cations, and the organic anions neutralize charges of the excess cations. Loss of H^+ causes the pH of the cell sap to rise slightly, even though acids are being produced. When anion absorption is in excess, organic acid anions, especially malate, disappear from the roots through respiration and perhaps also by translocation, helping to prevent extreme buildup of negative charges in the roots.

5.9 Electropotentials Across Roots Modify Ion Absorption

In spite of the exchanges of H^+, OH^- and HCO_3^- during absorption of nutrient ions, electrical neutrality is not exactly maintained. Instead, both the cytoplasm and vacuole appear to be slightly negatively charged relative to the surrounding soil solution, and usually there is very little charge difference between cytoplasm and vacuole (Higinbotham, 1973; Higinbotham and Anderson, 1974). These electropotential differences across the plasma membrane are measured by inserting tiny electrodes directly into the vacuole or cytoplasm, although it is often difficult to tell which compartment the electrode tip is in. Values between -70 and -150 millivolts relative to the outside medium are quite commonly observed. They are believed to arise largely from an ATP-dependent pumping of

H^+ outwardly (or OH^- inwardly) not fully compensated by uptake of nutrient cations.

These electropotential differences strongly influence ion uptake, just as do concentration differences. If potentials are more negative inside cells, cations will tend to diffuse in, restoring neutrality. Thus, we must know both the electropotential and the chemical potential (essentially concentration) gradients across the membrane before we can calculate the energetics of the absorption process. These two gradients constitute what is usually called the **electrochemical potential gradient.**

Many physiologists attempt to classify uptake of ions and uncharged molecules into one of two categories: active or passive. **Active uptake** is usually defined as absorption against an electrochemical potential gradient, not just a chemical potential gradient. **Passive uptake** is therefore absorption down an electrochemical potential gradient, not just a chemical potential gradient. Absorption of anions leading to accumulation is considered almost always to be active because of the negative internal electropotential, yet cation absorption against a concentration gradient (i.e., when accumulation ratios are greater than unity) could often be passive, because the negative electropotential of the cytoplasm attracts cations and overcomes the otherwise unfavorable chemical potential gradient.

Even though cation uptake often occurs passively, this does not negate the requirement for aerobic respiration and an energy supply to maintain the conditions that make the electropotential gradient favorable. We shall see that energy, usually in the form of ATP, is necessary to drive uptake of both anions and cations, even though from a thermodynamic standpoint one may be entering actively, while the other is dragged along passively.

5.10 Ion Accumulation by Roots Requires Respiration and ATP Production

As we saw in Fig. 5-13, roots must be provided with O_2 and actively respiring to rapidly absorb and accumulate ions and other solutes. The same is true of other plant cells. There are three major evidences for this conclusion. *First,* if O_2 is not readily available, the absorption rate is greatly reduced. Figure 5-14 illustrates how important this is for K^+ uptake by excised barley roots. In this and many other examples, uptake was saturated by a bubbling gas mixture containing about 3 percent O_2, a technique that also saturates the respiratory rate. In contrast, excised rice roots showed less dependence on O_2, which is consistent with the fact that rice is commonly grown in waterlogged, O_2-deficient soils. Probably all plants have an intercellular space system extending from shoots to roots that is filled with air and through which O_2 can diffuse (Greenwood and Goodman, 1971). This system is particularly well developed in rice, no doubt contributing to its relatively greater ability to absorb solutes from wet soils deficient in O_2. Such plants function with a sort of snorkel system.

Studying Water, Minerals, and Roots   **Paul J. Kramer**

F.B.S. 1965

Paul J. Kramer, one of the "deans" of plant-water relations, has pondered the roles of water and minerals in plants since 1931 in his laboratory at Duke University, where he is Professor of Botany, producing many significant data and ideas, not to mention graduate students who are now carrying on the work all over the world.

My entrance into the field of plant water relations research was somewhat accidental. In 1928, when I was a young graduate student at Ohio State University searching for a thesis topic, Professor E. N. Transeau showed me a paper by Burton E. Livingston in which it was claimed that osmosis plays a negligible role in water absorption. Transeau suggested that, as Livingston presented little evidence for his view, I might investigate the problem, and I have been working in that field ever since.

Most textbook writers of those days assumed that osmosis was somehow involved in water absorption, but their discussions were so vague that it was quite impossible to understand how absorption really occurred. My early research was done very simply with T-tubes, pipettes, rubber tubing, a vacuum pump, a pressure chamber built from pipe fittings, tomato and sunflower root systems, and papaya petioles. It was intended to test ideas of Atkins, Renner, and other early workers, and the results indicated that root systems can function as osmometers, and transpiring shoots can absorb water through dead root systems. This research led me to support the neglected view of Renner that two mechanisms are involved in water absorption. According to this view, the root systems of plants growing in moist, well-aerated soil function as osmometers when transpiration is slow, resulting in the development of root pressure and the occurrence of guttation. When rapid transpiration lowers the pressure or produces tension in the xylem sap, however, water is pulled in through the roots, and osmotic movement is negligible. Thus, in transpiring plants, most or all of the water enters passively and the roots act merely as absorbing surfaces. Further research demonstrated that factors such as cold soil and deficient aeration reduce absorption by increasing the resistance to water flow through roots, rather than by inhibiting some mysterious kind of active absorption mechanism. This early research contributed to the development of a relatively clear explanation of how water is absorbed and of how certain environmental factors affect the rate of absorption. Looking back, those early experiments seem prosaic, but at the time they were very exciting.

During the 1950s, I began to realize that many of the contradictory reports in the literature concerning the relationship between soil moisture and plant growth resulted from the fact that one cannot accurately predict plant water stress from measurement of soil water stress. Holger Brix, John Boyer, and others began working as students in my laboratory with the recently introduced thermocouple psychrometers, and I started a campaign to educate plant scientists about the necessity of measuring plant water potential. As they were finally beginning to realize the importance of water in relation to plant growth, this idea was generally accepted. I notice, however, that even today workers in water relations sometimes reduce the value of their research by failing to measure the degree of water stress to which their plants are subjected.

My work on water absorption naturally led to an interest in roots as absorbing organs. With the aid of a zoologist colleague, Karl Wilbur, I measured uptake of phosphorus by mycorrhizal roots. Then Herman Wiebe, as a graduate student in my laboratory, did some interesting experiments with radioactive tracers, which showed that the region of most rapid salt absorption and translocation to shoots often is several centimeters behind the root tip, rather than near it, as previously claimed. This conclusion was viewed with considerable skepticism at first, but we have had the pleasure of seeing it verified by more recent research, including work in R. Scott Russell's laboratory at Letcombe, England. I also concluded that considerable salt and water are absorbed by mass flow through suberized roots. This view has not yet been generally accepted, though evidence supporting it has been available for over 30 years. Perhaps this illustrates the difficulty of getting well-entrenched old ideas replaced by new ones.

Recently, Dr. Edwin Fiscus and other investigators in my laboratory have been studying various anomalies in water conduction, especially the reports of decrease in resistance with increase in rate of flow. Most of these reports involve the strict application of Ohm's law (flow being proportional to driving force and inversely proportional to resistance) without taking into account such anatomical and physiological facts as the role of phyllotaxy (leaf arrangement on the stem) in controlling the water supply to individual leaves or the fact that, as the rate of water flow through roots increases, the driving force changes from primarily osmotic to primarily mass flow. Thus, a simple application of Ohm's law is inadequate to explain water absorption without consideration of the driving forces involved.

I am glad that I entered plant physiology at a time when it was still possible to be a generalist. Thus, in addition to my work on water relations, salt absorption, and root systems, I was also able to carry on research on the physiology of woody plants. To me, research in whole plant physiology on the borderline between physiology and ecology, in what perhaps can be called environmental physiology, has been extremely interesting because of the variety of problems that it has presented. I also think that it has enabled me to contribute more to a general understanding of how plants live and grow than I could have contributed by concentration in a narrow area of research. In any event, there has never been any danger of boredom!

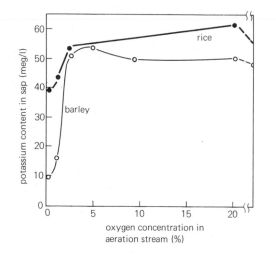

Figure 5-14 Influence of oxygen content in gas mixture used to aerate nutrient solutions upon potassium absorption by excised rice and barley roots. Plants were grown 4 weeks in a complete nutrient solution, and the roots were then excised and kept 24 hours in a 5 m*M* KBr solution aerated with various mixtures of oxygen and nitrogen. (From Vlamis, 1944.)

Second, there is a correlation between the inhibitory effects of cold temperatures upon rates of ion absorption and rates of respiration. The rate of salt absorption by crop plants usually increases with temperature from the melting point of water up to about 40 C, where it then begins to decrease. The slow growth of such plants in cold soils in the spring is partly due to the limited rate of ion uptake under these conditions. Increased absorption with rising temperatures is caused partly by the increased rate of diffusion of salts to the roots but mainly by the increased respiration of the plants at higher temperatures.

Third, several compounds that inhibit respiration also lower the rates of ion absorption. Three of the more effective inhibitors are malonic acid (a structural analogue of the Krebs cycle intermediate succinic acid; see Section 12.5), the azide ion (N_3^-), and the cyanide ion (CN^-). The latter two substances interfere with the mitochondrial cytochrome electron transport system, as discussed in Section 12.5.

To understand why respiration is essential for ion uptake, we might logically ask what substance is formed during respiration that is essential for uptake processes. The principal products of respiration are CO_2, H_2O, and ATP. HCO_3^- arising from CO_2 could exchange with incoming anions, thereby stimulating ion uptake, yet the most important product of respiration for ion absorption is ATP, the compound that provides the actual energy for most cellular processes. Much evidence supports such an involvement of ATP. For example, 2,4-dinitrophenol is a potent inhibitor of ion uptake even at concentrations so low that respiration is not inhibited or is even stimulated. How can dinitrophenol inhibit ion uptake without inhibiting respiration? It does this by interfering with ATP synthesis, even though electron transport, O_2 uptake, and CO_2 release may even occur faster in its presence. (The mechanism by which ATP synthesis and respiratory electron transport might become "uncoupled" by dinitrophenol is discussed in Special Topic 9.1.) Other compounds, such as the antibiotic oligomycin, also inhibit mitochondrial ATP production and ion uptake with little or no effect on respiratory gas exchange. So it is clear that ATP production is more closely related to ion uptake than are the other phases of respiration.

An additional evidence that ATP is the principal energy source for ion uptake in roots comes from studies on the relation of the ability of plasma membranes to hydrolyze ATP (thus apparently harvesting its energy) and their ability to absorb cations. Both the plasma membrane, the tonoplast, and other membranes contain **ATPase enzymes** capable of hydrolyzing ATP to ADP plus phosphate (Hodges, 1973; 1976). The activity of ATPases from plasma membranes of cereal grain roots is stimulated by Mg^{+2} (or Mn^{+2}) and additionally by K^+ (or Rb^+). There is excellent correlation between the enhancing effects of various concentrations of K^+ upon ATPase activities and the uptake rates of K^+ by the roots at these same concentrations. It appears that these ATPases are involved in utilizing energy from ATP to drive cation influx and H^+ efflux. More recently, plasma membrane preparations were found by Hodges to contain ATPases activated by monovalent anions. Perhaps they function in anion absorption.

Even though we are not yet certain how ATP is involved in ion accumulation, the fact that it is and that it is produced by respiration allows us better to understand how certain environmental factors influence growth. Roots of plants growing in soils that are cold, waterlogged, or excessively compacted (poorly aerated) probably have retarded respiration rates. Absorption of ions essential for growth of both roots and shoots will be limited under these conditions. We might also predict that bright, sunny days would lead to enhanced sugar production by photosynthesis, faster sugar translocation to the roots, more rapid respiration of these sugars by the roots with faster ATP production and, therefore, greater rates of ion absorption. Measurements have indeed shown that respiration rates of roots are closely correlated with translocation of sugars from the shoot (Hatrick and Bowling, 1973). Other data showed that ion uptake by rapidly respiring plants was about twice as fast as for those having low rates of translocation and respiration caused by prior exposure to darkness. We may conclude that this mechanism is one of the ways that growth of roots and shoots is correlated.

6

Mineral Nutrition

What elements must a plant absorb to live and grow? What "food" does a plant need? Can a plant grow when provided only with elements in inorganic form? Or do plants, like animals, require vitamins? If only minerals, then which minerals? Can we tell when a plant is lacking some essential element? How can we best apply the limiting element to overcome its deficiency? What are the functions of these elements in the plant?

These are the questions of plant mineral nutrition, an important subscience of plant physiology. Since we must properly "feed" plants before we can feed ourselves, these are important questions. Answers obtained so far have revolutionized agriculture during the past century and a half, but we are now trying to refine the answers, because an overuse of fertilizers may damage the environment. The questions of mineral nutrition also have profound implications for basic understanding of plants. In one sense, plant growth is the incorporation of various elements into plant substance.

This chapter is concerned with the elements necessary for proper plant nutrition, some of their functions, and deficiency symptoms occurring in their absence. We shall also mention some elements that might eventually prove essential for certain species.

6.1 The Elements in Dry Matter

An important approach to gaining at least partial answers to the questions of mineral nutrition is to analyze healthy plants for the minerals they contain. When freshly harvested plants or plant parts are heated to approximately 80 C (176 F) for several hours, nearly all the water is driven off, and the remainder is the so-called dry matter. This dry matter represents about 10 to 20 percent of the original fresh weight of many leaves and roots, although a much higher percentage of woody tissues is dry matter, and most of the weight of mature seeds is dry matter. The principal components of dry matter are cell wall materials such as carbohydrates and lignin, plus such protoplasmic components as proteins, lipids, amino acids, organic acids, and certain elements that form no essential part of any compounds.

An analysis of the principal elements present in the dried shoot system of a corn (maize) plant, published in 1924, is given in Table 6-1. Note that oxygen and carbon are by far the most abundant elements (on a weight basis), and that hydrogen ranks third. This is approximately the same distribution of elements found in carbohydrates, including cellulose, the most abundant compound in wood. Still smaller amounts of nitrogen are found, followed by several other elements that exist in even lower concentrations. Had a complete elemental analysis been made, trace amounts of numerous other elements could have been found, some of which are essential to corn and other plants.

Table 6-1 also includes two elements, aluminum and silicon, that are believed to be nonessential for growth of higher plants. More will be said of these elements later, but a point to be emphasized is that plants will absorb and accumulate numerous elements from the soil solution that they do not require. At least 60 elements have been found in plants, including gold, lead, mercury, arsenic, and uranium. Even the

Table 6-1 Elemental Analysis of the Stem, Leaves, Cob, and Grain of a Mature Maize Plant ("Pride of Saline"), Based on Average Values for Five Plants[a]

Element	Weight (g)	Percentage of Total Dry Weight
Carbon	364.19	43.569
Oxygen	371.42	44.431
Hydrogen	52.17	6.244
Nitrogen	12.19	1.459
Sulfur	1.416	0.167
Phosphorus	1.697	0.203
Calcium	1.893	0.227
Potassium	7.679	0.921
Magnesium	1.525	0.179
Iron	0.714	0.083
Manganese	0.269	0.035
Silicon	9.756	1.172
Aluminum	0.894	0.107
Chlorine	1.216	0.143
Undetermined	7.8	0.933

[a]Data of Latshaw and Miller, 1924, Journal of Agricultural Research 27:854.

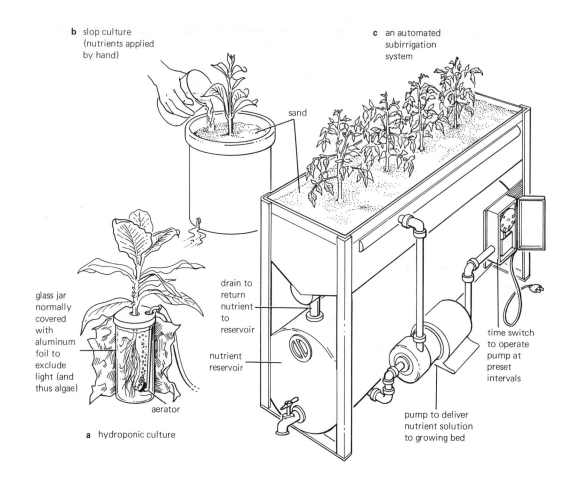

Figure 6-1 Three methods for growing plants with nutrient solutions. The container should be large compared to the volume of roots so that the solution composition does not change too rapidly. (**a**) Hydroponic culture is used in laboratory studies that require the most accurate control of nutrients, but slop culture (**b**) requires no aeration or special plant support; it is adequate for many studies. (**c**) The automated subirrigation system (which is hypothetical) could be used in a commercial operation or for a laboratory study requiring many plants.

b slop culture (nutrients applied by hand)

c an automated subirrigation system

sand

glass jar normally covered with aluminum foil to exclude light (and thus algae)

aerator

a hydroponic culture

drain to return nutrient to reservoir

nutrient reservoir

time switch to operate pump at preset intervals

pump to deliver nutrient solution to growing bed

"man-made" element plutonium, which has not been detected in the natural environment, can be absorbed by plants.

Soils are composed largely of aluminum, oxygen, and silicon, yet plants by no means reflect this composition. Part of the reason they do not is because they absorb carbon and much of their oxygen from the air. Another reason is that most of the oxygen, silicon, and aluminum present in soil minerals is not readily soluble in the soil water. Finally, roots exhibit considerable selection over the kinds of elements absorbed and the rates at which these are absorbed, as discussed in Chapter 5.

6.2 Methods of Studying Plant Nutrition: Hydroponics

Beginning in about 1804, scientists began to appreciate the fact that plants require certain mineral elements: calcium, potassium, sulfur, phosphorus, and iron. Then, during the years around 1860, two German plant physiologists, Julius Sachs and W. Knop, independently recognized the problems of determining the kinds and amounts of elements essential to plants growing in a medium as complex as the soil. They therefore grew plants by immersing their roots in an aqueous nutrient solution, the chemical composition of which was limited only by the purity of available chemicals. Growing plants in this way is referred to as **hydroponic culture** (Fig 6-1). Although Sachs and Knop apparently did not realize it, other investigators later showed that many plants grew much better if the roots were aerated, as shown in Fig. 6-1 (see also Fig. 5-12).

As techniques became available to purify the salts and the water used for hydroponic culture, more exact control was possible over the elements available to the plants. This proved to be especially important with several elements that are required only in very small amounts. Furthermore, a requirement for certain nutrients, such as molybdenum, copper, zinc, and boron, is often difficult to demonstrate for certain species having large seeds, because these seeds contain sufficient amounts of these elements for the growth of mature plants derived from them. In such cases, deficiency symptoms are more easily observed in the second generation when seeds taken from parents grown without the added nutrients are used.

In spite of the advantages of liquid hydroponic cultures in mineral nutrition studies, the technique has some disadvantages. The need for aeration is one, and it is usually necessary to replace the solution every day or two for maximum growth. This is because the solution composition changes continuously as certain ions are absorbed more rapidly than others. This preferential uptake not only depletes certain ions but also causes undesirable pH changes, as discussed in Section 5.8. To avoid some of the problems with liquid cultures, many investigators use washed white quartz sand or a mineral called *perlite* (expanded pumice) as a medium for the roots (Fig. 6-1). Nutrient solutions are simply poured or allowed to drip on top of these media at suitable intervals in excess amounts to insure leaching of the old solution through holes in the bottom of each pot. The technique is convenient but may not be suitable for some detailed studies. Such hydroponic cultures, however, are used by some greenhouse owners to grow commercial crops such as tomatoes, especially in winter (see the boxed essay on p. 86).

Numerous useful formulations of nutrient solutions have been devised from studies in which the proper concentrations of various elements have been investigated. Two such recipes are listed in Table 6-2, one by Dennis R. Hoagland, who pioneered mineral nutrition work in the United States, and a slightly more modern one. Both have the necessary elements in the proper amounts to allow good growth of many higher plants, but an ideal solution for one species is seldom exactly ideal for another. Also note that the Evans solution in Table 6-2 contains all its nitrogen in the form of the nitrate ion, but the rapidity with which this ion is absorbed frequently causes rapid rises in the nutrient solution pH. This problem can be minimized by supplying part of the nitrogen as an ammonium salt—that is, $NH_4H_2PO_4$, as in the Hoagland solution of Table 6-2.

Nearly all nutrient solutions are more concentrated than the solutions in soils. For example, the phosphorus concentration of the Evans solution in Table 6-2 is 500 micromolar (μM), while about three fourths of the determinations in one survey of 149 soils gave phosphorus levels less than 1.5 μM

(Reisenauer, 1966). Over half of these soil solutions had potassium concentrations less than 1.25 millimolar (mM), while the Evans solution is 5.5 mM in K^+. As we have seen (Section 5.3), many minerals in the soil are not in solution but adsorbed on clay surfaces and organic matter or precipitated as insoluble salts, going into the soil solution only slowly as they are removed by plants or by leaching. Many plants grow well in solutions having concentrations of essential elements as low as those dissolved in the soil solution, provided the solutions are replenished often enough to maintain such concentrations. The higher concentrations typically employed avoid the necessity to change solutions more than once a day or once every few days, depending on growth rates. Of course, the concentration must be low enough to prevent plasmolysis of the root cells. Most solutions have osmotic potentials no more negative than −1 bar, so this is not a problem.

6.3 The Essential Elements

The nutrient solutions of Table 6-2 contain 13 mineral elements believed essential for all angiosperms and gymnosperms, although in fact the nutrient requirements of most higher plant species have not been investigated. Adding H_2O and CO_2 to the 13 minerals, 16 elements are considered essential. With these and sunlight, plants can apparently synthesize all the compounds that they require. But is it possible that higher plants required some organic molecules (vitamins) synthesized by microorganisms that grow on plant roots? Apparently not, since plants have now been grown under sterile conditions in plastic or glass enclosures from which all microorganisms were excluded. Higher plants really are autotropic.

There are two principal criteria by which an element may be judged essential or nonessential to any plant (Epstein, 1972): *First*, an element is essential if the plant cannot complete

Table 6-2 Two Nutrient Solutions for Hydroponic Culture

| Hoagland's Solution[a] | | | Evans' Modified Shive's Solution[b] | | |
Salt	Molarity	mg/l (ppm)	Salt	Molarity	mg/l (ppm))
KNO_3	0.010		$Ca(NO_3)_2 \cdot 4H_2O$	0.005	
$Ca(NO_3)_2$	0.003		K_2SO_4	0.0025	
$NH_4H_2PO_4$	0.230		KH_2PO_4	0.0005	
$MgSO_4 \cdot 7H_2O$	0.490		$MgSO_4 \cdot 7H_2O$	0.002	
Mixture of 0.5% $FeSO_4$ and 0.4% tartaric acid:			Fe-versenate		0.5 Fe
0.6 ml/l added 3 times/week			KCl		9.0 Cl
$MnCl_2 \cdot 4H_2O$		0.5 Mn; 6.5 Cl	$MnSO_4$		0.25 Mn
H_3BO_3		0.5 B	H_3BO_3		0.25 B
$ZnSO_4 \cdot 7H_2O$		0.05 Zn	$ZnSO_4$		0.25 Zn
$CuSO_4 \cdot 5H_2O$		0.02 Cu	$CuSO_4$		0.02 Cu
$H_2MoO_4 \cdot H_2O$		0.05 Mo	Na_2MoO_4		0.02 Mo

[a]From D. R. Hoagland and D. I. Arnon, 1938, University of California Agricultural Experimental Station Circular # 347.

[b]From H. J. Evans and A. Nason, 1953, Plant Physiology 28:233-254.

its life cycle (i.e., form viable seeds) in the total absence of the element. Second, an element is essential if one can show that it forms part of any molecule or constituent of the plant that is itself essential (for example: N in proteins, Mg in chlorophyll). Either criterion is sufficient to demonstrate essentiality, and for most in our list of 16, both have been met. Historically, the first criterion has been the principal one employed.

It is usually easier to state positively that an element is essential than that it is not. Experimenters therefore often state that if an element in question is necessary, it is required only in concentrations less than the sensitivity limits of their detecting instruments. For example, it was reported that if silicon is essential for tomato plants, the amount needed is less than 0.2 μg per gram of dry tissue. Because of such problems, it is possible that a few more nutrient elements needed in barely detectable amounts will eventually be added to our list.

Table 6-3 lists the 16 elements presently believed essential to all higher plants, the chemical form most readily available, the approximate concentration in the plant considered adequate, and the approximate number of atoms of each element needed compared to molybdenum. The table shows that about 60 million times as many atoms of hydrogen than of molybdenum are required, a dramatic difference. The first seven listed are often referred to as the **trace elements, minor elements,** or **micronutrients** (needed in tissue concentrations equal to or less than 100 μg/g dry matter), and the last nine as **macronutrients** or **major elements** (needed in concentrations of 1,000 μg/g dry matter or more). The internal concentrations listed as adequate should be considered as useful but not totally accurate guidelines, especially for calcium and magnesium and probably several of the micronutrient cations, such as copper, zinc, manganese, and iron. Some results show that good growth of corn and tobacco can be obtained at much lower tissue concentrations of calcium and magnesium if the levels of certain micronutrient cations in the nutrient solution are also decreased. This suggests that much of the calcium in plants is needed only to prevent injury from excesses of chemically similar elements.

Besides the 16 essential elements of Table 6-3, some species require others. For example, there is good evidence that sodium is required by certain desert species such as *Atriplex vesicaria,* common to dry inland pastures of Australia, and *Halogeton glomeratus,* a common introduced weed of salty arid soils in the western United States. Peter F. Brownell and C. J. Crossland (1972) investigated the sodium nutrition of 32 species and concluded that those having the C-4 photosynthetic pathway (Section 10.3) probably do require Na^+. Growth enhancement due to Na^+ is shown in Fig. 6-2 for three C-4 pathway plants. Furthermore, certain species that can fix CO_2 in photosynthesis via the so-called crassulacean acid metabolism pathway (Section 10.5) also grow faster with sodium, but only when this pathway is stimulated by long nights and short days (Figure 6-3). As explained in Chapter 10, there are several similarities between the chemical reactions of the crassulacean acid metabolism pathway and the C-4 pathway, but what Na^+ has to do with this is not yet known.

Silicon is another element that enhances growth of some plants when added to the culture solution, although it has not been proved essential for any angiosperm or gymnosperm

Table 6-3 Essential Elements for Most Higher Plants and Internal Concentrations Considered Adequate

Element	Chemical Symbol	Form Available To Plants[b]	Atomic Wt	Concentration in Dry Tissue		Relative No. of Atoms Compared to Molybdenum
				ppm	Percent	
Molybdenum	Mo	$MoO_4^=$	95.95	0.1	0.00001	1
Copper	Cu	Cu^+, $\mathbf{Cu^{+2}}$	63.54	6	0.0006	100
Zinc	Zn	Zn^{+2}	65.38	20	0.0020	300
Manganese	Mn	Mn^{+2}	54.94	50	0.0050	1,000
Boron	B	H_3BO_3	10.82	20	0.002	2,000
Iron	Fe	Fe^{+3}, Fe^{+2}	55.85	100	0.010	2,000
Chlorine	Cl	Cl^-	35.46	100	0.010	3,000
Sulfur	S	$SO_4^=$	32.07	1,000	0.1	30,000
Phosphorus	P	$\mathbf{H_2PO_4^-}$, $HPO_4^=$	30.98	2,000	0.2	60,000
Magnesium	Mg	Mg^{+2}	24.32	2,000	0.2	80,000
Calcium	Ca	Ca^{+2}	40.08	5,000	0.5	125,000
Potassium	K	K^+	39.10	10,000	1.0	250,000
Nitrogen	N	$\mathbf{NO_3^-}$, NH_4^+	14.01	15,000	1.5	1,000,000
Oxygen	O	O_2, H_2O	16.00	450,000	45	30,000,000
Carbon	C	CO_2	12.01	450,000	45	35,000,000
Hydrogen	H	H_2O	1.01	60,000	6	60,000,000

[a]Modified after P. R. Stout, 1961, Proceedings of the Ninth Annual California Fertilizer Conference, 21–23.

[b]If one of two forms is more common than the other, this is indicated by boldface type.

Figure 6-2 Comparisons between plants of *Echinochloa utilis* (left), *Amaranthus tricolor* (center), and *Kochia childsii* (right), which received no addition (−Na) and 0.10 meq/liter sodium chloride (+Na). (From P. F. Brownell and C. J. Crossland, 1972, Plant Physiology 49:794–797.)

Figure 6-3 Comparisons between plants of *Bryophyllum tubiflorum* receiving the following treatments. Left to right: no addition, 0.1 meq/l NaCl (16-hr light period; overall temperature range, 15 to 38 C); no addition, 0.1 meq/l NaCl (8-hr light period; light temperature 33 C, dark temperature 13 C). (From P. F. Brownell and C. J. Crossland, 1974, Plant Physiology 54:416–417.)

(Lewin and Reimann, 1969). Corn (see Table 6-1) and several other members of the grass family accumulate the element to the extent of 1 to 4 percent of their dry weight, while rice and *Equisetum arvense* (a "horsetail" or "scouring rush") contain up to 16 percent silicon. Silicon is believed to be essential for *Equisetum*. For certain algae (diatoms and some flagellate Chrysophyceae) that are surrounded by a silica-rich sheath, the element is certainly essential.

Silicon exists in the soil solution as undissociated silicic acid, H_4SiO_4, and is absorbed in this form. It accumulates largely as hydrated amorphous silica ($SiO_2 \cdot nH_2O$) most abundantly in walls of epidermal cells, but also in primary and secondary walls of other cells of roots, stems, and leaves, and in grass inflorescences. When sheep and cattle feed on pasture grasses in which silica is abundant, it is largely excreted in the urine, but under some conditions it forms kidney stones. Silica is also blamed for causing excessive wear on the teeth of sheep.

Cobalt is believed to be essential for blue-green algae and many bacteria, but not for other algae or higher plants. It is required for nitrogen fixation by microorganisms in root nodules on legumes and many (perhaps all) nonlegumes (see Section 13.2), but not for normal functions of the plant itself. Figure 6-4 illustrates the growth of soybean in the presence and absence of very low concentrations of cobalt when the only source of nitrogen was that absorbed from the atmosphere and fixed in the root nodules. Cobalt concentrations as low as 0.1 μg/l (0.1 ppb) were high enough for good growth, and neither vanadium, germanium, nickel, nor aluminum could substitute for cobalt. When the host plant is given nitrate nitrogen, no cobalt is required for its normal growth. Free-living bacteria and blue-green algae that fix nitrogen apart from any symbiotic relationship also require cobalt. Organisms requiring this element, including many animals, probably need it principally because it is present in the vitamin B_{12} they require. Plants seem to contain no vitamin B_{12}. Of course, animals obtain their cobalt by eating plants that contain this element, even if such plants don't require it.

In addition to the 16 elements of Table 6-3, higher animals require sodium, iodine, cobalt, selenium, and apparently

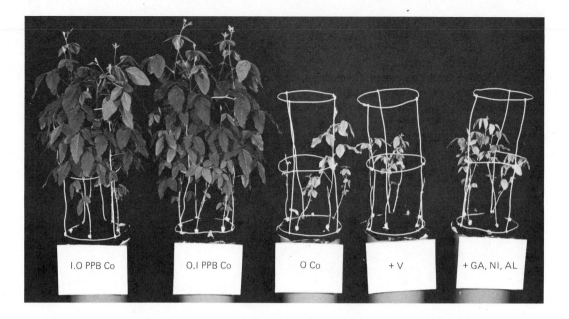

Figure 6-4 Specific cobalt requirement for nitrogen fixation in soybean. (From S. Ahmed and H. J. Evans, 1960, Soil Science 90:205, with the permission of the publisher. Copyright © 1960 by Williams & Wilkins Company, Baltimore, Maryland.)

also nickel, silicon, chromium, tin, vanadium, and fluorine, but not boron. Some plants may also prove to require nickel, because this metal is a tightly bound part of the enzyme urease (Dixon et al., 1976), but we are not yet certain that urease itself is essential to plants. Another element perhaps essential for a few species is selenium.

6.4 Quantitative Requirements and Tissue Analysis

In Table 6-3 we listed the concentrations of essential elements in plant tissues that appear to be essential to prevent deficiency symptoms from occurring. Such values provide useful guides to plant physiologists and also to foresters, orchard managers, and to farmers, because determinations of the amounts of elements actually in the tissues (especially in *selected* leaves) are much more indicative than soil tests of whether the plant would grow faster if more of a given element were added to the soil. Figure 6-5 shows an idealized plot of growth rate as a function of the concentration of any particular element in the tissue. In the range of concentrations called the **deficient zone,** as we provide more of the element and increase its concentration in the plant, the growth rate is stimulated dramatically. After the **critical concentration** (minimal tissue concentration giving almost maximal growth), increases in concentration (fertilizations) do not appreciably affect the growth rate (**adequate zone**). The adequate zone represents **luxury consumption** of the element. It is fairly wide for most elements but is narrower for micronutrients such as boron, zinc, iron, copper, manganese, and molybdenum. Continued increases of these elements usually lead to toxicities and a reduced growth rate

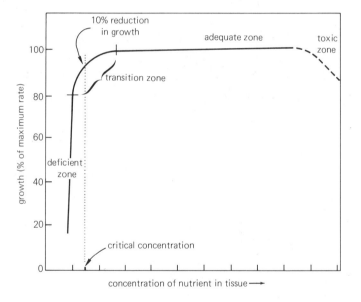

Figure 6-5 Generalized plot of growth as a function of the concentration of a nutrient in plant tissue. (After Epstein, 1972.)

(**toxic zone**). With increasing environmental pollution, toxic effects of both essential and nonessential elements are receiving much more attention than in the past.

Figure 6-6 shows average responses to calcium for 18 dicots and 11 monocots. The separate curves for these two groups provide evidence that the critical calcium levels for dicots (about 0.2 percent, dry weight basis) are significantly

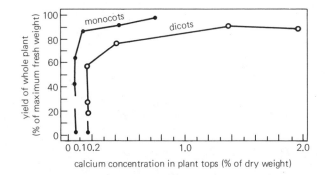

Figure 6-6 The relation between calcium concentration in tops and relative yields of eighteen dicots and eleven monocots after 17 to 19 days' growth in constant calcium concentrations in solution. Each point represents the average values for all species in each plant group given a single Ca^{++} treatment (0.3, 0.8, 2.5, 10, 100, or 1,000 μM). (From J. F. Loneragan, 1968, Nature 220:1307–1308.)

Table 6-4 Percentage of Calcium, Potassium, Magnesium, Nitrogen, and Phosphorus in the Tops of Several Species of Plants Grown in a Greenhouse in an Alberta Loam Soil[a]

| Species | Percent of Dry Weight | | | | |
	Ca	K	Mg	N	P
Sunflower	1.68	3.47	0.730	1.47	0.080
Bean	1.46	1.19	0.570	1.48	0.053
Wheat	0.46	4.16	0.225	2.26	0.058
Barley	0.68	4.04	0.292	1.94	0.125

[a]Data of Newton, 1928, Soil Science, 26:86.

higher than for monocots (less than 0.1 percent) and emphasize the approximate nature of the tissue concentrations listed as adequate in Table 6-3. Table 6-4 also contrasts calcium absorption by two dicots and two monocots growing on a common soil and lists data for other elements these species contained. The dicots, sunflower and bean, not only absorbed more calcium but also more magnesium than did the monocots, wheat and barley. In contrast, the bean (a legume) absorbed much less potassium than did the monocots or sunflower, a nonlegume dicot. The high percentages of K^+ in wheat, barley, and sunflower in Table 6-4 illustrate luxury consumption of this element. In general, members of the grass family absorb more potassium than do legumes.

Data presented in Fig. 6-6 and Table 6-4 emphasize the concept that the nutrient requirements of some species are different from those of others. Furthermore, graphs similar to that of Fig. 6-6 have been used effectively to plan the most efficient uses of fertilizers for crop plants and forest trees (van den Driessche, 1974). In the past, the primary reason for not fertilizing soils with concentrations of nitrogen, phosphorus, or potassium beyond those required to reach the critical concentrations has been the diminishing yield return (cost ratio), but now we are aware that some of the excess nitrate and phosphate ions not absorbed by plants are leached from the soil and ultimately appear in lakes and streams. Here they have been blamed for causing excessive growth of algae, which leads to eutrophication problems. Furthermore, nitrogen fertilizers require considerable energy to manufacture and may be among the most energy-expensive components in modern intensive agriculture. Therefore, fertilizer practices must consider not only the attainment of high yield but also the pollution of water and world energy requirements.

The ability of plants to obtain the essential nutrients from the soil is, along with the ability to obtain light, water, and a suitable temperature, very important in determining where they grow. Although we know much about mineral nutrition of crop plants, far too little is known about wild species, including forest trees. It is generally recognized that, except for carefully fertilized orchards, most trees grow on rather unfertile soils, and their nutrient requirements are lower than those of crop species bred to respond to fertilizers. Nevertheless, thousands of acres of forest trees in the northwestern United States are now being fertilized with nitrogen. Of course, the fall of leaves from deciduous trees in autumn returns some of the absorbed nutrients to the soil, while in crop plants such as alfalfa, the nutrients in the shoot are removed with the crop. Furthermore, most of the mobile nutrient elements move out of tree leaves into the twigs and branches before leaf fall. These are used in new growth the next season. Perennial range grasses similarly conserve minerals by translocation to lower stem tissues making up the crown. As mentioned in Chapter 5, mycorrhizae greatly facilitate the absorption of nutrients by many tree species, but forest growth is still frequently limited by insufficient nutrients (Schmidtling, 1973). Unfortunately, it is seldom known which elements are in shortest supply or how to provide them economically.

6.5 Chelating Agents

The micronutrient cations iron, zinc, manganese, and copper are relatively insoluble in nutrient solutions when provided as common inorganic salts, and they are also nearly insoluble in many soil solutions. This insolubility is especially marked if the pH is above 5, as it is in nearly all agricultural soils of the western United States. Under these conditions, the micronutrient cations react with hydroxyl ions, precipitating out the hydrous metal oxides. An example in which the ferric form of iron yields the reddish-brown oxide (rust) is shown in reaction R6-1:

$$2\ Fe^{+3} + 6\ OH^- \rightarrow 2\ Fe(OH)_3 \rightarrow Fe_2O_3 \cdot 3\ H_2O \qquad \text{(R6-1)}$$

Because of these and other reactions contributing to insolubility, certain plants cannot absorb enough of the metals, especially iron and zinc. One way to overcome this deficiency problem is to provide the elements as metal **chelates** (Greek,

Hydroponics and organic gardening are two agricultural practices that have their theoretical roots firmly planted in the mineral nutrition subscience of plant physiology. It is difficult to imagine two more diametrically opposed concepts, yet both seem to work. The hydroponic gardener feeds his plants only minerals and water and attempts to have total control over their nutrient environment; the organic gardener feeds his plants almost any organic matter and has virtually no chemical information about what they are actually receiving. There are advocates and opponents on both sides of both practices, and the issues are probably diffuse enough that any statement will invariably contain opinion. But the practices can be evaluated in terms of the principles discussed in this chapter. Let me do so—no doubt incorporating some of my own opinion.

Hydroponics is grounded on solid theory, being derived from the early experiments of Sachs and Knop and all the developments in the field of mineral nutrition since then. We may not know for sure about the best conceivable nutrient medium for each species, but this is not a serious limitation, since most commercially grown species do beautifully when provided with available commercial nutrient solutions. Tomatoes, lettuce, cucumbers, and a few other vegetables are grown this way in greenhouses during the winter. Solutions may be circulated over sand or gravel, but open solutions are also used. Because the medium is so well controlled, there are advantages in disease and weed control, and soil problems of salts, clays, hardpans, and so on do not exist. Plants may grow extremely well by hydroponics and are as nutritious and healthful for human consumption as any other plants. Given suitable nutrients and other conditions, plants synthesize all the vitamins, proteins, and other nutrients required by man.

Those are the facts on the plus side, but I remain unconvinced that the practice is economically sound. The nutrient solutions (prepared by dissolving purchased mixtures of salts in water) and equipment required to circulate them are expensive. Nutrients soon get out of balance, so solutions must be replenished and are usually discarded after two weeks. And plants are quite capable of growing luxuriantly and well in ordinary soil. True, the soil may be deficient in certain elements, but these can be applied much more easily than by hydroponics. We have learned much about agriculture, and if this knowledge is applied, say, in a greenhouse designed to produce winter tomatoes, a crop can be produced much more cheaply by conventional methods than by hydroponics. The same control over temperatures, insects, and other factors must be applied in both cases. There may be situations where hydroponics is appropriate (e.g., a South Sea island without soil), but several people have lost their life savings in hydroponic greenhouse ventures that failed.

It is difficult for a plant physiologist to remain objective and coolheaded about organic gardening. Whereas hydroponics is firmly founded on valid principles of plant physiology (if not economics), organic gardening seems to be founded in mysticism and the antiscience feeling that is currently abroad in the land. The organic gardener applies only large quantities of organic matter to the soil so that plants will be provided with the "vitamins" that they need and so that they will not be harmed by nasty old "chemicals." As a matter of fact, the organic gardener seems to think that chemicals are artificial rather than "natural" and, therefore, harmful to man and to be avoided like the plague.* These concepts are applied to insecticides or herbicides as well as to mineral fertilizers.

To modern plant physiologists, these principles seem patently absurd. We see no distinction between "natural" and "artificial" when it comes to chemicals. We are engaged in studying the chemicals that make up living organisms, and most of these chemicals in organic matter are converted by decay to the same form in which they are provided as mineral fertilizers. Furthermore, we know that plants are truly autotrophic, in that they need no special organic compounds ("vitamins") to grow well. And we are annoyed by the organic gardener's use of the word "organic," which we understand as meaning "compounds containing carbon," and don't see what that has to do with gardening! So in principle, organic gardening seems to be about as wrong as can be.

In practice, however, the organic gardener is certainly right. Usually his vegetables and other plants are as luxuriant and beautiful as any produced by any agricultural method. Though he may have the theory garbled, it is difficult to go wrong by adding large quantities of organic matter to the soil (although insect, weed, and disease problems may be accentuated). Nutrients are provided, and certainly in an excellent form, one in which they will be released gradually by decay, maintaining optimal soil levels over long intervals of time. Furthermore, some of the organic substances released by decay may act as chelates that help in the absorption of iron and other trace elements that are often immobilized in alkaline soils. And residues of decaying organic matter, consisting largely of lignin and other materials and collectively referred to as **humus**, provide an excellent subtrate for plant growth. The water-holding capacity is high, and the material may even greatly increase the cation exchange capacity of the soil. No wonder the organic gardener does well!

It also makes sense to recycle organic wastes by incorporating them into the soil. Certainly these wastes are one of the most serious sociological problems of our time. Why not utilize them in this way? Many studies are being carried out to answer that question. Except for home gardens, however, it appears that economics are again not always favorable. It is cheaper to use concentrated mineral fertilizers than to collect, distribute, and incorporate large quantities of organic wastes. Furthermore, these wastes may sometimes contain heavy metal ions or other toxins. Nevertheless, the economic questions hinge on the cost of producing mineral fertilizers, especially nitrogen, in a world in which energy is becoming increasingly expensive at the same time that waste disposal is becoming an increasingly serious problem. It is not unlikely that organic matter (e.g., from sewage disposal plants) will be used on an ever-widening scale in the near future, although avoiding herbicides, insecticides, and other chemicals is another matter.

Let's watch developments during the coming years to see to what extent the principles of hydroponics and the practices of organic gardening become an important part of our agricultural systems.

---

*Admittedly, not all organic gardners subscribe to the extreme views I have described.

"clawlike"). A chelate is the soluble and rather stable product formed when certain atoms in an organic molecule called a **chelating agent** or **ligand** donate electrons to the metal cation. Negatively charged carboxyl groups and many nitrogen atoms possess available electrons that can be shared in this way. One of the best known synthetic ligands provides both carboxyl groups and nitrogen atoms. This is **ethylene-diaminetetraacetic acid,** abbreviated **EDTA** and often sold in garden shops under the trade name "versene." The molecular structure of the zinc-EDTA chelate is shown in Fig. 6-7a.

Ligands such as EDTA are now commonly used to prevent or correct deficiency symptoms of iron in many parts of the western United States where the high pH of the calcareous soils renders iron unavailable to some species. The same problem occurs in certain western areas with zinc, and studies indicate that iron or zinc chelates can often correct these deficiencies, especially when applied to the foliage. In the acidic soils of parts of the eastern United States, especially Florida, iron deficiency can result from an interaction between iron and high amounts of copper and aluminum. Here, too, foliar applications of Fe-chelates prevent iron deficiencies in certain sensitive fruit trees and other crops. Iron chelates added to the soils of house plants often give an improved green color and faster growth.

Good chelating agents used in soils have at least two important properties. *First,* they should be resistant to microbial attack in the soil, and *second,* they should form stable chelates with the micronutrient ions but not with the far more abundant competitive cations such as calcium or magnesium. EDTA has a high affinity for calcium ions and is thus a poor chelating agent for calcareous soils. Here it has been replaced with **Fe-EDDHA (ethylenediamine di(o-hydroxyphenyl acetic acid),** a much better source of iron for calcareous soils (Fig. 6-7b). Several other synthetic ligands are used for certain purposes.

Naturally occurring chelates of micronutrient cations also exist in the soil, maintaining higher availability of these elements than would otherwise be the case. Although the chelating agents have not all been identified, it is suspected that several compounds present in the soil organic matter are capable of this, including certain phenolic compounds, proteins, amino acids, and organic acids. It is believed that chelating agents usually function as effective micronutrient sources simply by keeping the elements in solution yet releasing part of them at or near the root surface (Lindsay, 1974). Occasionally the intact metal chelate is absorbed, as evidenced by the characteristic color of synthetic chelates in the cells, then is broken down by enzymes in the cells.

Once absorbed by the plant, the divalent metals are apparently kept soluble by chelation with certain cellular constituents. Anions of organic acids, especially citric acid, appear to be most important as chelating agents for transport through the xylem, but amino acids may also participate. Ultimately, much of the iron, zinc, manganese, copper, and molybdenum are bound to proteins. In this form they facilitate electron transport processes or promote the catalytic

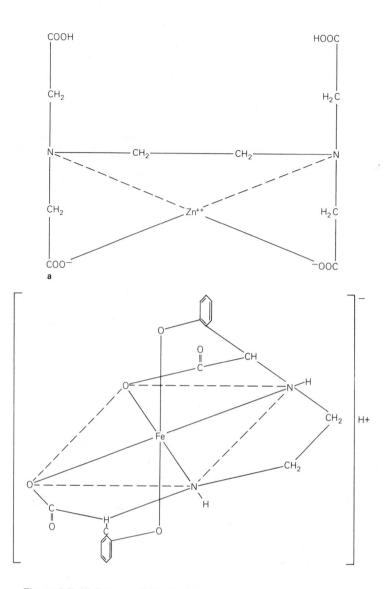

Figure 6-7 (**a**) The zinc-EDTA chelate. (**b**) The iron chelate Fe-EDDHA, a bright red compound sold under the trade names Chel 138 HFe and Sequestrene 138 Fe, is an excellent iron chelate, especially useful on calcareous soils. It contains 6 percent iron by weight and has a molecular weight of 413.

activity of enzymes. Monovalent cations such as K^+ and Na^+ do not form stable chelates, but even these are associated loosely by ionic attractions with organic acid anions and with proteins.

6.6 Functions of Essential Elements: Some Principles

The functions of essential elements have sometimes been classified into two groups, (1) a role in the **structure** of an important compound, and (2) a requirement to stimulate the rate at which a chemical reaction occurs (**enzyme-activating role**) (Rains, 1976). Frequently, however, there is no sharp distinction between these functions, because several elements form structural parts of enzymes and help catalyze the chemical reaction in which the enzyme participates. Carbon, oxygen, and hydrogen are perhaps the most obvious elements performing both functions, although nitrogen and sulfur, also found in enzymes, are equally important in this regard. Another example of an element with both roles is magnesium; it is an essential part of chlorophyll molecules and also activates many enzymes.

Some ions perform still another function by contributing to the osmotic potentials of plant cells, thus allowing the buildup of turgor pressure necessary to maintain form and to allow certain pressure-dependent movements (i.e., stomatal opening, Section 3.4; and "sleep" movements of leaves, Section 18.4). Potassium ions are especially important in this regard. Potassium and chloride, both monovalent ions, are good examples of elements necessary because they temporarily combine with and activate certain enzymes; no permanent structural roles that would make these elements essential are known.

6.7 Nutrient Deficiency Symptoms

Plants usually respond to an inadequate supply of essential elements by the formation of characteristic symptoms. Such visually observed symptoms are of interest in that they often aid in determining the necessary functions of the element in the plant, and some of these are of real practical importance because they are used by agriculturists and foresters to determine how and when to fertilize their crops. Several such symptoms are described in this section and are illustrated in books by Sprague (1964), Gauch (1972), and Hewitt and Smith (1975).

It should be pointed out that most of the symptoms described are those appearing on the shoot of the plant and are easily observed. The roots are, of course, also injured. Unless the plants are grown hydroponically, the condition of the roots cannot easily be seen without removing them from the soil, and root deficiency symptoms have been generally less well described. Furthermore, the symptoms differ to some extent depending on the species and the severity of the problem.

The deficiency symptoms for any element depend primarily upon two factors:

1. The function or functions of that element

2. Whether or not the element is readily translocated from old leaves to younger leaves

A good example emphasizing both factors is the **chlorosis** (yellowing caused by restricted chlorophyll synthesis) that results from magnesium deficiency. As mentioned previously, magnesium is an essential part of chlorophyll molecules, so chlorophyll is not formed in its absence. Furthermore, the chlorosis of the lower, older leaves becomes more severe than that of the younger leaves. This difference illustrates an important principle: **Young parts of the plant have a pronounced ability to withdraw mobile nutrients from older parts.** We do not understand this withdrawing power, but hormonal relations are involved. In Chapter 17, we shall emphasize the important role that cytokinin hormones have in this regard.

Whether withdrawal is successful, as with magnesium, depends upon the mobility of the element. As discussed in Chapter 7, some elements readily move through the phloem from one leaf to another, including nitrogen, phosphorus, potassium, sodium, magnesium, chlorine, and to some extent sulfur. Others such as boron, iron, and calcium are quite immobile, while zinc, manganese, copper, and molybdenum are somewhat intermediate. If the element is mobile, symptoms appear earliest and most pronounced in the older leaves, while symptoms resulting from a lack of a relatively immobile element such as calcium or iron appear first in the younger leaves. A key to the deficiency symptoms, emphasizing the mobility principle, is given in Table 6-5.

Nitrogen Most soils are more commonly deficient in nitrogen than any other element, partly because soil parent materials usually contain little or no nitrogen, and because this element is rather easily lost by leaching of nitrate ions or by conversion to volatile N_2 by microorganisms. Nitrogen is usually absorbed as NO_3^- or NH_4^+.

Because nitrogen is present in so many essential compounds, it is not surprising that plants grown without added nitrogen grow very slowly. Those containing enough to attain limited growth exhibit deficiency symptoms consisting of a general chlorosis, especially in the older leaves. In severe cases, these leaves become completely yellow and eventually fall off the plant. The younger leaves remain green for longer periods of time, because they receive soluble forms of nitrogen transported from the older leaves. Some plants, including tomato and certain varieties of corn, exhibit a purplish coloration in the stems, petioles, and lower leaf surfaces due to the accumulation of anthocyanin pigments.

Plants grown with excessive amounts of nitrogen usually have dark green leaves and show an abundance of foliage, usually with a poorly developed root system and therefore a high shoot-to-root ratio. Potato plants grown with superabundant nitrogen show an excessive shoot growth but form only small tubers. The reason for this is not yet known, but undoubtedly sugar translocation to the tubers is affected in some way, due perhaps to a hormone imbalance. Excess nitrogen also causes tomato fruits to split as they ripen. Flowering and formation of seeds of several agricultural crops are retarded by excess nitrogen. Flowering is not affected in some species, however, and still others that flower only in favorable daylength conditions (especially short days) do so faster with abundant nitrogen.

Phosphorus Second to nitrogen, phosphorus is the most often limiting element in soils. It is absorbed by plants primarily as the monovalent phosphate ion ($H_2PO_4^-$) and less rapidly as the divalent ($HPO_4^=$) phosphate ion. The soil pH controls the relative abundance of these two forms, $H_2PO_4^-$ being favored below pH 7 and $HPO_4^=$ above pH 7. Much phosphate is converted into organic forms upon entry into the root or after it has been transported through the xylem into the shoot. Phosphorus-deficient plants are stunted in appearance and, in contrast to those lacking nitrogen, are often a rather dark green in color. Anthocyanin pigments again sometimes accumulate. Maturity of the plants is often delayed, while those containing abundant phosphate usually mature rather early. There is apparently a close interaction between phosphorus and nitrogen in this respect, excess nitrogen delaying maturity and abundant phosphorus speeding maturity. Furthermore, if excess phosphorus is provided, root growth is often stimulated relative to shoot growth. This, in contrast to effects with excess nitrogen, causes low shoot-to-root ratios (i.e., extensive root systems).

Phosphate is easily redistributed in most plants from one organ to another and is lost from older leaves, accumulating in younger leaves and in developing flowers and seeds. As a result, deficiency symptoms occur first in more mature leaves.

Table 6-5 A Key to Plant Nutrient Deficiency Symptoms[a]

Symptoms	Element Deficient
Older or lower leaves of plant mostly affected; effects localized or generalized.	
Effects mostly generalized over whole plant; more or less drying or firing of lower leaves; plant light or dark green.	
Plant light green; lower leaves yellow, drying to light brown color; stalks short and slender if element is deficient in later stages of growth	Nitrogen
Plant dark green, often developing red and purple colors; stalks short and slender if element is deficient in later stages of growth	Phosphorus
Effects mostly localized; mottling or chlorosis with or without spots of dead tissue on lower leaves; little or no drying up of lower leaves.	
Mottled or chlorotic leaves, typically may redden, as with cotton; sometimes with dead spots; tips and margins turned or cupped upward; stalks slender	Magnesium
Mottled or chlorotic leaves with large or small spots of dead tissue.	
Spots of dead tissue small, usually at tips and between veins, more marked at margins of leaves, stalks slender	Potassium
Spots generalized, rapidly enlarging, generally involving areas between veins and eventually involving secondary and even primary veins; leaves thick; stalks with shortened internodes	Zinc
Newer or bud leaves affected; symptoms localized.	
Terminal bud dies, following appearance of distortions at tips or bases of young leaves.	
Young leaves of terminal bud at first typically hooked, finally dying back at tips and margins, so that later growth is characterized by a cut-out appearance at these points; stalk finally dies at terminal bud	Calcium
Young leaves of terminal bud becoming light green at bases, with final breakdown here; in later growth, leaves become twisted; stalk finally dies back at terminal bud.	Boron
Terminal bud commonly remains alive: wilting or chlorosis of younger or bud leaves with or without spots of dead tissue; veins light or dark green.	
Young leaves permanently wilted (wither-tip effect) without spotting or marked chlorosis; twig or stalk just below tip and seedhead often unable to stand erect in later stages when shortage is acute.	Copper
Young leaves not wilted; chlorosis present with or without spots of dead tissue scattered over the leaf.	
Spots of dead tissue scattered over the leaf; smallest veins tend to remain green, producing a checkered or reticulating effect.	Manganese
Dead spots not commonly present; chlorosis may or may not involve veins, making them light or dark green in color.	
Young leaves with veins and tissue between veins light green in color.	Sulfur
Young leaves chlorotic, principal veins typically green; stalks short and slender.	Iron

[a]*Source:* McMurtrey, 1950, *Diagnostic Techniques for Soils and Crops.* American Potash Institute.

Potassium After nitrogen and phosphorus, soils are usually most deficient in potassium. Because of the importance of these three elements, commercial fertilizers list the percentages of N, P, and K they contain, but the last two are actually expressed as equivalent percents of P_2O_5 and K_2O. Like nitrogen and phosphorus, K^+ is easily redistributed from mature to younger organs, so deficiency symptoms are first observable on the older, lower leaves. In dicots, these leaves initially become somewhat chlorotic, but scattered, dark-colored **necrotic lesions** (dead or dying spots) soon develop. In many monocots, such as the cereal crops, the cells at the tips and margins of the leaves die first, and this necrosis spreads basipetally toward the younger, lower parts at the base of the leaf. Potassium-deficient corn and other cereal grains develop weak stalks, and the roots become more easily infected with root-rotting organisms. These factors cause the plants to be easily bent to the ground (lodged) by wind or rain. Potassium is seldom deficient in soils of the western United States, which have relatively high pH values; acidic soils, on the other hand, are often deficient in potassium.

Sulfur Sulfur is absorbed as divalent sulfate ions ($SO_4^=$). It is metabolized by roots only to the extent that they require it, and most of the absorbed sulfate is translocated unchanged to the shoots. Because sufficient sulfate is present in most soils, sulfur-deficient plants are rare in nature, but they have been observed in Australian pine plantations, in cotton fields of the southeastern United States, and on most crops in Teton Valley of southeastern Idaho, where the second author of this textbook was reared. The symptoms consist of a general yellowing throughout the entire leaves. Sulfur is not easily redistributed from mature tissues in some species, so deficiencies are usually noted first in the younger leaves. Sometimes, however, most of the leaves become chlorotic at the same time.

Sulfur can also be absorbed by leaves through the stomates as gaseous sulfur dioxide (SO_2), an environmental pollutant released primarily from burning coal and wood. SO_2 is converted to bisulfite (HSO_3^-) when it reacts with water in the cells, and in this form it inhibits photosynthesis and causes chlorophyll destruction. Plants in the eastern United States are damaged by the "acid rain" caused by SO_2 released from power plants, but this same pollutant may act as a fertilizer for crops in the midwest.

Magnesium Magnesium is absorbed as the Mg^{+2} ion. In its absence, a chlorosis of the older leaves is the first symptom, as already emphasized. This chlorosis is usually interveinal, because the vascular bundles retain chlorophyll for longer periods than the parenchyma cells between them. Magnesium is almost never limiting to plant growth in soils.

Calcium Calcium is absorbed as the divalent Ca^{+2} ion but, in contrast to Mg^{+2}, it is almost immobile. As a result of its immo-

bility, the deficiency symptoms are always most pronounced in young tissues. Meristematic zones of roots, stems, and leaves where cell divisions are occurring are most susceptible, perhaps because of a calcium requirement for formation of a new middle lamella in the cell plate arising between daughter cells or because of a necessity for calcium in the mitotic spindle apparatus. Twisted and deformed tissues result from calcium deficiency, and the meristematic areas die early. Calcium is also essential for membrane integrity in all living cells. Most soils contain sufficient calcium for plant growth, but acidic soils where high rainfall occurs are frequently fertilized with lime (a mixture of CaO and $CaCO_3$) to raise the pH.

Iron Iron-deficient plants are characterized by development of a pronounced interveinal chlorosis similar to that caused by magnesium deficiency but occurring on the younger leaves. The reason that iron deficiency results in a rapid inhibition of chlorophyll formation is incompletely known, even though this problem has been studied for several decades. Iron is relatively immobile in the plant, as in the soil, perhaps because it is internally precipitated as an insoluble oxide or in the form of inorganic or organic ferric phosphate compounds. Direct evidence that such precipitates are formed is weak, and perhaps other unknown but similarly insoluble compounds are formed. One abundant and stable form of iron in leaves is stored in chloroplasts as an iron-protein complex called **phytoferritin.** The entry of iron into the phloem transport stream is probably minimized by formation of such insoluble compounds. In any case, once it is taken into an organ from the soil through the xylem (as Fe^{+2} or sometimes Fe^{+3}), its redistribution is severely limited.

Iron deficiencies are often found in particularly sensitive species in the rose family, including both shrubs and fruit trees. In soils of the western United States, the high pH and the presence of bicarbonates contribute to iron deficiency, while in acidic soils aluminum is more abundant and restricts iron absorption.

Chlorine Chlorine is absorbed from the soil as the chloride ion (Cl^-) and probably remains in this form in the plant without becoming a structural part of organic molecules. One function is to stimulate photosynthesis, but one or more other functions probably also exist. Deficiency symptoms, discovered first in 1953, consist of wilted leaves that then become chlorotic and necrotic, eventually attaining a bronze color. Roots become stunted in length but thickened, or club-shaped, near the tips. Chlorine is rarely if ever deficient in plants growing in nature, because of its high solubility and availability in soils and because it is also transported in dust or in tiny moisture droplets by wind and rain to the leaves where absorption occurs. Because of its presence in human skin, it was necessary for investigators demonstrating its essentiality to wear rubber gloves.

Manganese Manganese exists in various oxidation states in the soil but is probably absorbed largely as the divalent manganous ion (Mn^{+2}). In this state, the element is probably most stable in the plant. Deficiencies of manganese are not common, although various disorders such as "gray speck" of oats, "marsh spot" of peas, and "speckled yellows" of sugar beets result when inadequate amounts are present. Initial symptoms are often an interveinal chlorosis on younger or older leaves, depending on the species, followed by, or associated with, necrotic lesions. Electron microscopy of chloroplasts from spinach leaves showed that the absence of manganese caused a disorganization of the membrane system extending through those bodies, but had little effect on the structure of nuclei and mitochondria. This and much biochemical work suggests that the element plays a structural role in the chloroplast membrane system and that one of its important roles is in photosynthesis.

Boron Boron is almost entirely absorbed from soils as undissociated boric acid, H_3BO_3 (Oertli and Grgurevic, 1975), and it is not known to what extent the plant alters this ion. Deficiencies are not common in most areas, yet several disorders related to disintegration of internal tissues such as "heart rot" of beets, "stem crack" of celery, "water core" of turnip, and "drought spot" of apples result from an inadequate boron supply. Plants deficient in boron show a wide variety of symptoms, depending upon the species, but research led by Luke Albert at the University of Rhode Island showed that one of the earliest symptoms is cessation of root tip elongation, probably caused by inhibited DNA synthesis, because of insufficient formation of the pyrimidines cytosine and thymine. This symptom is illustrated for tomato roots in Fig. 6-8. Boron also plays an undetermined but essential role in pollen tube germination.

Zinc Zinc is absorbed largely or entirely as the divalent Zn^{+2} ion. Disorders caused by zinc deficiency include "little leaf" and "rosette" of apples, peaches, and pecans, resulting from reduction in the size of the leaves and the length of the internodes. Leaf margins are often distorted and puckered in appearance. Interveinal chloroses often occur in the leaves of cereal grains and fruit trees, suggesting that zinc somehow participates in chlorophyll formation. The retardation of stem growth in its absence may result partly from its probable requirement to produce a growth hormone, indoleacetic acid (auxin).

Copper Plants are rarely deficient in copper. Since it is available in nearly all soils, deficiency symptoms are largely known

Figure 6-8 Roots from tomato plants grown hydroponically (**a**) in a complete nutrient solution containing 0.1 mg/liter boron (controls) and (**b**) in the absence of added boron for 3 days, before a 6-day treatment with adequate boron, during which recovery did not occur (boron deficient). (From L. S. Albert and C. M. Wilson, 1961, Plant Physiology 36:244–251.)

from nutrient solution studies. Without copper, the young leaves often become dark green in color and are twisted or otherwise misshapen, often exhibiting necrotic spots. Citrus orchards are occasionally deficient, in which case the dying young leaves led to the name "die back disease" for this deficiency. Copper is largely absorbed as the divalent cupric or monovalent cuprous ion. Because such small amounts are needed by plants (Table 6-3), it readily becomes toxic in hydroponic studies unless the amounts are carefully controlled.

Molybdenum Molybdenum exists to a large extent in soils in the form of molybdate (MoO_4^-) salts and also as MoS_2. It is not known whether the element can be absorbed in both oxidation states. Most plants require less molybdenum than any other element, yet molybdenum deficiencies are geographically widespread. Examples of disorders caused by inadequate molybdenum include "whiptail" of cauliflower and broccoli, found in certain areas of the eastern United States. Symptoms often consist of an interveinal chlorosis occurring first on the older or mid-stem leaves, then progressing to the youngest leaves. Sometimes, as in the "whiptail disease," the plants may not become chlorotic but may develop severely twisted young leaves, which eventually die. In acidic soils, adding lime increases availability of molybdenum and may eliminate or reduce the severity of such disorders.

Table 6-6 Metabolic Roles of the Essential Elements[a]

Nitrogen: An essential part of all amino acids, proteins (including enzymes), coenzymes (p. 114), chlorophyll molecules (Fig. 9-4), nucleotides and nucleic acids, and many other plant components.

Phosphorus: Part of ATP and other nucleotides, nucleic acids, certain proteins, several coenzymes, membrane phospholipids (Secs. 5.5 and 14.2), and attached to many different sugars that are important in photosynthesis (Chs. 10 and 11) and respiration (Ch. 12); inorganic and organic phosphates both act as buffers to help maintain constant cellular pH.

Sulfur: Present in the essential amino acids cysteine and methionine (Fig. 8-2) and thus a part of nearly all proteins; part of a membrane sulfolipid (Sec. 14.2), the vitamin coenzymes thiamine and biotin, and coenzyme A, which contains cysteine and is involved in respiration (Sec. 12.5) and in synthesis of cytochromes, fatty acids, sterols, and gibberellin and abscisic acid hormones (none of which contain sulfur); and part of S-adenosylmethionine; which donates methyl groups during formation of such compounds as sterols and lignins.

Potassium (K$^+$): Important in stomate movements (Ch. 3) and activates many enzymes (see lists of Evans and Sorger, 1966[b], and Suelter, 1970[c]). Two examples are pyruvate kinase (an enzyme participating in glycolysis, Sec. 12.3) and starch synthetase (Sec. 10.6). K$^+$ activates protein synthesis in plants and other organisms, but the specific K-dependent enzyme remains to be identified. In many cases (e.g., starch and protein synthesis), high concentrations (0.05M) of K$^+$ are required for full enzyme activation. Another almost unique function of K$^+$ is inferred from much indirect evidence: It apparently acts as the primary charge-balancing cation during transport of anions (primarily NO_3^-, but also $SO_4^=$, phosphates, and organic acid anions) from one plant part to another. K$^+$ moves readily through xylem and phloem, whereas Ca^{+2} and Mg^{+2} move readily in xylem but hardly at all in phloem. Considerable K$^+$ is probably required for this translocation function and for maintaining turgor and activating enzymes; hence, plants require much potassium (ca. 1% of dry weight; see Table 6-3).

Magnesium (Mg^{+2}): Chelated in and essential to activity of all chlorophyll molecules (Fig. 9-4); also essential for maximum rates of most and perhaps all of hundreds of enzymatic reactions involving ATP (chelates with ATP or ADP, p. 114). Most Mg^{+2} is apparently bound to ATP, ADP, other nucleotides, or organic acids. Mg^{+2} also enhances activity of several other respiratory enzymes (Secs. 12.3 and 12.5) and is essential for full activity of the two principal CO_2-fixing enzymes, ribulose diphosphate carboxylase (p. 137) and phosphoenolypyruvate carboxylase (p. 140). Mg^{+2} is also crucial to protein synthesis (holds two ribosome subunits together and facilitates peptide bond synthesis).

Calcium (Ca^{+2}): Exists to a great extent in central vacuoles in insoluble crystals of calcium oxalate and $CaCO_3$ (sometimes $CaSO_4$ and $Ca_3(PO_4)_2$), thus keeping possibly toxic oxalates out of the cytoplasm (yet we lack good evidence that failure to do so would harm plants). Some Ca^{+2} also occurs in the middle lamella between adjacent cell walls, where it is bound to carboxyl groups of pectins; hence, may be important in "cementing" walls together, although more evidence is needed (Gauch, 1972). Ca^{+2} may also confer stability upon the mitotic spindle apparatus essential to cell division, and Ca^{+2} also conditions the normal selective properties of membranes (p. 74). A few enzymes seem to be activated much more by Ca^{+2} than by other divalent cations such as Mg^{+2}, Mn^{+2}, or Zn^{+2}. One is alpha amylase (digests starch; Sec. 12.2), which apparently contains calcium.

Iron: Exists in divalent or trivalent forms in cytochromes and ferredoxin (proteins essential to light-driven reactions of photosynthesis; p. 134), in cytochromes involved in the mitochondrial electron transport system of respiration (Sec. 12.6), in catalase and peroxidase (enzymes that cata-lyze the breakdown of toxic H_2O_2 into H_2O and O_2), and in nitrite reductase and nitrate reductase (essential to nitrogen metabolism; Sec. 13.3).

Chlorine (Cl$^-$): Clearly important, but only one function presently well recognized, the stimulation of photosynthesis by somehow enhancing electron transfer from H_2O to chlorophyll in photosystem II (p. 132). The club-shaped root tips of chloride-deficient plants suggests other undiscovered role(s).

Manganese: Exists primarily as Mn^{+2} in plants but can apparently undergo oxidation (during photosynthesis) to the Mn^{+3} valence state; like Cl$^-$, enhances electron transfer from H_2O to chlorophyll (p. 132); also activates certain enzymes involved in fatty acid synthesis (Sec. 14.1), nucleotide synthesis (i.e., orotidine-5′-phosphate decarboxylase), respiration (Fig. 12-6), and others. Often activation by Mn^{+2} is also accomplished by the more abundant Mg^{+2}; Mn^{+2} chelates with phosphate groups in such nucleotides as ADP and ATP, as does Mg^{+2}.

Boron: Essential for higher plants, which grow slowly and will not complete their life cycles without it; but its biochemical roles are uncertain. Boric acid (H_3BO_3) forms strong complexes with oxygen atoms of vicinal hydroxyl groups present in sugars and polysaccharides; much boron probably exists in plants in such combinations. Boron is apparently not essential for several species of green algae, fungi, bacteria, and higher animals, although blue-green algae and diatoms do require it. Monocots so far investigated generally need less than half as much boron as dicots. Suggested essential roles include: in translocation of sugar (via the phloem, Ch. 7), in synthesis of pyrimidine bases (thus, RNA and DNA), and in synthesis of certain flavonoids (Sec. 14.5). Gauch (1972) has a lengthy review of possible boron functions in plants.

Zinc: Few known important functions in higher plants, one a possible role in synthesis of the amino acid tryptophan and of the hormone auxin (IAA, derived from tryptophan; Sec. 16.1), but more research is needed. In microorganisms and animals and probably also in plants, Zn^{+2} is essential for activity of the enzymes alcohol dehydrogenase and lactic acid dehydrogenase (function in anaerobic respiration; Sec. 12.3). Glutamic acid dehydrogenase (crucial to nitrogen metabolism; Sec. 13.4) seems to contain essential zinc; also a part of carbonic anhydrase, a chloroplast enzyme that catalyzes the reaction between CO_2 and H_2O to form H_2CO_3. Zinc in DNA polymerase (catalyzes replication of DNA prior to mitosis and meiosis) is essential in various animals, bacteria, and possibly plants. Other enzymes activated by Zn^{+2} are usually activated equally well by such divalent cations as Mn^{+2}, Cu^{+2}, or Ca^{+2}. An effect of Zn^{+2} not duplicated by Mg^{+2} (although Mg^{+2} also participates) or other divalent cations is to maintain the structure of ribosomes, thus allowing protein synthesis.

Copper: Essential to photosynthesis, existing in the protein plastocyanin, an electron carrier in the light reactions (p. 134); essential to respiration (part of cytochrome oxidase; Sec. 12.6), and a necessary component of polyphenoloxidase (an enzyme that oxidizes numerous phenolic compounds) and, apparently, of ascorbic acid oxidase and various amine oxidases such as one that converts tryptamine to the growth hormone indoleacetic acid. No doubt also part of other plant enzymes.

Molybdenum: Required in the smallest quantities by plants, its only known functions being in nitrate reduction (an essential part of nitrate reductase, acting as an electron carrier by undergoing oxidation and reduction between the Mo^{+5} and Mo^{+6} states; Sec. 13.3), and nitrogen fixation (an integral part of one of two proteins making up the nitrogenase complex; Sec. 13.2). Plants can be grown in the absence of molybdenum if they are provided with reduced nitrogen (NH_4^+); hence no other functions may exist (Hewitt and Gundry, 1970[d]).

[a] Numbers in parentheses refer to pages, sections, chapters, and figures in this text. For more details consult Gauch (1972), Hewitt and Smith (1975), and Rains (1976).
[b] H. J. Evans and G. J. Sorger, 1966, Annual review of plant physiology 12:91–112.
[c] C. H. Suelter, 1970, Science 168:789–795.
[d] E. J. Hewitt and C. S. Gundry, 1970, Journal of horticultural science 45:351–358.

7

Phloem Transport

The plant, thrusting its roots into the soil and its leaves into the atmosphere and sunlight, is a complex of specialized organs. To function properly, a balanced and integrated transfer of materials must take place within this complex structure. Some of the materials absorbed by the roots must be moved to the leaves for assimilation. We have considered the movement of water and inorganic salts in this direction in the xylem, but minerals may be redistributed from the leaves in the phloem. As photosynthesis proceeds, its products and other metabolic products must move out of the leaves to other parts of the plant. Roots, stems, and young leaves will require these materials, and developing fruits and flowers will either metabolize or store them. Certain hormones are synthesized only in specific parts of the plant and then are moved to other parts.

Plant physiologists have devoted considerable effort to the study of this movement or **translocation** of dissolved materials throughout the plant. To begin with, we need descriptive information. In which tissues does the movement take place? How fast? In what form? Then we can ask: What are the mechanisms of translocation and what are the controls that coordinate supply with demand?

Interestingly enough, though the problem has been studied intensively for over a century, there is no consensus among plant physiologists that we have the answers. Some think that one proposed mechanism or control is supported by sufficient evidence to be acceptable; others are not so sure. This situation is in sharp contrast to that of animals. The concept of blood circulation has hardly been in doubt since William Harvey lectured on it in the early 1600s. The heart as a pump is much easier to comprehend than the subtle cellular mechanisms responsible for translocation in plants.

At this point we arrive at a level of plant function about as complex as any basically nonbiochemical plant process. Although substances are sometimes biochemically changed as they enter or leave the transport system and even along the way, the net result remains essentially a physical process and not a chemical reaction: Substances are moved from one position in space to another.

7.1 Surgical Experiments

As early as 1675, the Italian anatomist and microscopist Marcello Malphighi **girdled** a tree by removing a strip of bark from a ring around its trunk. The experiment was repeated by Stephen Hales in 1727. These were some of the first experiments performed in the field of plant physiology. Such experiments are still used as demonstrations, and in their most sophisticated recent development they have been combined with radioactive tracers. Bark may be surgically separated from wood, or wood may even be removed, leaving the bark virtually intact. Three broad conclusions and many more specific ones may be derived from these surgical experiments.

In the earliest studies, it became apparent that the bark (containing the phloem) could be removed from the stem or trunk (the xylem) with no immediate effects on growth of the shoot or on transpiration from the leaves. In more advanced experiments, water with tritium replacing the hydrogen or containing some radioactive solute is observed to move readily through the wood in the transpiration stream, even though the bark has been removed. Analysis of the xylem sap shows that it contains mostly dissolved minerals from the soil plus small amounts of various organic compounds, including sugars and amino acids (Bollard, 1960). Hence our first conclusion about translocation: **Water with its dissolved minerals moves primarily upward in the plant through xylem tissues.**

In Malphighi's and Hales's experiment, the bark below the girdle dried up and eventually died, while the bark above the girdle swelled somewhat and remained healthy (Fig. 7-1). It seems clear that the foodstuff products of metabolism or assimilation (**assimilates**) in leaves, including the products of photosynthesis (**photosynthates**), are necessary for growth of plant parts that cannot photosynthesize and even for some parts that photosynthesize only at low levels, such as some stems and fruits. These materials move through the bark in the phloem. Indeed, it is possible to kill a tree by girdling the trunk, making the roots dependent upon their stored food, which runs out after a few weeks to a few years, depending on the species.

Phloem in angiosperms consists of four principal cell types: **sieve elements** (which form **sieve tubes** consisting of rows of sieve elements with perforated **sieve plates** between each pair of elements), **companion cells, fibers,** and **phloem parenchyma.** We shall discuss the anatomy of phloem in a later section, but detailed studies have shown that assimilates move through the sieve tubes, which contain living cytoplasm but no nuclei. Hence the second conclusion: **Assimilates, including photosynthates, move primarily through sieve tubes in the phloem. This is phloem transport.**

Figure 7-1 The effects of girdling the trunk of a tree by removing bark from around the circumference. Note the swelled bark above the girdle compared to that below. The trunk has been cut to reveal the annual growth rings. Note that an entire year's growth was laid down above the girdle but not below.

Various parts of the plant besides the main stem have been girdled. It is possible, for example, to remove the bark between a leafy branch and developing fruit. Again, sugars accumulate in the bark on the side of the leafy branch. Or the girdle may be placed below the shoot tip between the leaves on the stem and the developing tip. Sugars still accumulate on the side of the leaves, at the *bottom* of the girdle. Gravity has nothing to do with movement of materials in the bark; the controlling relationship is the relative positions of the **source** and the **sink.** The leaves with their photosynthetic capacity typically constitute the source, but an exporting storage organ such as a beet or a carrot root in the spring of its second year would also be a source. Any growing, storing, or metabolizing tissue might be a sink. Developing young leaves, fruits, roots, corms, tubers, flowers, or growing stem and root tips are the common examples. Since the roots are such an important sink, we tend to think of movement in the phloem as *downward* from the leaves to the roots. But materials also move in the *phloem* upward from the leaves to such sinks as the growing stem tips, young leaves, seeds, or fruits. Hence the third important point: **Assimilates move from source to sink.**

7.2 The Use of Tracers

It was long the aim of plant physiologists to measure translocation directly by following the actual movement of marked materials in the transport system. Early investigators used dyes and, indeed, the dye fluorescein moves readily in phloem cells and can be used as an effective tracer. Viruses and herbicides have also been used, but by far the most important tracers are the radioactive nuclides made available after World War II. Radioactive phosphorus, sulfur, chlorine, calcium, strontium, rubidium, potassium, hydrogen (tritium), and carbon have been used in these studies. More recently, heavy stable isotopes such as those of carbon or oxygen are coming into wider user.

Tracers may be applied by the **reverse flap technique,** in which a flap including a vein is cut from a leaf, as shown in Figure 7-2. Another widely used approach exposes the leaf in a closed container to carbon dioxide labeled with carbon-14 (^{14}C). The $^{14}CO_2$ is incorporated into assimilates by photosynthesis, and in this form it is exported from the leaf in the translocation stream. It is important to know what happens chemically to the tracer in the plant. As a rule, such inorganic ions as phosphate, sulfate, potassium, or rubidium remain unchanged, and most of the ^{14}C in $^{14}CO_2$ is incorporated into sucrose or some other sugar, but radioactive $^{14}CO_2$ is incorporated into every organic compound in the plant, given sufficient time.

The radioactive tracer may be detected after it has been transported by bringing a Geiger tube into contact with the plant stem or other part, but a much more widely used method is **autoradiography** (Fig. 7-3). The plant is placed in contact with a sheet of X-ray film, sometimes for several months, and the film is then developed to show the location of the radioactivity. The immediate problem is to immobilize the radioactivity in the plant after harvest. Plants may be placed between blocks of dry ice and then, while still frozen, subjected to a vacuum, allowing the water to sublime away **(freeze drying** or **lyophilizing).** Another way of guarding against

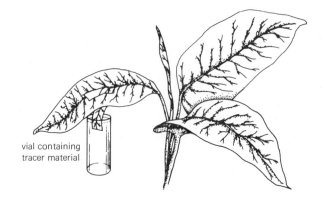

vial containing tracer material

Figure 7-2 The reverse-flap technique for applying material to leaves.

Figure 7-3 Results of an experiment in which autoradiography is used. The purpose of the experiment was to observe the effect of wilting on translocation. The first true leaf on the right of each soybean plant was held for 1 hour in an illuminated chamber containing $^{14}CO_2$, which was converted by photosynthesis into radioactively labeled assimilates. After 6 hours, the plants were harvested, dried, pressed (left, in each pair), and placed in tight contact with X-ray film. After 2 weeks of exposure, the film was developed (right, in each pair). The dark areas show where the most ^{14}C was located. In both cases, the leaf exposed to the $^{14}CO_2$ has by far the most tracer, but more is moved from the turgid plant (left) than the wilted one (right). (Specimens and films courtesy Herman H. Wiebe; see Wiebe, 1962.)

movement is to dismember the plant before making the autoradiograms.

Several elegant autoradiographic techniques have been developed. Orlin Biddulph and S. R. Cory at Washington State University, for example, distinguished between radioactive phosphorus and radioactive carbon by placing a film of aluminum foil between the plant and the X-ray film for the first exposure. The strong beta particles (electrons) from ^{32}P penetrate the aluminum, but the weak beta particles from ^{14}C do not. After several months the ^{32}P, with a half-life of only 14.3 days, had decayed away. They removed the aluminum foil and made a second autoradiogram showing the position of the carbon, which has a half-life of 5,730 years. In this study, the carbon and phosphate usually moved together in the assimilate stream. The conclusion of such studies: **Tracers provide an extremely valuable tool for study of phloem transport.**

7.3 In What Form?

In what form do materials move in the phloem? In 1944, Alden S. Crafts and O. Lorenz at the University of California at

Davis showed that the exudate from cut phloem could not be representative of the material actually accumulating in pumpkin fruits from sieve tubes. The dried exudate contained a higher percentage of nitrogen than the pumpkins to which it was being transported. Cutting the phloem probably caused some of the protein in phloem cells to be torn from the cell, mixing into the exudate. Furthermore, phloem exudate is usually more dilute than intact phloem sap. Releasing the pressure in the phloem by cutting lowers the water potential, and water then moves in by osmosis. These effects of cutting on composition and concentration are more noticeable in some species than in others (little effect on composition in many trees, for example), but they are complications to be reckoned with.

A technique that apparently overcomes many of these problems utilizes aphids (Fig. 7-4), which insert their stylets selectively into a single sieve element. Although they are called "sucking" insects, pressure in the sieve tubes actually forces the sap into the insect during the feeding process. To a degree, the insect strikes an "artesian well"; fluid forced into its body is secreted on the body surface as honeydew. The honeydew can be collected and analyzed, although its composition may be changed by passing through the insect's

Figure 7-4 Study of translocation in the phloem by the use of aphids. (**a**) An aphid hanging upside down on a branch of a tree. Note the droplet of honeydew being exuded from the insect. (**b**) A cross section of the tree showing an aphid stylet that has penetrated to a sieve-tube element. (From Martin H. Zimmermann, 1961, Science 133:73–79.)

body. This may be unimportant with radioactive mineral tracers. Or the insect may be anesthetized in a gentle stream of carbon dioxide (in which case honeydew secretion stops), and then it may be cut from the stylet with a thin sliver of a razor blade. Exudate then flows sometimes for days through the cut stylet, and it seems that this should be closely similar to intact phloem sap. The stylet is a far better hypodermic syringe than man has devised.

Nine tenths or more of the material translocated in phloem consists of carbohydrates, mostly nonreducing sugars, at rather high concentrations of 10 to 25 percent (Zimmermann, 1960). Plants contain both reducing and nonreducing sugars. The **reducing sugars** have an exposed ketone or aldehyde group on each molecule that will reduce Cu^{+2} to Cu^+ in Benedict's or Fehling's solution. In the **nonreducing** sugars, or related compounds, this group has been changed to a nonreducing alcohol, or it is blocked by combining with a similar group from another molecule. For example, combination of the six-carbon reducing sugar glucose, attached through its aldehyde group to the ketone of the six-carbon reducing sugar fructose, forms sucrose, which is nonreducing. Sucrose is by far the most common sugar in plants, and in many, if not most, species it is the only sugar found in phloem sap. Examples include grapes, sugar beets, and many trees. (Some-

times glucose and fructose are found in phloem exudate, but it has been shown that these are breakdown products of sucrose and are not themselves translocated.) In a few trees, such as white ash and lilac, other nonreducing sugars also occur in the phloem. Raffinose, which consists of sucrose with one D-galactose unit attached, is one of these, and others present are often related to raffinose, having more than one D-galactose unit attached to sucrose. Sugar alcohols also sometimes appear in phloem sap. A few species, including ash, contain mannitol, and many species in the rose family, including apple and apricot, transport mainly sorbitol. It is highly significant that only nonreducing sugars are translocated in phloem and that reducing sugars and their phosphate derivatives are not. Although the reasons for this are not known, nonreducing sugars are less labile to enzymatic destruction in sieve elements. Indeed, for perhaps the same reason, reducing sugars are rare in any plant cell. Glucose and fructose, for example, usually occur in cells as their phosphate derivatives.

Phloem sap also contains nitrogenous substances, especially amino acids and amides, at concentrations of 0.03 to 0.4 percent. Although these are always present, they increase 2 to 10 times in the autumn when leaves are aging and about to fall. This transport of a nitrogenous compound from leaves before they fall conserves much of a plant's nitrogen. Proteins and other materials are broken down, and their substances are exported for storage in stems, roots, or rarely fruits. Still other substances occur in phloem sap, including mineral ions and hormones. In any case, here is another conclusion: **The bulk of material translocated in the phloem consists of carbohydrate in the form of sucrose and other nonreducing sugars, although other important assimilates are also translocated.**

7.4 Where and When?

Which parts of the plant are exporting the most material via the phloem system? And to which parts of the plant is this material going? When is translocation of assimilates most active?

Early studies measured dry weights of different plant parts at different times. To correct for photosynthesis and respiration, plants were sometimes dismembered and kept moist in light or darkness. In more recent experiments, radioactive tracers were applied to different plant parts, and quantities translocated to other parts were noted as a function of time.

All these studies agree that mature but still healthy leaves export by far the most material. Young, developing leaves act as sinks to which material is exported by mature leaves. As a leaf matures, the time comes when it begins to export rather than import assimilates, often when it is one-third to one-half full size. As we've seen, old leaves often export their breakdown products. They may also import very small quantities of assimilates from healthy leaves, but this is uncommon. Roots and stems are more important sinks than young or old leaves. In one study of a vegetative tomato plant, for example, about half the material exported from mature leaves went to the roots, a third to the stem, and the remainder to the young developing leaves. When a plant becomes reproductive, these proportions may change radically. The developing flowers act as sinks, and when fruits begin to form, they may become the most important sinks. When the kernels of wheat are developing, for example, materials may be mobilized from virtually all the plant including the roots.

The distribution pattern is influenced by the pattern of vascular tissue in the plant. Sometimes, phloem strands from one leaf will bypass a nearby importing organ, which then can import only from more distant leaves. A number of such strange anomalies have been described, but in general, the lower leaves export to roots and the upper leaves to stem tips and young leaves, as might be expected.

As might also be expected, most export from mature leaves occurs during the day, when photosynthesis is most active. Not all photosynthate is exported at the time it is produced, however; considerable amounts are temporarily stored in the chloroplasts in the form of starch grains. At night, these may be broken down into soluble sugars and exported through the phloem system. The conclusion of such studies: **The source-to-sink relationship clearly holds both in the light and in the dark, and the greatest quantities of photosynthate are typically translocated from mature leaves during the daytime when they are being produced.**

7.5 How Fast?

In 1944, Crafts and Lorenz published another study. They measured growth of 39 pumpkins from August 5 to September 7, estimating that individual fruits gained an average of 482 g of dry material during their 792 hours of growth. Thus, material moved into each fruit through its peduncle at an average rate of 0.61 g per hour. Sections were taken from the peduncles, projected onto cardboard, and the cellular outlines traced. After suitable calibration, the outlined phloem tissues were cut from the cardboard and weighed; sieve tube cross sections were then cut from the phloem sections and also weighed. Crafts and Lorenz estimated that each peduncle included an average cross section of about 0.186 cm^2 of phloem tissue, about 20 percent of which consisted of sieve tube cross sections (0.0372 cm^2). Thus, material was moving into the fruits with a **mass transfer rate** of 0.61 g/hr $\div$ 0.0372 cm^2 = 16.4 g/cm^2/hr. Thus the mass transfer rate indicates the quantity of material passing through a given cross section of sieve tubes per unit of time.

What about the **velocity** of movement, which is a measure of the linear distance transversed by an assimilate molecule per unit of time? Crafts and Lorenz assumed that the dry material had a specific gravity or density of about 1.5 g/cm^2. This figure divided into the rate gives a velocity of about 11 cm/hr. Of course, the material does not move in dry form but as a solute in water. If the solute concentration is 10 percent, its velocity will be about 10 times (100 $\div$ 10) that calculated for the dry material, or 110 cm/hr. Phloem exudates do often consist of 10-percent solutions, but we've seen that they are usually more dilute than phloem sap. Osmotic potentials of -20 to -30 bars are common in intact sieve tubes. These values are approximately equivalent to 20- to 30-percent sucrose solutions, so if the assimilates are moving in a 20-percent solution, then they are moving 5 times (100 $\div$ 20) faster than the calculated figure for the dry material, or about 55 cm/hr. If they are moving only in sieve tube "protoplasm" (presumably a layer pressed against the cell walls) with a much smaller cross section, Crafts and Lorenz calculated that they could be moving 50 to 100 times the calculated velocity for dry material.

Numerous measurements similar to those of Crafts and Lorenz have been made. Other techniques have also been applied, especially using tracers. If a wave of radioactivity is measured as it passes two points along a stem, linear velocities should be indicated, although there are several difficulties that we shall not consider here. The conclusion of all these studies: **The bulk of photosynthates move in the phloem at peak velocities of 30 to 150 cm/hr.**

It is difficult to appreciate these velocities. If a lone sieve element in an angiosperm is 0.5 mm long (gymnosperm sieve cells are longer, about 1.4 mm), at translocation velocities of 90 cm/hr an entire element would be emptied and refilled in two seconds. At 300 × magnification, an entire phloem element could not be seen at one time in a microscope field, so if one could watch movement of particles of tracer dye through the microscope in sieve elements, velocities would be too rapid to follow conveniently with the eye. Slow-motion moving pictures would be necessary! Without magnification, movement at 100 cm/hr is equivalent to the tip of a 16-cm (6.3-in.) minute hand on a clock. Such movement can easily be detected with the unaided eye.

7.6 Phloem Loading

In 1949, Brunhild Roeckl determined the osmotic potentials of photosynthetic cells and sieve sap of *Robinia pseudoacacia* (black locust) using plasmolysis, refractometry, and cryoscopic techniques (Section 2.6). Such measurements have since been repeated by several others. Typically, the mesophyll cells have an osmotic potential of −13 to −18 bars, whereas sieve elements in leaves have an osmotic potential of about −20 to −30 bars. Since most of the osmotic potential is due to the presence of sugars in both kinds of cells, it is clear that the sugar concentration is approximately 1.5 to 2 times as high in the sieve elements as in the surrounding mesophyll cells. The process in which sugars are raised to high concentrations in phloem cells close to a source such as photosynthesizing leaf cells is called **phloem loading.**

How does it work? To find out, leaves are allowed to photosynthesize in the presence of $^{14}CO_2$, sampled at different times, and examined by high resolution autoradiography in which a thin layer of dry photographic emulsion is pressed against the tissue to detect the exact locations of the radioactivity (Fig. 7-5). As would be expected, the label first appears in the photosynthetic cells, but the most careful studies indicate that it is then secreted into the cell walls surrounding the photosynthesizing cells rather than passing through plasmodesmata directly into the phloem cells. That is, assimilate is secreted from the symplast into the apoplast or free space (cell walls) before entering the symplast in the phloem! When it moves into the sieve elements, it may be via companion cells or **transfer cells,** special cells with involuted cell walls that serve greatly to increase their membrane surface area (see later). The veins in many leaves are surrounded by a layer of **bundle sheath cells.** Some anatomists have suggested that these might act almost like an endodermis, controlling the entrance or exit of materials into or from both phloem and xylem, especially in the minor veins. Thus, they could well be important in phloem loading, but so far tracer studies do not suggest any special role for these cells. They may prove to be important in plants that photosynthesize by the C-4 pathway (see Section 10.3).

Specific sugars or other compounds can be labeled with ^{14}C and applied to leaves in various ways such as the reverse flap technique, or by **vacuum infiltration,** in which a leaf is placed in a solution, vacuum is applied causing air to bubble out of the leaf through the stomates, and vacuum is released, causing solution to immediately penetrate the leaf. Another method involves **abrasion,** in which some rough material is rubbed on the leaf surface, destroying the cuticle but rupturing only a few epidermal cells, after which the solution of the labeled compound can be applied. High-resolution autoradiography confirms in both cases the route of movement from the apoplast through transfer and companion cells into sieve elements, although sieve elements may concentrate the labeled material directly from the apoplast.

These techniques allow us to study the energy requirement and selectivity of the phloem-loading process. A. L. Kursanov and his coworkers in Russia (late 1950s and early 1960s) infiltrated solutions of ATP (the energy currency of cells) into leaves, which greatly promoted loading. Thus, phloem loading is a typical energy-requiring accumulation process such as was discussed in Chapter 5. Sugars and other compounds are moved across membranes into the sieve elements against concentration gradients and by the expenditure of metabolic energy.

The loading process is highly selective. Only certain compounds are loaded, accounting for the observation that only nonreducing sugars, specific ones depending upon the plant species, are transported in phloem sap. Donald R. Geiger and his coworkers at the University of Dayton in Ohio applied various labeled sugars to abraded sugar beet leaves and studied their uptake into small veins (phloem tissues). Sucrose was loaded much more rapidly than the other sugars, and sucrose is the sugar normally translocated by this plant. The ^{14}C from sugars that are readily converted to sucrose was also loaded,

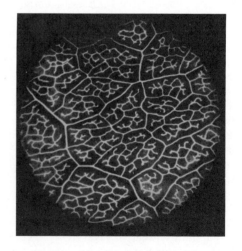

Figure 7-5 An enlarged positive print of an autoradiograph showing phloem loading in sugar beet leaves. Light areas are the minor veins that have accumulated the radioactive sucrose. (Courtesy Donald Geiger.)

although not as rapidly, but sugars that are not metabolized were not taken up at all. In other studies, some amino acids were readily loaded into the phloem, while others were not. This is true of minerals, also; those that are readily transported in phloem (nitrogen, phosphorus, potassium) are readily loaded; those that are not transported (calcium, iron, boron) are not (Chapter 6). This may be true even for such synthetic compounds as the herbicides 2,3,4-T (relatively immobile in phloem), 2,4-D (intermediate), and maleic hydrazide (most mobile), although they may enter the phloem by passively *permeating* rather than by being actively loaded. **So movement in the phloem is apparently dependent on the loading or other entry process rather than on mobility in phloem per se.** It seems that virtually any ion or molecule that can enter either phloem or xylem can move through those tissues. As we continue to increase our understanding of these matters, we may better select chemicals to apply for insect or disease control and to control the development of the plant itself.

7.7 An Introduction to Phloem Transport Theories

Just how do assimilates move through the phloem? The question brings up a number of interesting points about how science works. The scientific process forms a sort of ascending spiral. Hypotheses are established, tests are performed and data gathered, and the hypotheses are then accepted, rejected, or (most likely) modified. Thus, experiments are carried out in response to the hypotheses, but the hypotheses must be evaluated on the basis of the experiments. Let's follow this approach for the remainder of this chapter. The observations and conclusions we have examined have led during the last century to several hypotheses, three of them since 1957! It is an indication of the depth of our ignorance that only one or two of the hypotheses can be eliminated on the basis of present observations.

The hypotheses fall logically into two broad categories: passive and active. In passive mechanisms, metabolic energy is used to establish differences, usually in pressure, at the source and the sink. Energy is not used along the pathway of movement to keep transport going, but the cells along the pathway may require metabolic energy to maintain a viable structure. Active theories postulate that energy is utilized along the pathway to drive transport itself. Theories might also be classified on the basis of whether assimilates move in solution in a bulk or massflow (see Section 1.5), or whether individual assimilates are moved more or less independently of the liquids in the system. If assimilates move by a bulk flow, then all would move in the same direction within the phloem cell, although it is conceivable that different assimilates might move at somewhat different velocities as they interact with the channel along the way or move laterally to other cells. Other mechanisms might permit a **true bidirectional movement,** in which different substances move in opposite directions within the same cell at the same time.

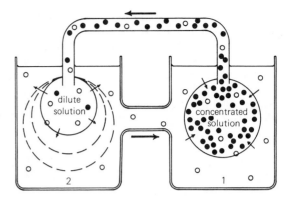

Figure 7-6 A model illustrating the pressure flow theory of solute translocation as proposed by Münch. Note that concentration of black particles will control the rate and direction of flow, but white particles (which are much more dilute) will move along in the resulting stream. Dashed lines on the left imply that flow may occur due to expansion (growth) of the second osmometer (tissue), as well as movement out through the membrane. (From Salisbury, 1966.)

Passive Theories of Phloem Transport. Diffusion is an example of a passive translocation mechanism, although it is several thousand times too slow to account for phloem transport. An **interfacial movement** would be another example. Substances, depending upon their polarity and electrical charge, will move rapidly along an interface between water and ether, for example. Could there be interfaces within the sieve tubes along which assimilates move by this mechanism? There is no convincing evidence for such interfaces.

By far the most widely studied hypothesis is that proposed by E. Münch in Germany in 1926. This **pressure flow hypothesis** is the ideal example of a passive phloem transport mechanism. Münch's model is simple, straightforward, and based upon a real model that can be built in the laboratory: two osmometers connected to each other with a tube (Fig. 7-6). Both osmometers may be immersed in the same solution or in different solutions, which may or may not be connected. The first osmometer contains a solution more concentrated than the surrounding solution, while the second osmometer contains a solution less concentrated than that in the first osmometer, but either more or less concentrated than the surrounding medium. Water moves into the first osmometer by osmosis, and pressure builds up. Since the two are connected, pressure is transferred from the first osmometer to the second (with the velocity of sound, which is basically a pressure transference phenomenon). Soon the increasing pressure in the second osmometer causes a more positive water potential than in the surrounding medium, so water molecules diffuse out by osmosis. Solute molecules are retained in the second osmometer by its membrane. The result is a movement of water into the first osmometer, movement of solution through the tube, and movement of water out the second osmometer. Solutes from the first osmometer are carried along in a bulk flow to the second osmometer. If the walls of the second osmometer stretch, as in the growth of a fruit or tuber, pressure is relieved even if no water moves out.

In Münch's model, mass flow ceases when enough solute has been moved from the first osmometer to the second to equalize osmotic potentials and pressure potentials in both osmometers. Münch suggested that the living plant contained a comparable system but with advantages (Fig. 7-7). Sieve elements near to source cells (usually leaf mesophyll cells) are analogous to the first osmometer, but assimilate concentration is maintained high in these sieve cells by photosynthesis and then phloem loading. (Münch was unaware of phloem loading.) Concentration of assimilates in the other end of the phloem system, near the sink cells, is probably kept low by phloem unloading, although less is known about that. Assimilates are rendered osmotically ineffective in the sink cells by metabolism, incorporation into protoplasm (growth), or storage as starch or fats. The connecting channel between source and sink is the phloem system with its sieve tubes; the surrounding dilute solutions are those of the apoplast, specifically those in cell walls and in the xylem. (Münch proposed the concepts of the symplast and the apoplast in his presentation of the mass flow hypothesis.)

Note that the system is passive according to this definition. Energy drives photosynthesis and loading at the source, establishing high osmotic concentrations there, and may also be used for unloading or for metabolic conversion to osmotically ineffective materials at the sink end. But movement itself is in response to the pressure gradient established osmotically and not to any pumping action along the way. Of course, energy is no doubt required to keep the channels in a functional condition. This mechanism would not permit bidirectional movement in the same transporting sieve tubes.

Active Theories of Phloem Transport Perhaps the simplest form of an active translocation mechanism would be one based upon **cyclosis** or **cytoplasmic streaming.** In many cells, the cytoplasm flows around the periphery of the cell, and metabolic energy (ATP) is required to maintain this cyclosis. As early as 1885, the Dutch botanist Hugo de Vries suggested that cyclosis might account for phloem transport. Wherever cytoplasmic streaming occurs, and wherever solutes diffuse from cell to cell, their rate of movement must be accelerated by the streaming (e.g., in parenchymous tissue, such as cortex). Such a mechanism would permit true bidirectional movement within a single cell (Fig. 7-8), but for three reasons it is no longer believed that cyclosis accounts for phloem transport: (1) It does not occur in functional sieve tube elements, (2) it is too slow to account for observed velocities, and (3) the diffusion step would greatly limit the process anyway.

Another idea was proposed by D. S. Fensom in 1957 and D. C. Spanner in 1958. Spanner continues to champion this **electro-osmosis hypothesis.** To simplify a rather complex idea: It suggests that potassium ions move through the protein fibers extending through the sieve plates, carrying water and other solutes along. The ions are returned passively via nearby

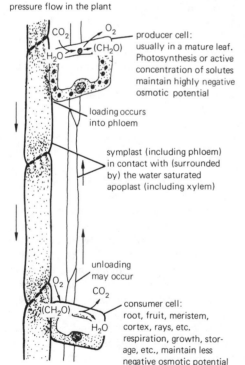

pressure flow in the plant

producer cell: usually in a mature leaf. Photosynthesis or active concentration of solutes maintain highly negative osmotic potential

loading occurs into phloem

symplast (including phloem) in contact with (surrounded by) the water saturated apoplast (including xylem)

unloading may occur

consumer cell: root, fruit, meristem, cortex, rays, etc. respiration, growth, storage, etc., maintain less negative osmotic potential

Figure 7-7 The pressure flow theory of Münch as it may be applied to the plant. (From Salisbury, 1966.)

Figure 7-8 A schematic illustration of the cytoplasmic streaming hypothesis of solute translocation. Note that diffusion of white or black particles across the sieve plate will occur in response only to their own concentration gradients. Movement of the two materials might occur simultaneously in opposite directions. Rate of movement is accelerated by streaming. (From Salisbury, 1966.)

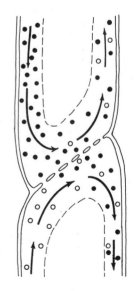

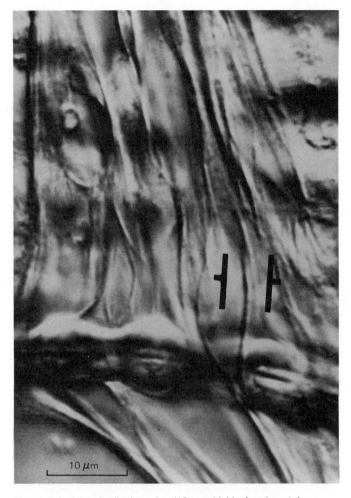

Figure 7-9 A longitudinal section (18 μm thick) of a sieve tube viewed in the Nomarski microscope and showing parallel strands of cytoplasm, one of which (-"-) is easily followed through the sieve pore. The bar represents 10 μm. (From Thaine and De Maria, 1973.)

companion cells or via the cell walls, and then are actively pumped (using the energy of ATP) back into the sieve tube element on the other side of the sieve plate. This establishes a strong K^+ gradient between each set of sieve tube elements, and K^+ then moves through the fibers along this gradient. There is some evidence for the electro-potential gradient (Bowling, 1969).

In 1961, Robert Thaine published a hypothesis of phloem transport based upon his observations with the light microscope of what appeared to be moving strands or particles within strands in living stems of *Primula obconica* (Fig. 7-9). He suggested that the strands extended through many sieve tube elements, passing through the sieve plate pores. This **transcellular strand theory** would allow bidirectional movement if strands in a cell could move in opposite directions. It now appears that there are three hypothetical mechanisms of movement: The entire strand moves like a thread through the sieve tubes (now considered unlikely); particles move within the hollow strands, pushing the contents along; or the hollow strands have waves of contraction (a **peristalsis**) that move the contents along. In all three variations, energy is required in the phloem cells to account for the movement.

Beginning in 1968, several investigators suggested that movement in the phloem might be produced by contractions or other activities of the ultramicroscopic protein fibers that are observed only in electron micrographs. Various mechanisms have been suggested, including a peristalsis of the fibers that would be similar to that of Thaine's transcellular strands but on a much smaller scale. Again, energy would be required in the pathway for activity of the fibers, and it is possible that movement might be occurring in different directions (different fibers) at the same time within the same cell.

Testing the Hypotheses How can we test the hypotheses? We need answers to at least the following three groups of questions:

1. What is the anatomy of the phloem tissue? Can we discern structures that will support or reject any of these ideas? What are the detailed structures and ultrastructures in *living* phloem tissues? What are the properties of molecules that occur in phloem?

2. What are the detailed patterns of phloem transport? Does true bidirectional movement ever occur? Can we recognize *apparent* bidirectional movement (see Section 7.9) for what it is? Does bulk flow occur? Do substances move at different rates within the same phloem cells?

3. What is the role of metabolism or metabolic energy in phloem transport? Is energy required to load and/or unload the phloem system, to maintain the system or, within the sieve tubes, to move assimilates?

7.8 Phloem Anatomy

A function can often be understood by understanding the structure where it occurs. An examination of the valves and chambers of the heart makes its function as a pump clear to anyone with some sense of mechanics. Can we discover the mechanism of phloem transport by studying the structure of phloem tissue? Perhaps, if we can understand the structure well enough. Certainly we cannot hope to understand the mechanism of phloem transport *without* understanding phloem anatomy.

Each transport theory requires a certain phloem structure. Yet it is a sobering thought that after more than a century of study with high-quality light microscopes and about two decades of study with electron microscopes, we are still unable to eliminate any of the principal theories (pressure flow, electro-osmosis, transcellular strands, or protein fibers) on the basis of anatomical studies, although cyclosis theories have been mostly eliminated. Regardless of the currently viable theory, its proponents can find support for it in the published results of anatomical studies.

What is the problem? For one thing, the scale of the functioning elements in the phloem transport system is extremely small compared to the valves and chambers of the heart. The

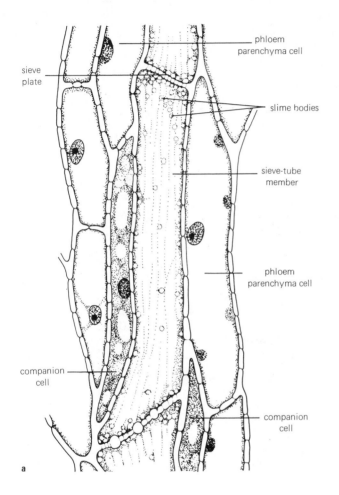

Figure 7-10 (a) A longitudinal view of a mature sieve-tube member and companion cell. (b) A face view of a sieve plate; the black areas are actually holes in the wall.

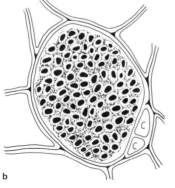

structures that result in phloem function are at least as small as the phloem cells and may be as small as the protein strands that may (or may not!) penetrate the sieve plates.

Furthermore (and this is probably the more important limitation), the functioning units in phloem transport are living cells buried below the plant surface. Any attempt to study a phloem cell may so change its structure that its function can no longer be discerned. For example, the minute bodies that are often seen plugging the sieve plate are probably broken free from the cell periphery when pressure is released by cutting to take a sample. Or the strands that we see passing through sieve pores in electron micrographs may have been produced by the fixative used to prepare the specimen for observation (that is, they may be **artifacts,** not uncommon in microscope work). This difficulty of observing functioning phloem cells remains the most serious limitation of our study of phloem transport.

Phloem Tissue Consider again the four kinds of cells in phloem tissue (Fig. 7-10). First are the **sieve elements,** which are the elongated cells in which phloem transport actually takes place. In angiosperms, they are connected end to end with pore-filled **sieve plates** between, forming long cellular aggregations called **sieve tubes.** In gymnosperms and lower vascular plants, sieve plates are not as clear; there are sieve areas with smaller pores on lateral walls and on slanted end walls. Hence, the units are called **sieve cells** instead of sieve elements. Second are the **companion cells** (in angiosperms) or **albuminous cells** (in gymnosperms), which are living cells closely associated with the sieve elements or sieve cells and having relatively dense cytoplasm and distinct nuclei. There are usually many plasmodesmata in the walls between sieve elements and their companion cells, with the plasmodesmata pores frequently being branched on the side of the companion cell. The exact function of the companion cells remains unknown, although they are always present and viable in functioning phloem and degrade in senescent phloem. In leaves, they might function in phloem loading. Third are the **phloem parenchyma cells,** which are thin-walled cells similar to other parenchyma cells throughout the plant except that some are more elongated. They may act in storage as well as lateral transport of solutes and water. Fourth are the **phloem fibers,** which, as in other tissues, are thick-walled cells that must provide strength to phloem tissue.

In recent years, plant anatomists have begun to appreciate special plant cells called **transfer cells.** Such cells are characterized by inwardly directed projections of cell wall material that greatly expand the interface between cell wall and cell membrane (Fig. 7-11). This increased surface for the plasmalemma could be of considerable importance in any process in which transfer of materials across the membrane between apoplast and symplast plays a central role, although few physiological studies have been performed on transfer cells. How important are these cells in the phloem? They are certainly not restricted to phloem but occur throughout the plant,

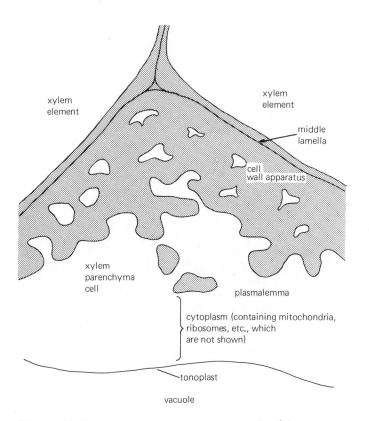

Figure 7-11 Drawing based on electron micrographs of the extensive cell wall ingrowths (called the cell wall apparatus) in a transfer cell.

such as in both xylem and phloem parenchyma of leaf nodes (where they are common) and in reproductive structures such as the interface between the gametophyte and the sporophyte of both lower and higher plants. In all vascular plants so far studied, they seem to be important in active transport processes occurring over short distances, such as in salt glands, nectaries, and the connections (haustoria) between a parasite and its host. But do they play a role in phloem loading? Only in certain species, it appears, and that has yet to be demonstrated. In a survey of over 1,000 species, it was found that phloem transfer cells associated with sieve elements in minor leaf veins were relatively rare, occurring only in certain herbaceous dicot families. They may aid phloem loading in these evolutionarily advanced groups, but they are not essential for this process in all higher plants.

Phloem Development Consider the development of a sieve element and its companion cell (Fig. 7-12). A single cambial cell divides twice to produce both a sieve element and its accompanying companion cell. The sieve element expands rapidly in diameter and becomes highly vacuolated, with a thin layer of cytoplasm pressed against the cell wall. Minute bodies appear in this cytoplasm. They are generally ovoid in

shape, some appearing rather amorphous, while others have a more fibrillar or stranded appearance. Traditionally, they have been called **slime bodies,** but they apparently consist of P-protein (see later). As they grow, their limits become less well defined, and they eventually fuse, often becoming completely dispersed in the cell. At about this time, the nucleus begins to degenerate. Eventually it disappears completely in most sieve elements, although there are a few exceptions in which degenerate or even whole nuclei may remain. The tonoplast membrane between the vacuole and the cytoplasm also disappears at about this stage, but the plasmalemma remains intact. Cytoplasmic streaming has been observed in some developing sieve elements, but when the elements are mature, this activity apparently ceases.

While these events are taking place, the sieve plate is developing. This process begins with small deposits of a special glucose polymer called **callose,** usually around plasmodesmata. The deposits increase in size until they assume the shape of the final pore. The middle lamella first disappears

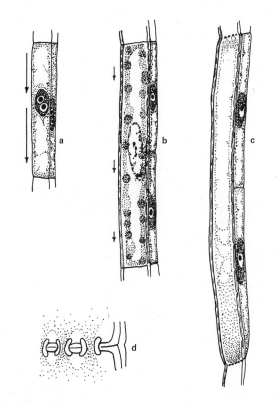

Figure 7-12 Some morphological aspects of sieve-tube elements. (**a**) Young sieve-tube element with companion cells. Long arrows indicate considerable cytoplasmic streaming. (**b**) Sieve tube of intermediate maturity. Slime bodies are evident, the nucleus is beginning to disappear, and cytoplasmic streaming is no longer readily evident. (**c**) Mature sieve-tube element. The nucleus, tonoplast, and cytoplasmic streaming are no longer evident. (**d**) Longitudinal section of sieve tube through a sieve plate, showing cytoplasmic connections through the pores. (From Salisbury, 1966.)

in the center of the deposit, and then the deposits on both sides of the wall fuse as the wall between disappears, so the pore is lined with callose resembling a grommet from the beginning. The plasmalemma extends through the pore so that it is continuous from cell to cell. The pores are usually 0.1 to 5.0 μm in diameter, much larger than any solute or even virus particle. The sieve elements are usually 20 to 40 μm in diameter, and 100 to 150 μm (0.10 to 0.5 mm) long. Of course, there are numerous exceptions to most of what has just been summarized, depending upon species. For instance, in gymnosperms the albuminous cells and the sieve cells do not arise from a single cambial cell.

Some sieve tubes remain functional for several years. This is probably most striking in perennial monocots, such as palm trees, which have little cambial tissue capable of producing secondary phloem. Some sieve elements laid down at the base of a palm tree, for example, apparently must remain functional throughout the life of the tree, which may be well over 100 years. Some new sieve elements are formed by a primary thickening meristem.

Phloem Ultrastructure Companion cells have unusually dense cytoplasm with small vacuoles. Mitochondria are abundant, as are dictyosomes and endoplasmic reticulum. The nucleus is well defined until the associated sieve element finally begins to become nonfunctional. Phloem parenchyma cells, in contrast to companion cells, contain large vacuoles and perhaps fewer organelles (although this may be only apparent due to relatively less cytoplasm). Chloroplasts are often conspicuous in these cells.

It is the ultrastructure of the sieve tube that is of most immediate interest. In addition to the slime bodies, earlier workers had often seen amorphous material in sieve elements in the **lumen;** that is, the part that was originally the vacuole before disintegration of the tonoplast. With the electron microscope, most, if not all, of this "slime" proved to be a fibrillar proteinaceous material. The diameter of the fibers is on the order of 70 to 240 A. There are clearly several different kinds of this protein, and some of these may be intraconvertible. The protein is referred to as **P-protein** (for phloem protein), and slime bodies may be called **P-protein bodies.**

Does P-protein play an active role in phloem transport? Or is its role only the secondary one of stopping "bleeding" following injury (see later)? Many workers have suggested that P-protein might use the energy of ATP to contract, this contraction providing the motive force for phloem transport. Contractile protein called **actin** is well known from animal muscle cells and is now known from plant cells as well. It is associated with cytoplasmic streaming in *Nitella* and *Chara,* for example. Is P-protein actin-like? Preliminary evidence for this was presented in 1974, but studies by Barry A. Palevitz and Peter K. Hepler (1975) seem to refute the idea. The test for actin is a highly characteristic association with heavy meromyosin from rabbit, a test that is clearly positive in the cases of *Nitella* and *Chara* but fails with P-protein. But nothing says

that only actin can contract. Does P-protein some way contract or otherwise act in an active phloem transport? Time and further research should tell.

Apparently, P-protein plays a critical role in stopping the loss of sieve tube contents when injured. Internal pressure is high, and surging occurs when a sieve element is cut, causing P-protein to flow to the sieve plate, blocking the pores. There is also evidence that P-protein coagulates when exposed to air, and in a few cases the sieve cell may collapse upon wounding. Without one or a combination of these mechanisms, plants might "bleed to death" when injured by grazing or other means. Yet these and other mechanisms make it extremely difficult to study and understand the uninjured sieve element. One approach to solving this problem is to freeze the tissue rapidly before fixation, but sieve elements contain highly concentrated sugar solutions and colloidal sols that are difficult to freeze.

These problems of tissue preparation have left us undecided about several questions of phloem anatomy. For example, what is the normal status of the sieve plate pores? If mass flow occurs through them, they must be at least partially open. If there are pumps, however, pores could be filled with P-protein or even with Thaine's strands. Electron micrographs of both "open" and "blocked" pores have been published in numerous papers (Figs. 7-13 and 7-14). In a recent paper of Donald Fisher (1975), sieve tubes known to be functional were reported to be open; blocked sieve tubes were known to be nonfunctional.

Virtually all electron micrographs do show *some* P-protein fibers extending through the sieve plate pores. In intermediate conditions of P-protein density, two workers may look at the same micrograph, one emphasizing the open areas between the fibers and the other emphasizing the fibers; to one, the pores appear "open," to the other, "blocked."

What about Thaine's transcellular strands? Thaine suggests that the P-protein is organized into the larger strands (1 to 7 μm = 10,000 to 70,000 A in diameter), extending through the sieve plate pores and throughout the length of the sieve tubes and containing moving particles. He has motion pictures of the strands and their particles in living tissue. Other workers say the strands do not exist as such but are only illusions caused by diffraction of the cell walls. Or perhaps they are standing waves caused by flowing sieve tube sap, which itself would move too rapidly to see. Thaine countered this by showing that the strands go in and out of focus as the microscope objective is focused up and down through a sieve element. In 1973, Thaine and Margaret E. De Maria published photographs taken with the differential diffraction microscope (Nomarski optics) showing strands penetrating a sieve plate pore (Fig. 7-9). Using a rapid-freeze technique, they also obtained electron micrographs showing structures that they interpreted as strands. Though freezing occurred in less than 1 second, parts of the strands were apparently destroyed. Thaine and De Maria suggest that this demonstrates their fragile nature and indicates why other workers have been unable to see them in electron micrographs.

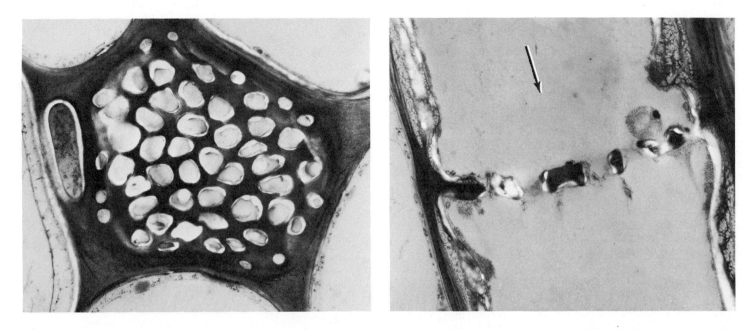

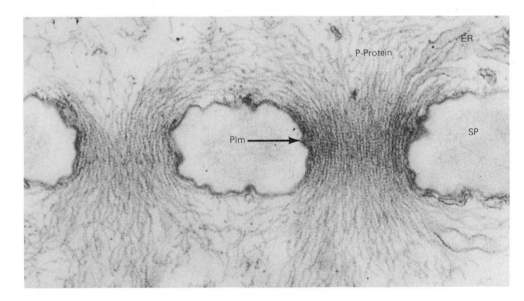

Figure 7-13 Electron micrographs of the sieve plate in soybean (*Glycine max*), showing rather open pores with relatively small amounts of P-protein fibers. Left, cross section; right, longitudinal section (both × 14,000). Arrow indicates probable direction of flow. (From Fisher, 1975.)

Figure 7-14 A highly magnified (× 50,000), longitudinal section through a sieve plate (SP) of Dutchman's-pipe, *Aristolochia brasiliensis*, showing sieve pores filled with P-protein fibers. The plasmalemma (Plm) can be seen lining the pores. (From Behnke, 1971.)

7.9 Bidirectional Phloem Transport

The mass flow and electro-osmosis hypotheses both require that everything within a given sieve tube be moving in the same direction at any given time. True bidirectional transport would be incompatible with these ideas. Strands, however, might be moving, or more probably contracting with peristaltic movements, in opposite directions within a given sieve tube, and this could also be true of some P-protein based mechanism. Does bidirectional movement actually occur in phloem transport?

Unfortunately, this question is much easier to ask than to answer. Numerous experimental observations seem to indicate that such movement does occur, but so far it has always been possible to devise alternative explanations. This has been so frustrating that M. J. P. Canny (1971) has described such studies as "will-o'-the-wisp experiments." Nevertheless, if we carefully examine the difficulties, it may someday be possible to devise an experiment that will avoid them.

As early as the 1930s, many workers studied the phloem transport of various substances, expecially viruses and herbicides. The flow of assimilates could be altered in various ways;

by darkening a mature leaf, for example, so that it became a sink instead of a source. In virtually all these experiments, the tracer substance moved with the assimilate stream, but the approaches were too limited to be certain that bidirectional movement *could not* occur under any circumstances. On the other hand, several workers have studied the movement of two different tracer substances from two different points on the plant. Fluorescein dye or radioactive tracers were applied at different points and their movement followed, much as in the experiment of Biddulph and Cory described earlier.

There is no question that two tracers, applied to different parts of a plant, often move bidirectionally. Tracers applied to young leaves may move basipetally (toward the base), while tracers applied to older leaves below may move acropetally (toward the tip), so that the two pass each other in the stem, moving in opposite directions. Such bidirectional movement may even occur in a leaf petiole. But do both tracers move in the same vascular bundle or the same sieve tube? Many studies have indicated that they do not. Carol A. Peterson and Herbert B. Currier (1969) abraded the stem surface of several species and applied fluorescein, which was absorbed into intact phloem tissue. After various time intervals, they cut sections from above and below the point of dye application. The tracer moved both up and down, but after fairly short time intervals it was never present in the same bundle both above and below a treated area; each bundle and sieve tube translocated the dye in only one direction. With longer time intervals, the dye could move to a node in one bundle, pass laterally to another bundle, and move back down the stem in an opposite direction.

Studies with aphids might settle the question of whether or not tracers can move in opposite directions in a single phloem cell. Figure 7-15 shows the setup for an aphid experiment as performed by Walter Eschrich in 1967. Fluorescein was applied to one leaf and $^{14}CO_2$ to another, with the honeydew from aphids between being collected on a slowly rotating cellophane disk. Much of the honeydew contained both tracers. In 1969, L. C. Ho and Alan J. Peel performed similar experiments with willow shoots, using ^{14}C, ^{3}H, and ^{32}P. They not only observed two tracers in the honeydew exudates, but of the 22 aphids employed, six produced double-labeled honeydew *from the first drop.*

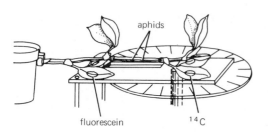

Figure 7-15 Experimental arrangement for the study of the simultaneous movement of fluorescein and ^{14}C-assimilates in *Vicia faba* plants. (From Eschrich, 1967.)

Are these experiments conclusive? Not completely; there are alternatives. A stylet inserted into a sieve tube might itself become a sink so that substances flow from both directions toward the stylet. This appears unlikely, however. It seems more likely that the direction of flow changes with time in a given sieve tube, perhaps as the source and sink roles change; or perhaps in response to more subtle hormone mechanisms. Thus, if the experiment lasts an hour or more (as is usually the case), then flow was possibly reversed during the experimental period so that some of the honeydew was produced from one source, while the rest was produced from another. Furthermore, there is considerable lateral transport between sieve tubes and even between the phloem and the xylem, especially at the nodes but in the internodes as well; in some species, sieve elements have lateral sieve pores. So it is possible that two tracers going in opposite directions but in different vascular bundles or sieve tubes, or even in xylem and phloem, become mixed by lateral transport. Thus, it is possible that true bidirectional transport does not occur. But it is also possible that it does occur, and a few plant physiologists feel that it does.

7.10 Pressure in the Phloem

Sieve tubes contain fluids under pressure, as indicated by the feeding habits of aphids. Cutting of phloem tissue often results in a brief exudation until sieve plates become plugged. There are a few cases, however, in which exudation occurs for several hours to many days. When the top of a sugar palm trunk or its influorescence is tapped, for example, as much as 10 liters of sugary sap may drip out of the cut sieve tubes in a day, and the Palmyra palm in India produces on the order of 11 liters of sap from cut phloem each day. This sap consists of about 10 percent sucrose and 0.25 percent mineral salts and has probably been diluted by water moving osmotically from the xylem (apoplast) into the phloem after pressure is released. There is little doubt that pressure exists in sieve tubes, as a pressure flow mechanism requires. Pressure may also be compatible with some of the other hypotheses.

The Münch hypothesis postulates not only that the sieve tube contents should be under pressure but that a pressure gradient should exist, and that this gradient should be sufficient to account for observed flow rates. Thus, it would be desirable to obtain measured values for pressures along the flow route. An osmotic gradient has been observed from source to sink, so if the fluids in the surrounding apoplast have approximately the same water potentials throughout the system,* the gradient in osmotic potentials would infer a gradient in pressures. Nevertheless, it would be satisfying to measure the pressures directly, and several workers have

---

*This cannot be strictly the case, since there is a gradient in water potentials throughout the xylem as we have seen in Chapter 4.

attempted to do this in recent years. One approach is to insert a hollow stainless steel needle attached to a glass sealed microcapillary directly into a bundle of sieve tubes, observing the extent to which the gas in the capillary is compressed (as in the pressure measurement of large algal cells; see Fig. 2-10). Another approach is to attach a pressure gauge to a cut palm shoot, and still another is to apply a pressure cuff similar to those used in measuring blood pressure, increasing pressure in the cuff until wound exudate stops. Pressures as high as 22 bars have been measured. At this time (1977), none of these methods has been completely satisfactory, but they could provide exciting new approaches to this hoary old problem.

7.11 Different Transport Velocities for Different Substances

The most simplistic view of the mass flow hypothesis might suggest that substances should move in the phloem not only in the same direction but at the same velocity. Thus, several workers (e.g., Biddulph and Cory, 1957, and Fensom, 1972) have attempted to measure the velocity of flow of different tracer substances (e.g., ^{14}C-sucrose along with ^{32}P and ^3H_2O). The different substances move at different velocities, but there are complications. Most studies, for example, have not considered the probability of different phloem-loading rates, which are known to vary for different substances. A few studies have avoided this difficulty by measuring velocities between two points along the transport system. This is somewhat more difficult, but results are similar: ^{14}C-sugars often move most rapidly, with ^{32}P-labeled phosphates moving more slowly, and ^3H_2O moving slowest of all. At first glance, it would seem that if the water moves more slowly than the solutes it is supposed to be carrying, a mass flow hypothesis could not be tenable.

There are two important complications, however: First, water and other substances exchange rapidly along the pathway. That is, much water passes out through sieve tube membranes into surrounding tissues, while much water from these tissues moves into the sieve tubes. Sucrose and phosphate do not pass as readily through the membranes along the way, so they might appear to move much faster than the carrier water molecules. Second, it is simplistic to imagine that the phloem transport system consists only of inert tubes. Sieve elements are alive and contain cytoplasm, P-protein, and other substances. Hence, solutes in the transport stream must interact with the matrix along the transport routes, much as solutes interact with the paper in a paper chromatogram, and therefore must move at different rates.

It appears, then, that measurements of velocities again fail to provide us with a definitive test of the various phloem transport hypotheses. Considering exchanges and interactions, all the proposed mechanisms might be expected to move different solutes at different velocities.

7.12 The Role of Metabolism

To study energy requirements during phloem transport, various inhibitors have been used: low temperatures, **anoxia** (withholding oxygen), and various chemical respiration inhibitors (cyanide, dinitrophenol, and others). Results of these experiments are difficult to interpret because of the inherent complexity of the plant. It is helpful to evaluate such inhibitory effects as they might apply to the source, the sink, or the pathway.

Source effects are perhaps the most difficult to evaluate, because the inhibitor might be inhibiting photosynthesis, phloem loading, or some other mechanism essential to phloem transport. In any case, all these inhibitors applied at the source do inhibit translocation. We have seen that ATP promotes loading, so anything that inhibits respiration (which produces ATP) would be expected to inhibit loading. Inhibitors applied at the sink also restrict translocation. Probably they inhibit active unloading, which otherwise prevents a pile-up of sugars at the end, thus steepening the osmotic and hence the pressure gradients.

Could effects upon the transport system itself help us decide between the various models? We find ourselves in the familiar position: All theories predict that inhibitors might have an effect upon movement through the pathway. Active theories suggest that any inhibition of metabolism would reduce the metabolic driving forces essential to phloem transport as well as maintenance of the pathway. Passive theories suggest that metabolism is essential only to maintain the integrity of the transport pathway. In any case, inhibitors applied along the pathway do restrict translocation. Furthermore, sieve elements prove to be highly active metabolically compared to parenchymous cells. They are rich in ATP, which even appears in phloem exudate.

Temperature effects could be instructive. Some plants (e.g., beans) are particularly sensitive to chilling, showing inhibited translocation at temperatures of 12 C, or below. Other plants (e.g., sugar beets) are more resistant, exhibiting little chilling response until temperatures approach 0 C. Microscopic examination has revealed distinct structural changes in the sensitive but not the resistant plants at and below the chilling threshold. Response to temperatures above the threshold can be accounted for on the basis of viscosity changes within the sieve tubes. So these studies are quite compatible with the mass flow hypothesis—as they are with the other hypotheses as well.

7.13 Phloem Transport Theories: A Final Consideration

So which hypothesis best accounts for phloem transport? Must we believe that only one mechanism can operate in the plant kingdom? Accelerated movement due to cytoplasmic streaming must be a fact of life for those cells in which streaming occurs. Pressure flow could work in some situations and must function if only on a limited scale. One or more of the other mechanisms could operate concurrently with pressure flow. But let's review the hypotheses:

Pressure Flow As we have seen, several observations support this idea, while several others can be reconciled with it but only after much careful evaluation of the logic of alternative explanations. The strongest argument in its favor seems to be its simplicity. When one builds a laboratory model of the Münch system (see Fig. 7-6), it is easy to make it work and to list the requisites for suitable function: (1) an osmotic gradient between the two osmometers, (2) membranes that will allow the establishment of a pressure gradient in response to the osmotic gradient, (3) a channel between the two osmometers that will allow flow, and (4) a surrounding medium having a higher water potential than that in the osmometer with the most negative osmotic potential. If these requisites are met in the plant, the system must function as it does in the model. All workers agree that the osmotic system (the symplast) with surrounding membranes exists in the plant, and pressures are observed in the transport system, although they still must be accurately measured to see if the gradient is sufficient to drive flow over long distances, as in trees. The surrounding medium with high water potential is the hydrated apoplast.

Problems arise on two grounds: First, anatomical studies of sieve tubes make us question whether a suitable channel exists for bulk flow between source and sink. The sieve pores seem too small, offering high resistance, and in many electron micrographs they appear to be partially to almost completely blocked with P-protein. Indeed, even the lumen of the sieve elements may contain such protein. It is easy to calculate that these protein strands would offer considerable resistance to flow, so that the calculated pressure differences between source and sink would not be sufficient to cause flow at the observed rates. On the other hand, blocking of sieve pores may be an artifact of preparation, and flow is actually observed from cut phloem tissues and from aphid stylets. If this is not a bulk flow through the sieve plates, then water must be moving into cut or tapped phloem by osmosis at rates much higher than seems possible. And where would the sugar come from? Second, the pressure flow mechanism comes in conflict with observations of apparent bidirectional movement. As we have seen, none of these experiments is conclusive, and all can be reconciled with pressure flow. In the late 1960s and early 1970s, with the increased use of aphids and other techniques and the reports of apparent bidirectional movement, many plant physiologists began to question seriously the Münch hypothesis. It is our impression that the tide is turning back to the Münch model at the time of this writing (1977). With a developing appreciation for phloem loading, pressure in the phloem, and transport velocities, the pressure flow idea has become stronger in the minds of many plant physiologists.

Electro-Osmosis Most observations can be made to fit an electro-osmosis theory, although calculations indicate that the energy required for the proposed circulation of potassium might be far too great, and present evidence favors open sieve plate pores, while this theory requires P-protein in the pores. In any case, there is little in favor of an electro-osmosis theory (it is as difficult to reconcile with bidirectional movement as is pressure flow), except that it is compatible with P-protein in sieve pores, and predicted electropotential gradients across sieve plates have been reported (Bowling, 1969). More data are required if this theory is to remain viable.

Transcellular Strands This hypothesis in its various forms would account for bidirectional movement, energy requirement along the pathway, and several other observations. Thaine's most recent observations are impressive enough to merit careful study and attempts by others at confirmation or rejection. A mathematical formulation, however, has recently rejected the model (Ferrier and Tyree, 1976).

Contraction of P-Protein Fibers This hypothesis might also account for bidirectional movement, pressure, velocities, and energy requirement in the pathway. It is not yet completely developed, so it is difficult to think of suitable tests, but clearly P-protein is not actin-like, and its primary role may be only to heal wounds. Its distribution in sieve tubes must be determined and reconciled with any hypothesis including mass flow, and we cannot yet eliminate the possibility that P-protein is directly involved in transport.

7.14 Some Final Problems

Although we have examined a wealth of detail about phloem transport, there are other items that we have not considered. For example, plant hormones and growth regulators may influence the direction and rate of transport of other substances. Minerals and assimilates tend to move toward points of growth regulator application, for example, as was observed in the 1930s with auxins. The most striking effects occur with application of cytokinins (Chapter 17), which will create sinks in leaves so that various substances move into a treated leaf.

Light has an interesting influence on translocation. Not only is the most sugar exported from photosynthesizing leaves during the light period, but most of it goes to the stem tip or other parts of the shoot rather than to the roots. In the dark, on the other hand, translocation to the root system is strongly favored over translocation to the stem tip. What controls direction, rates, and velocities in phloem transport? The source-to-sink relationship is important, but other factors must also play a role.

There are further intriguing problems in the field of translocation. Why does the flowering hormone (Chapter 22) move so much more slowly than anything else? And how can we understand the movement of the excitation stimulus in *Mimosa*, the sensitive plant? This stimulus is chemical; yet its velocity is extremely high compared to other substances.

Section Two

Plant Biochemistry

An apparatus used to measure photosynthesis of a single leaf.

8

Enzymes, Proteins, and Amino Acids

The functions of living organisms are largely determined by the kinds of chemical reactions occurring inside their cells. By acting as powerful catalysts, the protein molecules that act as enzymes determine which reactions proceed. From the products of a series of reactions, each catalyzed by a different enzyme, arise a large number of complex molecules that are integrated into subcellular structures such as cell walls, membranes, ribosomes, and starch grains. In this way, enzymes also influence the diverse forms attained by different plants and by different parts of the same plant. Thus, broad variations in morphology in the plant kingdom probably arise largely from differences in the kinds, quantities, and catalytic activities of enzymes the plants contain. As we shall see in Chapter 18, the genotype of the organism determines the *kinds* of enzymes it can produce, but the environment may strongly control the *amounts* of each that are actually formed and also their catalytic activities.

8.1 Enzymes in Cells

Enzymes are not uniformly mixed throughout cells. The enzymes responsible for photosynthesis are located in the chloroplasts, many of those essential for aerobic respiration occur exclusively in the mitochondria, while still other respiratory enzymes exist somewhere in the cytoplasm surrounding the mitochondria. Similarly, most of the enzymes essential to DNA and RNA synthesis and mitosis occur in nuclei. Enzymes governing the steps in a sequence of reactions (a **metabolic pathway**) are sometimes arranged in cellular membrane systems so that a kind of assembly-line production process occurs. Thus, the product of one reaction is released at a site where it can immediately be converted to a related compound by the next enzyme involved in the pathway, and so on until the metabolic pathway is completed and a quite different compound is formed.

For two reasons, these methods of compartmentalization almost surely increase the efficiency of numerous cellular processes: First, they help insure that the concentrations of reactants are kept relatively high at the sites where the enzymes acting upon them are located. Second, they help insure that a given compound will be directed toward the final useful product and not diverted into some other pathway by action

of another competing enzyme elsewhere in the cell that can also act upon it. This compartmentalization, however, is often not absolute, nor, apparently, should it be. For example, the membranes surrounding chloroplasts allow the outward passage of certain sugar phosphates produced during photosynthesis. These compounds are then acted upon by numerous enzymes outside the plastids that are involved in cell wall synthesis and respiration essential to growth and maintenance of the plant.

8.2 Properties of Enzymes

Specificity and Nomenclature One of the most important properties of enzymes is their specificity. Each commonly acts on only a single **substrate** (reactant) or a small group of closely related substrates that themselves have virtually identical functional groups capable of undergoing reaction. With some enzymes, the specificity appears to be absolute, but with others, there is simply a gradation between their abilities to convert related compounds to products. As we shall see, specificity results from specific combinations between the enzyme and substrate in a lock-and-key process.

Approximately 2,000 different enzymes have so far been discovered in living organisms, and many more will be added to the list as research continues. Each known enzyme has been named according to a standardized system and has also been given a usually simpler common or trivial name. In both systems, the names commonly end in the suffix *-ase* and usually describe the compound or kind of compound acted upon and the type of reaction catalyzed. For example, **cytochrome oxidase,** an important respiratory enzyme, oxidizes (removes an electron from) a cytochrome molecule. **Malic acid dehydrogenase** removes two hydrogen atoms from malic acid. These are the common names for the enzymes and, although short, neither gives sufficiently complete information about the reaction catalyzed. For example, neither tells the acceptor of the removed electron or hydrogen atoms.

The International Enzyme Commission lists longer but more descriptive and standardized names for all well-characterized enzymes (Dixon and Webb, 1964). As an example, cytochrome oxidase is named cytochrome *c*: O_2 oxidoreductase, indicating that the particular cytochrome from which electrons are removed is the c type and that oxygen molecules are the electron acceptors. Malic acid dehydrogenase is called L-malate: NAD oxidoreductase, indicating that the enzyme is specific for the ionized L form of malic acid (L-malate) and that a coenzyme abbreviated as **NAD** is the hydrogen atom acceptor.

Reversibility Enzymes catalyze the rate at which chemical equilibrium is established among products and reactants. From the standpoint of equilibrium, the assignment of the terms "reactants" and "products" is rather arbitrary and depends upon our point of view. Under normal physiological conditions, the enzyme has no influence on the relative equilibrium quantities (ratios) of products to reactants that would otherwise eventually be reached. Thus, if the equilibrium is unfavorable for formation of a component, the enzyme cannot change this.

The equilibrium constant depends upon the chemical potential of all compounds involved in the reaction. If the chemical potential of reactants is high compared to the products, the reaction may proceed essentially only in one direction. Most of the decarboxylations, in which CO_2 is split out of a molecule, are examples of such reactions, because the CO_2 can escape; its concentration, and hence its chemical potential, remains low. Hydrolytic reactions involving water, such as the breakdown of starch by amylases and the splitting of phosphate from various molecules by phosphatases, are other almost irreversible processes. Other enzymes using different substrates with higher energy levels carry out starch synthesis and the addition of phosphate to molecules. In fact, large molecules such as fats, proteins, starch, nucleic acids, and even certain sugars are synthesized by one series of enzymes and degraded by another. Synthetic and degradative enzymes are often kept separate from each other by membranes or are formed by the plant at different times, so that competition between degradation and synthesis is minimized.

Chemical Composition Every known enzyme consists of a protein as a major part of its structure, and many contain nothing other than protein. Some proteins, however, appear to have no catalytic function and are not classified as enzymes. For example, microtubule proteins and some of the proteins in membranes seem to perform more of a structural than a catalytic function, although future studies might show enzymatic functions for all of these. Numerous proteins, such as cytochromes, that are involved in transport of electrons during photosynthesis and respiration are not called enzymes by some workers, but rather "electron carriers." Furthermore, several so-called seed storage proteins that accumulate in the endosperm of cereal grains, the cotyledon or perisperm storage cells of dicots, or the female gametophyte of gymnosperms, are also not presently known to have enzymatic functions. The major role of seed storage proteins is to act as a reservoir of amino acids for the seedling after germination.

Here again, however, continued research may show that some storage proteins are catalysts.

Proteins consist of one or more long chains (**polypeptide chains**) of tens to hundreds of amino acids. The composition and size of each protein depends upon the kind and number of its amino acid subunits. Commonly, 17 to 20 *different* kinds of amino acids are present, but some proteins have slightly more or less. The total number of amino acids present varies greatly among different proteins, and so protein molecular weights also vary. Most plant proteins so far characterized have molecular weights of at least 40,000 g/mol, yet that of **ferredoxin,** one of the proteins involved in photosynthesis, is only about 11,500 and that of **ribulose diphosphate carboxylase,** another photosynthetic enzyme and probably the most abundant enzyme on earth, is over 500,000. The latter is apparently composed of eight small and identical polypeptide chains and eight larger polypeptide chains that are also identical to each other. The numerous chains of such complex enzymes are frequently held together by noncovalent bonds, often ionic and hydrogen bonds, and can be separated *in vitro* (see Section 8.3). Nevertheless, if care is taken to prevent chain separation during extraction, even complex enzymes such as ribulose diphosphate carboxylase can be isolated as homogenous crystals (Fig. 8-1).

Amino acids may be represented by the general formula:

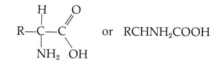

or $\quad RCHNH_2COOH$

The $—NH_2$ is the amino group and the $—COOH$ is the carboxyl group. These two groups are common to all amino acids, with

Figure 8-1 Crystals of pure ribulose diphosphate carboxylase. Note the dodecahedron shape; the largest crystals are more than 0.2 mm in diameter. (Courtesy S. G. Wildman.)

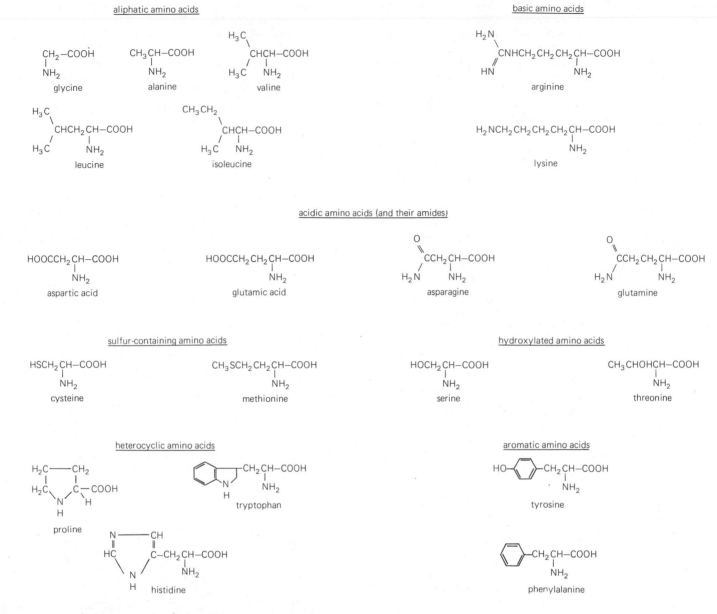

Figure 8-2 Molecular structures of 20 amino acids present in most proteins.

slight modification of the amino group in proline. *R* denotes the remainder of the molecule, which is different for each amino acid. Figure 8-2 shows the structures of the 20 amino acids commonly found in proteins. The *R* groups cause the amino acids to differ greatly in physical properties, such as water solubility. The aliphatic types in the upper left of the figure and the aromatic types at the lower right are much less soluble in water (more **hydrophobic** or water-fearing) than the more **hydrophilic** (water-loving) basic, acidic, and hydroxylated types.

Structures of two amides, **glutamine** and **asparagine,** that occur in many proteins are included in Fig. 8-2. These amides are formed from the two amino acids, **glutamic acid** and **aspartic acid,** each of which has an additional carboxyl group as part of *R*. The amides are important structural parts of most proteins. They also represent important forms in which nitrogen is transported from one part of the plant to another and in which surplus nitrogen may be stored (see Section 13.4).

The union of various amino acids and amides in proteins occurs through **peptide bonds** involving the carboxyl group on one and the amino of the next, as summarized in an over-simplified form in R8-1, the arrow indicating the peptide bond:

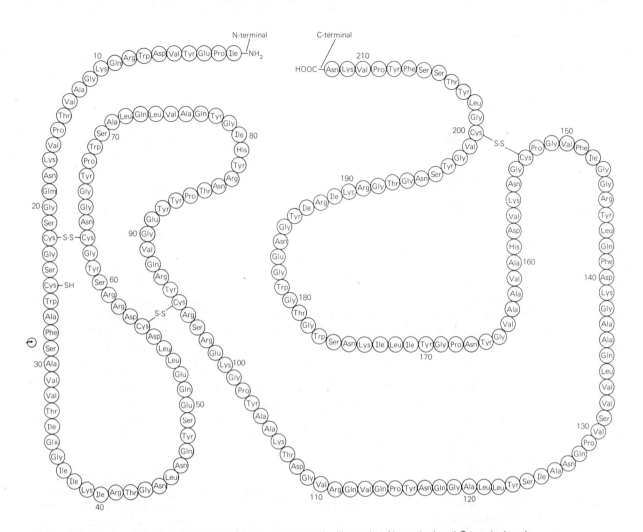

Figure 8-3 The single polypeptide chain of the enzyme papain, illustrating N-terminal and C-terminal ends. Papain is a protease enzyme that hydrolyzes peptide bonds in other proteins, releasing free amino acids. The disulfide bridges between cysteine molecules (S—S) at positions 22–63, 56–95, and 153–200 are strong covalent bonds and help confer a particular three-dimensional structure on the protein. Gln and asn represent the amides glutamine and asparagine, while other abbreviations can be determined from Fig. 8-2. (From Tetrahedon, 1976, 32:292, fig. 1.)

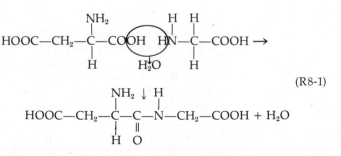

(R8-1)

The actual mechanism by which peptide bonds are produced to form polypeptide chains of proteins is a complex process involving various kinds of RNA molecules and enzymes, but an important fact may be noted here. The first peptide bond always involves the carboxyl group of the first amino acid at the starting end of the chain and the amino group of the second amino acid joined to it. The amino group of the first amino acid remains free. Therefore, a completed polypeptide chain always has a free amino group at the starting, or N-terminal, end and a free carboxyl group at the finishing, or C-terminal, end as in Figure 8-3.

When aspartic and glutamic acids, each of which has two carboxyl groups, form peptide bonds with other amino acids, only the carboxyl group adjacent to the amino group can participate. The other carboxyl group always remains free and gives acidic properties to the protein. When lysine and arginine, each of which has two amino groups, form peptide bonds, an additional amino group some distance from the carboxyl group is always free. The nitrogen atom of each of these groups possesses two electrons that can be shared by H^+ in the cells; as a result, H^+ are attracted to these basic nitrogen atoms, causing them to become positively charged.

Proteins that are rich in aspartic and glutamic acids usually have net negative charges in the cells because each loses an H^+ during dissociation of its carboxyl group not involved in a peptide bond. Alternatively, those rich in lysine and arginine usually have net positive charges. This is important, because whether an enzyme is catalytically active or whether it is bound to another cellular component frequently depends upon whether a given free functional amino or carboxyl group is charged or uncharged. For example, chromosomes contain five major kinds of positively charged proteins rich in lysine or arginine called **histones;** these histones are held to the negatively charged DNA by ionic bonds, and these bonds are important in controlling the structure and genetic activity of the chromosomes. Furthermore, the different net charges on enzymes often allow us to separate them from one another by various chemical and physical processes (see Special Topic 2.1). Their functions and properties can then be studied without interference by other enzymes.

Prosthetic Groups, Coenzymes, and Vitamins: Enzyme Helpers In addition to the protein parts of enzymes, some contain a much smaller, organic nonprotein portion called a **prosthetic group.** Prosthetic groups are usually attached tightly to the protein part by covalent bonds, and are generally essential to catalytic activity. An example is found among some of the dehydrogenase enzymes involved in respiration and fatty acid degradation. Here a yellow-colored pigment called a **flavin** is attached to the protein. The flavin is essential to enzyme activity because of its ability to accept and then transfer hydrogen atoms during the course of the reaction. Some enzymes also contain organic prosthetic groups to which is attached a metal ion (e.g., cytochrome oxidase; see Chapter 12). Other proteins, **glycoproteins** (from the Greek *glykys,* "sweet"), contain one or a group of connected sugars attached to their protein parts (Sharon, 1974). Such attached carbohydrates might contribute to enzymatic action or to protection of the enzyme against temperature extremes or against internal destructive agents such as proteases, although actual demonstrations of such protection have been rare.

Many enzymes without strongly attached prosthetic groups require for activity the participation of another organic compound or a metal ion, or both. These additional substances are usually called **coenzymes,** although some people refer to the metal ions simply as **metal activators.** Coenzymes and metal activators are generally not tightly held to the enzymes, but often no sharp distinction between coenzymes and prosthetic groups occurs. *Many vitamins synthesized by plants act*

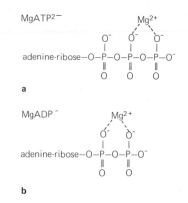

Figure 8-4 The Mg^{2+} chelate of (**a**) ATP and (**b**) ADP.

as coenzymes or prosthetic groups for important enzymes in both plants and animals, and this explains their essentiality for life.

The magnesium ion acts as a metal activator for most enzymes that use ATP or other nucleoside di- or triphosphates as a substrate. Magnesium does not usually truly activate the enzyme but instead combines with the ATP or other nucleotide, making it more susceptible to catalytic action. A stable chelate between the ATP and Mg^{2+} is formed, probably having the structure shown in Fig. 8-4a. The enzyme-substrate complex is then a Mg-ATP-enzyme complex. Mg^{2+} also combines with ADP, as shown in Fig. 8-4b. Furthermore, Mn^{2+} can combine with ADP or ATP in a similar way, forming a chelate that is often as active as that formed with Mg^{2+}. This situation in which cations combine directly with the substrate rather than the enzyme may prove to be a widespread phenomenon with other cations in plants, but direct combination of certain enzymes with manganese, iron, zinc, copper, calcium, potassium, and probably other cations also occurs.

Amino Acid Sequences The number of different ways in which amino acids could theoretically be arranged in proteins is a staggering figure. Consider as a simple example only one small catalytically active protein with a molecular weight of 12,500, consisting of 100 amino acids with an average molecular weight of 125. If 20 *different* amino acids were present, the number of theoretically possible ways to arrange them would be nearly 20^{100} (10^{130}). Of course, all but a few of these arrangements would not possess the enzymatic activity of our protein, and many of them would not be likely to catalyze any reaction. The number of different proteins of all sizes and kinds shown in nature does not even begin to approach such a figure,* although estimates of the existence of as many as 10,000 unique proteins in plants have been made. Consider-

---

*If each of the 10^{130} possible kinds of proteins with 100 amino acids occupied 10^6 cubic angstroms, all 10^{130} taken together would fill the entire known universe.

ing this, how various forms of life can be so different is no longer so puzzling. We know that the arrangements of nucleotides in genes that code for these proteins determine their amino acid sequences. The science of genetics is partly devoted to studies of gene variation and its importance in morphological characteristics (phenotypes).

The methods used to determine the amino acid sequence in any protein are tedious and require the availability of a pure protein. As a result, we presently know the complete sequences for relatively few proteins. Such studies are important, however, because the results are necessary for us to learn how enzymes catalyze reactions. Furthermore, comparison of sequences in proteins having the same function in different organisms provides a powerful tool in evolutionary studies. For example, ferredoxins from angiosperms, a green alga, and certain photosynthetic and nonphotosynthetic bacteria have been sequenced. The similarities and differences have been used, with the help of computers, to construct tentative family trees (Cammack et al., 1971). The same kinds of comparisons have been made with cytochrome c molecules from 38 organisms, including primitive and advanced plants and animals (Boulter et al., 1972). Results from hemoglobins have already provided several clues to animal evolution. Comparison of one of the histone molecules (histone IV, also called F2A1) found in nuclei of both calves and peas showed that the same amino acids occur in 100 of the 102 amino acid positions present in each. These results suggest that many mutations in genes controlling this protein have been eliminated by natural selection during the last 1.5 billion years or so, and therefore that this protein and each amino acid in it play an important, even though still poorly understood, role in the lives of various organisms (Spiker, 1975).

Three-Dimensional Structures of Enzymes and Other Proteins Although proteins consist of one or more long polypeptide chains (Fig. 8-3), results from numerous techniques such as X-ray diffraction, viscosity measurements, and centrifugation studies show that these chains are usually coiled and twisted to form somewhat spherical or globular molecules. An example of a rather simple globular protein is shown in Fig. 8-5. The three-dimensional structure of a protein is apparently usually determined by its spontaneous attainment of the lowest free energy configuration consistent with its amino acid composition and sequence at the normal cellular conditions of pH, temperature, ionic strength, and so on. This configuration is usually attained as soon as it is synthesized on the ribosomes and might begin to occur even before synthesis is complete. The more hydrophobic amino acids such as valine, leucine, isoleucine, methionine, phenylalanine, and often tyrosine become concentrated on the inside of the structure where they are partly or largely shielded from water, while the more hydrophilic amino acids such as serine, glutamic acid, glutamine, aspartic acid, asparagine, lysine, histidine, and arginine are more commonly exposed to the surfaces where they are in contact with water. The folding or coiling of a given polypeptide chain so that some portions of it are in contact with other portions takes place in part because of stabilizing attractive forces between certain R groups some distance apart in the chain. Another factor affecting folding is

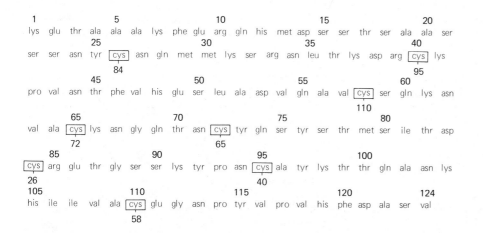

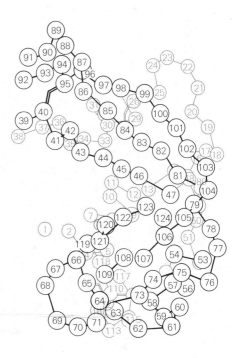

Figure 8-5 Three-dimensional configuration of ribonuclease A. The active site of the enzyme is in the cleft at the left center of the molecule. The positions of individual amino acid residues are marked by the numbered circles; these correspond to the sequence list above. As in Fig. 8-3, gln and asn represent glutamine and asparagine. Disulfide bridges are indicated by a bent double line. Cysteine residues joined by disulfide linkages are enclosed in a box in the sequence list. The number indicated by the arrow at the bottom of each box shows the amino acid residue to which the disulfide linkage is joined. (From R. E. Dickerson and I. Geis, 1969, The structure and action of proteins, W. A. Benjamin, Inc., Menlo Park, Calif. Copyright© 1969 by Dickerson and Geis.)

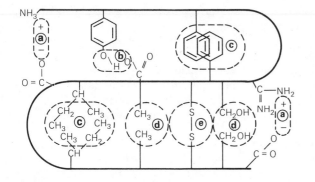

Figure 8-6 Probable types of bonding responsible for holding one polypeptide chain close to another. (**a**) Electrostatic attraction. (**b**) Hydrogen bonding. (**c**) Interaction of nonpolar side chain groups caused by repulsion of each by water. (**d**) Van der Waals attractions. (**e**) Disulfide bonding between former —SH groups of cysteine molecules (Modified after C. B. Anfinsen, 1959.)

repulsion of hydrophobic R groups by water so that these hydrophobic groups are brought close together. Some of the important kinds of intrachain bonding resulting from these factors are shown in Fig. 8-6.

An added feature of complexity in proteins and enzymes is that *many, perhaps most, consist of more than one polypeptide chain,* just as we saw for ribulose diphosphate carboxylase. Bonds holding the chains together are probably similar to those holding separate parts of the same chain together (Fig. 8-6), although **disulfide bonds** (S—S) seem to be rarer. In some, the chains are identical, in which case **homopolymers** are made, but usually they are different and **heteropolymers** result. When different, each polypeptide chain in the heteropolymer is coded for by a different gene. Thus, heteropolymers having two or more different polypeptide chains require the presence of two or more different corresponding genes in the same cell. In ribulose diphosphate carboxylase, the gene coding for the smaller polypeptide chain is a nuclear gene, and either allele of diploid plants contains genetic information for a functional subunit of this type. However, the larger polypeptide chain of ribulose diphosphate carboxylase is coded for by a chloroplast gene, so inheritance for this subunit is strictly maternal in most species. (Maternal inheritance of chloroplast genes occurs because plastids are present in the egg but usually not in the sperm, and subsequent divisions of chloroplasts in the zygote and developing plant involve reproduction of those genes.)

How would enzymes having two or more polypeptide chains behave if each chain could catalyze the same reaction? First, we might expect to find that if each chain could act as effectively when independent as when combined with another one (whether it is identical or different) there would be no selective advantage to their combination. The fact that many combinations do exist suggests an advantage to this, and the advantage may be related to the ability of both homopolymers and heteropolymers to undergo allosteric transitions (structural changes altering catalytic activity; see Section 8.6).

Isozymes During the 1950s, a technique called **electrophoresis** was developed to separate proteins, and this technique led to an important new discovery about many enzymes. Electrophoresis is essentially the separation of proteins or other charged molecules in an electrical field. A mixture of enzymes can be placed on an inert medium such as a layer of starch gel or a column or slab of polyacrylamide gel that is wetted with a buffer at a controlled pH. The various R groups in each enzyme ionize to an extent controlled by their chemical nature and the pH. For example, if the pH is 7, enzymes rich in aspartic and glutamic acid will attain a net negative charge because of their dissociated carboxyl groups. Alternatively, enzymes rich in lysine or arginine are more likely to be positively charged at this pH.

Each enzyme in the mixture attains a different charge, and, if these differences are great enough, the enzymes can be separated by forcing electrical current from a negative electrode inserted at one end of the gel to a positive electrode at the other end. The enzymes migrate different distances in the electrical field, depending upon their net charge and size. After they have migrated, their positions in the gel can be detected by various procedures; for example, by incubating the gel with the proper substrate, then chemically staining the reaction product. When a particular enzyme is investigated in this way, it is often found that more than one stained zone appears in the gel, indicating the presence of more than one enzyme that can act on the same substrate and convert it to the same product. Such enzymes are referred to as **isozymes** or **isoenzymes.** Isozymes usually have very similar amino acid sequences, but the small sequence differences often allow separation by electrophoresis or by various chromatographic procedures. These differences frequently result from the fact that a different gene codes for each isozyme or, if it is a heteropolymer, for each of the polypeptide chains in it. If only one polypeptide chain is present, two isozymes could result from genetic coding by each of the two allelic genes derived from a different parent. Other possibilities resulting from gene duplication followed by a mutation in only one of the two genes also exist. Since a single mutation in a gene usually causes the insertion of a single different amino acid in the enzyme, this is not likely to change the enzyme enough to allow its separation from another almost identical isozyme coded by a nonmutated gene. Therefore, there must be many isozymes that will prove almost impossible to detect. Thus far, most isozymes have proven to be homopolymers or heteropolymers rather than monomeric proteins.

The importance to a plant in having different isozymes capable of catalyzing the same reaction apparently is that these isozymes differ somewhat in their responses to various environmental factors. Sometimes one isozyme will exist in one tissue and another in a different tissue. Different isozymes can

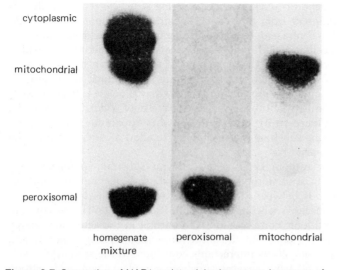

Figure 8-7 Separation of NAD⁺-malate dehydrogenase isozymes of spinach leaves by starch gel electrophoresis. In the homogenate mixture at left, three isozymes are detectable. Two of these correspond to isozymes extracted from isolated peroxisomes and mitochondria, while the third resides in the cytoplasm. (From I. P. Ting, I. Führ, R. Curry, and W. C. Zschoche, 1975, in Markert, 1975, pp. 369–383.)

cytoplasmic

mitochondrial

peroxisomal

homegenate mixture   peroxisomal   mitochondrial

8.3 Mechanisms of Enzyme Action

Only the most energetic molecules are usually able to undergo changes during chemical reactions. Such molecules temporarily become more energetic than others of the same kind by being subjected to different numbers and types of collisions. If we could analyze the energies in a population of such molecules, statistical predictions indicate that a distribution similar to the hypothetical values of Fig. 1-1 (p. 10) would be found. The curves in that figure show that an increase in temperature can greatly increase the number of molecules having relatively high energies. Note that the higher temperature curve is skewed more dramatically to the right than the lower is. If we assume that only those molecules represented by the shaded areas at the top of the figure have high enough energies to react in the absence of enzymes, the area under the higher temperature curve is about twice as large as that under the lower temperature curve. Therefore, twice as many molecules will react in a given period of time at the higher temperature.

But how do enzymes increase reaction rates? Do they similarly cause a shift in the frequency distribution curve as seen with a temperature increase? The answer is no, but to understand this we must first consider another aspect of the problem. Figure 8-8 shows that as substrates react to form products, an energy barrier must be overcome. This barrier is called the **energy of activation.** A temperature rise would simply increase the number of molecules having energies

sometimes even be found within the same cell. Figure 8-7 illustrates the separation of three isozymes of malate dehydrogenase from various organelles in spinach leaf cells, one isozyme from the mitochondria, one from the peroxisomes and one from the cytoplasm. Each isozyme is exposed to a different chemical environment within the cell, and each participates in a different sequence of reactions. The environmental factors influencing growth and development of cells containing some isozymic types may influence other cells with other isozymes differently.

Even if an isozyme in one tissue has the same amino acid sequence as another isozyme in a different tissue, they might differ in charge, catalytic ability, or response to environmental changes. These surprising differences result from subtle but often crucial modifications in structure occurring after the isozymes are synthesized (Trewavas, 1976). The differences are independent of the genes coding for the isozymes. Several kinds of modifications are now known, including **phosphorylation** (in which a phosphate is esterified to the hydroxyl group of one or more serine or threonines present), **glycosylation** (in which one or more sugars is attached to the enzyme), and **methylation** or **acetylation** (in which one or more methyl or acetyl groups is added). Such modifications can probably even occur in the same cell at different stages in its development. They seem to allow another kind of control over the kind of reactions occurring in a given cell or part of a cell at a particular time in its life.

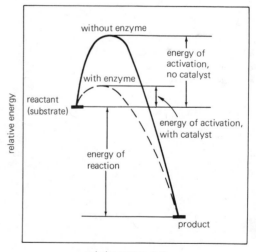

Figure 8-8 Energy diagram for a metabolic reaction occurring in the presence and absence of an enzyme. Reacting substrate molecules must pass over an "energy hump" (accumulate activation energy) to allow formulation of new chemical bonds present in the product, even though the product may be at a lower free energy level than the substrate. A catalyst, such as an enzyme, lowers the activation energy required, thus increasing the fraction of molecules that can react in a given time.

equal to that of the average base level energy plus the activation energy. In contrast, enzymes lower the activation energy so that a much greater fraction of the substrate molecules have sufficient energy to react. In other words, enzymes do not shift the curves of Fig. 1-1, but in their presence the shaded area of that figure would be displaced greatly to the left.

Just how the activation energy is decreased by an enzyme is generally not well understood, although it is known that the enzyme and substrate or substrates combine temporarily during the course of the reaction to form an **enzyme-substrate complex.** This complex somehow causes certain bonds of the substrate to become broken, then rearranged to form products much more rapidly than in the absence of the enzyme.

The enzyme-substrate complex first hypothesized by the great organic chemist Emil Fischer, about 1884, assumed a rigid lock-and-key union between the two (Fig. 8-9a). The portion of the enzyme to which the substrate (or substrates) combines as it undergoes conversion to a product is called the **active site.** But if the active site were rigid and specific for a given substrate, it is difficult to understand how reversibility of the

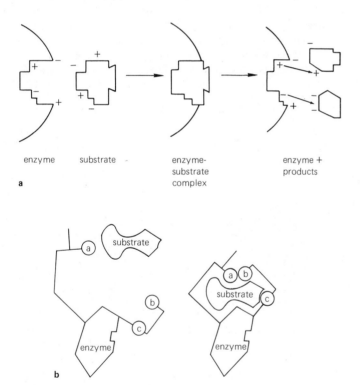

Figure 8-9 (**a**) The lock-and-key model of the active site, as hypothesized by Emil Fischer. The active site is considered to be a rigid arrangement of charged groups that precisely matches complementary groups of the substrate. (**b**) A modified conception of the active site, as advanced by D. E. Koshland. Here the catalytic groups A and B must be aligned, but the orientation of these groups is altered by the approaching substrate, resulting in a better fit. (From Wolfe, 1972.)

reaction could occur, because the structure of the product is different from that of the substrate and should not fit well. As contrasted to a rigidly arranged active site, Daniel E. Koshland (1973) at the University of California provided evidence that the active site of enzymes can be induced by close approach of the substrate (or product) to undergo a change in conformation (shape) that allows a better combination between the two. This idea is now widely known as the **induced-fit hypothesis** (Fig. 8-9b). Apparently, the structure of the substrate is also changed to allow a better fit during the process.

Although only a few enzyme-substrate complexes have been well studied, available data indicate that the kinds of bonds between them can be covalent, ionic, hydrogen, and van der Waals bonds (Bell and Koshland, 1971). The covalent and ionic bonds are most important in the activation energy for reaction, but the more numerous hydrogen bonds and van der Waals interactions contribute to the structural orientation of the enzyme-substrate complex. Even when strong covalent bonds are formed, they are usually very rapidly broken to release new product molecules. Both covalent and noncovalent bonds are usually formed between parts of the R portions of amino acid residues, not between those atoms involved in the peptide bonds. This fact stresses the importance of enzymes having the proper amino acids (composition) and having them in the right places (amino acid sequence).

8.4 Denaturation

The previous discussion of enzyme-substrate complexes and three-dimensional structures of proteins suggests that if the structure of an enzyme is altered so that the substrate no longer can bind in the proper places with it, catalytic activity will be eliminated. Numerous factors cause such alterations and are said to cause enzyme **denaturation.** In many cases, denaturation is irreversible. High temperatures easily break hydrogen bonds and commonly cause irreversible denaturation. You can't unboil an egg! We have said that free energy relations largely determine the enzyme structures, so why doesn't cooling a heated protein bring it back to its original stable configuration? Presumably, extreme heating causes the formation of new covalent bonds between different polypeptide chains or between parts of the same chain, and these bonds are so stable that they break at almost negligible rates.

Cold temperatures are nearly always maintained during extraction and purification of enzymes to prevent heat denaturation. This is true even though the enzymes normally exist undenatured in the cells at higher temperatures. We do not fully understand why purification of enzymes at temperatures identical to those at which the cells normally exist causes denaturation, but we suspect that extraction and purification procedures remove or dilute substances that normally protect the enzymes. Alternatively, homogenization of cells often releases and allows exposure of enzymes to denaturing substances from one subcellular compartment (e.g., vacuoles) that *in vivo* are prevented by membranes from contacting such enzymes. Interestingly, a few enzymes are known to be inactivated by *low* temperatures during purification. Again, a change in structure is no doubt the cause.

Oxygen or unnatural chemical oxidizing agents also denature numerous enzymes, often by causing disulfide bridges to be formed in chains at which SH groups of cysteine are normally present. Reducing agents can denature for the opposite reason, that is, break essential disulfide bridges to form two SH groups. Heavy metal cations such as Ag^+, Hg^{2+}, Hg^+, or Pb^{2+} can denature enzymes, and it is for this reason that great concern about the presence of these in the environment has developed (Chisolm, 1971; Goldwater, 1971).

When enzymes are in a dry state, they are much less susceptible to heat denaturation than when water is abundant. This is mainly why dry seeds and dry fungal or bacterial spores can resist high temperatures and why the presence of steam in autoclaves used for sterilization increases the effectiveness of the heat treatment above that obtained in a dry oven at the same temperature. The dry state also prevents enzyme denaturation by cold temperatures during the winter.

8.5 Factors Influencing Rates of Enzymatic Reactions

Enzyme and Substrate Concentrations: Either Can Be Limiting Catalysis is carried out only if the enzyme and substrate combine to form a transient complex. The rate of reaction is thus dependent upon the number of successful collisions of the two, which in turn is dependent upon the concentration of each. If enough substrate is present, doubling the enzyme concentration usually causes a twofold increase in the rate of the reaction (Fig. 8-10a). Upon adding still more enzyme, the rate begins to level out because the substrate becomes limiting.

Figure 8-10b shows the effect of substrate concentration on the reaction rate when the enzyme concentration is held constant. There is usually an apparent direct proportionality between rate and substrate concentration until the enzyme concentration becomes limiting. Further addition of substrate will eventually cause no further rise in the reaction rate, because nearly all of the enzyme molecules are then combined with substrate. When this occurs, there are no more free enzyme active sites to cause catalysis. To increase the speed of reaction then requires addition of more enzyme.

Figure 8-10b illustrates another useful fact about enzymes that can be obtained from such graphs. This is the substrate concentration required to cause half the maximal reaction rate, a value named the **Michaelis-Menten constant** (K_m). K_m values are more or less constants independent of the amount of enzyme present, at least within reasonable limits. The values do vary somewhat with the pH, temperature, and ionic strength, and also with the kind or amounts of coenzymes present when these are required. Most enzymes so far studied have K_m values between 10^{-3} and $10^{-7} M$, although exceptions exist. If an enzyme catalyzes a reaction between two or more different substrates, it will have a different K_m value for each.

Certain advantages result from knowing the K_m value for an enzyme of interest. First, if we can measure the concentration of the substrate in that part of the cell in which the enzyme resides, we can predict whether the cell would need more enzyme or more substrate to speed up the reaction. Some such studies indicate that respiration enzymes are usually not saturated by substrates. The same is probably often true with photosynthetic enzymes in leaves exposed to bright light, because CO_2 fixation can be stimulated by increasing the concentration of this gas around such leaves. Second, K_m values represent approximate inverse measures of the affinity of the enzyme for a given substrate; thus, the lower the K_m, the more stable the enzyme-substrate complex. Using this argument, it has frequently been possible to say that for an enzyme that can catalyze reactions with two similar but different substrates (e.g., glucose and fructose), the substrate for which the enzyme has the lower K_m is the one more frequently acted upon in the cell.

p**H** The pH of the medium influences enzyme activity in various ways. Usually there is an optimum pH at which

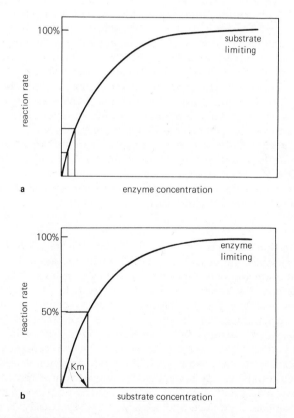

Figure 8-10 (**a**) Effects of enzyme concentration on rate of reaction when substrate concentration is held constant. (**b**) Effect of substrate concentration on reaction rate when enzyme concentration is held constant.

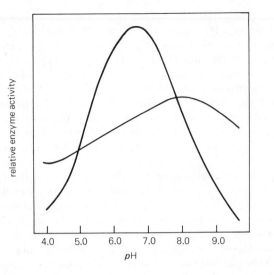

Figure 8-11 Influence of *p*H upon activity of two different enzymes. The *p*H optimum and shape of the curve vary greatly among enzymes and depend somewhat upon reaction conditions.

enzymes function, with decreases in activity at higher or lower *p*H values. Sometimes a plot of activity versus *p*H, as in Fig. 8-11, gives an almost bell-shaped curve, while with another enzyme the curve may be almost flat. The optimum *p*H is often between 6 and 8 but may be somewhat higher or lower. Extremes of the *p*H scale usually cause denaturation.

Apart from the denaturation effect, the *p*H can influence reaction rates in at least two ways. First, enzyme activity often depends upon the presence of free amino or carboxyl groups. These may be either charged or uncharged, but only one form is presumed effective in a given case. If an uncharged amino group is essential, the *p*H optimum will be relatively high, while a neutral carboxyl group requires a low *p*H. Second, the *p*H controls the ionization of many substrates, some of which must be ionized for the reaction to proceed.

Reaction Products The rate of an enzymatic reaction can be determined by measuring the rate of disappearance of substrates or the rate of product appearance, or both. By either method, the reaction is usually observed to proceed more slowly as time passes. This rate decrease is sometimes caused by denaturation of the enzyme while the reaction is being measured, but other factors are also involved. One of the most important factors is the continuous decrease in concentration of substrate or substrates. As the products accumulate, their concentrations may become high enough to cause appreciable reversibility of the reaction, provided that the relative chemical potentials of products and reactants allow reversibility. Finally, the forward reaction catalyzed by a few enzymes is markedly reduced by the formation of products, because these molecules combine with the enzyme in such a way that formation of the complex between enzyme and reactant is inhibited.

Inhibitors Many substances block the catalytic effects of enzymes. These may be either inorganic, such as many heavy metal cations, or they may be organic. Both groups are usually classified according to whether their effect is competitive or noncompetitive with the substrate. **Competitive inhibitors** commonly have structures sufficiently similar to the substrate that they compete for the active site of the enzyme. When such a combination of enzyme and inhibitor is formed, the concentration of effective enzyme molecules is lowered, decreasing the reaction rate. The inhibitor itself may or may not undergo a change caused by the enzyme. Addition of more of the natural substrate overcomes the effect of a competitive inhibitor. A classic example of competitive inhibition is caused by malonate ($^-$OOC—CH_2—COO$^-$), the doubly charged anion of malonic acid, on the action of succinate dehydrogenase. This enzyme functions in mitochondria to carry out an essential reaction of the Krebs cycle (see Chapter 12). It removes two H atoms from succinate and adds these to its covalently bound prosthetic group, **flavin adenine dinucleotide**, abbreviated **FAD,** forming fumarate and enzyme-bound FADH$_2$:

$$
\begin{array}{lll}
\text{COO}^- & & \text{COO}^- \\
| & & | \\
\text{CH}_2 & + \text{ enzyme-FAD} \rightleftharpoons & \text{CH} \quad\quad + \text{ enzyme-FADH}_2 \\
| & & \| \\
\text{CH}_2 & & \text{HC} \\
| & & | \\
\text{COO}^- & & \text{COO}^- \\
\text{succinate} & & \text{fumarate} \quad\quad\quad \text{(R8-2)}
\end{array}
$$

Malonate combines reversibly with the enzyme in place of succinate, but since hydrogen removal cannot occur, no reaction takes place. In this way, succinate dehydrogenase molecules bound to malonate are unable to catalyze the normal dehydrogenation of succinate, and respiration is poisoned. Interestingly, beans and certain other legumes contain unusually high concentrations of malonate, probably in the central vacuole where it cannot affect respiration.

Noncompetitive inhibitors also combine with the enzyme, but at a location different from the active site. This effect is not overcome by simply raising the substrate concentration. Noncompetitive inhibitors generally show less structural resemblance to the substrate than do the competitive type. Toxic metal ions and compounds that combine with or destroy essential sulfhydryl groups often are noncompetitive inhibitors. For example, excess O_2 can oxidize −SH groups close to each other, removing the H atom from each and forming new disulfide bridges, thus changing the structure of the enzyme so that its active site no longer can combine well with the substrate or substrates. Heavy metal ions such as Hg^{2+} and Ag$^+$ can replace the H atom on a sulfhydryl group, forming heavy mercaptides that are often insoluble.

Most poisons affect plants and animals because they inhibit enzymes. Certain of these poisons will be discussed later in relation to the specific processes that are affected. Enzymes may also be inhibited noncompetitively by any protein denaturant, such as strong acids or bases, or by high concentrations of urea, which break hydrogen bonds.

Plant Proteins and Human Nutrition

Humans depend, directly or indirectly, upon photosynthetic plants as a source of many of their own amino acids, so the composition of seed, leaf, and stem proteins is important in our diets. We and other animals use these amino acids to build our own proteins and as a food (energy) source. Although adult humans can synthesize most of the amino acids they require from carbohydrates and various organic nitrogen compounds, present evidence indicates that eight amino acids must be provided in the diet. These are leucine, isoleucine, valine, lysine, methionine, tryptophan, phenylalanine, and threonine. Furthermore, adequate amounts of the sulfur-containing amino acid cysteine can apparently only be formed when sufficient methionine (another S-amino acid) is provided, and we use phenylalanine to synthesize tyrosine, which otherwise would be essential.

Most of the proteins in the human diet come from seed proteins, especially those of the cereal grains rice, wheat, and corn (maize). Approximately two thirds of the world's population depends on wheat or rice as the principal source of calories and protein. Corn is equally important in many tropical and subtropical parts of Central and South America. A smaller but still important contribution is made by legume seeds such as beans, peas, and soybeans. Soybeans are an unusually rich, fairly well-balanced protein source; about 40 percent of their dry weight is protein compared to about 10 percent for most cereal grains (Table 8-1).

Compared to most animal proteins, cereal grain proteins are low in lysine, while legume seeds are low in methionine. For example, the lysine content of total protein from seeds of 12,561 wheat varieties averaged 3.14 percent on a weight basis compared to 6.4 percent in whole egg protein (Table 8-1). Bean seed proteins averaged only 1.0 percent methionine compared to 3.1 percent in egg proteins. Plant breeders are making some progress in introducing new varieties or hybrid species with increased protein contents and increased percentages of essential amino acids. Examples include the opaque-2 and floury-2 varieties of corn, both of which are considerably richer in both lysine and tryptophan than are commonly grown varieties (Harpstead, 1971; Sylvester-Bradley and Folkes, 1976).

Table 8-1 Protein Content and Amino Acid Composition of Selected Food Legumes and Cereals[a]

Food	Protein %	Amino Acid Composition. % of Total Protein								
		Lysine	Methioniné	Threonine	Tryptophan	Isoleucine	Leucine	Tyrosine	Phenylalanine	Valine
Soybean	40.5	6.9	1.5	4.3	1.5	5.9	8.4	3.5	5.4	5.7
Peas	23.8	7.3	1.2	3.9	1.1	5.6	8.3	4.0	5.0	5.6
Beans	21.4	7.4	1.0	4.3	0.9	5.7	8.6	3.9	5.5	6.1
Oats	14.2	3.7	1.5	3.3	1.3	5.2	7.5	3.7	5.3	6.0
Barley	12.8	3.4	1.4	3.4	1.3	4.3	6.9	3.6	5.2	5.0
Wheat	12.3	3.1	1.5	2.9	1.2	4.3	6.7	3.7	4.9	4.6
Rye	12.1	4.1	1.6	3.7	1.1	4.3	6.7	3.2	4.7	5.2
Sorghum	11.0	2.7	1.7	3.6	1.1	5.4	16.1	2.8	5.0	5.7
Maize	10.0	2.9	1.9	4.0	0.6	4.6	13.0	6.1	4.5	5.1
Rice	7.5	4.0	1.8	3.9	1.1	4.7	8.6	4.6	5.0	7.0
Whole egg	12.8	6.4	3.1	5.0	1.7	6.6	8.8	4.3	5.8	7.4

[a]Data from Orr and Watt, 1957 and Johnson and Lay, 1974.

8.6 Allosteric Proteins and Feedback Control

We have seen that numerous *foreign* ions or molecules can inhibit enzymatic action, in most cases by altering the configuration of the enzyme so that it cannot effectively form a complex with the substrate. Several enzymes can also be altered by *normal* cellular constituents, with resulting decreases or increases in their functions. Such effects usually appear to be important mechanisms for homeostatic control at the metabolic level. The more common case is inhibition of a particular reaction by a metabolite that is chemically unrelated to the substrate with which the enzyme reacts.

To understand this, consider an example in which a compound A is converted by a series of enzymatic reactions

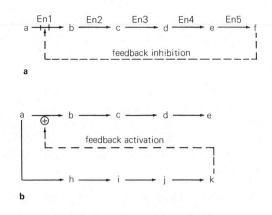

Figure 8-12 (**a**) Feedback inhibition, (**b**) feedback activation.

via intermediates B, C, D, and E to an essential product F (Fig. 8-12a). After this number of reactions, compound F no longer bears much structural resemblance to A. Nevertheless, F can sometimes reversibly combine with the first enzyme, thereby inhibiting its combination with A. This is an example of **feedback inhibition** or **end-product inhibition**. Its advantage is that it provides a very rapid and sensitive mechanism to prevent oversynthesis of compound F, because the process occurs only after F has built up to a level apparently sufficient for cellular needs. Later, when the amount of F in the cell has been reduced (say, by incorporation into a structural component of the cell), fewer F molecules are bound to enzyme 1, and it becomes active again. Known cases of feedback inhibition nearly always involve action of a metabolic pathway product acting upon the first enzyme of that pathway. A well-studied example of feedback inhibition in plants occurs in formation of the nucleotide uridine monophosphate (UMP), beginning with aspartic acid and carbamyl phosphate. The pathway requires five enzyme-catalyzed steps, but only the first enzyme, aspartic transcarbamylase, is susceptible to feedback control by UMP, and no other reaction is blocked in a similar way by other reactants and products in the pathway.

To understand cases where a metabolic product increases activity of an enzyme, consider the situation in which another compound, *a*, may be converted by an enzyme to compound *b* in a series of reactions that lead to *e*, yet *a* may also be acted upon by a competing enzyme to initiate a second reaction leading to product *k* (Fig. 8-12b). Here the cellular levels of *e* and *k* depend upon the relative activities of the first enzymes unique to their pathways of formation. We find that oversynthesis of *k* may be prevented by its activation of the competing enzyme converting *a* to *b*. Other more complicated

kinds of feedback loops are also known, but primarily in bacteria.

Enzymes that combine with and respond, either negatively or positively, to small molecules such as F or *k* are called **allosteric enzymes.** The sites at which combination with the smaller molecules occurs are called **allosteric sites** (*allo* = "other," i.e., different from the active site). Sometimes an allosteric site is on a polypeptide chain different from the chain containing the active site. The small molecules undergoing reversible binding to allosteric sites are called **allosteric effectors.** The result of allosteric binding is apparently to lock the enzyme into a different configuration so that its K_m for the normal substrate is either increased or decreased. Figure 8-13 shows an oversimplified diagram of how a single enzyme with two different allosteric sites could be activated by one allosteric effector and inhibited by another. The activation case is another example of induced fit, but here it is an allosteric effector instead of the substrate that alters the shape of the protein so that it now combines more readily with the substrate. It is likely that many plant hormones act as allosteric effectors, but in no case has this yet been demonstrated. Since an allosteric effector is not altered permanently by the enzyme, the hormone could be used in small amounts again and again to activate many enzyme molecules.

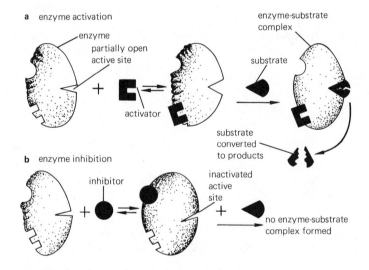

Figure 8-13 A hypothetical model illustrating how the presence of activators and repressors might influence allosteric enzymes, thus affecting reaction rates. In (**a**), attachment of the activator to the enzyme opens the active site, allowing it to combine with the substrate. In (**b**), attachment of the inhibitor closes the active site, preventing attachment of the substrate.

9

Photosynthesis

Photosynthesis is essentially the only mechanism of energy input into the living world. Organisms that can obtain energy by oxidizing compounds not synthesized by photosynthesis are known, but they are of little quantitative significance in the overall energy budget. These include chemosynthetic bacteria that obtain energy by oxidizing inorganic substrates such as ferrous ions and sulfur, and the continued presence of such substrates depends indirectly on photosynthesis. Because of its importance to life, therefore, we shall devote three chapters to photosynthesis.

Like energy-yielding oxidation reactions upon which all life depends, photosynthesis involves oxidation and reduction. The overall process is an oxidation of water (removal of electrons with release of O_2 as a byproduct) and a reduction of CO_2 to form organic compounds such as carbohydrates. You realize that the reverse of this process, the combustion or oxidation of gasoline or carbohydrates in wood to form CO_2 and H_2O, is a spontaneous process with a negative ΔG. It is the similar yet effectively controlled process of respiration, discussed in Chapter 12, that keeps all organisms alive. During combustion or respiration, electrons are removed from carbon compounds and passed downhill, energetically speaking, where they and H^+ combine with a willing electron acceptor, O_2, to make stable H_2O. Considered in this way, photosynthesis uses light energy to drive electrons uphill away from H_2O to a less-willing electron acceptor, CO_2. These relations are summarized in Fig. 9-1. In this chapter, we shall emphasize the light-harvesting apparatus of plants, the chloroplast, and the way in which it accomplishes this uphill transport of electrons.

9.1 Historical Summary of Photosynthesis

Before the early eighteenth century, scientists believed that plants obtained all of their elements from the soil. In 1727, Stephen Hales suggested that part of their nourishment came from the atmosphere and even that light participated somehow in this process. It was not known then that air contains several different gaseous elements. In 1771, Joseph Priestley, an English clergyman and chemist, implicated O_2 (though it was yet to be named and proven to be a molecule) when he

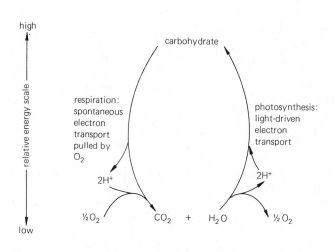

Figure 9-1 Contrasting energy relations of photosynthesis and respiration.

found that green plants could renew air made bad by breathing of animals. Then a Dutch physician, Jan Ingenhousz, demonstrated that light was necessary for this purification of air. He found that plants, too, made bad air in darkness. This caused him to recommend (unnecessarily) that plants be removed from houses during the night to avoid the possibility of poisoning the occupants.

In 1782, Jean Senebier showed that the presence of the noxious gas produced by animals and plants in darkness (CO_2) stimulated production of purified air (O_2) by plants in the light. So by this time, the participation of two gases in photosynthesis had been demonstrated. Work of Lavoisier and others made it apparent that these gases were CO_2 and O_2. Water was implicated by Nicholas de Saussure when, in 1804, he made the first quantitative measurements of photosynthesis. He found that plants gained more dry weight during photosynthesis than could be accounted for by the amount by which the weight of CO_2 absorbed exceeded the weight of O_2 released. He attributed the difference in weight to an uptake of H_2O. He also noted that approximately equal volumes of CO_2 and O_2 were exchanged during photosynthesis.

The nature of the other product of photosynthesis, organic matter, was demonstrated by Julius Sachs in 1864 when he observed the growth of starch grains in illuminated chloroplasts. The starch is detected only in areas of the leaf exposed to the light. Thus, the overall reaction of photosynthesis was demonstrated to be as follows:

$$nCO_2 + nH_2O + light \rightarrow (CH_2O)_n + nO_2 \qquad (R9\text{-}1)$$

In this reaction, (CH_2O) is simply an abbreviation for starch or other carbohydrates with this empirical formula. We now realize that starch is not the only important organic matter product with this approximate empirical formula. The disaccharide sucrose and polymers made from fructose, the fructosans, are also abundant examples of (CH_2O). Free hexose sugars such as glucose and fructose are not important products of photosynthesis.

A further important discovery was that of C. B. van Niel, who, in the early 1930s, pointed out the similarity between the overall photosynthetic process in green plants and that in certain bacteria. Various bacteria were known to reduce CO_2 using light energy and an electron source different from H_2O. Some of these use organic acids, such as acetic or succinic acid, as electron sources, while those to which van Niel gave primary attention use H_2S and deposit sulfur as a byproduct. The overall photosynthetic equation for these bacteria proved to be as follows:

$$nCO_2 + 2nH_2S + light \rightarrow (CH_2O)_n + nH_2O + 2nS \qquad (R9\text{-}2)$$

When R9-2 is compared to R9-1 above for green plants, an analogy can be seen between the role of H_2S and H_2O, and of O_2 and sulfur. This suggested to van Niel that the O_2 released by plants is derived from water, not from CO_2. This idea was supported in the late 1930s by work of R. Hill and R. Scarisbrick in England, which showed that isolated chloroplasts and chloroplast fragments could release O_2 in the light if they were given a suitable acceptor for the electrons being taken from H_2O. Certain ferric (Fe^{+3}) salts were the earliest electron acceptors provided, and they became reduced to the ferrous (Fe^{+2}) form. This light-driven split of H_2O in the absence of CO_2 fixation became known as the **Hill reaction.** It showed that whole cells were not necessary for at least some of the reactions of photosynthesis and that the light-driven O_2 release is not mandatorily tied to reduction of CO_2.

More convincing evidence that the O_2 released is derived from H_2O came in 1941 from results of Samuel Ruben and his associates. They supplied the green alga Chlorella with H_2O containing ^{18}O, a heavy, nonradioactive oxygen isotope that was detected with a mass spectrometer. The photosynthetic O_2 became labeled with ^{18}O, thus supporting van Niel's hypothesis. Nevertheless, even this experiment was not conclusive, because in solution the ^{18}O in H_2O interchanges somewhat with oxygen atoms in HCO_3^- ions, with which CO_2 is in equilibrium. A chloroplast enzyme, **carbonic anhydrase,** catalyzes the reaction shown by the first arrow in R9-3:

$$CO_2 + H_2{}^{18}O \xrightarrow{\text{carbonic anhydrase}} H_2C^{18}O_3 \rightarrow HC^{18}O_3^- + H^+ \quad (R9\text{-}3)$$

Both reactions will proceed without an enzyme, but carbonic anhydrase greatly speeds the first one. Since the reactions are reversible, H_2O, CO_2, and HCO_3^- all contain ^{18}O after some time has passed. Occasional criticisms of Ruben's techniques prompted Alan Stemler and Richard Radmer, in 1975, to repeat and extend Ruben's experiments using broken chloroplast fragments in which carbonic anhydrase could be essentially eliminated and in which the pH could be better controlled. Their data apparently prove that O_2 comes from H_2O, not from CO_2 or HCO_3^-. We must therefore modify the summary equation for photosynthesis given in R9-1 to include two H_2O molecules as reactants:

$$nCO_2 + 2nH_2O^* + light \xrightarrow{\text{chloroplasts}} (CH_2O)_n$$
$$+ nO_2{}^* + nH_2O \qquad (R9\text{-}4)$$

In 1951, it was found that a natural plant constituent, the vitamin B (niacin or nicotinamide)-containing coenzyme called **nicotinamide adenine dinucleotide phosphate** (commonly abbreviated **NADP$^+$**), could also act as a Hill reagent by accepting electrons from H_2O in reactions occurring in isolated chloroplasts. This discovery again stimulated photosynthesis research, because it was already known that the reduced form of NADP$^+$, **NADPH,** could transfer electrons to a number of plant compounds, and it was immediately suspected that its normal role in the chloroplasts was the reduction of CO_2. This suspicion proved correct. Thus, *one of two essential functions of light in photosynthesis is to provide electrons for reduction of NADP$^+$ to NADPH. The other function is to provide energy to form ATP from ADP and $H_2PO_4^-$.*

This conversion of ADP and $H_2PO_4^-$ to ATP in chloroplasts was discovered in the laboratory of Daniel Arnon, at the University of California, in 1954. Before that time, the only important known mechanism to form ATP was in respiration, especially those reactions occurring in the mitochondria called oxidative phosphorylation (see Chapter 12). Arnon found that ATP was synthesized in isolated chloroplasts only during light, and the process became known as **photosynthetic phosphorylation,** or simply **photophosphorylation.** This process of ATP formation by photophosphorylation can be summarized by reaction R9-5:

$$ADP^{-2} + H_2PO_4^- + light \xrightarrow{\text{chloroplasts}} ATP^{-3} + H_2O \quad (R9\text{-}5)$$

Photophosphorylation in chloroplasts accounts for much more ATP formation in leaves during the light than does oxidative phosphorylation in the mitochondria of those leaves, so it is clearly of great quantitative significance. Notice, however, that our summary equation for photosynthesis (R9-4) says nothing about ATP, NADPH, or NADP$^+$. The reason for this is that once ATP and NADPH are formed, their energy is used up in the process of CO_2 reduction and carbohydrate synthesis, and ADP, $H_2PO_4^-$, and NADP$^+$ are again released. So ADP and $H_2PO_4^-$ are rapidly converted to ATP by light energy, and the ATP is just as rapidly broken down when photosynthesis is occurring at a constant rate. How ATP and NADPH are used to help fix CO_2 is a subject of Chapter 10, but in the remainder of this chapter we shall be concerned with chloroplasts and the intricate processes by which they trap light energy and use it to form ATP and NADPH.

9.2 Chloroplasts: Structures, Pigments, and Development

Structures and Photosynthetic Pigments Chloroplasts of many shapes and sizes are found in various kinds of plants. Most are easily seen with the light microscope, but their fine structure can only be discovered by electron microscopy. True

chloroplasts are surrounded by a double membrane system or envelope, the inner membrane of which controls molecular traffic into and out of them. Assuming that organisms can logically be segregated into five kingdoms, chloroplasts occur in all members of the Kingdom *Plantae* and in certain algae or algal-like members of the Kingdom *Protista*. Blue-green algae and photosynthetic bacterial members of the Kingdom *Monera* have no chloroplasts, but they still have photosynthetic pigments embedded in specialized membranes, just as do the other examples mentioned. No chloroplasts occur in membranes of the *Fungi* or *Animalia*. We shall discuss only the chloroplasts typical of vascular plants, bryophytes, and many green algae in the Kingdom *Plantae*.

The structure of such a chloroplast from an oat leaf is

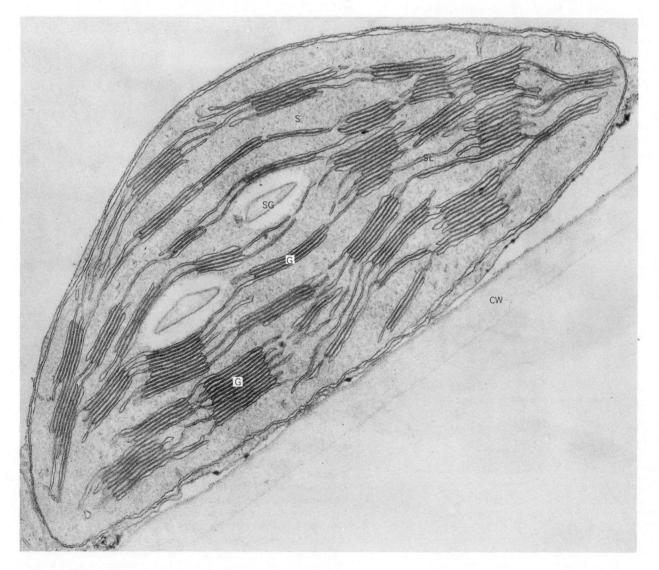

Figure 9-2 Oat leaf chlorophast. S, stroma; SL, stroma lamella; G, granum; SG, starch grain; CW, cell wall. (Courtesy P. J. Hanchey-Bauer.)

shown in Fig. 9-2. Each of the internal membranes containing the photosynthetic pigments seems to be the wall of a flattened tube or sac, sometimes called a **lamella** (plural **lamellae**). In certain regions, the lamellae are stacked to form **grana** (single stack = **granum**). Each granum contains several short lamellae called **grana lamellae.** The longer lamella that connect one granum to another extend through the chloroplast matrix called the **stroma,** so these membranes are usually referred to as **stroma lamellae.** Collectively, grana and stroma lamellae are often called **thylakoids** (Greek *thylakos,* "sac" or "pouch"). Stroma lamellae can often be seen to extend into and make up part of one or more grana, and in those locations there is no apparent distinction between them and the grana lamellae. Figure 9-3 illustrates a three-dimensional interpretation of the relation between grana and stroma lamellae.

The pigments present in thylakoids consist largely of two kinds of **chlorophylls**, chlorophyll *a* and the very similar chlorophyll *b*. The content of chlorophyll *a* is usually two to three times that of chlorophyll *b* in most green plants. Also present are yellow to orange pigments classified as **carotenoids**. There are two kinds of carotenoids, the pure hydrocarbon **carotenes** and the oxygen-containing **xanthophylls**. Carotenoids also exist in the chloroplast envelope, giving it a yellowish color, while chlorophylls are not found in the envelopes. The structures of chlorophylls *a* and *b* and some carotenoids are shown in Fig. 9-4.

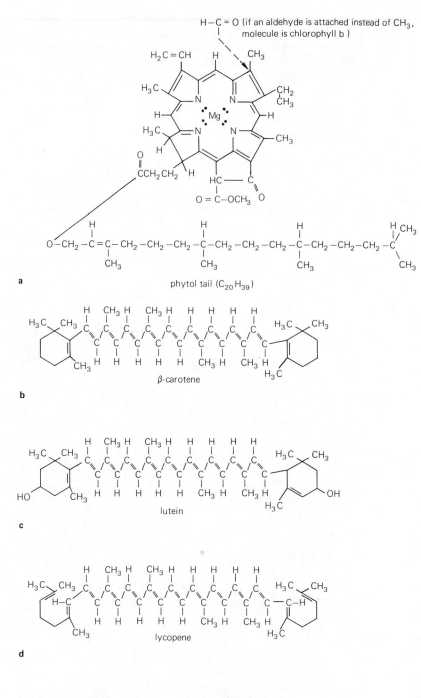

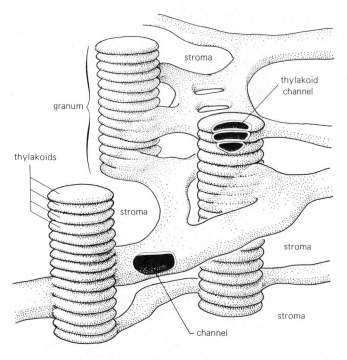

Figure 9-3 A three-dimensional interpretation of the arrangement of the internal membranes of a chloroplast, emphasizing the relation between stroma lamellae and grana. Note the channels in both kinds of thylakoids. (Redrawn from an original by T. E. Weir.)

Figure 9-4 Structures of some chlorophyll and carotenoid pigments (**a**) Structure of chlorophyll *a* and its relation to chlorophyll *b*. The tetrapyrrole ring structure at left gives the green color, while the hydrophobic $C_{20}H_{39}$ phytol tail common to both chlorophylls probably extends into the interior of the lamellae. Chlorophyll *a* is blue-green; chlorophyll *b*, yellow-green. (**b**) β-carotene, a yellow carotenoid with the empirical formula $C_{40}H_{56}$. (**c**) Lutein, a yellow xanthophyll with the empirical formula $C_{40}H_{56}O_2$. (**d**) Lycopene, a reddish carotene with the empirical formula $C_{40}H_{56}$. Lycopene is not found in chloroplasts but gives the red color to tomato fruits.

The electron microscope gives us no information about how these pigments are arranged in thylakoids, but other techniques show that chlorophylls are bound to and perhaps even embedded within protein molecules. The carotenoid pigments are quite hydrophobic, especially β-**carotene** (Fig. 9-4b) and other carotenes, and they probably extend toward the membrane interior.

Chloroplast Development: One Plastid Leads to Another The number of chloroplasts in different plant cells is highly variable. Certain algae have only one per cell, and each time a cell divides, so does its chloroplast. Angiosperms and gymnosperms have numerous chloroplasts in most photosynthetic cells, and they have several colorless plastids **(leucoplasts)** in root cells and certain other nonphotosynthetic cells. Each plastid seems to arise by division of another, the whole process starting with division of the fertilized egg. In most species, all plastids in the zygote come from the egg, the sperm contributing none (Kirk and Tilney-Bassett, 1967).

Consider a young leaf as it grows to maturity. Much of the early growth is accompanied by cell divisions in certain meristematic areas, but usually the last half or two thirds of the enlargement is caused only by expansion of pre-existing cells. A few of the plastids present in an immature leaf occur as small **proplastids** with a poorly developed thylakoid system. These are apparently formed by division of similar proplastids arising originally in the zygote. Such proplastids apparently either develop into mature chloroplasts or divide by pinching to form other proplastids.

The number of chloroplasts per cell increases greatly as the cells grow. In spinach, for example, the number in both palisade parenchyma and spongy mesophyll cells increases from about 50 to 500 per cell as a leaf expands from 1 or 2 cm long to full size. Most of this increase results from division of chloroplasts, not of proplastids. This division process is controlled by light; in spinach there is a direct relation between the daily amount of light provided and the chloroplast number. The red and blue wavelengths absorbed by the chlorophylls are the most effective wavelengths in inducing chloroplast division.

If chloroplasts divide, what controls their division? We do not know, but many chloroplast properties are determined by nuclear genes, while others are determined by plastid genes. Chloroplasts contain DNA that can be detected with the electron microscope as small fibrils. This DNA can be extracted and, when done carefully, it exists as a continuous circle about 45 μm long. An electron micrograph of one such circular molecule obtained from a spinach chloroplast is shown in Fig. 9-5. An average-sized chloroplast contains about two dozen such circular molecules, all of which are apparently identical, but the number present increases as the plastid grows and develops prior to the next division (Possingham and Rose, 1976). As you know, the function of DNA molecules is to code for production of RNA molecules, most of which in turn code for proteins. Each circle of chloroplast DNA has

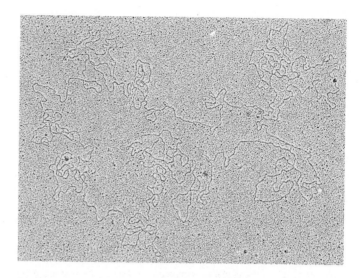

Figure 9-5 Circular DNA molecule isolated from a spinach chloroplast. This molecule was 44.7 μm long. (From R. Herrmann, H. Bohnert, R. V. Kowallik, and J. M. Schmidt, 1975, Biochimica et Biophysica Acta, 378:305–317.)

approximately enough genetic information to code for 100 different proteins, yet most chloroplast proteins are coded for by nuclear genes (Kirk, 1972). Chloroplasts are therefore clearly not autonomous (independent) but are dependent upon their host cells.

The Role of Light in Chloroplast Development When seeds of angiosperms are germinated in the dark or the plants are transferred from light to darkness, chlorophyll synthesis is prevented, although the conifers, mosses, some ferns, and most algae can form chlorophylls in darkness. Restricted chlorophyll production in angiosperms results partly from a lowered rate of chloroplast division, as described earlier, but two more primary factors are involved. In darkness, trace amounts of a green precursor molecule that can be converted into chlorophyll a is formed. This substance, called **protochlorophyllide a,** differs from chlorophyll a only by the absence of the phytol tail and two hydrogen atoms. Upon illumination, protochlorophyllide a is rapidly hydrogenated (reduced) to **chlorophyllide a.** A phytol tail is then esterified onto the chlorophyllide a, and chlorophyll a results. Thus, **one of the main reasons that light stimulates chlorophyll synthesis in angiosperms is that it is needed for reduction of protochlorophyllide a.**

A second way that light controls chlorophyll synthesis is at an early step in the production of protochlorophyllide a. A small molecule from which each of the four nitrogen-containing **pyrrole** rings in protochlorophyllide a and in all chlorophylls originates is **delta-aminolevulinic acid,**

$$NH_2-CH_2-\underset{\underset{O}{\|}}{C}-CH_2-CH_2-COOH,$$

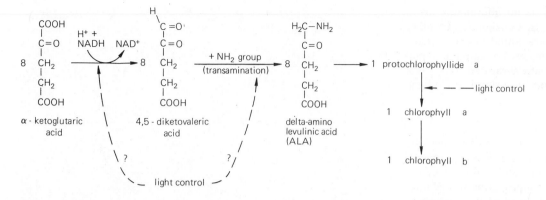

Figure 9-6 Some reactions of chlorophyll synthesis in relation to promotion by light. In the first step, the upper carboxyl group is reduced to an aldehyde group by two electrons donated by a coenzyme with reducing properties called NADH. (The structure of NADH is in Fig. 9-12.) In the second step, an amino group is added to the aldehyde group of 4,5-diketovaleric acid. This is a transamination reaction discussed more generally in Chapter 13. Light somehow promotes one or both of the first two reactions. Conversion of eight molecules of ALA to one molecule of protochlorophyllide *a* represents numerous identified reactions independent of light. Conversion of protochlorophyllide *a* to chlorophyll *a* is light dependent, while subsequent conversion of chlorophyll *a* to chlorophyll *b* occurs in either light or darkness.

commonly abbreviated **ALA.** Now ALA can be rapidly converted to protochlorophyllide *a* in darkness, yet ALA cannot be formed in darkness. To find out why light is needed for ALA synthesis, we must learn the chemical reactions and the enzymes responsible for this synthesis. The organic acid *α*-**ketoglutaric acid,** a member of the Krebs cycle (see Chapter 12), is the most direct precursor of ALA so far detected (Lohr and Friedmann, 1976; Beale et al., 1975). Although we still do not understand the mechanism of light action, it seems clear that **a second reason light stimulates chlorophyll synthesis is because it enhances formation of ALA, a precursor to both protochlorophyllide a and chlorophyll a.** Chlorophyll *b* is formed from chlorophyll *a,* and this occurs readily even in darkness. These relations are summarized in Fig. 9-6.

In the absence of chlorophyll, dark-grown angiosperms usually appear yellow, because most carotenoids continue to be formed in the dark. These carotenoids accumulate in the envelopes surrounding the plastids and in the stroma lamellae, if such lamellae are formed in darkness. Plastids present in dark-grown seedlings are called **etioplasts** (French *etioler,* "to grow pale or weak"). Etiolated corn and bean seedlings having many such etioplasts are compared with normal light-grown seedlings in Fig. 19-1, p. 290.

The structure of a fully developed etioplast is shown in Fig. 9-7. Etioplasts have an interesting system of lamellae arranged in a regular lattice called the **prolamellar body.** Upon reaching maturity, etioplasts of some species produce a considerable stroma lamellae system from the prolamellar body. After exposure to light, when chlorophyll synthesis begins, the prolamellar body disappears and grana are formed from the stroma lamellae.

Prolamellar bodies also develop in proplastids of young leaves of plants that are not etiolated but exposed to normal

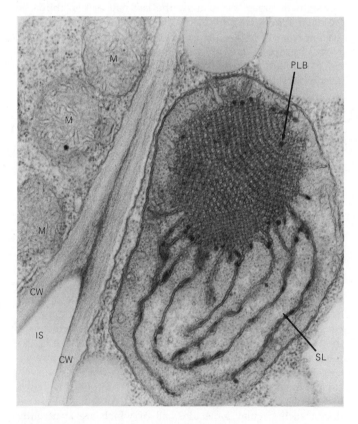

Figure 9-7 Etioplast from a cotyledon of a dark-grown radish seedling, illustrating the prolamellar body (PLB) and stroma lamellae (SL) radiating from it. Also shown are two adjacent cell walls (CW), an intercellular space (IS) between the walls, and mitochondria (M) in the lower cell to the left. (Courtesy Nicholas Carpita.)

day-night conditions. Functional chloroplasts formed from such immature plastids can divide to produce many other chloroplasts, as discussed earlier.

9.3 The Role of Light in Photosynthesis

Some Principles of Light Absorption by Plants To find out how light causes photosynthesis, we must learn something about its properties. As discussed in Special Topic 3.1, light has a **wave nature** and a **particulate nature.** Light represents that part of radiant energy having wavelengths visible to the human eye (approximately 390 to 760 nm). This is a very narrow region of the electromagnetic spectrum.

The particulate nature of light is usually expressed in statements that light comes in **quanta** or **photons,** discrete packets of energy, each having a specific associated wavelength. The energy in each photon is inversely proportional to the wavelength, so the violet and blue wavelengths are more energetic than the longer orange and red ones. One mole (6.02×10^{23}) of photons is called an **einstein.** Quantitative aspects of these relations are in Special Topic 3.1.

A fundamental principle of light absorption, often called **Einstein's law,** is that any pigment (colored molecule) can absorb only one photon at a time and that this photon causes the excitation of one electron. Specific valence (bonding) electrons in stable ground state orbitals are those usually excited, and each electron can be driven away from the positively charged nucleus for a distance corresponding to an energy exactly equal to the energy of the photon absorbed (Fig. 9-8). The pigment molecule is then in an **excited state,** and it is this excitation energy that is used in photosynthesis.

Chlorophylls and other pigments can remain in an excited state only for short periods, usually a billionth (10^{-9}) of a second or even much less. As shown in Fig. 9-8, the excitation energy can be totally lost by heat release as the electron moves back to the ground state. This is what is now happening to electrons in the ink of the words you are reading. A second way that some pigments, including chlorophyll, can lose excitation energy is by a combination of heat loss and **fluorescence** (light production accompanying rapid decay of electronic energy). Chlorophyll fluorescence produces only red light, and this deep red color is easily seen when a concentrated solution of either chlorophyll a or b or a mixture of chloroplast pigments is illuminated, especially with ultraviolet or blue radiation. In the leaf, fluorescence is greatly minimized because the excitation energy can be used in photosynthesis.

Figure 9-8 helps explain why blue light is always less efficient than red in photosynthesis: After excitation with a blue photon, the electron always decays extremely rapidly by heat release to a lower energy level, a level that red light produces without heat loss when it is absorbed. From this lower level, either additional heat loss, fluorescence, or photosynthesis can occur.

Photosynthesis requires that energy in excited electrons of various pigments be transferred to an energy collecting

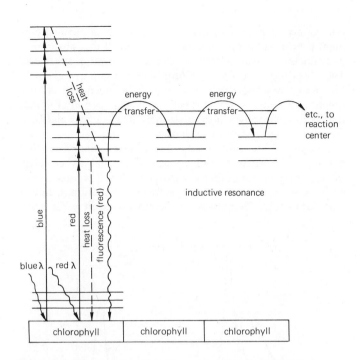

Figure 9-8 Simplified model to explain how light energy striking a chlorophyll molecule is given up. Note that excitation by either blue or red light leads to the same final energy level (often called the first excited singlet). From here, the energy can be lost by decay back to the ground state (heat loss or fluorescence of red light) or can be transferred to an adjacent pigment by inductive resonance. Each time a pigment transfers its excitation energy to an adjacent pigment, the excited electron in the first pigment returns to the ground state.

pigment, or **reaction center.** We shall explain later that there are two kinds of reaction centers in thylakoids, both of which consist of chlorophyll a molecules that are made special by their association with particular proteins and other membrane components. Figure 9-8 illustrates that the energy in an excited pigment can be transferred to an adjacent pigment, and from it to another pigment, and so on until the energy finally arrives at the reaction center. Various theories to explain energy migration within a group of neighboring pigments are available, one of which is **inductive resonance.** We shall not discuss this theory, but we emphasize that excitation of any one of numerous pigment molecules in a thylakoid allows momentary collection of the light energy in a chlorophyll a reaction center.

Leaves of most species absorb more than 90 percent of the violet and blue wavelengths that strike them and almost as high a percentage of the orange and red wavelengths (see Fig. 3-12). Essentially all this absorption is by the chloroplasts. In the thylakoids, each photon can excite an electron in a carotenoid or chlorophyll. Since chlorophylls are green, they absorb green wavelengths relatively ineffectively and instead reflect or transmit these wavelengths. We can measure the relative absorbance of various wavelengths by a purified pigment with a spectrophotometer. A graph of this absorption as a function of wavelength is called an **absorption spectrum.** The absorption spectra of chlorophylls *a* and *b* are given in Fig. 9-9a. They show that very little of the green and yellow-green light between 500 and 600 nm is absorbed *in vitro* and that both chlorophylls absorb strongly in the violet and blue and the orange and red wavelengths.

Some of the carotenoids also transfer their excitation energy to the same reaction centers as do the vast majority of chlorophylls, although most carotenoids are photosynthetically inactive. The absorption spectra of two abundant chloroplast carotenoids, *β*-**carotene** and **lutein,** are given in Fig. 9-9b.

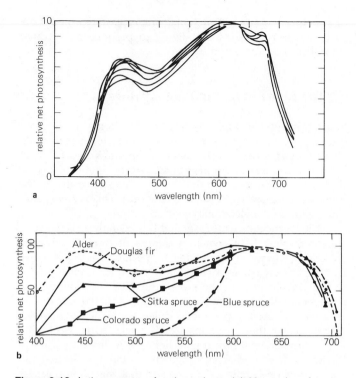

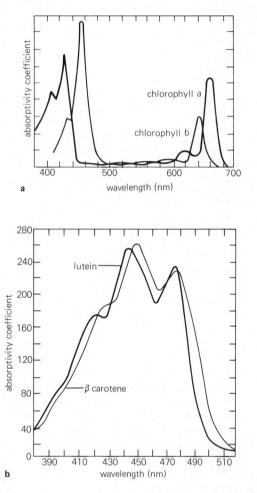

Figure 9-9 (**a**) Absorption spectra of chlorophylls *a* and *b* dissolved in diethyl ether. The absorptivity coefficient used here is equal to the absorbance (optical density) given by a solution at a concentration of 1 g/l with a thickness (light path length) of 1 cm. (From F. Zscheile and C. Comar, 1941, Botanical Gazette 102:463.) (**b**) Absorption spectra of β-carotene in hexane and of lutein (a xanthophyll) in ethanol. The absorptivity coefficient used is the same as that described in 9-9a. (Data from F. Zscheile et al., 1942, Plant Physiology 17:331.)

Figure 9-10 Action spectra of various plants. (**a**) 22 species of crop plants. (From K. J. McCree, 1972, Agricultural Meteorology 9:191–216.) (**b**) Some trees. (Drawn from data of J. B. Clark and G. R. Lister, 1975, Plant Physiology 55:401–406.)

These yellow pigments absorb only the blue and violet wavelengths *in vitro*. They reflect and transmit green, yellow, orange, and red wavelengths, and this combination appears yellow to our eyes.

When we compare the effects of different wavelengths on the rate of photosynthesis, always making sure not to add so much energy of any wavelength that the process becomes saturated, we obtain an **action spectrum.** Action spectra for photosynthesis and other photobiological processes help identify the pigment involved, because these spectra often closely match the absorption spectrum of any pigment suspected to participate. Figure 9-10 plots the relative rates of photosynthesis for several species as a function of the number of photons of each wavelength striking the leaf.*

Figure 9-10a compares action spectra for 22 herbaceous dicot and monocot crop plants, and Fig. 9-10b shows data from five trees. All the plants show a major peak in the red region and a distinct lower peak or at least a shoulder (the

---

*Slightly different action spectra with lower blue peaks are obtained when we plot the data as a function of the energy in each applied wavelength. This is because, although a blue photon absorbed by photosynthetically active pigment is as effective as any other photon, more of its energy is wasted as heat. Therefore, blue light can be as *effective* in photosynthesis as red but never as *efficient*.

conifers) in the blue, both of which result mainly from light absorption by chlorophylls. The conifers show less response in the blue because their waxy blue-green needles reflect more of the blue light and also because they contain high amounts of carotenoids, most of which are not photosynthetically active but which still absorb some of the unreflected blue light in competition with chlorophylls. Such carotenoids are said to screen the blue light.

Compared to the absorption spectra of purified chlorophylls and carotenoids, the action of green and yellow light in photosynthesis and the absorption of these wavelengths by leaves is surprisingly high for all species mentioned. Nevertheless, carotenoids and chlorophylls are apparently the only pigments absorbing light.* The main reason the action spectra are higher than the absorption spectra for yellow and green wavelengths is that although the chance of any such wavelength being absorbed is small, those that are unabsorbed are repeatedly reflected from chloroplast to chloroplast in the complex network of photosynthetic cells. With each such reflection, a small additional percentage of these is absorbed, until finally half or more of them are absorbed by most leaves. Furthermore, the *in vitro* absorption by these pigments in an organic solvent occurs at shorter wavelengths than when they are present in the chloroplast lamellae. When *in vivo*, the absorption of the carotenoids shifts from the blue into the green, and much photosynthesis in the green at about 500 nm results from absorption by certain active carotenoids.

The chlorophylls show only small shifts in the blue region, but chlorophyll *a* shows several shifts in the red. The association of chlorophyll *a* with itself to form dimers and trimers and with thylakoid proteins causes additional peaks to occur in the red region (Brown, 1972). We are interested in two of these minor peaks, those at about 682 and 703 nm, because they result from special chlorophyll *a* molecules acting as reaction-center pigments. These pigments are abbreviated **P680** and **P700.** Their functions are discussed in more detail in the following section.

9.4 The Emerson Enhancement Effect: Cooperating Photosystems

In the 1950s, Robert Emerson at the University of Illinois was interested in why red light of wavelengths longer than about 680 nm was so inefficient in causing photosynthesis (Fig. 9-10), even though it was absorbed *in vivo* by chlorophyll *a*. His research group found that if light of shorter wavelengths was provided at the same time as the longer red wavelengths, the rate of photosynthesis was even greater than could be

---

*In many algae, carotenoids and phycobilin pigments (red phycoerythrin or blue phycocyanin) absorb light, causing photosynthesis. The action spectra of these algae are very different from those of green plants. For example, green light is most effective in many red algae, and this apparently allows some red algae species to live at depths to which only green light penetrates.

expected from adding the rates found when either color was provided alone. This synergism or enhancement became known as the **Emerson enhancement effect.** We can think of this enhancement as the long red wavelengths helping out the shorter red, orange, yellow, green, and blue, or as the short group helping the long red. We now realize that two groups of functioning pigments called **photosystems** cooperatively cause this enhancement. Photosystem I is enriched in chlorophyll *a* relative to photosystem II and contains less chlorophyll *b*. Photosystem I is the only photosystem to absorb wavelengths above about 680 nm, but both it and photosystem II can absorb shorter wavelengths. More recently, a group of chlorophylls *a* and *b* that can transfer excitation energy to either photosystem was discovered (Satoh et al., 1976).

The reaction center for photosystem II is P680 and for photosystem I is P700. Apparently, any photon absorbed by a light-harvesting chlorophyll within either photosystem has a high probability of transferring its excitation energy to the corresponding reaction center that acts as a temporary energy trap, even though P680 and P700 represent only 1 percent or less of the total chlorophyll molecules present. Such energy transfer from certain carotenoids is also important. What do P680 and P700 do with this energy? The Emerson enhancement effect and other results prove that the two centers cooperate to drive photosynthesis, but how?

9.5 Transport of Electrons From H_2O to $NADP^+$

Functions of P680 and P700 In the introduction to this chapter, we emphasized that light acts in photosynthesis to drive electrons away from water to $NADP^+$, after which the NADPH is used to reduce CO_2. Obviously, this is an indirect process, since it is P680 and P700, not H_2O itself, that become excited after transfer of light energy from surrounding pigments. In their excited states, P680 and P700 have very different properties than they do in their ground states. They can, if a suitable primary electron acceptor molecule is present, transfer an electron to that acceptor. This initiates and drives the energetically uphill process of electron transport, because the primary acceptor of either photosystem can then spontaneously donate its electron in an energetically downhill process to other membrane components capable of being reduced and oxidized. Another way of saying this is that excited P680 and P700 molecules can reduce certain primary electron acceptors that have negative reduction potentials (are energetically difficult to reduce), and each reduced acceptor can readily transfer an electron to another molecule with a more positive reduction potential.

In thylakoids, several kinds of proteins and a few other kinds of molecules are arranged in the two photosystems to form an efficient electron transport system that moves electrons from P680 and P700 to $NADP^+$. Once oxidized, P680 is such a strong electron acceptor that it now attracts electrons from H_2O, again by a process involving at least one other redox protein. We might compare these electron transport components to a "bucket brigade." Just as people in the bucket brigade have the purpose of rapidly moving a bucket of water toward a fire, so the function of P680, P700, and the other components is to move electrons rapidly from H_2O to $NADP^+$. The reduction potentials of these components and their arrangement in the thylakoids insure that electron transport occurs from H_2O to P680 in photosystem II, then, upon photoexcitation of P700, to the redox components of photosystem I, and finally to $NADP^+$. This process of electron transport is accompanied by a separation of H^+ and OH^- across the thylakoid lamellae. This separation of charges provides the potential energy used in photosynthetic phosphorylation.

A Model for the Electron Transport System Let us now investigate some of the components involved in electron transport and see how they cooperate. Another look at thylakoid structures shown in Fig. 9-3 might be helpful. Notice that there is a space or channel between the lamellae of each thylakoid. This space contains H_2O and dissolved solutes and is important to both electron transport and photosynthetic phosphorylation.

Our tentative and simplified model is presented in Fig. 9-11. It represents a compromise between Z-scheme models shown in many beginning textbooks and the structural aspects emphasized in more advanced reviews (Trebst, 1974; Anderson, 1975). An attempt is made to integrate relative energy changes and postional aspects of the electron transport system, but neither is shown accurately. First, observe the overall aspects and note that electrons move from H_2O, the ultimate donor, at the lower right to $NADP^+$ at the upper left. From an energy standpoint, this represents an uphill process driven by light. To transfer each electron from H_2O to $NADP^+$ via this model requires a *net* change of about 1.14 electron-volts (eV), although most of the steps are energetically downhill, and only the two light-driven steps (from P680 and P700) are uphill. This requires about 26,300 cal/mol* or, since four electrons are removed from the two H_2O required, about 105,200 cal per 2 mol of H_2O oxidized.

Also note that H_2O is oxidized inside the thylakoid channel. The release of H^+ there decreases the *p*H and raises the number of positive charges, causing a more positive electropotential relative to the outside (stroma side) of the lamella. These H^+ attract OH^- into the channel, while the excess OH^- on the stroma side attract H^+. These attractive forces have the overall

result of removing H_2O from a site where ATP and H_2O are being formed from ADP and $H_2PO_4^-$, thereby favoring ATP formation (for more details, see Special Topic 9.1).

Our model represents ATP synthesis only at the top in connection with an ATPase known to bound loosely to the stroma side of the membrane where it operates to form ATP. Nevertheless, modern investigations indicate that *H^+ release during the oxidation of H_2O is one of two primary "sites" of noncyclic photosynthetic phosphorylation.* The term "noncyclic" comes from the failure of the electrons to cycle back to O_2 to form H_2O.

As electrons are drawn from H_2O toward photooxidized P680 in photosystem II, they are accepted by an unidentified protein or group of proteins thought to contain at least four manganese ions. Certainly, Mn^{+2} bound to protein is somehow required for this process, and this is one reason it is essential to plants. The process is also stimulated by chloride, and this helps explain the essentiality of this ion.

Excitation of four electrons in photosystem II requires four photons. Results of Pierre Joliot and Bessel Kok in the early 1970s indicated that the electron donor to P680 must donate four electrons and accumulate four positive charges before one O_2 molecule can be released. Perhaps these four charges represent a group of four Mn^{+3} ions, produced from Mn^{+2} as an electron moves from each such Mn^{+2} to P680.

The primary acceptor to which P680 transfers an electron is still not identified. It is often called **Q,** because of its ability to quench fluorescence of chlorophyll *a* under certain conditions. Q might prove to be a quinone; if so, its name would be especially appropriate.

A group of **quinones** exist in chloroplasts, including a few of the vitamin K type (naphthoquinones), a few vitamins E (tocopherylquinones), and some **plastoquinones (PQ)**. One of the most abundant plastoquinones is **plastoquinone A (PQA)**. It apparently accepts electrons directly from Q, two such electrons being required for its complete reduction to **$PQAH_2$**. Reduction of PQA also requires $2H^+$:

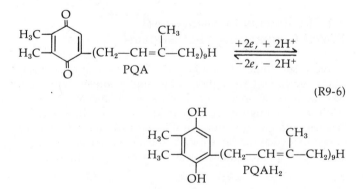

$$(R9\text{-}6)$$

---

*The Gibbs free energy change (ΔG) for oxidation-reduction reactions can be calculated from the standard reduction potential in volts at *p*H 7 (ΔE) by the following equation:

$$\Delta G = -nF\Delta E$$

where n = number of equivalents of electrons transferred
F = Faraday constant (23,062 cal/electron volt equivalents)

In our example, $\Delta G = -(4\,\text{equivalents})\,(23{,}062\,\text{cal/electron volt equivalents})\,(1.14\,\text{electron volts}) = -105{,}200\,\text{cal.}$

*The term "site" is commonly used by biochemists in a special way: It refers to the place in the electron transport process at which energy release needed to power ATP synthesis occurs, but it does not represent the physical site at which such synthesis takes place.

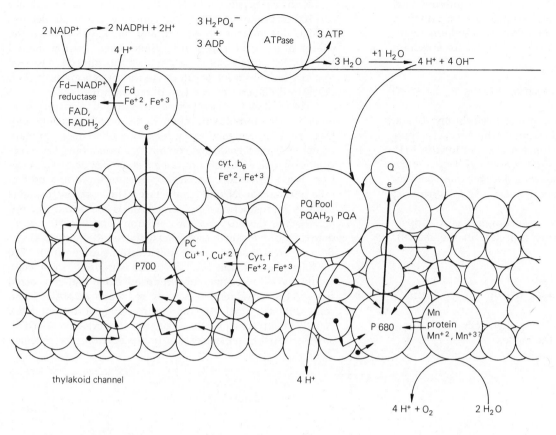

stroma side of lamella

Photosystem I

Photosystem II

thylakoid channel

Figure 9-11 A model for the light-stimulated transport of electrons and H$^+$ in chloroplasts. Two H$_2$O (lower right) are oxidized by cooperation of both photosystems. Smallest circles represent chlorophyll and carotenoid molecules that extend completely across the lamellae. Each solid dot represents absorption of one photon, while connecting arrows indicate energy transfer among pigments and ultimately to the reaction center P680 or P700; transfer of all four available electrons in two H$_2$O requires absorption of eight photons, four by each photosystem. Heavy arrows upward from P680 and P700 represent electron excitation and transfer to the corresponding unidentified primary electron acceptors, simplified as Q and Fd. Standard reduction potentials for identified reactions are: 1/2 O$_2$ + 2e + 2H$^+$ = 2H$_2$O, + 0.815 volt (V); PQA + 2e + 2H$^+$ = PQAH$_2$, + 0.1 V; cyt. f(Fe^{+3}) + e = cyt. f(Fe^{+2}), + 0.36 V; PC(Cu^{+2}) + e = PC(Cu^{+1}), + 0.32 V; Fd(Fe^{+3}) + e = Fd(Fe^{+2}), − 0.43 V; NADP$^+$ + 2e + H$^+$ = NADPH, − 0.32 V; cyt. b_6(Fe^{+3}) + e = cyt. b_6(Fe^{+2}), − 0.14 V. Fd, Fd-NADP$^+$ reductase, and the ATPase proteins are rather loosely bound on or in the lamella near the stroma side, while the other proteins are buried more deeply. The ATPase catalyzes phosphorylation (ATP formation) when the other product of this reaction, H$_2$O, is removed; this removal apparently occurs by use of H$^+$ in the reduction of PQA. When PQAH$_2$ is oxidized, the H$^+$ are transferred into the thylakoid channel.

Since our model is based upon transfer of four electrons from 2H$_2$O, 4H$^+$ are shown entering the PQ pool from the stroma side of the lamella at the top of Fig. 9-11.

Two herbicides (weed killers), 3-(3,4-dichlorophenyl)-1,1-dimethylurea, usually abbreviated DCMU (common name, **diuron**) and a closely related analogue, monuron, block the electron transport system between Q and PQA. This is their primary mechanism of action as weed killers (Black, 1976). The triazine herbicides atrazine and simazine and certain substituted uracil herbicides such as bromacil and isocil also inhibit at, or very near, this electron transport site.

Three iron-containing proteins called cytochromes, **cytochrome f, cytochrome b$_6$,** and **cytochrome b$_3$** exist in chloroplasts. All contain iron in the Fe^{+2} or Fe^{+3} state as part of a heme prosthetic group. Of these, only b_6 and f, both of which exist in photosystem I, are widely accepted as functioning in the electron transport system. The role of b_3 is still unknown. In all cytochromes, the iron atom is responsible for accepting and donating one electron:

$$\text{cytochrome (Fe}^{+3}) \underset{-1e}{\overset{+1e}{\rightleftharpoons}} \text{cytochrome (Fe}^{+2}) \qquad (R9\text{-}7)$$

As $PQAH_2$ transfers an electron to cytochrome f, one H^+ is released to the thylakoid channel. This process is presumably repeated rapidly four times to account for the 4 H^+ shown in Fig. 9-11. *The uptake of H^+ from the stroma side of the lamella shown at the upper right and release into the channel probably represents the second primary "site" of noncyclic photosynthetic phosphorylation.*

The electron acceptor from cytochrome f is a copper-containing protein called **plastocyanin** (**PC** in the model). It functions as in R9-9:

$$\begin{aligned} \text{cytochrome } f \text{ (Fe}^{+2}) + PC(Cu^{+2}) &\rightarrow \\ \text{cytochrome } f \text{ (Fe}^{+3}) + PC \text{ (Cu}^{+1}) \end{aligned} \qquad (R9\text{-}8)$$

This electron transfer process is repeated four times to account for the four electrons in the model.

Reduced plastocyanin is the immediate electron donor to P700 in photosystem I. P700 can only accept an electron if it has just lost one of its own, and this happens when the energy of a photon collected in photosystem I is transmitted to P700. Repetition of such a photoact four times is essential to allow four electrons to fall down the energy gradient from PQH_2 to P700 between the two photosystems.

The first well-characterized acceptor of electrons from excited P700 is **ferredoxin**, shown in the model as Fd, although there is good evidence for at least one other more primary acceptor. Plant ferredoxins contain two iron atoms per molecule, only one of which (in its Fe^{+3} state) can accept an electron. From ferredoxin, an electron is transferred to a protein called **ferredoxin-NADP$^+$ reductase.** This enzyme contains FAD as a prosthetic group, and FAD is the actual electron acceptor for the enzyme. Two electrons and two H^+ are required for conversion of each FAD to the reduced form, $FADH_2$. Because we are dealing with four electrons driven from P700 by four photons, the FAD in ferredoxin-NADP$^+$ reductase would have to become fully reduced twice and reoxidized twice. This reoxidation occurs by reduction of NADP$^+$ and represents the final step of electron transport from H_2O. NADP$^+$ accepts two electrons as it is converted to NADPH, so two NADPH molecules are produced from the repeated

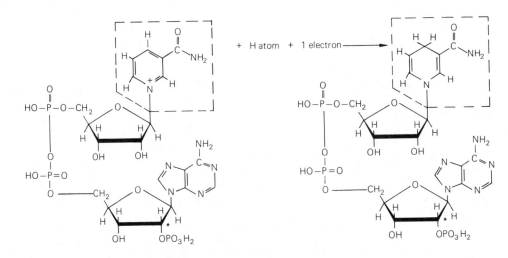

Figure 9-12 Structures of NADP$^+$ (left) and NADPH (right). The part of the NADP$^+$ molecule undergoing reduction, the nicotinamide ring, is enclosed by a dotted line. One electron is added to the nitrogen atom of the nicotinamide, neutralizing its positive charge, while the second electron is added as part of an H atom to its uppermost carbon atom. The relatively complex NADP$^+$ is a combination of two nucleotides, adenosine monophosphate (AMP) and nicotinamide mononucleotide. All nucleotides are made of three major parts: (1) a heterocyclic ring, in this case nicotinamide but in other nucleotides a purine or pyrimidine base, (2) the pentose sugar ribose, and (3) phosphate. Phosphate is esterified to the C-5 position of the ribose unit. The two nucleotides in NADP$^+$ are connected in an anhydride linkage between the C-5 phosphate group of each ribose moiety. Notice also that NADP$^+$ contains another phosphate group esterified to the OH group at the C-2 position (see asterisk) of that ribose moiety belonging to AMP. The presence of this additional phosphate is the only way in which NADP$^+$ and NADPH differ from another important electron-carrying coenzyme called **NAD$^+$** (**nicotinamide adenine dinucleotide**). NAD$^+$ and its reduced form, **NADH**, are much less abundant than NADP$^+$ and NADPH in chloroplasts, but they are involved in electron transport during several reactions of respiration (Chapter 12), nitrogen metabolism (Chapter 13), fat breakdown (Chapter 14) and even of a few photosynthesis reactions (Chapter 10).

oxidation of $FADH_2$ in the reductase enzyme. Fig. 9-12 shows the structures of $NADP^+$ and NADPH.

An interesting question concerns what happens if nearly all of the $NADP^+$ is converted to NADPH by these light-driven reactions, but CO_2 cannot penetrate the leaf because, for example, the stomates are closed. NADPH should accumulate under these conditions, because insufficient CO_2 is available to accept its electrons and convert it back to $NADP^+$. There are two control mechanisms that seem to prevent this excessive conversion of $NADP^+$ to NADPH in intense light. The best-known mechanism involves an alternate pathway for electrons to cycle from ferredoxin back to P700. Cytochrome b_6 is thought to participate in this pathway, as shown in the model. One photon is required for each electron cycling through. This light-driven alternate pathway of electron transport from P700 to ferredoxin and then back again is called **cyclic electron transport**, as contrasted to the **noncyclic electron transport** involving a permanent loss of electrons when they flow from H_2O to form NADPH. Cyclic electron transport also requires H^+ uptake at the PQ step, so a pH gradient across the lamella is again formed, and conversion of ADP + $H_2PO_4^-$ to ATP occurs (**cyclic photophosphorylation**). Because it leads to production of ATP but not NADPH, cyclic photophosphorylation is thought to help regulate the ratios of these two energy-rich compounds.

A second mechanism (not shown in our model) by which the relative amounts of ATP and NADPH might be controlled is by **pseudocyclic photophosphorylation** (Simonis and Urbach, 1973; Egneus et al., 1975). This process is similar to cyclic photophosphorylation in that reduced ferredoxin molecules do not transfer electrons to $NADP^+$, but instead they transfer the electrons to O_2, eventually forming H_2O. ATP is produced by the regular steps of noncyclic photophosphorylation, but no NADPH is formed. How important cyclic and pseudocyclic photophosphorylation are to ATP synthesis during photosynthesis is still controversial, but it is clear that O_2 can compete with CO_2 for electrons removed from H_2O by light energy (Radmer and Kok, 1976).

Some Quantitative Aspects of Photosynthetic Light Reactions According to our model, eight photons of light are required to oxidize two H_2O molecules and reduce two $NADP^+$ to two NADPH. Four of these photons are used in photosystem II and four in photosystem I, each photon being required to push one electron up part of the energy gradient. Since two H_2O are required to release one O_2 and reduce one CO_2 (R9-4), our model predicts that eight photons are required to fix one CO_2 molecule. Most measurements of the actual photon or quantum requirement of photosynthesis indicate values between 8 and 12, so the model is fairly consistent with the data. In the next chapter, we shall show that each CO_2 fixed in the Calvin cycle requires two NADPH molecules, entirely consistent with the model, and that three ATP molecules are required to help convert each CO_2 to a sugar phosphate. If at least three ATP molecules are formed during noncyclic photophosphorylation for each O_2 released, as indicated at the top of the model, adequate energy is provided for photosynthetic reduction of CO_2 without cyclic or pseudocyclic photophosphorylation and additional photons. Assuming this is true, we may now write a probable summary equation for the light reactions of photosynthesis:

$$2H_2O + 2NADP^+ + 3ADP^{-2} + 3H_2PO_4^- \qquad \text{(R9-9)}$$
$$+ \text{ 8-12 photons} \rightarrow O_2 + 2NADPH + 2H^+$$
$$+ 3ATP^{-3} + 3H_2O$$

Carbon Dioxide Fixation and Carbohydrate Synthesis

The sequence of reactions leading from CO_2 to the more complex organic products produced in photosynthesis was solved only after radioactive carbon 14 became available about 1949. Carbon dioxide containing ^{14}C was then prepared, and all molecules produced from it during photosynthetic experiments were labeled. Paper chromatography was developed at about the same time, making separation of many of the photosynthetic compounds practical.

Labeled molecules present on the chromatograms were detected by autoradiography (Section 7.2). In this technique, X-ray film is placed in tight contact with the paper chromatograms in darkness for a few days to a few months (depending upon the amount of radioactivity present), while the radioactivity exposes the film. When the film is developed, dark spots that indicate the locations of the radioactive compounds are seen. The identity of the compounds and the amount of radioactivity present in each can be determined after cutting out the areas of paper that correspond to the dark spots on the film. Geiger-Müller tubes or the more sensitive **liquid scintillation counters** are used to measure radioactivity, while direct chemical analyses and rechromatography of unknown substances with known compounds **(cochromatography)** are used to identify radioactive photosynthetic products. These and other techniques have given us a fairly complete understanding of the pathways of CO_2 fixation and carbohydrate synthesis.

10.1 Products of Carbon Dioxide Fixation

The First Product Chromatographic procedures in conjunction with the use of $^{14}CO_2$ were first applied to the problem of photosynthesis by Melvin Calvin, Andrew A. Benson, James A. Bassham, and others at the University of California in Berkeley in 1949 (see the personal essays by Calvin and Bassham near the end of this chapter). They allowed green algae such as *Chlorella* to attain a steady rate of photosynthesis, then introduced radioactive $^{14}CO_2$ (as $H^{14}CO_3^-$) into the solution in which the algae were growing. By opening a valve on the bottom of the container at various times after introduction of the isotope, the algae were dropped into boiling 80-percent ethanol to kill them rapidly and to extract any metabolites. This extract was chromatographed, and autoradiograms were made.

It was found that even after photosynthesizing only 7 seconds in the presence of $^{14}CO_2$, the algae had formed as many as 12 radioactive compounds. After 60 seconds, many dark spots on the film were detected, as shown in Fig. 10-1. By this time, both amino acids and organic acids had become radioactive. The compounds containing most of the ^{14}C were phosphorylated sugars, shown at the lower right of the figure. To identify the first product formed from CO_2, the time periods were shortened to as little as 2 seconds. When this was done, most of the ^{14}C was found in a phosphorylated three-carbon acid called **3-phosphoglyceric acid,** abbreviated **3-PGA.** This acid, then, was the first detectable product of photosynthetic CO_2 fixation in these algae. Similar results were found with these general techniques using leaves of certain higher plants or chloroplasts isolated from them.

The Compound Combining with CO_2 The search next began for a two-carbon substance with which CO_2 could react to form 3-PGA. Such a two-carbon unit was not found. When $^{14}CO_2$ was fed to the algae for some time and the supply was then suddenly removed, it was expected that the compound with which it normally combines would accumulate. The substance which did accumulate was a five-carbon sugar, phosphorylated at each end of the molecule, **ribulose-1,5-diphosphate.** At the same time, there was a rapid drop in the level of labeled 3-PGA. This suggested that ribulose diphosphate is the normal substrate to which CO_2 is added, leading to the formation of 3-PGA.

The Reaction of CO_2 Fixation Soon an enzyme was found that catalyzes the combination of CO_2 with ribulose diphosphate to form two molecules of 3-PGA. Considerable evidence now indicates that an unstable intermediate product is formed (shown in brackets in R10-1) that splits into two 3-PGA molecules with the addition of H_2O. Thus, if $^{14}CO_2$ is reacting, one of the two 3-PGA molecules becomes labeled with ^{14}C and one remains unlabeled during the reaction.

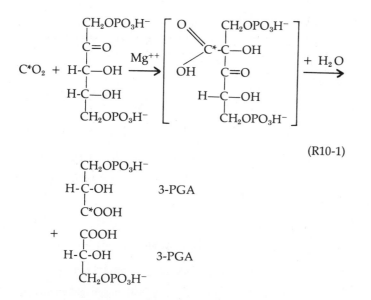

$$C^*O_2 + \begin{array}{c} CH_2OPO_3H^- \\ | \\ C=O \\ | \\ H\text{-}C\text{—OH} \\ | \\ H\text{-}C\text{—OH} \\ | \\ CH_2OPO_3H^- \end{array} \xrightarrow{Mg^{++}} \left[\begin{array}{c} O \\ \| \\ C \\ OH \end{array} \begin{array}{c} CH_2OPO_3H^- \\ | \\ C^*\text{-}C\text{—OH} \\ | \\ C=O \\ | \\ H\text{—C}\text{—OH} \\ | \\ CH_2OPO_3H^- \end{array} \right] + H_2O \longrightarrow$$

(R10-1)

$$\begin{array}{c} CH_2OPO_3H^- \\ | \\ H\text{-}C\text{-OH} \\ | \\ C^*OOH \end{array} \quad 3\text{-PGA}$$

$$+ \quad \begin{array}{c} COOH \\ | \\ H\text{-}C\text{-OH} \\ | \\ CH_2OPO_3H^- \end{array} \quad 3\text{-PGA}$$

The enzyme catalyzing this reaction was first named **carboxydismutase** but is now more commonly called **ribulose diphosphate carboxylase.** It is apparently functional in all photosynthetic organisms and is important not only because of the essential reaction it catalyzes, but also because it is probably by far the most abundant protein in nature. Chloroplasts contain approximately half of the total protein in leaves, and about one fourth to one half of their total protein is made up of this enzyme, so roughly one eighth to one fourth of leaf protein commonly exists as this enzyme. It is therefore very important in the diets of animals, including humans.

CO_2 fixation by ribulose diphosphate carboxylase is not reversible, and most of the CO_2 absorbed by the majority of plants is first combined with it rather than with other CO_2-fixing enzymes. Nevertheless, another especially important CO_2-fixing reaction is known; it will be discussed in Section 10.3.

10.2 The Calvin Cycle

Further investigations of radioactive compounds formed from $^{14}CO_2$ showed that other sugar phosphates containing five, six, and even seven carbon atoms were produced. These compounds included the pentose (five-carbon) phosphates ribose-5-phosphate, xylulose-5-phosphate, and ribulose-5-phosphate; the hexose (six-carbon) phosphates fructose-6-phosphate, fructose-1,6-diphosphate, and glucose-6-phosphate; and the heptose (seven-carbon) phosphates sedoheptulose-7-phosphate, and sedoheptulose-1,7-diphosphate. By noting time sequences in which these became labeled from $^{14}CO_2$ and by degrading the molecules to determine which atoms contained the ^{14}C, it was possible to predict a metabolic pathway relating them.

When 3-PGA was degraded, most of the radioactivity proved to be in the carboxyl carbon, as shown by asterisks

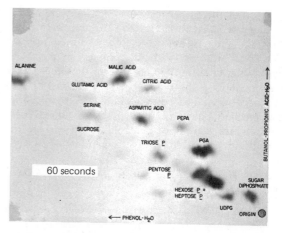

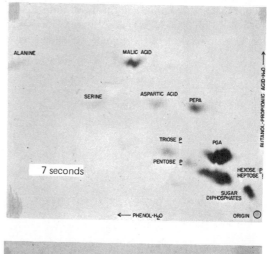

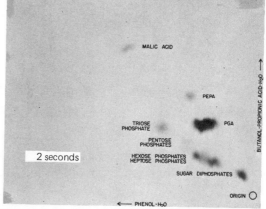

Figure 10-1 Autoradiograms showing the products of photosynthesis in the alga *Chlorella pyrenoidosa* after various times of exposure to $^{14}CO_2$. (Top) 60 seconds, (middle) 7 seconds, (bottom) 2 seconds. Note the increasing importance of 3-PGA and other sugar phosphates as the exposure time is shortened. (From J. A. Bassham, 1965, in J. Bonner and J. E. Varner, eds., Plant Biochemistry, Academic Press, Inc., New York, pp. 883–884.)

in R10-1, but the alpha and beta carbons were also labeled. This suggested that the latter two carbons were not derived from $^{14}CO_2$ directly but were rather formed from the carboxyl carbon atom of 3-PGA by some cyclic process. A cyclic pathway that uses the 3-PGA to form the other sugar phosphates mentioned and that also converts some of the carbon atoms back to ribulose diphosphate was soon worked out. These reactions have collectively been named the **Calvin cycle.** Calvin was awarded a Nobel Prize in 1961 for this work.

The Calvin cycle consists of four principal parts:

1. *CO_2 and H_2O are added to ribulose diphosphate to form two molecules of 3-PGA.*

2. *3-PGA is reduced to 3-phosphoglyceraldehyde using electrons provided by NADPH and with energy provided by ATP.* The use of NADPH and ATP, both of which are produced from the light reactions of photosynthesis, to reduce 3-PGA to 3-phosphoglyceraldehyde is shown in reaction R10-2. Note that carboxylic acids such as 3-PGA are energetically

difficult to reduce and that the plant avoids direct reduction of carboxyl group of 3-PGA by forming an acid anhydride, 1,3-diphosphoglyceric acid, with phosphate derived from ATP.

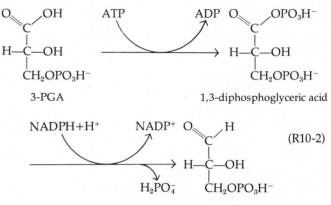

(R10-2)

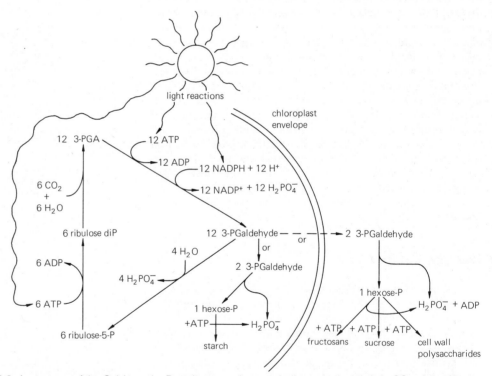

Figure 10-2 A summary of the Calvin cycle. Reactions are shown to indicate the input of six CO_2 molecules (upper left) and the net output of two three-carbon molecules (3-phosphoglyceraldehyde, abbreviated 3-PGaldehyde), or of one hexose phosphate. This hexose-P can be either fructose-6-P or glucose-6-P, as discussed in Section 10.6. Although starch accumulates as a product of photosynthesis in most chloroplasts, sucrose and other polysaccharides are formed outside the plastids. Only a few compounds involved in the cycle can readily move out of plastids to form those polysaccharides. 3-PGaldehyde can be readily transported across the chloroplast envelope (Heber, 1974), so the model shows the probable transport and use of two such molecules. Ten other 3-PGaldehyde molecules are needed to regenerate ribulose-5-P (lower left). Production of starch, fructosans, sucrose, and cell wall polysaccharides is discussed later in this chapter. To form each glycosidic bond in such polymers requires an additional ATP, shown in this model.

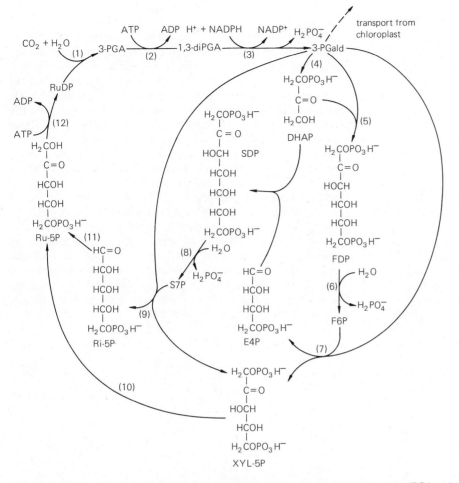

Figure 10-3 The Calvin cycle. Abbreviations: 3-PGA, 3-phosphoglyceric acid; 1,3-diPGA, 1,3-diphosphoglyceric acid; DHAP, dihydroxyacetonephosphate; E4P, erythrose-4-phosphate; FDP, fructose-1, 6-diphosphate; F6P, fructose-6-phosphate; SDP, sedoheptulose-1, 7-diphosphate; S7P, sedoheptulose-7-phosphate; Ri-5P, ribose-5-phosphate; xyl-5P, xylulose-5-phosphate; Ru-5P, ribulose-5-phosphate; RuDP, ribulose-1,5-diphosphate. Structures of RuDP, 3-PGA, 1,3-diPGA, and 3-PGald are given in text reactions R10-1 and R10-2. Names of enzymes catalyzing individual reactions are: (1) ribulose diphosphate carboxylase, (2) 3-phosphoglyceric acid kinase, (3) 3-phosphoglyceraldehyde dehydrogenase, (4) triose phosphate isomerase, (5) aldolase, (6) fructose-1,6 diphosphatase, (7) aldolase, (8) sedoheptulose-1,7-diphosphatase, (9) transketolase, (10) ribulose phosphate epimerase, (11) ribose phosphate isomerase, (12) ribulose phosphate kinase. Note that aldolase and transketolase have dual reaction specificities. Transketolase transfers a two-carbon fragment (attached to a coenzyme, thiamine pyrophosphate, not shown) only from sugar phosphates having a keto group at C-2 and the specific arrangement of OH and H about C-3 that occurs in S7P, xyl-5P, and F6P. All reactions are physiologically reversible except (1) and the phosphatase reactions (6) and (8).

3. *Some of the 3-phosphoglyceraldehyde molecules are converted to fructose diphosphate, part of which is then converted to xylulose-5-phosphate. Other 3-phosphoglyceraldehyde molecules unite with sedoheptulose-7-phosphate, yielding ribose-5-phosphate and xylulose-5-phosphate. Ribulose-5-phosphate is produced directly from either of these two pentose phosphates.*

4. *Ribulose-5-diphosphate is phosphorylated by ATP to form ribulose diphosphate, which can then accept CO_2 to continue the cycle.*

The Calvin cycle accounts for regeneration of ribulose diphosphate and also allows for the net storage of carbohydrate molecules, since some of the 3-phosphoglyceraldehyde is used for the synthesis of sucrose, starch, cellulose, pectins, and other polysaccharides, instead of being entirely converted to ribulose-5-phosphate. With six turns of the cycle, six CO_2 molecules are fixed and one hexose phosphate is synthesized. These relationships are summarized in Fig. 10-2. The complete Calvin cycle is shown in Fig. 10-3. We emphasize that the cycle predicts a requirement of two NADPH molecules (one for each 3-PGA formed) and three ATP molecules per CO_2 fixed. One ATP is required for reduction of each of the two 3-PGA molecules (R10-2) and a third ATP to convert ribulose-5-P back to ribulose-1,5-diP (part 4 earlier). The summary reaction whereby these ATP and NADPH molecules are produced in the light-driven reactions is shown in Chapter 9, R9-10.

10.3 The C-4 Dicarboxylic Acid Pathway: Some Species Fix CO_2 Differently

Reactions of the Calvin cycle were thought to have solved the mechanism of CO_2 fixation and reduction in plants, but a new era in photosynthesis research was stimulated by a discovery of Hugo P. Kortschak, Constance E. Hartt, and George O. Burr, made in Hawaii in 1965. They found that sugar cane leaves, in which photosynthesis is unusually rapid and efficient, fix most CO_2 into carbon-4 of **malic** and **aspartic** acids. After approximately 1 second of photosynthesis in $^{14}CO_2$, 80 percent of the ^{14}C recovered from labeled compounds was present in these two acids, and only 10 percent was found in PGA, indicating that in this plant 3-PGA is not the first product of photosynthesis. These results were soon confirmed by M. D. Hatch and C. R. Slack in Australia, who found that some tropical species of the grass family, including corn, displayed similar labeling patterns after fixing $^{14}CO_2$. Other grasses, such as wheat, oat, rice, and bamboo, not closely related taxonomically to most of the others, showed 3-PGA as the predominant fixation product.

It is thus clear that the primary carboxylation reaction of some species is different from that involving ribulose diphosphate. Species that show four-carbon acids as the primary initial CO_2 fixation products are now commonly referred to as **C-4 species;** those fixing CO_2 largely into 3-PGA are called **C-3 species.** Some representative species having the C-4 pathway of CO_2 fixation are listed in Table 10-1. Most C-4 species are monocots, but a few are dicots. The pathway has been found in certain members of at least 13 families, 117 genera, and 485 species of angiosperms. Extensive lists of C-4 species are given in Krenzer et al. (1975) and Downton (1975). At least 11 genera include species of both the C-4 and C-3 type. All gymnosperms, pteridophytes, bryophytes, and algae studied are C-3 plants.

The reaction (R10-3) by which CO_2 (actually HCO_3^-) is converted into carbon-4 of malic and aspartic acids is through its initial combination with **phosphoenolpyruvic acid (PEP)** to form **oxaloacetic acid** and $H_2PO_4^-$. Oxaloacetic acid is not usually a detectable product of photosynthesis, but it can

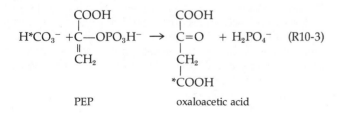

(R10-3)

PEP   oxaloacetic acid

be found when special precautions are taken to prevent its destruction in the isolation and chromatography procedures. This reaction, along with that catalyzed by ribulose diphosphate carboxylase, is one of the few carboxylations occurring in plants with a decrease in free energy. **Phosphoenolpyruvate carboxylase,** an enzyme widely distributed in plants, is the catalyst involved. This enzyme, often called **PEP carboxylase,** exists in various isozymic forms and is probably present in all living plant cells. The reason it is of special importance in leaves of C-4 species is because of its high activity there and because a cyclic pathway that maintains a constant and relatively plentiful supply of PEP is also present. In leaves of C-3 species and in root, fruit, and other cells lacking chlorophyll, regardless of the species, other isozymes of PEP carboxylase are also present. Here the enzyme's major functions seem to be to help replace Krebs cycle acids used in synthetic reactions (see Chapter 12) and to help form malate ions needed in charge balancing functions (see Section 5.8). Reactions converting oxaloacetate to malic and aspartic acids in C-4 plants are shown in R10-4 and R10-5.

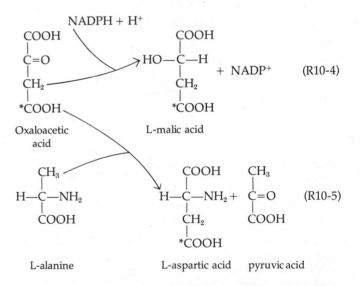

The formation of malic acid in R10-4 is catalyzed by **malic acid dehydrogenase,** with the necessary electrons being provided by NADPH. The formation of aspartic acid requires another amino acid such as alanine as the source of an amino group. This type of reaction is referred to as a **transamination,** since a transfer of an amino group is involved (see Chapter 13).

It became evident that there is a division of labor between different kinds of photosynthetic cells in C-4 species. Both kinds of cells are important for the production of sugars and starch. In the C-4 species, one or occasionally two distinct layers of tightly packed, often thick-walled photosynthetic cells (a bundle sheath) almost always surrounds the leaf vascular bundles. This concentric arrangement of photosynthetic cells is described as **Kranz** (German, ''halo'' or ''wreath'') **anatomy** (Laetsch, 1974). Compared to C-3 plants, in which a much less distinct bundle sheath is sometimes also present,

Table 10-1 Selected Examples of C-4 and C-3 Plants

C-4 monocots

Cyperaceae, sedge family
  Cyperus esculentus L.
  Cyperus rotundus L.
Gramineae (*Poaceae*), grass family
  Aristidoideae
    Aristida purpurea Nutt., purple threeawn
    Aristida hamulosa Hen. Y., Havard threeawn
  Eragrostoideae (Chloridoideae)
    Bouteloua curtipendula (Michx.) Torr., sideoats grama
    Bouteloua gracilis (H.B.K.) Lag., bluegrama
    Buchloë dactyloides (Nutt.) Engelm, buffalograss
    Chloris cucullata Bisch, hooded windmillgrass
    Chloris distichophylla Lag., weeping windmillgrass
    Chloris gayana Kunth, Rhodesgrass
    Cynodon dactylon (L.) Pers., Bermudagrass
    Eragrostis chloromelas Steud., Boer lovegrass
    Eragrostis pilosa (L.) Beauv., India lovegrass
    Leptochloa dubia (H.B.K.) Nees., green sprangletop
    Muhlenbergia schreberi J. F. Gmel., nimblewill
    Sporobolus cryptandrus (Torr.) A. Gray, sand dropseed
    Zoysia japonica Steud., Japanese lawngrass
  Panicoideae
    Andropogon scoparius Michx., little bluestem
    Cenchrus pauciflorus Benth, mat sandbur
    Digitaria sanguinalis (L.) Scop., crabgrass
    Echinochloa crus-galli (L.) Beauv., barnyardgrass
    Euchlaena mexicana Schrad., Mexican teosinte
    Panicum antidotale Retz., blue panicum
    Panicum capillare L., common witchgrass
    Panicum halli Vasey, Halls panicum
    Paspalum notatum Flügge, Bahiagrass
    Pennisetum purpureum Schum., Napiergrass
    Saccharum officinarum L., sugar cane
    Setaria italica (L.) Beauv., foxtail millet
    Setaria lutescens (Weigel) Hubb., yellow bristlegrass
    Sorghum bicolor (L.) Moench, gooseneck sorghum
    Sorghum sudanense (Piper) Stapf., Sudangrass
    Tripsacum dactyloides L., Eastern gamagrass
    Zea mays L., corn or maize

C-3 monocots

Cyperaceae, sedge family
  Cyperus alternifolium gracilis L., a sedge
Gramineae, grass family
  Agropyron repens (L.) Beauv., quackgrass
  Agrostis alba L., redtop
  Avena sativa L., oat
  Dactylis glomerata L., orchardgrass

Festuca arundinacea Schreb., meadow festuca
Hordeum vulgare L., barley
Lolium multiflorum Lam., Italian ryegrass
Oryza sativa L., rice
Panicum commutatum Schult., witchgrass
Phalaris canariensis L., canarygrass
Poa pratensis L., bluegrass
Triticum aestivum L., wheat

C-4 dicots

Amaranthaceae, amaranth family
  Amaranthus albus L., white pigweed
  Amaranthus retroflexus L., redroot pigweed
  Atriplex rosea L., tumbling orach
  Atriplex semibaccata R. Br., Australian saltbrush
  Froelichia gracilis (Hook) Mog., froelichia
  Gomphrena globosa L., common globeamaranth
  Tidestromia oblongifolia Wats (Standl.), honeysweet tidestromia
Chenopodiaceae, goosefoot family
  Kochia childsii L., Childs summer cypress
  Salsola kali L., common Russian thistle
Euphorbiaceae
  Euphorbia maculata L., spotted euphorbia
Portulacaceae, portulaca family
  Portulaca oleracea L., common purslane

C-3 dicots

Chenopodiaceae, goosefoot family
  Atriplex hastata L., fat-hen saltbrush
  Beta vulgaris L., beet
  Chenopodium album L., lambquarters goosefoot
  Spinacea oleracea L., spinach
Cruciferae, mustard family
  Brassica nigra (L.) Koch, black mustard
Compositae, sunflower family
  Helianthus annuus L., sunflower
  Lactuca sativa L., lettuce
  Xanthium strumarium L., cocklebur
Leguminosae, pea family
  Arachis hypogea L., peanut
  Glycine max Merrill, soybean
  Phaseolus vulgaris L., bean
Malvaceae, mallow family
  Gossypium hirsutum L., cotton
Solanaceae, nightshade family
  Datura stramonium L., jimson weed
Umbelliferae, parsley family
  Daucus carota L., carrot

the bundle sheath cells of many C-4 plants contain far more chloroplasts, mitochondria, and other organelles, and smaller central vacuoles. A comparison of leaf anatomy in representative C-3 and C-4 monocots is shown in Fig. 10-4. Chloroplasts of the bundle sheath cells frequently contain nearly all of the leaf starch, with little present in chloroplasts of the surrounding, more loosely arranged mesophyll cells. Isolation of the bundle sheath and mesophyll cells confirmed earlier suggestions that malic and aspartic acids are formed in the mesophyll cells and that 3-PGA, sucrose, and starch are produced mainly in the bundle sheath cells. Ribulose diphosphate carboxylase exists almost entirely in the bundle sheath cells,

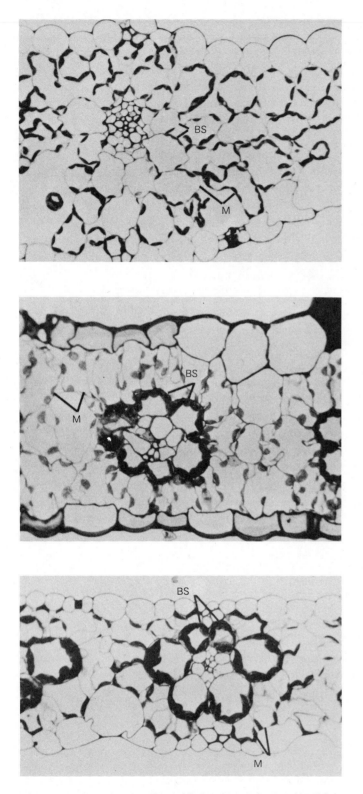

Figure 10-4 Leaf cross sections of C-3 monocots (oat, top) and C-4 monocots (corn, middle, and Rhodesgrass, bottom). (From S. E. Frederick and E. H. Newcomb, 1971, Planta 96:152–174.)

and certain other enzymes of the Calvin cycle exist primarily there. As a result, the complete Calvin cycle can occur in bundle sheath cells. Alternatively, PEP carboxylase is almost entirely restricted to the mesophyll cells. *Thus, C-4 species really use both kinds of CO_2-fixing mechanisms.*

The reason that CO_2 first appears in C-4 acids seems to be that it is the mesophyll cells into which CO_2 first penetrates after stomatal entry and that activities of PEP carboxylase are high there. Ultimately, a small fraction of CO_2 will also enter the bundle sheath cells and be fixed into 3-PGA, but the surface area exposed to the air space in the leaf is much smaller for bundle sheath than for mesophyll cells. Most of the CO_2 recently fixed in the carboxyl groups of malic and aspartic acids is rapidly transferred, perhaps via abundant plasmodesmata (see arrows in Fig. 10-5), into the bundle sheath cells. Here the acids undergo decarboxylation with release of CO_2 that is then fixed by ribulose diphosphate carboxylase into 3-PGA. The principal source of CO_2 for the bundle sheath cells is therefore from C-4 acids formed in the mesophyll. Sucrose and starch are ultimately formed from 3-PGA in the bundle sheath cells using Calvin cycle reactions and other reactions not yet mentioned. In summary, the division of labor referred to in the preceding paragraph seems to involve trapping of CO_2 into C-4 acids by the mesophyll cells, then, after transfer of these acids to the bundle sheath cells, decarboxylation and refixation of the CO_2. The 3-carbon acid (thought to be pyruvate or alanine) resulting from decarboxylation is then returned to the mesophyll cells where it can again be carboxylated to keep the cycle going. The C-4 acids formed in the mesophyll cells seem to be just carriers of the CO_2 to the bundle sheath cells. This idea is illustrated in our model of the C-4 pathway and its relation to the Calvin cycle in Fig. 10-6.

Two additional aspects of this model require explanation. First, by what mechanisms are aspartate and malate decarboxylated in the bundle sheath cells? Three such mechanisms occur, depending upon the species (Gutierrez et al., 1974a; Hatch and Kagawa, 1976), but we shall mention only two almost identical ones. As aspartate moves into the bundle sheath cell, it is metabolized in the mitochondria, first to oxaloacetate by a transaminase. Next, the resulting oxaloacetate is reduced to malate by an NADH-dependent malate dehydrogenase isozyme. (NADH is a coenzyme capable of transferring two electrons and is nearly identical in structure to NADPH, as explained in Fig. 9-12. The oxidized form of NADH is NAD^+. These two coenzymes operate in mitochondria, while $NADP^+$ and NADPH are much more important in chloroplasts.) The malate is then oxidatively decarboxylated by a **malic enzyme** isozyme, using NAD^+ as the electron acceptor. Pyruvate (CH_3—$\overset{\overset{\textstyle O}{\|}}{C}$—$COO^-$), CO_2, and NADH are the products. The pyruvate is probably then converted to alanine by transamination. As alanine moves back to the mesophyll cells, the nitrogen in it replaces that lost when aspartate was transported to the bundle sheath cells.

The malate transferred directly into the bundle sheath

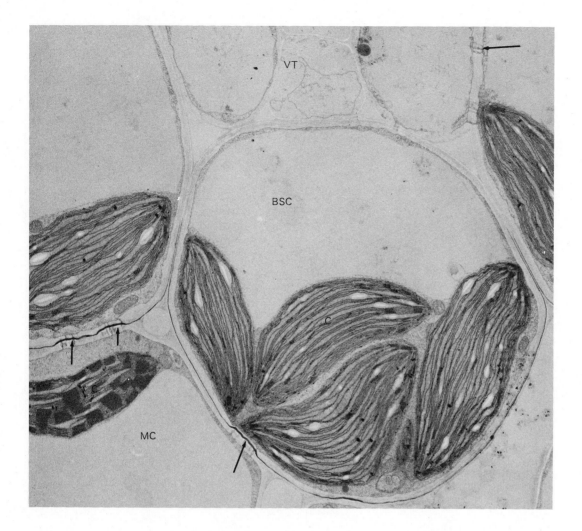

Figure 10-5 Electron micrograph of adjacent mesophyll cell (MC) and bundle sheath (BSC) in the C-4 plant crabgrass (*Digitaria sanguinalis*). Note abundant grana and lack of starch in the mesophyll cell chloroplast, but absence of grana and presence of several small starch granules in bundle sheath chloroplasts. Arrows mark plasmodesmata where passage of organic acids is suspected to occur. Vascular tissue (VT) is shown at top. (From C. C. Black et al., 1973.)

cells is decarboxylated in a similar way, but in the chloroplasts rather than in the mitochondria and with a different isozyme of the malic enzyme. This isozyme is $NADP^+$-dependent and therefore yields NADPH in addition to pyruvate and CO_2. This NADPH helps reduce 3-PGA in the Calvin cycle, as shown in the model. Such NADPH is apparently crucial to the cycle, because those species transporting mainly malate rather than aspartate from the mesophyll cells (corn, crabgrass, sorghum, sugar cane, and others) seem unable to generate enough NADPH by the light reactions of photosynthesis in the bundle sheath cells. The reason for this is that their bundle sheath chloroplasts have very few grana (Fig. 10-5). Photosystem II, upon which $NADP^+$ conversion to NADPH by the light reactions is dependent, is located mainly in the grana. Photosystem I is located in both grana and stroma lamellae but is unable to form NADPH without cooperation of Photosystem II (see Fig. 9-11).

A second important aspect of the diagram requiring emphasis is the way in which pyruvate or alanine transported back to the mesophyll cells is used to regenerate PEP. The alanine is apparently reconverted to pyruvate, then a chloroplast enzyme named **pyruvate, phosphate dikinase** converts pyruvate to PEP. This reaction (R10-6) requires ATP as an energy source but is unusual in that $H_2PO_4^-$ is also a reactant:

$$
\begin{array}{ccc}
\text{COOH} & & \text{COOH} \\
| & & | \\
\text{C=O} + \text{ATP} + \text{H}_2\text{PO}_4^- \;\rightleftharpoons\; & & \text{C—OPO}_3\text{H}^- \\
| & & \| \\
\text{CH}_3 & & \text{CH}_2 \\
\text{pyruvic acid} & & \text{PEP}
\end{array}
$$

(R10-6)

$$
+ \text{AMP} + \text{HO—}\underset{\underset{\text{O}}{\|}}{\overset{\overset{\text{O}^-}{|}}{\text{P}}}\text{—O—}\underset{\underset{\text{O}}{\|}}{\overset{\overset{\text{O}^-}{|}}{\text{P}}}\text{—OH}
$$

pyrophosphate

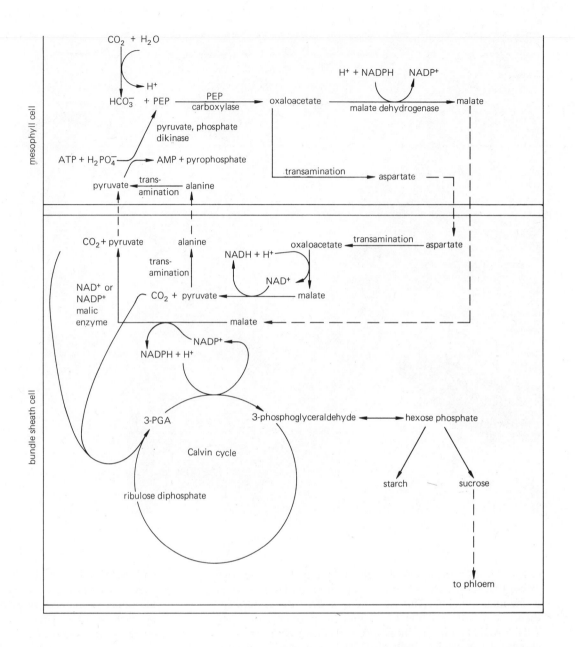

Figure 10-6 A summary of metabolic division of labor in mesophyll and bundle sheath cells of C-4 plants. CO_2 is initially fixed into C-4 acids in the mesophyll, then these acids move into the bundle sheath (probably as K^+ salts) where they are decarboxylated. The CO_2 thus released is fixed via the Calvin cycle in the bundle sheath chloroplasts. Sucrose and starch are common products, as shown. After decarboxylation of the C-4 acids, three-carbon molecules such as pyruvate and alanine move back to the mesophyll cells where they are converted to PEP so that CO_2 fixation can continue there.

In C-4 plants, pyruvate, phosphate dikinase has been found only in the mesophyll cells, and it has not been detected in C-3 plants, so it appears to represent an enzyme that evolved for a rather unique task only in certain species.

Calculations of the energy required to operate the C-4 pathway with the Calvin cycle indicate that for each CO_2 fixed by this scheme, two ATP molecules besides those three needed in the Calvin cycle are required. These additional ATP molecules are necessary to regenerate ATP from AMP, so that PEP synthesis can be maintained. There is no additional NADPH requirement, because for each NADPH used to reduce oxaloacetate in the mesophyll cells, one is also regained during malic enzyme action in the bundle sheath

cells. In spite of the apparent inefficiency of ATP utilization in C-4 species, these plants almost always show more rapid rates of photosynthesis on a leaf area basis than do C-3 species *when both are exposed to high light intensities and warm temperatures.* Under these conditions, ATP formation should not limit photosynthetic rates. Instead, CO_2 may then be limiting or, if water is limiting, the stomates might be partially or completely closed, also restricting CO_2 absorption. The C-4 species are adapted to and apparently evolved in areas of periodic drought, such as tropical savannas. When the temperatures are in the range of 25 to 35 C and the light intensity is high, C-4 plants are at least twice as efficient as C-3 plants in converting the sun's energy into dry matter production. Ecological and other physiological aspects of photosynthesis are discussed in the next chapter, but it is useful to mention now that the greater efficiency of C-4 species results from a light-enhanced *loss* of part of the CO_2 fixed in C-3 plants but little or no such waste in the C-4 plants.

10.4 Control of Photosynthetic Enzymes by Light

We have emphasized the role of light in providing ATP and NADPH needed for CO_2 fixation and reduction, yet light has another important role as a regulator of the activity of several chloroplast enzymes involved in photosynthesis. Many such enzymes apparently exist in two interconvertible forms, an active form in light and an inactive or much less active form in darkness. Carbohydrate production from CO_2 is therefore shut off tightly at night because of enzyme inactivity, closed stomates, and a deficiency of ATP and NADPH.

In C-3 species, five Calvin cycle enzymes are controlled by light, including ribulose diphosphate carboxylase, 3-phosphoglyceraldehyde dehydrogenase, fructose-1,6-diphosphate phosphatase, sedoheptulose-1,7-diphosphate phosphatase, and ribulose-5-phosphate kinase. The function of each enzyme can be determined from the legend to the Calvin cycle in Fig. 10-3. In C-4 species, additional enzymes in mesophyll cells (Fig. 10-6) are subject to similar control, including PEP carboxylase, $NADP^+$-malate dehydrogenase, and pyruvate, phosphate dikinase (Wolf, 1972; Preiss and Kosuge, 1976).

Most of these enzymes can be converted from the inactive to the active form *in vitro* by adding **dithiothreitol**, a compound with two sulfhydryl groups having the structure $HS—CH_2—(CHOH)_2—CH_2—SH$. Dithiothreitol activates such enzymes by reducing each sulfur of disulfide (—S—S—) bridges in them to the —SH form. Inactivation occurs when the sulfhydryl groups are oxidized back to the disulfide form. *In vivo*, activation in light and inactivation in darkness occur by the same mechanisms, except that dithiothreitol is not the source of the essential sulfhydryl groups. When light causes electron transport through photosystem I (see Fig. 9-11), ferredoxin and other unidentified proteins in this photosystem not shown in Fig. 9-11 become reduced. This reduction involves conversion of disulfide groups in such proteins to sulfhydryl groups. These sulfhydryls then somehow promote reduction of the photosynthetic enzymes, thereby activating them. In darkness, the sulfhydryls of photosystem I proteins become oxidized to disulfides, presumably by O_2 or by $NADP^+$, and these disulfides then oxidize the sulfhydryls in the photosynthetic enzymes to disulfides, inactivating such enzymes (Anderson and Avron, 1976).

The enzymes apparently not subject to this kind of control are ribulose diphosphate carboxylase and PEP carboxylase. They are influenced by light through its effects on the pH and Mg^{++} concentration of the stroma in which they exist and by allosteric feedback inhibition by certain molecules formed during photosynthesis (Chu and Bassham, 1975). The pH and

Mg^{++} effects result from light-driven movement of H^+ from the stroma into the chloroplast lamellae channels (see Fig. 9-11) and the movement of Mg^{++} in the reverse direction in exchange for these H^+. When exposed to light, the pH in the stroma rises from about 7 to 8, and the Mg^{++} concentration in the stroma rises from 1 or 2 mM to about 10 mM, both effects apparently increasing the activity of each carboxylating enzyme. Each enzyme requires Mg^{++} for maximal activity and functions best at about pH 8. Formation of the active form of ribulose diphosphate carboxylase from the inactive form also occurs more extensively in the presence of CO_2, an effect independent of the role of CO_2 as a substrate. In darkness, these changes in pH and magnesium Mg^{++}, and HCO_3^- are reversed, and the carboxylases become less active. Other examples of positive and negative allosteric control of these two enzymes by metabolites are also known.

10.5 CO_2 Fixation in Succulent Species

Numerous species living in arid climates have fleshy leaves with relatively low surface to volume ratios and accompanying low transpiration rates (Fig. 10-7). Such species are frequently referred to as **succulents.** As discussed in Chapter 3, many of these species open their stomates primarily at night and fix CO_2 into organic acids, especially malic acid, then. Figure 10-8 illustrates the daily cycle of transpiration and CO_2 fixation in one such species, *Agave americana*. Their metabolism of CO_2 is unusual and, because it was first investigated in members of the Crassulaceae, it is commonly called **crassulacean acid metabolism,** often abbreviated **CAM.** CAM has been found in 18 families, including the familiar Cactaceae, Orchidaceae, Bromeliaceae, Liliaceae, and Euphorbiaceae. It should be noted that not all CAM plants are succulents, and that some succulents (e.g., the halophytes, "salt lovers") do not possess CAM. Species with CAM usually lack a well-developed palisade layer of cells, and most of the leaf or stem cells are spongy mesophyll. Bundle sheath cells are present but, in contrast to those of C-4 plants, are quite similar to the mesophyll cells.

It was found several years ago that the first stable product of CO_2 fixation in CAM plants is malic acid, and that citric, isocitric, and other acids are derived from it, probably by reactions of the Krebs cycle. More recently, PEP carboxylase was found responsible for most of the CO_2 fixation at night, in contrast to its low activity in C-4 plants in darkness, and ribulose diphosphate carboxylase becomes active during daylight. The role of ribulose diphosphate carboxylase seems to be identical to its function in bundle sheath cells of C-4 species; that is, refixation of CO_2 lost from organic acids such as malic. A model consistent with what we know about CO_2 fixation in CAM plants is shown in Fig. 10-9.

Figure 10-9 shows that during darkness, starch breaks down by reactions of glycolysis (Chapter 12) as far as PEP. CO_2 is fixed into oxaloacetic acid by a PEP carboxylase, then this acid is reduced to malate by an NADH-dependent malate

Figure 10-7 Miscellaneous succulents. From left to right, *Opuntia*, *Aloe obscura*, *Echeveria corderoyi*, *Crassula argentea*, *Agave horrida*. (New York Botanical Garden photograph courtesy Arthur Cronquist.)

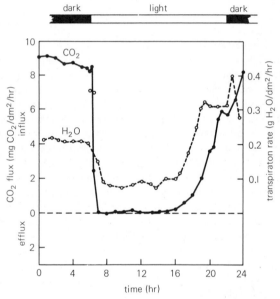

Figure 10-8 CO_2 fixation and transpiration rates of the CAM plant *Agave americana* during alternate light and dark periods. (From T. F. Neales, A. A. Patterson, and V. J. Hartney, 1968, Nature 219:469–472.)

The CO_2 released in the PEP carboxykinase reaction is trapped by ribulose diphosphate carboxylase, presumably activated by light, to form two 3-PGA molecules. These are subsequently reduced to sugars and starch with the help of NADPH and ATP formed in the light-driven reactions of photosynthesis discussed in Chapter 9. The PEP arising during the PEP carboxykinase reaction is also converted in part to sugars and starch, although some of it is respired and some converted to proteins, nucleic acids, fats, and so on.

In summary, CAM plants, like C-4 species, use both PEP carboxylase and ribulose diphosphate carboxylase in the overall process of CO_2 fixation. In C-4 plants, there is a spatial separation of the reactions, while in CAM plants, there is a time separation coupled to protection against drought, a remarkable physiological adaptation.

Although the ability of a plant to perform CAM is genetically determined, it is also environmentally controlled. In general, CAM is favored by hot days with high light intensities, cool nights, and dry soils, all of which predominate in deserts. Several species switch to a greater rate of CO_2 fixation in daylight by a C-3 photosynthetic mode after a rainstorm or when the night temperatures become higher. The stomates then remain open longer during the morning daylight hours (Neales, 1975; Ting, 1970).

10.6 Formation of Sucrose and Starch

In each of the three major pathways by which CO_2 is fixed, the principal end products accumulating in the light are usually sucrose and starch. Free hexose sugars such as glucose and fructose normally are much less abundant than sucrose in photosynthetic cells, although both glucose and fructose are present in combined form in sucrose, and starch contains only glucose units. In many of the grass species (especially those that originated in temperate zones, including the Hor-

dehydrogenase. Much of the malic acid is stored in the vacuole until daybreak, but some of it is converted to other acids by mitochondrial Krebs cycle reactions not shown. During daylight, these acids come out of the vacuoles, and malate is then dehydrogenated to form oxaloacetate again. Oxaloacetate is now decarboxylated to form CO_2 and PEP. This reaction is not, however, a simple reversal of the reaction in which CO_2 was originally fixed during the previous night. Instead, another enzyme called **PEP carboxykinase** catalyzes the reaction (Daly et al., 1977).

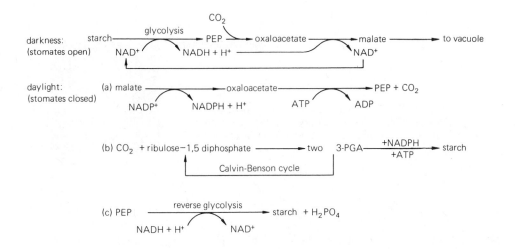

darkness: (stomates open)

daylight: (stomates closed)

(a) malate $\longrightarrow$ oxaloacetate $\longrightarrow$ PEP + CO_2

(b) CO_2 + ribulose–1,5 diphosphate $\longrightarrow$ two 3-PGA $\xrightarrow{\substack{+NADPH \\ +ATP}}$ starch

Calvin-Benson cycle

(c) PEP $\xrightarrow{\text{reverse glycolysis}}$ starch + H_2PO_4

Figure 10-9 A summary of CO_2 fixation in CAM plants.

deae, Aveneae, and Festuceae tribes) and also in a few dicots, starch is not a major accumulating product of photosynthesis, but sucrose and polysaccharides much smaller than starch and composed almost entirely of fructose units (usually 3 to 35), called **fructosans** (or, by some authors, **fructans**), predominate. In this section we will summarize the mechanisms by which sucrose and starch are formed from photosynthetic products. The structures and utilization of fructosans are discussed in Chapter 12.

Synthesis of Sucrose Reactions of sucrose synthesis are especially important, because this sugar is such a common and abundant carbohydrate food source in plants. It is also of great commercial importance, because of its unusual abundance in sugar beet roots and sugar cane stems. Free (unphosphorylated) hexoses usually occur in much lower concentrations than does sucrose, although levels of phosphorylated hexoses such as glucose-6-phosphate are sometimes quite high. As we explained in Chapter 7, sucrose is by far the most abundant sugar transported in the phloem sieve tubes of most plants, so a complete understanding of translocation necessitates knowledge of how and where it is synthesized. The Calvin cycle forms an abundant supply of 3-phosphoglyceraldehyde, some of which is transported out of the chloroplast, as indicated in the summary diagram of Fig. 10-2 (Heber, 1974; Bamberger et al., 1975). Some of these molecules are converted by reactions not shown into glucose-6-phosphate, which can be rearranged to yield either glucose-1-phosphate or fructose-6-phosphate. Fructose-6-phosphate and glucose-1-phosphate are two basic hexose units needed to form sucrose, but they cannot combine directly to form this sugar. Instead, a small fraction of the ATP molecules produced in the light-driven reactions of photosynthesis indirectly provides the energy for the combination reaction that occurs during sucrose formation outside the chloroplast.

The process begins with activation of the glucose unit in glucose-1-phosphate. Energy for this is not provided di-

rectly by ATP, but by a corresponding "high-energy" nucleotide triphosphate called **uridine triphosphate (UTP)**, as shown in R10-7. The products are pyrophosphate and a nucleotide-sugar named **uridine diphosphoglucose (UDPG)**. The pyrophosphate is then hydrolyzed by a pyrophosphatase enzyme (R10-8).

$$UTP + \text{glucose-1-phosphate} \rightleftharpoons UDPG + \text{pyrophosphate}$$
(R10-7)

$$\text{pyrophosphate} + H_2O \longrightarrow 2\ H_2PO_4^- \qquad (R10-8)$$

Enzymes capable of transferring glucose from UDPG to a large number of acceptor molecules exist in plants. Furthermore, an analogous nucleotide-sugar called **adenosine diphosphoglucose (ADPG)** also contains glucose that can be used for sucrose formation. In the reversible reaction by which sucrose is formed from UDPG (R10-9), frustose-6-phosphate is the acceptor molecule, and sucrose-6-phosphate is formed. The enzyme catalyzing this reaction is named **sucrose phosphate synthetase.** A glycosidic bond is formed between C-1 of glucose and C-2 of fructose-6-phosphate; these are the carbons that allow these two hexoses to reduce Cu^{++} to Cu^+ in Benedict's or Fehling's solutions, but sucrose is a nonreducing sugar because these carbons are bridged by oxygen (Fig. 10-10).

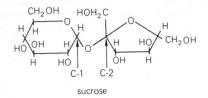

sucrose

Figure 10-10 The structure of sucrose, a disaccharide.

Exploring the Path of Carbon in Photosynthesis (I)

James A. Bassham

James A. Bassham, whom most of his friends call "Al," participated as a graduate student in elucidating the pathway of carbon in photosynthesis and is still working with Melvin Calvin as Deputy Director of the Laboratory of Chemical Dynamics at the University of California, Berkeley, where he is also Adjunct Professor of Biochemistry. He tells us his story:

When I think back to my days as a graduate student 30 years ago, the first two impressions that enter my mind are those of good fortune and excitement. The good fortune had to do with my becoming involved in that particular research at that time and place.

The chain of events that led me there began with my having to fill out a form for entry to the University of California at Berkeley and deciding to put down chemistry in the appropriate blank space because I had received a chemistry prize in high school. Then, during my freshman chemistry class at Cal one spring morning, our section instructor, Professor Sam Ruben, chose to put aside his usual laboratory lecture notes and tell us a little about his research. He and other scientists had discovered a new radioisotope of carbon, called carbon-14, and were putting it to many uses, of which the most interesting to me was as a tracer to map the pathway of carbon fixation by green plants making sugar and using sunlight. This was certainly more interesting than precipitating various sulfides with hydrogen sulfide on the porch of the freshman chemistry laboratory as part of my training in qualitative analysis. The advent of World War II, however, and three years of service in the U.S. Navy were to put such thoughts from my mind until a later date.

The third link in this chain of circumstances followed my return to the University at Berkeley as a graduate student in chemistry, after the close of the war. Since I had been an undergraduate student at Berkeley, the Chemistry Department was reluctant to admit me as a graduate student, it being their wise policy to encourage former undergraduates to go elsewhere for graduate training. After much discussion, they allowed me to enter for course work only. By some miracle, after the first semester had passed, the Dean invited me to stay on and work for my doctorate degree.

As is usual in such cases, I was directed to interview several professors of organic chemistry so that, hopefully, I would find one willing to accept me to do research on some project for my thesis. The first person I spoke to was a young professor named Melvin Calvin. By coincidence, the first topic he mentioned to me was the mapping of the path of carbon reduction during photosynthesis using carbon-14 as a label. I soon learned that Dr. Ruben had lost his life in an unfortunate laboratory accident during the war, and that the work on carbon-14 was now being carried forward in the newly formed

BioOrganic Chemistry Group of the University of California. Professor Ernest Lawrence, director of the Radiation Laboratory (now the Lawrence Berkeley Laboratory) had invited Melvin Calvin to form a division to explore the uses of carbon-14 in investigations of biochemistry and organic reaction mechanisms. The work was already under way.

Professor Calvin then proceeded to tell me of several projects involving organic synthesis and reaction mechanisms, making use of this radioisotope, but he might have saved himself the trouble because my mind was already made up. If at all possible, and if acceptable to him, I was very eager to get going on the work of using carbon-14 to study photosynthesis. He agreed to this and before long escorted me from the old red brick chemistry building across the court to an even older and shabbier wooden frame building that had been constructed as a "temporary" building about half a century before. As most university colleagues will know, temporary buildings built on university campuses are usually good for at least 50 years.

This old building turned out to be ideal for research in many ways. Its most recent former occupant had been Ernest O. Lawrence and, in fact, he still maintained an office in the building at that time. The 37-inch cyclotron still occupied a part of the building, along with mechanical and glass-blowing shops. With the subsequent removal of the cyclotron, a large uncluttered space became ours and was suitable for the kind of open laboratory interdisciplinary group that developed and has been maintained during the intervening years. Since the building was old and "temporary," we could make holes and run pipes and wires pretty much where we chose to, making any other modifications that suited our purposes. For example, when we found that stray radiation from a cyclotron in the Crocker Laboratory across the alley from us was interfering with our work by causing variable background in our counting chambers, we simply cut a hole in the floor, burrowed into the earth below, lined it with concrete and put our counters down there, where the background was much lower. Imagine the red tape and delay that would be involved in doing such a thing with a regular university building, with appropriate approvals by duly constituted committees!

In this building, called the Old Radiation Laboratory (ORL), I soon became acquainted with a small group of people who were to prove instrumental in the work, leading to the mapping of the path of carbon in photosynthesis. Of key importance among these was Andrew A. Benson, a young postdoctoral scientist, working with Professor Calvin on the photosynthesis project. Andy was an excellent experimentalist and taught me a great deal about how to devise apparatus and techniques for solving new kinds of problems in the laboratory. Of course, as those familiar with the mapping of the path of carbon in photosynthesis know, he played a very important role in the identification of various sugar phosphate intermediates, which were to turn out to be the essential compounds in the photosynthetic pathway. There were others who played important roles, visiting scientists, staff scientists, and students, but space will not permit me to describe them all.

Thinking about these former colleagues and the old laboratory brings to me my second main impression of that time: a sense of great excitement. The Old Radiation Laboratory was an exciting place to work, the research project was fascinating, and the people were all, without exception, enthusiastic about what they were doing. I suppose, to some extent, this is because we had a new and, at that time, almost exclusive technique and an important problem to which to apply it. It went far beyond that, however, and, to a large extent, stemmed directly from the personality of Melvin Calvin. He was in the laboratory every morning, or as soon as he could get free from his teaching duties, asking questions about the latest experiments, and laying out a program of new experiments to pursue. Never mind the fact that just the day before he had outlined an experiment which, in our opinion, would take a month or so, he would be in the next morning to ask the progress we had made with it. As we all know, this is an excellent procedure for stimulating graduate students. I don't mean to imply that he was a taskmaster; he was simply motivated by a tremendous enthusiasm for the project and the results we were getting. When we were fortunate to get a new experimental result, it always took on a new and greater importance after he had examined it.

This sense of achievement was further heightened when we developed techniques for analyzing the radioactive products of photosynthesis by using two-dimensional paper chromatography and radioautography. It was a beautiful analytical tool but suffered from one drawback: It required about two weeks from the original experiment until the films were developed to locate and count the radioactive spots. Of course, we had to learn to go on doing other experiments without waiting for the results of the previous one. Thus, there was a considerable sense of suspense when the X-ray films were developing in the dark room and the spots first began to appear before our eyes.

Some other phases of the research were even more painstaking and drawn out. For example, the degradation of molecules following short periods of photosynthesis to locate the radiocarbon within the molecule could take several weeks, or even months, before the final product was obtained.

Driven by our excitement to overcome these difficulties, we tended to work long hours and long weeks. When this reached an intolerable point and we felt the need for some mental rejuvenation, Andy Benson would organize an expedition to the high Sierra, and several of us would dash off to the mountains for a strenuous weekend of climbing 14,000-foot peaks. After such excursions, we returned to the laboratory physically exhausted but mentally refreshed. I suspect that Melvin Calvin was always relieved to see us return to the laboratory after these mountain and rock-climbing expeditions with no more serious infirmities than a few sore muscles or blisters on our feet.

About midway through the time when the path of carbon in photosynthesis was being mapped, I had done enough work on my part of the project to justify writing it up as a doctoral thesis. My main reluctance in doing this was the thought that I might then have to depart from the interesting project and find a job somewhere else doing something which was sure to be far less exciting. Fortunately for me, I was invited to stay on in a postdoctoral status and was able to remain a part of the team while the path of carbon in photosynthesis was fully mapped.

The techniques that we developed during that period proved to be extremely valuable for studies of metabolic regulation in plant cells and even in animal cells, and have shaped my whole scientific career since that time. To a large extent, we have been able to maintain in our laboratory a sense of excitement and cooperation over the years, in spite of two moves and growth from a dozen people to over a hundred. The person who has been most responsible for maintaining that sense of excitement and purpose is Melvin Calvin.

In R10-10, the phosphate is irreversibly hydrolyzed from sucrose-6-phosphate by a phosphatase enzyme to yield sucrose. Finally, in R10-12, the energy of ATP is used to regenerate a UTP molecule so that the process can continue. ADP is converted back to ATP by photosynthetic phosphorylation.

$$\text{UDPG} + \text{fructose-6-phosphate} \rightarrow \text{sucrose-6-phosphate} + \text{UDP} \quad \text{(R10-9)}$$

$$\text{sucrose-6-phosphate} + H_2O \rightarrow \text{sucrose} + H_2PO_4^- \quad \text{(R10-10)}$$

$$\text{UDP} + \text{ATP} \rightarrow \text{UTP} + \text{ADP} \quad \text{(R10-11)}$$

Another reaction (not shown) by which free fructose can accept glucose from UDPG, thus forming sucrose directly, is also known. It is catalyzed by **sucrose synthetase.** In photosynthetic cells, however, most sucrose is formed by action of sucrose phosphate synthetase and a phosphatase (R10-9 and R10-10).

A calculation of the energy cost to the plant in producing sucrose from two hexose phosphates can be calculated by adding the necessary reactions (R10-7 through R10-11). The net equation (R10-12) shows that one ATP is required for each sucrose:

$$\text{glucose-1-phosphate} + \text{fructose-6-phosphate} + 2\ H_2O$$
$$+ \text{ATP} \rightarrow \text{sucrose} + 3\ H_2PO_4^- + \text{ADP} \quad \text{(R10-12)}$$

Considering that three ATP molecules are required in the Calvin cycle for each carbon in each hexose (36 total ATPs), the one additional ATP needed to form the glycosidic bond in sucrose is a small energy requirement.

Formation of Starch The major food stored in most plant cells is starch. It is always found inside plastids, where it is formed by photosynthesis in most leaf chloroplasts, or in leucoplasts (amyloplasts) of storage organs where it is synthesized following sugar translocation from the leaves. Starch normally builds up in chloroplasts during daylight when photosynthesis is faster than translocation, then some of it disappears at night because of respiration and continued translocation.

You are familiar with many starch-rich plant organs, such as potato tubers, banana fruits, and the seeds of cereal grains so common in our diets. The crops from which these foods come have been selected for hundreds of years by man for their food content, but most of these could no longer survive in nature. Most perennial plants store starch before and during the dormant period, and this is used in regrowth the next growing season. In deciduous trees and shrubs, it is stored largely in the younger twigs, both in the bark and in living xylem parenchyma cells. Herbaceous perennials store starch in the roots, the base of the stem (the crown), or in underground bulbs or tubers. The cortex and pith cells are frequent sites of starch storage, both in annual and perennial stems. Maple sugar trees store starch in the xylem of its twigs and trunk during late summer and early fall, then convert this to sucrose and other sugars in early spring when it can be collected as maple sugar.

Two types of starch are present in most plastids, **amylose** and **amylopectin,** both of which are composed of D-glucose units connected in α-1,4-linkages (Fig. 10-11a). The α-1,4-linkages cause the chains to coil into a helix (Fig. 10-12). Amylopectin consists of branched molecules, the branches occurring between C-6 of a glucose in the main chain and C-1 of the first glucose in the branch chain. The number of glucose units present in various amylopectins is believed to range from 2,000 upward to 200,000. Amylose molecules are almost entirely unbranched and also contain a few thousand sugar units. Amylose becomes purple or blue when stained with an iodine-potassium iodide solution. Amylopectin reacts much less intensely with this reagent, and it exhibits a purple to red

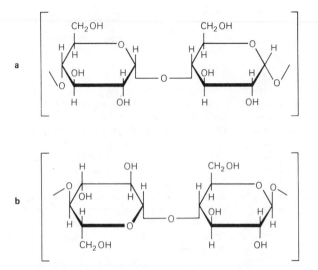

Figure 10-11 (a) The alpha (α) linkage between glucose residues, as in starch. (b) The beta (β) linkage between glucose residues, as in cellulose.

color. The iodine test is often used to determine the presence of starch in cells.

The percentage of amylose in starch from most species varies from 0 to about 40 percent (Akazawa, 1976). Potato tubers contain about 22 percent amylose and 78 percent amylopectin in their starch granules. These ratios are similar for starches of banana fruits and the seeds of wheat, rice, and field corn. The amylose content is genetically controlled, and some waxy varieties of cereal grains have 99 percent or more of amylopectin in their starch. The inheritance of waxy grains is controlled by a recessive gene, and only those homozygous for the gene produce amylose-free starch. In a few other plants, such as wrinkled peas, the starch is composed largely of amylose.

Formation of starch in chloroplasts from products of the Calvin cycle first involves the formation of glucose-1-P, just as is required for sucrose synthesis. Three enzymes are capable of forming the α-1,4 bonds in both amylose and amylopectin, but we are not certain which is usually most important. One of these enzymes, **starch phosphorylase,** catalyzes the reversible reaction R10-13:

$$\text{glucose-1-phosphate} \rightleftharpoons \text{starch} + H_2PO_4^- \quad \text{(R10-13)}$$

This reaction has an equilibrium constant in the forward direction of only about 3 at pH 7, a pH value about 1 unit below that of the stroma in illuminated chloroplasts. At pH 5, the K_{eq} is about 11, so starch formation is energetically more favorable at pH values considerably below those existing in chloroplasts when starch is actually formed. Whether starch formation or starch phosphorolysis is favored in the plastids also depends upon the existing ratio of $H_2PO_4^-$ to glucose-1-phosphate. These ratios are usually greater than 100 for the

AMYLOSE

AMYLOPECTIN

Figure 10-12 A schematic representation of starch molecules. Amylose and amylopectin are similar, except that amylopectin is branched.

entire cells and are suspected to be quite high in the plastids, too. Such high ratios make starch formation thermodynamically unfavorable by starch phosphorylase. The function of this enzyme is more likely to be that of starch degradation, although amylase enzymes might prove even more important in most cases of starch breakdown in plastids (see Section 12.2).

Two enzymes catalyzing essentially irreversible reactions also lead to α-1,4 bond formation in starch. These enzymes use an activated form of glucose rather than glucose-1-phosphate directly. One, **ADPG-starch transglucosylase,** transfers the glucose in ADPG, while the other, **UDPG-starch transglucosylase,** catalyzes a similar transfer of glucose from UDPG. Both enzymes are activated by K^+ ions, especially the ADPG-dependent one, and this is one of the reasons potassium is essential to plants. The generalized reaction for the transglucosylase is as follows:

$$\text{ADPG (or UDPG)} \rightarrow \text{starch} + \text{ADP (or UDP)} \quad \text{(R10-14)}$$

ADPG is apparently much more important than UDPG for starch synthesis in chloroplasts, and ADPG is also preferred in developing seeds. The same may be true in other storage tissues where starch accumulates in amyloplasts.

UDPG or ADPG can be formed directly from sucrose by analogous reactions summarized in R10-15.

$$\text{UDP (or ADP)} + \text{sucrose} \rightleftharpoons \text{UDPG (or ADPG)} + \text{fructose} \quad \text{(R10-15)}$$

The fructose released in this reaction is also efficiently converted (via glucose-1-phosphate) to starch, so both hexose units of sucrose are used for starch synthesis. A diagram of reactions by which starch is formed from sucrose is shown in Fig. 10-13. The net reaction can be summarized as in R10-16:

$$\text{sucrose} + 2\ \text{ATP} + H_2O \rightarrow \text{starch (2 glucose units)} + 2\ \text{ADP} + 2\ H_2PO_4^- \quad \text{(R10-16)}$$

This shows that the energy cost is two ATPs for each sucrose, or one for each glycosyl bond formed when one glucose is added to the growing chain of starch.

10.7 Formation of Cell Wall Polysaccharides

In growing plants, one group of unusually abundant products of photosynthesis is the cell wall polysaccharides. Before

Exploring the Path of Carbon in Photosynthesis (II)   **Melvin Calvin**

Melvin Calvin is a busy man. While working for the AEC (Atomic Energy Commission, now ERDA, the Energy Research and Development Administration) in 1974, I [F.B.S.] visited the laboratory he directs on two or three occasions, since it was one of the national laboratories supported largely by the AEC. We had a formal inspection of the lab that year—and found it to be administered in a unique way based on close trust and cooperation among the directing scientists but designed to frustrate thoroughly a Washington bureaucrat (which I was at that time)!

I told Professors Calvin and Bassham that we were planning to revise our text and asked if they would write essays for us, telling about the work on what is now called the Calvin cycle. They agreed, but Calvin was unable to find the time. However, on January 4, 1977, I recorded the following telephone conversation with him:

Frank B. Salisbury: How did you get into science?

Melvin Calvin: Well, it was a very practical consideration. When I was still in grade school, I was concerned with what my future livelihood would be. I looked around and decided that almost everything that I had contact with (for example, in a food store, the food itself and its processing, the cans and the making of them, the paper and the making of it, the dye stuff on the paper), everything had to do with chemistry. This prompted me to say: "I'll try to understand how the food gets made, how it's processed, and how it eventually reaches the grocery store." I realized that how the food got there was an essential activity for human survival, and that I had a good chance of finding a job if I knew anything about it.

F.B.S.: So it really was a practical consideration. But what aimed you at biology, especially plant physiology?

M.C.: Well, partly it was the food thing, and partly (much later) as I learned more of chemistry, it was the mysteriousness of how a plant could use sunshine to make food.

F.B.S.: So the practical approach became a bit of an intellectual challenge as well?

M.C.: Yes, exactly. You've got to make a living, but you want to do something interesting; if you can do both, you really have it made.

F.B.S.: What got you from the grocery store to Ernest Lawrence?

M.C.: That is a long, long trip, that one! I was an undergraduate at Michigan Tech, a graduate student at Minnesota, and then did postdoctorate work in England with Michael Polanyi. It was there that the real interest in photosynthesis, which came long ago, began to be executed. That's when I started working on chlorophyll and chlorophyll analogues: How do they work electronically? What kinds of things were they? Thus began the marriage of the practicality of food production and the intellectual challenge of energy conversion. So when I came to Berkeley, I started working on synthetic analogues of chlorophyll and heme. I was living in an environment in which radioactive isotopes were being turned up every day, and the idea of having the radioactive isotope of carbon was fairly rampant around the place. Work on the chemistry of the production of sugars wasn't something that happened because we had the carbon; the carbon was something we wanted, and we knew what we wanted it for long before we had it. Then Ruben and Kamen came along with both the carbons, carbon-11 and carbon-14. I was busy with porphyrin chemistry during the early war years, and my background in this and metal complexes led to my association with the Manhattan District for the development of methods of uranium purification and plutonium isolation. That's how I got to Ernest. And it came from chlorophyll, whether you like it or not! By that time, Sam Ruben had been killed, and Ernest said, "Well, you ought to do something with radiocarbon." I said, "Yes, I ought to," and I knew exactly what I had to do.

F.B.S.: What were your essential insights or acts of creativity?

M.C.: You mean in terms of the carbon cycle? Those came after a good deal of work. We knew the experiments that had to be done, and the mechanics of doing them were obvious. It was just a matter of having the material and the time, and with the end of the war we had both. Carbon-14 had appeared just before the war, so the moment we had the time and the opportunity to generate the radiocarbon (which we did in 1944–1945) the work began in earnest, I mean in a serious way! We made carbon-14 at the Hanford and Oakridge reactors. Everybody else did, too; we had lots of it. The experiments were easy to design. We did several, and by 1951 we had already mapped a good many of the early compounds on the way to sugar, although actual delineation of the whole sequence didn't occur until somewhat later.

F.B.S.: Putting it all together must have been the creative part.

M.C.: Yes. The first thing we saw were 3s [three-carbon intermediates], then we saw 5s and 6s and 7s. We didn't see the 4s until toward the end of the line. They were an essential part of the puzzle when we did find them. Putting it all together occurred in the early and middle 1950s. It came in bits and pieces and fits and starts. The first important step was the recognition of phosphoglyceric acid as the first product. We kept looking for a 2-C piece because the first thing we saw was 3, and it was logical that CO_2 should add to a 2. But we never did find a 2. As you know, there wasn't one! The reaction was a 5 + 1 that gives two 3s.

I saw that. I can remember being at home and reading some papers in JACS about the mechanism of decarboxylation of β-keto acids and of dicarboxylic acids (malonic acid and acetoacetic acid), completely mechanistic studies in organic chemistry. I then recognized how a CO_2 would add to a β-keto sugar. It was the recognition that ribulose diphosphate was the acceptor of the CO_2 that allowed me to finally draw the whole thing out. It was done almost all at once, because the pieces and parts had been accumulating for several years.

F.B.S.: That was the white heat of creation?

M.C.: Yes. The actual drawing of the reaction was on a scrap of paper right beside me where I was reading. The article prompted my thinking of the reverse decarboxylation reaction and what I had to do it with. The ribulose diphosphate was the only thing I had to do it with. I tried it with fructose diphosphate, but you come out with one carbon too many. So it had to be a five-carbon and not a six-carbon keto sugar. We had already found the five-carbon sugar, but I didn't know what it was doing there. I know which chair I was sitting in when I put it together! Once you hit it, you know it's right. Everything fits when you hit the right key. And I know when that happened! It was quite an exciting few minutes. The next step was to come back in the lab and pick up the missing pieces, which we did.

F.B.S.: You were the director of the lab with several graduate students and postdocs working with you. How did that interaction work?

M.C.: Oh, that was wonderful! I enjoyed that immensely. That was the best part of the thing, you know. Every day I would come in and say, "What's new?" We would review and then see what the next experiment had to be. Usually there were anywhere from one to six people involved at any one time.

F.B.S.: Were you involved in the laboratory part of the work?

M.C.: Oh, yes. It was a joint effort, the design of the experiments. Some of them I would carry out myself. Some were easy! For example, the identification of phosphoglyceric acid I did personally. And I did it by the way the stuff behaved on an ion exchange column a year or two before we had paper chromatography. One of the graduate students designed a transient experiment with CO_2. Alec Wilson was a boy from New Zealand and did a beautiful job! The experiment pointed the finger at ribulose, because when you shut off the CO_2, the ribulose diphosphate rises. Shutting the light on or off was an easy experiment. But to shut off or turn on the radioactive CO_2 wasn't so easy. We had to build a very special apparatus, and he did it. It was a beautiful technical accomplishment.

F.B.S.: Clearly, this was one of the classical examples in our time, of which there must be many now, of a group interaction, with a bunch of you working together in a moment of excitement.

M.C.: Yes. On a daily basis. Every day and every night. A graduate student would be there three years, an undergraduate maybe one, a postdoc one or two. So there was a constant change of people, a replenishment of the enthusiasm of the lab. I guess that's part of the reason why I enjoy this kind of life. You'll understand that.

F.B.S.: That was the beginning of your development of a "laboratory in the round," which I saw when I visited you. Tell me about that.

M.C.: That's exactly right. We started the lab in what was called the Old Radiation Laboratory, where Ernest Lawrence had originally built the 37-inch cyclotron. He had used an old wooden building and taken all the walls out, so when the cyclotron was moved out right after the war, there was a big old wooden building with no walls in it. We put in chemical benches, a counting room, and things like that. This led to a large laboratory in which there were no rooms that separated people from each other. There were 20 people in one large room! The interaction between them was a perfectly natural one and one that, when it came time for that building to be destroyed, we wanted to recreate in the new building. That's how our round building was born. We moved in 1962.

F.B.S.: Tell me about your Thursday lunch system of administration.

M.C.: Every week we have lunch together with the five to seven senior staff, each one of which has a research activity of his own and is responsible for some function of the lab. Those luncheon meetings are where the normal administrative decisions are made, with respect to how many people we can take next year, how much money we can spend for a piece of equipment, and so on. Of course, you can't

help but get into science, but predominantly it's an effort to deal with administrative problems of running a lab with about 120 people in it. This system helps you to make decisions that you shouldn't make alone anyway. It works. Everybody in the building knows what's going on, both scientifically and in terms of the budget. I find that some of our federal agencies are beginning, by the way they operate, to destroy this approach. Only AEC and ERDA, so far, have found it possible to make a grant to the lab as a whole rather than to individuals. You can get institutional grants from some agencies, but it is a big thing, which requires almost an act of Congress.

F.B.S.: Your lab was the only one administered by the AEC that received only four large grants to cover your multitude of projects. Other national laboratories had to break their budgets into individual projects.

M.C.: We keep fighting that, because the pressure is now greater than ever to have each individual apply directly to the federal source. That means that there is no unity in the operation of the facility. It becomes no more than just a collection of individuals. I'm beginning to see that now, from farther away; if I were younger I would fight it much harder than I can now. Fighting takes so much out of you!

F.B.S.: Two final questions: Just what is the nature of creativity in science? What price do you have to pay, and what are the rewards?

M.C.: I have a phrase in response to that to my students. I tell them that it's no trick to get the right answer when you have all the data. A computer can do that. The real creative trick is to get the right answer when you only have half enough data, half of what you have is wrong, and you don't know which half is wrong! When you get the right answer under those circumstances, you are doing something creative! The students learn after a while that this is not a joke. You must be able to sift out the critical points and put it all together in the right way, ignoring what doesn't seem to fit right. If you ignore the right things, you come out with the right answer! But if you pay attention to the wrong things, you don't.

F.B.S.: It's not a mechanical thing; if you have only half the data, you've got to sense or feel your way to the right answer.

M.C.: One which fits your concept of the physical world. The process involves intuition. You can't do it on a part-time basis, obviously. It is done day and night, winter and summer, under all circumstances, at any given moment. You can't tell where you will be when you are doing it. It does intrude on what some people call their private lives. In fact, I don't see how it can work otherwise. So, you ask for price. The price usually is paid by the person's family in the form of neglect and competition. I don't see any way to beat that.

F.B.S.: So what are the rewards? Are they worth it?

M.C.: The satisfaction of guessing correctly and then showing that it is correct is really great! You can usually tell when you are right, even before you've proved it; that's when the satisfaction is the greatest. When something is born and you *know* it is right; then you set about for the next 10 years trying to show it's right. It sounds as though you're making the discoveries over a period of 10 years; you're really not. You make it all at once, and then you spend a long time showing that this is the way it is. Even today, I have some ideas about how photoelectric charge separation occurs—a conversion of solar energy—and I'm beginning to see physically how to do it. We will simulate the thing in a synthetic system within a decade, I would say, probably less. How soon it will be economic in terms of building useful gadgets is another matter. But it's on its way. That's the great sport.

a leaf is about half grown, nearly all of the carbohydrate it contains is either converted to starch in the chloroplasts or used to form cell wall materials and protoplasmic constituents. During the last half of its growth, the leaf exports sucrose or other translocatable carbohydrates via the phloem to feed younger leaves, stems, roots, flowers, seeds, and fruits. In the cells of these organs, much of the imported carbohydrate is converted to cell wall materials. Let us first investigate the principal components in walls of growing cells (primary walls), then we shall summarize what we know and don't know about how their synthesis occurs. Relations of wall structure to growth will be deferred until Chapter 15, while secondary walls are discussed briefly in Chapter 14.

Early chemical analyses of primary walls indicated that three major fractions existed, based primarily on relative solubilities in acids and bases. These fractions were identified as **pectins** (especially abundant in the middle lamella), **hemicelluloses,** and **cellulose.** Modern, improved techniques show that only cellulose is a fairly discrete chemical entity, while the other fractions contain rather complex mixtures of polymers of sugars and sugar derivatives that are now being identified and named. Let us examine these sugars that cell walls are made of.

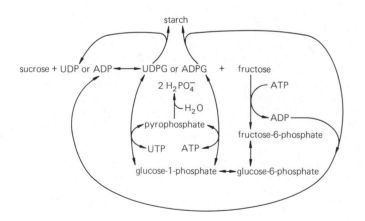

Figure 10-13 Summary reactions by which sucrose is converted into starch.

The Sugar Building Blocks of Primary Cell Walls In all primary walls of angiosperms and gymnosperms for which reliable analyses are available, polymers of only five sugars and five sugar derivatives occur in appreciable quantities. Thus, there are fewer different subunits in cell wall polysaccharides than in proteins (18 to 20 different amino acids). The number of ways in which these sugars might be connected to each other is large, because each hydroxyl group represents a potential branch point, yet the actual complexity of cell wall polysaccharides is not nearly as great as one might expect.

The five sugars are D-**glucose,** D-**mannose,** and D-**galactose,** all hexoses, and two pentoses, D-**xylose** and L-**arabinose.** No ribose, deoxyribose, ribulose, or other sugars have been detected. The sugar derivatives include two uronic acids, D-**glucuronic acid** and D-**galacturonic acid** in which the carbon-6 of glucose and galactose has been oxidized to a carboxyl group, and two deoxy sugars, L-**rhamnose** (6-deoxy-L-mannose) and L-**fucose** (6-deoxy-L-galactose), in which carbon-6 has been reduced. One important additional complexity is also evident: The carboxyls of many of the D-galacturonic acid residues in the classic pectin fraction are esterified with methyl groups to form the fifth sugar derivative shown at the lower right in Fig. 10-14. The nonesterified carboxyl groups of D-glucuronic acid and D-galacturonic acid are largely ionized in the walls and are associated with cations, especially calcium and magnesium but presumably also potassium. Except for glucose, these sugars or the derivatives mentioned are almost never found in the free or uncombined form in plant cells.

Although proteins were not considered to be part of primary cell walls by earlier investigators, modern chemical procedures employed by Peter Albersheim at the University of Colorado (see his personal essay at the end of this chapter) and by others show that glycoproteins are always present. Proteins comprise at least 10 percent of the weight of most primary walls. The functions of these wall proteins are not yet understood. They might represent structural proteins or enzymes involved in assembling polysaccharides into the wall framework.

Polysaccharides of Primary Walls The least complex wall polysaccharide is **cellulose.** Cellulose molecules are excep-

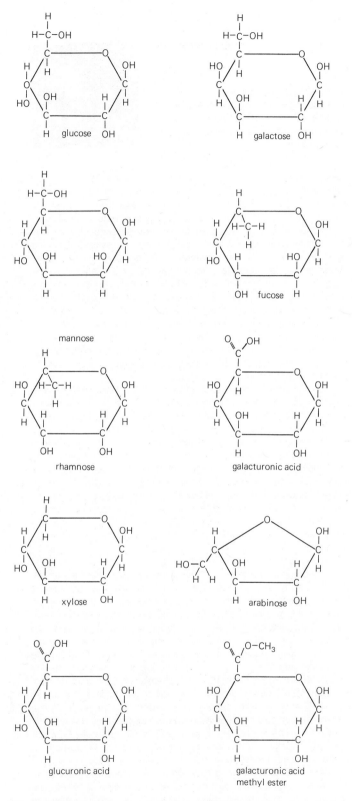

Figure 10-14 The sugars and sugar derivatives found in primary wall polysaccharides.

tionally long, unbranched, and not coiled, and are assembled from a variable number of D-glucose units, frequently 2,000 or more in primary walls (and at least 14,000 in some secondary walls). These glucose units are attached between C-1 of one glucose and C-4 of the next. The linkage is of the β type, as illustrated in Fig. 10-11b.

Approximately 40 cellulose chains, each of which apparently contains a pair of cellulose "molecules," are held to each other by hydrogen bonds and packed together along their long axes to form **microfibrils** (Fig. 10-15). These microfibrils are highly ordered and behave enough like crystals so that X-ray diffraction and other techniques of crystallography can give information about them. Each of the 80 or so cellulose "molecules" making up a given microfibril behaves somewhat like an individual strand in a steel cable and is as strong, for its weight, although the cellulose molecules are not wound around each other as cable strands are. Together, they strongly resist stretching. The importance of microfibril orientation in the wall to growth is discussed in Chapter 15. Some microfibrils extend for long distances threaded over and under others, oriented either parallel to the long axis of the cell, more nearly around the cell like hoops around a barrel, or at some intermediate orientation. Embedded between the microfibrils and forming a more nearly amorphous matrix lie most of the other polysaccharides, the classical pectin and hemicellulose fractions, and the wall protein.

One of the best-studied walls exists in the dicotyledon *Acer pseudoplatanus* (sometimes called sycamore-maple). It consists of five major structural components: 23 percent cellulose, 21 percent **xyloglucan** (a xylose-glucose hemicellulose polymer), 20 percent **arabinogalactan** (an arabinose-galactose hemicellulose polymer), 16 percent **rhamnogalacturonan** (a rhamnose-galacturonic acid pectin polymer) and 10 percent of a protein rich in the unusual amino acid **hydroxyproline.** The single gymnosperm so far analyzed (Douglas fir) has about as much cellulose (22 percent) as sycamore-maple, while several members of the grass family contain only from 9 to 13 percent cellulose. These monocots have compositions quite similar to one another, but quite different from the dicots. Their wall proteins have only traces of hydroxyproline.

The chemical analyses described here were performed in Albersheim's laboratory, and he has suggested a model (1976) to explain how the various constituents of primary walls are assembled to form the whole wall. The xyloglucan appears to be hydrogen bonded to hydroxyl groups along the cellulose chains, while all other constituents are covalently bonded to each other. In a simplified form, the model is as follows:

> -cellulose-xyloglucan-arabinogalactan-
> rhamnogalacturonan-arabinogalactan-protein-
> arabinogalactan-rhamnogalacturonan-
> arabinogalactan-xyloglucan-cellulose-

Each cellulose microfibril is therefore hydrogen bonded on either side to a xyloglucan, so an entire network or "giant molecule" is produced.

Try a Little Diversity in Your Science

Peter Albersheim

F.B.S. 1975

It is often said that a scientist learns more and more about less and less until he knows everything about nothing. (An administrator takes an opposite approach until he knows nothing about everything!) Actually, a successful modern scientist usually achieves his success because of his breadth, not his narrowness. Peter Albersheim, Professor of Chemistry at the University of Colorado, widely recognized in at least two fields (cell walls, plant pathology), got to thinking about these matters and produced for us the following essay:

While we were riding the double chair lift at the Keystone, Colorado ski resort, Frank Salisbury asked me to prepare a short essay for this textbook. How could I refuse? Indeed, I was honored by the realization that I would be joining the authors of such essays in Frank's earlier botany book: my former mentors, James Bonner and Frits Went, and such leaders in the field of plant physiology during the period of my graduate studies as Sterling Hendricks, Johannes van Overbeek, and Kenneth Thimann. In considering these individuals, it became clear to me that I was in the next generation of scientists, and I asked myself what was different between this generation and the last and what were the factors that allowed our laboratory to make contributions that must have led Frank to ask me to write one of these essays. The answer becomes quite clear in my own mind: The breadth of background in our laboratory has permitted us to make successful attacks on interesting problems in plant physiology.

The diversity in our laboratory must have developed from my own training. I have a degree in agriculture from Cornell University, but I elected to take sufficient chemistry courses to have the equivalent of a chemistry major. My formal training for the Ph.D. degree at Cal Tech was routine enough: I majored in biochemistry with a minor in chemistry. But the whole emphasis of Cal Tech is to consider every possible area of science, and the interaction of biochemists, geneticists, ecologists, behavioralists, chemists, and the like was impressive indeed. In addition, I lived for two years in Cal Tech's equivalent of a faculty club, where on most evenings I ate dinner at one of the round tables with seven scientists representative of the entire spectrum of the Cal Tech community. As a postdoctoral fellow in Switzerland, I published papers on electron microscopy in one botany department, on plant pathology in a second botany department, and on biochemistry in a department of agricultural chemistry. At Harvard I was forced to learn about really new areas of science, for I participated in the teaching of general biology to 350 freshmen. I was asked to lecture on such topics as animal embryology and endocrinology. When it came time to select a more permanent position, I deliberately selected the Department of Chemistry at Colorado over more biologically oriented departments at other institutions. I selected this department because I felt that this was a critical area of my own background where I was most lacking. However, I kept my formal contacts with biology by holding, as well, an appointment in the Department of Molecular, Cellular and Developmental Biology at the same university. For that department, I was charged with developing and presenting the first semester's lectures for their year-long general introductory course. Finally, I have also had the experience of teaching molecular genetics and a wide variety of biochemistry courses.

The breadth of my own background is not sufficient to make me an expert in all of the areas required to solve the plant science problems that our laboratory is attacking. Diverse training, however, does allow me to talk with ease on a meaningful scientific level to such a range of specialists as plant pathologists, ecologists, geneticists, and theoretical chemists. It is undoubtedly my ability to interact with these individuals that has attracted a number of them to my laboratory.

Our laboratory is best known for our contribution to the deciphering of the structure of plant cell walls. The major breakthrough in this area came from a three-year intensive collaborative effort between myself and three graduate students, Wolfgang D. Bauer, Kenneth Keegstra, and Kenneth Talmadge. Dr. Bauer had already spent four years as a graduate student in physical chemistry and electronics. Dr. Keegstra had done independent research in organic chemistry, and Dr. Talmadge had had professional experience in analytical chemistry and was enrolled as an analytical chemistry graduate student when he joined our laboratory. It took the combined talents of all of us to develop the methods and technology necessary to solve this important structural problem.

The success of the collaborative efforts of these students was made evident by their winning the 1973 George Olmstead Award of the American Paper Institute. These men have obviously taken the emphasis on diversity to heart. Dietz Bauer, the physical chemist turned biochemist, obtained postdoctoral training in plant pathology and today is an independent scientist at a research institute working on pathology problems. Kenneth Keegstra, the organic chemist, is today an assistant professor in a microbiology department in a medical school working on the cell surfaces of viruses and virally transformed animal cells. Kenneth Talmadge, the analytical chemist, is carrying out independent research in a hospital on the effects of female hormones on the nuclear proteins of the receptor cells of these hormones.

Today, my own laboratory retains its diverse character. In addition to a group of graduate students whose backgrounds are as different as those of Bauer, Keegstra, and Talmadge, the postdoctoral fellows presently in the laboratory have Ph.D.'s in plant physiology, food technology, plant pathology, biochemistry, and physical chemistry. The physical chemist has a background in electrical engineering equivalent to that of a Ph.D. and has had experience in computer programming in industry, experience that we are taking full advantage of. Another important member of our group is a research chemist whose background combines organic chemistry and enzyme kinetics. Our interests are being further widened by collaboration with a variety of investigators in other laboratories. These include an analytical chemist and an organic chemist in our own department, an immunologist in a medical school, and three plant pathologists and a botanist-plant physiologist at four other universities. The excitement generated by these projects is catalytic and wonderful to experience. The diverse personnel has enabled us to broaden our own research interests to include not only cell wall structure and the mechanisms of cell elongation, but also to launch a major attack on the molecular basis of

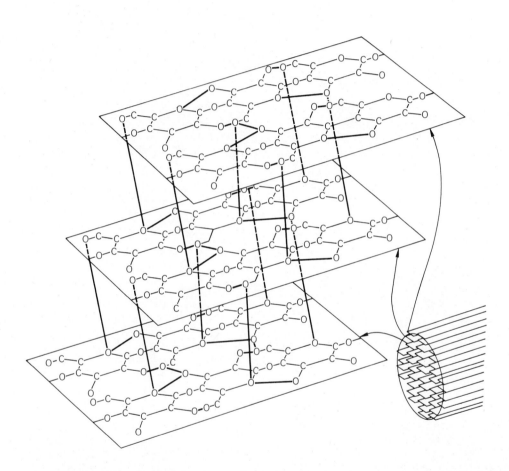

Figure 10-15 Model to explain the arrangement of cellulose molecules in a microfibril. Each molecule in such a sheetlike pair is held to its partner and to similar molecules above and below by hydrogen bonds (heavy lines). For simplicity, no H atoms are shown. About 40 such sheetlike arrays occur in each microfibril (lower right). The pattern of H-bonding is tentative. (From "The Walls of Growing Plant Cells" by P. Albersheim. Copyright © 1975 by Scientific American, Inc. All rights reserved.)

It is interesting that *cellulose is actually a minor constituent,* representing one fourth or less of the weight of the primary wall of these species. This is to be contrasted to secondary walls, which contain approximately one half cellulose and one half lignin. It is because of its high concentration in secondary walls of xylem (wood) that cellulose is the most abundant organic compound on earth. It is perhaps unfortunate that humans cannot digest cellulose, because this would help solve the world's food problem. Plants have cellulose-hydrolyzing enzymes called **cellulases,** but these are not active enough to allow cellulose to act as a food, so it remains intact for long periods and is not used by the plant as an energy source. Certain bacteria and fungi more readily digest cellulose and the other wall polymers when leaves and other dead parts of plants fall to the ground. These organisms do use wall polysaccharides as a food source, thereby returning CO_2 to the atmosphere. Attempts are being made to use bacterial cellulases to hydrolyze cellulose from sawdust and other wood products into nutritionally available glucose molecules. Such glucose can be fermented to ethanol, which might someday compete with gasoline as a fuel.

Perspectives in Cell Wall Formation Now that we know something about the composition of primary cell walls, we should be able to learn how they are synthesized from earlier products of photosynthesis. This problem will not be easy to solve, however, because of the plasma membrane barrier

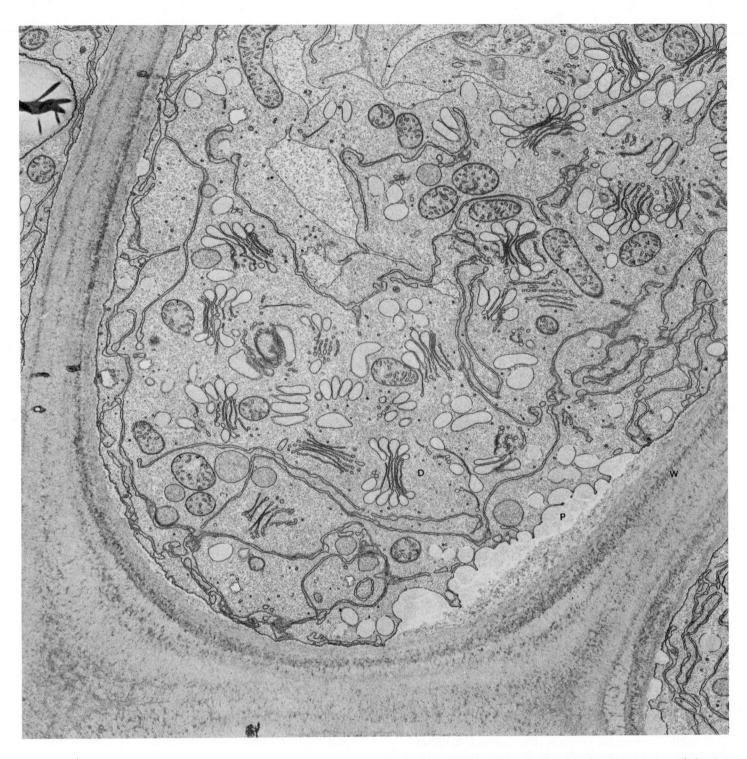

Figure 10-16 Electronmicrograph of a corn root cap cell showing fusion of secretory products (P) pinched off dictyosomes (D). The plasma membrane and cell wall (W) are also shown. (From D. J. Morre, D. D. Jones, and H. H. Mollenhauer, 1967, Planta 74:286–301.)

between the cytoplasm, where polysaccharide precursors are formed, and the final site of assembly of these precursors. The noncellulosic polysaccharides are not synthesized in the wall but in **dictyosomes** of the **Golgi apparatus**. Tiny vesicles (membrane-bound sacs) pinch off the ends of the dictyosomes and somehow move to the plasma membrane where the membrane of each vesicle fuses with the plasma membrane, accompanied by secretion of the polysaccharide contents into the wall. Figure 10-16 illustrates this process for the root cap cells of corn.

This activity of dictyosomes presumably increases the area of the plasma membrane and supplies polysaccharides that are then assembled into the wall. This secretion of pectins and hemicelluloses is unusually active in root cap cells. In these cells, some of the polysaccharides become attached to the outside of the walls in a slimy layer, while some move on into the soil where microorganisms presumably thrive upon them.

Although noncellulosic polysaccharides are clearly present in dictyosome vesicles and transported therein to the walls, this is apparently not true for cellulose. Where it is synthesized and how it is assembled in the wall remains unsolved (Northcote, 1974; Karr, 1976).

An important generalization concerning synthesis of wall polysaccharides is that all of the sugars incorporated into them must first be combined with a nucleotide such as UDP or guanosine diphosphate (GDP) before they can be polymerized. Thus, UDPG and a similar nucleotide-sugar called **guanosine diphosphoglucose (GDPG)** both act as effective glucose donors for synthesis of cellulose-like polymers in studies carried out *in vitro*. UDP-galactose, UDP-galacturonic acid, UDP-glucuronic acid, UDP-xylose, UDP-L-rhamnose, and UDP-L-arabinose all occur in plant cells, and these are effective sugar donors for *in vitro* production of certain polysaccharides. Reactions by which most of these nucleotide-sugars can be synthesized from UDP-glucose are well known (Gander, 1976).

Unfortunately, we are still a long way from understanding just how these reactions occur in the cell, and especially what controls the kinds of polymers produced in different species. How proteins get into the walls and what they do there are also interesting problems yet to be solved. Finally, as we shall discuss in Chapter 18, it is essential that we learn what dictates the pattern of cellulose microfibril deposition, because this pattern controls the direction of cell growth and thus influences the final shapes of cells, organs, and plants.

Photosynthesis: Environmental and Agricultural Aspects

The atmospheres of Jupiter, Saturn, and Uranus contain vast amounts of hydrogen, helium, ammonium, and methane; they are reducing atmospheres. The earth's atmosphere, on the other hand, with its 21 percent O_2, is a highly oxidizing atmosphere. This not only makes animal life as we know it possible, but it strongly influences almost everything on the earth's surface. The rocks and the soil are highly oxidized. Furthermore, O_2 high in the stratosphere is converted by ultraviolet light to ozone (O_3), which then absorbs the ultraviolet radiation that would otherwise be highly damaging or lethal to many of the earth's organisms. What is the origin of the O_2 that makes our kind of life possible? A small amount is produced by the photolysis of water vapor into O_2 and H_2, followed by the subsequent escape of H_2 into space, but most of the O_2 in our atmosphere is produced by photosynthesis (Cloud and Gibor, 1970).

The amount of photosynthesis presently occurring on earth is a staggering figure. Estimates vary greatly, but a value of about 70 billion metric tons of carbon (170 billion tons of dry matter with an empirical formula close to CH_2O) fixed

Table 11-1 Estimates of Net Primary Productivity and Plant Biomass for Various Ecosystems[a,b]

Ecosystem Type	Area (10^6 km²)	Net Primary Production, per Unit Area (g/m²/yr)		World Net Primary Production (10^9 t/yr)	Biomass per Unit Area (kg/m²)		World Biomass (10^9 t)
		Normal Range	Mean		Normal Range	Mean	
Tropical rain forest	17.0	1000–3500	2,200	37.4	6–80	45	765
Tropical seasonal forest	7.5	1000–2500	1,600	12.0	6–60	35	260
Temperate evergreen forest	5.0	600–2500	1,300	6.5	6–200	35	175
Temperate deciduous forest	7.0	600–2500	1,200	8.4	6–60	30	210
Boreal forest	12.0	400–2000	800	9.6	6–40	20	240
Woodland and shrubland	8.5	250–1200	700	6.0	2–20	6	50
Savanna	15.0	200–2000	900	13.5	0.2–15	4	60
Temperate grassland	9.0	200–1500	600	5.4	0.2–5	1.6	14
Tundra and alpine	8.0	10–400	140	1.1	0.1–3	0.6	5
Desert and semidesert scrub	18.0	10–250	90	1.6	0.1–4	0.7	13
Extreme desert, rock, sand, and ice	24.0	0–10	3	0.07	0–0.2	0.02	0.5
Cultivated land	14.0	100–3500	650	9.1	0.4–12	1	14
Swamp and marsh	2.0	800–3500	2,000	4.0	3–50	15	30
Lake and stream	2.0	100–1500	250	0.5	0–0.1	0.02	0.05
Total continental	149		773	115		12.3	1,837
Open ocean	332.0	2–400	125	41.5	0–0.005	0.003	1.0
Upwelling zones	0.4	400–1,000	500	0.2	0.005–0.1	0.02	0.008
Continental shelf	26.6	200–600	360	9.6	0.001–0.04	0.01	0.27
Algal beds and reefs	0.6	500–4,000	2,500	1.6	0.04–4	2	1.2
Estuaries	1.4	200–3,500	1,500	2.1	0.01–6	1	1.4
Total marine	361		152	55.0		0.01	3.9
Full total	510		333	170.		3.6	1,841

[a]From Whittaker, 1975.

[b]Units are square kilometers, dry grams or kilograms per square meter, and dry metric tons (t) of organic matter. One metric ton equals 1.1023 English tons.

per year was obtained relatively recently. Table 11-1 lists estimates of photosynthetic productivity for specific types of ecosystems. Roughly two thirds of this productivity occurs on land and only one third in the oceans. This vast productivity occurs in spite of the low CO_2 concentration in the atmosphere (only about 0.032 percent by volume, or 320 ppm). Most of the CO_2 is ultimately converted into cellulose, the major component of wood. At present, the major portion of known carbon is in the oceanic carbonate deposits, including sedimentary limestones and dolomites. Oil shale, coal, and petroleum are also important reservoirs. The air has less carbon in it than the other sources, but is still estimated to be equal to 700 billion metric tons (Bolin, 1970). Thus, the equivalent of 10 percent of the carbon is used in photosynthesis each year.

The amount of CO_2 in the air has increased slowly over the last hundred or so years (apparently since significant burning of fossil fuels began) and is presently rising at the rate of about 0.7 ppm per year. These small increases show either that return of the gas to the air is closely balanced by its use in photosynthesis or that its concentration is otherwise maintained fairly constant. Important return agencies are the respiration of plants, microorganisms, and animals and the activity of volcanoes, factories, and automobiles (Fig. 11-1), but even more important in helping maintain a fairly constant atmospheric CO_2 level are the oceans. Here the dissolved carbonates are in equilibrium with CO_2, a change in one eventually affecting the other, so the concentration of CO_2 on a worldwide scale is buffered by the oceans. Nevertheless, the CO_2 content may fall to 260 ppm or less during daylight hours around the growing plants in a corn field, for example, while the content there may reach 400 ppm during darkness as a result of respiration by plants and soil microbes.

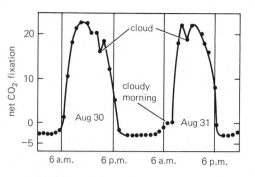

Figure 11-2 Photosynthesis in an alfalfa plot over a two-day period in late summer. The effect of periods of cloud cover can be noted. Negative CO_2 fixation values during hours of darkness indicate the respiration rates. (From data of M. Thomas and G. Hill, 1949, in J. Franck and W. E. Loomis, eds., Photosynthesis in plants, Iowa State University Press, p. 35.)

11.1 The Daily Course of Photosynthesis

Figure 11-2 illustrates the rather typical (except for CAM plants) daily course of CO_2 fixation by photosynthesis and its release in respiration at night in a plot of alfalfa. Certain interesting facts can be derived from this graph that will receive further attention in this chapter. First, the maximum rate of CO_2 fixation occurs about noon, when the light intensity is highest. That light is often limiting photosynthesis is also shown by the reduced rates when the plants were briefly exposed to cloud shadows. The figure also shows the relative magnitudes of photosynthesis and dark respiration; in this example the photosynthetic rate reached a maximum of about eight times the night respiratory rate.

11.2 Photosynthetic Rates in Various Species

The photosynthetic rates of various species living in such diverse conditions as arid deserts, high mountains, and tropical rain forests differ greatly (Larcher, 1969; Black, 1973; Cooper, 1975). This is partly due to differences in environmental conditions such as light, temperature, and CO_2 availability, but individual species show remarkable differences under specific conditions optimum for each. Species possessing the C-4 pathway of CO_2 fixation have the highest photosynthetic rates so far measured, while slow-growing desert succulents exhibiting crassulacean acid metabolism (CAM) have among the slowest rates. Table 11-2 summarizes the approximate range of maximum values for a few major groups of plant types representing many different species. Relatively few data are available for CAM plants.

Another interesting group of plants, for which even fewer results are available, includes numerous perennial

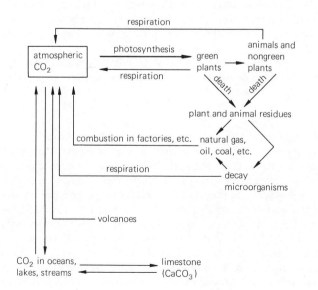

Figure 11-1 The carbon cycle in nature.

Table 11-2 Maximum Photosynthetic Rates of Major Plant Types Under Natural Conditions

Type of Plant	Example	Maximum Photosynthesis (mg CO_2/dm²/hr)[a]
CAM	*Agave americana* (century plant)	1–4
Tropical, subtropical and Mediterranean evergreen trees and shrubs; temperate zone evergreen conifers	*Pinus sylvestris* (Scotch pine)	5–15
Temperate zone deciduous trees and shrubs	*Fagus sylvatica* (European beech)	5–20
Temperate zone herbs and C-3 pathway crop plants	*Glycine max* (soybean)	15–30
Tropical grasses, dicots, and sedges with C-4 pathway	*Zea mays* (corn or maize)	35–70

[a]Values are calculated on the basis of one surface of the leaf; for conifers, data are for the optical projection of needles. Other common units include g CO_2/hr/m² leaf area and mg or μmol CO_2/mg chlorophyll/hr. It is frequently difficult to convert data from a leaf area basis to a chlorophyll basis or vice versa, but approximate conversions can be made. For many broad leaves, one dm² of leaf surface corresponds to about 2 g fresh weight of tissue, and this frequently contains about 3 mg of total chlorophyll (1.5 mg chlorophyll/g fresh weight). Whenever possible, photosynthetic data in this text will be expressed as mg CO_2/dm²/hr. An extensive list of tree photosynthetic rates was compiled by Larcher, 1969.

alpine and arctic species. These usually have short growing seasons, the alpine species with moderate daylengths and high light intensities and the arctic species with long daylengths and low light intensities. Their photosynthesis exceeds respiration so much that they can double in dry weight within a month or less, so carbohydrate accumulation is apparently not a survival problem.

11.3 Environmental Factors Affecting Photosynthesis

Clearly, many factors enter into the photosynthetic equations and might be expected to influence photosynthesis: H_2O, CO_2, light, nutrients, and temperature, as well as the age and genetics of the plant. Which most limits photosynthesis in natural or agricultural ecosystems? From Table 11-1, we may conclude that higher plants apparently are most limited by availability of water. Deserts are extremely unproductive, while marshes, estuaries, and tropical rain forests are by far the most productive ecosystems, with the exception of certain suitably irrigated agricultural crops. As we have already seen (Section 3.6), when water potentials become more negative (that is, when water becomes limiting), cellular expansion is first retarded so that plant growth is reduced. With only a little more water stress, stomates begin to close, CO_2 uptake is restricted, and photosynthesis is limited. These relationships are examined in more detail in Chapter 24.

Table 11-2 suggests that two other factors are important in natural ecosystems: First, alpine and arctic tundras have low productivity, mainly because of the low temperatures and short growing seasons of these habitats, as discussed later. Second, the oceans are also low in productivity on a unit

area basis, although there are portions of the oceans that are highly productive. What limits productivity in the open oceans? Temperatures are suitable, and there is ample water, sunlight, and dissolved CO_2. Mineral nutrients are the limiting factor. In open oceans, many organisms settle to the bottom when they die, taking their minerals with them. Thus, the surface waters become impoverished in phosphates, nitrates, and other essential nutrients, while the deep waters are rich in these nutrients. When special conditions bring deep waters to the surface, such an upwelling results in a profuse bloom of phytoplankton. But most of the oceans are "nutrient deserts" and thus have a much lower productivity than was thought several years ago. In land plants, phosphorus and nitrogen are the most important nutrients limiting productivity of both native and crop plants.

With this general view of planetary productivity, it is now appropriate to examine certain factors influencing photosynthesis in more detail.

Light Effects The alfalfa data in Fig. 11-2 indicate that, as expected, reduced light intensity lowers photosynthetic rates, but we wish to understand the relationship on a more quantitative basis. Let us first examine how much light energy sunlight provides. At the upper boundary of the atmosphere and at the earth's mean distance from the sun, the radiant flux (intensity) of total solar radiation is 1.94 cal/cm²/min (the **solar constant**). As this radiation passes through the atmosphere to the earth's surface, much energy is lost by absorption and scattering caused by water vapor, dust, CO_2, and ozone, so that only about 1.3 cal/cm²/min remain, depending on elevation, time of day, latitude, and other factors. Of this, about 52 percent is in the infrared region, about 4 percent in the ultraviolet, and about 44 percent (0.6 cal/cm²/min) has wavelengths between 400 and 700 nm capable of causing photosynthesis.* The actual percentage of radiant energy in the 400 to 700 nm range varies with atmospheric conditions, apparently depending on absorption of infrared radiation by atmospheric water vapor on cloudy days.

The value 0.6 cal/cm²/min corresponds to about 10,000 foot candles (ft-c) or 108,000 lux, units that many people are more accustomed to. However, footcandle and lux units are not energy units but are measures of illuminance, a subjective description of the ability of the human eye to perceive light (see Special Topic 3.1). Wherever possible, we shall use cal/cm²/min as a measure of intensity and calories as a measure of total energy. During the summer in the United States, each cm² of land surface receives some 400 to 700 calories of photosynthetically active light per day, depending on cloud cover, location, and date. About 85 percent of this active light (400 to 700 nm) might be absorbed by a representative leaf, although this value varies considerably with leaf structure and age. The remainder is transmitted to lower leaves or the ground below or is reflected to the surroundings. Of that absorbed and potentially capable of causing photosynthesis, more than 95 percent is usually lost as heat, while less than 5 percent is usually captured during photosynthesis.

Let us now investigate how varying the light intensity affects photosynthetic rates when *single leaves* are exposed to normal air with about 320 ppm CO_2. There is, of course, no net CO_2 fixed in darkness except in CAM plants, and even in dim light the respiratory loss of CO_2 may exceed that used in photosynthesis. The light intensity at which photosynthesis just balances respiration (net CO_2 exchange is zero) is called the **light compensation point** (Fig. 11-3). This point varies with the species, with the light intensity during growth, and to some extent with the temperature at which measurements are made and the CO_2 concentration, but it is usually less than 2 percent of maximum sunlight (approximately the intensity in a well-lighted classroom). Only above the light compensation point can dry weight increases occur. Differences in light compensation points are caused primarily by differences in respiration rates. When respiration is slow, the leaf requires less light to photosynthesize rapidly enough to balance the CO_2 being lost, so the light compensation point is then also low.

Figure 11-3 shows responses to light intensity changes exhibited by single leaves of three dicot species after growth in their native habitats. The upper curve is for a C-4 shrub species, *Tidestromia oblongifolia,* that grows under unusually hot and arid conditions at high light intensities in Death Valley, California; the middle curve is for *Atriplex patula* subspecies *hastata,* a C-3 plant that grows along the Pacific coast

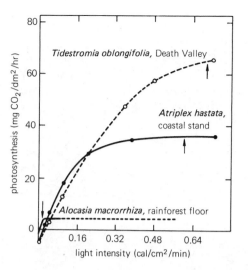

Figure 11-3 Influence of light on photosynthetic rates in single, attached leaves of three species native to different habitats. Maximum light intensities to which the plants are normally exposed are indicated by arrows. The light compensation points are indicated on the graph where the lines cross the abscissa. (Redrawn from Berry, 1975.)

of the western United States; and the lowest curve is for *Alocasia macrorrhiza,* which grows on the floor of a rain forest in Queensland, Australia. The energy of 400 to 700 nm light received each day by the first and last species during their growth varies 300-fold.

The *Alocasia* responses are typical of many species accustomed to shade (**shade plants**), including most house plants. First, these species exhibit much lower photosynthetic rates under bright sunlight than do crop plants or other species grown in open areas. Second, their photosynthetic responses are saturated at much lower intensities than are those of other species. Third, they usually photosynthesize at higher rates under very low light intensities than do other species, and fourth, their light compensation points are unusually low. These characteristics cause them to grow slowly in their natural habitats, yet they can survive where species with higher light compensation points would die.

Leaves on the shaded side of plants growing in bright light (**shade leaves**) have characteristics more like those of leaves on true shade plants than do the **sun leaves** on the sunny side. These differences result from anatomical and biochemical adaptations during leaf development, although they have definite genetic limits. *Tidestromia oblongifolia* cannot adapt to the strong shade conditions where *Alocasia macrorrhiza* grows, and the same is probably true of most of our crop plants. Shade leaves are often larger in area and thinner than sun leaves. Sun leaves are typically thicker than shade leaves because they form longer palisade parenchyma cells or an additional layer of such cells (Fig. 11-4). On a weight basis, shade leaves also generally have considerably more chlorophyll, mainly because each chloroplast has more grana than

---

*Although much of the information in this chapter was collected from data of researchers who assumed that 400 to 700 nm represented the useful energy for photosynthesis, later results showed that wavelengths as short as 350 nm and as long as 750 nm are effective (see the action spectrum data of Fig. 9-10a).

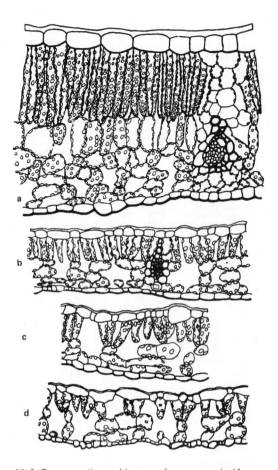

Figure 11-4 Cross sections of leaves of sugar maple (*Acer saccharum*), an unusually shade-tolerant tree, exposed to different light intensities during growth. (**a**) Leaf from south side of isolated tree. Note thick cuticle over the upper epidermis and long palisade parenchyma cells. (**b**) Leaf from center of crown of an isolated tree. (**c, d**) Leaves from base of two forest trees. All trees were growing near Minneapolis, Minnesota. (From H. C. Hanson, 1917, American Journal of Botany 4:533–560.)

do those of sun leaves. On the other hand, chloroplasts of shade leaves have less total stroma protein, including ribulose diphosphate carboxylase, and less lamellar electron transport protein than do sun leaves (Anderson et al., 1973; Berry, 1975). Those results indicate that shade leaves invest more of their energy in producing light-harvesting pigments that allow use of essentially all the limited amount of light striking them. Furthermore, the chloroplasts in leaves exposed to deep shade become arranged by phototaxis within the cells in patterns that maximize light absorption (see Section 19.7). The petioles also respond to the direction and intensity of

light by bending (see phototropism, Section 18.4), causing the leaf blades to move into less shaded regions, forming leaf mosaics.

The photosynthetic light responses of *Tidestromia oblongifolia* shown in Fig. 11-3 are typical of C-4 species native to sunny habitats and C-4 crop species such as corn, sorghum, and sugar cane. They show no rate saturation up to and even beyond full sunlight and have maximum photosynthetic rates at least twice those of most C-3 species. For such C-4 crops, rates as high as 50 to 60 mg CO_2 fixed/dm²/hr are not uncommon. The *Atriplex hastata* responses in Fig. 11-3 are representative of many C-3 crop species, such as potatoes, sugar beets, soybeans, cotton, alfalfa, tomatoes, and orchard grass. These species show rate saturation at intensities one-third to one-half full sunlight. Peanut and sunflower, however, are two C-3 species that do not become rate saturated until nearly full sunlight and that show maximum rates almost as high as those of C-4 crops. Most trees show maxima intermediate between those of typical C-3 crops and shade plants (Table 11-1), and they are often saturated by intensities as low as one-fourth full sunlight.

The more rapid photosynthetic rates of C-4 than C-3 species under high intensity light result in a lower water requirement per gram of dry matter produced for the C-4 species, although CAM plants have much lower requirements than either of these. Table 11-3 compares this and several other photosynthetic characteristics of C-3, C-4, and CAM plants, most of which are discussed in Chapter 10 or in the following sections.

Photorespiration Otto Warburg, a famous German biochemist who devoted part of his attention to photosynthesis, recorded in 1920 that photosynthesis in algae is inhibited by O_2. This inhibition occurs in all C-3 species studied since and is termed the **Warburg effect.** Figure 11-6 illustrates this effect for soybean leaves exposed to different CO_2 concentrations. Note that even the normal O_2 levels of 21 percent are quite inhibitory compared to zero O_2. Furthermore, the percentage of inhibition is greater at the lower CO_2 concentration, although the actual rate is much greater at the higher CO_2 level. In contrast to these results, photosynthesis in leaves of C-4 species is hardly affected by varying O_2 concentrations, although the process is inhibited in isolated bundle sheath cells.

To understand these differences between C-3 and C-4 species, remember that net CO_2 fixation is the amount by which photosynthesis exceeds respiration. From Fig. 11-2, we saw that respiration during darkness is small compared to photosynthetic rates, but is it possible that respiration rates in light are faster than in darkness and that O_2 inhibition of photosynthesis results from a faster release of CO_2 by respiration occurring in the light? Data indicating that this is so were obtained with an infrared CO_2 analyzer and first published by John P. Decker in the 1950s, but very few plant physiologists were then willing to accept his evidence for a

Adaptation to Light Intensity Changes

Adaptation of sun leaves or sun plants to shade is important in the "foliage plant" industry, where decorative plants are grown in well-lit greenhouses, then transferred to dimly lit homes or offices. Sudden transfers often cause death of older leaves or even of the entire plant, while gradual exposures to decreasing light intensity over a period of several days lowers the light compensation point and eventually allows slow growth in shade.

The reverse adaptation from shade to sun conditions is also possible, especially for the younger leaves. Mature leaves do not readily adapt to bright light, so shade plants usually cannot be moved to direct sunlight without fairly rapid death of the older leaves. Even seedlings of some species are sensitive to excess light. A dramatic example of this is represented by Englemann spruce (*Picea engelmannii* Parry) in the central and southern Rocky Mountains. When seedlings of this species are transplanted in the open during reforestation work, they usually become chlorotic and die. These symptoms result from a phenomenon known as **solarization,** a light-dependent absorption of O_2 and release of CO_2 that causes bleaching of the chloroplast pigments, especially those of photosystem II. A major function of certain carotenoid pigments is apparently to protect against solarization by absorbing light energy that is released as heat instead of being transferred to chlorophylls. In Englemann spruce and some other species, this protection is insufficient. If the spruce seedlings are shaded with logs, stumps, or brush, their survival rate is much higher (Fig. 11-5). Light sensitivity is an important factor in plant succession, because unusually sensitive species never become established except in the shade of others.

Figure 11-5 One method of shading spruce seedlings to prevent solarization during reforestation planting. (Courtesy Frank Ronco.)

Table 11-3 Some Photosynthetic Characteristics of Three Major Plant Groups[a]

Characteristic	C-3	C-4	CAM
Leaf anatomy	No distinct bundle sheath of photosynthetic cells	Well-organized bundle sheath, rich in organelles	Usually nc palisade cells, large vacuoles in mesophyll cells
Carboxylating enzyme	Ribulose diphosphate carboxylase	PEP carboxylase, then ribulose diphosphate carboxylase	Darkness: PEP carboxylase. Light: mainly ribulose diphosphate carboxylase
Theoretical energy requirement (CO_2:ATP:NADPH)	1:3:2	1:5:2	1:6.5:2
Transpiration ratio (g H_2O/g dry weight increase)	450–950	250–350	50–55
Leaf chlorophyll *a* to *b* ratio	2.8 ± 0.4	3.9 ± 0.6	2.5 to 3.0
Requirement for Na^+ as a micronutrient	No	Yes	Unknown
CO_2 compensation point (ppm CO_2)	30–70	0–10	0–5 in dark
Photosynthesis inhibited by 21% O_2?	Yes	No	Yes
Photorespiration detectable?	Yes	Only in bundle sheath	Difficult to detect
Optimum temperature for photosynthesis	15–25 C	30–40 C	$\simeq$ 35 C
Dry matter production (tons/hectare/year)	22 ± 0.3	39 ± 17	Low and highly variable

[a]Slightly modified from more extensive table of Black, 1973.

more rapid light respiration. We now know that respiration in leaves of C-3 species is often two or three times as rapid in light as in darkness and that it can be half as rapid as the net photosynthetic CO_2 fixation rate occurring simultaneously. It is certainly large enough to influence the curves of Figs. 11-2, 11-3, and 11-6. Respiration in illuminated photosynthetic organs occurs by two processes, those that occur in all plant parts during darkness, and by an additional much more rapid process known as **photorespiration.** The two processes are spatially separated within the cells, normal respiration occurring in the cytoplasm and in mitochondria and photorespiration occurring in chloroplasts, peroxisomes, and mitochondria in a cooperative way. Loss of CO_2 by photorespiration in C-4 species is almost undetectable, and this is the principal reason that they show much higher net photosynthetic rates at high light intensities than do C-3 species.

To understand why photorespiratory rates are so different in the two groups, we must first understand the chemical reactions of photorespiration. As we shall see, photorespira-

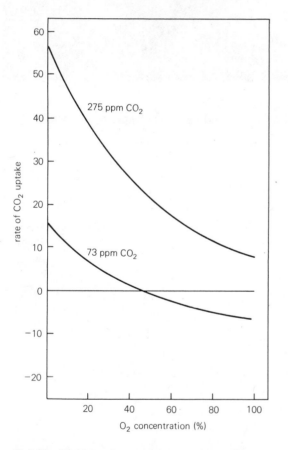

rate of CO_2 uptake

275 ppm CO_2

73 ppm CO_2

O_2 concentration (%)

Figure 11-6 The Warburg effect: inhibition of photosynthesis in soybean (C-3) plants by O_2. The light intensity was equal to about one-sixth maximum sunlight; the temperature was 22.5 C. Negative values represent a net loss of CO_2 by respiration. (From M. L. Forrester, G. Krotkov, and C. D. Nelson, 1966, Plant Physiology 41:422–427.)

tion is stimulated by three additional factors besides high light intensity: high O_2 levels, low CO_2 levels, and high temperatures.

Separation of bundle sheath cells from mesophyll cells in C-4 plants shows that only the bundle sheath cells photorespire. In the leaf, however, the CO_2 released is efficiently recaptured by ribulose diphosphate carboxylase or by PEP carboxylase before it escapes into the atmosphere. All photosynthetic cells of C-3 plants apparently photorespire. A major difference is that these mesophyll cells cannot produce a two-carbon acid called **glycolic acid** (HO—CH_2—$COOH$), while the bundle sheath cells can. The importance of glycolic acid is that its carboxyl group is the main source of CO_2 actually lost in photorespiration.

In 1971, W. L. Ogren and George Bowes at the University of Illinois theorized from various data that carbons 1 and 2 of ribulose diphosphate were the precursors to glycolic acid. They showed experimentally for the first time that O_2 could inhibit CO_2 fixation by ribulose diphosphate carboxylase, thus apparently explaining the Warburg effect. They also showed that ribulose diphosphate carboxylase catalyzes an oxidation of ribulose diphosphate by O_2. *Thus, ribulose diphosphate carboxylase is also an oxygenase.* This O_2 is thought to represent about two thirds of the total O_2 absorbed during photorespiration (Björkman, 1973; Chollet and Ogren, 1975).

The two products formed from ribulose diphosphate and O_2 are 3-PGA and **phosphoglycolic acid,** a two-carbon acid. Using heavy oxygen ($^{18}O_2$), it was shown that only one of the O_2 atoms is incorporated into phosphoglycolic acid, the other apparently being converted to water (OH^- ion):

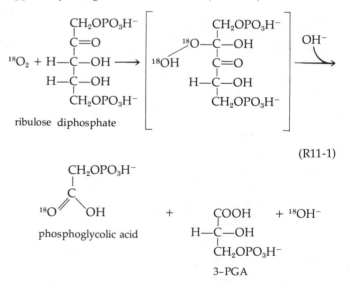

(R11-1)

Oxygen and CO_2 thus compete for the same enzyme and for the same ribulose diphosphate substrate. This competition explains the greater inhibition of photosynthesis in C-3 plants at low than at higher CO_2 levels, shown in Fig. 11-6. It is thought that the CO_2 released when malate and aspartate are decarboxylated in the bundle sheath cells of C-4 plants (see

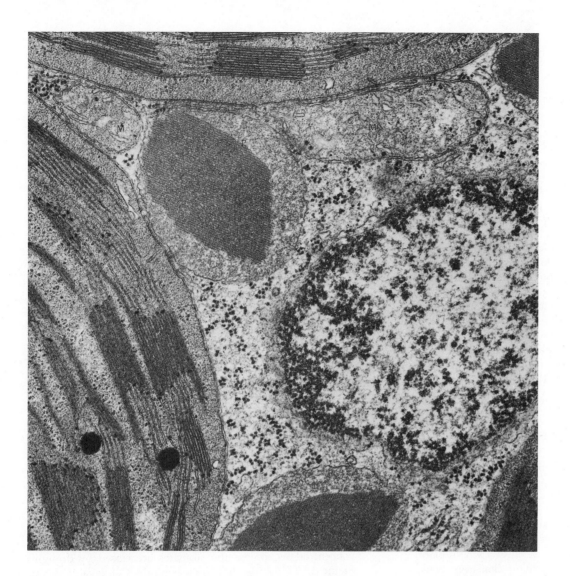

Figure 11-7 Close association of chloroplasts, peroxisomes (P), and mitochondria (M) in a leaf cell. The crystalline-like matrix in these peroxisomes is due to the enzyme catalase, but many peroxisomes containing catalase show no such matrix. (Courtesy Eugene Vigil.)

Section 10.3) accumulates there and minimizes O_2 competition, thereby inhibiting photorespiration. When O_2 is fixed in bundle sheath cells of C-4 plants or any photosynthetic cell of C-3 plants, a precursor to the principal photorespiratory source of evolved CO_2 is formed. This phosphoglycolic acid is dephosphorylated by a specific phosphatase, releasing free glycolic acid.

Now we can understand how light stimulates O_2 fixation, glycolate production and therefore photorespiration: It provides ATP and NADPH needed to regenerate ribulose diphosphate from 3-PGA by the Calvin cycle. This enhanced rate of production of ribulose diphosphate with increased light therefore increases both CO_2 fixation and O_2 fixation. Up to a certain intensity, increasing light stimulates CO_2 fixation more than it does O_2 fixation, so the net rate of photosynthesis rises. At still higher intensities, photorespiration apparently balances the increment of CO_2 fixation, so the net

rate of photosynthesis becomes saturated (as shown in Fig. 11-3 for C-3 species). For every ribulose diphosphate undergoing a reaction, its combination with CO_2 rather than O_2 leads to one more CO_2 fixed, one less O_2 molecule absorbed at this step, and potentially one less CO_2 molecule lost by oxidation of glycolic acid. *Most of the apparent photorespiratory loss of CO_2 in C-3 plants is not really a release of CO_2 from glycolic acid but simply the failure of CO_2 to be fixed because of O_2 competition.*

How is glycolic acid oxidized to account for production of CO_2 that is actually lost? It is clear that this process does not occur in the chloroplasts. The first step is movement of glycolic acid out of the chloroplast into adjacent **peroxisomes.** Peroxisomes are small organelles that contain several oxidative enzymes (Tolbert, 1971, 1973; Vigil, 1973; Breidenbach, 1976). They exist almost exclusively in photosynthetic tissues and often appear in electron micrographs in direct contact with chloroplasts (Fig. 11-7). In the peroxisomes, glycolic acid is oxidized to **glyoxylic acid** by **glycolic acid oxidase,** an enzyme containing riboflavin (vitamin B_2) as part of an essential prosthetic group:

$$CH_2-COOH + O_2 \rightarrow HC-COOH + H_2O_2 \qquad (R11\text{-}2)$$
$$|\qquad\qquad\qquad\quad \|$$
$$OH \qquad\qquad\qquad\quad O$$

Glycolic acid oxidase transfers electrons (H atoms) from glycolic acid to O_2, reducing the O_2 to H_2O_2 (hydrogen peroxide). Much of this H_2O_2 is then broken down by **catalase** (another peroxisomal enzyme) to water and O_2:

$$2\ H_2O_2 \rightarrow 2\ H_2O + O_2 \qquad (R11\text{-}3)$$

The ultimate fate of glyoxylic acid is not well understood. Some of it might be oxidized to CO_2 and formic acid (HCOOH) by H_2O_2 not removed by catalase, but most of it is converted to glycine. Then, after transport of glycine into mitochondria (Fig. 11-7), two molecules of it are converted to one molecule of serine and one molecule of CO_2. *This reaction is the major source of CO_2 truly released in photorespiration.*

Serine can be converted to 3-phosphoglyceric acid by a series of reactions that involve loss of the amino group and gain of a phosphate group from ATP. The series of reactions by which glycolate is converted to 3-PGA via glyoxylate, glycine, serine, and other three-carbon acids is known as the **glycolate pathway.** The 3-PGA can be converted to sucrose and starch in the chloroplasts. This pathway therefore represents a means for conserving some of the carbons that are split off ribulose diphosphate (as glycolate) when O_2 reacts with it.

Light Effects in a Plant Canopy The curves of Fig. 11-3 show how photosynthesis in single leaves changes with light intensity, but we may reasonably ask whether a whole plant, an entire crop, or a forest would exhibit similar response curves. Figure 11-8 shows that stands of C-4 crop plants respond in almost a linear fashion to increases in light intensity measured at the top of the canopy. This curve is somewhat like that for a single leaf of corn or sugar cane (Fig. 11-3), except that it shows even less indication of light saturation. Figure 11-8 also lists typical results for two C-3 crops, cotton and wheat. For them we see a distinct tendency toward light saturation, but only at much higher light intensities than in the single leaf C-3 example (*Atriplex hastata*) of Fig. 11-3.

The principal reason for differences in single leaves and whole plants or groups of plants is that the upper leaves absorb much of the incident light, allowing less for the lower leaves. In this situation, exposure to a higher intensity may saturate the upper leaves, but more light is then transmitted and reflected toward the shaded leaves below that are not saturated. As a result, single plants, crops, or forests as a whole probably very seldom receive enough light to maximize the photosynthetic rate. This is consistent with the alfalfa data of Fig. 11-2, where temporary cloudiness decreased photosynthesis.

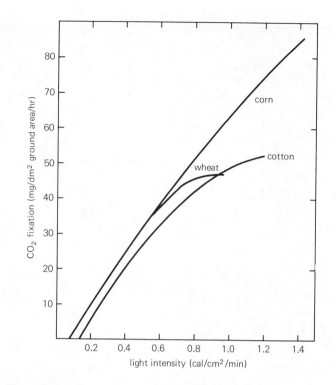

Figure 11-8 Effect of total solar radiation intensity at the top of the canopy on net photosynthetic rates in corn, wheat, and cotton plants. (Drawn from data of [corn] D. N. Baker and R. B. Musgrave, 1964, Crop Science 4:127–131; [wheat] D. W. Pukridge, 1968, Australian Journal of Agricultural Research 20:623–634; [cotton] D. N. Baker, 1965, Crop Science 5:53–56.)

Availability of CO_2 Not only are photosynthetic rates enhanced by increased light intensities but also by higher CO_2 concentrations, unless the stomates are closed by drought. Figure 11-9 illustrates how increasing the CO_2 level in the air stimulates photosynthesis in a C-3 plant at three different light intensities. Here the additional CO_2 decreases photorespiration by increasing the ratio of CO_2 compared to O_2 reacting with ribulose diphosphate carboxylase. This, of course, will lead to faster rates of net photosynthesis. Note that at high CO_2 concentrations, the higher light intensities increase the rate of photosynthesis and that to saturate this rate, a higher CO_2 concentration is required at the higher light intensities than at the lower. C-4 species respond to increased CO_2 in much the same manner, except that their rates are often saturated by lower CO_2 levels.

A more distinct difference between C-4 and C-3 species can be observed if the CO_2 levels are decreased well below normal atmospheric levels. If the light intensity is above the light compensation point, net photosynthesis of C-3 species usually reaches zero at CO_2 concentrations between 50 and 100 ppm, while C-4 plants continue net CO_2 fixation at levels between 0 and 5 ppm. This CO_2 concentration where photosynthetic fixation just balances respiratory loss is called the

CO₂ **compensation point,** a few examples of which are illustrated in Fig. 11-10. Note that the value for corn appears to be 0 ppm, while that for the C-3 species sunflower, red clover, and maple is about 50 ppm.

The difference in CO_2 compensation points for C-4 and C-3 species is exhibited dramatically by the contrasting responses when a plant of each type is placed in a common sealed chamber in which photosynthesis can occur (Moss and Smith, 1972). (The plants must be grown hydroponically in a soilless medium such as sand, perlite, or vermiculite to avoid CO_2 release by soil microorganisms.) Both plants fix CO_2 until the compensation point of the C-3 is reached; since the C-4 plant will photosynthesize at still lower CO_2 concentrations, this occurs with CO_2 lost by respiration, including photorespiration, from the C-3 plant. As a result, the C-3 plant will usually die within a week or so, while the C-4 plant will continue to grow for some time. Here there is a net transfer of CO_2 from one plant to the other (Fig. 11-11).

The lower CO_2 compensation points in C-4 than in C-3 species arise from the much lower photorespiratory release of CO_2 by C-4 plants. The difference in compensation points essentially disappears if the O_2 concentration to which the plants are exposed is decreased from the normal 21 percent down to about 2 percent. Here the CO_2 compensation points of C-3 species also approach zero, because insufficient O_2 is present to compete with CO_2 for ribulose diphosphate.

During the summer growing season, a lack of CO_2 is a

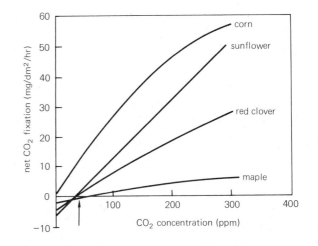

Figure 11-10 Influence of reduced CO_2 concentrations on photosynthetic rate in C-4 (corn) and C-3 plants. Artificial lights providing approximately the same energy as sunlight in the 400–700 nm region were used. (From J. D. Hesketh, 1963, Crop Science 3:107–110, 493–496.)

Figure 11-11 Left, a C-3 plant (wheat) and right, a C-4 plant (corn) grown in a soilless medium of perlite inside an airtight chamber in which the CO_2 supply became depleted. The corn remained green except at the leaf tips, while the wheat leaves were brown and apparently dead. (Courtesy Dale N. Moss.)

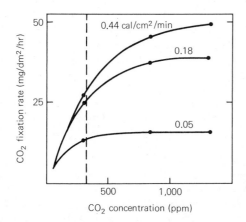

Figure 11-9 Effects of atmospheric CO_2 enrichment on CO_2 fixation in sugar beet leaves. Intact, fully developed leaves from young plants were used. Fixation rates for three different light intensities are shown: 0.05, 0.18, and 0.44 cal/cm²/min of 400–700 nm light. The dashed line represents the present atmospheric CO_2 concentration. Higher CO_2 levels stimulated CO_2 fixation more at increasing light intensities. At the highest intensity, which is only slightly less than that obtained from full sunlight, the highest CO_2 concentration nearly saturated the fixation rate, but at lower intensities, this rate was saturated by lower CO_2 concentrations. Leaf temperatures were between 21 and 24 C. (Redrawn from data of P. Gaastra, 1959, Mededelinger van de Landbouwhogeschool Te Wageningen, The Netherlands.)

common cause of suboptimal photosynthetic rates, especially for leaves exposed to bright light. Even slight breezes can enhance photosynthesis by replacing CO_2-depleted air in the boundary layer around a leaf with normal air. Students sometimes ask which is the usual limiting factor in plant growth, CO_2 or light. The answer is that both can be limiting, and both usually are. The upper, more illuminated leaves will usually respond to increases in CO_2, while the lower leaves may be CO_2 saturated but will respond to additional light. Thus, an increase in either factor increases CO_2 fixation of a whole plant or crop.

Greenhouse crops usually lack enough CO_2 for maximal growth, and this is especially serious in winter, when the

CO₂, O₂, and Stomatal Control

We mentioned in Chapter 3 that somates might be considered disadvantageous to plants because they allow loss of water by transpiration, but that they are essential for gas exchange. Calculations of Herman Wiebe at Utah State University emphasize that only the exchange of CO_2 is controlled in an important way by stomates, while O_2 can diffuse through cuticles of most species rapidly enough to allow undiminished respiration. Roots, for example, apparently respire quite well with the O_2 absorbed from the soil through a layer of suberin, although the air-filled intercellular space system continuous with the shoot may provide some of this oxygen. Stomates are also absent in numerous flower petals, underground storage organs such as potato tubers (stems), and in many fruits—that is, generally in organs that never photosynthesize. Oxygen penetrates these organs readily enough, just as it penetrates leaves at night when the stomates are tightly closed or as it penetrates rapidly growing and respiring immature leaves with undeveloped stomates.

Wiebe emphasized three reasons that open stomates are essential for uptake of CO_2 but not O_2. First, plants in light generally respire only about 1/10 as fast as they photosynthesize, so they use up far more CO_2 than O_2. Second, if plants became deficient in O_2 by using all of the O_2 inside them, the diffusion gradient would be extremely large, because the concentration in the external air is about 209,000 ppm by volume. This gradient is 654 times as large as the 320 ppm difference in CO_2 concentration that could exist, so diffusion of O_2 would be much faster than that of CO_2 through any pathway unless the permeability of this pathway to CO_2 was much greater. Third, the respiratory enzyme that uses O_2 (cytochrome oxidase, see Section 12.6) has a very high affinity (low K_m) for O_2, so it is usually provided with much more O_2 than it requires for maximum activity. On the contrary, both ribulose diphosphate carboxylase and PEP carboxylase have K_m values for CO_2 that approximate the CO_2 concentration in normal air. Thus, plants usually are provided with more O_2 and less CO_2 than they need for maximum respiration and photosynthesis.

---

greenhouses must be kept closed. Some growers fertilize the air with CO_2 released from high-pressure tanks or other sources, thereby obtaining increased yields of many ornamental and food crops during the winter months (Wittwer and Robb, 1964). CO_2 levels are usually not allowed to exceed 1,000 ppm under these conditions, because such concentrations are frequently toxic or cause stomatal closure, nullifying any increased photosynthesis. In the summer months, greenhouses are usually cooled with evaporative cooling systems in which outside air is drawn across wet pads in the greenhouse walls. Increased plant growth in such greenhouses must result in part from the increased CO_2 levels that are caused by the incoming fresh air. (See boxed essay.)

Temperature The temperature range over which plants can photosynthesize is surprisingly large. Certain bacteria and blue-green algae photosynthesize at temperatures at least as high as 70 C, while conifers can photosynthesize extremely slowly at −6 C or below. In some antarctic lichens, photosynthesis occurs at −18 C, with an optimum near 0 C! In many higher plant species exposed to bright sunlight on a hot summer day, leaf temperatures often reach 35 C or higher, with photosynthesis continuing.

The effect of temperature on photosynthesis depends upon the species, the environmental conditions under which it was grown, and the environmental conditions during measurement. In general, desert species have higher temperature optima than arctic or alpine species, and annual desert plants that develop during the hot summer months (mostly C-4 species) have higher optima than those that grow there only during the winter and spring (mostly C-3 species). Crops such as corn, sorghum, cotton, and soybeans that grow well in warm climates usually have higher optima than do crops such as potatoes, peas, wheat, oats, and barley that are cultivated in cooler regions. In general, optimum temperatures for photosynthesis are similar to the daytime temperatures at which the plants normally grow. Figure 11-12 illustrates this relationship for two grasses native to the Great Plains in Wyoming, *Spartina pectinata* (prairie cordgrass), a C-4 plant, and *Leucopoa kingii* (king's fescue), a C-3 plant. *Spartina* grows at a lower elevation (Wheatland site, Fig. 11-12a) than does *Leucopoa* (Pole mountain site), and the mean day temperatures are higher for the Wheatland site. Figure 11-12b shows that the optimum photosynthetic temperature for Spartina is near 35 C, compared to about 25 C for Leucopoa.

Although there are many exceptions, C-4 plants generally have higher optima than do C-3 plants, and this difference is often controlled by differences in photorespiration. Normal temperature increases have little influence on the light-driven split of H_2O or the diffusion of CO_2 into the leaf, but they more markedly influence the biochemical reactions of CO_2 fixation and reduction. Thus, increases in temperature usually stimulate photosynthetic rates until the stomates close or enzyme denaturation begins to occur. However, respiratory CO_2 loss also increases with temperature, and this is especially pronounced for the chemical reactions of photorespiration, primarily because a temperature rise raises the ratio of dissolved O_2 compared to CO_2 (Ku and Edwards, 1977). As a result of O_2 competition, net CO_2 fixation in C-3 plants is not promoted by increased temperature nearly as much as one might expect. The stimulating effect of a temperature rise is nearly balanced by increased respiration and photorespiration over much of the temperature range at which C-3 plants

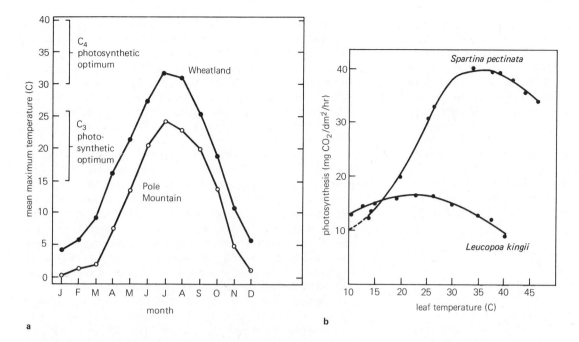

Figure 11-12 Effect of temperature on photosynthesis in grasses native to the northern Great Plains. (**a**) Mean maximum temperatures during various months at two Wyoming sites. The Wheatland site on the Laramie River has an elevation of 1470 m, while the Pole Mountain site near Laramie is at 2600 m. Grasses native to the Wheatland site are primarily C-4 species; those at Pole Mountain are mainly C-3 species. The temperature optima for the C-4 species are about 15 C higher than for the C-3 species. (**b**) Temperature-photosynthesis response curves for two species native to the sites in Fig. 11-12a. *Spartina*, a C-4 plant, has a higher temperature optimum and greater photosynthetic rates at most temperatures than does *Leucopoa*, a C-3 plant. All measurements were made on whole plants in the field on cloudless days near noon. Stomates of both species remained open at temperatures up to 40 C, but partially closed at higher temperatures. (Unpublished data of A. T. Harrison.)

normally grow, so a rather flat and broad temperature response curve between 15 and 30 C often occurs. Since photorespiration is of less significance in C-4 plants, they often exhibit optima in the 30 to 40 C range (Table 11-3).

The most dramatic example of photosynthesis at high temperatures in an angiosperm was found in the C-4 dicot *Tidestromia oblongifolia* by Olle Björkman and his colleagues at the Carnegie Institution in Stanford, California. As mentioned before, this plant grows in Death Valley, California, and does so under the hottest natural environment in the western hemisphere. As contrasted to most species that grow in this area only in winter or spring, *Tidestromia* grows in the hot summer months. It has a remarkable photosynthetic optimum of 47 C (117 F) and, as expected, proved to be a C-4 plant.

Leaves can adapt somewhat to temperature if they are exposed for a few days to different temperatures, and this is apparently advantageous in allowing plants to adjust to seasonal changes (Berry, 1975; Chabot and Lewis, 1976; Doley and Yates, 1976). As mentioned in Section 10.5, CAM plants adjust to increasing night temperatures by keeping their stomates open longer in the morning and fixing a greater

fraction of CO_2 during daylight than they do when exposed to cool nights.

Leaf Age As individual leaves on a plant grow and develop, their ability to photosynthesize increases for a time and then, often even before the leaf reaches maturity, the rate of photosynthesis begins to decrease. Old, senescent leaves eventually become yellow and are unable to photosynthesize because of chlorophyll breakdown and loss of functional chloroplasts. However, even apparently healthy leaves of conifers that often persist for several years usually show gradually decreasing photosynthetic rates during successive summers.

Carbohydrate Translocation Another interesting internal control of photosynthesis concerns the rate at which photosynthetic products such as sucrose can be translocated from the leaves to various sink organs. It is often found that removal of developing tubers, seeds, or fruits (excellent sinks) significantly inhibits photosynthesis after a few days, especially in adjacent leaves that normally translocate to these organs

(Neales and Incoll, 1968). Furthermore, species having high photosynthetic rates also have relatively high translocation rates, consistent with the idea that effective removal of photosynthetic products maintains rapid CO_2 fixation. Severe infection of leaves by pathogens often so severely inhibits photosynthetic rates that these leaves become sugar importers rather than exporters; the adjacent healthy leaves then gradually attain marked increases in photosynthetic rates, suggesting that enhanced translocation has removed some governor over CO_2 fixation. We do not yet understand the mechanism of these relations, but we can speculate that if enough effective sinks could be produced on plants, perhaps by increasing the number of ears on a corn plant, fruits on an apple tree, or tubers on a potato plant, photosynthetic yields could be significantly increased.

11.4 Photosynthetic Rates and Crop Production

Many crop physiologists, ecologists, and plant breeders are concerned with the problem of how certain environmental factors and the plant itself can be altered to obtain maximal crop yields (Wittwer, 1974; Loomis et al., 1971; Cooper, 1975; Kriedemann, 1971). Photosynthetic efficiencies appear never to exceed about 22 percent of the absorbed radiation energy in the 400 to 700 nm region, and there is no known way by which this can be increased. In fact, this maximum efficiency is obtained only at relatively low light intensities, not in brightest sunlight when yields are optimum. Considering the supply of light to available land area on which a crop is growing, we find that overall yield efficiencies are always much below 22 percent. Many crops, including forest trees and herbaceous species, convert about 1 to 2 percent of the photosynthetically active energy striking the field during their growing season into the energy stored in carbohydrates. The efficiencies are generally better for C-4 than for C-3 species under high light intensities, mainly because of lack of photorespiration in the former, but such factors as insufficient CO_2, nutrients, or water, and unfavorable temperatures account for much of the difference between observed and theoretical efficiencies. Also, much of the light incident on the field is not absorbed by plants when they are small. Finally, much of the energy in photosynthetic products is used in respiration during darkness. Respiration is essential to provide energy for carbohydrate translocation, ion absorption, reduction of nitrate and sulfate ions, and for synthesis of polysaccharides, proteins, chlorophyll, and other large molecules (Penning de Vries, 1975; Penning de Vries et al., 1974).

Agriculturists are much more concerned with total productivity than with efficiency, so one goal of researchers has been to increase the amount of light absorbed by crops. Varieties of annual crop species that produce extensive leaf cover early in the season are being sought, because these will intercept more light than varieties that produce relatively more early stem or root growth. Much research has concerned the architecture of plant canopies in relation to productivity. A term called the **leaf area index** (**LAI**) is widely used to indicate the ratio of leaf area (one surface only) of a crop to the ground area upon which the plants grow. LAI values of up to 8 are common for many mature crop communities, depending on species and planting density. Deciduous forests also have LAI values approaching 8, and many of the lower leaves are then extensively shaded, receiving perhaps less than 1 percent of full sunlight.

Productivity rates increase with LAI up to a certain point because of more effective light interception, but still larger LAI values often cause no more increases and then even decreases on a ground area basis (Fig. 11-13). The rate decreases are partly due to respiration losses in stems and in the shaded leaves that do not receive enough light to overbalance their respiratory carbon losses at high LAI values. In fact, at low light intensities the LAI values do not become as large as at high intensities, partly because the lower leaves become senescent and fall off. It is interesting that such shaded leaves have much lower dark respiration rates than more exposed leaves on the same plant and, for C-3 species, we have already emphasized that exposure to lower light intensities will decrease their photorespiratory losses. Were it not for low rates of respiration, they might be rather serious welfare cases on the more exposed leaves. Unusually low rates of respiration also appear to be characteristic of leaves of plants growing in the deep shade of forests; this allows such plants to grow at very low light intensities and low photosynthetic rates.

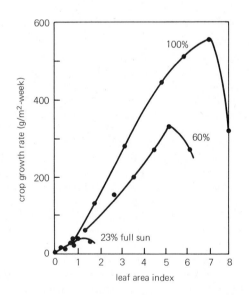

Figure 11-13 Growth of sunflower plant communities (100 plants/m²) at various leaf area indices and light intensities given as percent of full sunlight. At full sunlight, the optimum LAI is 7; the optimum at 60 percent full sunlight is only 5; and at 23 percent full sunlight, it is only 1.5. (From Leopold and Kriedemann, 1975.)

Increased stem elongation is often an advantage for plants competing in nature for light, but in a uniformly growing cereal crop no such advantage occurs, and increased grain yields can be obtained with dwarf or semidwarf varieties that allocate relatively more photosynthate to grain than to stems. Plant breeders are also providing varieties of several crops that have other alterations in canopy structure that increase yields. For example, computer results predicted that at LAI values above 2 or 3, depending on the species, varieties having erect leaves near the top of the canopy should photosynthesize more rapidly than those having more horizontally oriented leaves in this position. Indeed, yields of the erect-leaf type have been significantly greater. It is expected that further cooperation between physiologists and geneticists will bring about increased yields of several other species. The elimination of photorespiration in future agricultural and forest crop plants presently appears to be a very worthwhile goal (Zelitch, 1975a; Wildner and Henkel, 1976).

12

Respiration

All active cells respire continuously, often absorbing O_2 and releasing CO_2 in nearly equal volumes. Yet, as you know, respiration is much more involved than a simple exchange of gases. The overall process is an oxidation-reduction in which foods are oxidized to CO_2, and the O_2 absorbed is reduced to form H_2O. Starch, fructosans, sucrose or other sugars, fats, organic acids and, under rare conditions, even proteins can serve as respiratory substrates. The common respiration of glucose, for example, can be written as in R12-1.

$$C_6H_{12}O_6 + 6\ O_2 \rightarrow 6\ CO_2 + 6\ H_2O + energy \qquad (R12\text{-}1)$$

Much of the energy (approx. 686 kcal/mol of glucose) released during respiration is heat. When temperatures are low, this heat might stimulate metabolism and be beneficial to growth of certain species, but usually it is just transferred to the atmosphere or soil with little consequence to the plant. Far more important than heat is the energy trapped in compounds that can be used later for the many essential processes of life, such as those involved in growth and ion accumulation. ATP is the most important of these compounds, while NADH and NADPH are also important because of their ability to transfer electrons.

The summary equation just given for respiration is misleading in a way, because respiration, like photosynthesis, is not a single reaction. It is rather a series of many component reactions, each catalyzed by a different enzyme. It is a "burning" that occurs in a water medium, near neutral pH, at moderate temperatures, and without smoke! This gradual, stepwise breakdown of large molecules provides a means for trapping energy in ATP, NADH, and NADPH. Furthermore, as the breakdown proceeds, carbon skeleton intermediates are provided for a large number of other essential plant products. These include amino acids for proteins, nucleotides for nucleic acids, and carbon precursors for porphyrin pigments (such as chlorophyll and cytochromes), fats, sterols, carotenoids, anthocyanins, and certain other aromatic compounds. Of course, when these are formed, conversion of the original substrates to CO_2 and H_2O is not complete. Usually only some of the respiratory substrates are fully oxidized to CO_2 and H_2O, while the rest are used in synthetic (**anabolic**) processes. The energy trapped during oxidation can be used to synthesize the large molecules required for growth. *When plants are growing rapidly, most of the disappearing sugars are diverted into such synthetic reactions and never appear as CO_2.* Whether most of the carbon atoms in the food being respired are converted to CO_2 or to any of the large molecules mentioned earlier depends on the kind of cell involved, its position in the plant, and whether or not the plant is rapidly growing.

12.1 The Respiratory Quotient

If carbohydrates such as sucrose, fructosans, or starch are serving as respiratory substrates, and if they are completely oxidized, the volume of O_2 taken up exactly balances the volume of CO_2 released from the cells. This ratio of CO_2/O_2, called the **respiratory quotient** or **RQ,** is often very near unity. For example, the RQ obtained from leaves of many different species averaged about 1.05. Germinating seeds of the cereal grains and many legumes such as peas and beans, which contain starch as the main reserve food, also exhibit RQ values of approximately 1.0. Seeds from many other species, however, contain much fat or oil. When fats and oils are oxidized during germination, the RQ is often as low as 0.7. Consider the oxidation of a common fatty acid, oleic acid:

$$C_{18}H_{34}O_2 + 25.5\ O_2 \rightarrow 18\ CO_2 + 17\ H_2O \qquad (R12\text{-}2)$$

The RQ for this reaction is $18/25.5 = 0.71$. By measuring the RQ for any seed or other plant part, information can be obtained about the type of compounds being oxidized. The problem is complicated, because at any time several different types of substances may be respired, so the measured RQ is only an average value, dependent upon the respiratory contribution of each substrate and its relative content of carbon, hydrogen, and oxygen. In this chapter, we shall emphasize respiration of polysaccharide foods. The utilization of fats is discussed in Chapter 14.

12.2 Formation of Hexose Sugars from Reserve Polysaccharides

Degradation of Starch As discussed in Section 10.6, starch is stored in plastids as water-insoluble granules or grains that consist of branched amylopectin molecules and essentially unbranched amyloses. Starch accumulated in the chloroplasts during photosynthesis is an important food reserve in leaves of most species. Starches that have been formed in amyloplasts of storage organs after translocation of sucrose or certain other sugars are also principal respiratory substrates for these organs at certain stages in their development. Parenchyma cells in both roots and stems commonly store starch; in perennial species, the starch stored there during the growing season is maintained during winter months and is used in new growth the following spring. Potato tubers are rich in starch-containing amyloplasts, and most of this starch disappears by respiration and translocation from tuber sections planted to obtain a new crop. The endosperm or cotyledon storage tissues of many monocot and dicot seeds contain abundant starch, and most of this also disappears during seedling development.

Figure 12-1a shows the relation of the starch-storing endosperm to the rest of the seed in corn, while a germinating corn seedling is shown in Fig. 12-1b. These pictures illustrate a situation in which respiration is incomplete, because although some of the carbon atoms in starch are being totally degraded to CO_2, others that are converted into sucrose molecules in the scutellum move into the growing root and shoot. Some of the carbons in this sucrose are respired to CO_2 and H_2O, while others are diverted into cell wall materials, proteins, and other substances needed for growth of the seedling.

Most steps in the degradation of starch to glucose can be catalyzed by three different enzymes, although still others are needed to complete the process. The first three include an **alpha amylase** (α-amylase), a **beta amylase** (β-amylase), and starch phosphorylase. Some studies show that of these, only alpha amylase can attack intact starch granules, so when β-amylase and starch phosphorylase are involved, they must act on the first products released by α-amylase (Dunn, 1974). Many α-amylases are activated by Ca^{2+}, thus providing one

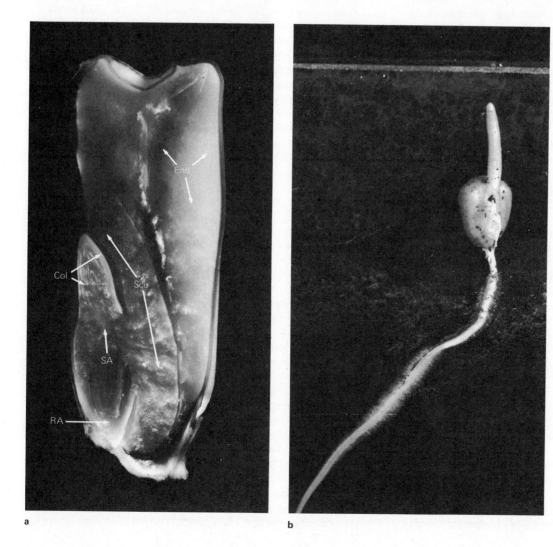

Figure 12-1 (**a**) Longitudinal section of a corn seed, showing the relation of the starch-storing endosperm (END) to the other seed parts. COL, coleoptile; SCU, scutellum or cotyledon; SA, shoot apex; RA, radicle. (From O'Brien and McCully, 1969.) (**b**) Corn seedling being nourished by the endosperm. (From Jensen and Salisbury, 1974.)

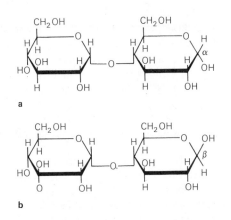

Figure 12-2 Alpha(α)- and beta(β)- maltose released from starch during action of α- and β-amylases.

explanation for the essentiality of this element. Alpha amylase randomly attacks 1,4-bonds throughout both amylose and amylopectin, at first releasing still large and complex products. Later, fragments containing about ten glucose units, the **dextrins,** are released, and eventually α-**maltose,** a disaccharide containing two glucose units (Fig. 12-2a), and glucose are produced from amylose. Alpha amylase cannot, however, attack the 1,6-bonds at the branch points in amylopectin (see Fig. 10-12), so amylopectin digestion stops when branched dextrins of very short chain lengths still remain. Beta amylase hydrolyzes starch into β-**maltose** (Fig. 12-2b), starting only from one end. The β-maltose is rapidly changed by mutarotation into the natural mixture of α- and β-isomers. Hydrolysis of amylose by β-amylase is nearly complete, but amylopectin breakdown is incomplete because the branch linkages are not attacked. Dextrins again remain.

Activity of both amylases involves the uptake of one H_2O for each bond cleaved. They are therefore classified as **hydrolytic enzymes.** Hydrolytic reactions are not reversible, and no starch synthesis by the amylases can be detected. *A general principle is that large molecules are usually synthesized by one series of reactions (pathway) and broken down by another* (Beevers, 1974). For example, we discussed in Section 10.6 how polysaccharide synthesis requires an activated form of a sugar such as ADPG, UDPG, or perhaps even glucose-1-phosphate.

The amylases are widespread in various tissues but are most active in germinating seeds high in starch. In leaves, α-amylase is probably of much greater importance than β-amylase is for starch hydrolysis. The α-amylase is located primarily inside the chloroplasts with apparent excellent access to the starch grains that it will attack. It apparently functions both day and night, although of course during daylight there is a net production of starch from photosynthesis.

Starch phosphorylase breaks down starch not by incorporating water into the products as the amylases do, but by incorporating phosphate. It is therefore a **phosphorolytic**

enzyme rather than a hydrolytic one, and the reaction that it catalyzes is reversible:

$$\text{starch} + H_2PO_4^- \rightleftharpoons \text{glucose-1-phosphate} \quad \text{(R12-3)}$$

As will become more apparent later, formation of glucose-1-phosphate avoids the need for an ATP molecule to convert glucose into a glucose phosphate during respiration.

Amylopectin is only partially degraded by starch phosphorylase, the reaction proceeding consecutively from the terminal end of each chain to within two or three glucose residues of the α-1,6 branch linkages and leaving dextrins. Amylose, almost entirely free of such branches, is completely hydrolyzed by repeated removal of glucose units beginning at one end of the chain. As with the amylases, starch phosphorylase is widespread in plants, and it is often difficult to decide which enzyme digests most starch in the cells concerned. The present theory is that α-amylase is always essential for initial attack, as mentioned, and for cereal grain seeds both amylases appear functional, but starch phosphorylase does not. For seeds of other species, for leaves, and for other tissues, starch phosphorylase apparently also contributes.

The 1,6-branch linkages in amylopectin that are not attacked by any of these enzymes are hydrolyzed by a **debranching enzyme** sometimes called the **R enzyme** and by **dextrinases.** Action of these enzymes therefore provides additional end groups in the undegraded dextrins for attack by the amylases or starch phosphorylase, and allows complete digestion of starch into glucose, maltose, or glucose-1-phosphate.

The maltose seldom accumulates to any appreciable extent in plants, and it appears to be slowly hydrolyzed to glucose by α-amylase or more rapidly by a **maltase** enzyme as in R12-4:

$$\text{maltose} + H_2O \rightarrow 2\alpha\text{-D-glucose} \quad \text{(R12-4)}$$

The resulting glucose units are now available for conversion into other polysaccharides as discussed in Chapter 10 or, as emphasized in the present chapter, for degradation by subsequent respiratory processes.

In summary, amylases hydrolyze amylose to glucose and maltose, while starch phosphorylase converts amylose to glucose-1-phosphate. The action of all three enzymes on amylopectin leaves a dextrin, the branch linkages in which must be hydrolyzed by a debranching enzyme and by dextrinases. Maltose is hydrolyzed to glucose by maltase.

There is an interesting effect of temperature on the ratio of starch to sugars in certain tissues. In the potato tuber, for example, storage at temperatures only slightly above freezing causes an accumulation of glucose, fructose, and sucrose, with an accompanying loss of 1 to 5 percent of the starch. This explains the sweet taste of potatoes stored at low temperatures. When the tubers have been rewarmed to above 10 C, the sugars are reconverted back to starch. It is important for the potato chip industry to use potatoes containing very little sugar, because sugary potatoes become brown or black during the cooking process. Thus, careful control of storage temperatures is important. Similar effects of low temperature on starch-to-sugar conversions have been observed in tulip bulbs, acorns, maple stems, and in roots of sweet potatoes and mullein (*Verbascum thapsus*), although the critical temperatures for starch breakdown vary with the tissues involved (Glier and Caruso, 1974). In banana fruits, low temperatures have the opposite effect, and sugars are converted to starch in the refrigerator.

The mechanism of these temperature effects has not been well explained. It is probable that temperature affects several reactions involving starch and the various sugars and sugar phosphates, not just a single reversible reaction. Because sugars are important substrates for respiration, the amount available, as influenced by temperature, affects the rates of respiration and of other metabolic processes.

Hydrolysis of Fructosans As mentioned in Section 10.6, the principal carbohydrate food reserve material in some species, most notably certain grasses of temperate regions and certain members of the Compositae (Asteraceae) family, is not starch. Instead, fructosans predominate. Compared to the starch molecules, fructosans are very small. The number of fructose units varies from a dozen or so in grass leaves to about 35 in members of the Compositae. Some fructosans are branched, while others are not. Many contain fructoses connected by β-links between C-1 of one molecule and C-2 of another. These are collectively often called **inulins,** as in dandelion roots, Jerusalem artichoke tubers, dahlia tubers, and iris bulbs (Fig. 12-3a). Most fructosans of grasses are connected by β-links between C-2 and C-6, as exemplified by the **phleins** from timothy roots (*Phleum pratense*) or the **levans** of ryegrass (*Lolium perenne*) and orchardgrass (*Dactylis glomerata*), as shown in Fig. 12-3b.

Surprisingly little is known about fructosan metabolism compared to its importance, although these molecules are hydrolyzed by β-fructofuranosidase enzymes having specificity for the particular β-2,1 or β-2,6 links involved. For example, one

such enzyme from the Jerusalem artichoke tuber successively cleaves fructose units from inulin until a mixture of fructose and the terminal sucrose unit remain:

$$\text{glucose-fructose-(fructose)}_n + n\,H_2O \rightarrow$$
$$\text{(fructosan)}$$

(R12-5)

$$n \text{ fructose } + \text{ glucose-fructose}$$
$$\text{(sucrose)}$$

Hydrolysis of Sucrose The sucrose produced from fructosans in reactions such as R12-5 and those sucrose molecules translocated by the phloem from leaves to various receiving (sink) cells must be degraded to glucose and fructose before respiratory breakdown can continue. In R10-15, we showed a mechanism for sucrose degradation occurring in cells in which most of the hexose units of translocated sucrose are converted to starch; that mechanism involved reaction of sucrose with UDP or ADP. But for cells in which starch production is not important and most of the hexose units are degraded to CO_2 and water, sucrose appears to be irreversibly hydrolyzed to free glucose and fructose by **invertases:**

$$\text{sucrose} + H_2O \rightarrow \text{glucose} + \text{fructose} \qquad \text{(R12-6)}$$

12.3 Glycolysis and Fermentation

There is a group of reactions, called **glycolysis,** in which glucose, glucose-1-P, or fructose set free by the reactions just discussed are converted to pyruvic acid in the cytoplasm outside any organelle. Glycolysis represents the first of three major phases of respiration and is followed by Krebs cycle and electron transport processes occurring in the mitochondria. Historically, the individual reactions of glycolysis, now believed to occur in all living organisms, were discovered between about 1912 and 1935 by German scientists interested

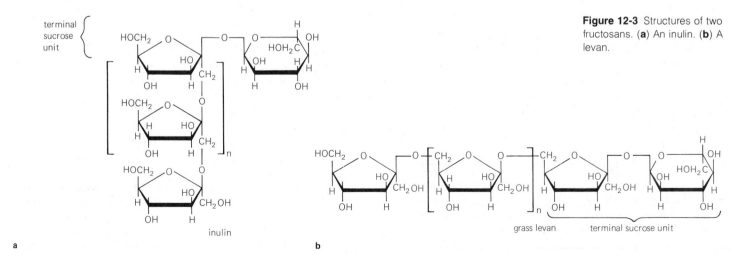

Figure 12-3 Structures of two fructosans. (**a**) An inulin. (**b**) A levan.

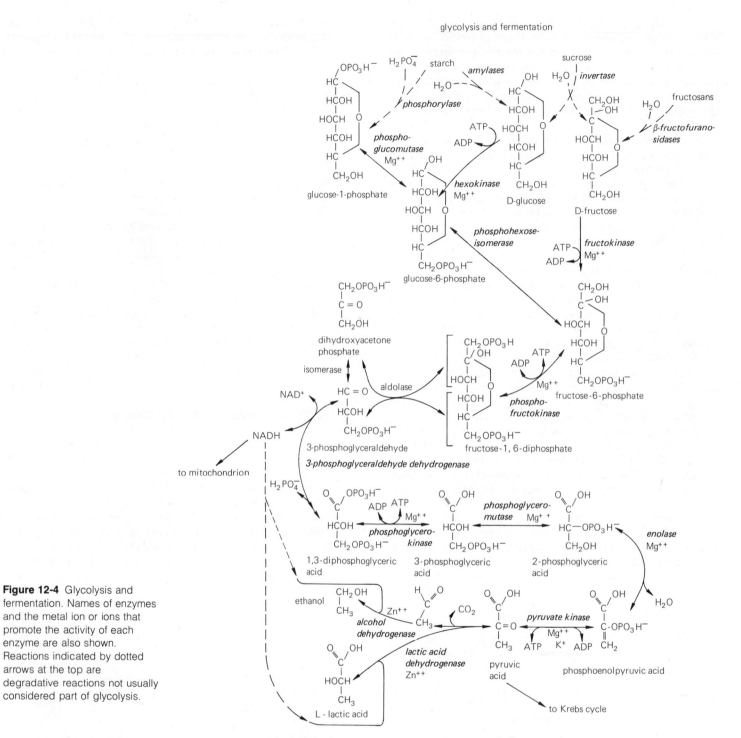

Figure 12-4 Glycolysis and fermentation. Names of enzymes and the metal ion or ions that promote the activity of each enzyme are also shown. Reactions indicated by dotted arrows at the top are degradative reactions not usually considered part of glycolysis.

in the production of alcohol by yeast and by other Germans concerned with the breakdown of animal starch (glycogen) to pyruvic acid in muscle cells (Lipmann, 1975). Another name for the pathway by which these substrates are degraded to pyruvic acid is the *Embden-Meyerhof-Parnas pathway*, in honor of three people who contributed most toward elucidation of

the reactions involved. As we shall see, all of the reactions of glycolysis are part of **fermentation,** the process by which sugars are converted to ethyl alcohol plus CO_2 or to lactic acid or malic acid.

The individual reactions of glycolysis and fermentation, the enzymes catalyzing these, and particular requirements of the enzymes for metal ion coenzymes are listed in Fig. 12-4. Inspection shows that the glycolytic production of pyruvic acid (Fig. 12-4 bottom, center) does not involve any reaction in which O_2 is absorbed or CO_2 is released. These reactions occur later.

An important function of glycolysis is the formation of molecules that can be removed from the pathway to synthesize several other constituents of which the plant is composed. This function is not apparent from Fig. 12-4, but it will be given special attention in Section 12.8 and Fig. 12-10.

A second essential function of glycolysis is production of some ATP, for even though the initial utilization of glucose or fructose requires input of two ATP molecules, two ATPs are later released for each three-carbon unit involved (in the reactions catalyzed by phosphoglycerokinase and pyruvate kinase). Thus, there is a total production of four ATPs per hexose used, and a net production of two ATPs per hexose. If glucose-1-P, glucose-6-P, or fructose-6-P is the substrate, one less ATP is required to start with, so the net ATP production is three per hexose phosphate. In photosynthetic cells, 3-PGA and 3-phosphoglyceraldehyde produced in the Calvin cycle and dihydroxyacetone phosphate produced from 3-phosphoglyceraldehyde become respiratory substrates when they are transported into the cytoplasm from the chloroplasts.

A third essential function of glycolysis is production of another molecule, NADH, that we may conveniently consider to be energy rich. NADH (or, in leaves, both NADH and NADPH) is formed by reduction of NAD^+ (or $NADP^+$) during the oxidation of 3-phosphoglyceraldehyde to 1,3-diPGA. Remember that this reaction is the reverse of that which occurs in photosynthesis (R10-2). In photosynthesis, there is sufficient NADPH available to drive that reaction toward aldehyde formation, but in respiration there is not. The energy in NADH may be used in either of two ways. First, NADH may enter the mitochondria where electron transport reactions oxidize it and convert the energy into the terminal phosphate bond energy of two ATPs (described later). If this happens, the two NADH molecules arising from each hexose may be considered equal to the readily available energy of four ATP molecules. Second, NADH (or NADPH) can be used as a source of electrons in numerous anabolic reactions that will be discussed later in this chapter and in other chapters.

Whether NADH is oxidized in the mitochondria or used to drive other reductive reactions is dependent on the internal O_2 concentration, because, as will soon become apparent, the mitochondria cannot oxidize NADH back to NAD^+ unless O_2 is present. In the absence of O_2, there is soon less NAD^+ because it is converted to NADH, and the pyruvic acid becomes reduced to form **ethyl alcohol** (**ethanol**) and CO_2, or to **lactic acid** (Fig. 12-4, bottom, left). Note that ethanol formation involves the loss of CO_2 from pyruvate and the subsequent reduction of the acetaldehyde intermediate by NADH. When yeast cells are grown under anaerobic conditions, mitochondrial development is greatly reduced, although the mitochondria of anaerobically grown rice coleoptiles seem rather normal (Vartapetian et al., 1975). The sugars provided

yeast cells are converted largely to ethanol and CO_2 by the fermentation process. Apparently, nearly all of the NADH formed during fermentation is used just as rapidly. ATP production is limited during fermentation, and this is one of the principal reasons for slow growth of anaerobic organisms.

Fermentation of glucose requiring use of the NADH molecules released in glycolysis may be summarized by R12-7:

$$\text{glucose} + 2\ ADP^{-2} + 2\ H_2PO_4^- \longrightarrow$$
$$2\ \text{ethanol} + 2\ CO_2 + 2\ ATP^{-3} + 2\ H_2O \qquad \text{(R12-7)}$$

Fermentation also occurs in the cells of many higher plants when O_2 is not available or cannot readily enter. In water-logged soils, the O_2 content and its rate of diffusion are reduced, so aerobic respiration in roots is slowed, fermentation occurs, and growth is slow. Roots of trees growing in swamps are often shallow for this reason. Poor entry of O_2 into seeds often occurs in the initial stages of germination, because the seed coats prevent adequate gas diffusion. Here, too, part of the pyruvic acid is converted to ethanol or to lactic acid. Lactic acid is less common and usually less abundant than ethanol in higher plants but is quite common in algae and fungi under anaerobic conditions. Overworked muscles of humans suffer from a lack of O_2 and also produce lactic acid from pyruvate, causing the familiar stiffness.

Because of the close relation between glycolysis and fermentation or lactic acid production, it is easy to overlook the fact that glycolysis occurs normally in aerobic organisms in the presence of O_2. Glycolysis (beginning with glucose) may be summarized by R12-8:

$$\text{glucose} + 2\ NAD^+ + 2\ ADP^{-2} + 2\ H_2PO_4^- \longrightarrow 2\ \text{pyruvate}$$
$$+ 2\ NADH + 2\ H^+ + 2\ ATP^{-3} + 2\ H_2O \qquad \text{(R12-8)}$$

The pyruvate formed in this reaction then moves into the mitochondria, where it is oxidized to CO_2 by reactions of the Krebs cycle. The electrons and protons arising in these oxidation steps are accepted by NAD^+ and FAD, forming NADH plus H^+ and $FADH_2$. Finally, these electrons are transported through a mitochondrial electron transport system to O_2, which, with H^+, forms H_2O. Before describing these reactions in detail, we shall first discuss some of the morphological properties of mitochondria.

12.4 Mitochondria

In some respects, mitochondria are similar to chloroplasts, although functionally they are quite different. Both contain circular DNA that has genetic information for production of a small percentage of their enzymes, both are formed mainly or entirely by division of the corresponding preexisting organelle, and both are surrounded by a double membrane or envelope with an extensive inner membrane system. Each plant cell contains many mitochondria, frequently about 200,

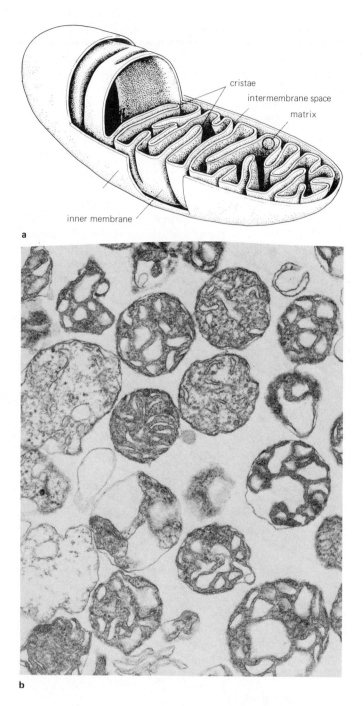

and estimates of the maximum number per cell have reached 2,000 for immature metaxylem vessel elements from young corn shoots (Malone et al., 1974).

Mitochondria are only a micron or so long, about the length of many bacteria, and although they can be seen with a light microscope, their fine structure is made clear only with the electron microscope. Their general morphology is shown by electron micrographs of earlier chapters, such as Fig. 11-7, and is illustrated in a highly idealized form in Fig. 12-5a. The inner membrane of the mitochondrial envelope is highly convoluted, protruding into the interior in sheetlike patterns in many places. Each such convolution is called a **crista** (plural **cristae**). In some plant mitochondria, sheetlike cristae are well developed (Bajrachara et al., 1976; Öpik, 1974), but this varies with the type of cell, its age, and its extent of development. In many, one crista is fused to another in the interior of the mitochondrion, forming a continuous sac-like intermembrane compartment between them (Fig. 12-5b), while other modifications are shown in electron micrographs and drawings by Malone et al. (1974). Regardless of their form, cristae contain most of the enzymes that catalyze steps of the electron transport system following the Krebs cycle, so the increased surface area they provide is of great importance. The Krebs cycle reactions occur in the protein-rich region between the cristae.

12.5 The Krebs Cycle

The **Krebs cycle** was named in honor of the English biochemist H. A. Krebs, who, in 1937, proposed a cycle of reactions to explain how pyruvate breakdown takes place in the breast muscle of pigeons. He called his proposed pathway the **citric acid cycle,** because citric acid is an important intermediate. Another common name for the same group of reactions is the **tricarboxylic acid (or TCA) cycle,** a term used because some of the acids involved have three carboxyl groups. It was not until the early 1950s that mitochondria capable of carrying out this cycle were isolated from plant cells.

The initial step leading to the Krebs cycle involves loss of CO_2 from pyruvate and combination of the remaining two-carbon acetate unit with a sulfur-containing compound, **coenzyme A** (abbreviated **CoA**), forming **acetyl CoA** (CH_3—C—SCoA). This reaction normally proceeds only in the for-
$\overset{\|}{O}$

ward direction. This and another comparable role in the Krebs cycle represent important reasons why sulfur is an essential element.

The reaction of pyruvate decarboxylation also involves a phosphorylated form of **thiamine (vitamin B_1)** as a prosthetic group. Participation of thiamine in this reaction partially explains the essential function of vitamin B_1 in plants and animals. Besides the loss of CO_2, two hydrogen atoms are removed from pyruvic acid during the formation of acetyl CoA. The enzyme catalyzing the complete reaction is called **pyruvic acid dehydrogenase,** but it is actually an organized

Figure 12-5 (**a**) An idealized drawing of a mitochondrion with cristae. (From Jensen and Salisbury, 1974.) (**b**) An electron micrograph of mitochondria isolated from corn shoots. The mitochondria have been cut in various planes. The unstained (white) areas correspond to the intermembrane spaces shown in Fig. 12-5(a). These spaces apparently end adjacent to the outer membrane in regions above and below the plane of sectioning. The matrix of Fig. 12-5(a) is shown here as the heavily stained regions between the cristae. (From C. Malone et al., 1974, Plant Physiology 53:920.)

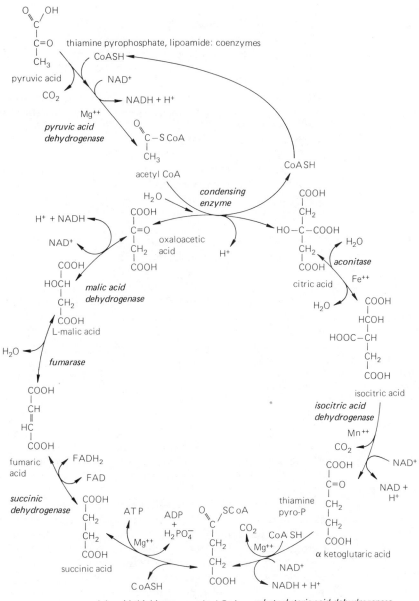

Figure 12-6 Reactions of the Krebs cycle, including enzymes and coenzymes.

complex containing numerous copies of three different enzymes (Reid et al., 1977). The hydrogen atoms removed are finally accepted by NAD^+, yielding NADH. This and other Krebs cycle reactions are described in Fig. 12-6.

The Krebs cycle accomplishes the removal of electrons from organic acid intermediates and transfer of these electrons to NAD^+ or FAD. Notice that none of the dehydrogenase enzymes of the cycle use $NADP^+$ as an electron acceptor. In fact, $NADP^+$ is usually undetectable in plant mitochondria, a situation opposite to that of chloroplasts where $NADP^+$ is abundant but where there is much less NAD^+. Not only are NADH and $FADH_2$ important products of the Krebs cycle,

but one molecule of ATP is formed from ADP and $H_2PO_4^-$ during the conversion of **succinyl coenzyme A** to **succinic acid.** Two additional CO_2 molecules are released in these Krebs cycle reactions, and there is a net loss of both carbon atoms from the incoming acetate of acetyl CoA. The release of CO_2 in the Krebs cycle accounts for the product CO_2 in the summary equation for respiration R12-1, but no O_2 is absorbed during any Krebs cycle reaction.

Primary functions of the Krebs cycle may be listed as follows:

1. Production of the electron donors NADH and $FADH_2$ that are subsequently oxidized

2. Direct synthesis of a limited amount of ATP (one ATP for each pyruvate oxidized, or half that formed in glycolysis)

3. Formation of carbon skeletons that can be used to synthesize certain amino acids that, in turn, are converted into larger molecules (see Section 12.8 and Fig. 12-10)

Considering that two pyruvates are produced in glycolysis from each glucose, the overall reaction for the Krebs cycle may be written as follows:

$$2 \text{ pyruvate} + 8 \text{ NAD}^+ + 2 \text{ FAD} + 2 \text{ ADP}^{-2} + 2 \text{ H}_2\text{PO}_4^- +$$
$$4 \text{ H}_2\text{O} \rightarrow 6 \text{ CO}_2 + 2 \text{ ATP}^{-3} + 8 \text{ NADH} + 8 \text{ H}^+ + 2 \text{ FADH}_2$$
$$(\text{R12-9})$$

12.6 The Electron Transport System and Oxidative Phosphorylation

The Electron Transport System When NADH and FADH$_2$ produced in the Krebs cycle or in glycolysis are oxidized, ATP is produced. Although this oxidation involves O$_2$ uptake and H$_2$O production, neither NADH nor FADH$_2$ can combine directly with O$_2$ to form H$_2$O. Rather, the electrons are transferred via several intermediate compounds, comparable to a bucket brigade, before H$_2$O is made. These electron carriers constitute what is often called the **electron transport system** or **cytochrome system** of the mitochondria. The transfer of electrons proceeds from carriers that are thermodynamically difficult to reduce (those with low reduction potentials) to those that have a greater tendency to accept electrons (have higher reduction potentials). Oxygen has the greatest tendency to accept electrons, and so it ultimately does so. Each member of the system usually accepts electrons only from the previous member close to it. They are thought to be arranged in an assembly-line fashion in the cristae, and there are several thousand electron transport systems in each mitochondrion. As in the chloroplast electron transport system involved in transfer of electrons *from* water molecules, the mitochondrial system involves cytochromes (apparently three of the *b* type and two of the *c* type) and a quinone (**ubiquinone**). Also present are several **flavoproteins** (riboflavin-containing proteins), some iron-sulfur proteins similar to ferredoxin, another cytochrome-containing substance called **cytochrome oxidase,** and a few other electron carriers not yet identified. The cytochromes and cytochrome oxidase both contain **iron** as part of a heme group.

An oversimplified model for the sequence by which electrons are transferred from NADH or FADH$_2$ through the transport chain, finally to be accepted by O$_2$ in a reaction catalyzed by cytochrome oxidase, is shown in Fig. 12-7. (We shall first emphasize the main-line pathway proceeding from left to right and from NADH in the figure, although we shall return later to the cyanide-resistant pathway shown at the top.) *The principal importance of this electron transport system is that the flow of electrons to O$_2$ drives ATP formation from ADP*

and H$_2$PO$_4^-$. We call this production of ATP in the mitochondria **oxidative phosphorylation.** Two ATP molecules are believed to be formed when both electrons from FADH$_2$ move through the system to O$_2$, while three such ATPs are thought to result when both electrons from any of the NADH molecules arising in the Krebs cycle are transferred. When each NADH molecule produced in glycolysis is oxidized, only two ATPs are produced (Palmer, 1976). Oxidative phosphorylation is thus a far more effective way of producing ATP than are glycolysis and fermentation; the advantage of O$_2$ to growing organisms should now be more apparent.

A summary reaction for the electron transport system in which both of the NADH molecules from glycolysis and eight NADH molecules and two FADH$_2$ molecules from the Krebs cycle are oxidized may be written as in R12-10:

$$10 \text{ NADH} + 10\text{H}^+ + 2\text{FADH}_2 + 32 \text{ ADP}^{-2} + 32 \text{ H}_2\text{PO}_4^- +$$
$$6 \text{ O}_2 \rightarrow 10 \text{ NAD}^+ + 2 \text{ FAD} + 32 \text{ ATP}^{-3} + 42 \text{ H}_2\text{O}$$
$$(\text{R12-10})$$

If we now add each of the summary reactions for the three major partial processes of respiration, including glycolysis (R12-8), the Krebs cycle (R12-9) and the electron transport system (R12-10), we arrive at the following net equation for respiration of glucose:

$$\text{glucose (C}_6\text{H}_{12}\text{O}_6) + 6 \text{ O}_2 + 36 \text{ ADP}^{-2} + 36 \text{ H}_2\text{PO}_4^- \rightarrow$$
$$6 \text{ CO}_2 + 36 \text{ ATP}^{-3} + 42 \text{ H}_2\text{O} \qquad (\text{R12-11})$$

Except for the inclusion of ADP, H$_2$PO$_4^-$, ATP, and the additional 36 H$_2$O molecules formed when ATP is formed from ADP and H$_2$PO$_4^-$, R12-11 is the same as the summary equation R12-1 at the beginning of the chapter. ATP represents the energy usable by the plant.

We can also estimate the efficiency of respiration in terms of how much energy in glucose can be trapped in the terminal phosphate bond of ATP. The standard Gibbs free energy change (ΔG) at *p*H 7 for one mol of glucose or fructose is $-686,000$ cal, so we shall use this as the energy in the reactants of R12-11. Among the products, only the energy in the terminal phosphate of ATP is additional useful energy. The standard ΔG for the terminal phosphate in each mol of ATP is about $-7,600$ cal/mol at *p*H 7, or $-273,600$ cal in 36 mol of ATP. Thus, the efficiency is about $-273,600/-686,000$, or 40 percent. The remaining 60 percent is lost as heat.

Coupling of Electron Transport and Oxidative Phosphorylation Since mitochondria can be isolated from cells (Fig. 12-5b), oxidative phosphorylation occurring *in vitro* can be studied. In fact, most of what we know about the electron transport sequence has been learned in this way. One of the limiting factors in O$_2$ uptake and electron transfer is the amount of ADP and H$_2$PO$_4^-$. If they are limiting, the rates of ATP production and O$_2$ uptake are reduced. These and other results indicate that the processes of oxidative phosphorylation and electron transport are closely dependent upon each other.

We call this interdependence **coupling.** It appears that in the living plant the two processes are usually also closely coupled, because experimentally imposing a work load on the plant, such as by providing an abundance of nutrient ions to be actively absorbed, causes an increase in O_2 absorption. This increased respiration is known as **salt respiration.** Salt respiration results from the provision of ADP and $H_2PO_4^-$ formed when the energy in the terminal phosphate bond of ATP is utilized in ion transport, as discussed in Section 5.10. It is a classic example of feedback control not involving inhibition or activation of enzymes.

Another way of stimulating respiration of plants is by adding proper concentrations of 2,4-dinitrophenol. Dinitrophenol prevents conversion of ADP and $H_2PO_4^-$ to ATP, yet it enhances flow of electrons and H^+ to O_2. It therefore acts as an *uncoupler* of oxidative phosphorylation, because a net production of ATP does not occur in its presence, even though oxidation continues rapidly because of ADP availability (Beevers, 1974). (How dinitrophenol apparently acts as an uncoupler through a ferryboat-type transport of H^+ across the membrane, thereby neutralizing charge separations according to the chemiosmotic theory, is discussed in Special Topic 9.1.)

The Mechanism of Oxidative Phosphorylation ATP synthesis in the electron transport system depends upon the formation of a pH gradient across the inner mitochondrial membrane in a manner analogous to photosynthetic phosphorylation in chloroplasts, except that the direction of H^+ movement is toward the outside of the membrane, rather than toward the inside as in the chloroplast lamellae. The Mitchell chemiosmotic theory discussed in Special Topic 9.1, primarily in relation to photosynthetic phosphorylation, may be consulted for details. The important principle is that as O_2 pulls electrons from NADH or $FADH_2$ to form H_2O, H^+ are removed from the inner matrix between the cristae and are released into the intermembrane space between the two membranes. During formation of ATP and H_2O from ADP and $H_2PO_4^-$, the OH^- in this H_2O are attracted toward the H^+ in the intermembrane space, while the H^+ in the H_2O are attracted toward the high pH region in the matrix. This removal of H_2O allows ATP formation to occur.

Cyanide-Resistant Respiration Aerobic respiration of most organisms, including some plants, is strongly inhibited by certain negative ions that combine with the iron in cytochrome

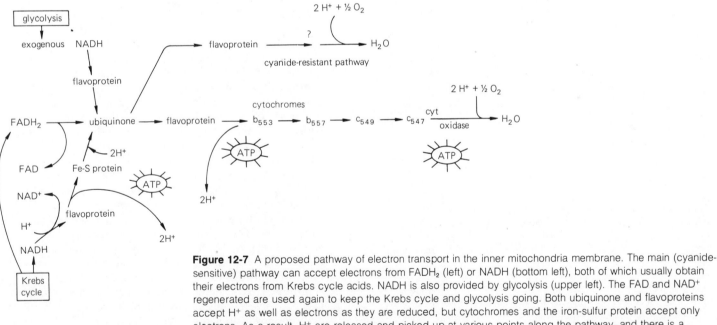

Figure 12-7 A proposed pathway of electron transport in the inner mitochondria membrane. The main (cyanide-sensitive) pathway can accept electrons from $FADH_2$ (left) or NADH (bottom left), both of which usually obtain their electrons from Krebs cycle acids. NADH is also provided by glycolysis (upper left). The FAD and NAD^+ regenerated are used again to keep the Krebs cycle and glycolysis going. Both ubiquinone and flavoproteins accept H^+ as well as electrons as they are reduced, but cytochromes and the iron-sulfur protein accept only electrons. As a result, H^+ are released and picked up at various points along the pathway, and there is a transport of H^+ toward the outer (cytoplasmic) side of the inner mitochondria membrane where these reactions occur. Production of ATP molecules from ADP and $H_2PO_4^-$ by oxidative phosphorylation neutralizes this H^+ gradient, thus facilitating ATP synthesis. Note that there are three "sites" of ATP formation (oxidative phosphorylation) for each NADH arising in the Krebs cycle but only two such sites for $FADH_2$ or for NADH arising exogenously, as in glycolysis. Cytochrome oxidase catalyzes the last step (O_2 absorption and H_2O production), and it is sensitive to cyanide, azide, and carbon monoxide. The oxidase in the cyanide-insensitive pathway (top) has not been identified.

oxidase. Two such ions, cyanide (CN^-) and azide (N_3^-), are particularly effective. Carbon monoxide (CO) also forms a strong complex with this iron, preventing electron transport and poisoning respiration. In many plant tissues, however, poisoning cytochrome oxidase by such inhibitors has only a minor effect on respiration. The respiration that continues in this situation is said to be **cyanide-resistant respiration.** Several fungi and algae and a few bacteria and animals are also resistant to cyanide, azide, and CO (Henry and Nyns, 1975).

The reason respiration can continue when cytochrome oxidase is blocked is that such mitochondria have an alternate branch in the electron transport pathway (see Fig. 12-7, top). This branch or alternate route also allows transport of electrons to oxygen. The terminal oxidase and most other components of this route have not yet been identified, but it is known that little or no oxidative phosphorylation is coupled to it; that is, it leads mainly to production of heat, not ATP. (This heat production is sometimes beneficial to certain plants, as in the pollination ecology of the arum lilies. See Fig. 12-8 and the personal essay by Bastiaan J. D. Meeuse on pp. 188–189.) Even though oxidative phosphorylation from the cyanide-resistant pathway itself seems not to occur, Fig. 12-7 shows that the dichotomy from the cytochrome oxidase pathway occurs only after the first oxidative phosphorylation "site" (point of energy production for ATP formation). Therefore, if electron transport through the cyanide-resistant pathway is fast enough, both heat and ATP will be formed at rapid rates, although at a considerable expense of reserve foods.

12.7 The Pentose Phosphate Pathway

After 1950, plant physiologists gradually became aware that glycolysis and the Krebs cycle were not the only reactions by which plants could obtain energy from the oxidation of sugars into carbon dioxide and water. Much of the research indicating that a different pathway also occurs in plants was performed in the 1950s by Martin Gibbs at Cornell University and by Bernard Axelrod and Harry Beevers at Purdue University. Because five-carbon sugar phosphates are intermediates, this series of reactions is often called the **pentose phosphate pathway,** sometimes abbreviated **PPP.** (It has also been called the Warburg-Dickens pathway, the hexose monophosphate shunt, and the phosphogluconate pathway.)

Several of the compounds of the PPP are also members of the Calvin cycle, in which sugar phosphates are synthesized in chloroplasts. (In fact, most reactions of the PPP and the Calvin cycle were discovered during the same period in the early 1950s.) The major difference between the Calvin cycle and the PPP is that in the PPP, sugar phosphates are oxidized rather than synthesized. In this respect, the reactions of the PPP are similar to those of glycolysis. In addition, glycolysis and the PPP have certain reactants in common and both occur in the cytoplasm outside any organelle, so the two pathways are interwoven to some extent. One important difference is that in the PPP, $NADP^+$ always accepts the electrons from the sugar phosphates, whereas in glycolysis, NAD^+ is the more common acceptor.

Reactions of PPP are outlined in Fig. 12-9. The first reaction involves glucose-6-phosphate, which can arise either from starch breakdown by starch phosphorylase followed by phosphoglucomutase action in glycolysis, or from the addition

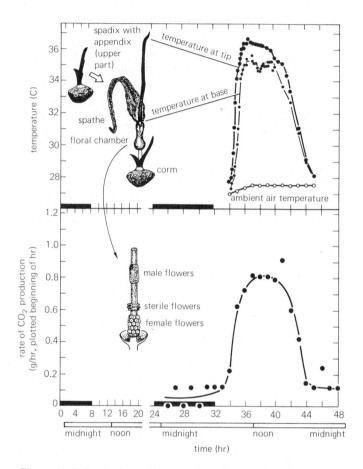

Figure 12-8 Respiration and temperature of a spadix of *Sauromatum guttatum* as a function of time. *Sauromatum* is a Pakistani and Indian genus in the Araceae family. Growth from the corm to a structure about 50 cm tall may occur in about 9 days (drawings at upper left), with a maximum growth rate of 7 to 10 cm/day. If this occurs in constant light, the spathe remains wrapped around the spadix, but after the "normal" time for flowering has passed, a single period of darkness, if it is long enough (bar on the abscissa—two 8-hour dark periods were given in this experiment), will initiate opening of the spathe and a burst in CO_2 production (note extremely large quantities) with a concurrent rise in temperature. The heat apparently serves to volatilize various compounds (especially amines and ammonia), which give an odor of rotting meat. Carrion flies and beetles are attracted and serve in pollination They enter the floral chamber (lower drawing, somewhat schematic). (Original data. Experiment performed for use in this text by B. J. D. Meeuse, R. C. Buggeln, and J. R. Klima of the University of Washington, Seattle.)

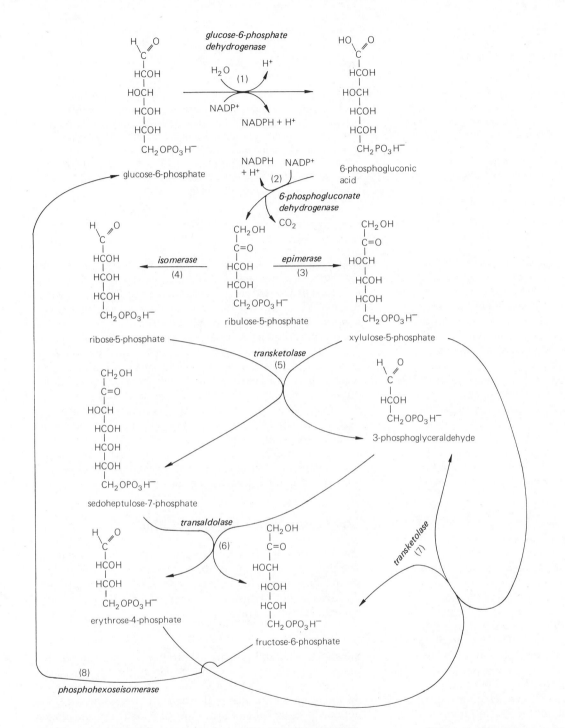

Figure 12-9 The pentose phosphate respiratory pathway, PPP. Reactions are numbered to make the sequence easier to follow. Three particularly significant products of the pathway are two molecules of NADPH (first two reactions at top), ribose-5-P (left center) used as a precursor for ribose in nucleotides and nucleic acids, and erythrose-4-P (bottom center), a precursor of several aromatic compounds.

of the terminal phosphate of ATP to glucose, or directly from photosynthetic reactions. It is immediately oxidized (dehydrogenated) by **glucose-6-phosphate dehydrogenase** to **6-phosphogluconic acid.** This is an example of oxidation of an aldehyde to an acid, and NADP$^+$ is the electron acceptor. The 6-phosphogluconate doesn't accumulate but is both dehydrogenated and decarboxylated rapidly by **6-phosphogluconic acid dehydrogenase,** yielding the five-carbon compound

ribulose-5-phosphate, NADPH, and CO_2. These two dehydrogenase reactions are not reversible. They represent the only oxidations in the PPP, and the CO_2 arising in the second step is the only CO_2 released in the entire pathway. The main function of the subsequent reactions is to cycle the ribulose-5-phosphate back into glucose-6-phosphate, which then undergoes dehydrogenation and decarboxylation by the two dehydrogenase enzymes mentioned.

Until the late 1960s, it was assumed that the PPP occurs only in the cytoplasm outside any organelle. It then became clear that it also occurs in chloroplasts, but only in darkness (Schnarrenberger et al., 1973). In the light, glucose-6-phosphate dehydrogenase is inactivated, so the entire pathway stops and the Calvin cycle begins to operate. The mechanism of inactivation of this dehydrogenase is closely related to that by which some of the Calvin cycle enzymes are activated (Section 10.4), except that sulfhydryl groups of the chloroplast photosystem have opposite effects.

The PPP is important because it acts as a mechanism for glucose breakdown, because it provides reduced NADPH for synthetic reactions, because it provides ribose-5-phosphate used in nucleotide and nucleic acid synthesis and, as will be seen in Chapter 14, because the erythrose-4-phosphate needed in synthesis of lignin and other aromatic compounds can arise from this pathway.

12.8 Respiratory Production of Molecules Used for Synthetic Processes

Near the beginning of this chapter, we stated that respiration is important to cells because many compounds are formed that can be diverted into other substances needed for growth. Many of these are large molecules, including lipids, proteins, and nucleic acids. To form them, the aid of the high transfer potential in the terminal phosphate bond of ATP is needed, and, in some of the reactions, the electrons present in NADH or NADPH are also required. Other processes requiring significant quantities of NADH or NADPH are the reductions of nitrate and sulfate ions (Chapter 13). We emphasized in the preceding section the importance of the PPP in producing NADPH, ribose-5-P, and erythrose-4-P for anabolic reactions. The role of glycolysis and the Krebs cycle in producing carbon skeletons for synthesis of larger molecules is summarized in Fig. 12-10. You should remember when studying this figure that if carbon skeletons are diverted from the respiratory pathway as shown, not all the carbons from the original respiratory substrate (e.g., starch) will be released as CO_2, and not all the electrons normally transferred by NADH or NADPH will combine with O_2 to form H_2O. Yet it is essential that some of the substrate molecules be totally oxidized, because use of diverted carbon skeletons to form larger molecules will usually be effective only when oxidative phosphorylation is producing an adequate supply of ATP.

Another important point is that when organic acids of the Krebs cycle are removed by conversion into aspartic acid, glutamic acid, chlorophyll, and cytochromes, for example, the regeneration of oxaloacetic acid will be prevented. Thus, diversion of organic acids from the cycle would soon cause it to stop if it were not for another mechanism to generate oxaloacetate. In all plants, both day and night, there is some fixation of CO_2 into oxaloacete by the reaction catalyzed by PEP carboxylase (see R10-3 and Fig. 12-10, left). This reaction is essential for growth processes because it replenishes organic acids converted into larger molecules.

12.9 Factors Affecting Respiration

Many environmental factors influence the respiratory activity. The previous discussion of the individual reactions involved should allow you to understand better how such factors affect the overall rate of respiration and its importance to plant maintenance and growth.

Substrate Availability Any respiration depends upon the presence of a food source, and it is evident that starved plants having low starch, fructosan, or sugar reserves respire at relatively low rates. Plants deficient in sugars often respire noticeably faster when sugars are provided. We mentioned in the previous chapter that the lower, shaded leaves usually respire slower than the upper leaves exposed to higher light intensities. If this were not true, the lower leaves would probably die sooner than they do. The difference in sugar contents resulting from unequal photosynthetic rates is probably responsible for the lower respiratory rates of shaded than illuminated leaves.

If starvation becomes extensive, even proteins can be

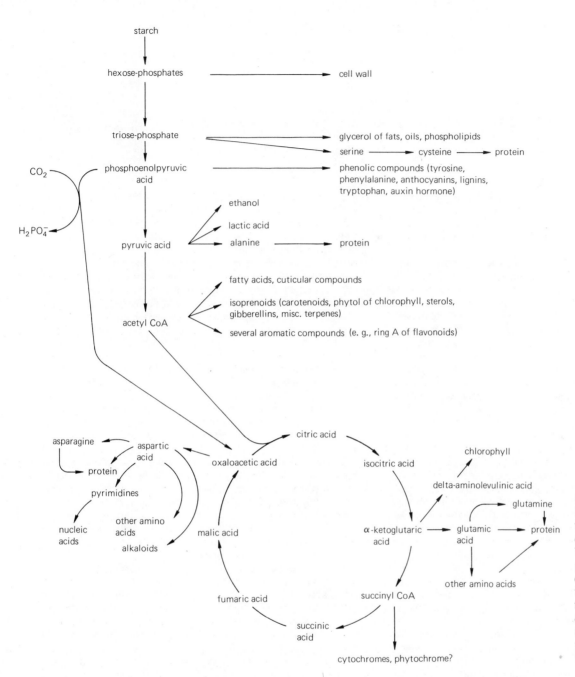

starch

hexose-phosphates → cell wall

triose-phosphate → glycerol of fats, oils, phospholipids

serine → cysteine → protein

phosphoenolpyruvic acid → phenolic compounds (tyrosine, phenylalanine, anthocyanins, lignins, tryptophan, auxin hormone)

CO_2

$H_2PO_4^-$

pyruvic acid → ethanol

lactic acid

alanine → protein

acetyl CoA → fatty acids, cuticular compounds

isoprenoids (carotenoids, phytol of chlorophyll, sterols, gibberellins, misc. terpenes)

several aromatic compounds (e. g., ring A of flavonoids)

asparagine

aspartic acid

protein

pyrimidines

nucleic acids

other amino acids

alkaloids

malic acid

oxaloacetic acid

citric acid

isocitric acid

chlorophyll

delta-aminolevulinic acid

glutamine

α-ketoglutaric acid → glutamic acid → protein

other amino acids

succinyl CoA

fumaric acid

succinic acid

cytochromes, phytochrome?

Figure 12-10 Glycolysis and Krebs cycle simplified to show their roles in formation of some other essential compounds. There are many unsolved problems of transport from one organelle to another apparent in this diagram, such as the transport of α-ketoglutarate (or δ-aminolevulinic acid) from mitochondria to chloroplasts, where it is used in chlorophyll synthesis.

oxidized. These proteins are hydrolyzed into their amino acid subunits, which are then **catabolized** (degraded) by glycolytic and Krebs cycle reactions. In the case of glutamic and aspartic acids, the relation to the Krebs cycle is especially clear, because they are converted to α-ketoglutaric and oxaloacetic acids, respectively (Fig. 12-10). Similarly, alanine is oxidized via pyruvic acid. As leaves become senescent and yellow, most of the protein and other nitrogenous compounds in the chloroplasts are broken down. Ammonium ions released from various amino acids are combined in glutamine and

asparagine (as amide groups) during this process, and this prevents ammonium toxicity. These processes will be discussed in the next chapter.

Available Oxygen The O_2 supply also influences respiration, but the magnitude of its influence differs greatly with various plant species and probably even with different organs of the same plant. Normal variations in O_2 content of the air are much too small to influence the respiratory rate of most leaves

The Hazards of Hubris

Bastiaan Jacob Dirk Meeuse

F.B.S. 1975

Bastiaan Jacob Dirk Meeuse, who often signs himself "Don Sebastian, Pooh of Puyallup," or other colorful aliases and pseudonyms, is a plant physiologist at the University of Washington in Seattle. He is, like Frits Went and Johannes van Overbeek, a botanist trained or produced by the Dutch East Indies. But what better introduction could he have than the following essay?

If luck is indeed the reward for virtue in a previous incarnation, I must in the dim grey past have been a pretty good boy. I was born and raised in the East Indies—in retrospect, quite a breeding ground for Dutch biologists! I lived in Buitenzorg (now Bogor) for four years. The town is famous for its botanical gardens and was, in my youth, also a place where much scientific research in biology was going on. Without realizing it, I was exposed to a whole array of subtle biological influences and ways of thinking—and, in addition, I was leading the free, outdoorsy life of a healthy boy. I developed a strong sense of identification with the plants and animals around me. For that reason, I shall never find it in my heart to see an organism only as a source of a particular enzyme! Also, I quickly learned that no two organisms are the same, and that it pays to get to know the different types. One does not want to confuse a poisonous snake with its imitators!

Getting settled emotionally in Holland was a traumatic experience, but going to Leiden to start my studies in biology (later to be continued in Delft) was like being reborn. Again, I was incredibly lucky to be guided by men with such diverse approaches and viewpoints as L. G. M. Baas Becking, the plant physiologist and ecologist who also was a specialist in orchids; A. J. Kluyver, the great microbiologist and biochemist; G. van Iterson, Jr., a multifaceted man of astounding thoroughness and grasp; and Niko Tinbergen, a pioneer in animal behavior studies. The generosity and understanding they showed toward me, their callow and awkward student, is simply beyond praise.

Another rebirth came on May 5, 1945, when Holland was liberated after five years of Nazi occupation, and still another when I moved to America in 1947. But alas, my enthusiasm for contemporary biology, as practiced by freshly trained "biologists," is on the wane. A few years ago, I took a young molecular biologist to a mountain stream to show him our dipper or water ouzels, explaining to him on the way up that these birds actually fly under water. He listened eagerly, then asked me quite sincerely: "Do they have gills?" Flabbergasted, it took me a few seconds before I could answer: "Yes . . . just like whales." He laughed sheepishly, and we shook hands. I don't think there were any hard feelings—but *what sort of biological training had this bright young fellow received?* At about the same time, a famous biochemist came over to talk to us about his research on biochemical evolution. As he explained it, almost flippantly, only after he had spent considerable time and effort (and money!) did he "discover" the basic fact that horseshoe crabs are entirely different from other crabs. (As an elementary zoology book could have told him, they are indeed farther apart than flies and butterflies.) The listeners laughed uproariously, but I feel that instead they should have keelhauled him immediately—in shark-infested waters! Quite obviously, he had never looked at his animals carefully.

Since that day, I have come to the sad conclusion that this gentle-man's attitude is not rare at all, so I am not whipping a dead horse when, in this essay, I break a lance for the age-old idea of patient, humble, and careful observation, without hubris*—*at least as a first step*. Few biological case histories can teach us the value of this attitude better than that of the arum lilies, such as *Arum maculatum,* whose traplike inflorescences develop heat and smell to attract the pollinators. In their behavior, these plants liberally drop clues that even a biochemist would do well to heed. Heat, normally an undesirable byproduct of metabolism in plants, has definite survival value here, for without it there would be insufficient evaporation of the odoriferous compounds (amines, ammonia, indole, skatole) that act as attractants. No heat would mean no fertilization and no sexually produced offspring. Heat generation in *Arum* and *Sauromatum* (the voodoo lily) takes place in the highly specialized appendix, the naked and sterile club- or finger-shaped upper part of the inflorescence, where large masses of starch are broken down in a respiration process so fierce that at its peak it compares favorably with that of a flying hummingbird. The result is a 10 to 15 C temperature difference with the environment, as first reported (for *Arum*) by the famous Lamarck in 1778. During the period of heat production (i.e., insect attraction), only the pistillate flowers at the base of the floral column are receptive, a situation that practically insures cross-pollination. The staminate flowers, placed higher up on the column, then shower the captive insects with their pollen until, many hours later, certain wilting phenomena begin to make escape from the trap possible. As a rule, escapees are soon trapped again by another inflorescence that is still in the smelly, receptive stage; there they will carry out an act of cross-pollination for a second time.

Two of the main points of interest in the situation are the thermogenic nature of the respiration process in the appendix and the perfect timing noticeable in the pollination events. The thermogenicity is due to the fact that the mitochondria of the appendix, in addition to having the "classical" respiratory chain for electron transport, which is coupled with the generation of ATP, possess an alternate pathway that is not so coupled. Functional replacement of the classical pathway by the alternate one, on the first day of flowering, leads to an "uncoupling"; that is, replacement leads to a considerable curtailment of ATP production and thus to a spectacular increase in heat production per unit of time. In addition, the total flux of electrons through glycolysis and the respiratory mitochondrial transport chain is spectacularly increased.

With the valuable help of a group of loyal collaborators (Richard Buggeln, James Chen, Conrad Hess, Tom Johnson, John Klima, Dennis Olason, Bruce Smith, and Sue Summer Thayer), I have (among

*Hubris: exaggerated pride or self-confidence.

and stems (Armstrong and Gaynard, 1976). Furthermore, the rate of O_2 penetration is usually sufficient to maintain normal O_2 uptake levels by the mitochondria.

One might suspect that in such bulky tissues as carrot roots, potato tubers, and other storage organs, the rate of O_2 penetration would be so low as to cause the respiration inside to be primarily anaerobic. Quantitative data on gas penetration into such organs is meager, but the measurements at hand show that rate of O_2 movement through them is certainly much less than in air. In pure water, O_2 diffusion is nearly 300,000 times slower than in air. However, the French physiologist H. Devaux showed, in 1890, that central regions of bulky plant tissues do respire aerobically. He demonstrated the importance of intercellular spaces for gaseous diffusion. We now know that these spaces may represent significant amounts of the total tissue volume. For example, in potato tubers approximately 1 percent of the volume is occupied by air spaces. Such intercellular air spaces extend from the stomata of leaves to every cell in the plant, aiding their aerobic respiration.

We mentioned in Chapter 5 that diffusion of O_2 through the intercellular space system from the leaves to the roots was probably important in providing O_2 needed for root respiration in waterlogged soils. Rice roots are particularly adapted to growing under anaerobic conditions. This adaptation consists in part of forming new roots with an unusually large volume of intercellular spaces, and it is accompanied by the gradual development of a greater capacity to absorb ions anaerobically. These facts emphasize that O_2 and other gases can move through plant tissues more rapidly than might have been expected for an organism with no lungs or hemoglobin in blood to help transport the gas. Nevertheless, we do not yet understand why perpetually flooded soils cause poor growth of most species. Such species generally cannot adapt as rice does, and an O_2 deficiency in the roots must be one important factor. Another factor is the accumulation of ethylene, as discussed in Chapter 17.

The influence of O_2 on respiration depends upon the way in which the rate is measured. Of course, if no O_2 is present, as when pure N_2 is used to replace it, none can be absorbed. Under this condition, the rates of sugar loss and CO_2 output are, however, not reduced to zero. In fact, they sometimes increase! This surprising result shows that O_2 can actually inhibit the utilization of sugars in respiration. The effect is on glycolysis and is known as the **Pasteur effect,** after Louis Pasteur, who first recorded the phenomenon in his studies with fermentation and winemaking over a century ago.

Figure 12-11 shows an example of the Pasteur effect in apple fruits. As O_2 is reduced, the CO_2 production gradually decreases but then begins to increase again at low O_2 levels. This increase results when the O_2 level is so low that it no longer interferes with glycolysis. Glycolysis is then anaerobic, and ethanol usually accumulates. The inhibition of glycolysis by O_2 in animal cells results largely from allosteric control of phosphofructokinase, the enzyme that converts fructose-6-phosphate into fructose-1,6-diphosphate with ATP (Fig. 12-4). **Phosphofructokinase** is also an important control site in plant cells but not the only one (Turner and Turner, 1975).

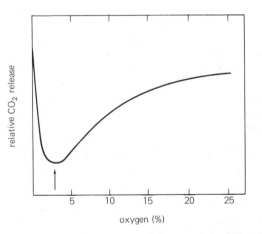

Figure 12-11 The influence of atmospheric oxygen concentration upon CO_2 production in apple fruits. On the right side of the arrow, increasing O_2 supply increases respiration because of stimulated Krebs cycle activity, yet anaerobic CO_2 release from pyruvate and accompanying ethanol release become insignificant in this region of the curve due to indirect inhibition effects of O_2 on glycolysis. At the left of the arrow, the O_2 concentration is low enough to allow a very rapid breakdown of sugars to ethanol and CO_2. (Redrawn from James, 1963.)

Although this enzyme requires ATP to carry out the reaction, it is inactivated by ATP when the levels of this nucleotide become too high. On the other hand, it is activated by $H_2PO_4^-$. When O_2 is abundant and ATP is being formed from ADP and $H_2PO_4^-$ by oxidative phosphorylation, the $ATP/H_2PO_4^-$ ratio rises, and phosphofructokinase is inactivated. When O_2 is deficient, the $ATP/H_2PO_4^-$ ratio decreases, and the enzyme is activated. This represents an important example of negative and positive feedback control, discussed in general in Section 8.6.

The Pasteur effect is probably important to the maintenance of food reserves in flooded roots and other examples of O_2 deficiency. It also has some practical importance in fruit storage, especially with apples. Here the object in storage is to prevent extensive sugar loss during overripening. This is done by carefully decreasing the O_2 to the point where aerobic respiration is at a minimum, but where sugar breakdown by anaerobic processes is not stimulated. Additional CO_2 is also added to the air, and the temperature is lowered, which further prevents overripening. As discussed in Chapter 17, CO_2 inhibits action of a fruit-ripening hormone, ethylene, and this is at least one explanation for its effectiveness in inhibiting overripening.

Temperature For most plant species and even for individual tissues, the Q_{10} for respiration between 5 and 25 C is usually near 2.0 to 2.5. (The Q_{10} is the ratio of the rate of a process at one temperature divided by the rate at a temperature 10 C lower; see Section 1.3.) With further increases in temperature up to 30 and 35 C, the respiration rate still increases, but less rapidly, so the Q_{10} begins to decrease. A possible explanation for the decrease is that the rate of O_2 penetration into the cells begins to limit respiration at these higher temperatures where chemical reactions could otherwise proceed rapidly. Diffusion of O_2 and CO_2 are also stimulated slightly by increased temperature, but the Q_{10} for these physical processes is less than half that of the Q_{10} for the chemical reactions of respiration.

With a further rise in temperature to 40 C or so, the rate of respiration is actually decreased, especially if the plants are maintained under such conditions for long periods. Apparently, the required enzymes begin to be denatured at a rapid rate, preventing a continued stimulation of metabolism. With pea seedlings it was found that, although increasing the temperature from 25 to 45 C initially caused a much more rapid respiration, within about two hours the rate was less than before. A probable explanation is that the two-hour period was long enough to partly denature the enzymes.

Type and Age of Plant Because there are large morphological differences among members of the plant kingdom, it is to be expected that differences in metabolism also exist. In general, bacteria and fungi respire considerably more rapidly than do higher plants. Various organs or tissues of higher plants also exhibit large variations in rates. One reason that bacteria and

fungi have so much higher values than higher plants, based on dry weight, is that they contain only small amounts of stored food reserves and have no nonmetabolic woody cells. Such dead woody cells contribute to the dry weight and strength of the plant, but not to respiration. Similarly, root tips and other regions containing meristematic cells with large protoplasm contents have high respiratory rates expressed on a dry weight basis. If comparisons are made on a protein basis, these differences are smaller. In general, there is a fairly good correlation between the rate of growth of a particular cell type and its respiration rate. This results from several factors, such as the use of ATP, NADPH, and NADH for synthesis of proteins, cell wall materials, membrane components, and nucleic acids. Inactive seeds and spores have the lowest (usually undetectable) respiration rates, but here the effect is not entirely due to a lack of growth. Rather, certain changes in the protoplasm, especially desiccation, shut off metabolism. Such seeds and spores generally contain abundant food reserves.

The age of intact plants influences their respiration to a large degree. Figure 12-12 shows how the rate changes in a whole sunflower from germination until after flowering. The rate is expressed as the amount of CO_2 released per amount of pre-existing dry weight. The curve is extrapolated to zero time to show the common initial large burst in respiratory activity as the dry seeds absorb water and germinate. Respiration remained high during the period of most rapid vegetative growth, but then fell before flowering. In this example, much of the respiration in the mature plants is carried out by the young leaves and roots and by the growing flowers.

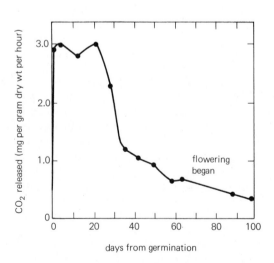

Figure 12-12 Respiration of whole sunflower plants from germination until maturity. The rate gradually declined after the 22nd day, even though the rate for individual parts, such as inflorescences, increased for a time after that. (Drawn from data of F. Kidd, et al., 1921, Proceedings of the Royal Society of London, series B 92:368.)

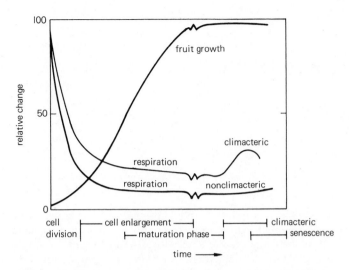

Figure 12-13 Stages in development and maturation of fruits that undergo the climacteric respiration increase and of those that do not. Discontinuities in the lines indicate that the time scale was changed to show differences in development rates of different fruits. The growth pattern may be single or double sigmoid (see Chapter 15). (From J. Biale, 1964, Science 146:880. Copyright 1964 by the American Association for the Advancement of Science.)

Changes in respiration also occur during the development of ripening fruits. In all fruits, the respiration rate is very high when they are young, while the cells are still rapidly dividing and growing (Fig. 12-13). The rate then gradually declines, even if the fruits are picked. In many species, however, of which the apple is a good example, the gradual decrease in respiration is reversed by a sharp increase, known as the **climacteric rise**. The climacteric usually coincides with full ripeness and flavor of the fruits, and its appearance is hastened by traces of ethylene, known to stimulate fruit ripening (Hulme, 1970). Further storage is accompanied by senescence and by decreases in respiration, which would approach zero in the absence of attack by microorganisms.

Some fruits do not show the climacteric rise, including the citrus fruits, grapes, pineapple, and strawberry (lowest curve, Fig. 12-13). Grapefruits, oranges, and lemons are allowed to ripen on the trees and, if removed sooner, their respiration simply continues at a gradually decreasing rate. The advantages and disadvantages to a climacteric rise are unknown. The biochemical basis for the climacteric respiratory rise is also unclear, but it is being actively investigated.

13

Assimilation of Nitrogen and Sulfur

The importance of nitrogen to plants is emphasized by the fact that only carbon, oxygen, and hydrogen are more abundant in them. Although nitrogen occurs in a vast number of plant constituents, most of it is in proteins. Sulfur is only about one tenth as abundant in plants as nitrogen, but again it occurs in many molecules, especially proteins. Both elements are usually absorbed from the soil in highly oxidized forms and must be reduced by energy-dependent processes before they are incorporated into proteins and other cellular constituents. Humans cannot duplicate this reduction, just as we cannot reduce CO_2. To describe the ways in which nitrate

and sulfate are reduced and subsequently combined with carbohydrate skeletons to form amino acids is an important task of this chapter. We shall also discuss fixation of atmospheric N_2 and interconversions of nitrogen compounds during various stages of plant development.

13.1 The Nitrogen Cycle

Nitrogen exists in several forms in our environment. The continuous interconversion of these forms by physical and

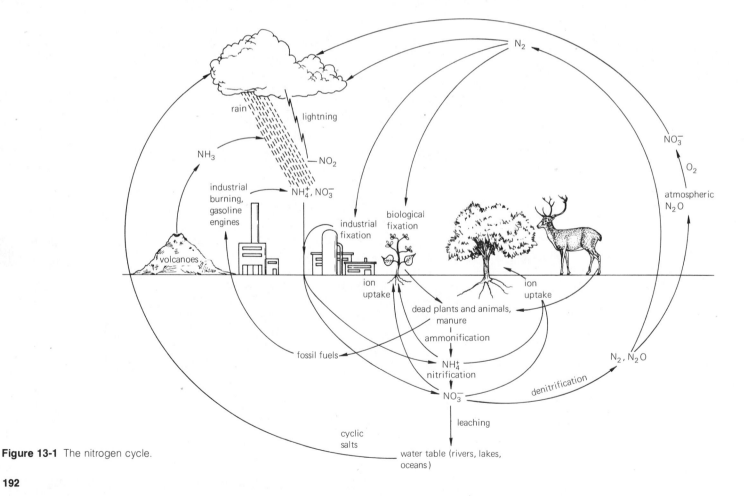

Figure 13-1 The nitrogen cycle.

Nitrogen Fertilizers and World Energy Problems

For most native species used by man, economics do not warrant application of nitrogen fertilizers. Growth of such plants depends upon soil nitrogen deposited by rainfall, upon decomposition of previous vegetation and, as will be described in the next section, upon nitrogen fixed by soil bacteria, by bacteria or actinomycete-like organisms in root nodules, and by other kinds of organisms. On the other hand, most high-yielding crops are fertilized with anhydrous ammonia, liquid ammonia, or $(NH_4)_2SO_4$. Nitrate salts are also used, but they are more expensive. Legume crops are seldom fertilized with nitrogen, because they contain root nodules in which nitrogen fixation occurs, and they usually do not yield more if fertilized.

The major source of nitrogen fertilizers is presently the petroleum-dependent industrial formation of ammonia by the Haber-Bosch process. This process involves reaction of N_2 and H_2 at high pressure and temperature. The ammonia can then be converted to $(NH_4)_2SO_4$, NO_3^-, or urea fertilizers. The Haber-Bosch process is expensive, and so are transportation, storage, and application of the fertilizers. Furthermore, crops generally recover only about half the applied nitrogen, mainly because of losses due to leaching of NO_3^- and to denitrification. Presently, of the total amount of energy required in production of a corn crop in the United States, about one third is needed to manufacture, transport, and apply the nitrogen fertilizer. Nevertheless, such fertilizers are widely used, and our food supply depends upon them (Hardy and Havelka, 1975).

R. W. F. Hardy, research director for E. I. duPont de Nemours and Co., compiled data showing a direct relation between the worldwide use of nitrogen fertilizers and cereal grain production from 1956 to 1971 (Fig. 13-2). This relation held for both more developed and less developed countries. Since most of the world's food comes from cereal grains, the continuing importance of such fertilizers is obvious.

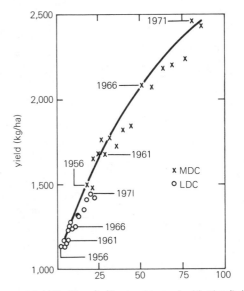

Figure 13-2 The relation between the use of nitrogen fertilizers and the yield of cereal grains from 1956 to 1971 in more developed countries (MDC) and less developed countries (LDC). Although yields leveled off late in this period, they were generally proportional to the amount of fertilizer applied per hectare. (From R. W. F. Hardy, 1975, *in* W. E. Newton and D. J. Nyman, eds., Symposium on dinitrogen fixation, Washington State University Press, Pullman, Wa., pp. 693–717.)

biological processes constitutes the nitrogen cycle, as summarized in Fig. 13-1 (Delwiche, 1970).

Vast amounts of N_2 occur in the atmosphere (78 percent by volume), yet it is energetically difficult for living organisms to obtain it. Although N_2 moves into leaf cells along with CO_2 through stomates, enzymes are available to reduce only the CO_2, so N_2 molecules move out as fast as in. Most of the N_2 in living organisms arrives there only after fixation (reduction) by numerous microorganisms, some of which exist in the roots of certain plants, or by industrial fixation to form fertilizers. Small amounts of nitrogen also move from the atmosphere to the soil as NH_4^+ and NO_3^- in rain and are then absorbed by roots. This NH_4^+ arises from industrial burning, volcanic activity, and forest fires, while NO_3^- arises from an oxidation of N_2 by O_2 or ozone caused by lightning and ultraviolet radiation. Another source of NO_3^- is from the oceans. Wind-whipped white caps produce minute droplets of water called aerosols, from which the water evaporates, leaving ocean salts suspended in the atmosphere. These salts can be brought to the land in rainwater. They are called **cyclic salts** and are more important near coastlines than farther inland.

Absorption of NO_3^- and NH_4^+ by plants allows them and the animals that eat them to form numerous nitrogenous compounds, most abundantly proteins. Shedding of dead leaves and roots, death of whole plants and animals, and manure represent important sources of nitrogen returned to the soil. Most of the soil nitrogen occurs as decaying microbial, plant, and animal matter, but the majority of this is insoluble and not immediately available for plant use. Several soils contain small but detectable amounts of various amino acids, the origin of which is probably from the action of microorganisms on decaying organic matter and from excretion by living roots. Such amino acids can be absorbed and metabolized by plants, but in nature these and other more complex nitrogen compounds contribute only negligibly to the plant's nitrogen nutrition in a direct way. They are, however, of great importance as a nitrogen reservoir from which NH_4^+ and NO_3^- arise.

Conversion of organic nitrogen to NH_4^+ by soil microbes is called **ammonification.** In warm, moist soils with near neutral pHs, NH_4^+ is further oxidized by bacteria to NO_3^- within a few days of its formation or its addition as a fertilizer.

This oxidation, called **nitrification,** provides energy for survival and growth of such microbes, just as does oxidation of more complex foods for other organisms. On many acidic soils, however, nitrifying bacteria are less abundant, so NH_4^+ becomes a more important nitrogen source than NO_3^-. Many forest trees absorb most of their nitrogen as NH_4^+, because of the low pH common to forest soils and perhaps because other factors contribute to slow rates of nitrification. Because of its positive charge, NH_4^+ is adsorbed to soil colloids, while NO_3^- is not adsorbed and is much more readily leached. Nitrate is also lost from soils by **denitrification,** the process by which N_2 or N_2O are formed from NO_3^- by anaerobic bacteria (Delwiche and Bryan, 1976). These bacteria use NO_3^- rather than O_2 as an electron acceptor during respiration, thus obtaining energy for survival. Denitrification occurs relatively deep in the soil where O_2 penetration is limited, in waterlogged or compacted soils, and in certain regions near the soil surface where the O_2 demand is high because of especially rapid aerobic oxidation of organic matter.

13.2 Nitrogen Fixation

The process by which N_2 is reduced to NH_4^+ is called **nitrogen fixation.** It is, so far as we know, always carried out by prokaryotic microorganisms. Principal N_2 fixers include certain free-living soil bacteria, free-living blue-green algae on soil surfaces or in water, blue-green algae in symbiotic associations with fungi in lichens or with ferns, mosses, and liverworts, and bacteria or other microbes associated symbiotically with roots, especially those of legumes. It is of great importance to the food chain in forests, deserts, freshwater and marine environments, and even arctic regions.

About 10 percent of the 12,000 or more species in the Leguminosae (Fabaceae) family have been examined for N_2 fixation, and approximately 90 percent of these possess this ability. Important nonlegumes fixing N_2 are primarily trees and shrubs, including members of the genera *Alnus* (alder) *Myrica* (such as *M. gale,* the bog myrtle), *Shepherdia, Coriaria, Hippophae, Ceanothus, Eleagnus, Casaurina,* and seven other known genera (Bond, 1976). They are typically pioneer plants on nitrogen-deficient soils, for example, *M. gale* on the bog soils of western Scotland and *Casaurina equisetifolia* on sand dunes of tropical islands. Figure 13-3 demonstrates the important role of root nodules in providing nitrogen to *Myrica gale.* Only the nodulated group on the left could survive in the nutrient solution lacking nitrogen salts.

Most of the microorganisms responsible for N_2 fixation in the roots of nonlegumes have not been identified. In some tropical trees, they are blue-green algae, but in most species, prokaryotic actinomycete-like organisms (filamentous bacteria) carry out this process. In the legumes, bacterial species of the genus *Rhizobium* are responsible. A different *Rhizobium* species is generally effective on each legume species.

The *Rhizobia* are aerobic bacteria that can persist saprophytically in the soil until they infect a root hair or a damaged

Figure 13-3 Bog myrtle *(Myrica gale)* plants cultured with (left) and without (right) nodules in a hydroponic solution lacking nitrogen salts. Plants were grown from seed for 5 months in the solutions. (From G. Bond, 1963, in P. S. Nutman and B. Mosse, eds., Symbiotic associations, Thirteenth Symposium of the Society for General Microbiology.)

epidermal cell. Root hairs of some legumes respond to invasion by surrounding the bacteria with a threadlike structure called the **infection thread,** although such a thread has not been detected in many legumes (Dart, 1975). The infection thread consists largely of an infolded and extended plasma membrane of the cell being invaded, along with new cellulose formed *inside* this membrane. The bacteria multiply extensively inside the thread, which extends inwardly and penetrates the cortex and, less commonly, even the pericycle. Here the bacteria are released into the cytoplasm of such cells and apparently stimulate some of them (especially tetraploid cells) to divide. These divisions lead to a proliferation of tissues, eventually forming a mature **root nodule** made up of tetraploid cells filled with bacteria and extending outside the root (Fig. 13-4). The bacteria usually occur in the cytoplasm in groups, each group surrounded by a membrane. Each enlarged, nonmotile bacterium is referred to as a **bacteroid.** Figure 13-5 shows an electron micrograph of groups of bacteroids in soybean root nodules.

Nitrogen fixation in legumes occurs by bacteroids. The legume provides the bacteroids with carbohydrates, which they oxidize. Some of the electrons and ATP obtained during this oxidation are used to reduce N_2 to NH_4^+. The NH_4^+ is then converted into the other nitrogenous compounds, most of which are absorbed by the surrounding plant cells and translocated to other organs. Thus, the plant benefits from the nitrogen compounds provided by the bacteroids, while the bacteroids require the carbohydrates and perhaps other compounds from the plant.

The Biochemistry and Physiology of Nitrogen Fixation Nitrogen fixation is dependent on ATP, a source of electrons that has not yet been identified, and an enzyme complex called **nitrogenase.** The number of ATP molecules needed for each N_2 molecule fixed is not yet known; estimates vary from 6 to

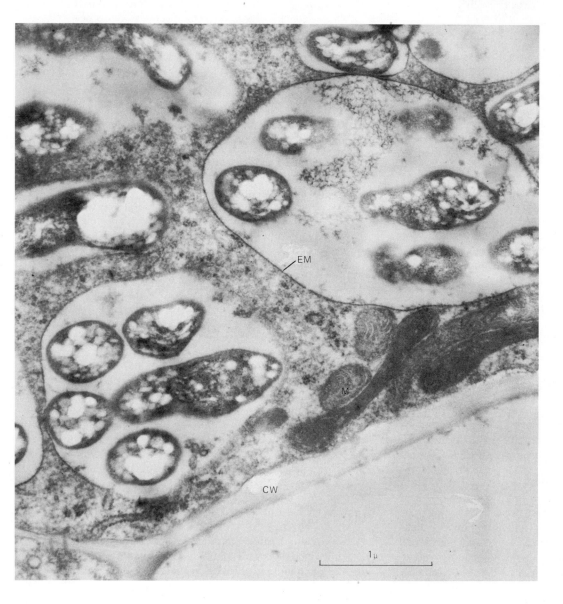

Figure 13-4 External appearance of soybean root nodules. Note the approximately spherical shape. Healthy nodules are pink because of leghemoglobin, a reddish protein. (From F. J. Bergersen and D. J. Goodchild, 1973, Australian Journal of Biological Sciences 26:729–740.)

Figure 13-5 Electron micrograph of part of a bacteroid-containing cell from a soybean root nodule. Bacteroids are in groups of four to six, each group present in a sac-like structure that is surrounded by an envelope membrane (EM). Cell wall (CW) and a few mitochondria (M) of the nodule cell are visible. Light areas in bacteroids are probably food reserves of poly-β-hydroxybutyric acid. (From D. J. Goodchild and F. J. Bergersen, 1966, Journal of Bacteriology 92:204–213.)

20. The original source of electrons in the plant is sucrose or other carbohydrate transported from leaves to roots through the phloem. Respiration in the bacteroids then provides the more immediate electron donor for the nitrogenase system. Ferredoxin is one important electron donor, and it probably obtains its electrons from NADH or NADPH formed during the oxidation of sucrose. Molybdenum and iron are essential for N_2 fixation, because they form part of a protein in some ways similar to ferredoxin (but about 20 times as large) called the **Mo-Fe protein.** Both the single molybdenum atom and one or more of the 20 or so atoms of iron in this protein probably undergo alternate oxidation and reduction during fixation. The other smaller **Fe protein** in the nitrogenase complex contains three or four iron atoms but no molybdenum. Both proteins are essential for fixation.

The NH_4^+ is not translocated out of the nodules, but is converted to glutamine, glutamic acid, and asparagine by reactions discussed in Section 13.4. Each of these amino acid or amide molecules can be converted into bacterial proteins or transported to host cells. The amides are especially efficient molecules in which to transport nitrogen because of the relatively high N:C ratio in them. Asparagine is the most abundant compound translocated in soybeans, alfalfa, and field peas, but in alder a less common amide, citrulline, proved most important in the transport system. These molecules move from the bacteroid-containing cells into the xylem cells of the numerous vascular bundles that surround the nodule proper and then into the xylem of the root and shoot to which the vascular bundles of the nodule are connected. Surprisingly, the root cells seem to receive appreciable nitrogen only after it has moved into the leaves. Apparently, transport from leaves back to roots occurs through the phloem. There is therefore a cycling process: from nodules upward to shoots in the xylem, then movement of leftover molecules downward to roots via the phloem.

For one species, field pea, an attempt to produce a balance sheet for carbon flow into and out of the nodule was made (Minchin and Pate, 1973). For every 100 carbon atoms fixed by photosynthesis during vegetative growth, about 32 moved to the nodule. Of these 32, 12 were lost as CO_2 during respiration, 5 were used in growth of the nodule, and the other 15 were transported back to the shoot as asparagine and other nitrogenous compounds.

Several factors influence the rate of N_2 fixation and probably also the carbon flow balance. In general, those factors that enhance photosynthesis, such as adequate moisture, warm temperature, bright sunlight, and high CO_2 levels, also stimulate N_2 fixation. Consistent with this, the rate of fixation in nodules is usually maximal in the early afternoon when translocation of sugars from the leaves is occurring rapidly. These factors are important for future research efforts to design crops that will fix N_2 faster (see box), because fixation is often limited by the rate of photosynthesis (Hardy and Havelka, 1975; Sinclair and deWit, 1975).

The stage of growth also influences N_2 fixation. Two important United States grain legumes, soybeans and peanuts, both show maximum fixation rates after flowering when the demand for nitrogen in the developing seeds and fruits increases. These species, as is common to legumes, contain seeds that are especially protein rich. In fact, soybean seeds contain 40 percent protein, the highest percentage of any plant known. Quantitatively, about 90 percent of the N_2 fixation in both species occurs during the period of reproductive development and only 10 percent during the first two months of vegetative growth. Somewhat surprisingly, N_2 fixation provides only one fourth of the total in mature soybean, peanut, and pea plants grown on soils of normal fertility; the remaining three fourths is absorbed as NO_2^- or NH_4^+ from the soil, mainly during the period of vegetative growth. Nevertheless, yields of these plants cannot usually be increased by fertilizing with NO_3^-, NH_4^+, or urea, because N_2 fixation is decreased in proportion to the added amount of fertilizer nitrogen absorbed. This decrease probably results from inhibition of nitrogenase synthesis and more rapid senescence of the nodules by NH_4^+ or organic nitrogen compounds formed from such fertilizers.

The amount of N_2 fixed by perennial native and legume crop species during various times in the growing season is unknown but is probably again greatest during reproductive development. The percentage of nitrogen in such species derived from fixed N_2 is likely to be much greater than that in annual grain legumes, because the nodules are perennial, and fixation should begin earlier than in annuals, where nodule development must start anew each year. Furthermore, native N_2-fixing plants often grow in relatively unfertile soils where the main input of nitrogen is from fixation. For alfalfa fields, from which the crop is removed as soon as it begins to bloom, the main supply of nitrogen is also from N_2 fixation.

13.3 Metabolism of Nitrate Ions

For the vast majority of plants that cannot fix N_2, the only available nitrogen sources are NO_3^- and NH_4^+. We shall emphasize first the assimilation of NO_3^- because of its relative abundance in the soil and because NH_4^+ is derived in the plant from it (Hewitt, 1975; Hewitt et al., 1976).

Sites of Nitrate Assimilation Since both roots and shoots require organic nitrogen compounds, in which of these organs is NO_3^- reduced and incorporated into organic compounds? Roots of some species can synthesize all of the organic nitrogen compounds they need from NO_3^-, while the roots of others rely on the shoots for reduced, organic nitrogen. Figure 13-6 shows the kinds of nitrogenous substances present in the xylem transport stream of several herbaceous species during vegetative growth with their roots in sand watered with a nitrate-containing nutrient solution. (The legumes did not contain root nodules.) None of these plants translocated detectable amounts of NH_4^+ to the shoots, but some transported large quantities of organic nitrogen compounds derived from NH_4^+. The cocklebur and lupine (a legume) represent extremes

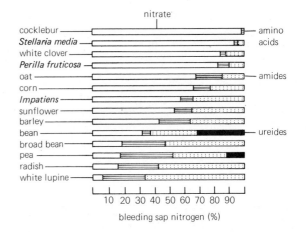

Figure 13-6 Relative amounts of nitrogen compounds in the xylem sap of various species. Plants were grown with their roots in sterile sand watered with a sterile nutrient solution containing 140 mg/1 nitrogen as nitrate (10 mM NO_3^-), then the stems were severed to collect xylem sap from the cut stumps. Species at the top of the figure transport primarily nitrate; those at the bottom, mainly amides and amino acids. Two legumes also transport ureides, especially citrulline. (From J. S. Pate, 1973, Soil Biology and Biochemistry 5:109–119.)

among those studied. Cocklebur roots have almost no ability to reduce NO_3^-, so they depend upon amino acids and amides translocated in the phloem from the leaves. In the lupine, nearly all of the NO_3^- was absorbed and converted into amino acids and amides in the roots. A number of woody species, most notably apple, behave like the lupine and translocate almost no NO_3^- to the shoots; the shoots of these species are provided a diet of organic nitrogen. Much more research is needed with conifers, other forest trees, and herbaceous plants grown in nature where the fungal hyphae in mycorrhizae may contribute organic nitrogen compounds to the root cells.

The relative amounts of NO_3^- and organic nitrogen in the xylem of various species depend upon environmental conditions. Even plants that normally do not translocate much NO_3^- do so if provided with excessive amounts in the soil, or if the roots are kept cold. Under these conditions, reduction of NO_3^- in the roots cannot keep pace with the transport process. Reduction will then occur in the leaves and stems, taking place at rapid rates under high light intensities and warm temperatures.

The Process of Nitrate Reduction The overall equation for reduction of NO_3^- to NH_4^+ is an energy-dependent process summarized in R13-1.

$$NO_3^- + 8 \text{ electrons} + 10 \text{ H}^+ \rightarrow NH_4^+ + 3 \text{ H}_2O \quad \text{(R13-1)}$$

The oxidation number of nitrogen changes from +5 to −3. The eight electrons required are provided in a number of separate partial reactions. As the H^+ are removed, a slight rise in *p*H causes PEP carboxylase to become more active, thus leading to malic acid formation (see R10-3 and R10-4). The resulting malic acid ionizes and replaces the H^+ used in NO_3^- reduction.

The only intermediate product detected during NO_3^- reduction is NO_2^-. Nitrite is produced by action of **nitrate reductase,** an enzyme occurring in the cytoplasm that transfers electrons from NADH or occasionally NADPH, depending on the species from which it is obtained, to NO_3^-. Nitrate reductase contains riboflavin (FAD) and molybdenum (Mo^{6+}), both of which accept and transfer electrons. The nitrate reductases of some fungi and green algae also contain a cytochrome component, and it seems reasonable that the enzymes in higher plants do also. The role of nitrate reductase is shown by R13-2, which represents only a partial reaction (addition of two electrons) of the overall process summarized in R13-1.

$$NADH + H^+ \rangle \langle FAD \rangle \overset{2\ H^+}{\underset{\downarrow}{\land}} 2\ Mo^{5+} \rangle \langle 2\ H^+ + NO_3^-$$
$$NAD^+ \swarrow \searrow FADH_2 \rangle \uparrow\uparrow \langle 2\ Mo^{6+} \swarrow \searrow NO_2^- + H_2O \quad (R13-2)$$

Synthesis of nitrate reductase can be stimulated in whole plants or excised plant parts by various factors, most important of which is NO_3^-. This is a specific case of the general phenomenon of **enzyme induction,** enhanced formation of an enzyme by a particular chemical added to the cells or formed by them under a particular environmental condition (Marcus, 1970; Filner et al., 1969). Enzyme induction occurs extensively in microorganisms but probably to a more limited extent in complex plants and animals. It is often caused by the substrate upon which the enzyme acts, and this is true for nitrate reductase. If a seedling is grown in darkness in a nutrient solution without NO_3^-, activity of nitrate reductase in the roots and shoots is very low or undetectable. If NO_3^- is then added, the enzyme begins to be synthesized within a few hours. If light is also supplied, synthesis in the leaves is much faster, partly because NO_3^- is translocated to the leaf cells more rapidly and partly because photosynthesis supplies carbohydrates that are oxidized and provide energy for enzyme production. Furthermore, light activates the phytochrome system (Chapter 19), which somehow causes greater activity of the ribosomal enzyme-synthesizing system in the leaves. Nitrate reductase can also be induced in excised cotyledons of some dicots by adding cytokinin hormones, but the presence of NO_3^- in such cells is probably also necessary. Thus, although various factors besides NO_3^- contribute to induction, nitrate reductase is the best known example of a substrate-inducible enzyme in plants.

Reduction of Nitrite to Ammonium Ions Nitrite arising in the cytoplasm from nitrate reductase action is transported into chloroplasts in leaves or into proplastids in roots, where subsequent reduction to NH_4^+ catalyzed by **nitrite reductase** occurs:

$$NO_2^- + 6\ \text{electrons} + 8\ H^+ \longrightarrow NH_4^+ + 2\ H_2O \quad (R13-3)$$

In leaves, reduction of NO_2^- to NH_4^+ is directly coupled to the chloroplast photosynthetic electron transport process. By this we mean that the light-driven split of H_2O provides the six electrons used to reduce NO_2^-, and that the use of these electrons oxidizes electron donors such as ferredoxin so that they can again accept electrons from H_2O. Although reduced ferredoxin is a normal donor of electrons to leaf nitrite reductase, the reducing substance of roots is unknown. When nitrite reductase is studied *in vitro,* it will not accept electrons from NADH, NADPH, or reduced flavin compounds such as $FADH_2$. Although reduced ferredoxin will provide electrons for isolated root nitrite reductases, it does not do this *in vivo,* because neither the proplastids nor other parts of roots have detectable amounts of ferredoxin.

Nitrate reductase is an unusual enzyme because it binds and holds NO_2^- and the subsequent intermediate reduction products, releasing only the final product NH_4^+. Therefore, most of these intermediates have not yet been identified.

Apparently, hydroxylamine, NH_2OH, is an intermediate, and it is less tightly bound than the others.

Nitrite reductase and ferredoxin both contain iron, and, since nitrate reductase probably contains a cytochrome, an essential role for iron in NO_3^- assimilation is well established.

Relation of Nitrate Reduction to Photosynthesis, Respiration, and Intracellular Traffic High light intensities and warm temperatures enhance the conversion of NO_3^- to NH_4^+. One reason for this is that nitrate reductase is then most active. A second reason is that photosynthesis and respiration then occur at rapid rates, providing the reduced NADH and ferredoxin for both nitrate reductase and nitrite reductase. Reduction of NO_3^- to NH_4^+ in leaves is competitive with CO_2 reduction, because both processes require electrons arising ultimately from the photosynthetic light reactions.

When we consider the details of what appears to be a rather simple relationship, the complexity of subcellular traffic of ions and molecules becomes apparent. Consider, for example, how light increases reduction of NO_3^-. Light acts in the chloroplast to form NADPH, but nitrate reductase is outside this organelle and usually requires NADH. Therefore, some reducing agent must be formed in chloroplasts and transported out where nitrate reductase has access to it. How has the plant solved this transport problem? One theory involves a shuttle in which oxaloacetate moves into the chloroplast in exchange for malate. The oxaloacetate is reduced to malate by NADPH, then the malate is shuttled out where it is oxidized by a cytoplasmic NAD^+-dependent malate dehydrogenase, regenerating oxaloacetate and providing NADH for reduction of NO_3^- to NO_2^-. In this way, reducing power is transported out of the plastid indirectly in the form of malate, not NADH or NADPH to which the chloroplast is impermeable.

Another theory is that 3-phosphoglyceraldehyde molecules formed during photosynthesis are transported by a shuttle system into the cytoplasm, where they are oxidized by NAD^+ and the enzyme 3-phosphoglyceralehyde dehydrogenase. This reaction of glycolysis forms 3-phosphoglyceric acid (3-PGA) and NADH used by nitrate reductase. Some of the 3-PGA is transported by the same shuttle system back into the chloroplast, where it can again be reduced by light-stimulated processes. According to this theory, 3-phosphoglyceraldehyde acts as an electron carrier from the chloroplasts, and 3-PGA is the other member of the shuttle system. The NO_2^- arising from nitrate reductase action then moves into the plastid, where it is further converted to NH_4^+ by electrons from reduced ferredoxin. These processes are illustrated in the upper left portion of Fig. 13-7. As we shall see, further metabolism of NH_4^+ to form certain organic nitrogen compounds illustrated in this figure also occurs in the chloroplasts of leaves and probably in the proplastids of roots.

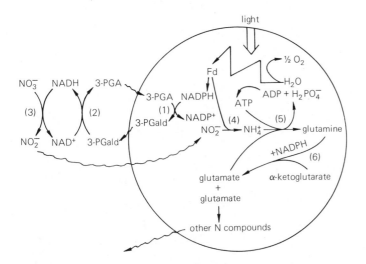

Figure 13-7 How photosynthetic production of ATP, NADPH, and reduced ferredoxin stimulate metabolism of NO_3^- and NO_2^- in leaves. In chloroplasts, NADPH is used to reduce 3-PGA (reaction 1), as described in R10-2. Some of the resulting molecules of 3-phosphoglyceraldehyde (3-PGald) move out of the chloroplasts where they are oxidized by an NAD^+-dependent dehydrogenase (reaction 2). NADH is then used by nitrate reductase to form NO_2^- (reaction 3). Nitrite moves into the chloroplast and is reduced to NH_4^+ with the aid of reduced ferredoxin [Fd (Fe^{+2})] and nitrite reductase in reaction 4. Then, in reaction 5, ATP provides energy for converting NH_4^+ and glutamate to glutamine. Next, the amide group of glutamine is transferred to α-ketoglutarate (source unknown) in the NADPH- or reduced ferredoxin-dependent reaction 6 catalyzed by glutamate synthase. One of the two glutamates produced must keep the cycle going (reaction 5), while the other is converted into proteins and other compounds of the plastid or could be transported out.

13.4 Assimilation of NH_4^+ into Organic Compounds

Whether NH_4^+ is absorbed from the soil or produced by energy-dependent reduction processes in roots, leaves, or stems from NO_3^-, it does not accumulate in the plant. Ammonium is, in fact, somewhat toxic, because it inhibits the production of ATP in both the mitochondrial and photosynthetic electron transport systems, especially the latter, by acting as a ferryboat-type uncoupler (see Special Topic 9.1). There are two possible ways to convert NH_4^+ into organic compounds. In roots, the first-discovered reaction involves the NADH-dependent addition of NH_4^+ to α-ketoglutaric acid, forming glutamic acid and NAD^+:

$$\begin{array}{l} COOH \\ | \\ C{=}O + NH_4^+ + NADH \\ | \\ CH_2 \\ | \\ CH_2 \\ | \\ COOH \\ \alpha\text{-ketoglutaric acid} \end{array} \rightleftharpoons \begin{array}{l} \quad\;\; COOH \\ \quad\;\; | \\ H{-}C{-}NH_2 + NAD^+ + H_2O \\ \quad\;\; | \\ \quad\;\; CH_2 \\ \quad\;\; | \\ \quad\;\; CH_2 \\ \quad\;\; | \\ \quad\;\; COOH \\ \text{glutamic acid} \end{array} \qquad \text{(R13-4)}$$

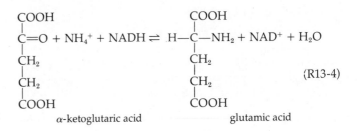

This reaction is catalyzed by **glutamate dehydrogenase,** a zinc-dependent enzyme. Glutamate dehydrogenase occurs in the mitochondria were NADH and α-ketoglutaric acid produced in the Krebs cycle allow its effectiveness (see Figs. 12-6 and 12-10). A different isozyme also exists in chloroplasts; its importance will be mentioned later. Note that removal of α-ketoglutarate from the mitochondria for glutamate production would stop the regeneration of oxaloacetate by Krebs cycle reactions, thereby preventing further α-ketoglutarate production, if another route to form Krebs cycle acids were not available. As mentioned in Section 12.8, the formation of oxaloacetate by PEP carboxylase apparently provides this route.

Although glutamate dehydrogenase is active in roots, there are two facts that suggest it might not be important in NH_4^+ assimilation. First, the reaction it catalyzes is freely reversible, suggesting that a more important function for the enzyme might be the conversion of glutamic acid (produced during degradation of proteins) into a substrate that can be oxidized by the Krebs cycle. Second, glutamate dehydrogenase has such a high K_m value for NH_4^+ that toxic levels of NH_4^+ would probably have to accumulate in the mitochondria before the enzyme could synthesize glutamate.

The other reaction by which NH_4^+ is converted into an organic compound is catalyzed by **glutamine synthetase.** In this reaction, the OH of the carboxyl group farthest from the alpha carbon of glutamate is replaced by an NH_2 group from NH_4^+, as shown in R13-5. ATP is essential to drive the reaction.

$$
\underset{\text{glutamic acid}}{
\begin{array}{l}
\text{COOH} \\
| \\
\text{H—C—NH}_2 \\
| \\
\text{CH}_2 \\
| \\
\text{CH}_2 \\
| \\
\text{COOH}
\end{array}}
+ NH_4^+ + ATP \xrightarrow{\text{Mg}^{2+}} \rightleftharpoons
$$

$$
\underset{\text{glutamine}}{
\begin{array}{l}
\text{COOH} \\
| \\
\text{H—C—NH}_2 \\
| \\
\text{CH}_2 \\
| \\
\text{CH}_2 \\
| \\
\text{C—NH}_2 \\
|| \\
\text{O}
\end{array}}
+ ADP + H_2PO_4^- + H^+
$$

(R13-5)

Conversion of NH_4^+ and glutamate to glutamine by glutamine synthetase occurs in both roots and leaves, probably mainly in proplastids and chloroplasts, respectively. Glutamine synthetase has a much lower K_m (much higher affinity) for NH_4^+ than does glutamate dehydrogenase, so it should begin to use NH_4^+ as fast as it is absorbed or formed

from NO_3^-. In leaves, the reaction catalyzed by glutamine synthetase is the principal mechanism for assimilating NH_4^+ into an organic compound, and the same may prove true for roots. Since this enzyme requires glutamate as a substrate, there must be some reaction to provide it.

In 1970, an enzyme other than glutamate dehydrogenase capable of synthesizing glutamate was discovered in bacteria. This enzyme, called **glutamate synthase,** was subsequently found in both roots and leaves of higher plants and in blue-green and green algae, but it seems to be absent in most fungi (Miflin and Lea, 1976). Like glutamine synthetase, glutamate synthase functions primarily in chloroplasts and proplastids of leaves and roots. It transfers irreversibly the amide group of glutamine to α-ketoglutaric acid, producing two molecules of glutamate. In chloroplasts, reduced ferredoxin is essential as an electron donor, while in roots either NADH or NADPH is required. R13-6 describes the process with NADPH as an example electron donor:

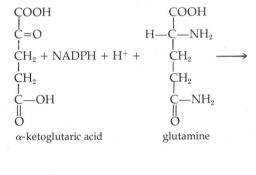

α-ketoglutaric acid   glutamine

(R13-6)

$$2 \quad$$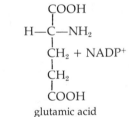

glutamic acid

The glutamate arising in this reaction can be converted into proteins or enter several other reactions inside the plastid. It also can be transported into the cytoplasm and metabolized there. Figure 13-7 summarizes the principal relations between photosynthesis and nitrate metabolism that we have described for leaves. Much evidence favors the importance of glutamine synthetase and glutamate synthase and the unimportance of glutamate dehydrogenase for NH_4^+ assimilation in higher plants.

Probably because of its high ratio of nitrogen to carbon compared to most other compounds, glutamine is an important storage form of nitrogen in plants. Storage organs such as potato tubers and the roots of beet, carrot, radish, and turnip, are especially rich in this amide. In mature leaves, glutamine is often formed from glutamic acid and NH_4^+ produced during protein degradation. It is then transported to younger leaves or to roots, flowers, seeds, or fruits where its nitrogen can be reused. Fertilization of grasses with NH_4^+ salts often leads to the appearance of a white crystalline material rich in glutamine on the tips of the leaves. The cause of this is guttation, presumably driven by root pressure.

The other principal plant amide, asparagine, probably performs some of the same functions as glutamine, especially in legumes (Atkins et al., 1975). It is formed when NH_4^+ is absorbed in excess or when it is released from proteins being hydrolyzed, as in senescing leaves or in cotyledons of young seedlings. As mentioned earlier, asparagine is usually the principal nitrogen compound transported from root nodules. Both asparagine and glutamine are also important forms of nitrogen translocated in the xylem from roots to shoots (for example, see radish and lupine data in Fig. 13-6). They can be incorporated directly into proteins or metabolized in other ways. Asparagine is formed in an ATP-dependent reaction catalyzed by **asparagine synthetase.** Aspartic acid and glutamine are the other reactants, as shown in R13-7:

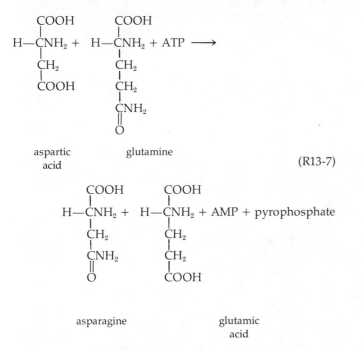

asparagine   glutamic
                       acid

13.5 Transamination Reactions

When NH_4^+ containing isotopic ^{15}N is fed to plants or to excised plant parts, glutamic and aspartic acids are usually the two amino acids that become most rapidly labeled with ^{15}N, as detected with a mass spectrograph. Subsequently, the ^{15}N appears in the other amino acids. The reason for this sequence of labeling is that glutamate transfers its amino group directly to a variety of α-keto acids in freely reversible transamination reactions. An excellent example of transamination occurs between glutamate and oxaloacetate, producing α-ketoglutarate and aspartate:

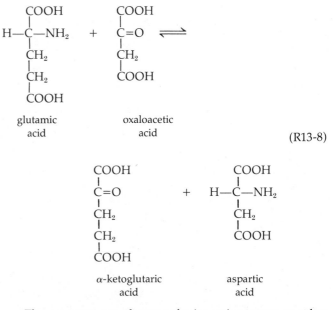

The aspartate can also transfer its amino group to other α-keto acids to form different amino acids. Transfer to pyruvate, for example, would yield alanine. **Pyridoxal phosphate,** a vitamin B_6-containing coenzyme, is essential for transaminase enzymes. The reactions by which amino acids other than aspartate are formed (Bryan, 1976) will not be discussed here. We emphasize, however, that the amino groups of nearly all amino acids in both plants and animals probably passed through glutamate on the route from NO_3^-, NH_4^+, or atmospheric N_2. Furthermore, plants synthesize amino acids that animals do not. Besides the 18 amino acids and 2 amides common in proteins, about 200 nonprotein amino acids have been identified in the plant kingdom. The functions of such amino acids in the plants containing them are generally unknown. Some are toxic to insects, mammals, and other plants, suggesting ecological defense roles (Fowden, 1973; Bell, 1976).

13.6 Nitrogen Transformations During Plant Development

Nitrogen Metabolism of Germinating Seeds In storage cells in all kinds of seeds, the reserve proteins are deposited in membrane-bound structures called **aleurone grains** or **protein bodies.** The latter term is now more commonly used, even though various nonprotein substances such as phosphate, potassium, calcium, and magnesium are also stored there. Protein bodies are also formed in cells of the embryonic axis of the seed, and Fig. 13-8a shows an electron micrograph of these darkly stained bodies in cells of the radicle (embryonic root) of the lettuce seed. In these cells, protein bodies and oleosomes (lipid or fat bodies) are almost the only visible subcellular structures.

Imbibition of water by a dry seed sets off a variety of chemical reactions that lead to germination (radicle protrusion)

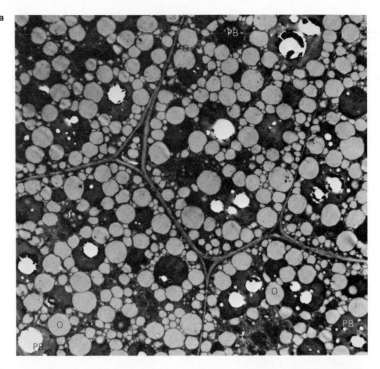

a

Figure 13-8 (**a**) Electron micrograph of a cortex cell in the radicle of an ungerminated Grand Rapids lettuce (*Lactuca sativa*) seed. The numerous unstained (light grey) structures are oleosomes (O) in which oils and fats are stored, while the larger darkly stained structures are protein bodies (PB). White areas in some of the protein bodies identify sites where most of the reserve phosphorus is stored as phytins. Phytins are calcium, magnesium, and potassium salts of phytic acid, myoinositol hexaphosphoric acid. (Courtesy Nicholas Carpita.) (**b**) Digestion of reserve protein in radicle cortex cells of recently germinated lettuce seeds. Protein bodies (PB) surrounding the nucleus were beginning to fuse to form the large vacuole, and most of the protein in them has disappeared. Numerous oleosomes (O) are still visible. (Courtesy Nicholas Carpita.)

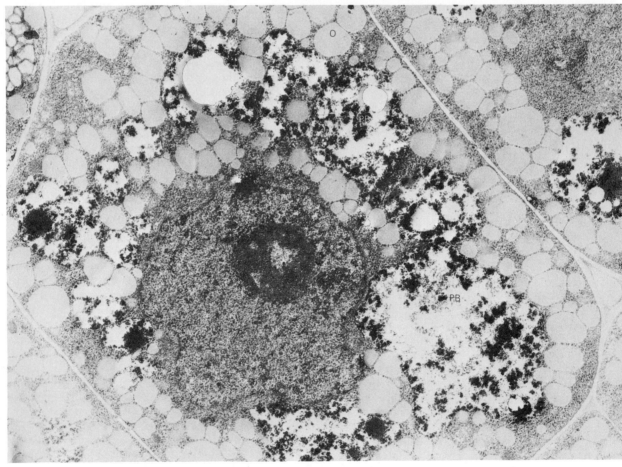

b

and subsequent seedling development. The proteins present in the protein bodies are hydrolyzed by **proteases** and **peptidases** to amino acids and amides (note disappearance of most of the protein in the protein bodies of Fig. 13-8b) (Ashton, 1976; Chrispeels et al., 1976). These amino acids and amides are then converted into proteins and other nitrogen compounds of the growing seedling.

Membranes surrounding disintegrating protein bodies are not degraded; rather, these membranes fuse to form the tonoplast around the central vacuole. In the radicle, the amino acids released during hydrolysis of reserve proteins are used by the same cells to form new enzymes and other proteins. The same is true of reserve proteins in the epicotyl and hypocotyl, but eventually the root and shoot require some of the amino acids and amides arising during breakdown of storage proteins in the cotyledons, endosperm, or female gametophyte, depending on the seed. Subsequently, the roots begin to absorb NO_3^- and NH_4^+, and nitrogen metabolism for another plant has begun.

Traffic of Nitrogen Compounds from One Organ to Another
The gross aspects of protein transformations in vegetative organs of an herbaceous plant during its development were demonstrated long ago (McKee, 1962). Many of these are illustrated in Fig. 13-9, in which the changes in amounts of total nitrogen in roots, stems, leaves, and seeds of a broad bean (*Vicia faba*) plant from the seedling stage until maturity are shown. These changes largely reflect degradation and synthesis of proteins, because most of the nitrogen in any plant part is in protein. For leaves, about half of this protein is in

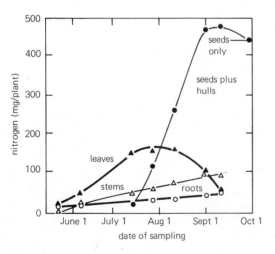

Figure 13-9 Changes in nitrogen content of various organs of the broad bean (*Vicia faba*) during growth. The extensive accumulation of nitrogen compounds in the fruits (seeds plus hulls) was accompanied by a loss from the leaves and a large uptake from the soil. (Data of A. Emmerling, 1880, Ladw. Versuchsstat 24:113.)

chloroplasts. Note that in the broad bean the leaves actually lost nitrogen during August and September, while the seeds were accumulating it. This transfer of nitrogenous compounds from leaves, especially those that are mature, to developing protein bodies in seeds or to fruits via the phloem is typical of plants. The principal organic compounds translocated are glutamine, asparagine, glutamate, and aspartate, just as in xylem, and neither NO_3^- nor NH_4^+ is translocated in significant amounts in the phloem. The amount of nitrogen transferred from vegetative organs to fruits in the broad bean was much less than that gained by the fruits during the same period (Fig. 13-9). The additional nitrogen demand of seeds in such legumes is usually satisfied by increasing rates of N_2 fixation during seed development. In perennial plants, much of the nitrogen and other elements that are mobile in the phloem move into the crown and roots after the seed demands are satisfied. As a result, these elements are available for the next season's growth, and the standing dead plant returns less to the soil than otherwise would have occurred.

In small grains and many other annuals, the transfer of nitrogen from vegetative parts to the seeds is sometimes more extensive than in legumes, even though their seeds contain lower percentages of protein than do legume seeds (Sinclair and deWit, 1975). Wheat leaves, for example, can lose up to 85 percent of their nitrogen (and an equal percentage of phosphate) before they die. This extensive reutilization is accompanied by a large *decrease* in the rate of uptake of soil nitrogen as reproductive growth begins. Thus, wheat and oats can absorb 90 percent of the nitrogen (and phosphate) needed for maturity before they are half grown. Causal factors relating flowering, decreased nitrogen absorption rates, and rapid transfer from vegetative organs represent important and challenging things to study.

In deciduous trees, much of the nitrogen moving out of the leaves in late summer and early autumn is stored in the twigs in storage proteins and other compounds that are available for new growth the next spring. Were it not for this conservation process, losses of nitrogen during leaf fall would probably cause productivity of forest trees typically growing on nitrogen-deficient soils to be even lower.

The loss of protein from leaves often begins even before they are full grown. Figure 13-10 illustrates this for the field pea (*Pisum arvense*), an annual legume. Loss of both protein and nonprotein amino acids and amides began 15 to 20 days before yellowing (chlorosis) of the leaves became apparent. In this leaf, the nitrogen that was accumulated in the first two weeks of development was retained only temporarily and was probably soon transported to younger leaves and from them eventually to seeds.

RNA molecules are also degraded in mature and senescing leaves and in seed storage tissues. Hydrolytic enzymes called ribonucleases are responsible for this degradation (Wilson, 1975). These enzymes generally release purine and pyrimidine nucleotides in which the phosphate group is attached either to carbon-3 or to carbon-5 of the ribose unit. Whether these nucleotides are translocated to other organs directly or after

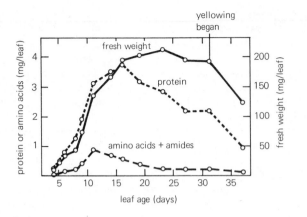

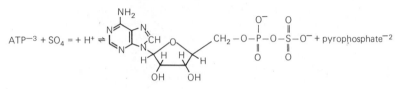

The first step in roots and leaves is reaction of $SO_4^=$ with ATP, producing **adenosine-5´-phosphosulfate (APS)** and pyrophosphate. It is catalyzed by **ATP sulfurylase.** The pyrophosphate is rapidly and irreversibly split into $H_2PO_4^-$ by a pyrophosphatase enzyme; the $H_2PO_4^-$ can then be used in

$ATP^{-3} + SO_4^= + H^+ \rightleftharpoons$ adenosine-5´-phosphosulfate (APS) $+ pyrophosphate^{-2}$

adenosine-5′-phosphosulfate (APS)

$pyrophosphate^{-2} + H_2O \longrightarrow 2\ H_2PO_4^-$

$APS^{-2} + 8\ ferredoxin\ (Fe^{+2}) + 5\ H^+ \longrightarrow AMP^{-1} + S^= + 8\ ferredoxin\ (Fe^{+3})$
$+\ 3\ H_2O$

Figure 13-11 A summary of reactions involved in sulfate reduction in chloroplasts. Processes in roots are similar, but NADPH or NADH rather than ferredoxin is presumably the electron donor necessary for the last reaction.

Figure 13-10 Changes in protein and soluble amino acids plus amides during development of a pea leaf (the fifth-formed leaf). Loss of chlorophyll was readily detectable after 1 month, as indicated by the arrow at the upper right. Note that there was much more protein than soluble amino acids and amides (weight basis) and that both of these fractions began to disappear from the leaf even before it was full grown. (Data of D. J. Carr and J. S. Pate, 1967, in H. H. Woolhouse, ed., Aspects of the biology of ageing, 21st Symposium of the Society for Experimental Biology, Cambridge University Press, pp. 589–599. Amino acid plus amide data were converted from a mol to a mg basis, assuming an average molecular weight of 130.)

degradation and conversion of the nitrogen to glutamate, aspartate, and their amides is unknown. Still less is known about DNA breakdown in plants, except that DNA is much more stable in leaves than RNA. Even quite senescent leaves from which all chlorophyll and most of the protein and RNA have disappeared still retain much of their DNA. This DNA remains in the leaf as it is shed, and its nitrogen is cycled back to the soil.

13.7 Metabolism of Sulfur

Except for small amounts of SO_2 absorbed by the shoots of plants growing near smokestacks, $SO_4^=$ ions absorbed by the roots provide the necessary sulfur for plant growth. Just as the reduction of NO_3^- and CO_2 are energy-dependent reduction processes, so is the conversion of sulfate to sulfide shown in R13-9:

$SO_4^= + 1\ ATP + 8\ electrons + 8\ H^+ \longrightarrow$
$\qquad S^= + 4H_2O + 1\ AMP + 1\ pyrophosphate \qquad$ (R13-9)

This process occurs in both roots and shoots of some species, but most of the sulfur transported to the leaves is in nonreduced $SO_4^=$. We know little about $SO_4^=$ reduction in roots, but the initial steps, at least, are apparently the same as those occurring in leaves. ATP is essential in both organs. In leaves, the entire process occurs in chloroplasts.

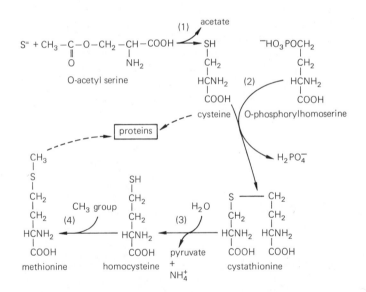

Figure 13-12 Reactions by which sulfide is converted to cysteine and methionine. Reaction 1, catalyzed by cysteine synthetase, involves replacement of the acetate group in O-acetyl serine by sulfide. Reaction 2, catalyzed by cystathionine synthetase, splits phosphate from O-phosphorylhomoserine and joins the sulfur atom in cysteine to the terminal CH_2 group of the homoserine residue. Reaction 3 is catalyzed by cystathionase, an enzyme that hydrolyzes cystathionine between the S and the carbon shown by the wavy line. Pyruvate and NH_4^+ are released; these products were formerly part of the cysteine molecule. Homocysteine is converted by methionine synthetase to methionine by receipt of a methyl group in reaction 4. N^5-methyltetrahydrofolic acid is the methyl donor for this reaction.

mitochondria or chloroplasts to regenerate ATP. These processes are shown in the first two reactions of Fig. 13-11.

The sulfur of APS is reduced in chloroplasts by electrons donated from reduced ferredoxin (Wilson and Reuveny, 1976; Schmidt, 1976). For roots, proplastids represent a reasonable but unproven comparable site of reduction, perhaps using NADPH as the electron donor. The oxidation number of sulfur changes from +6 to −2 during APS reduction, yet no intermediate compounds are normally released from the reductase enzyme. The reaction is complicated, and we shall only emphasize that in chloroplasts, eight reduced ferredoxin molecules are needed to form sulfide and AMP from one molecule of APS (Fig. 13-11, last reaction).

The sulfide resulting from reduction of APS does not accumulate, because it is rapidly converted into organic compounds, especially **cysteine** and **methionine.** Reactions by which these two amino acids are formed are shown in Fig. 13-12. Most cysteine and methionine molecules are converted into proteins, although small amounts of cysteine are incorporated into coenzyme A, and traces of methionine are used to form S-adenosyl methionine, a coenzyme that transfers the methyl groups of the methionine in it to lignins, pectins, chlorophyll, and flavonoids. In some species, especially onion, garlic, and cabbage and its close relatives, mercaptans (R—SH) such as methyl mercaptan and n-propyl mercaptan, sulfides (R—S—R), or sulfoxides (R—S—R)

$$\underset{\text{O}}{\overset{\parallel}{}}$$

are significant products of sulfur metabolism (Richmond, 1973). Most such compounds have offensive odors. Furthermore, small amounts of H_2S have been detected by gas chromatography as volatile sulfide products of leaves.

Although plants, bacteria, and fungi generally reduce and convert sulfur into cysteine, methionine, and other essential sulfur compounds, mammals cannot. Because of this, we and the animals we eat depend on plants for reduced sulfur and particularly for the essential amino acids cysteine and methionine.

Lipids and Aromatic Compounds

In the preceding six chapters, we have emphasized that plants contain an imposing variety of carbohydrates and nitrogenous and sulfur-containing compounds. Many of the reactions by which these substances are formed have been explained, especially in relation to the importance of light in providing energy to drive these reactions in the shoot system. We have seen that light energy is used to drive the reduction of CO_2, NO_3^-, and $SO_4^=$, processes that human beings and other animals cannot accomplish. In this chapter, we shall discuss the properties and functions of many other compounds that plants require for growth or survival. Some of these, such as fats and oils, are important food reserves that are deposited in specialized tissues and cells only at certain times in the life cycle. Others, such as waxes and the components of cutin and suberin, are protective coats over the plant's exterior, while still others aid in perpetuation of the species by facilitating pollination or by performing a defense role against other competitive organisms. Besides these, certain plants produce many compounds such as rubber for which no function is presently known. In a way, plants can be compared to sophisticated organic-chemical laboratories in which several thousand kinds of molecules can be synthesized, many of which remain to be discovered.

We shall begin by discussing the **lipids.** These are a group of fatlike substances, rich in carbon and hydrogen, that dissolve in organic solvents such as chloroform, acetone, ethers, certain alcohols, and benzene, but that do not dissolve in water. Among these are fats and oils, phospholipids and glycolipids, waxes, and many of the components of cutin and suberin.

14.1 Fats and Oils

Chemically, fats and oils are very similar compounds, but fats are solids at room temperatures while oils are liquids. Both compounds are composed of long-chain **fatty acids** esterified by their single carboxyl group to a hydroxyl of the three-carbon alcohol **glycerol.** In fats and oils, all three hydroxyl groups of glycerol are esterified, so these compounds are often called **triglycerides.** Except where an important distinction is to be made between fat or oil triglycerides, we shall refer to them as fats. The general formula for a fat is given in Fig. 14-1.

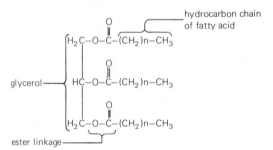

Figure 14-1 The general structure of a fat or oil, both of which are triglycerides.

The melting points and other physical properties of fats are determined by the kinds of fatty acids they contain. A given fat usually contains three different fatty acids, although occasionally two are identical. These acids almost always have an even number of carbon atoms, usually 16 or 18, and they may or may not be **unsaturated** (contain double bonds). The melting point increases with the length of the fatty acid and with the extent of its saturation, so solid fats usually have saturated fatty acids. In oils, one to three double bonds are often present in each fatty acid. Examples of commercially important plant oils are those from the seeds of cotton, corn, peanuts, and soybeans. All of these oils principally contain fatty acids with 18 carbon atoms, including **oleic acid,** with one double bond, and **linoleic acid,** with two double bonds. In fact, these two acids, in the order named, are the most abundant fatty acids existing in seeds, while **linolenic acid,** an 18-carbon acid with three double bonds, is by far the most abundant fatty acid of most leaves (Hitchcock and Nichols, 1971; Stumpf, 1976).

Table 14-1 contains a list of some of the important fatty acids occurring in plants, with the number of carbon atoms, structure, degree of unsaturation, and position of double bonds listed for each. The most abundant saturated fatty acids are **palmitic acid,** with 16 carbons, and **stearic acid,** with 18 carbons. These acids are important constituents of fats from both plants and animals. Coconut fat is also a rich source of **lauric acid,** a saturated acid with only 12 carbons. The seeds of many plants contain a high percentage of fatty acids not

Table 14-1 Fatty Acids Abundant or Common in Various Plants

Name	No. Carbons: No. Double Bonds	Structure
Lauric	12:0	$CH_3(CH_2)_{10}COOH$
Myristic	14:0	$CH_3(CH_2)_{12}COOH$
Palmitic	16:0	$CH_3(CH_2)_{14}COOH$
Stearic	18:0	$CH_3(CH_2)_{16}COOH$
Oleic	18:1 at C-9,10	$CH_3(CH_2)_7\overset{\displaystyle H}{C}{=}\overset{\displaystyle H}{C}{-}(CH_2)_7COOH$ 10 9
Linoleic	18:2 at C-9,10;12,13	$CH_3(CH_2)_4C{=}C{-}CH_2C{=}C{-}C{-}(CH_2)_7{-}COOH$ 13 12 10 9
Linolenic	18:3 at C-9,10;12,13;15,16	$CH_3CH_2C{=}C{-}CH_2C{=}C{-}CH_2C{=}C{-}(CH_2)_7{-}COOH$

important in fats of their vegetative organs. Castor beans (*Ricinus communis*), for example, contain **ricinoleic acid** (12-hydroxyoleic acid), which makes up between 80 and 90 percent of the fatty acids in castor oil, but which is absent from castor bean leaves and is rare in other species.

The kinds of fatty acids in fats varies in an interesting way with the temperature at which the plant is grown. At lower temperatures, a higher percentage of unsaturated fatty acids such as linoleic and linolenic acids having lower melting points is produced, while higher temperatures favor more solid fats and a greater proportion of oleic acid. The mechanism causing this is only partly understood. One factor involved is the greater solubility of O_2 in the cells at lower temperatures; this O_2 is essential for introduction of double bonds in fatty acids such as those in linoleic and linolenic. These two unsaturated acids also predominate in the membrane lipids of plants grown under relatively cool conditions, and the ability of these lipids to remain liquid or semiliquid when the temperature drops is important to normal membrane function in extreme latitudes and at high elevations. This subject is discussed in relation to chilling damage in certain species in Chapter 24.

Distribution and Importance of Plant Fats Fats occur only in small amounts in leaves, stems, and roots but are abundant in many seeds and some fruits. Fats are usually concentrated in the endosperm or cotyledon storage tissues of seeds, but they also occur in the embryonic axis. Compared to carbohydrates, fats contain larger amounts of carbon and hydrogen and less oxygen, and, when they are respired, relatively large amounts of O_2 are used. As a result, considerable ATP is formed, demonstrating that greater amounts of energy can be stored in a small volume as fats than as carbohydrates.

Perhaps because of this, small seeds nearly always contain fats as the primary storage materials. When these fats are respired, sufficient amounts of energy are released to allow establishment of the seedling, and yet the small weight of such seeds often allows them to be scattered effectively by wind. Larger seeds, especially those such as pea, bean, and corn selected by man for agriculture, often contain much starch and only small amounts of fats, but seeds of conifers and those in nuts are usually fat rich (Table 14-2).

Fats are always stored in specialized bodies within the cells (see Fig. 13-8), and there are often hundreds or thousands of such bodies in each storage cell. These bodies have been called **lipid bodies, spherosomes,** or **oleosomes** (Latin *oleo,* "oil"). We prefer the term "oleosome," suggested by Yatsu et al. (1971), because it indicates that they contain oil and distinguishes them from peroxisomes (Fig. 11-7) and glyoxysomes (see later, this section), which are also rather spherical and about the same size (approx. 1 μm diameter).

Oleosomes can be isolated from seeds in fairly pure form, allowing analysis of their composition. Considerable disagreement about whether each oleosome is surrounded by a typical unit membrane arose from electron microscope studies. The failure of oleosomes to fuse into one large lipid droplet suggests that a membrane is present, but it frequently cannot be seen in electron micrographs, and when it is visible it appears to be only about half as thick (approx. 3 nm) as a typical unit membrane (approx. 8 nm). Apparently, oleosome membranes are indeed half-membranes whose polar, hydrophilic surfaces are exposed to the exterior and whose nonpolar, hydrophobic surfaces face the fats at the inside. If correct, this interpretation could mean that fats are synthesized and inserted between the two halves of a unit membrane, probably of the endoplasmic reticulum, forcing these halves apart and ultimately leading to a full-grown oleosome.

Table 14-2 The Chemical Composition of Some Seeds of Economic Importance.[a,b]

Species	Family	Nature of Reserve Tissue	Percent Content		
			Carbohydrate	Protein	Lipid
Maize (*Zea mays*)	Gramineae	Endosperm	51–74	10	5
Wheat (*Triticum vulgare*)	Gramineae	Endosperm	60–75	13	2
Pea (*Pisum sativum*)	Leguminosae	Cotyledons	34–46	20	2
Peanut (*Arachis hypogaea*)	Leguminosae	Cotyledons	12–33	20–30	40–50
Soybean (*Glycine* sp.)	Leguminosae	Cotyledons	14	37	17
Brazil nut (*Bertholletia excelsa*)	Lecythidaceae	Hypocotyl	4	14	62
Castor bean (*Ricinus communis*)	Euphorbiaceae	Endosperm	0	18	64
Date palm (*Phoenix dactylifera*)	Palmae	Endosperm	57	6	10
Sunflower (*Helianthus annuus*)	Compositae	Cotyledons	2	25	45–50
Oak (*Quercus robur*)	Fagaceae	Cotyledons	47	3	3
Douglas fir (*Pseudotsuga menziesii*)	Pinaceae	Gametophyte	2	30	36

[a]The percentages are based on the fresh (air-dry) weights of the seeds, except for date palm, where the percentages are expressed on a dry weight basis.

[b]From Street and Öpik, 1970. A longer list of analyses of whole seeds is given by T. R. Sinclair and C. T. de Wit, 1975.

Synthesis of Fats The abundance of fats in certain seeds and fruits has stimulated efforts to learn how they are formed during development. It is now clear that fats are not transported from leaves but are synthesized from sucrose or other translocated sugar. This failure of fats to be translocated in the vascular tissues also means that when reserve fats are utilized during seed germination, they must first be converted to a substance that can be translocated to the elongating root and shoot axes. As we shall see, this substance is also most commonly sucrose.

Conversion of sucrose or other sugars to fats requires production of the fatty acids and of the glycerol backbone to which the fatty acids become esterified. The glycerol unit (**L-α-glycerophosphate**) arises by reduction of dihydroxyacetone phosphate produced in glycolysis (Section 12.3). The fatty acids are formed by multiple condensations of acetate units in acetyl CoA. These reactions apparently occur in or on several membrane systems, including those of the endoplasmic reticulum and chloroplast envelopes. The vitamin biotin and Mn^{2+} ions are coenzymes in the process. A summary of the multistep condensations of acetate units worked out for plants primarily by P. K. Stumpf at the University of California,

Davis, is shown for the 16-carbon saturated acid palmitic acid (actually its CoA ester) in R14-1:

$$8 \text{ acetyl CoA} + 7 \text{ ATP}^{-3} + 14 \text{ NADPH} + 14 \text{ H}^+ \rightarrow$$
$$\text{palmityl CoA} + 7 \text{ CoA} + 7 \text{ ADP}^{-2} + 7 \text{ H}_2\text{PO}_4^-$$
$$+ 14 \text{ NADP}^+ + 7 \text{ H}_2\text{O} \qquad \text{(R14-1)}$$

Subsequently, CoA is removed when the fatty acid is used in formation of fats or membrane lipids.

This summary emphasizes that all of the carbon atoms in the fatty acid are derived from acetyl groups of acetyl CoA and that the process requires almost two pairs of electrons (two NADPH molecules) and one ATP molecule for each acetyl group incorporated. The high NADPH and ATP requirement means that fat synthesis is an energy-expensive process. The source of the necessary NADPH is uncertain; presumably the pentose phosphate pathway of respiration (Section 12.7) provides part of it.

Neither free fatty acids nor their CoA esters accumulate during fatty acid formation. Instead, they rapidly react with α-glycerophosphate to form triglycerides deposited within oleosomes or to yield diglycerides such as the glycolipids and phospholipids of membranes described later. When triglycerides are produced, $H_2PO_4^-$ is released from carbon-3 of the α-glycerophosphate molecule just before the third fatty acid is attached to that position.

Degradation of Fats The catabolism of fats begins with the action of the **lipases,** which hydrolyze the ester bonds, releasing glycerol and the three fatty acids, as shown in R14-2:

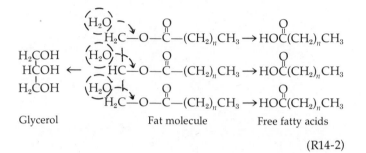

$$(R14-2)$$

The glycerol resulting from lipase action is converted with ATP to α-glycerophosphate; this molecule is then oxidized by NAD^+ to dihydroxyacetone phosphate, which is either respired by glycolytic and Krebs cycle reactions or converted to sucrose by reversal of glycolysis.

The fatty acids resulting from the action of lipase upon triglycerides can be oxidized to CO_2 and H_2O, accompanied by the production of large amounts of ATP, NADH, and $FADH_2$. The fat stored in some seeds provides an energy supply needed for germination or early seedling development, and thus for survival of the species. Most of this energy is released during oxidation of the fatty acids, although the oxidation of glycerol should not be ignored. Fatty acid oxidation occurs primarily by the β-**oxidation** pathway, which is well known in animal cells also. An additional α-**oxidation** pathway contributes to some extent in certain plant tissues. The process of β-oxidation releases two-carbon acetate units as acetyl CoA and forms $FADH_2$ and NADH, both valuable oxidizable substances. Details of β-oxidation will not be described here, although the summary equation for β-oxidation of palmitic acid is presented in R14-3:

$$\text{palmitate} + ATP^{-3} + 7\,NAD^+ + 7\,FAD + 7\,H_2O + 8\,CoASH$$
$$\rightarrow 8\,\text{acetyl CoA} + AMP^{-1} + \text{pyrophosphate}^{-2}$$
$$+ 7\,NADH + 7\,H^+ + 7\,FADH_2 \quad (R14-3)$$

The acetate units of acetyl CoA are oxidized in some cells to CO_2 via the Krebs cycle, thereby forming even more NADH and $FADH_2$. The latter compounds can be oxidized by the mitochondrial electron transport system to produce ATP. The α-oxidation pathway is much different from β-oxidation; its contribution is minor and may be primarily important in providing propionic acid needed for coenzyme A formation

and odd-numbered fatty acids present in suberin and in cuticular waxes (Stumpf, 1976).

Conversion of Fats to Sucrose: The Glyoxylate Cycle During germination of seeds that store fats as the important storage material, such fats disappear, and sugars, especially sucrose, accumulate. Sucrose is then translocated to the developing embryonic axis where it is used for growth, as illustrated for the peanut in Fig. 14-2.

The mechanism for conversion of fats to sugars was long in doubt. Even after it was learned that fatty acids are broken down into acetyl CoA, problems still existed, because acetyl CoA cannot be converted back to pyruvic acid and then into sugars by a reversal of glycolysis. Such a conversion cannot occur because the Krebs cycle reaction that normally produces acetyl CoA and CO_2 from pyruvic acid is irreversible (Fig. 12-6). This problem was solved in the late 1950s and early 1960s largely by Harry Beevers, then at Purdue University, through his demonstrations that the **glyoxylate cycle** functions in young seedlings derived from fat-rich seeds of many species. This cycle had previously been found in certain microorganisms that can exist on acetate as the only source of carbon. These microorganisms derive all their energy from the oxidation of acetate to CO_2 and H_2O via acetyl CoA, and they also convert acetate to sugars and all other cellular materials required for their growth.

During seedling development, fatty acids arising from reserve fats are oxidized to acetate units of acetyl CoA almost entirely in bodies that Beevers discovered and named **glyoxysomes.** This was surprising at the time, because it was thought that β-oxidation occurred in the mitochondria. Morphologically, glyoxysomes (Fig. 14-3) are indistinguishable from peroxisomes in photosynthetic cells, and both are frequently referred to simply as **microbodies** (Breidenbach, 1976; Vigil, 1973). Studies with isolated glyoxysomes showed that most of the enzymes they contain are identical to those of peroxisomes, yet only glyoxysomes have the full enzyme complement necessary to carry out both β-oxidation and the glyoxylate cycle.

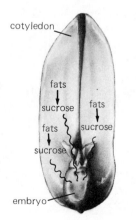

Figure 14-2 Fats and oils stored in the cytoplasm of cells such as those of the cotyledons of peanut seeds are broken down to acetyl CoA by β-oxidation, then the acetyl CoA is converted to sucrose through the glyoxylate cycle and reverse glycolysis. The sucrose can be transported within the embryonic axis, while the larger, insoluble fat molecules are not transportable.

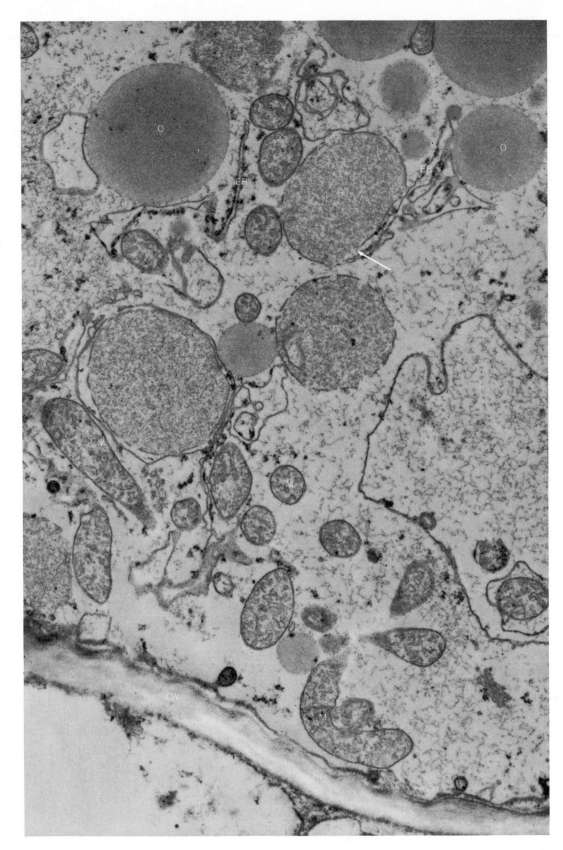

Figure 14-3 Part of a fat-storing cell in the megagametophyte of a ponderosa pine seed germinated 7 days, showing glyoxysomes (G), fat-storing oleosomes (O), endoplasmic reticulum (ER), and mitochondria (M). Most of the fat has been converted to sugars. Some of the glyoxysomes appear to be connected to the ER, from which they arise; note arrow. (From T. M. Ching, 1970, Plant Physiology 46:475–482.)

Both glyoxysomes and peroxisomes originate from the endoplasmic reticulum. Apparently, the ends of the ER envelopes swell, followed by a pinching-off process, much as vesicles are pinched off dictyosomes (see Fig. 10-16). Glyoxysomes persist only a few days while the storage fats are converted to sugars, then they disappear. The plant forms these organelles and the enzymes they contain to perform a specific task, then destroys them when the task is complete.

Reactions of the glyoxylate cycle are shown in Fig. 14-4. Acetyl CoA derived from fatty acids by β-oxidation (upper right) reacts with oxaloacetic acid to form citric acid, just as in the Krebs cycle. After isocitric acid (six carbons) is formed, it undergoes cleavage by an enzyme unique to this cycle called **isocitrate lyase.** Succinate (four carbons) and glyoxylate (two carbons) are produced. The succinate moves out of the glyoxysome into the mitochondria, where it is sequentially converted to fumarate, malate, and oxaloacetate by Krebs cycle reactions. The glyoxylate reacts with another acetyl CoA to form malate and free coenzyme A (CoASH). This reaction is catalyzed by another enzyme restricted to the glyoxylate pathway called **malate synthetase.** This malate is also oxidized to oxaloacetic acid.

The oxaloacetate formed in the glyoxysome reacts with another acetyl CoA to keep the glyoxylate cycle going, as shown. The other oxaloacetate produced in a mitochondrion presumably has no acetyl CoA to react with, because glycolysis is not functioning to form pyruvate, from which acetyl CoA is derived, and because the acetyl CoA produced by β-oxidation is restricted to the glyoxysomes. This oxaloacetate is instead converted to phosphoenolpyruvate (PEP) and CO_2, as shown in Fig. 14-4 (upper left), by **PEP carboxykinase.** This conversion is an ATP-dependent reaction, just as are certain others required to convert PEP into sucrose by reverse glycolysis. The formation of CO_2 accounts for loss of one fourth of the carbon atoms in fatty acids during conversion of fats to sugars. The glycerol arising from fat hydrolysis is effectively salvaged by conversion to dihydroxyacetone phosphate, an intermediate of glycolysis, which is then also converted to sugars (Fig. 14-4, top).

The energy required to convert oxaloacetate via phosphoenolpyruvate to sucrose by reversing glycolysis presumably arises as follows: First, some of the NADH essential to convert 3-phosphoglyceric acid to 3-phosphoglyceraldehyde is formed in the glyoxylate cycle when malate is oxidized to oxaloacetate. Still more NADH arises at various steps in β-oxidation (R14-3). Second, some of these NADH molecules are oxidized by the mitochondria, and this produces ATP by oxidative phosphorylation. This ATP is presumably available for driving oxaloacetate to PEP, 3-PGA to 3-phosphoglyceraldehyde, and 3-phosphoglyceraldehyde via the hexose phosphates and UDPG to sucrose. As the glyoxylate pathway continues to utilize acetyl CoA, the concentrations of PEP and other intermediates of glycolysis probably build up, helping to make sucrose formation thermodynamically feasible. Sucrose is then translocated from the storage tissues of the seed. *In summary, the glyoxylate cycle provides seedlings a mechanism for conversion*

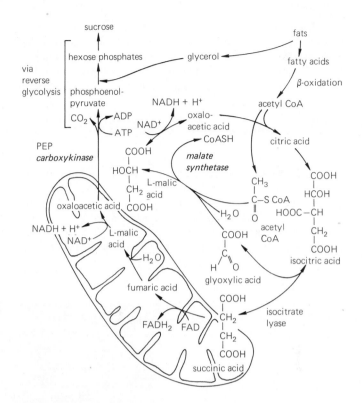

Figure 14-4 Conversion of fatty acids to sucrose through the glyoxylate cycle. Fatty acids arise from triglycerides in oleosomes (lipid bodies) during action of at least two lipases, one of which is in the oleosome and one in the glyoxysome membrane. β-oxidation of fatty acids occurs in the glyoxysome; it releases $FADH_2$ and NADH and provides acetyl CoA for the glyoxylate cycle. Succinic acid is not converted to oxaloacetate in the glyoxysomes but instead moves into the mitochondria, where those reactions of the Krebs cycle occur. The glyoxysomes do, however, convert malic acid to oxaloacetate so that the glyoxylate cycle can continue.

of fats, which cannot be transported to the embryo, into sucrose, which can be transported.

We mentioned previously that there is an apparent advantage to small, wind-blown seeds in storing fats as a food reserve, because these molecules contain more energy per unit weight than does starch (about 2.5 times as much). But what advantage do numerous other seeds and many vegetative organs, including leaves, stems, roots, bulbs, corms, and tubers, obtain from storing starch? First, starch is economical to store, because its production requires little energy besides that already present in the sugar phosphate products of photosynthesis. Starch utilization also requires only a few enzymes, so it is readily and efficiently converted back to sucrose. Synthesis of fatty acids from sucrose involves loss of one third of the carbon atoms as CO_2 during formation of acetyl CoA from pyruvate. Furthermore, reconversion of a fatty acid to sucrose during seedling development involves loss of about one fourth of its carbon atoms as CO_2, suggesting that storage of fats might be quite inefficient. Calculations indicate, however, that on an energy basis the formation of fats from sugars, followed by the reverse transformation, has an overall efficiency as high as 75 percent.

14.2 Phospholipids and Glycolipids

These lipids are universal constituents of plant membranes and are generally not important to the plant as a food source. Chemically, they are similar to triglycerides except that only two fatty acids, usually unsaturated, are esterified to glycerol, while the third hydroxyl group of glycerol is connected to one of several possible constituents, depending on the lipid. Phospholipids always have a phosphate group esterified to this position of the glycerol molecule, and differences among them arise from the type of fatty acids attached to the other two positions and from the kind of substituent connected to the phosphate. One of the major phospholipids found in most plants is **phosphatidyl choline**, a **lecithin**, in which the phosphate is esterified to **choline:**

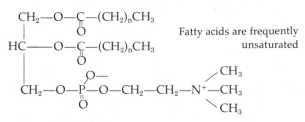

Fatty acids are frequently unsaturated

Phosphatidyl choline

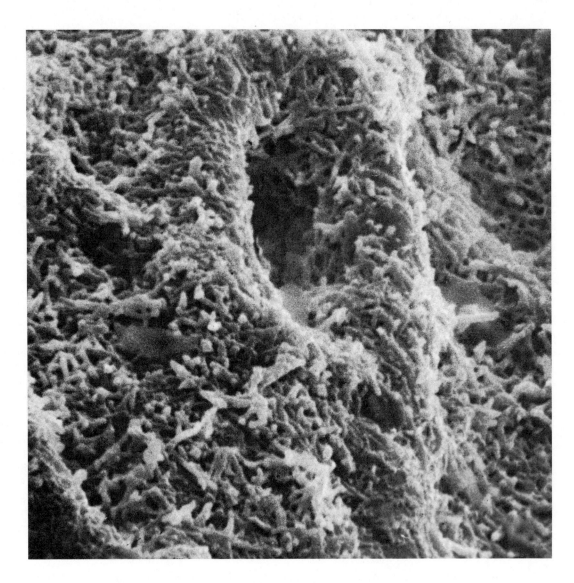

Figure 14-5 Wax on the leaf surface of a carnation. Carnation (*Dianthus* sp.) is a common plant with a prolific layer of wax on the cuticle. The structure of wax on the surface can be as thin flakes, plates, rodlets, or rods. When the wax is in the form of rodlets or rods, it is visible to the naked eye as a bluish "bloom," which can be easily rubbed off the leaf. (From Troughton and Donaldson, 1972.)

Other principal phospholipids include **phosphatidyl glycerol,** in which glycerol rather than choline is esterified to the phosphate, **phosphatidyl inositol,** in which the sugar alcohol inositol is esterified to the phosphate, and **phosphatidyl ethanolamine** (a **cephalin**), in which ethanolamine, HO—CH_2—CH_2—NH_2, is similarly esterified.

Glycolipids have no phosphate but instead are usually connected to a sugar. The three most abundant glycolipids are **monogalactosyl diglyceride,** in which a single galactose (a hexose sugar) is connected to the glycerol position not esterified with a fatty acid, **digalactosyl diglyceride,** in which a disaccharide of two galactose units is attached at this position, and the plant sulfolipid, **sulfoquinovosyl diglyceride,** universal in thylakoid membranes of chloroplasts:

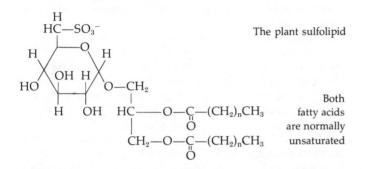

The plant sulfolipid

Both fatty acids are normally unsaturated

These three glycolipids comprise about 80 percent of the lipid fraction of higher plant chloroplasts (excluding lipid-like pigments such as carotenoids and chlorophylls), and many of the reactions involved in their formation occur in the chloroplast envelope. Of the relatively few phospholipids in thylakoids, phosphatidyl glycerol appears to be the most abundant. The hydrophobic fatty acids of both phospholipids and glycolipids protrude toward the interior of the membrane in which they occur, while the hydrophilic phosphate or sugar parts are associated with either surface of the membrane.

14.3 Waxes, Cutin, and Suberin: Plant Protective Coats

The entire shoot system of an herbaceous plant is covered by a waxy **cuticle** that slows water loss from all of its parts, including leaves, stems, flowers, fruits, and seeds. A scanning electron micrograph of the cuticle on a carnation leaf illustrates the fine structure of this layer for numerous plants (Fig. 14-5). Without this protective cover, transpiration of most plants would probably be so rapid that they would die. The cuticle also provides protection against some plant pathogens and against minor mechanical damage. It is of special interest in agriculture, because it repels water used in various sprays containing fungicides, herbicides, insecticides, or growth regulators. Because of its hydrophobic nature, most spray formulations contain a detergent to reduce surface tension and allow water to spread on the foliage.

Some 50 to 90 percent of the cuticle is composed of a heterogenous mixture of components called **cutin,** while much of the remainder consists of overlaying waxes. Cutin is a heterogenous polymer consisting largely of various combinations of members in two groups of fatty acids, a group having 16 carbons and one having 18 carbons. Most of these fatty acids have one or more hydroxyl groups, analogous to ricinoleic acid mentioned in Section 14.1. Small amounts of phenolic compounds are also present in cutin. The waxes include a variety of long-chain hydrocarbons that also have little oxygen. Many waxes contain long-chain fatty acids esterified with long-chain monohydric alcohols, but they also contain free long-chain alcohols, aldehydes, and ketones ranging from 22 to 32 carbon atoms, and even true hydrocarbons containing up to 37 carbons. Cutins and waxes are synthesized by the epidermis and are then somehow secreted onto the surface. The waxes lie mostly outside the cuticle but are also mixed with it. The waxes accumulate in various patterns, one of which is the rodlike pattern shown in Fig. 14-5.

A frequently less distinct protective coating over underground plant parts is generally called **suberin.** Suberin also covers the cork cells formed in tree bark and many cells produced as scar tissue after wounding, such as after leaf abscission and on potato tubers cut for planting. Recently, modern techniques of gas chromatography and mass spectrometry have been used to identify several of the components in suberins from several root crop vegetables (Kolattukudy, 1975). A lipid portion (up to half of the total suberin) is a complex mixture of long-chain fatty acids, hydroxylated fatty acids, dicarboxylic acids, and long-chain alcohols. Nearly all members of these groups have more than 16 carbon atoms. The remaining half to three fourths of the suberin contains phenolic compounds, of which ferulic acid (see Fig. 14-11) is a major component. Thus, suberin is similar to cutin in having an important lipid fraction but differs in having a much more abundant phenolic fraction and in the kinds of fatty acids present.

14.4 The Isoprenoid Compounds

Numerous plant products having some of the general properties of lipids form a diverse group of compounds with a common structural unit. These are classified as **isoprenoids.** Included are hormones such as the gibberellins and abscisic acid (see Chapters 16 and 17), farnesol (a stomatal regulator in sorghum, Section 3.5), sterols, carotenoids, turpentine, rubber, and the phytol tail of chlorophyll. Hundreds of isoprenoids have been found, and the actual number existing in the plant kingdom is probably in the thousands. Many of these are of interest because of their commercial uses and because they illustrate the ability of plants to synthesize a vast complex of compounds not usually formed by animals. For most isoprenoids, no function in the plant is presently known. Nevertheless, many influence other plant or animal species with resulting benefit to the species containing them.

Such chemicals (exclusive of foods) that influence another species are sometimes called **allelochemics** (see Section 23.2.10 and Whittaker and Feeny, 1971). A few examples of allelochemical action will be given.

The isoprenoids are dimers, trimers, or polymers of **isoprene units,** in which these units are joined in a head-to-tail fashion:

$$\text{(head)} \quad -CH_2-\overset{\overset{\displaystyle CH_3}{|}}{C}=CH-CH- \quad \text{(tail)}$$

isoprene unit

The isoprene unit is synthesized entirely from acetyl CoA. Three acetyl CoA molecules provide the five carbons for one isoprene unit, while the sixth carbon is lost as CO_2. These reactions are described in most biochemistry books but will not be discussed here. A description of a few of the isoprenoids follows.

Sterols The best known sterols include **stigmasterol, β-sitosterol, ergosterol, campesterol,** and **cholesterol.** Cholesterol was thought to be present only in animals until 1958, when it was discovered in red algae. In the 1960s, it was found in potato and other species, but always in small amounts relative to most of the other sterols present. Stigmasterol, β-sitosterol, and campesterol are the most abundant angiosperm sterols (Grunwald, 1975; Heftmann, 1973).

The structures of these four sterols are given in Fig. 14-6. They and most other plant sterols contain 27 to 29 carbon atoms derived from six isoprene units. They exist in the free form but are also present as glycosides, in which various sugars are attached to the hydroxyl group, or as esters, in which the hydroxyl group is connected to a fatty acid. Their primary significance is probably to act as structural and functional components of membranes. They are widely distributed and probably universal in the plant kingdom but are apparently absent in most blue-green algae and bacteria (procaryotes).

Besides their membrane functions, certain sterols influence the kinds of insects and higher animals that feed on the plants containing them (Roeske et al., 1976). For example, glycoside derivatives of certain sterols found in foxglove (*Digitalis purpurea*) cause convulsive heart attacks in vertebrates. Plants containing sterol glycosides such as **digitoxigenin** have been used since prehistoric times as sources of arrow poisons.

Sterols are of importance to man because of their use as starting materials for the chemical synthesis of certain synthetic animal hormones, including the female ovarian hormone progesterone. Ergosterol is a minor component of plant sterols and seems to be abundant only in certain fungi. It is of special importance to humans because it is converted by ultraviolet radiation to vitamin D_2 (calciferol). The final sterol listed in Fig. 14-6 is another fungal sterol, **antheridiol.** Antheridiol is a sex hormone in the aquatic fungus *Achlya bisexualis*. It is secreted by female strains of the fungus and induces formation of antheridial hyphae. Some people have speculated that sterols also act as hormones that induce flowering in plants, but there is little evidence to support this idea.

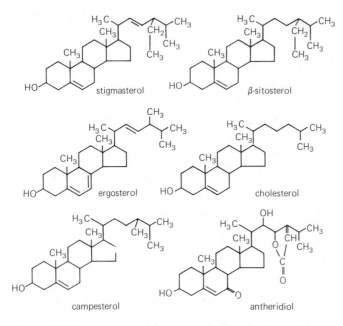

Figure 14-6 Some plant and fungal sterols. Ergosterol probably occurs in trace amounts in some plants, but antheridiol has been found only in certain fungi.

The Carotenoids The carotenoids represent a group of isoprenoids discussed in Chapter 9 in relation to their functions in photosynthesis. They are yellow, orange, or red pigments that exist in various kinds of colored plastids (**chromoplasts**) in roots, stems, leaves, flowers, and fruits of various plants (Goodwin, 1973; 1976). The carotenes are pure hydrocarbons, while the xanthophylls also contain oxygen. Both groups generally contain 40 carbon atoms. Neither group is water soluble, but both dissolve readily in petroleum ether, 80 percent acetone, and many other organic solvents.

About 80 different carotenoids are known in the plant kingdom, although only a few are found in any given species. The most abundant carotenoid found in higher plants is β-carotene, the compound imparting the orange color to carrot roots. **Lutein,** an example of the oxygen-containing xanthophylls, is apparently present in all plants and is the predominant xanthophyll of most leaves. **Lycopene** is the compound giving the reddish color to tomato fruits. The structures of β-carotene, lutein, and lycopene shown in Fig. 9-4 are typical of several carotenoids.

The functions of the carotenoids are still not completely clear. As discussed in Chapters 9 and 11, some of those in chloroplasts participate in photosynthesis, and others prevent photooxidation of chlorophylls. It has been postulated that flower carotenoids might benefit certain plants by attracting insects that carry pollen from one plant to another. None of these ideas, of course, explains the reason why there is so much β-carotene in carrot roots! That is probably because carrots have been cultivated and selected by man, and the carotene they contain is attractive and also useful to humans, because our livers convert it to two vitamin A molecules.

Miscellaneous Isoprenoids and Essential Oils Numerous miscellaneous isoprenoid compounds are present in various amounts among certain members of the plant kingdom. Some of these are referred to simply as **terpenes.** In these, the isoprene unit is condensed into ring compounds containing 10, 15, 20, 30, and even more carbon atoms. Many of the terpenes containing 10 or 15 carbons are called **essential oils,** because they are volatile and contribute to the odor or essence of certain species. Many volatile hydrocarbons released from plants also contribute to smog and other forms of air pollution. Frits Went (1974) estimated that as much as 1.4 billion tons of volatile plant products, mostly terpenes, are released by plants each year, especially over tropical forests. The Blue Mountains in Australia and the Blue Ridge and Smoky Mountains of Virginia and North Carolina in the United States were probably named because of atmospheric scattering of blue light by tiny particles derived from terpenes.

One of the best-known essential oils is turpentine, present in certain specialized cells of members of the genus *Pinus.* The turpentine of some species consists largely of **n-heptane** although 10-carbon terpenes such as α-**pinene,** β-**pinene,** and **camphene** (Fig. 14-7) are also present. The essential oils sometimes contain hydroxyl groups or are chemically modified in other ways. The structures of two such 10-carbon compounds, **menthol** and **menthone,** both components of mint oils, and of **1:8 cineole,** the major constituent of eucalyptus oil, are also shown in Fig. 14-7. Camphene and α-pinene are reportedly volatilized from healthy Douglas fir trees on warm days, or from severed resin canals of injured trees. These molecules attract flying female bark beetles, which then usually attack the damaged trees or, during epidemics, even healthy trees (Rudinsky, 1970). Cineole apparently performs an important function in pollination of orchids by male euglossine bees. Such bees are attracted by the fragrance of the orchid flowers, an important constituent of which is 1:8 cineole (Dodson et al., 1969). This molecule, along with camphor (Fig. 14-11), was shown by Muller (see Muller and Chou, 1972) to be released from *Salvia leucophylla* into the California desert soil in which it grows. Both compounds inhibit root growth of various annual herbs around the *Salvia* shrubs, because a ring forms around the shrubs that is devoid of the herbs. Cineole and camphor are thus phytotoxins, and their effects are said to be **alleopathic.**

Essential oils often contain compounds other than isoprenoids. The essential oils of orange and jasmine flowers, for example, contain indole, a nitrogen-containing compound structurally related to the amino acid tryptophan.

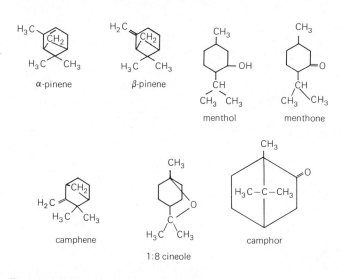

Figure 14-7 Structures of some C-10 terpenes.

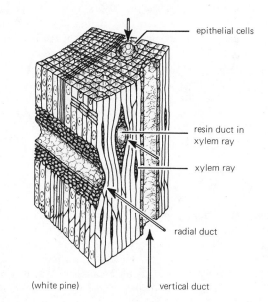

Figure 14-8 A three-dimensional schematic drawing of a piece of pine wood (highly magnified), drawn to show vertical and horizontal (radial) resin ducts with their lining of epithelial cells. These ducts extend throughout the plant.

Complex mixtures of terpenes containing 10 to 30 carbon atoms make up the **resins,** which are common in coniferous trees of North America and in several angiosperm trees of the tropics. Resins and related materials are formed in leaves by specialized epithelial cells, which line the resin ducts, and are then secreted inside the ducts where they accumulate. Similar resin ducts also occur in the vascular tissues of other organs, as shown in Fig. 14-8. It is speculated that resins protect such trees against many insects.

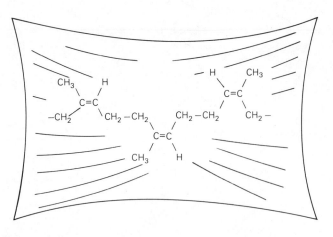

Figure 14-9 Three isoprene units in rubber.

Rubber Rubber is also an isoprenoid compound, the largest of all. It contains some 3,000 to 6,000 isoprene units linked together in very long, unbranched chains (Fig. 14-9). Most natural rubber is commercially obtained from the latex of the tropical plant *Hevea brasiliensis,* a member of the Euphorbiaceae family. About one third of this latex is pure rubber. It has been reported, however, that over 2,000 plant species form rubber in varying amounts. Various *Taraxacum* (dandelion) species are among the most well-known North American species possessing this ability.

14.5 Phenolic Compounds and Their Relatives

Flowering plants, ferns, mosses, liverworts, and many microorganisms contain various kinds and amounts of phenolic compounds. With certain important exceptions, the functions of most of these substances are obscure. Many presently appear to be simply byproducts of metabolism, but this probably reflects to a great extent our ignorance of alleochemical interactions.

All the compounds to be discussed possess a benzene ring that contains various attached substituent groups, such as hydroxyl, carboxyl, and methoxyl (—O—CH₃) groups, and often other nonaromatic ring structures. The phenolics differ from the lipids in being more soluble in water and less soluble in nonpolar organic solvents. Some, however, are fairly soluble in ether, especially when the *p*H is low enough to prevent ionization of the carboxyl and hydroxyl groups that are often present.

The Aromatic Amino Acids Phenylalanine, tyrosine, and tryptophan are aromatic amino acids that are formed by a route common to many of the phenolic compounds.

Two respiratory intermediates are precursors of these amino acids and of many other phenolic compounds as well. These two are PEP from the glycolytic pathway and erythrose-4-phosphate from the pentose phosphate respiratory pathway. These two molecules combine, producing a seven-carbon compound that forms a ring structure, which is then converted by several reactions into a rather stable compound called **shikimic acid.** These steps make up what has come to be known as the **shikimic acid pathway.** Reactions of this pathway and others summarizing formation of several aromatic compounds are in Fig. 14-10.

Miscellaneous Simple Phenols and Related Compounds An extensive number of other relatively simple compounds also arise from the shikimic acid pathway and subsequent reactions. Among these are the acids **cinnamic, p-coumaric, caffeic, ferulic, protocatechuic, chlorogenic** (Fig. 14-11), and **gallic** (Fig 14-10, upper right). The first four are derived entirely from phenylalanine and tyrosine. They are important not because they are abundant in uncombined (free) form, but because they are converted into several significant derivatives. These derivatives include phytoalexins, coumarins, lignin, and various flavonoids such as the anthocyanins, all of which will be described shortly.

An important reaction in formation of these derivatives is the conversion of phenylalanine to cinnamic acid (R14-4). This is a deamination in which ammonia is split out of phenylalanine and is catalyzed by **phenylalanine ammonia lyase,** formerly called phenylalanine deaminase.

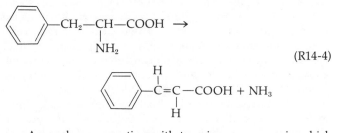

(R14-4)

An analogous reaction with tyrosine can occur, in which **tyrosine ammonia lyase** converts tyrosine to ammonia and p-coumaric acid. This acid can also be formed readily in a wide variety of species by addition of one atom of oxygen from O_2 and an H atom from NADPH directly to the *para* position of cinnamic acid. Subsequent addition of another hydroxyl group adjacent to the OH group of p-coumarate by a similar reaction forms caffeic acid. Addition of a methyl group from S-adenosyl methionine to an OH group of caffeic acid yields ferulic acid. Caffeic acid forms an ester with an alcohol group in still another acid formed in the shikimic acid pathway, **quinic acid,** thus producing chlorogenic acid.

Protocatechuic and chlorogenic acids probably have special functions in disease resistance of certain plants. Protocatechuic acid is one of the compounds preventing smudge in certain colored varieties of onions, a disease caused by the fungus *Colletotrichum circinans.* This acid occurs in the scales

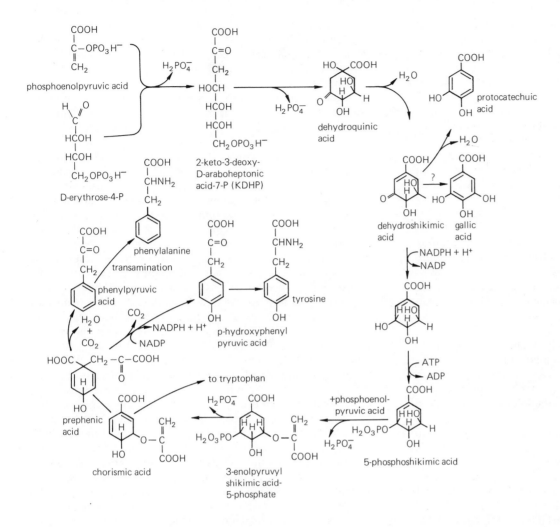

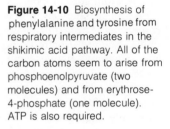

Figure 14-10 Biosynthesis of phenylalanine and tyrosine from respiratory intermediates in the shikimic acid pathway. All of the carbon atoms seem to arise from phosphoenolpyruvate (two molecules) and from erythrose-4-phosphate (one molecule). ATP is also required.

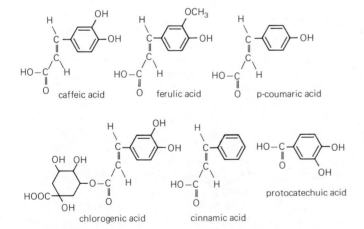

Figure 14-11 Structures of phenolic acids often found in plants. All are shown in the *trans* form. Chlorogenic acid is an ester formed from caffeic and quinic acids.

of the neck of colored onions that are resistant to the pathogen, but it is absent from susceptible white varities. When extracted from colored onions, it strongly inhibits the growth of the smudge fungus and of other fungi.

High amounts of chlorogenic acid might similarly prevent certain diseases in resistant varieties; suggested examples are the scabs of potato tubers and of apple leaves. Chlorogenic acid is widely distributed in various parts of many plants and usually occurs in easily detectable quantities. It is formed in relatively high amounts in many potato tubers, and its oxidation followed by a free-radical polymerization causes formation of large, uncharacterized quinones responsible for the darkening of freshly cut tubers. Enzymes called **polyphenol oxidases** catalyze this reaction, using O_2 as the electron acceptor. It is thought that chlorogenic acid and certain other related compounds can be readily formed and oxidized into potent fungistatic quinones by certain disease resistant varieties but less readily by susceptible ones. In this way, the infection is well localized in the resistant plants. Ferulic acid and its derivatives also play a role in plant protection, since they form part of the phenolic fraction of suberin.

Gallic acid is important because of its conversion to **gallotannins,** which are heterogenous polymers containing numerous gallic acid molecules connected in various ways to one another and to sugars that are also present. Many gallotannins greatly inhibit growth (Green and Corcoran, 1975), and tolerance may involve sequestering them in the vacuoles where they cannot denature the cytoplasmic enzymes. Gallotannins and other tannins are used commercially to tan leather because they act as protein denaturants. Gallotannins were found by Elroy Rice at Oklahoma University to act as alleopathic agents, inhibiting growth of other species around those plants that form and release them. Free cinnamic acid and some of its esters may function in allelopathic responses of guayule (*Parthenium argentatum*) plants. Similarly, p-coumaric, ferulic, and caffeic acids leached from leaves of various deciduous trees or released from decaying leaves reportedly inhibit germination and development of certain grasses that would otherwise grow in the relatively bare areas under such trees (Lodhi, 1976). Numerous other phenolic acids derived from the shikimic acid pathway are probably involved in this relatively unstudied but interesting area of physiological ecology (Rice, 1974).

A group of compounds closely related to the phenolic acids and also derived from the shikimic acid pathway are the **coumarins.** Approximately 50 coumarins are known, although only a few are usually found in any particular plant family. The structures of two coumarins, **scopoletin** and **coumarin** itself, are given in Fig. 14-12.

Coumarin is a volatile compound that is formed from nonvolatile derivatives upon senescence or injury. This is especially significant in alfalfa and sweet clover, where coumarin causes the characteristic odor of recently mown hay. Certain sweet clover strains have been developed that contain low amounts of coumarin and others that contain it in a bound form. These are of economic importance, because free coumarin can be converted to a toxic product, **dicumarol,** if the clover becomes spoiled during storage. Dicumarol is a hemorrhagic and anticoagulant agent responsible for sweet clover disease (a bleeding disease) in ruminant animals that are fed plants that contain it. It is also claimed that coumarin in sweet clover acts as both a feeding attractant and a feeding deterrent for adult vegetable weevils. Such weevils are attracted to the clover, but they reportedly are repelled by the first bite (Lindstedt, 1971).

Scopoletin is a toxic coumarin widespread in plants and often found in seed coats. In certain seeds, scopoletin apparently prevents germination, causing a dormancy that exists until the chemical is leached out (for example, by a rainstorm heavy enough to provide sufficient moisture for seedling establishment). It might thus function as a natural inhibitor of seed germination. It appears to be formed from coumarin, which, in turn, arises from phenylalanine and cinnamic acid.

Also illustrated in Fig. 14-12 are the structures of two coumarin-like compounds named **preocenes** isolated in 1976 from the plant *Ageratum houstonianum.* They cause premature metamorphosis in several insect species by decreasing the level of insect juvenile hormone, thereby causing formation of sterile adults. Such compounds appear promising as insecticides that have no influence on species other than the target ones.

Certain toxic coumarin derivatives are synthesized in response to parasite invasion and might play a role in disease resistance. Since about 1960, various other antifungal compounds that also appear to be synthesized by the plant only when it is infected by certain fungi have been discovered (Deverall, 1977). These compounds were hypothesized to act in a way comparable to the antibodies of animal cells. They are collectively referred to as **phytoalexins** (from the Greek *phyton,* "plant," and *alexin,* a warding-off substance). Compounds that act as phytoalexins include **pisatin,** in pea pods, and **phaseolin,** in bean pods. Others that apparently are produced only when the plant is invaded and that thus seem to be phytoalexins include **orchinol,** from orchid tubers, **trifolirhizin,** from red clover root, and an **isocoumarin,** from carrot roots. It appears that nonpathogenic fungi often induce such high, toxic levels of phytoalexins in the host that their establishment is prevented, while pathogenic fungi are successful parasites because they induce only nontoxic phytoalexin levels. The importance of phytoalexins in disease resistance is still controversial (see Special Topic 23.2).

14.6 Lignin

Lignin is a strengthening material that occurs together with cellulose and other polysaccharides in the cell walls of all higher plants. It occurs in largest amounts in woody plants where it accumulates in the middle lamella, primary walls, and secondary walls of the xylem elements. It usually occurs between the cellulose microfibrils, where it serves to resist compression forces. Resistance to tension (stretching) is primarily a function of the cellulose. Lignin may easily be detected by the bright red color it gives when stained with a mixture of phloroglucinol and hydrochloric acid. It is normally a brown, amorphous solid.

Lignin is not readily soluble in most solvents, primarily because it, like cellulose, has a high molecular weight (probably more than 10,000), and because in the native state it is chemically united to the cellulose at various points by ether linkages through the cellulose hydroxyl groups. It is possible

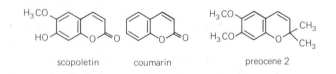

scopoletin   coumarin   preocene 2

Figure 14-12 Structures of two coumarins and of preocene 2, a plant compound that acts as an insect antijuvenile hormone substance. Preocene 1, another antijuvenile hormone substance, also occurs in plants; it lacks the upper methoxyl group on the benzene ring (see W. S. Bowers et al., 1976, Science 193:542–547).

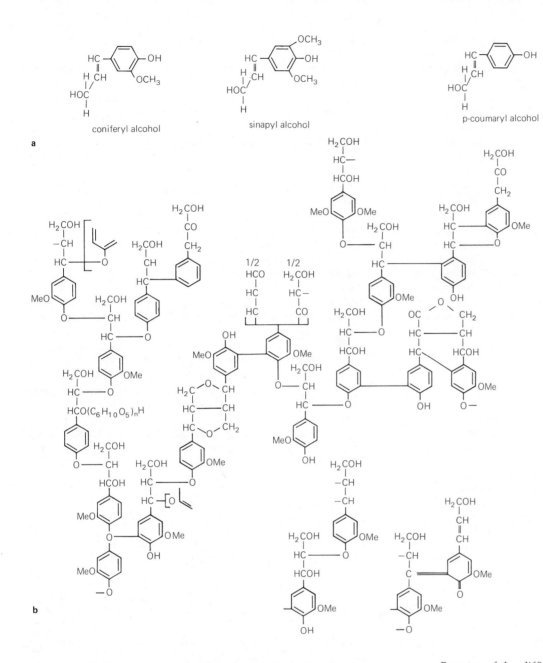

a

b

coniferyl alcohol

sinapyl alcohol

p-coumaryl alcohol

Figure 14-13 (a) Common phenolic subunits found in lignins. **(b)** Constitutional model of spruce lignin, showing possible mechanisms by which alcohol subunits may be connected. The ratio of alcohols in this lignin is about 14 coumaryl:80 coniferyl:6 sinapyl. Beechwood, a typical hardwood lignin, may have ratios about 5:49:46. Methyl groups are represented as Me. (From K. Freudenberg, 1965, Science 148:595. Copyright 1965 by the American Association for the Advancement of Science.)

to remove lignin from wood by dissolving it in sodium bisulfite, which, however, breaks it down into small water-soluble derivatives. The treatment with sodium bisulfite, called the **sulfite process,** is one of two primary methods used in the paper industry to remove lignin from wood (Hall, 1974). The other method, called the **kraft process,** involves cooking the wood chips in a mixture of sodium sulfide, sodium carbonate, and sodium hydroxide, a highly destructive method that releases malodorous organic sulfur compounds. If not removed, lignin causes paper made from wood to become yellow. Lignins remain almost entirely a waste product in the pulp and paper industry. Canadian industries alone were estimated to discard over 3 million tons annually.

Because of the difficulty in dissolving lignin without destroying it and some of its subunits, its exact chemical structure has been impossible to ascertain. Much of what we know about lignin structure has been determined by analyzing several large molecules that are intermediates in its synthesis. This is contrasted with our knowledge of polysaccharides, proteins, and nucleic acids, the structures of which were largely determined by analyzing degradation products. In general, lignins contain three aromatic alcohols, **coniferyl alcohol** (which predominates in the conifers), **sinapyl alcohol,** and **p-coumaryl alcohol.** Grass and dicot lignins also contain large amounts of phenolic acids, such as p-coumaric and ferulic acid, that are esterified to alcohol groups of each other and to sinapyl alcohol, p-coumaryl alcohol, and so on. The structures of these alcohols are given in Fig. 14-13, along with some proposed models indicating the ways they might be connected in lignin.

Several reactions by which lignin is formed are known, even though we do not fully understand its structure. Phenylalanine is converted to aromatic acids such as coumaric and ferulic acid, then these are converted to CoA esters. The esters are reduced by NADPH to the phenolic alcohols mentioned in the preceding paragraph (Wengenmayer et al., 1976). These alcohols are polymerized into lignin. An iron-containing enzyme called **peroxidase** catalyzes the first reaction involved in polymerization. Peroxidase exists in many isozyme forms, some of which exist in the cell walls. These isozymes function by removing H atoms from alcohol groups, apparently combining two of these with hydrogen peroxide (H_2O_2), supplied from a still unknown source, to form H_2O molecules. The remaining part of the phenolic alcohol is now a free radical, and several kinds of electronic shifts allow migration of the unpaired electron to other parts of the molecule. Many such free radicals combine in various ways to form bonds such as those proposed in Fig. 14-13b, so lignins presumably always have variable structures.

14.7 The Flavonoids

The flavonoids are 15-carbon compounds that are generally distributed throughout the plant kingdom (Harborne, 1976; Swain, 1976). The basic flavonoid skeleton, shown in the following diagram, is usually modified in such a way that more double bonds are present, causing the compounds to absorb visible light and thus giving them color. The two carbon rings at the left and right ends of the molecule are designated the A and B rings, respectively.

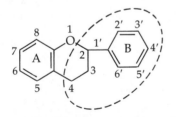

The dashed lines around the B ring and the three carbons of the central ring indicate that part of flavonoids derived from the shikimic acid pathway. This part may be compared with cinnamic acid (Fig. 14-11), which is a precursor to it. Ring A and the oxygen of the central ring are derived entirely from acetate units provided by acetyl CoA. Hydroxyl groups are nearly always present in the flavonoids, especially attached to ring B in the 3' and 4' positions (compare p-coumaric and caffeic acids of Fig. 14-11), to the 5 and 7 positions of ring A, or to the 3 position of the central ring. These hydroxyl groups serve as points of attachment for various sugars that increase the water solubility of flavonoids. Many flavonoids accumulate

in the central vacuole, although many of them are apparently synthesized in chloroplasts.

Three groups of flavonoids are of particular interest in plant physiology. These are the **anthocyanins,** the **flavonols,** and the **flavones.** The anthocyanins (from the Greek *anthos,* "flower," and *kyanos,* "dark blue") are colored pigments that commonly occur in red, purple, and blue flowers. They are also present in various other plant parts, such as certain fruits, stems, leaves, and even roots. Most fruits and many flowers owe their colors to anthocyanins, although some, such as tomato fruits and several yellow flowers, are colored by carotenoids. The bright colors of autumn leaves are due largely to anthocyanin accumulation on bright, cool days, although yellow carotenoids are the predominant pigments in autumn leaves of some species.

Anthocyanins seem generally absent in the liverworts, algae, and other lower plants, although some anthocyanins and other flavonoids occur in certain mosses. They have only rarely been demonstrated in gymnosperms. Several different anthocyanins exist in higher plants, and often more than one is present in a particular flower or other organ. They are nearly always present as glycosides, containing most commonly one or two glucose or galactose units attached to the hydroxyl group in the central ring or to that on the 5 position of ring A, as described in Fig. 14-14. When the sugars are removed, the remaining parts of the molecules, which are still colored, are called **anthocyanidins.**

Anthocyanidins are usually named after the particular plant from which they were first obtained. The most common anthocyanidin is **cyanidin,** which was first isolated from the

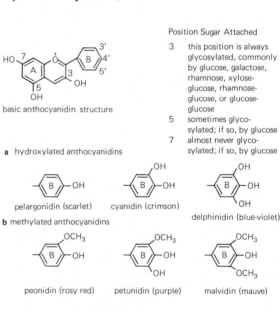

Figure 14-14 The basic anthocyanidin ring, showing variations of *B* ring by hydroxylation and methylation to produce various anthocyanins. Anthocyanins are produced by attachment of sugars (glycosylation) to the 3-hydroxyl position of the anthocyanidin, and sometimes also to the 5 or 7 position.

blue cornflower, *Centaurea cyanus.* Another, **pelargonidin,** was named after a bright red geranium of the genus *Pelargonium.* A third, **delphinidin,** obtained its name from the genus *Delphinium* (blue larkspur). These anthocyanidins differ only in the number of hydroxyl groups attached to the B ring of the basic flavonoid structure. Other important anthocyanidins include the reddish **peonidin** (present in peonies), the purple **petunidin** (in petunias) and the mauve-colored (purplish) pigment **malvidin,** first found in a member of the Malvaceae, the mallow family.

The exact color of the anthocyanins depends first upon the substituent groups present on the B ring. When methyl groups are present, as in peonidin, they cause a reddening effect. Second, the anthocyanins are sometimes associated with other phenolic compounds, and this causes them to become blue. Finally, the *p*H of the cell sap influences their color. Most anthocyanins are reddish in acid solution but become purple and blue as the *p*H is raised. This probably results from ionization of OH groups on the B ring, followed by electron shifts in this ring. These changes are fully reversible. Because of these properties and the common presence of more than one anthocyanin, there is wide variation in the hues of flower colors in higher plants.

The functions that anthocyanins perform have presented interesting topics for discussion ever since their discovery. One of their useful functions in flowers is apparently the attraction of birds and bees that carry pollen from one plant to another, thus facilitating pollination (Harborne, 1976). Charles Darwin long ago suggested that a fruit's beauty serves as a guide to birds and beasts in such a way that the fruit may be eaten and its seeds widely disseminated in the manure. Presumably, anthocyanins contribute to this beauty. Anthocyanins may also play a role in disease resistance, although the evidence for this is weak. Their abundance certainly suggests some functions that have favored their evolution (McClure, 1975).

Anthocyanins and other flavonoids are of particular interest to many plant geneticists, because it is possible to correlate many morphological differences among closely related species in a particular genus, for example, with changes in the type of flavonoids they contain. A knowledge of the flavonoids present in related species of the same genus gives information that can be used by taxonomists to classify and determine the lines of evolution of these plants.

The **flavonols** and **flavones** are closely related to the anthocyanins, except that they differ in the central oxygen-containing ring structure, as follows:

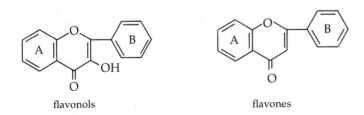

flavonols   flavones

Most of the flavones and flavonols are yellowish or ivory-colored pigments, and like the anthocyanins, they often contribute to the color of flowers. Even those flavones and flavonols that are not colored nevertheless absorb ultraviolet wavelengths and therefore affect the spectrum of radiation visible to bees or other insects that are attracted to flowers containing them. These molecules are also widely distributed in leaves, especially in the chloroplasts, where they are apparently synthesized (Harborne, 1976). They apparently function there as feeding deterrents and, since they absorb ultraviolet radiation, as a protection against ultraviolet rays.

Light stimulates the formation of flavonoids in many plants (Smith, 1972; Wong, 1976). The anthocyanins have been studied most in this respect. It has probably been known for centuries that the reddest apples are found on the sunny side of the tree. This is because anthocyanins accumulate in these fruits, a process increased by light. The most effective wavelengths are blue and red. The pigment or pigments that absorb this light are more fully discussed in Chapter 19. The nutritional status of a plant also affects its production of anthocyanins. A deficiency of nitrogen, phosphorus, or sulfur leads to accumulations of anthocyanins in certain plants, as mentioned in Chapter 6. Low temperatures also increase anthocyanin formation in some species, as in the coloration of certain autumn leaves.

Certain species also synthesize one or more **isoflavones,** which differ from flavones in that ring B is attached to the carbon atom of the central ring adjacent to the point of attachment in flavones. The functions of isoflavones are generally unknown, but they may act as allelochemics. Thus, their structures resemble those of animal estrogens such as estradiol, and they cause infertility in female livestock, especially sheep (Shutt, 1976). Various legumes, particularly subterranean clover, accumulate especially high levels of isoflavones. These compounds cause the "clover disease" of sheep, first noted as a decline in fertility in western Australia in the 1960s. They are also suspected to be a factor controlling rodent populations in certain regions.

14.8 Betalains

The red pigment of beets is a **betacyanin,** one of a group of red and yellow **betalain** pigments that were long thought to be related to the anthocyanins, even though they contain nitrogen. It is now known that neither red betacyanins nor the other kind of betalain pigments, the yellow **betaxanthins,** are at all structurally related to the anthocyanins, and that anthocyanins and betalains do not occur together in the same plant. Betalains seem restricted to 10 plant families, all of which are members of the order Caryophyllales (Centrospermae). Betalains do not undergo the characteristic extensive changes in color with *p*H that the anthocyanins do.

Like the anthocyanins and most other flavonoids, betalains can be hydrolyzed into a sugar and a remaining colored portion. The most extensively studied member of this group is **betanin,** crystallized from red beet roots. Betanin can be hydrolyzed into glucose and **betanidin,** a reddish pigment with the following structure:

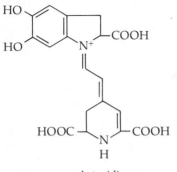

betanidin

Little is yet known of the metabolism or functions of betacyanins, but a role in pollination comparable to that of anthocyanins in other species seems likely (Piatelli, 1976).

14.9 The Alkaloids

Many plants contain aromatic nitrogenous compounds called **alkaloids** that appear to have no important function in those species producing them. The alkaloids usually contain nitrogen in a heterocyclic ring. This nitrogen will act as a base, since it has the ability to accept hydrogen ions, so most solutions of alkaloids are usually slightly basic, as their name indicates. Most alkaloids are white crystalline compounds and are only slightly water soluble.

More than 3,000 alkaloids have been found in some 4,000 species of plants, although any given species typically contains only a few such compounds. The first alkaloid to be isolated and crystallized was the drug **morphine,** isolated in 1805 from the opium poppy, *Papaver somniferum.* Other well-known alkaloids include **nicotine,** present in cultivated varieties of tobacco (*Nicotiana tabacum*); **quinine,** from cuprea bark; **caffeine,** from coffee beans and tea leaves; **strychnine,** from the seeds of *Strychnos nuxvomica;* **theobromine,** from cocoa beans; **atropine,** from the poisonous black nightshade (*Atropa belladonna*); **colchicine,** from *Colchicum byzantinum;* **mescaline,** a hallucinogenic and euphoric drug from flowering heads of the cactus *Lophophora williamsii;* and **lycoctonine,** a toxic alkaloid in *Delphinium barbeyi* (larkspur). The structures of several alkaloids are shown in Fig. 14-15.

Many other alkaloids have far more complex chemical structures than either nicotine or caffeine. They are often synthesized only in plant shoots, but nicotine appears to be produced only in the roots of tobacco. The chemical reactions involved in formation of alkaloids are largely unknown, although the precursor compounds providing the carbon skele-

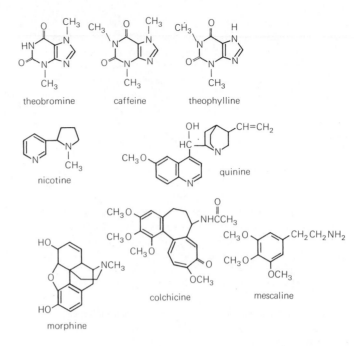

Figure 14-15 Structures of some representative alkaloids.

tons and nitrogen atoms of a few have been established. Synthesis of nicotine has received greatest attention, initially because of its commercial importance, and now because of concern about harmful effects of smoking. Nicotinic acid (niacin), present in NAD and NADP molecules, is a precursor of nicotine. The nitrogen and carbon atoms of nicotinic acid, in turn, arise from a product obtained when aspartic acid and 3-phosphoglyceraldehyde are combined. Other amino acids are precursors of various alkaloids, tryptophan probably providing several of the atoms of quinine; phenylalanine, of colchicine; lysine, of anabasine; and tyrosine, of morphine. Very few of the thousands of enzymes that must be necessary for production of the various alkaloids have been demonstrated in the plant kingdom. Furthermore, Hughes and Genest (1973) speculated that there are many thousands of alkaloids yet to be discovered in plants.

The physiological role of most alkaloids is unknown, and it has been suggested that they perform no important metabolic function, being merely byproducts of other more important pathways. Yet it is reasonable that they could be of ecological importance, providing some survival value to the plant. Plants containing such toxic substances might be avoided by grazing animals and leaf-feeding insects, for example. Nevertheless, larkspur is not avoided by cattle, even when other forage is available, and it accounts for more cattle deaths in the United States than any other poisonous plant (Keeler, 1975).

Section Three

Plant Development

Phototropism in oat, soybean, and pea seedlings.

15

Growth and Development

Again think of your favorite plant as you did at the beginning of the Prologue. You probably now see that plant somewhat differently. You may visualize water moving into and across various root cells, up through the xylem into living leaf cells, where hydrogen bonds are broken and the water molecules evaporate into air spaces of the leaf and finally out through the open stomates into the atmosphere. With a little more reflection, you may imagine CO_2 molecules diffusing through those stomates into chloroplasts of photosynthetic cells, being fixed into carbohydrates in chemical reactions energized by ATP and NADPH arising from light-dependent reactions. Think, too, of certain photosynthetic products being loaded into phloem sieve tubes and moved to specific sinks. Ions are being selectively and actively absorbed, some of which are assimilated into organic compounds and some of which act as coenzymes. Though no one can explain all that is going on in cells, at least your favorite plant no longer seems like such a static object. It is a well-organized living thing capable of processing many of the things in its environment and maintaining a relatively low entropy.

Yet in all this, we have overlooked one of the most fascinating things about living organisms, their ability to grow and develop. The continuous synthesis of large, complex molecules from the smaller ions and molecules that are the raw materials for growth leads not only to larger cells, but often to more complex ones. Furthermore, not all cells grow and develop in the same way, so a mature plant consists of numerous cell types. The process by which cells become specialized is called **differentiation,** and the process of growth and differentiation of individual cells into recognizable tissues, organs, and organisms is often called **development.** Another useful term for this process is **morphogenesis** (Greek *morpho,* "form," and *genesis,* "origin" or "beginning").

We know that genes govern the synthesis of enzymes, which in turn control the chemistry of cells, and that all this must some way account for growth and development. We do not know, however, exactly what it is that governs which genes should be transcribed in which cells at a given time. This lack of understanding provides one of the most challenging problems for modern biologists. Yet we know a great deal about what happens during the growth and development of a plant. We know that chemical messengers called hormones somehow act as intermediaries in many growth processes.

Study of such substances has been an important thrust in plant physiology since shortly after the beginning of this century. The two chapters following this introductory one are devoted to a summary of what has been learned, but much of the discussion in subsequent chapters also concerns hormones. In Chapter 18 we will examine some specific studies of differentiation, certainly an important key to eventual understanding of development.

As the science of plant physiology has evolved, it has become apparent that development can often be strongly modified by environment. Light, which plays an important role, is considered in Chapter 19. Often, light effects are exhibited within constraints placed by an internal timing mechanism that exists in plants and animals: the biological clock, which is the subject of Chapter 20. Plants also respond strongly to temperature changes, and there are some especially interesting responses to low temperature, subjects of Chapter 21.

Perhaps the most striking example of morphogenesis in plants is the conversion from the vegetative to the reproductive stage. Cells must differentiate in radically new ways, and hormones are involved. Light, particularly the relative lengths of day and night, often modifies or controls flower initiation, and so do temperature changes. We shall examine these things in Chapter 22.

15.1 What Is Meant by Growth?

To most people, growth probably means an increase in size. As organisms grow from the zygote, they increase in weight, cell number, amount of protoplasm, and complexity. For many purposes, we must be able to measure growth. In theory, we could measure any one of the growth features just mentioned, but the two principal methods of measurement involve determination of either the increase in volume or weight. Volume (size) increases are often approximated by measuring expansion in only one or two directions, such as length, height, width, diameter, or area. Weight increases can be determined by simply harvesting the entire plant or that part we wish to measure and weighing it rapidly before too much water evaporates from it. This gives us the **fresh weight,** which is a somewhat variable quantity dependent upon the plant's water status. A leaf, for example, may have a greater fresh weight in the morning than it does at midafternoon simply because of transpiration. Because of this problem, many people, particularly those interested in crop productivity, prefer to use the increase in **dry weight** of a plant or plant part as a measure of its growth. The dry weight is commonly obtained by drying the freshly harvested plant material for 24 to 48 hours at 70 to 80 C. The leaf that has a lower fresh weight at midafternoon will probably have a greater dry weight, because it photosynthesized and absorbed mineral salts from the soil during the preceding morning. Here dry weight may be a more valid estimate of what we mean by growth than fresh weight.

Sometimes dry weights give inadequate indication of important growth increases. For example, when a seed germinates and develops into a seedling in total darkness, provided only with water, the size and fresh weight increase greatly, but the dry weight decreases slightly due to respiratory loss of CO_2 (Fig. 15-1). Although the total dry weight of such dark-grown seedlings is less than that of the original seed, the growing parts of the stem and roots do increase in dry weight. This increase occurs because assimilates are translocated from nongrowing parts to the growing regions, so growth again involves a dry weight increase.

Early stages in seedling development involve production of some new cells by **mitosis** (nuclear division) and subsequent **cytokinesis** (cell division), but normal-appearing seedlings can be produced from seeds of some species in the absence of mitosis or cell division. When seeds of lettuce and wheat

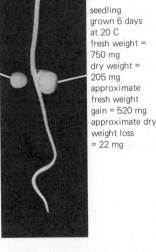

seed
fresh (air dry)
weight = 230 mg
dry weight (after
48 hr at 70 C)
= 227 mg

seedling
grown 6 days
at 20 C
fresh weight =
750 mg
dry weight =
205 mg
approximate
fresh weight
gain = 520 mg
approximate dry
weight loss
= 22 mg

Figure 15-1 Changes in fresh and dry weight of a pea seed as it develops into a seedling in darkness. The fresh weight increases greatly because of water uptake, but the dry weight decreases slightly because of respiration. (Photo by C. W. Ross.)

are irradiated with gamma rays from a cobalt-60 source at intensities high enough to stop DNA synthesis, mitosis, and cell division, germination occurs, and growth continues until seedlings with giant cells are produced. These seedlings, called **gamma plantlets,** can survive up to three weeks but then die, presumably because new cells are eventually necessary. They illustrate that even if we could conveniently measure the increase in cell number during growth, this number is sometimes a poor measure of growth. Many other examples of growth without cell division are known. There are also a few examples of cell division without increase in overall size, as in the maturation of the embryo sac. We shall use an increase in size as the fundamental criterion of growth, even though there are problems in measuring it.

15.2 Growth in Relation to Time (Growth Kinetics)

Many investigators have measured quantitative changes in sizes of organisms as a function of time. Julius Sachs, who helped establish the science of plant physiology, pioneered these measurements in Germany in the nineteenth century. It was hoped that growth curves could be explained by mathematical models that would then help explain the factors controlling growth. Such attempts have generally been unsuccessful, although to a limited extent we can do the reverse, that is, explain the shapes of the curves with our knowledge of plant physiology and anatomy.

An idealized **S-shaped (sigmoid) growth curve** exhibited by numerous annual plants and individual parts of both annual and perennial plants is illustrated in Fig. 15-2a. Three primary phases can usually be detected: a **logarithmic phase,** a **linear phase,** and a **senescence phase** (Sinnott, 1960; Richards, 1969). In the logarithmic phase, the growth rate (the increase in size per unit of time) is slow at first (Fig. 15-2b), apparently because the germinating seed has fewer cells capable of growth, but the rate continuously increases as more cells are formed. Mathematicians have pointed out the analogy between the logarithmic phase and the growth of money invested to draw compound interest: The accumulated interest also draws interest, so the total principal grows logarithmically. A logarithmic phase is also exhibited by cultures of bacteria and other single-celled organisms, in which each product of division is capable of growth and division.

In the linear phase, increase in size continues at a constant, usually maximum rate for some time (Fig. 15-2b). We do not understand exactly why the growth rate should be constant in this phase, but one reason might be that stems and roots grow by meristems that produce cells that grow mainly only in length. As opposed to a bacterial culture free to proliferate in every direction of the medium, growth in the plant shoot and root becomes mainly one dimensional and linear. The final senescence phase is characterized by a decreasing growth rate (note drop in rate curve in Fig. 15-2b) as the plant reaches maturity and begins to senesce.

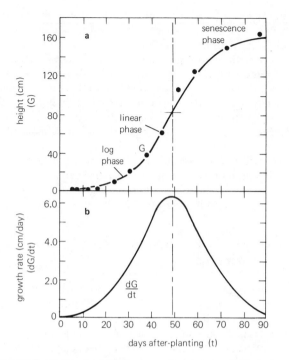

Figure 15-2 Idealized growth and growth rate curves. The rate curves in (**b**) are obtained from the first derivative of the growth curve in (**a**). (Drawn from data of W. G. Whaley, 1961, *in* W. Ruhland, ed., Encyclopedia of plant physiology, vol. 14, Springer-Verlag, Berlin, pp. 71–112.)

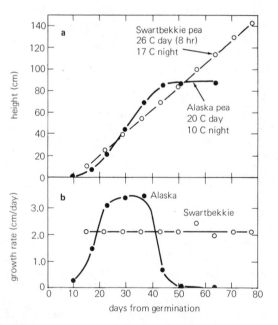

Figure 15-3 (**a**) Growth curves for two tall pea varieties, indicating departure from the typical sigmoid curve in Swartbekkie (note extended linear phase shown by open circles). (**b**) Growth rate curves derived from data in (**a**), comparing the somewhat more bell-shaped curve for the Alaska variety with the constant growth rate for Swartbekkie. (From Went, 1957.)

Although the curves of Fig. 15-2 are representative of many species, the measured growth curves often only approximate those. Sometimes the linear phase is hardly detectable, and then the logarithmic and senescence phases are almost continuous. Even more commonly, the linear phase is extended, as shown for the Swartbekkie pea in Fig. 15-3. The growth rate was constant at just over 2 cm height increase per day for nearly two months. (Senescence phase is not shown, although it occurred later.) The graphs also illustrate results with the Alaska pea, another tall variety, which showed a more sigmoid-like growth curve and a bell-shaped rate curve flattened on top because of the extended linear phase.

Growth of commercial fruits has been studied considerably, no doubt in part because of their economic importance, but there is little information about smaller, less fleshy fruits, even though these predominate (Coombe, 1976). Growth curves of apple, pear, tomato, banana, strawberry, date, cucumber, orange, avocado, melon, and pineapple fruits are sigmoid, while raspberry, grape, blueberry, fig, currant, olive, and all stonefruits (peach, apricot, cherry, and plum) show interesting double-sigmoid growth curves where a first "senescence" phase is followed by another logarithmic phase leading to the second sigmoid part of the curve.

Fewer data are available for perennial species, especially trees, but sigmoid curves would probably be produced, usually with important flat portions due to winter or dry periods. For specific seasons, data for shoots of trees are readily available, and modified sigmoid curves are indeed observed. Figure 15-4 shows such curves for both pines and deciduous hardwood trees. Note that there were important differences in actual height growth and in lengths of the growing period among the various species. Among the hardwoods, all except the poplar essentially stopped elongating in August. Among the pines, the nonnative (transplanted) red and white species stopped elongating in late spring, while the native species grew taller during a longer time period.

It is common for trees to cease height growth temporarily in late summer, when temperatures are still warm and days are long (Kramer and Kozlowski, 1960; Zimmermann and Brown, 1971; Perry, 1971). Sometimes growth resumes again before winter dormancy, a deeper dormancy that results in part from the increasing night lengths and decreasing day lengths and in part from the low temperatures of autumn (Chapter 21). Growth in stem diameter (due to expansion of cells produced by the vascular cambium) continues, at a decreasing rate, until well after height growth stops. The smaller widths of xylem cells produced in summer than in spring are responsible for the contrasting summer and spring wood of the annual rings, which allow estimates of age in many tree species. Photosynthesis continues until the leaves become senescent and yellow, as in deciduous trees, or until temperatures become too cold, as in evergreens. Because of this, increases in dry weight and radial growth can continue several weeks after stem elongation ceases. Root growth can continue as long as water and nutrients are available and the soil temperature remains high enough; that is, dormancy such

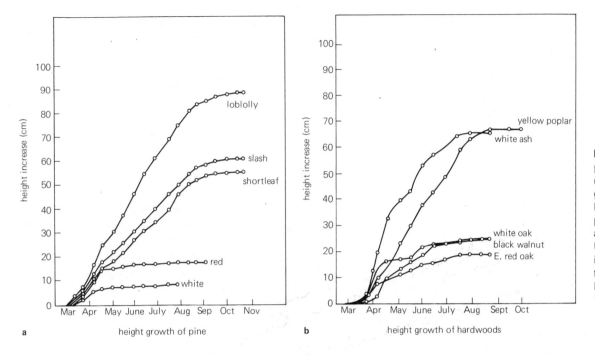

Figure 15-4 Shoot elongation in pines (**a**) and deciduous trees (**b**) in North Carolina during one growing season (1938). All trees had been planted a few years previously in the same plot. Not all are native to the southeastern United States. Note differences in growth rates and in lengths of the growing season. (From P. J. Kramer, 1943, Plant Physiology 18:239–251.)

as that found in shoots does not occur in roots so far examined. Because of this, orchard trees often benefit by continued root growth from late summer fertilization and water treatments.

15.3 Some Features of Plant Growth

Growth in the plant is not uniformly distributed throughout but is restricted to certain zones containing cells recently produced in a **meristem.** The principal meristematic zones are found near the root and the shoot tips (apices), in the vascular cambium, near the nodes of monocots, and in certain parts of young leaves. The root and shoot apical meristems are formed during embryo development as the seed forms, but the vascular cambium and meristematic areas of leaves are not distinguishable until after germination.

Some plant structures are determinate; others are indeterminate. A **determinate** structure grows to a certain size and then stops, eventually undergoing senescence and death. Leaves, flowers, and fruits are excellent examples of determinate structures, and the great majority of animals also grow in a determinate way. On the other hand, the vegetative stem and the root are **indeterminate** structures. They grow by meristems that continually replenish themselves, remaining youthful. A bristlecone pine that has been growing for 4,000 years could yield a cutting that would form roots at its base, producing another tree that might live for another 4,000 years. At the end of that time, another cutting might be taken, and so on, potentially forever. Some fruit trees have already been propagated this way for centuries. Although a meristem may

be killed, in one sense it is immortal, while a determinate structure is subject to senescence and death.

Although there are borderline cases, entire plants are in a sense either determinate or indeterminate. We use different terms, however: **Monocarpic species** (Greek *mono*, "single"; *carp*, "fruit") flower only once and then die; **polycarpic species** (*poly*, "many") flower, return to a vegetative mode of growth, and flower at least once more before dying. Most monocarpic species are **annuals,** but there are variations on the theme. Many germinate from seed in the spring, grow during the summer and autumn and die before winter, perpetuating themselves only as seeds. Spring wheat and rye are commercial annuals that are planted in the spring, but *winter* wheat or rye seeds germinate in the fall, overwintering as seedlings beneath the snow and flowering the next spring.

Typical **biennials,** such as beet (*Beta vulgaris*) and henbane (*Hyoscyamus niger*) germinate in the spring and spend the first season as a vegetative rosette of leaves that dies back in late fall. Such a plant overwinters as a root with a shoot apical meristem surrounded by some remaining protective dead leaves (called a **perennating bud**). During the second summer, the apical meristem forms stem cells that elongate into a flowering stalk (**bolting**).

The century plant (*Agave americana*) may exist for a decade or more before flowering once and dying. Though a monocarpic species, it would be called a **perennial,** because it lives for more than two growing seasons. It and many bamboos (*Bambusa* species), which may live more than half a century before flowering once and dying, are excellent examples of the extreme monocarpic growth habit. We can think of monocarpic plants as those that convert all of their indeterminate meristems to determinate flower-forming meristems. With no indeterminate meristems to continue vegetative growth, death becomes inevitable. Actually, there are other reasons for the senescence of monocarpic plants after flowering, as we shall see in subsequent sections.

Polycarpic plants do not convert all of their vegetative meristems to determinate reproductive ones. All are perennials. Woody perennials (shrubs and trees) may utilize only some of their axillary buds for the formation of flowers, keeping the terminal buds vegetative; alternatively, terminal buds may flower while axillary buds remain vegetative for next season's growth. Sometimes a single meristem forms only one flower, as in tulip, while single grass or Compositae (Asteraceae) meristems form an inflorescence or head of flowers. The bottle brush (*Callistemon* sp.) seems to form a terminal spike of flowers, but the apical meristem remains vegetative and continues to grow the next season, producing leaves and a woody stem. Woody perennials often become reproductive only after they are several years old. Herbaceous perennial dicots such as field bindweed (*Convolvulus arvensis*) or Canada thistle (*Cirsium arvense*) and perennial grasses die back each year except for one or more perennating buds close to the soil. Some form bulbs, corms, tubers, rhizomes, or other underground structures. We will encounter several important differences between monocarpic and polycarpic plants.

The seed contains a miniature plant telescoped into a tiny package: The embryonic root, shoot, and some of the primordial leaves form the **embryo.** The gamma plantlets discussed in Section 15.1 can develop as far as they do because of embryo differentiation during seed formation. Normally, the meristematic cells of the root and shoot apices give rise to other cells that will divide to form branch roots, still more leaves, axillary buds, and stem and root tissues, including the vascular cambium. Eventually, apical and axillary meristems may form flowers. We shall not trace these developmental processes here but shall discuss growth of roots, stems, leaves, and flowers in a general way, beginning with germination of a seed.

15.4 Growth of Roots

Organization of the Young Root In the great majority of species, seed germination begins with **radicle** (embryonic root) rather than **epicotyl** (shoot) protrusion through the seed coat (Berlyn, 1972). In some species, including sugar pine (*Pinus lambertiana*) and probably cherry, cytokinesis occurs in the radicle before germination is complete. In others (corn, barley, broad bean, lima bean, lettuce), few if any mitoses occur before radicle protrusion, so elongation is due to growth of cells formed when the seed was developing on the mother plant. Continued growth of the primary root in the seedling and of branch roots derived from it requires activity of apical meristems. A root tip is illustrated in Fig. 15-5. You should review

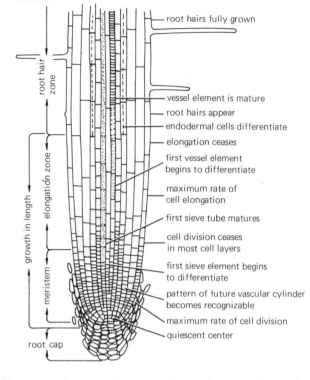

Figure 15-5 Simplified diagram of the growing zone of a root, in longitudinal section. The number of cells in a living root is normally much greater than is shown in this diagram. (From The living plant, Second Edition by Peter Martin Ray. Copyright © 1972, by Holt, Rinehart and Winston, Inc. Reprinted by permission of Holt, Rinehart and Winston.)

this, since it is typical of organization in most angiosperm and gymnosperm root tips.

The oldest cells of the cap are in the **distal** part (that part farthest from the point of attachment to the rest of the plant, i.e., the tip). In a more **proximal** position (closer to the point of attachment of the root cap to the rest of the root) are the young cells being formed from the apical meristem. The root cap protects the apical meristem as it is pushed through the soil and acts as the site of gravity perception for roots (see geotropism in Section 18.4). Furthermore, it secretes a polysaccharide-rich slime or **mucigel** over its outer surface that might lubricate the root as it slides through the soil. This requires activity of dictyosome vesicles, as shown in Fig. 10-16. As the root grows, slightly older parts become coated with the mucigel, presumably derived from the root cap. This mucigel harbors microorganisms and probably influences formation of mycorrhizae, root nodules, and ion uptake in unknown ways (Barlow, 1975).

Cells produced by divisions in the apical meristem develop into the epidermis, cortex, endodermis, pericycle, phloem, and xylem. Microscopists can observe where cell division is occurring, i.e., where the meristem is, by observation of cells in any of the mitotic stages. Another clever method is to determine where DNA synthesis is occurring, because a doubling of the DNA content usually means mitosis and cyto-kinesis will follow. The technique of providing cells with radio-active thymidine that is incorporated into DNA, followed by autoradiography to detect DNA synthesis, is often used. Just proximal to the root cap, there is a small zone where divisions do not occur, extremely common in roots, called the **quiescent center** (Clowes, 1975). If the meristem or the root cap is dam-aged, the quiescent center then becomes active and can regen-erate either of these parts.

Root Elongation Only a few of the cells shown in Fig. 15-5 had grown much, except for the oldest ones in the extreme tip of the root cap. Typically, many cells in the apical meristem are about 10 μm in each dimension, but slightly more proximal to the tip begins a **zone of elongation** that can extend as far as 5 mm from the tip. In this zone most of the cells elongate at least 15-fold, while increasing in diameter perhaps only two- or threefold.

Sachs studied the distribution of growth in roots a century ago by marking them at equal intervals with India ink. After the root has grown a day or two, the marks that were originally close to the tip are considerably farther apart, but those orig-inally a few millimeters from the tip exhibit no elongation. Figure 15-6 shows data obtained microscopically for pea seed-lings. Growth in root cell volume was painstakingly deter-mined by measuring both widths and lengths of the cells. Most of the volume increase (about 18-fold) was caused by elonga-tion rather than radial expansion, and almost no growth oc-curred farther than 4 or 5 mm from the tip. Elongation in a zone close to the root tip can push the root through the soil,

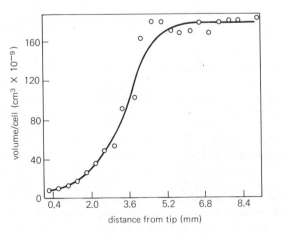

Figure 15-6 Average volume of pea root cells at increasing distances from the root tip: (From R. Brown, W. S. Reith, and E. Robinson, 1952, Symposium of the Society for Experimental Biology 6:329–347.)

while elongation farther from the tip might cause the root to bend and possibly break. Just proximal to the elongating region is the **root hair zone.** If elongation occurred in that zone, root hairs that wrap around or adhere to soil particles would be broken off.

Formation of Lateral Roots The frequency and distribution of lateral root formation partly controls the overall shape of the root system, and hence which zones of the soil are explored. Lateral or branch roots generally begin development several millimeters to a few centimeters proximal to the root tip. They originate in the pericycle, usually opposite the protoxylem points, growing outwardly through the cortex and epidermis, as illustrated in Fig. 15-7. This growth probably involves secretion by the branch root of hydrolytic enzymes that digest the walls of the cortex and epidermis, although apparently none of the postulated hydrolytic enzymes have been identified.

What causes certain fully differentiated pericycle cells to dedifferentiate and then develop into the apical meristem of a lateral root? As for so many questions about development, no answer is known. In some species, the plant hormone **auxin** (indoleacetic acid, Chapter 16) stimulates formation of lateral roots, although other hormones may also be involved. Cyto-kinin hormones (Chapter 17) synthesized in or near the root tip apparently retard formation of such roots near the tip. If the root tip is cut off, lateral roots develop from the cut surface.

Radial Growth of Roots The roots of gymnosperms and most dicots develop a vascular cambium from procambial cells located between the primary phloem and primary xylem near or in the root hair zone. This cambium is indirectly responsible for most of the increased width of these roots, because it forms expanding new xylem cells (toward the inside) and phloem cells (toward the outside). Auxin is one of the important agents causing activation of these procambial cells into a vascular cambium. This hormone is synthesized in young leaves and moves in a **basipetal** (base-seeking) direction down the stem and into the root (now in an **acropetal** or apex-seeking direc-tion). Most monocots do not form a vascular cambium, and the small radial enlargement they undergo is caused mainly by increases in diameter of nonmeristematic cells.

After the vascular cambium initiates secondary growth, a **cork cambium (phellogen)** arises in the pericycle. This be-comes a complete cylinder that forms cork (**phellem**) toward the outside and, later, some secondary cortex (**phelloderm**) inside (Cutter, 1969). The epidermis, the original cortex, and the endodermis are sloughed off, so the mature root consists of xylem at the center, vascular cambium, phloem, secondary cortex, cork cambium, and finally cork cells. Water-repellent suberin (Section 14.1) is deposited in the cork cell walls.

As the root system grows, more of it becomes suberized. For example, in mature loblolly pine (*Pinus taeda*) and yellow poplar or tuliptree (*Liriodendron tulipifera*), the surface area of unsuberized roots during the growing season is almost always

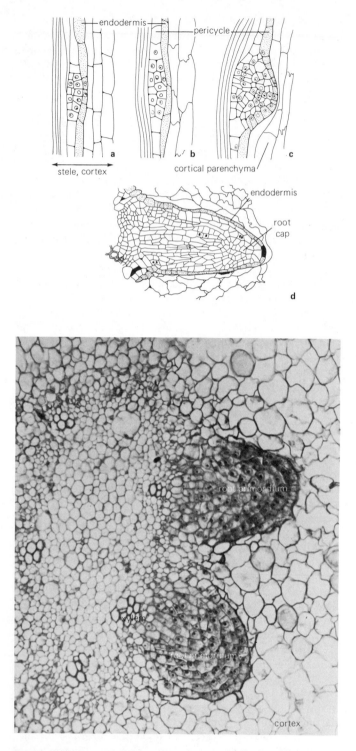

Figure 15-7 The origin of a secondary root starts with cell divisions in the pericycle that result in the establishment of a small mass of cells. These become the root primordium, which grows outwardly through the cortex. Frequently, the endodermis divides in pace with growth of the branch root, covering it, as in (**d**), until it breaks out of the main root. (From Jensen and Salisbury, 1972.)

less than 5 percent of the total. Apparently, suberized roots absorb water and mineral salts through lenticels, through tiny crevices formed by penetration of branch roots, and through holes left when branch roots die.

15.5 Growth of Stems

Stem Elongation The apical meristem of the shoot forms in the embryo and is the place where new leaves, branches, and floral parts originate. The basic shoot tip structure is similar in most higher plants, both angiosperms and gymnosperms. Figure 15-8 shows a photomicrograph of the terminal shoot of a representative dicot and of a monocot.

In growing stems, cell division occurs in regions much farther from the tip than occurs in roots (Sachs, 1965). In many gymnosperms and dicots, some cells divide and elongate several centimeters below the tip. In grasses, growth also occurs far down the stem, but it is restricted to specific, periodic regions. Near the shoot tip of young monocots, the leaf primordia are very close together, and the internodes are formed later by division and growth of cells between these primordia. At first, these divisions occur throughout the length of the young internode, but later the meristematic activity becomes restricted to the region at the base of each internode and just above the node itself. These periodic meristematic regions are called **intercalary meristems,** because they are intercalated (inserted) between regions of older, nondividing cells. Each internode consists of older cells at the top and younger cells derived at the base from the intercalary meristem.

In stems, auxin normally promotes elongation. Moving basipetally down the stem, it induces the elongation of cells that it encounters. Certain gibberellins, hormones that also stimulate stem growth, are abundant in young leaves and are probably synthesized there. They enhance cell division in the apical meristem and also promote growth of the cells produced. These important effects are discussed in Chapter 16.

Radial Growth of Stems Increased stem diameter in gymnosperms and most dicots results from radial expansion of cells produced by the vascular cambium, just as in roots. Again, auxin and gibberellins are apparently involved. In woody plants, renewed cambial activity in spring begins near the stem tips and progresses downwardly into the roots. If the buds containing young leaves are removed from a woody twig in late winter or early spring and the twig is then exposed to water and a suitable temperature, no cambial growth occurs. However, if a dab of lanolin paste containing auxin or a gibberellin is added to the apical cut surface from which the bud was removed, normal cambial activity begins. The two hormones together are especially effective, but lanolin alone is inactive. Such studies suggest that when new leaves begin to grow in the spring, they awaken the vascular cambium via the auxin and perhaps gibberellins they release. This is an excel-

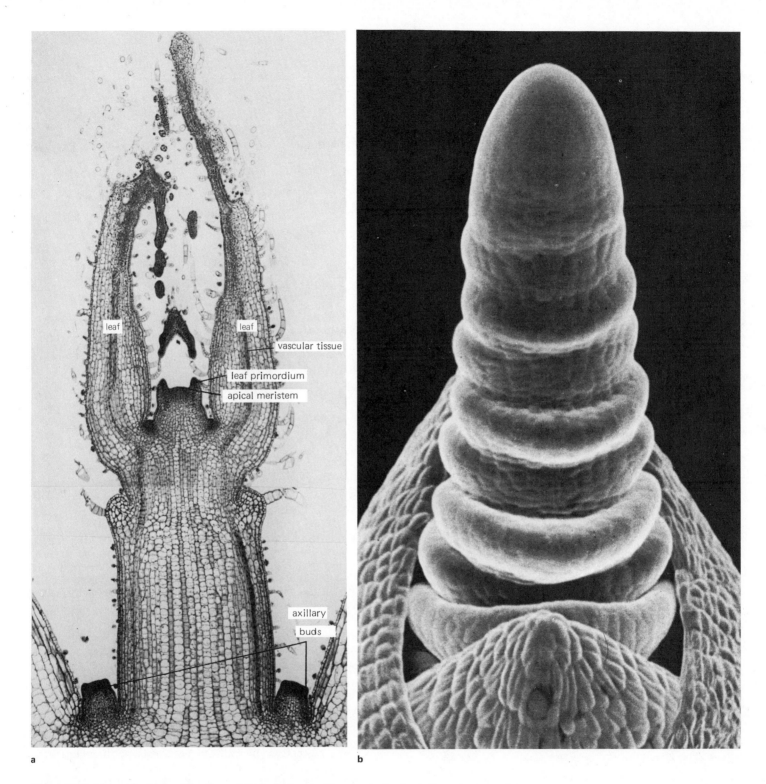

Figure 15-8 (**a**) Longitudinal section through the apical part of the shoot of a dicot. (From Jensen and Salisbury, 1972.) (**b**) A scanning electron micrograph of the apical meristem of wheat at a late vegetative stage. Leaf primordia in grasses are formed as ridges around the shoot axis. (From Troughton and Donaldson, 1972.)

lent example of a **correlative growth effect,** in which one part of a plant influences growth of another part, usually by transmission of a chemical signal, a hormone or a nutrient such as sucrose, from one organ or tissue to another. We shall encounter other examples.

15.6 Leaf Growth

The earliest sign of leaf development in both gymnosperms and angiosperms usually consists of divisions in one of the three outermost layers of cells near the surface of the shoot apex. In the first divisions, the new cell wall established between daughter cells is almost always in a plane approximately parallel to the closest surface of the apex; it is said to be **periclinal.** If the new wall is formed perpendicular to the closest surface, such a division is **anticlinal** (Fig. 15-9). Periclinal divisions followed by growth of the daughter cells cause a protuberance that is the leaf primordium, while anticlinal divisions increase the surface area of the primordium. Both kinds of divisions are important for further development of leaves and for growth in other parts of the plant.

Leaf primordia do not develop randomly around the circumference of the shoot apex. Rather, each species usually has a characteristic arrangement, or **phyllotaxis,** causing opposite or alternate leaves (Richards and Schwabe, 1969). No one knows why a given leaf primordium develops where it does. One theory holds that an already developing primordium sends out an inhibitor, preventing a new primordium from arising within the inhibitor's field of influence. Another theory suggests that primordia compete for space, so a new primordium can be initiated only when adequate space becomes available.

Note that the shape of the leaf primordium is controlled to a great extent by the planes of cell division. If most of the early divisions are periclinal, the primordium will be long and narrow, while relatively more anticlinal divisions will make a shorter, wider organ. *That the plane of cell division is an important factor controlling shapes of all multicellular organs is an important developmental principle;* we shall return to it in Chapter 18.

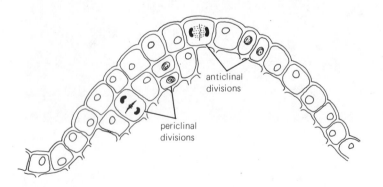

Figure 15-9 The relation of anticlinal and periclinal divisions at the shoot apex. (From Spratt, 1971.)

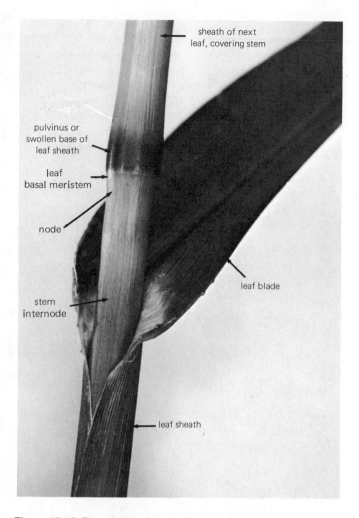

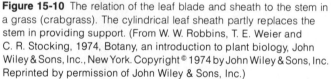

Figure 15-10 The relation of the leaf blade and sheath to the stem in a grass (crabgrass). The cylindrical leaf sheath partly replaces the stem in providing support. (From W. W. Robbins, T. E. Weier and C. R. Stocking, 1974, Botany, an introduction to plant biology, John Wiley & Sons, Inc., New York. Copyright © 1974 by John Wiley & Sons, Inc. Reprinted by permission of John Wiley & Sons, Inc.)

Subsequent leaf development is highly variable, as shown by the wide variety of leaf shapes. Continued extension outwardly occurs by both periclinal and anticlinal divisions at the primordium tip (apex). Later, often when the leaf is only a millimeter or so long, meristematic activity begins throughout its length. In grass leaves and conifer needles, this activity ceases first at the distal end (tip) and finally resides at the leaf base. An increase in width of the leaf blade in angiosperms results from meristems along each margin of the leaf axis, but these cease activity well before the leaf matures. In grasses, the basal meristem is an intercalary one that remains potentially active for long periods, even after leaf maturity. It can be stimulated by defoliation caused, for example, by a grazing animal or a lawnmower.

Figure 15-10 shows a crabgrass (*Digitaria sanguinalis*) leaf with its base encircling the stem, an encirclement that results from periclinal divisions in the primordium all the way around the shoot apex (see also Fig. 15-8b). The basal meristem in a grass leaf sheath often lies immediately external to an intercalary meristem of the system.

In dicot leaves, most cell divisions stop well before the leaf is full grown, frequently when it is half or less of its final size. In a bean primary leaf, cell division is complete when it has reached slightly less than one fifth its final area, so the last 80 percent of leaf expansion is caused solely by the growth of preformed cells. This growth occurs over the entire leaf area, although not uniformly; the same is true for many dicots. Cells in the young leaf are relatively compact. As leaf expansion occurs, the mesophyll cells stop growing before the epidermal cells do, so the expanding epidermis then pulls the mesophyll cells apart and causes development of an extensive intercellular space system in the mesophyll shown in Fig. 3-4.

A few leaf primordia and sometimes even floral primordia are usually detectable near the shoot apex of the embryo in the seed, but most primordia (especially in perennial species) are formed after germination. In conifers and deciduous trees, the early rapid growth of spring usually involves expansion of leaf primordia formed during the previous season and extension of the internodes between these primordia, while only in late summer are new primordia formed. These new primordia form part of the bud that is usually dormant during the winter or during a long dry period.

Since young leaves provide auxin and perhaps also gibberellins to the stem and root, what effect have these hormones on growth and development of the leaves in which they are formed? This proves to be a question almost impossible to answer. Neither hormone, when applied to a young leaf, influences the final leaf size (area) very much. Gibberellins frequently produce longer, narrower, and paler leaves. Cytokinins frequently cause formation of broad, fleshy leaves, but again, the effect is usually relatively small. Remember, however, that a leaf might normally contain nearly the right amount of each hormone needed for its growth, so that further addition would be either innocuous or inhibitory.

15.7 Growth and Development of Flowers

After establishment of roots, stems, and leaves, flowers and then fruits and seeds form, perpetuating the species and completing the life cycle. The time from germination to flower development varies greatly among different species. We saw that some monocarpic species flower in a few weeks after germination, while others (century plant and bamboos) may live for decades before flowering and death occur. Polycarpic species often live several years before they form flowers.

Most angiosperm species produce bisexual (**perfect**) flowers containing functional female and male parts, while others such as spinach, cottonwoods, willows, maples, and date palms are **dioecious**, containing **imperfect** staminate (male) and pistillate (female) flowers on different individual plants. **Monoecious** species such as corn, cocklebur, squash, pumpkins, and many hardwood trees form staminate and pistillate flowers at different positions along a single stem. The reproductive structures of conifers develop in unisexual cones (strobili). Most conifers are monoecious, although junipers and certain others are dioecious.

In monoecious squash plants, only male flowers are formed at first, then both, then only female. The ratio of male to female flowers can be increased by extending the daylength with supplementary light in both corn and squash, but in spinach and certain other species, long days cause an increase in femaleness. Other examples of daylength effects on development are described in later chapters. Temperature also influences the ratio of male to female flowers, but the direction of the response varies with species. Applied auxins frequently increase the percent of female flowers; gibberellins often increase maleness.

Anthesis, the opening of flowers with parts available for pollination, is sometimes a spectacular phenomenon, usually associated with full development of color and scent. While many flowers remain open from anthesis until **abscission** (falling off), others such as tulips open and close at certain times of the day for several days. Opening is usually caused by faster growth of the inner compared to the outer parts of the petals, but continued opening and closing is probably a response to temporary changes in turgor pressure across the two sides. Opening and closing are influenced by temperature and atmospheric vapor pressure, but the major factor is often an internal clock set by the daily dawn and/or dusk signal (see Chapter 20). For example, evening primrose flowers (*Oenothera* species) usually open in the evening about 12 hours before dawn, but they can be rephased to open in the morning by artificially reversing the light-dark cycles. The light influencing this response is absorbed by the flowers themselves.

After anthesis and pollination, the petals eventually wither, die, and abscise. In some species, withering follows anthesis rapidly. For example, in *Portulaca grandifolia* and many morning glories, including *Ipomea tricolor* and *Pharbitis nil,* opening of the flower occurs in the morning, and the corolla withers in late afternoon. Such withering is commonly associated with extensive transport of solutes from the flowers to other plant parts, often to the ovary, and with rapid water loss. There is an accelerated breakdown of protein and RNA from petals and sepals during withering, and hydrolytic enzymes such as proteases and ribonucleases are apparently activated by hormonal changes to cause such breakdown. Nitrogenous products such as amino acids and amides are then transported to seeds and other tissues where growth is occurring, so nutrients are conserved. Although withering and color fading is common, certain rose and *Dahlia* species lose petals that are still turgid and that contain most of their original protein.

One hormone causing flower withering is **ethylene** (C_2H_4), a simple hydrocarbon gas. In *Ipomea tricolor,* the flower

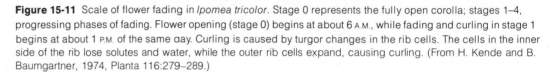

Figure 15-11 Scale of flower fading in *Ipomea tricolor*. Stage 0 represents the fully open corolla; stages 1–4, progressing phases of fading. Flower opening (stage 0) begins at about 6 A.M., while fading and curling in stage 1 begins at about 1 P.M. of the same day. Curling is caused by turgor changes in the rib cells. The cells in the inner side of the rib lose solutes and water, while the outer rib cells expand, causing curling. (From H. Kende and B. Baumgartner, 1974, Planta 116:279–289.)

produces ethylene during a short phase in late morning, and this ethylene then triggers a more extensive production of that gas in early afternoon, causing the petals to change from blue to purple, curl up, and wither (Fig. 15-11). If we could delay ethylene production or inhibit its action in valuable flowers such as orchids, we might extend their freshness (Beyer, 1976).

15.8 Seed and Fruit Development

Chemical Changes in Growing Seeds and Fruits The zygote, embryo sac, and ovule develop into the seed, while the surrounding ovary develops into the fruit. Numerous anatomical and chemical changes occur; we shall emphasize only the chemical changes. Frequently, sucrose, glucose, and fructose accumulate in the ovules until the endosperm nuclei become surrounded by cell walls, then the concentrations of these sugars decrease as they are used in cell wall formation and starch or fat synthesis. These sugars arise largely from sucrose, stachyose, or sorbitol transported through the phloem into the young seeds and fruits (Chapter 7). Most of the nitrogen of immature seeds and fruits is present in proteins, amino acids, and the amides glutamine and aspargine. The amino acids and amides decrease in concentration as storage proteins are formed in the protein bodies.

The roles of enzymes and nucleic acids in developing seeds are important to seed longevity. For a mature seed to remain alive for long periods before germination, it must either possess all of the enzymes necessary for germination and seedling establishment or have the genetic information available to synthesize them. Some of the enzymes essential to germination are produced in a stable form during seed development, others are translated from stable messenger RNA, transfer RNA, and ribosomal RNA molecules synthesized during seed maturation, while still others are formed from newly transcribed RNA molecules only after the seed is planted (Dure, 1975). Thus, different seeds control enzyme production in various ways, and there are different mechanisms even in the same seed for control of specific enzymes. Loss of water during

seed maturation is critical, leading to important but poorly understood changes in the physical and chemical properties of the cytoplasm. As a result, dry seeds respire extremely slowly and remain alive through extended drought or cold periods.

The chemical composition of edible fruits and the transformation of carbohydrates during ripening have been studied, but little is known about development of nonfleshy, less economically important fruits (Hulme, 1970; Coombe, 1976). In apples, the concentration of starch increases to a maximum and then decreases somewhat until harvest as it is converted to sugars. In apples and pears, fructose is often the most abundant sugar, while lesser amounts of sucrose, glucose, and sugar alcohols are also present. Grapes and cherries contain about equal amounts of glucose and fructose, but sucrose is often undetectable. The hexose concentration of grapes can reach unusually high values. Concentrations of glucose and of fructose in some varieties reach $0.6\,M$ for each sugar, giving the mature fruits an unusually negative osmotic potential and a sweet taste. During ripening of oranges, grapes, grapefruits, pineapples, and various berries, the organic acids (principally malic, citric, and isocitric) decrease and the sugars increase, so the fruits become sweeter. In lemons, however, the acids continue to increase during ripening so that the pH decreases and the fruits remain sour. Lemon fruits contain virtually no starch at any time during development.

Numerous other changes in fruit composition have been studied, including transformation of chloroplasts to carotenoid-rich chromoplasts, accumulation of anthocyanin pigments, and accumulation of flavoring components. The use of gas chromatography has allowed identification of hundreds of volatile substances such as aliphatic or aromatic esters, aldehydes, ketones, and alcohols contributing to the flavor and aroma of strawberries and other fruits (Nurnsten, 1970). This provides a basis for improvement of fruit flavors by plant breeding and for development of artificial flavoring substances.

Importance of Seeds for Fruit Growth Development of fruits is usually dependent upon germination of pollen grains on the stigma or upon this plus subsequent fertilization. Yet extracts of pollen grains added to certain flowers will simulate natural pollination and fertilization by causing ovary growth and wilting and abscission of the petals.

Developing seeds are usually essential for normal fruit growth. If seeds are present only in one side of a young apple fruit, only that side of the fruit will develop, and seeds are essential for normal strawberries (Fig. 15-12). There is good evidence that fruit development requires auxin normally produced by seeds: First, young seeds are rich sources of auxin, as are germinating pollen grains; second, the auxin content in the various tissues of the pistil sometimes increases progressively as the pollen tube grows through it; and third, the retarding effect of seed removal upon growth of some fruits can be overcome by adding an auxin to the ovaries (Nitsch, 1970).

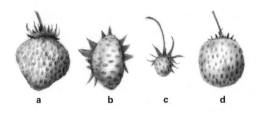

Figure 15-12 The influences of seeds (achenes, which are dry, true fruits) and an auxin on the development of the edible portion of strawberries (tissues derived from the receptacle). (**a**) Normal "fruit," (**b**) disk-shaped "fruit" resulting from removal of achenes on two sides early in development; (**c**) "fruit" from which all achenes were removed; (**d**) "fruit" from which all achenes were removed but which was treated with a lanolin paste containing an auxin. (Redrawn from J. P. Nitsch, 1950, American Journal of Botany 37:211.)

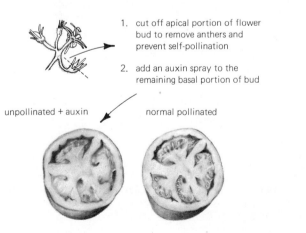

Figure 15-13 Production of seedless fruits by removal of anthers and application of an auxin to ovaries of tomato plants. (Fruits redrawn from Carl L. Wilson and Walter E. Loomis, Botany, 4th ed., Holt, Rinehart and Winston, New York. Copyright 1952, 1957, 1962, 1967 by Holt, Rinehart and Winston.)

Normal production of fruits lacking seeds is called **parthenocarpic fruit development.** It is especially common among fruits that produce many immature ovules, such as bananas, melons, figs, and pineapple. Parthenocarpy may result from ovary development without pollination (citrus, banana, pineapple), from fruit growth stimulated by pollination but without fertilization (certain orchids), or from fertilization followed by abortion of the embryos (grapes, peaches, cherries).

F. G. Gustafson first demonstrated in 1936 that seedless fruits could be produced by adding an auxin to the ovaries of flowers in which pollination was prevented (Fig. 15-13). Many species, especially members of the Solanaceae and Cucurbitaceae families, will produce parthenocarpic fruits in response to applied auxins. But auxins fail to stimulate parthenocarpic fruit development in most species, and in some of these (grapes, apples, pears, cherries, apricots, peaches), gibberellins are more effective. Gibberellins are also effective with certain species quite responsive to auxins (e.g., tomatoes and figs). That many seeds are unusually high in gibberellins also suggests that these hormones may be among those causing fruit enlargement; the presence of cytokinins in some fruits may also implicate these hormones. A proper balance among the auxins, gibberellins, cytokinins, and perhaps other undiscovered hormones is probably responsible for normal fruit development (Nitsch, 1970).

15.9 Relations Between Vegetative and Reproductive Growth

Gardeners have long practiced the technique of removing flower buds from certain plants to maintain vegetative growth. A commercial example is the "topping" (removal of flowers and fruits) of tobacco plants, which encourages leaf production. Such an effect on soybeans is shown in Fig. 15-14.

There is a competition for nutrients among vegetative and reproductive organs. Developing flowers and fruits, especially young fruits, possess a large but unexplained "drawing power" for mineral salts, sugars, and amino acids. During the accumulation of these substances by the reproductive organs, there is often an approximately corresponding decrease in the amounts present in the leaves. Studies with radioactive tracers show that nutrient accumulation in developing flowers, fruits, or tubers occurs largely at the expense of materials in nearby leaves. Nevertheless, the situation is more complex than a simple competition for nutrients. In the cocklebur (*Xanthium strumarium*), induction of flowering by long nights causes leaf senescence just as rapidly when flower buds are removed as when they are allowed to develop normally. In other cases, perhaps some inhibitor that is transported into the vegetative organs and causes their death prematurely is involved.

There is usually a competition among individual fruits of the same plant for nutrients. For example, fruit size decreases with increasing number of fruits allowed to form on tomato plants or apple trees. The mechanism by which fruits can divert nutrients out of leaves and into their own tissues, sometimes against apparent concentration gradients, is not understood. Various hormones, especially cytokinins (see Chapter 17), are probably involved.

In general, factors that stimulate shoot growth retard flower, tuber, and fruit development. High nitrogen fertilization causes luxuriant stem and leaf growth of tomatoes but reduces fruit development, while lower nitrogen levels lead to less stem and leaf growth but more fruit development. Similarly, excess nitrogen stimulates leaf growth but inhibits growth of potato tubers or apples and reduces sugar content in sugar beet roots.

Do processes interfering with vegetative growth also stimulate flower development? Sometimes they do. Heavy pruning, tying branches to the ground, or various other mutilation procedures often stimulate flowering. Furthermore, such commercial growth retardants as Phosphon D, CCC, Amo-1618 (see Chapter 16) and B995 (N-dimethylamino succinamic acid) inhibit growth of stems. This stunting is sometimes accompanied by a more rapid initiation of flower buds or a greater number of flowers per plant. These chemicals are used in commercial chrysanthemum production, for example, but they inhibit flowering in some other species.

15.10 Growth at the Cellular Level

Cytokinesis After mitosis is complete, the daughter nuclei are usually separated by formation of a new cell wall. This is begun by production of a **cell plate,** which arises by the fusion of hundreds of tiny vesicles, most of which are pinched off from the ends of dictyosome vesicles containing noncellulosic-polysaccharides such as pectins (Section 10.8). As the vesicles fuse, the pectin-rich **middle lamella** is formed. It is bounded by membranes that formerly were part of the dictyosome vesicles but now become the plasma membranes of each dividing daughter cell (Fig. 15-15). Subsequent formation of the new primary wall of each daughter cell also occurs, in part, by fusion of dictyosome vesicles containing other noncellulosic-polysaccharides.

What guides the movement of dictyosome vesicles to the equator of the cell during cytokinesis? One theory is that the vesicles migrate along tiny rodlike **microtubules** that extend toward opposite poles of the dividing cell (Hepler and Palevitz, 1974; Hepler 1976). Figure 15-15 shows numerous microtubules oriented with their long axes perpendicular to the equator. If formation of these microtubules is prevented by antimitotic drugs (see the boxed essay), dictyosome vesicles do not move to the equator of the cell at anaphase. If the drugs are added after anaphase is nearly complete, the cell plate cannot form, cytokinesis cannot occur, so a binucleate cell is produced.

Figure 15-14 Delay of senescence in soybean plants by daily removal of flower buds. (From Leopold and Kriedemann, 1975.)

Microtubules and Chromosome Separation

Microtubules making up the spindle fibers are somehow involved in chromosome movements during mitosis, but how they function is not established. One evidence for their involvement comes from use of certain antimitotic compounds such as **colchicine, podophyllotoxin, vinblastine** (all unusual drugs isolated from specific plants), **colcemid,** an analog of colchicine, and **trifluralin,** an herbicide (weed killer). All these compounds combine with the tubulin proteins composing the microtubules and prevent them from polymerizing into microtubules. At the growing end of a microtubule, even a single drug molecule might combine with the most recently added tubulin and prevent other tubulins from adding on. Without microtubules, no spindle apparatus and no chromosome-to-pole spindle fibers can be formed, so the chromosomes duplicated earlier do not separate. A nuclear envelope forms around these chromosomes, and the undivided cell now has a doubled chromosome number. Because of this, colchicine is sometimes used by plant breeders and others to increase the ploidy level of certain plants. One important application of colchicine was in development of *Triticale*, a rye-wheat hybrid.

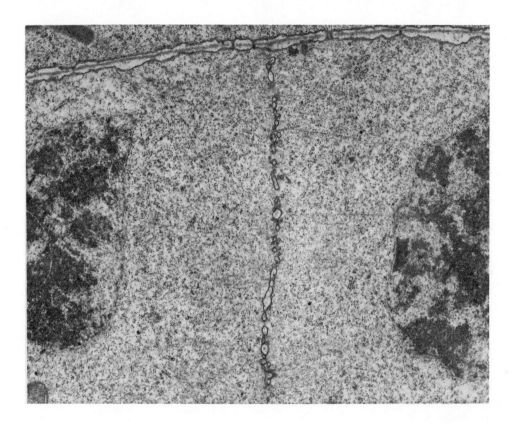

Figure 15-15 Formation of the cell plate during cytokinesis in a cotton root tip. Pectin-rich vesicles pinched off dictyosomes fuse at the equator to form the new middle lamella and the two plasma membranes in contact with it. Subsequent formation of a primary wall involves noncellulosic polysaccharides secreted from each cell into and onto the middle lamella in additional dictyosome vesicles, while cellulose appears to be formed in each plasma membrane without dictyosome vesicle involvement (Chapter 10). Narrow rodlike microtubules oriented perpendicular to the cell plate might function in guiding dictyosome vesicles to this plate. Formation of the nuclear envelope, probably from the endoplasmic reticulum, around each daughter nucleus is nearly complete. Numerous ribosomes (tiny dots) are also visible. (Courtesy Dan Hess.)

The Water Relations of Growth After cytokinesis, the daughter cells that do not remain meristematic grow extensively. This growth, primarily caused by an uptake of water, involves synthesis of new cell wall and membrane materials, since neither wall, plasma membrane, nor tonoplast is simply stretched as growth occurs. Reversible volume changes accompanying plasmolysis and deplasmolysis are accompanied by contraction and stretching of these membranes and, to a lesser extent, of the wall, but such changes in volume are small compared to growth changes. Wall synthesis has received much attention because of its suspected importance in controlling the growth rate (Cleland, 1971). Unfortunately, little is known about the rate of production of membrane lipids and proteins, or how membrane assembly and disassembly is controlled. It has usually been assumed that the growth rate of a cell depends mainly upon how rapidly the wall yields to allow water uptake, even though extension and growth of the wall and the plasma membrane are well coordinated.

What causes a cell to enlarge? An old hypothesis was that the wall and plasma membrane were extended in small stepwise increments by metabolic activities of the cell, then water entered at each step to fill the void. *The opposite and probably correct interpretation is that water pressure drives growth by forcing the wall and membranes to expand.* The rate of water movement into a cell is governed by two factors, the water potential gradient and the permeability of the membrane to water.

Membranes are always highly permeable to water, so to begin with, at least, we need only evaluate things that influence the water potential gradient. This may appear a simple task, but many interacting factors are involved.

Let us first assume that the cell is surrounded by water at a water potential (ψ) close to zero. The cell's water potential consists of its osmotic potential (ψ_π) plus its pressure potential (ψ_p):

$$\psi = \psi_\pi + \psi_p \qquad \text{(equation 5, p. 20)}$$

The ψ of the cell is often not measurably different from that of the solution outside, although there must be some gradient if water is to diffuse in. Assuming that membrane permeability is unchanged, the rate of water movement into the cell can be increased by lowering either ψ_π or ψ_p or both. The ψ_π is made more negative by solute uptake or by breakdown of macromolecules into their constituent subunits. Breakdown involves hydrolysis of starch to glucose; fructosans to fructose; proteins to amino acids; sucrose, raffinose, and stachyose to hexoses; and conversion of fats to sugars. Some environmental and internal factors affect ψ_π by changing the rate of solute absorption, while others change the rate of macromolecule breakdown.

Auxins stimulate growth primarily by decreasing ψ_p. They do this by somehow loosening the wall, allowing it to be stretched more readily in an irreversible manner, which releases pressure and leads to water uptake (Ray et al., 1972). This irreversible wall extension is called **plastic deformation** (a stretching such as that shown by chewing gum), while the reversible (rubber-band type) wall expansion is called **elastic deformation.** Stomatal opening and closing involve elastic changes in the walls of the mature guard cells, as do plasmolysis and deplasmolysis. Growth involves increases both in elastic and, especially, plastic components. Actually, auxins increase both elasticity and plasticity of primary cell walls. Gibberellins sometimes also cause extensive wall loosening, but they also increase the production of solutes, often sugars (see Chapter 16).

A **threshold turgor** must be exceeded before any growth occurs. In other words, although ψ_P may already be at a threshold value of 5 bars, additional pressure is needed for growth (Hsiao, 1973). Hence, the growth rate is very sensitive to the extent by which ψ_P is raised above the turgor threshold, and growth stops well before ψ_P reaches zero and before the tissues are wilted.

Young cells are usually able to grow under lower ψ_P values than are nearly full-grown cells, apparently because their walls are less rigid. Why does the wall become more rigid as growth nears completion? When secondary walls are produced, it is easy to imagine that the cell could not generate enough pressure to stretch them. In cells retaining only relatively thin primary walls, an accumulation of the amino acid hydroxyproline in wall proteins sometimes accompanies growth cessation, but if or how hydroxyproline-rich proteins might make the wall more rigid is unknown.

Although changes in ψ_π or ψ_P in the cells control growth of well-watered plants, decreases in either factor are partly nullified by water uptake. Suppose, in a cell at osmotic equilibrium with its surrounding solution, a few K^+ are absorbed or a few sucrose molecules are hydrolyzed into twice that number of hexoses. Either process momentarily lowers the cell's ψ_π (and therefore ψ), so water immediately enters and, by dilution, forces ψ_π and ψ back up toward their original values. Or say that auxin causes the wall to loosen and stretch irreversibly so that ψ_P, and thus ψ, is lowered. Water immediately enters, building the pressure back toward its original value. Even when water enters in response to increased solutes, the developing pressure causes the wall to yield so that ψ_P decreases toward its original value. Thus, cell growth normally occurs for considerable periods without appreciable changes in ψ_P, ψ_π, or ψ. The near constancy of ψ_π is illustrated in Fig. 15-16. Here the cells expanded in length 15-fold, yet their solute *concentrations* remained essentially constant as the increased *content* of solutes closely matched cell size increases. The solutes later proved to be mostly glucose, fructose, and K^+ salts, arising from sucrose and K^+ translocated from the cotyledon cells (McNeil, 1976).

What would happen in a tissue growing in water with no access to a solute supply such as mineral salts from a soil solution or sugars derived from photosynthesis? Water uptake would dilute the existing solutes, and ψ_π would rise toward

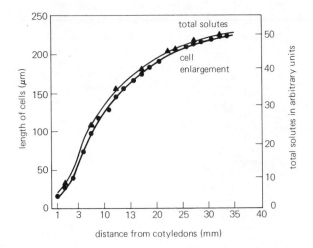

Figure 15-16 Relation between cell size and solute content in epidermal cells of various regions of sunflower hypocotyls. Etiolated seedlings 90 hr old and 45 to 50 mm high were used. Lengths and widths of the growing epidermal cells at various distances from the cotyledons were measured microscopically. Solute concentrations were measured from incipient plasmolysis values, then converted to arbitrary units by multiplying by the cell length (cell diameters remained almost constant during growth). Plants were grown in a peat-sand mixture and were watered only with tap water. (From W. A. Beck, 1941, Plant Physiology 16:637–641.)

zero. Since ψ remains immeasurably different across the plasma membrane, ψ_P must decrease, but growth will eventually stop when the turgor threshold is reached, unless the threshold decreases. Usually, growth stops in the absence of a solute supply, apparently because the wall either retains its rigidity or becomes even less plastic, as discussed previously. Therefore, *a plant requires water as the driving force for growth, but continued water uptake almost always requires mineral salt absorption or sugars and other organic solutes provided by translocation or photosynthesis.* This fact (along with the essential functions of mineral elements, sugars, and other organic solutes in metabolic processes) is essential to understand how the environment influences growth rates.

Primary Wall Changes during Growth In Section 10.8, we explained that primary walls of growing cells consist largely of an amorphous matrix of noncellulosic polysaccharides and some protein, through which run microfibrils of cellulose. It is these microfibrils, each behaving as a multistranded cable, that minimize extension in the direction of their long axes. Wall growth can, however, effectively occur in a direction that allows many of the microfibrils to slide past each other, somewhat analogous to extension of a chain-link fence. Furthermore, when new cellulose molecules are formed during growth, existing microfibrils are apparently lengthened, allowing some extension parallel to their lengths.

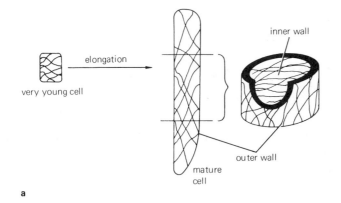

Figure 15-17 (**a**) Changes in orientation of cellulose microfibrils during cell elongation. In the young cell, the microfibrils are oriented almost randomly, but expansion occurs longitudinally because the newly deposited microfibrils on the inner surface of the wall are oriented perpendicular to the cell's long axis. The older microfibrils on the outside of the wall become oriented in the direction of elongation during growth. (**b**) Orientation of cellulose microfibrils in the inner (younger) and outer (older) part of the primary wall. Shown is a leaf hair cell of *Juncus effusus*, a sedge. Note that the microfibrils on the interior of the wall (IW) are perpendicular to the long axis of the cell, those on the exterior of the wall (EW) are parallel to the long axis of the cell, while those in between are intermediate in orientation. The direction of cell elongation is given by the long arrow. (From Jensen and Park, 1967.)

In meristematic cells, the cellulose microfibrils are rather randomly oriented relative to the length and width of the surface of each wall, so ψ_p (exerted equally in all directions) should cause uniform expansion in three dimensions and yield isodiametric cells. Why, then, are not all cells isodiametric? The answer is apparently twofold. First, the mirofibril orientation is not entirely random in some very young cells, so growth begins faster in a direction parallel to one axis (Fig. 15-17). Second, as growth occurs, new microfibrils are deposited into the wall adjacent to the plasma membrane in characteristic patterns. In this way, the wall retains a near uniform thickness during growth. If the orientation of these new microfibrils is random, growth is likely to be three-dimensional (as in fleshy fruits or leaf spongy mesophyll cells), but if they are deposited preferentially around one axis of the cell like hoops in a barrel, growth will begin to be more two dimensional and in directions perpendicular to the long axes of the microfibrils (as in elongating roots, stems, and petioles). Figure 15-17 illustrates this effect.

If the pattern of cellulose microfibril deposition is so important in controlling final cell shape, what controls that orientation? No answer is yet available, but there are consistent observations that in growing cells, microtubules close to the wall become oriented the same as the microfibrils. When colchicine, a drug that prevents microtubule formation, is added to cells, new cellulose microfibrils are randomly rather than transversely oriented. Removal of colchicine allows renewed production of transversely oriented microtubules and microfibrils. If microtubules control microfibril arrangement, we must learn what controls microtubule patterns.

The ability of cellulose microfibrils to slide past one another and past noncellulosic polysaccharides is important in determining growth rates. If this sliding is easy, we say that the cell wall is quite plastic. If it is not, growth has a rather high ψ_p threshold, and continued growth is slow. Much effort is presently directed at learning how auxins affect chemical bonds to increase wall plasticity. We shall return to this problem in the next chapter.

16

Hormones and Growth Regulators: Auxins and Gibberellins

In the previous chapter, we reviewed some of the growth regions of plants and introduced a few of the numerous effects of certain plant hormones on growth and development. In this and the next chapter, we shall summarize knowledge about such hormones and related growth regulators. Subsequent chapters present more details about the roles of these compounds in specific developmental processes. There are five groups of well-accepted hormones, including at least one **auxin** (indoleacetic acid, IAA), many **gibberellins,** several **cytokinins, abscisic acid** and related inhibitors, and **ethylene** (Thimann, 1972).

What is a plant hormone? Most plant physiologists accept a definition similar to that developed for animal hormones: *an organic compound synthesized in one part of a plant and translocated to another part where, in very low concentrations, it causes a physiological response.* The response in the target organ need not be promotive, because processes such as growth or differentiation are sometimes inhibited by hormones, especially abscisic acid. Since the hormone must be synthesized by the plant, such inorganic ions as K^+ that cause important responses are not hormones. Neither are organic growth regulators synthesized by organic chemists (e.g., 2,4-D, an auxin) or synthesized in organisms other than plants. The definition also states that a hormone must be translocated in the plant, but nothing is said about how or how far, nor does this mean that the hormone will not cause a response in the cell in which it is synthesized. Sucrose is not considered a hormone, even though it is synthesized and translocated by plants, because it causes growth only at relatively high concentrations. Hormones are usually effective at internal concentrations of $1~\mu M$ or less, while sugars, amino acids, organic acids, and other metabolites necessary for growth and development are usually present at concentrations of 1 mM to 50 mM.

In the last century, the famous German botanist Julius Sachs spoke of the possibility of specific organ-forming substances, supposing that one substance caused stem growth and others, leaf, root, flower, or fruit growth. The principle that development is influenced by chemicals in the plant is valid, but Sachs's explanation has proved too simple: Normal development of a single plant part involves interaction of several known hormones and probably others yet to be discovered, and a single hormone typically influences development of several or all plant parts.

16.1 The Auxins

The term "auxin" was first used by Frits Went, who as a graduate student in Holland in 1926 discovered that some unidentified compound caused bending of oat coleoptiles toward light (see his personal essay on pages 250–251). This bending phenomenon, called phototropism, is discussed in Chapter 18. The compound Went found is especially abundant in coleoptile tips, and Fig. 16-1 indicates how he showed its existence. The critical demonstration was that a substance present in the tips could diffuse from them into a tiny block of agar. The activity of this auxin substance was detected by bending of the coleoptile caused by enhanced growth on the side to which the agar block was applied.

Went's auxin is now known to be IAA (Fig. 16-2), and some experts believe that IAA represents the only true auxin

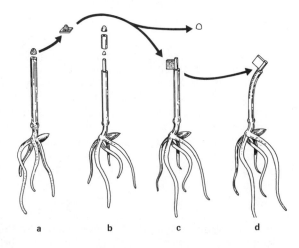

a   b   c   d

Figure 16-1 The demonstration by Went of auxin in the *Avena* coleoptile tip. Auxin is indicated by stippling. (**a**) The tip was removed and placed on a block of gelatin. (**b**) Another seedling was prepared by removing the tip, waiting a period of time, and removing the tip again (a new "physiological tip" sometimes forms). (**c**) The leaf inside the coleoptile was pulled out, and the gelatin block containing the auxin was placed against it. (**d**) Auxin moved into the coleoptile on one side, causing it to bend. (From Salisbury and Park, 1964.)

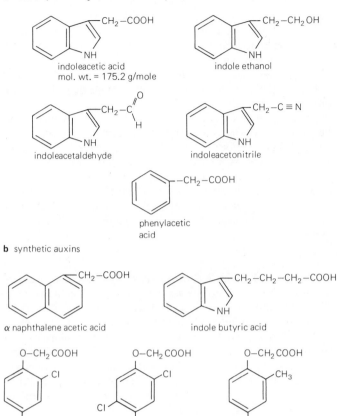

indoleacetic acid
mol. wt. = 175.2 g/mole

indole ethanol

indoleacetaldehyde

indoleacetonitrile

phenylacetic
acid

b synthetic auxins

α naphthalene acetic acid

indole butyric acid

2,4-D

2,4,5-T

MCPA

Figure 16-2 (**a**) Structures of some naturally occurring compounds having auxin activity, and (**b**) structures of other compounds that are only synthetic auxins.

hormone (Thimann, 1972). Nevertheless, there have been occasional demonstrations that plants contain other compounds that cause many of the same responses as IAA and that should perhaps be considered as auxins (Schneider and Wightman, 1974). Most of these remain unidentified, including some steroid-like compounds from *Coleus* plants and a nonindolic compound from citrus fruits. Another, **phenylacetic acid** (Fig. 16-2), is widespread among plants and is frequently more abundant than IAA, although it is less active. Little is known about its transport characteristics and whether it normally functions as an auxin.

Three additional compounds found in many plants have considerable auxin activity, but only because they are converted to IAA. These include **indole ethanol, indoleacetalde-**

hyde, and **indoleacetonitrite.** Each has a structure similar to IAA, but each lacks the carboxyl group (Fig. 16-2). They are oxidized to IAA by various enzymes.

Similar compounds synthesized by chemists also cause many of the physiological responses common to IAA and are generally considered as auxins. Of these, **naphthaleneacetic acid (NAA)**, **indolebutyric acid (IBA)**, **2,4-dichlorophenoxyacetic acid (2,4-D)**, **2,4,5-trichlorophenoxyacetic acid (2,4,5-T)**, and **2-methyl-4-chlorophenoxyacetic acid (MCPA)** (Fig. 16-2), are the best known. Since these are not synthesized by plants, they are not hormones.

Synthesis and Degradation of IAA IAA is structurally similar to the amino acid tryptophan and is synthesized from it. Two mechanisms for synthesis are known (Fig. 16-3), both of which involve removal of the amino group and the terminal carboxyl group from the side-chain of tryptophan. The preferred pathway for most species involves donation of the amino group to another α-keto acid by a transamination reaction to form **indolepyruvic acid,** then decarboxylation of indolepyruvate to form **indoleacetaldehyde** (Schneider and Wightman, 1974). Finally, indoleacetaldehyde is oxidized to IAA. The enzymes necessary for tryptophan conversion to IAA are most active in young developing tissues, such as shoot meristems and young leaves and fruit. In these tissues the auxin contents are also highest, suggesting that IAA is synthesized there.

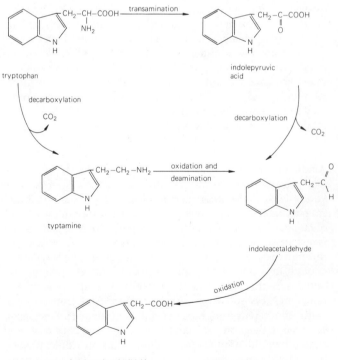

tryptophan

indolepyruvic
acid

decarboxylation

CO_2

decarboxylation

CO_2

typtamine

oxidation and
deamination

indoleacetaldehyde

oxidation

indoleacetic acid (IAA)

Figure 16-3 Possible mechanisms of formation of IAA in plant tissues.

It seems logical that the plant should have some mechanism to get rid of hormones as potent as IAA when they are no longer needed, and two general processes by which IAA is metabolized have indeed been found. In one process, it is combined with other molecules to form certain derivatives called **bound auxins.** Numerous bound auxins are known, including the peptide indoleacetyl aspartic acid and the esters IAA-inositol and IAA-glucoside, all of which involve bond formation between the carboxyl carbon of IAA and an amino or hydroxyl group of the other molecule. Other bound auxins are complex products containing glucose, protein, and phenolic compounds (Percival and Bandurski, 1976). The other process for IAA removal is degradative and involves a Mn^{2+}-dependent oxidation by O_2 and loss of the carboxyl group as CO_2. The products are variable, but **3-methyleneoxindole** is usually a principal one. The enzyme catalyzing this reaction is called **IAA oxidase.** Several IAA oxidase isozymes exist, and all or nearly all are identical to the peroxidases involved in early steps of lignin formation (Section 14.6). In beech and horseradish, for example, all 20 peroxidase isozymes had IAA oxidase activity (Gove and Hoyle, 1975). Synthetic auxins are not destroyed by such oxidases and therefore persist in the plant much longer than does IAA.

Certain phenolic compounds are suspected to influence the levels of IAA through effects on IAA oxidase activity (Schneider and Wightman, 1974). Ferulic and p-coumaric acids (Fig. 14-11) and kaempferol, a flavonoid, each having only one hydroxyl group (monophenols) act as coenzymes for IAA oxidase *in vitro,* and each inhibits stem growth. On the other hand, such o-diphenols as caffeic acid, chlorogenic acid (Fig. 14-11), and quercetin, another flavonoid, inhibit IAA oxidase and occasionally stimulate growth when applied to various plant parts. Several plants contain complex polymers of phenolics called **auxin protectors** that inhibit IAA oxidase. Although all of these compounds can be extracted from plants, they might normally exist in the vacuole and have little influence on cytoplasmic processes.

Auxin Transport A surprising thing about the ability of IAA to act as a hormone is the way in which it is transported from young leaves and other meristematic areas of the shoot and from nonmeristematic coleoptile tips. In contrast to movement of sugars, ions, and certain other solutes, IAA usually is not translocated through the phloem sieve tubes. It will move through these cells if it is applied to the surface of a leaf mature enough to export sugars, but normal transport in stems and petioles is through other living cells, perhaps phloem parenchyma (Section 7.8), cortex, or pith. This transport has features different from those of phloem transport. First, *auxin movement is slow,* only 0.5 cm/hr to 1.5 cm/hr in both roots and stems, but still 10 times faster than diffusion would predict. Second, *its transport is polar,* always occurring in stems preferentially in a basipetal (base-seeking) direction. Transport in roots is also polar but preferentially in an acropetal (apex-seeking) direction. Third, *its movement is an active process,* as

evidenced by transport against a concentration gradient and by the ability of ATP synthesis inhibitors or lack of oxygen to block it. Another powerful inhibitor of auxin translocation is **2,3,5-triiodobenzoic acid** (TIBA), but TIBA acts through some unknown mechanism distinct from ATP synthesis. Even synthetic auxins are moved by this polar mechanism, although transport of 2,4-D is usually slower than that of the others.

The mechanism of polar transport of auxins is unknown, but such transport down the stem from young leaves or meristematic cells of the shoot tip is important for control of such processes as renewed vascular cambium activity in woody plants during spring (Section 15.5), for growth of stem cells, and probably for inhibition of lateral bud development. It is this transport to coleoptile cells directly below the agar block in Fig. 16-1 that causes their elongation, resulting in bending.

Extraction and Identification of IAA How could you determine whether a given plant part has IAA, and if so how much? The problem is not easy, partly because IAA concentrations are so low, frequently only 0.1 to 1 μM (about 1 to 20 million IAA molecules, compared to about 10^{15} H_2O molecules in a hypothetical cell 100 μm long and 20 μm thick). Many of the procedures apply to other hormones, so we shall mention some general techniques. The first step is to remove the hormone from the tissue, and one popular procedure for auxins and gibberellins is to allow them to diffuse into an agar block. By this **diffusion technique** we obtain molecules that might normally be moving as hormones. More commonly, to extract auxins the tissues are first homogenized in such solvents as ethanol, cold ethyl ether, or chloroform, in which the relatively nonpolar IAA is soluble but in which most hydrophilic compounds found in plants are not. The extract is concentrated by solvent evaporation and then is usually subjected to some form of chromatography (often paper) to separate hormones from contaminating compounds. Positions of known hormones on paper chromatograms are established by chromatographing purified hormones obtained commercially in the same solvent system at the same time but on parts of the paper adjacent to the plant extract. That part of the paper corresponding to the section containing the known hormone is then cut out and treated with a solvent to dissolve any hormone originally present in the plant. The hormone in the resulting solution can be further purified and then identified and quantified by various analytical chemical methods such as infrared or mass spectroscopy and nuclear magnetic resonance.

Another technique that often must be used for hormone measurement, especially when the identity of the molecule is unknown or too little is available to obtain an infrared or mass spectrum, is a **bioassay.** Bioassays take advantage of the extreme sensitivity of certain plant parts to a particular hormone. Many are specific for one particular hormone type; for an auxin, for example, as opposed to a cytokinin or gibberellin.

One of Went's most significant contributions was his development of an auxin bioassay, the **coleoptile curvature test** (Fig. 16-1), in which semipurified extracts suspected to contain an auxin are incorporated into an agar block of about 10 mm³ volume, and the block is placed on one side of a coleoptile stump. The degree of bending is proportional to the IAA concentration within a certain range, and concentrations as low as 0.05 mg/l (0.2 μM) can be measured (Fig. 16-4). This limit of detection corresponds to about one billionth of a gram of IAA applied to one coleoptile. This bioassay is both sensitive and specific for auxins, but like all bioassays, it is subject to inhibition from certain other compounds, such as abscisic acid, if they contaminate the sample. Other bioassays are also listed by Audus (1972).

A less sensitive and less specific but easier auxin bioassay is the **straight growth test** (Fig. 16-5). Coleoptile sections or sections cut from the elongating portions of various dicots can be used. A popular dicot is the pea plant. It is grown in darkness (which increases its sensitivity to auxin) for about a week. Then a section 5 to 10 mm long is cut from the third (upper) internode and allowed to grow in buffered solutions with sucrose and various concentrations of the auxin-containing solution to be assayed. Most of the auxin-induced growth is caused by elongation, so the lengths of the sections can be measured later and compared to a standard curve obtained similarly with a known auxin. Straight growth is also stimulated somewhat by gibberellins and even potassium ions, although abscisic acid and ethylene are inhibitory, and cytokinins are either inhibitory or have little effect, depending on the species used.

The buffer (often 5 mM to 20 mM potassium phosphate at a pH near 6) and sucrose (about 0.1 M) are important for continued elongation of stem or coleoptile sections. Although auxins will stimulate growth in distilled water, this growth rate is not maintained nearly so long as when solutes are present in the medium. The sucrose provides energy through respiration that helps drive growth. Furthermore, both sucrose and the absorbed potassium and phosphate ions contribute to the cells' osmotic potentials, thereby preventing dilution of the cell contents by absorbed water and preventing a drop in turgor pressure. When growth occurs only in water, the coleoptile sections become less turgid because the absorbed water stretches the cells, and the cells' pressure potentials decrease because of expansion. This point is illustrated for oat coleoptile sections in Fig. 16-6. The growth rate without sucrose (lower curve) soon decreased, while sections with sucrose grew at a constant rate for 18 hours. The water potentials for both were only slightly less negative than that of the solution in which they were immersed (-2 bars), but the final ψ_π of those with sucrose was -12 bars, compared to only -8 bars without. Therefore there was a final ψ_p of about 10 bars in those with sucrose and only 6 bars without it.

Why should we not simply treat an intact stem with auxin and measure its growth response as a bioassay? Apparently, enough endogenous auxin is usually supplied to stems of intact plants from tissues above, so that additional auxin does

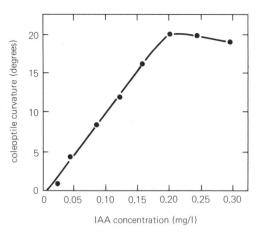

Figure 16-4 Curvature of decapitated oat coleoptiles caused by unilateral applications of various amounts of IAA upon the cut stump. (From Went and Thimann, 1937.)

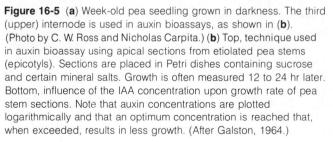

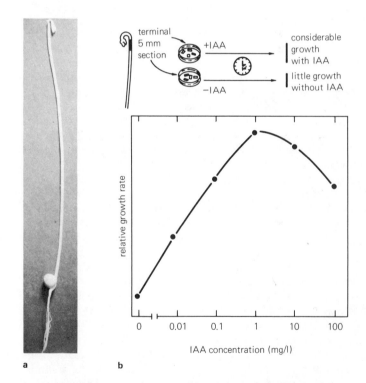

a　　　　b

Figure 16-5 (**a**) Week-old pea seedling grown in darkness. The third (upper) internode is used in auxin bioassays, as shown in (**b**). (Photo by C. W. Ross and Nicholas Carpita.) (**b**) Top, technique used in auxin bioassay using apical sections from etiolated pea stems (epicotyls). Sections are placed in Petri dishes containing sucrose and certain mineral salts. Growth is often measured 12 to 24 hr later. Bottom, influence of the IAA concentration upon growth rate of pea stem sections. Note that auxin concentrations are plotted logarithmically and that an optimum concentration is reached that, when exceeded, results in less growth. (After Galston, 1964.)

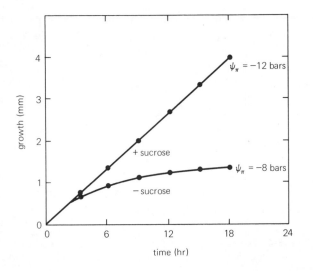

Figure 16-6 Progress curves for growth of oat coleoptile sections in IAA, $2 \times 10^{-5}\,M$, in presence or absence of $0.08\,M$ sucrose. (From J. Bonner, 1961, in R. M. Klein, ed. Plant growth regulation, Iowa State University Press, Ames, pp. 307–326.)

not stimulate growth. Most intact plants will not grow faster regardless of how the auxin is applied or which auxin is used, although cucumber stems and certain flower stalks (scapes) do respond positively to added auxins. Auxin bioassays such as the coleoptile curvature test and the straight growth test depend upon removal of the responding part from the normal auxin supply.

Effects of Auxins on Roots The role of IAA in stem elongation seems well established, and IAA is probably also necessary for growth of fruits (Section 15.8). It is commonly assumed that IAA is essential for growth of leaves and flowers, but there is little evidence for this. Roots have been studied somewhat more intensively and require special discussion.

IAA exists in roots at concentrations similar to those in many other plant parts (often 0.1 to 1 μM). As first shown in the 1930s, auxins will stimulate elongation of excised roots of many species (Batra et al., 1975; Scott, 1972) but only at extremely low concentrations (10^{-7} to $10^{-13}\,M$). At higher concentrations, total elongation is usually inhibited. The assumption is that root cells usually contain enough or almost enough auxin for normal elongation. Indeed, many excised roots grown for long periods *in vitro* do not require an added auxin, indicating that any requirement they have for this hormone is satisfied by their ability to synthesize it. An exogenous supply often causes inhibition of root growth. Part of this inhibition is caused by ethylene, because *auxins of all types stimulate many kinds of cells to produce ethylene*, especially when relatively large amounts of auxin are added. Ethylene retards elongation of both roots and stems (Section 17.2).

Where does the IAA in roots come from? Most of it probably arrives from young leaves or meristematic cells of the shoot. Young leaves not only have higher concentrations of IAA than older leaves, but the absolute amounts are greater in young leaves. As endogenous IAA moves basipetally, it stimulates elongation of shoots and presumably of roots, and it also increases the rate of cambial division in the stems and development of the procambium into a vascular cambium in the roots (Section 15.4).

Although continued root elongation is usually inhibited by auxin concentrations greater than 1 μM, early development of lateral or branch roots from the pericycle is stimulated by such treatments (Section 15.4; Torrey, 1976). Auxins also stimulate adventitious root development on stems. Many woody species (e.g., apples, most willows, Lombardy poplar) have preformed adventitious root primordia on their stems that remain dormant for some time unless stimulated by an auxin (Haissig, 1974). These primordia are often at nodes or on the lower sides of branches between nodes. Apple burr knots on the stems contain up to 100 root primordia each. Even plants without preformed root primordia in the stems will form adventitious roots. These result from division of an outer layer of phloem.

In many fruits and ornamentals, it is important to maintain genetic purity of specially developed hybrids by asexual reproduction. The practice is to propagate plants from excised stem sections called **cuttings.** Success depends on adventitious root development at or near the base of the cutting. Frequently, a mass of unspecialized cells called a **callus** forms at the base of the cutting, and roots develop from these as well as from adventitious root primordia in the stem. Thus, the primary role of auxin in root initiation seems to be a stimulation of cell divisions, which is consistent with auxin promotion of cambial activity.

Sachs obtained evidence in the 1880s that young leaves and active buds stimulate root initiation, and he suggested that a transmissible substance (hormone) was involved. In 1935, Went and Kenneth V. Thimann showed that IAA stimulates root initiation, and from this developed the first practical use of auxins. Synthetic auxins such as NAA and IBA (Fig. 16-2) are usually more effective than IAA, apparently because they are not destroyed by IAA oxidase or other enzymes and therefore persist longer. Commercial powders in which cut ends of stems are dipped to facilitate root production usually contain IBA or NAA mixed with talcum powder. Figure 16-7 illustrates this technique.

With most species, adventitious root development always occurs at the physiological base. Thus, whether the stem section is oriented normally or turned upside down, the roots nearly always form at the position that was originally more basal, and the young shoots develop from the originally more acropetal part nearer the shoot apex. This development of roots is entirely consistent with the polarized basipetal transport of IAA in stems, but some woody species, especially gymnosperms and some fruit trees (apples, pears), will not readily form adventitious roots even with auxin treatment.

Went suggested that other hormones were involved, and he named these still unidentified substances **rhizocalines.**

Auxin Effects on Lateral Bud Development Some species such as the spruces grow considerably more in height than in breadth, while others such as oaks show much less difference. This depends on the extent of branching, which results from development of lateral (axillary) buds in the leaf axils. In most species, the apical bud exerts an inhibitory influence (**apical dominance**) upon the lateral buds, preventing their development. This extra production of undeveloped buds has definite survival value, for if the apical bud is damaged or removed by a grazing animal or a windstorm, a lateral bud will then grow out and become the leader shoot. Apical dominance is widespread. It also occurs in bryophytes and pteridophytes and in roots. Another dominant effect of the shoot apex is to cause branches below to grow out somewhat horizontally.

Gardeners have long practiced the technique of removing the apical bud and young leaves to stimulate branching. This technique also allows the branches to grow more vertically, especially the uppermost one. In many species, continual removal of the youngest visible leaves is as effective as removing the entire shoot apex, suggesting that dominance resides in them. If an auxin is added to the cut stump after the shoot apex is discarded, lateral bud development and vertical orientation of existing branches are again retarded in many plants. This replacement of the bud or young leaves with an auxin suggests that the inhibitory compound they produce is IAA, but in many species (e.g., cocklebur) application of an auxin will not replace a bud or young leaves. Other compounds may be involved.

Even in species where auxins are effective, IAA itself may not be the inhibitory compound in lateral buds, but IAA might stimulate formation of another inhibitor, perhaps ethylene. Another idea to explain indirect inhibition of lateral bud growth by an auxin is that the auxin somehow causes lack of some nutrient or stimulatory hormone such as a cytokinin in these buds. There are still other hypotheses with even less supporting evidence (Phillips, 1975).

Auxins as Herbicides (Weed Killers) In the 1940s, work at the Boyce Thompson Institute in New York established that 2,4-D has auxin activity. Later work there and in England showed that 2,4-D, NAA, and certain other related compounds are effective **herbicides** or plant killers. Four of the most widely used auxin herbicides are 2,4-D, 2,4,5-T, MCPA (Fig. 16-2) and derivatives of picolinic acid such as **picloram** (sold under the trade name Tordon). Their popularity derives from their toxicity, their relatively low cost, and their property of affecting dicots much more than monocots (Ashton and Crafts, 1973). Because of this selectivity, they are often used to kill broadleaf dicot weeds in cereal grains and lawns. For grass pastures and rangelands in which woody perennials such as sagebrush and mesquite are often a problem, 2,4,5-T is particularly effective. Several derivatives of benzoic acid such as **dicamba** also have auxin activity and are more effective than the others against deep-rooted perennial weeds, including field bindweed or wild morning glory (*Convolvulis arvensis*), Canada thistle (*Cirsium arvense*), and dandelions (see Special Topic 16-1 by John O. Evans.)

These herbicides are formulated as salts of weak bases such as ammonia (amines), as emulsifiable acids, or as esters, and are mixed with an oil or detergent to facilitate spreading and penetration of the waxy cuticle after spraying on the foliage. They are absorbed through the cuticle and are translocated throughout the plant largely in the phloem. Therefore, it is usually best to spray them in the morning of a warm, sunny day so that they will be absorbed and translocated along with other photosynthetic products. Windy days are avoided to keep the herbicide from drifting to susceptible neighbor crops.

In spite of much research to determine how they kill only certain weeds, the mechanism of action of auxin herbicides is unknown. Part of their selectivity against broadleaf

Figure 16-7 Promoting root growth from cuttings by treatment with an auxin. (From E. J. Kormondy et al., 1977.)

weeds is caused by greater absorption and translocation than in grasses, but other factors are also involved. It is sometimes stated that they cause the plant to "grow itself to death," but this is misleading. Admittedly, certain parts of some organs do grow faster than other parts, so we see twisted and deformed leaf blades, petioles, and stems due to unequal growth. Much of this results from epinastic effects (Section 17.2) that arise from the common property of all auxins to enhance ethylene production, because ethylene is notorious for causing epinasty. But unequal growth causing twisting could result from inhibition of one part, not promotion of the other. Overall growth of the plant is definitely retarded and eventually stopped if enough herbicide is absorbed and translocated. Modern theories suggest that these compounds alter DNA transcription and RNA translation in such a way that the proper enzymes needed for coordinated growth are not produced properly, but how this is accomplished is unknown.

Other important auxin functions are in phototropism and geotropism (Chapter 18) and in delaying abscission of leaves, flowers, and fruits. We shall discuss them later.

Possible Mechanisms of Auxin Action How can auxins cause so many responses if a common initial chemical reaction is involved? Clearly, the response depends upon poorly understood physiological and anatomical conditions of the affected cells, for not all are target cells. Early research was guided by the idea that auxins and other hormones might act as coenzymes for specific enzymes. One reason for suspecting a catalytic activity is that such small quantities of hormones cause such marked effects. A few million molecules of IAA per cell may seem like a lot, but that would be only about one IAA molecule for each 100 million to 1 billion molecules of water. And one IAA molecule will ultimately affect thousands to millions of other molecules, so a tremendous *amplification* of the initial triggering response of this and other hormones is involved. If a hormone acted as a coenzyme, this amplification could be explained, but none of the hormone groups discussed in these two chapters is known to act in a definite coenzyme way, nor do we have evidence that they cause allosteric changes in any enzymes. Both possibilities still exist, however, because many chromosomal enzymes associated with DNA and RNA synthesis, many enzymes and ribosomal proteins responsible for protein synthesis, and nearly all membrane proteins have not been isolated and characterized. It is these enzymes and proteins that are receiving most attention in theories of hormone action.

The first evidence that plant hormones could affect nucleic acid synthesis came from work of Folke Skoog and his colleagues at the University of Wisconsin in 1953. They found that growth of pith tissue cut from tobacco cells and cultured *in vitro* is stimulated by IAA and that this growth is preceded by an increase in RNA matching the growth response. Subsequent extensive work by Joe L. Key, L. D. Noodén, and others showed that protein synthesis inhibitors such as the antibiotics **cycloheximide** (also known as **actidione**), **puromycin,**

and **chloramphenicol** are inhibitory to auxin-induced growth. So are RNA synthesis inhibitors such as **actinomycin D, 6-azauracil, 6-methylpurine,** and **3-deoxyadenosine (cordycepin)**, but compounds that block only DNA synthesis are not inhibitory to hormone-stimulated growth except where cell division is necessary. Key et al. (1967) demonstrated an excellent correlation between inhibition of 2,4-D-induced elongation of soybean hypocotyl sections and inhibition of RNA synthesis by actinomycin D. Inhibition of elongation and of protein synthesis by cycloheximide were also well correlated.

The conclusion from this and much work of others is that RNA molecules needed for synthesis of enzymes or other proteins are produced faster in the presence of auxins, and that the inhibitors prevent this. Ribosomal RNA synthesis is especially stimulated by auxins, and nucleoli responsible for this synthesis also increase in size after auxin treatment. In soybean hypocotyls, the increased ribosomal RNA results from enhanced activity of RNA polymerase I, the responsible enzyme (Guilfoyle et al., 1975). Synthesis of growth-limiting proteins at the ribosomal level then presumably occurs faster. As yet, no specific nuclear receptor proteins for auxins or other plant hormones that function in the way steroid hormone receptors do in mammals (O'Malley and Schrader, 1976) have been found, but the search continues (Kende and Gardner, 1976).

Although continued synthesis of RNA and protein is necessary for growth and development, cell elongation usually begins too fast to allow synthesis of different proteins. Auxins promote cytoplasmic streaming within a minute or two and stimulate growth of coleoptile and stem sections within 10 to 15 minutes. Michael L. Evans and Peter M. Ray (1969) developed a technique by which one can determine how fast growth is stimulated by auxins. They threaded 12 or more hollow coleoptile sections on a cotton thread and immersed these in an open glass cell through which a solution continuously flowed. By shining a light beam near one end of the sections, they could measure the combined elongation rate of all sections by a shadowgraph technique. As the section grows, a shadow is cast upon slowly moving photographic paper behind them, and the growth rate is thereby recorded on the paper.

An important conclusion from the work of Evans and Ray illustrated in Fig. 16-8 is that growth begins as early as 15 minutes after auxin is added to the flowing solution (note vertical white line indicating time of IAA addition). Such lag times are considered too short to allow auxins to act via protein synthesis (Evans, 1974). Then how can we reconcile this evidence with that derived from long-term growth responses of many hours indicating that synthesis of RNA and proteins is essential? One interpretation is that auxins initially stimulate wall loosening by an effect at the plasma membrane, but that this response requires some protein or enzyme that is destroyed within a few hours and must be resynthesized by RNA-dependent reactions. Auxin stimulation of RNA and protein synthesis might require that the hormone, or a molecule formed in response to hormonal action at the plasma

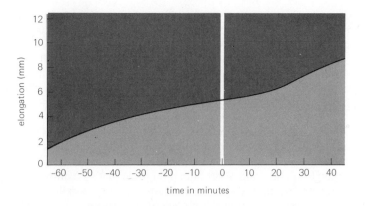

Figure 16-8 Representation of shadowgraph record of growth of excised oat (*Avena sativa* vr. Victory) coleoptile sections. The incubation medium was changed from water to 3 μg/ml IAA at the time corresponding to the vertical white line. The rapid elongation at the beginning of the record is the result of tactile stimulation of the coleoptiles during experimental manipulations. (From M. L. Evans and P. M. Ray, 1969, Journal of General Physiology 53:1–20.)

membrane, would move into the nucleus to affect transcription. In this way, nuclear activities might be controlled by events at the cell surface.

Is there any evidence that auxins do affect reactions at the plasma membrane rather than elsewhere in the cell? Some, but it is not overwhelming. Auxins stimulate H^+ release from coleoptile and stem cells, and both the magnitude and the timing of this release correlate nicely with the extent and timing of growth promotion in short experiments. The protons at first accumulate in the walls, causing the pH to drop from approximately 6 to about 5, but if the cuticle is peeled off, the protons also appear in the growth medium. Some of these protons are exchanged for incoming K^+, as we discussed in Chapter 5 for H^+ secretion accompanying K^+ uptake in roots. The assumption has been that auxins stimulate an ATP-dependent pump in the plasma membrane that catalyzes this apparently universal exchange process, but so far this has been impossible to prove. In 1977, Ray showed that the rough endoplasmic reticulum (ER) contains a binding site with high affinity for the auxin NAA. He speculated that auxin binding causes H^+ transport from the cytoplasm into the cavity of the ER, presumably with the aid of an ATPase. He suggested that the H^+ are next transferred into dictyosomes, then across the plasma membrane and into the primary wall by dictyosome vesicles. An action at the ER presumably explains the several-minute lag time necessary for growth promotion, whereas action at the plasma membrane should enhance growth almost immediately.

Auxin-promoted proton secretion is presumably essential for long-term growth stimulation (Rayle and Cleland, 1977), although there is evidence against this concept (Vanderhoef et al., 1977). Slightly acidic growth media stimulate stem or coleoptile elongation, especially if the cuticle is removed to allow protons to move into the cell walls. Solutions of low pH also greatly enhance root elongation. Carbon dioxide, which forms carbonic acid (H_2CO_3) in solution and in the cell walls, is a potent growth stimulator. Low pH values caused by acidic buffers or CO_2 cause detectable growth with a lag time of only a minute or two. These and numerous other facts suggest that auxins promote growth by causing proton secretion into the cell walls (see the personal essay by Rayle on pages 252–253). These protons somehow cause wall loosening. To maintain this growth clearly requires continued synthesis of proteins, perhaps ATPase enzymes of the plasma membrane or, according to Ray's hypothesis, proteins normally synthesized and transported by the rough ER along with the growth-promoting H^+. If protons formed at the plasma membrane are indeed "second messengers" for auxins during cell elongation, how do they stimulate wall loosening? A popular hypothesis is that protons decrease the wall pH just enough to allow hydrolytic enzymes to begin breaking covalent bonds in the wall, thereby loosening it. Certain of the many enzymes causing polysaccharide hydrolysis are indeed more active at slightly acidic pH values, but the one primarily responsible for wall loosening has not yet been identified.

16.2 The Gibberellins

Gibberellins were first discovered in Japan in studies with diseased rice plants that grew excessively tall. These plants often could not support themselves and eventually died from combined weakness and parasite damage. As early as the 1890s, the Japanese called this the *bakanae* ("foolish seedling") disease. It is caused by the fungus *Gibberella fujikuroi* (asexual or imperfect stage is *Fusarium moniliforme*). E. Kurosawa, a plant pathologist, found in 1926 that extracts of the fungus applied to the rice plants caused the same symptoms as the fungus itself, demonstrating that a definite chemical substance is responsible for the disease. In the 1930s, T. Yabuta and T. Hayashi were able to isolate and identify an active compound from the fungus, which they named **gibberellin.** Thus, the first gibberellin was discovered as early as IAA, yet because of preoccupation with IAA and synthetic auxins, lack of early contact with the Japanese, and then World War II, the Western Hemisphere did not become interested in gibberellin effects until the early 1950s. It is also perhaps significant that the Japanese scientists were primarily interested in *pathology* rather than *physiology*.

At least 50 gibberellins have now been discovered in fungi and plants (Pharis and Kuo, 1977). The structures of five of these are in Fig. 16-9. All gibberellins have either 19 or 20 carbon atoms grouped in a total of either four or five ring systems, and all have one or more carboxyl groups. They are abbreviated GA, with a subscript such as GA_1, GA_2, GA_3, and so on, to distinguish them. All could properly be called gibberellic acids, but GA_3 has been studied much more than the others (because of its availability), and it is often referred to as

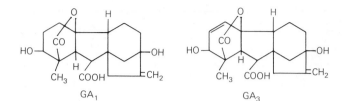

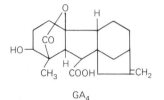

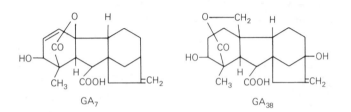

Figure 16-9 Structures of representative gibberellins. Most experiments have been performed with GA₃, gibberellic acid. Of those shown, all have 19 carbon atoms except GA₃₈, which has 20.

gibberellic acid. Gibberellins exist in angiosperms, gymnosperms, ferns, algae, and fungi, but apparently not in bacteria. Only in vascular plants have well-documented physiological roles for gibberellins been established.

Gibberellin-Promoted Growth of Intact Plants *Gibberellins have the unique ability among plant hormones to stimulate extensive growth of intact plants.* They generally enhance elongation of intact stems much more than of excised stem sections, so their effects are the opposite of those of auxins in this respect. An early demonstration of elongation caused by an ether-soluble substance extracted from bean seeds was made by John W. Mitchell and his colleagues in 1951 (Fig. 16-10). They did not know what caused this unusual stimulation, but they demonstrated that IAA was not responsible. Now we know that seeds of beans and many other dicots are rich sources of gibberellins, and that the symptoms Mitchell observed are identical to those caused by several purified gibberellins.

Most dicots and some monocots respond by growing faster when treated with gibberellins, but several species in the Pinaceae family show little or no elongation responses to gibberellins, perhaps because they already contain sufficient amounts. Cabbages in the rosette form with very short internodes may grow 2 m tall and then flower after GA₃ application, while untreated plants remain short and vegetative.

Short bush beans become climbing pole beans, and genetic dwarf mutants of rice, corn, peas, watermelons, squash, and cucumbers exhibit phenotypically tall characteristics of normal varieties when treated with GA₃ or certain other gibberellins. Dwarf meteor peas are sensitive to as little as one billionth of a gram of GA₃, so this plant is used as the basis for a gibberellin bioassay.

Five different dwarf mutants of corn grow as tall as their normal counterparts after proper gibberellin application, although at least two other mutants do not respond. Each of these responding dwarfs contains a mutation on a single different gene, and each mutation presumably controls a different enzyme needed in the pathway of gibberellin synthesis. Growth of normal varieties of corn is not appreciably stimulated by gibberellins.

Although numerous dwarfs respond to added gibberellins, we cannot conclude that all dwarfs are deficient in these hormones. Dwarf Japanese morning glory (*Pharbitis nil*) and pea, which grow tall when treated with exogenous gibberellins, seem to have as much gibberellin as their normal counter-

Figure 16-10 Growth stimulation of *Phaseolus vulgaris* L. by a gibberellin-containing extract prepared from seeds of the same variety (Black Valentine). An ether extract of seeds was evaporated, and 125 μg of the residue was mixed with lanolin and applied as a band around the first internode of the plant on the right. Plants were photographed 3 weeks after treatment. Plant on the left was untreated. (From J. W. Mitchell, D. P. Skaggs, and W. P. Anderson, 1951 (August), Science 114:159–161.)

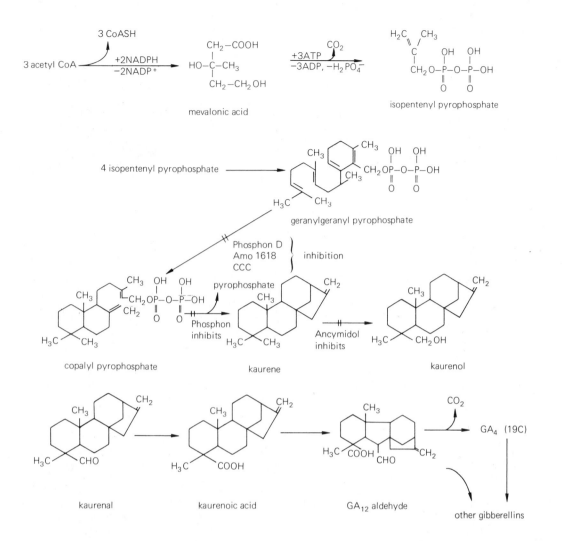

Figure 16-11 Some reactions of gibberellin biosynthesis. Many steps indicated as single arrows actually involve more than one enzyme-catalyzed reaction, especially those before kaurene.

parts, although there is no assurance that the gibberellins found in them are in active forms or in the proper subcellular locations. The same is true of several semidwarf, stiff-strawed wheat varieties that respond well to fertilizers and are being used as genetic stock for plant breeding experiments. Dwarfing could also be due to low auxin contents, to an inability to respond to endogenous gibberellins or auxins, or to high concentrations of unknown inhibitors.

Promotion of growth of various dwarf plants by gibberellins is used as a common gibberellin bioassay. The first 38 gibberellins to be discovered have been compared in dwarf pea, dwarf rice, and three other bioassays (Reeve and Crozier, 1974). In general, those that are highly active in one bioassay are also active in others, but there are important exceptions. Growth stimulation caused by gibberellins in many species is not accompanied by increases in dry weight but is caused only by enhanced water uptake, especially of the shoot system. In some species, however, there is an increase even in dry weight, resulting from greater leaf development and more photosynthesis (Stuart and Cathey, 1961).

Metabolism of the Gibberellins As mentioned in Section 14.4, gibberellins are examples of isoprenoid compounds. They are diterpenes synthesized from acetate units of acetyl coenzyme A. Geranylgeranyl pyrophosphate (Fig. 16-11), a 20-carbon compound, serves as the donor for all gibberellin carbon atoms. It is converted to copalyl pyrophosphate, which has two ring systems, then to kaurene. Kaurene has some gibberellin-like activity, probably because it is effectively converted in the plant to gibberellins (Fig. 16-11). Some of these conversion steps are oxidations occurring in the endoplasmic reticulum membranes (Hasson and West, 1976). They involve the intermediate compounds kaurenol (an alcohol), kaurenal (an aldehyde), and kaurenoic acid. The first compound with a true gibberellane ring system is the aldehyde of GA_{12}, a 20-carbon molecule. From it presumably arise both C-20 and C-19 gibberellins by as yet undetermined enzyme reactions. In leaves, chloroplasts represent a major site of interconversion of gibberellins, although reactions of the pathway up to kaurenoic acid probably occur outside the plastids.

Why a Biologist? Some Reflections*

Frits W. Went

F.B.S. 1966

With the discovery of auxin, Frits W. Went's fame was assured. In this essay, he tells how that happened while he was a young student working in his father's laboratory at the University of Utrecht in the Netherlands. After completing the doctoral degree, he spent five years in Java, a Dutch possession, and almost 20 years at the California Institute of Technology, where he continued hormone work and developed interests in desert ecology. He moved in 1958 to St. Louis, Missouri, and then in 1964 to the Desert Biology Laboratory at the University of Nevada, where he has continued his desert studies.

Years ago, I tried to discover what made a biologist become a biologist. Soon I found that there are about as many different motivations as biologists, but a few were more prevalent than others. Every human being is born with an enormous amount of intellectual curiosity. If this has not been blunted through unfortunate experiences in youth, then an inspiring teacher may guide this urge for understanding toward biological or other problems. Early acquaintance with plants or animals can direct an inquisitive mind—even without teacher direction—toward the mysteries of life: the problems of growth, form, function, environment, and heredity. The orderly mind may be attracted toward taxonomy or biophysics, whereas the mind intrigued by complexity may select ecology, and the mind trying to understand interrelationships may become a physiologist. The mechanically inclined person might attack biological problems with delicate instruments, whereas the artist might try to solve problems of shape and color in nature.

There is an equal plethora of methodological approaches to the solution of biological problems. Since Francis Bacon in the seventeenth century pointed out the inevitability of cause and effect, the experimental or inductive approach has taken precedence over the old Aristotelian deductive approach, which through pure reasoning forced life into a straitjacket of axioms or preconceived ideas.

Not a few bright minds have been misled by their or others' sophisticated hypotheses. Such complicated but artificial fabrications as the phlogiston theory of the eighteenth century and the ether theory of the nineteenth century have held back real understanding. It is difficult for us to say which present-day ideas may be discarded in the centuries to come, but the Second Law of Thermodynamics might well be one of them.

An important aspect of the motivation to become a biologist may be humanitarian—the urge of the person to take a positive part in the well-being of his fellow man. Most disciplines in biology, particularly agriculture and medicine, have an input to society, and let us not forget teaching, one of the most important motivations of all. This desire to transmit the knowledge gathered by our culture to the generations to come is quite the contrary of the attitude of a professor 100 years ago, who admonished his successor never to tell the students everything he knew!

In my own case, ever since I chose botany over chemistry or engineering as my life's endeavor, I have been intrigued with the form and function of a plant and with its place and role in the environment. When walking in the country, I am forever wondering why a particular plant grows in a particular spot; why it has its particular shape; why it does not grow 100 m farther on; why only some plants grow in the

*See also F. W. Went, 1974, Reflections and speculations, Annual Review of Plant Physiology 25:1–26.

desert or tropics; why a limited number of plants have become weeds; why some plants are closely similar to others that lived 200 million years ago, whereas most are recently evolved; why some trees can be tapped for sugar, whereas most cannot. And I want to look inside the plant to find how it grows and functions, why it branches the way it does, and how it responds to its environment in such a precise way.

Some of these questions have already been answered, at least partially. Yet in many cases the answers don't satisfy me; they are either not general enough, too simple, or blatantly anthropocentric. This means that for me nature is still full of interesting problems.

One of the early problems that attracted me as a student was phototropism. Many of my fellow students thought that our predecessors in my father's laboratory—Blaauw, Arisz, Bremekamp, and Koningsberger, with their doctoral theses on phototropism—had preempted this subject. But such others as Dolk, Dillewyn, and Gorter were fascinated by the unsolved problems of plant responses to environment, and we had almost nightly bull sessions. I had to fulfill my military service obligations, which left only evenings and nights free for more productive projects. We discussed the newest publications of Paal, Seubert, Nielsen, and Stark—dissecting, interpreting, or repeating them.

This was in early 1926, an exciting time, with the growth substance concept just around the corner. Our discussions were usually based on Paal's theory, that the stem tip normally produces a growth-promoting factor. The hottest point of debate was whether or not in phototropism this factor was destroyed by the light. To wind up the argument, I asserted that I would "prove that the growth regulator from the tip was light stable." Consequently, I had to extract it from seedlings and then expose it to light. For this, I prepared a tiny cube of gelatin, stuck it on a needle, and placed cut stem tips on it all the way around. When I removed the tips after an hour and placed the gelatin cube on one side of the seedling stump, nothing happened at first. But in the course of the night, the stump started to curve away from the gelatin block. It had acquired the capacity of the stem tips to grow! At 3:00 A.M. on April 17, 1926, I ran home to my parents' house nearby, burst into their bedroom, and said excitedly: "Father, come and see, I've got the growth substance."

My father (who was also my major professor) sleepily turned around and said: "Fine. Repeat the experiment tomorrow [which was my day off from military service]; if it is any good, it will work again, and then I can see it."

Then followed an exciting time. I lived for my nights in the laboratory. Every experiment seemed to work, and I learned a lot about the behavior of the growth substance in and outside the stem tip. Of course, I chose this subject for my thesis work. But then, with completion of my military obligation, something unexpected happened.

Although I improved my whole experimental procedure, rebuilt the temperature and humidity controls of the lab room, grew much better seedlings, and worked much more cleanly, the growth substance seemed to have disappeared, for none of my test plants responded any more—until I found that bacteria lurking in the gelatin ate all of the growth substance overnight! My procedure had changed, because I prepared the blocks during the day and left them overnight. When I pressed an icebox into service (refrigerators were at that time unavailable), the bacteria ceased their activity and all experiments succeeded again.

To begin with, my approach to scientific problems was rather naïve. I mentioned already that I set out early to *prove* that the growth substance was light stable. I soon learned that experiments cannot prove anything; they can only *test* an hypothesis. Thus, my experiments to test whether or not auxin is moved along a gradient by decrement unexpectedly showed that its transport was polar. And my tests on the behavior of auxin inside the seedling in unilateral light showed that this deflected the strictly downward auxin stream laterally, thus laying a solid basis for the Cholodny-Went theory of phototropism. Further work suggested that other growth factors were involved in the action of auxin. In an unaccountable way, I was later accused of promulgating theories, whilst I was only presenting experimental facts (which were finally, although many years later, accepted as true).

After receiving my Ph.D., I was faced by a completely new challenge. The working conditions for a tropical botanist in Java—before the blessing of air conditioning—were harder. There was less equipment, which operated less well in the oppressive moist heat, so my main efforts went to ecology or, rather, applied physiology. I found tropical plants admirably suited for work on root initiation, but only after moving to California, where I could again devote full time to physiological problems, could I work out a proper test to study root formation.

Whereas all my auxin experiments were based on more or less logical sequences of deductions leading to the crucial experiments, my ecological work was mostly a set of questions asked of nature, after observations of nature had posed the problems. We constructed a phytotron, in which such environmental factors as temperature, light, humidity, wind, or rain could be controlled. With this new tool I could establish, for example, that the profuse flowering of the desert in certain years depended upon the precise germination response of seeds from desert annuals to temperature and rain and not upon a mystic "survival of the fittest" or a "struggle for existence." In the last 30 years, phytotrons have helped make ecology an experimental rather than a descriptive science; now the extreme complexity of the organism in its total environment can be reduced to experimentally manageable subunits. It is satisfying when laboratory experiments give us a better insight into the mechanism of life, but for me the greatest thrill comes when these new insights help me understand what goes on in nature, when laboratory knowledge is applicable in the field. Thus, nature not only provides the inspiration; it is also the ultimate arbiter. The laboratory is only an interlude between perceiving and understanding.

Certain commercial growth retardants that inhibit stem elongation and cause overall stunting do so in part because they inhibit gibberellin synthesis. These include **Phosphon D, Amo-1618, CCC** or **Cycocel,** and **Ancymidol.** The first two block the conversion of geranylgeranyl pyrophosphate to copalyl pyrophosphate (Frost and West, 1977). Phosphon D seems also to inhibit the subsequent formation of kaurene, while Ancymidol inhibits a reaction between kaurene and kaurenol (Coolbaugh and Hamilton, 1976). Growth inhibition by each can be completely overcome in many plants by GA_3, which first suggested that their only effects are to inhibit gibberellin synthesis. However, Phosphon D, Amo-1618, and CCC inhibit sterol synthesis in tobacco, in which the stunting effects of Amo-1618 and CCC are overcome not only by GA_3 but also by three sterols, β-sitosterol, stigmasterol, and cholesterol (Douglas and Paleg, 1974). Such results are difficult to interpret if we assume that they block only gibberellin synthesis.

Once formed, gibberellins appear to be only slowly degraded, but they can be readily converted to bound forms that are probably inactive. In bound forms they might be stored or translocated before release at the proper time and place. Known bound forms include **glucosides,** in which glucose is connected in an ether bond to one of the OH groups or in an ester bond to a carboxyl group of the gibberellin. Other unidentified bound forms also exist, some of which appear to be stable protein-gibberellin conjugates. Another important metabolic process is conversion of highly active gibberellins to others less active. For example, Douglas fir shoots, which show little response to most applied gibberellins, can effectively convert GA_4 to the much less active GA_{34} by simply adding an additional hydroxyl group.

Location of Gibberellin Synthesis Which parts of the plant synthesize gibberellins? Clearly, if we find these hormones in a plant organ, they might have been either synthesized there or translocated there. The diffusion technique of Jones and Phillips (1964) has been of great help in distinguishing these two possibilities. An organ is excised, the cut surface placed in contact with agar (as in early studies with IAA), and gibberellins are allowed to diffuse into the agar for several hours. The amounts collected are estimated by bioassays and by more sophisticated chemical methods mentioned for auxin analysis. These amounts are compared with quantities extractable from similar organs or tissues both before and after diffusion. If the diffusible amounts are significantly greater than the extractable amounts, as they often are, the cells must have been synthesizing gibberellins or converting them from an inactive form (glucoside?) to an active form.

The diffusion technique shows that young leaves are major sites of active gibberellin synthesis, just as they are for IAA. This is consistent with the fact that if the stem apex containing young leaves is excised, and the cut stump then

Discovering the Wall-Loosening Factor

David L. Rayle

F.B.S. 1976

The thrill of scientific discovery can strike at any time during a scientist's active research life, but during a postdoctoral fellowship is a logical and common time. One is fresh and enthusiastic from one's doctoral studies and not burdened down with teaching and administrative duties. David L. Rayle, who is now Chairman and Professor of Botany at San Diego State University in California, tells us how, four decades after Frits Went discovered auxin, Rayle experienced the sweet taste of discovery in learning what auxin does.

The process of cell enlargement and the mechanism by which enlargement is controlled by plant growth hormones has attracted considerable attention and effort over the last 50 years, perhaps more than any other single aspect of plant development. You are perhaps curious why a particular problem such as the mechanism of hormone action in cell enlargement has generated such disproportionate attention and, of course, you should question whether the effort has produced meaningful results. Let me explain why, as a young scientist, I chose to enter such a competitive field and how I became involved in the development of the acid growth theory of auxin action.

As a personal scientific preference, I have always chosen to work with plant responses that were large and easily measurable. Cell enlargement is such a phenomenon. Plant cells can elongate to many times their initial length as they mature, and to a great extent this expansion determines the final size and shape of the mature organism. A second and very attractive aspect is that it is under tight hormonal control. That is, one can make enlargement in plant sections start or stop simply by adding or removing auxin. Lastly, I was attracted by the fact that hormone-induced cell enlargement is one of the few hormone-mediated responses in plants that is detectable in minutes rather than hours or days. This made me more confident that any biochemical change uncovered shortly after hormone addition is associated with a primary event induced by the hormone rather than a secondary or side event associated with or caused by an earlier event.

When I arrived in Robert Cleland's laboratory in Seattle as a postdoctoral student in 1968, ideas concerning the biochemical mechanism of auxin action were passing through a low point of excitement; in general, the field seemed to lack any clear-cut direction. Nevertheless, enough was known about cell enlargement for us to formulate the following model: First, when a plant cell enlarges, most of the increase in volume is due to the uptake of water into an expanding, centrally located vacuole. Second, we knew from studies by Bob Cleland and others that auxin enhances cell enlargement by increasing cell wall extensibility (cell wall loosening) rather than by making the osmotic potential of the cell more negative. This wall loosening leads to a transient decrease in cell water potential; water then enters the cell, causing the wall to stretch. The synthesis of new cell wall meanwhile "fixes" the expanded wall and repairs it so that further wall loosening can occur. Lastly, we knew that auxin did not act directly on the wall but rather in the cytoplasm or at the plasma membrane, so we reasoned that there must be some form of communication between the cytoplasm and wall. That is, auxin must initiate the release of a wall-loosening factor (WLF). The critical questions were then rather obvious: How does auxin initiate the release of WLF? What is the nature of WLF, and how does WLF initiate wall loosening? While these questions were simple and straightforward, the correct approach in solving them

was not at all obvious and, as I will explain, good fortune rather than any special insight on my part played an important role.

In Seattle, I began to reinvestigate and more completely characterize a phenomenon that had been discovered in the early 1930s: acid-induced growth. My reasoning for pursuing this seemingly odd and nearly forgotten growth response was that it might provide a simple model system for studying cell wall loosening, a model that one could then eventually apply to auxin-induced wall loosening. Interestingly, even when I was well into this work, I didn't have the slightest idea that acid-induced growth was in any way directly related to auxin-induced growth or the "normal" mode of wall loosening. Indeed, the final realization that the two responses were related was the most pleasant surprise of my scientific life.

Characterizing acid growth progressed rapidly during 1968. We found that the optimal pH for elongation in *Avena* coleoptiles was 3.0 and that the extension response was indeed mediated by wall loosening. Further, the H^+ response, unlike the auxin response, was unaffected by virtually all metabolic poisons and could even be demonstrated using isolated wall matrices if tension was applied to replace turgor. In spite of these rather obvious differences between H^+- and IAA-induced growth, I began to discover some striking similarities: Both agents at optimal levels induced the same growth rate; both yielded responses that were similarly affected by changes in temperature (Q_{10}); and both seemed to loosen the cell wall in a similar manner. These similarities as well as differences led Bob Cleland and me to begin suspecting that we might have something more than just a model system. They eventually led us to propose that H^+ and auxin might be altering the same bonds in the cell wall. However, because the optimal pH for extension was so physiologically unrealistic and eventually caused the cells to die, we were still unconvinced that the hydrogen ion was the natural WLF. Rather, we felt it was more probable that simply by coincidence H^+ and the natural WLF happen to loosen similar bonds within the cell wall.

One afternoon, while working in Meinhart Zenk's laboratory at the Ruhr University in West Germany, my thinking on this matter suddenly changed. I discovered, much to my embarrassment, that I had misinterpreted the pH optimum for acid-induced growth by nearly two orders of magnitude! It was pH 4.8 rather than 3.0! The key to this discovery was the simple realization that the cuticle is a potent barrier to H^+ flux and that all of our earlier work was performed with segments that had an intact cuticle. Indeed, the very simple step of rendering coleoptile segments permeable to H^+ completely changed my thinking about acid growth and made it physiologically reasonable to propose H^+ as the elusive WLF.

In the meantime, other investigators, chiefly Acchen Hagar in Münster and Erasmo Marré in Milan, were independently arriving at similar conclusions. However, one important piece of critical evidence treated with a gibberellin or auxin, stem elongation is promoted relative to stems similarly cut off but otherwise untreated. Mature leaves have little ability to synthesize either hormone type.

Roots also synthesize gibberellins in significant quantities, as shown by the diffusion technique, yet gibberellins have little direct effect on root growth, and they inhibit adventitious root formation. These hormones can be detected in the xylem exudates of roots and stems when these organs are excised and root pressure forces the xylem sap out. Inhibitors of gibberellin synthesis decrease amounts of gibberellins in these exudates. Repeated excision of part of the root system causes marked decreases in the concentrations of gibberellins in the shoot, suggesting that much of the shoot's gibberellin supply arises from the roots via the xylem, perhaps partially explaining why root pruning inhibits shoot growth. A gibberellin glucoside has also been found in xylem sap of maple and elm trees, and this raises the unsolved question of how effectively various gibberellins and their glucosides can be circulated and interconverted in the plant.

Gibberellins are present in embryos, seeds, and fruits, and they probably stimulate growth there. Nevertheless, we do not yet know whether those organs are generally capable of synthesizing gibberellins. In pea fruits, synthesis does indeed occur, as evidenced by two facts: the content in the seeds increases 300-fold when development occurs in excised pods, and Amo-1618 inhibits this increase.

Gibberellin Transport Gibberellins stored in abundance in seeds either free or in bound forms are transported out through the phloem into the developing seedling. From roots, gibberellins move through the xylem, as mentioned earlier. The transport pathway from young leaves into the stem below is still unknown. It almost surely does not involve vascular tissues, because young leaves import but hardly export through either xylem or phloem. Presumably, as with auxin, cortex or pith is involved. But in general, the direction of gibberellin flow is not nearly so polarized as it is for auxins. Gibberellins seem to move much more extensively than auxins and therefore act over long distances to control various responses. Certain other responses, not all of which involve long distance transport, are discussed in the next four sections.

Gibberellins Promote Germination of Dormant Seeds and Growth of Dormant Buds The buds of evergreens and de-ciduous trees and shrubs growing in temperate zones usually become dormant in late summer or early fall. Dormant buds are relatively hardy during cold winters and drought. Seeds of many noncultivated species are also dormant when first shed and will not sprout even when exposed to adequate moisture, temperature, and oxygen. Dormancy of buds and seeds is often overcome (broken) by extended winter cold periods, allowing growth in the spring when conditions are favorable (see Chapter 21). For some species, bud dormancy can also be overcome by increasing daylengths that occur in late winter, and for seeds of some species, dormancy is broken by brief periods of red light when they are moist (see Chapter 19). *Gibberellins overcome both kinds of seed dormancy and both kinds of bud dormancy in many species, acting as a substitute for low temperatures, long days, or red light.* In seeds, the principal gibberellin effect is to enhance cell elongation so the radicle can push through the endosperm, seed coat, or fruit coat that restricts its growth. Buds have been less carefully investigated, and whether a stimulation of cell division in addition to elongation is necessary is not known.

Flowering As we shall discuss in Chapter 22, the time at which plants form flowers is dependent upon several factors, including their age and certain properties of the environment. For example, the relative durations of light and darkness have key controlling influences upon several plants. Some species flower only if the light period exceeds a critical length, while others flower only if this period is shorter than some critical length. *Gibberellins can substitute for the long-day requirement in some species, again showing an interaction with light.*

Gibberellins Also Overcome the Need Some Species Have for an Inductive Cold Period to Flower (Vernalization). This is primarily true for biennial species such as cabbage and carrots, which normally form a rosette during the first year's growth. It appears that the formation of flowers caused either by long days or by cold periods might normally depend upon the buildup of endogenous gibberellins during these periods, because the gibberellin content of some affected plants increases following these treatments. This topic is discussed in more detail in Chapter 21.

Gibberellins Stimulate the Mobilization of Foods and Mineral Elements in Seed Storage Cells Soon after a seed germinates, the young root and shoot systems must begin to use mineral nutrients, fats, starch, fructosans, and proteins present in storage cells of the seed. The young seedling depends upon these substances before the time mineral salts can be absorbed from the soil and before extension of the shoot system into the light. The mineral salts are readily translocated via the phloem into and throughout the young roots and shoots, if they are mobile. The seedling has a problem with fats, polysaccharides, and proteins, because these molecules are not translocated. How is the problem solved? We discussed this briefly in Chapters 10, 13, and 14, explaining how foods are converted into sucrose and into mobile amino acids or amides. *Gibberellins greatly stimulate these conversions, especially in cereal grains* (Briggs, 1973; Jones, 1973; Varner and Ho, 1976).

The embryo (germ) of seeds of cereal grains and other grasses is surrounded by food reserves present in the essentially nonliving cells of the endosperm, which in turn is surrounded by a thin, living layer two to four cells thick called the **aleurone layer** (Fig. 16-12). After germination occurs, primarily in response to increased moisture, the aleurone cells provide the hydrolytic enzymes that digest the starch, proteins, phytin, RNA, and certain cell wall materials present in the endosperm cells. Some of the necessary enzymes for these digestion processes include β-amylase (already present in the seed), several α-amylase isozymes (synthesized in the aleurone layer), ribonuclease, phytase, and various proteases (some of which are also newly synthesized upon germination). If the embryo is removed from a barley seed, the aleurone cells do not produce and hardly secrete most hydrolytic enzymes, especially α-amylase. This suggested that the barley embryo

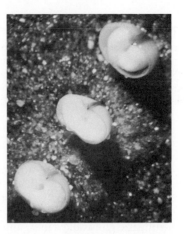

Figure 16-13 Gibberellin-stimulated digestion of the endosperm in barley leaf seeds. The embryo half of each seed was removed before treatment (top to bottom) with 5 μl of 0.1 μM GA$_3$, 0.001 μM GA$_3$, or H$_2$O. (Courtesy J. E. Varner.)

normally provides some hormone to the aleurone layer and that this hormone stimulates aleurone cells to manufacture these hydrolytic enzymes. *This hormone, which proved to be a gibberellin, also stimulates secretion of the hydrolytic enzymes into the endosperm, where they digest food reserves and cell walls.* Reserve mineral elements also become more readily available as a result of gibberellin action. Figure 16-13 illustrates endosperm degradation in barley half-seeds (from which the embryo was removed) in response to as little as nine trillionths of a gram of GA$_3$. The increase in α-amylase in aleurone layers of such half-seeds is frequently used as a gibberellin bioassay.

In grass seeds, gibberellins are synthesized in the scutellum (cotyledon) and perhaps also in other parts of the embryo. The kind of gibberellin probably depends upon the species, but in barley, GA$_1$ and GA$_3$ seem most important. This is a good example of a correlative phenomenon that doesn't cause growth of the affected cells but does stimulate growth of the seedling. Nevertheless, although barley, wheat, and wild oats (*Avena fatua*) aleurone layers respond to added GA$_3$ or certain other gibberellins by synthesizing α-amylase and certain other enzymes, some cultivated oat and most corn cultivars do not. Perhaps in oat and corn an untried gibberellin is responsible. There is considerable genetic variability of gibberellin responses among cereal grains, much of which probably depends on how much gibberellin already exists in the aleurone layer or the endosperm. Gibberellins generally have little or no effect on mobilization of food reserves in dicots and gymnosperms.

Other Gibberellin Effects We mentioned in Chapter 15 that gibberellins cause parthenocarpic (seedless) fruit development in some species, suggesting a normal function in fruit growth,

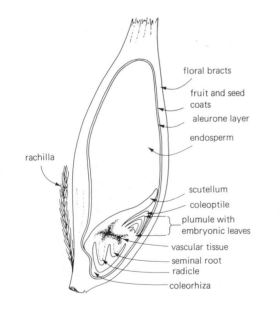

floral bracts

fruit and seed coats

aleurone layer

endosperm

rachilla

scutellum

coleoptile

plumule with embryonic leaves

vascular tissue

seminal root

radicle

coleorhiza

Figure 16-12 Barley seed sectioned to illustrate major tissues. (Original drawing by Arnold Larsen, Colorado State University Seed Laboratory.)

Commercial Uses of Gibberellins

Considering the numerous effects of gibberellins, it seems logical that they would be used commercially. Major limiting factors have been their cost and their frequent promotion of fresh weights but not of dry weights, especially regarding the obvious possible application in growth of pastures and hay crops. We still must rely on the *Gibberella* fungus to synthesize them. They are used extensively in the Central and Imperial Valleys of California to increase the size of Thompson seedless grape berries and the distance between them (Fig. 16-14). When applied at the right time and with the proper concentration, gibberellins cause the bunches to elongate so that they are less tightly packed and less susceptible to fungal infections. They are also used by some breweries to increase the rate of malting through their enhancing effects on starch digestion. Gibberellins have a dis-advantage for some malting processes by causing amino acid accumulation resulting from increased protein hydrolysis, but they are commonly used outside the United States in making distiller's malt for grain spirit production. Celery plants, which are valued for the lengths and crispness of their stalks, respond favorably to gibberellins, but the poor storage qualities of such stalks prevent wide use of these hormones in the celery industry. Gibberellins have also been sprayed on fruits and leaves of navel orange trees (when the fruits have lost more of their green color) to prevent several rind disorders that appear during storage. Here the hormones delay senescence and maintain firmer rinds. Other occasional and potential uses are listed by Weaver (1972).

Figure 16-14 Effects of gibberellin and girdling on growth of Thompson seedling grapes. Dates of treatment: blossom spray, June 13; shatter spray, June 23; girdling, June 23. Harvest, August 3. (Courtesy J. LaMar Anderson, Utah State University, Logan.)

and that gibberellins formed in young leaves might renew activity of the vascular cambium in woody plants. Another important effect is the delay of aging (senescence) in leaves and in citrus fruits. In Chapter 18, we shall discuss the ability of gibberellins to influence leaf shapes, a response that is especially apparent in leaves showing heterophylly or phase changes. Gibberellins may also participate, along with auxins and perhaps ethylene and abscisic acid, in geotropism and phototropism (see Chapter 18). Another effect they have in some plants is to retard the development of cold hardiness (see boxed essay).

Possible Mechanisms of Gibberellin Action The many effects of gibberellins suggest that they have more than one important primary site of action. Even a single effect on whole plants, such as enhanced stem elongation, results from at least three contributing events: *First, cell division is stimulated in the shoot apex,* especially in the more basal meristematic cells from which develop the long files of cortex and pith cells (Sachs, 1965). Careful work of Peter B. W. Liu and J. Brent Loy (1976) showed that gibberellins promote cell division because they stimulate cells in G_1 phase to enter S phase and because they also shorten the S phase.* The mechanism of

---

*In the cell cycle, G_1 is the "gap" phase following mitosis, S is the synthesis phase during which labeled precursors are incorporated into DNA, and G_2 is the second "gap" between synthesis and subsequent mitosis. The DNA content of a normal diploid cell equals 2 C during G_1 and 4 C during G_2.

this has not been elucidated, but it is reasonable that gibberellins increase the number of sites on the chromosome where DNA and RNA synthesis can begin. This might result from their ability to combine with chromosomal proteins and thereby unmask a covered initiation site for DNA and RNA synthesis, but this is only an hypothesis. Regardless, the increased number of cells will lead to more rapid stem growth, because each of these cells can then grow.

Second, gibberellins stimulate cell growth because they increase hydrolysis of starch, fructosans, and sucrose into glucose and fructose molecules. These hexoses provide energy via respiration, contribute to cell wall formation, and also make the cell's water potential momentarily more negative. As a result of the decrease in water potential, water enters more rapidly and dilutes the sugars, causing cell expansion. In sugar cane stems, gibberellin-promoted growth results in part from increased synthesis of invertase enzymes that hydrolyze incoming sucrose (Glasziou, 1969). In dwarf peas, the activities of both invertase and amylase enzymes rise in parallel with growth (Broughton and McComb, 1971). The same is true for amylase in dwarf corn. Less quantitative work with other species indicates that gibberellin-induced stem growth is associated with rises in amylase activity in small water plants and in certain trees, suggesting that the phenomenon is general, but we know of no results for conifers (see boxed essay).

Third, gibberellins sometimes increase wall plasticity. An excellent example of this occurs in internodes of oat, in which growth promotion of young cells derived from the intercalary meristem is unusually dramatic. Here no stimulation of cell division occurs. Elongation due to GA_3 is 15 times as great as in the untreated sections (Fig. 16-15), provided that sucrose and mineral salts are present to provide energy and prevent excessive dilution of the cell contents (i.e., prevent a rise in ψ_π). A significant increase in wall plasticity occurs (Adams et al., 1975), and a similar phenomenon apparently explains gibberellin-promoted growth in lettuce hypocotyls (Stuart and Jones, 1977).

Not only is stem elongation promoted by gibberellins, but so is growth of the whole plant, including leaves and roots.

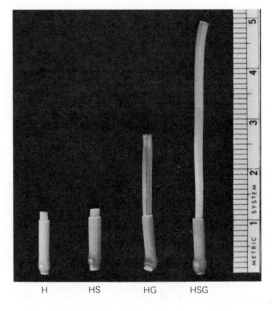

H   HS   HG   HSG

Figure 16-15 Effect of GA and sucrose on the growth of 1 cm oat stem segments. The segments are shown after 60 hr of treatment in Hoagland's nutrient solution (H), Hoagland's + 0.1 *M* sucrose (HS), Hoagland's + 30 μM GA (HG), and Hoagland's + sucrose + GA (HSG). A centimeter ruler denotes actual size. Elongation of the leaf sheaths did not occur, but growth of cells derived from the intercalary meristem (see Fig. 15-12) accounts for the stem elongation illustrated. (From P. A. Adams, P. Kaufman, and H. Ikuma, 1973, Plant Physiology 51:1102–1108.)

We stated that application of gibberellins directly to leaves may stimulate their growth slightly and may influence their shapes, yet direct application to roots usually has almost no effect on the roots themselves. But if the gibberellin is applied in any manner whereby it can move into the shoot apex, increased cell division and cell growth apparently lead to increased elongation of the stem and, in some species, to increased development of the young leaves. In species where faster leaf development occurs, enhanced photosynthetic rates then stimulate growth of the whole plant, including the roots.

How might gibberellins loosen cell walls and also increase formation of hydrolytic enzymes leading to stem elongation? We have no evidence about wall-loosening mechanisms, except that, in contrast to auxins, H^+ are apparently not involved. In the oat internodes, there is a lag of nearly 1 hour before promotion of elongation can be detected. This would probably allow sufficient time for gibberellins to increase synthesis of RNA or enzymes needed to amplify growth responses. Perhaps hydrolases that attack cell wall polysaccharides are synthesized faster or simply become more active in gibberellin-treated cells. There is much evidence regarding hydrolytic enzyme formation, most of it with barley aleurone layers and α-amylase, but still no clear-cut primary action of gibberellins can be described (Mann, 1975).

Hormones and Growth Regulators:

Cytokinins, Ethylene, and Abscisic Acid

The more we learn about growth and development, the more complex these processes seem to become. In the last chapter, we explained that both processes depend upon IAA and gibberellins, but that these hormones generally influence different parts of the plant in different ways. Thus, whereas IAA promotes cell division leading to production of lateral roots and adventitious roots, gibberellins enhance cell division in the shoot apical meristem leading to stem elongation. Both promote elongation of stem cells but apparently by different primary mechanisms. Both stimulate division and growth of fruit cells causing parthenocarpic fruits, yet this effect varies considerably with the species. In spite of the complexities, we now realize that both hormone types must be considered in understanding growth, and we know generally how to approach the problems of their interactions. In this chapter, we shall discuss the other three kinds of hormones that are presently known (cytokinins, ethylene, and abscisic acid), emphasizing that although each has distinct effects, growth and development normally involve an interplay between all known hormones and probably others yet undiscovered.

17.1 The Cytokinins

About 1920, G. Haberlandt discovered in Austria that an unknown compound present in vascular tissues of various plants stimulated cell division, causing cork cambium formation and wound healing in cut potato tubers. He also deduced that the wounded parenchyma cells produced another compound that participated in the division and healing process. This was the first demonstration that plants contain compounds, now called **cytokinins,** that stimulate cytokinesis. In the 1940s, J. van Overbeek found that the milky endosperm from immature coconuts is also rich in compounds that stimulate cytokinesis. In the early 1950s, Folke Skoog and his colleagues, who were then interested in auxin stimulation of plant tissue cultures, found that cells in pith sections from tobacco stems divided much more rapidly if a piece of vascular tissue was placed on top of the pith, verifying Haberlandt's results. They tried to identify the chemical factor from the vascular tissues, using growth of tobacco pith cells as a bioassay system. These cells were cultured on agar media con-

taining known sugars, mineral salts, vitamins, amino acids, and IAA. IAA itself stimulated growth slightly, mainly because enormous cells were formed, but these cells did not divide. Many of these cells were polyploids with several nuclei. In seeking substances that stimulated cell division, they found a purine-like compound in yeast extracts that was highly active. This led to investigations of the ability of DNA to promote cytokinesis and to the discovery, in 1954, of a very active compound formed by partial breakdown of aged or autoclaved herring sperm DNA. They named this compound **kinetin.**

Although kinetin itself has not been found in plants and is not the active substance found by Haberlandt in phloem, related cytokinins are present in most plants. F. C. Steward, also using tissue culture techniques in the 1950s, found several cytokinins in coconut milk that stimulate cell division in carrot root tissues. The most active of these were later shown by D. S. Letham (1974) and J. van Staden and S. E. Drewes (1975) to be compounds previously given the common name **zeatin** and **zeatin riboside.** Zeatin had first been identified in New Zealand by Letham in 1964 and almost simultaneously by Carlos Miller in Indiana, both of whom used the milky endosperm of corn (*Zea mays*) as a source.

Since then, other cytokinins with adenine-like structures similar to kinetin, zeatin, and zeatin riboside (Fig. 17-1) have been identified in numerous plant parts. None of these are constituents of DNA nor are they breakdown products of DNA, but some occur as odd bases in certain transfer RNA (tRNA) molecules of higher plants, bacteria, and even primates, and some exist as unbound, free cytokinins. Only about one tenth of the tRNAs (those having corresponding messenger RNA codons beginning with a uracil base) have cytokinins, and then only one cytokinin base is present in each tRNA molecule. It is always attached next to the 3' end of the anticodon. Cytokinins exist in tRNA as nucleotides (base + ribose + phosphate). The cytokinin bases include zeatin, **isopentenyl adenine** (Fig. 17-1), **2-methylthiozeatin,** and **2-methylthio-isopentenyl adenine** (Hall, 1973; Burrows, 1975). The methylthio derivatives have not been found as free cytokinins, but all the others mentioned have. These tRNA-bound cytokinins might be necessary for normal binding of tRNA to messenger RNA by anticodon-codon interactions during protein synthesis on the ribosomes, but it is free cytokinins not bound

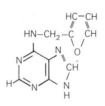

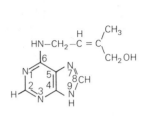

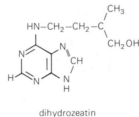

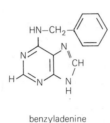

zeatin

mol. wt. = 219.2 g/mole

zeatin riboside or
ribosyl zeatin

kinetin

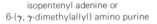

isopentenyl adenine or
6-(γ, γ-dimethylallyl) amino purine

dihydrozeatin

benzyladenine

Figure 17-1 Structures of common natural and synthetic cytokinins. All these are adenine derivatives in which the purine ring is numbered as shown for zeatin (upper left). Zeatin and zeatin riboside can exist with groups arranged about the side chain double bond either in the *trans* (as shown) or *cis* (with CH_3 and CH_2OH groups interchanged) configuration. The *cis* form predominates in tRNA-bound cytokinins, but the *trans* form exists in free zeatin and zeatin riboside.

in tRNA that apparently cause the known physiological responses.

A half dozen other free cytokinins that are adenine derivatives having a side chain on the N^6 position have since been identified in angiosperms, and more are sure to come. It is likely that all could exist as the free base, the nucleoside, or even the nucleotide forms, just as adenine does. So far, free cytokinins have been identified in only a few gymnosperms and apparently not at all in mosses and ferns, but they are probably present in all plants both in the free and tRNA forms. Free cytokinins exist in brown and red algae and apparently also in diatoms, and these hormones sometimes promote algal growth. Certain bacteria and fungi contain cytokinins that might influence disease processes caused by those microbes. Many ectomycorrhizal fungi (see Chapter 5) contain zeatin and zeatin riboside, both of which might stimulate (along with auxins) mycorrhizal formation.

Several hundred synthetic cytokinins were tested by Skoog, Letham, and others in various bioassays. Many are active, but those most commonly used by physiologists include kinetin and **benzyladenine,** also shown in Fig. 17-1.

Cytokinin Metabolism We know very little about how any cytokinin is synthesized, except for the adenine part of the molecule. Once the adenine moiety is formed from glycine, CO_2, and other small molecules, the side chain essential for activity could be attached to the N^6 position using isopentenyl pyrophosphate ($CH_2{=}\overset{\overset{\displaystyle CH_3}{|}}{C}{-}CH_2{-}CH_2{-}OP_2O_6^{-2}$, see Fig.

16-11), with the accompanying loss of pyrophosphate. Modification of the resulting five-carbon side chain might lead to the other cytokinins known. Indeed, for cytokinins in tRNA, this is how the process probably occurs. Addition of the sulfur atom and methyl group in each methylthiocytokinin in tRNA also takes place after the tRNA is already formed. Whether or not the side chain of free cytokinins is formed from isopentenyl pyrophosphate is unknown. Some cytokinins found by Letham (1973) in corn endosperm have only three or four carbon atoms in the side chain, and a cytokinin in poplar leaves has a benzene ring as the major part of the side chain (Horgan et al., 1975).

More is known about the destruction of cytokinins than about their synthesis. Destruction is probably important to regulate cytokinin levels. Removal of the side chain is one reaction that occurs. This is a direct inactivation process, because the adenine or adenosine products are essentially inactive as cytokinins. Another mechanism that might represent part of a useful storage phenomenon is formation of sugar derivatives (glucosides), just as occurs with IAA and gibberellins. **Raphanatin** (glucosylzeatin), first found in radish (*Raphanus sativus*), is an abundant glucoside in this species. In raphanatin, the C-1 of glucose is attached directly to the nitrogen atom at the 7 position of the purine ring of zeatin (see numbering system in Fig. 17-1). Raphanatin is metabolized very slowly in the radish root. In other species, glucose can also be attached to cytokinin bases at the nitrogen atom occupying position 9 of the purine ring, the position at which ribose is attached in nucleoside cytokinins, and to the hydroxyl group on the side chain of zeatin. Therefore, zeatin can be converted to raphanatin, to the 9-glucosyl derivative, to zeatin riboside (the 9-ribosyl derivative), and to the side chain glucoside. We cannot be sure that these derivatives represent only storage products, because all might be active by themselves. Zeatin riboside has usually been assumed to be active itself, yet there is no proof that it doesn't have to be converted to zeatin first.

Sites of Cytokinin Synthesis and Transport Cytokinins are relatively abundant in young fruits and seeds, in young leaves, and in root tips. It seems logical that they are synthesized there, but we cannot dismiss the possibility of transport from some other site. For roots, synthesis is probably involved, because if the roots or stems are cut off and xylem exudates are collected from the lower portions remaining, cytokinins continue to be exuded up to four days (Skene, 1975). A conclusion is that root tips synthesize cytokinins and transport them through the xylem to all parts of the plant. This might explain their accumulation in young leaves, fruits, and seeds into which xylem transport is effective.

Little direct evidence for cytokinin formation in organs except roots exists. Transport of these hormones out of young leaves, seeds, and fruits is quite restricted, and movement through nonvascular parts of stems is also very slow. If, for example, we add a cytokinin to a lateral bud subject to dominance from IAA or other substances arriving from the shoot apex, almost none of this cytokinin moves out. Cytokinins, especially as glucosides, can be translocated from leaves and through stems if they get into the phloem transport system, as evidenced by their occurrence in honeydew of aphids. Furthermore, if a mature leaf is cut off and the petiole kept moist, cytokinins move to the base of the petiole, probably through the phloem. This implies that mature leaves normally contribute cytokinins to young leaves, but this is uncertain. Since conduction of solutes from young leaves or fruits to older leaves is minimal or nonexistent, cytokinins formed in young cells usually remain there (exceptions being root tips).

Cytokinins Promote Cell Division and Organ Formation We have seen that a major function of cytokinins is to stimulate cytokinesis. If the pith from tobacco, soybean, and many other dicot stems is cut out and cultured on an agar medium containing an auxin and the proper nutrients, a mass of large, unspecialized, loosely arranged, and typically polyploid cells called a **callus** is formed. If a cytokinin is also added, cytokinesis is greatly promoted, as already mentioned. The amount of growth serves as a sensitive and highly specific bioassay.

Skoog also found that if the cytokinin-to-auxin ratio is maintained high, certain cells that subsequently develop into buds, leaves, and stems are produced in the callus. If the tissues are kept in the light, leaf development is promoted. But if the cytokinin-to-auxin ratio is lowered, root formation is favored. By choosing the proper ratio, callus from many species can be made to develop into an entire new plant (Section 18.2).

Cytokinins Delay Senescence When a mature but still active leaf is cut off, it begins to lose chlorophyll, RNA, proteins, and lipids from the chloroplast membranes more rapidly than if it were still attached, even if it is provided with mineral salts and water through the cut end (Beevers, 1977). This premature aging or senescence, a yellowing of leaves, occurs especially fast if the leaves are kept in darkness. Sometimes adventitious roots form at the base of the petiole in dicot leaves, and then senescence of the blade is greatly delayed. Something is apparently provided by the roots to the leaf that keeps it physiologically young. This something almost surely consists in part of a cytokinin provided through the xylem, because any cytokinin will partially replace the need for roots in delaying senescence. Various bioassays in which retardation of chlorophyll loss is measured take advantage of this principle. In sunflowers, the cytokinin content of the xylem sap increases during the period of rapid growth, then decreases greatly when growth stops and flowering begins. This suggests that a reduction in cytokinin transport from roots to shoots might promote senescence (Skene, 1975).

The delay of senescence by cytokinins appears to be a natural, partially root-controlled phenomenon and is. associated with other interesting phenomena. Cytokinins cause transport of many solutes from older parts of the leaf and even from older leaves into the treated zone. A dramatic illustration of this is shown in Fig. 17-2. Here the oldest (primary) leaves of the bean plant were painted at 4-day intervals with the synthetic cytokinin benzyladenine. Normally, these leaves

Figure 17-2 Senescence of a trifoliate bean leaf caused by treating the primary leaves of cuttings with synthetic cytokinin benzyladenine (30 μg/ml) at 4-day intervals. (From A. C. Leopold and M. Kawase, 1964, American Journal of Botany 51:294–298.)

become senescent before the trifoliate leaves above, but in this example the senescence pattern was reversed. The treated primary leaves apparently withdrew nutrients from the adjacent trifoliate, causing it to senesce first. Other studies with many dicots and monocots show that if only one part of a leaf is treated, radioactive metabolites added to another part of the same leaf or to an adjacent leaf migrate into and accumulate in the treated zone. The implication is that young leaves can remove nutrients from older ones partly because they are rich in cytokinins. Whether or not these hormones are involved in the normal transport of mobile nutrients into twigs and larger branches of woody plants before leaf fall in autumn is an interesting question. That cytokinins in reproductive structures might have survival value by stimulating movement of sugars, amino acids, and other solutes from mature leaves into seeds, flowers, and fruits is also an interesting hypothesis.

When certain fungi causing rust diseases infect leaves, necrotic areas of dead and dying cells are produced. These areas are often immediately surrounded by several green and starch-rich cells, even when the rest of the leaf has become yellow and senescent. These "green islands" are rich in cytokinins, presumably synthesized by the fungus. The cytokinins presumably help maintain food reserves for the fungus and influence further progress of the disease symptoms.

The ability of cytokinins to retard senescence also applies to certain cut flowers and fresh vegetables. The concentration of cytokinins in rose petals decreases as aging occurs, and applied cytokinins slow this aging process. For most cut flowers, however, exogenously supplied cytokinins cannot overcome the senescence-stimulating effects of ethylene produced by the flowers. The storage lives of Brussels sprouts and celery are often increased by relatively inexpensive commercial cytokinins such as benzyladenine, but this is not permitted for foods sold in the United States, even though we are constantly exposed to natural cytokinins in foods from plants.

Cytokinins Promote Lateral Bud Development If a cytokinin is added to an inactive lateral bud dominated by the shoot apex above it, the lateral bud often begins to grow. In many species, the lateral branch doesn't continue to grow, even with repeated cytokinin applications, but in others, such as tobacco, this growth continues dramatically. The ability of cytokinins to overcome apical dominance probably involves stimulated cell division in the bud and the ability of the bud to act as a sink into which nutrients from other parts of the plant are drawn.

Enhanced lateral branching also occurs in two diseases where the pathogen synthesizes cytokinins. One is a fasciation* disease caused by *Corynebacterium fascians* occurring in various dicots such as chrysanthemum, garden pea, and sweet

*In fasciation of plants, normally round stems become flattened and/or numerous lateral buds develop into branches, often forming a broom-like bundle of stems.

pea. This pathogen also causes certain kinds of witches'-brooms in trees, a phenomenon involving multiple bud and branch production. Two other pathogens (*Exobasidium* species) that cause witches'-brooms also synthesize cytokinins. Logical speculation is that cytokinin synthesis by the pathogen causes the symptom of branching.

As mentioned in Section 15.4, cytokinins repress branch root formation, at least in some species. Dominance of the root tip where cytokinin synthesis occurs is exhibited by retarded development of branch or lateral roots. If the tip is cut off, branch roots develop more rapidly, but addition of cytokinins to the cut surface again inhibits growth of the laterals.

Cytokinins Increase Expansion of Cotyledons and Leaves in Some Dicots When seeds of certain dicots are germinated in darkness, the cotyledons remain yellow and relatively small. If they are exposed to light, growth increases greatly, even if the light energy provided is too low to allow photosynthesis. This is a photomorphogenetic effect controlled by phytochrome, as discussed in Chapter 19, but cytokinins are probably also involved. If the cotyledons are excised and incubated with a cytokinin, growth is enhanced relative to controls without the hormone. This is true for at least a dozen species, including radish, lettuce, sunflower, cocklebur, white mustard, squash, cucumber, pumpkin, and muskmelon. Most (perhaps all) of these species contain fats as the major food reserve in the cotyledons. Furthermore, the cotyledons normally emerge above ground and become photosynthetic in each species. No response has been found in cotyledons of starch-rich species such as the pea, which remain underground after germination, or those of the bean, which emerge but do not become leafy. Figure 17-3 shows the promotive effects of zeatin on radish cotyledon enlargement in both light and darkness and also illustrates that light is stimulatory in the absence of zeatin. Auxins do not promote growth of cotyledons, and gibberellins also have little effect when the cotyledons are cultured in water, so this response provides a useful bioassay for cytokinins (Letham, 1971; Narain and Laloraya, 1974).

Does zeatin stimulate cotyledon growth only by increasing cell expansion, or is cell division also increased? In radish, cytokinesis is increased somewhat whether the cotyledons are cultured in light or darkness, but the principal effect is increased cell size (Gordon and Letham, 1975). The same is true for flax cotyledons incubated in light. Here the cell volumes are increased almost fivefold by benzyladenine.

Partly because of cotyledon growth, cytokinins stimulate germination of lettuce and a few other seeds that otherwise require light for germination. This dormancy-breaking effect is discussed in Section 19.3. Swelling of the cotyledons seems to help force the radicle through the restrictive seed coat or endosperm barriers, accomplishing the act of germination.

Since cotyledons in which growth is promoted by cytokinins become photosynthetic organs, we might ask whether

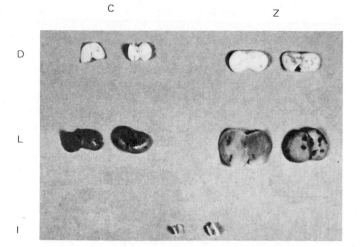

Figure 17-3 Stimulation of excised radish cotyledon enlargement by zeatin and light. Cotyledons at the bottom labeled I represent initial cotyledons excised from 2-day-old, dark-grown seedlings before growth studies. Excised cotyledons were incubated for 4 days on filter papers held in Petri dishes containing 2 mM potassium phosphate (pH 6.4) alone (controls, C) or also with 2.5 μM zeatin (Z). Cotyledons exposed to light (L) received continuous fluorescent radiation at an intensity near the photosynthetic light compensation point. Cotyledons incubated 4 days in darkness (D). (Unpublished results of A. K. Huff and C. W. Ross.)

true leaves also require cytokinins for growth. Definite promotive effects on intact dicot leaves of some species occur after repeated applications, but the effects are usually small and may arise indirectly through attraction of metabolites from other organs. If disks are cut from dicot leaves with a cork borer and kept moist, cytokinins stimulate expansion (through enhanced cell growth), again suggesting a normal function of cytokinins from some other organ in leaf growth.

Cytokinins Promote Chloroplast Development If we remove a plant part from its normal supply of cytokinins, we can test whether adding a cytokinin to it has any effect on chloroplast development. One example is callus tissue, the cells of which have the genetic potential to synthesize functional chloroplasts in light. If this tissue is kept in darkness and no cytokinin is provided, only etioplasts devoid of chlorophyll with no grana and almost no stroma lamellae are formed. Addition of a cytokinin promotes stroma lamellae formation, but still no chlorophyll and grana are formed in darkness. In light and with a cytokinin, stroma lamellae are produced, and grana and chlorophyll also appear. If no cytokinin is added to light-grown callus, development of stroma lamellae, grana, and chlorophyll is minimal (Stetler and Laetsch, 1965).

Another example of enhanced chloroplast development occurs in radish and squash cotyledons excised from dark-

grown seedlings. Prolamellar bodies (see Fig. 9-7) are present in the etioplasts, and from these develop some stroma lamellae in darkness if a cytokinin is added. When transferred to light, the previous cytokinin treatment speeds formation of grana and chlorophyll (Harvey et al., 1974). These facts suggest that cytokinins present in attached leaves are important for chloroplast development during normal leaf growth.

Effects of Cytokinins on Overall Plant Growth Normal growth no doubt requires free cytokinins, but the endogenous amounts are seldom limiting. Therefore, treatment of plants with cytokinins fails to stimulate overall growth. Suppose, however, that we stop delivery of cytokinins (and gibberellins) from roots to shoots by removing the roots. Can we now add cytokinins and gibberellins and restore shoot growth? In sunflower and pea, no, but in soybean seedlings, yes. In derooted soybean seedlings, in which the stem is at first only a hypocotyl, elongation during a 24-hour period is totally restored by GA$_3$ plus a synthetic cytokinin, both absorbed through the cut hypocotyl base. The main effect of cytokinins is to promote cell divisions in the uppermost, young hypocotyl cells, while the principal influence of GA$_3$ is to increase elongation of the growing cells below. These results suggest that cytokinins and gibberellins are required for normal stem elongation, but that other substances can also be limiting (Skene, 1975). The hypothesis that cytokinins are required for normal shoot growth is also supported by results with waterlogged plants exhibiting flooding injury. Delivery of both cytokinins and gibberellins from waterlogged roots is retarded, but in tomato plants benzyladenine alone relieves most of the flooding symptoms (chlorosis, adventitious root production, and inhibited stem elongation) when sprayed on the foliage (Railton and Reid, 1973).

Although elongation of stems of waterlogged plants or stems of some rootless plants can be promoted by cytokinins, *these hormones are potent inhibitors of elongation in certain excised stem and root sections.* In fact, kinetin, zeatin, and benzyladenine are as potent inhibitors of auxin-stimulated elongation in soybean hypocotyl sections as actinomycin D or cycloheximide (Vanderhoef and Stahl, 1975). Figure 17-4 shows the effect upon elongation of such sections when 4 μM kinetin was added to auxin-treated sections at various times. In all cases, elongation rates were eventually reduced to that of the control level without auxin (lowest curve).

In stems and roots, added cytokinins usually increase radial enlargement of the cells. This causes short, thick roots and stems that may weigh as much or more than untreated organs. As we shall describe later, ethylene also causes the same response in roots and stems. Could cytokinins, similar to *excesses* of auxins, inhibit elongation because they increase ethylene production? Cytokinins sometimes do increase ethylene production, but that is not the answer for stems, at least. They might change the orientation of newly synthesized cellulose microfibrils in the walls to a direction more nearly parallel with the cell's long axis (Apelbaum and Burg, 1971).

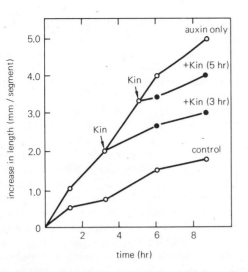

Figure 17-4 Inhibition of auxin-induced elongation in soybean hypocotyl sections by 4 μM kinetin added at different times of incubation (arrows). (From L. N. Vanderhoef et al., 1973, Physiologia Plantarum 29:22–27.)

This would inhibit longitudinal growth but allow radial expansion. Cytokinins stimulate vascular cambium activity in radish roots, and this, combined with inhibited elongation and promoted thickening of parenchyma cells, presumably causes a short, thick storage root. There is also evidence that cytokinins normally promote potato tuber formation, because treatment of the stolon sections with kinetin causes swelling at the ends and development of small tubers (Sections 21.7; 22.10; Palmer and Barker, 1973).

How Do Cytokinins Act? The variability of cytokinin effects suggests that they might have different mechanisms of action in different tissues, yet the simpler view is that a common primary effect is followed by numerous secondary effects dependent on the physiological state of the target cells. As with other hormones, amplification of the initial effect must occur, because cytokinins are present in such low concentrations (0.01 to 1 μM). Some effect of cytokinins on RNA and enzyme synthesis is suspected, partly because their effects are blocked by inhibitors of protein synthesis.

When cytokinins promote cytokinesis, there is an increase in nucleolar size in both flax cotyledons and tobacco pith. This suggests an increase in ribosomal RNA synthesis, a primary function of nucleoli, but this effect occurs only after several hours and is no doubt a secondary response. The hormones might stimulate formation of some protein needed for cytokinesis but not needed for DNA synthesis, since the latter process occurs unimpeded in callus tissue without added cytokinins. You'll remember that the giant, auxin-treated, pith-callus cells have high ploidy levels.

Numerous enzymes increase in activity after cytokinin treatment, including photosynthetic enzymes characteristic of well-developed plastids. In mature leaves where senescence is delayed, the rate of protein and RNA synthesis increases, but the less rapid loss of protein and RNA seems mainly due to a decreased rate of hydrolysis by proteases and ribonucleases (Kende, 1971). All known changes in enzyme patterns require several hours.

No wall-loosening effects caused by cytokinins have been found, and in soybean hypocotyl sections in which elongation is inhibited, the walls become stiffer. If wall stiffening is a general effect, how is cell size increased in excised, fat-rich cotyledons and certain leaves? Solute production must be increased, but what solutes are involved? Data in Fig. 17-5 suggest that reducing sugars (glucose, fructose, or both) might be responsible. Here radish cotyledons from 2-day-old, dark-grown seedlings were excised and incubated in the presence or absence of zeatin for various periods in weak light. There is a high correlation between growth measured by fresh weight increase and the *amount* of reducing sugars present. When water uptake is prevented by adding mannitol as an osmoticum to the growth medium, enhanced sugar production by zeatin still occurs (Huff and Ross, 1975). This shows that

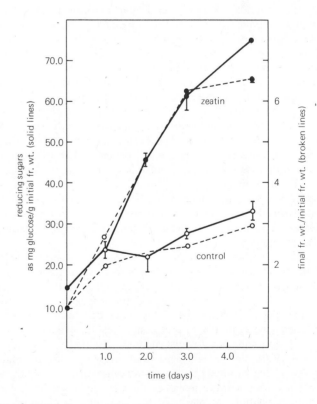

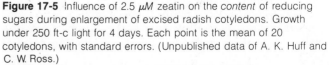

Figure 17-5 Influence of 2.5 μM zeatin on the *content* of reducing sugars during enlargement of excised radish cotyledons. Growth under 250 ft-c light for 4 days. Each point is the mean of 20 cotyledons, with standard errors. (Unpublished data of A. K. Huff and C. W. Ross.)

growth is not required for sugar production, so the correlation presumably means that cytokinins promote growth partly because they increase the number of sugar molecules in the cells. The *concentration* of sugars remains constant during growth, as expected, because water uptake balances solute production. The sugars probably arise from fat reserves, because oleosomes are abundant but starch grains are quite sparse.

More recent discoveries of considerable potential significance are that cytokinins bind tightly and rather specifically to certain ribosomal proteins (Fox and Erion, 1975; Takegami and Yoshida, 1975). This suggests that the primary site of action is on protein synthesis at the ribosome, and that changing enzyme patterns result from this effect. Since growth and development depend on enzymes, the kind of response would depend partly on the nature of ribosomal protein receptors in the target cells.

17.2 Ethylene, a Volatile Hormone

The ability of certain gases to stimulate fruit ripening has been known for many years. Even the ancient Chinese knew that their picked fruits would ripen more quickly in a room with burning incense. In 1910, an annual report by H. H. Cousins to the Jamaican Agricultural Department mentioned that oranges should not be stored with bananas on ships, because some emanation from the oranges caused the bananas to ripen prematurely. This was apparently the first suggestion that fruits release a gas that stimulates ripening, but it was not until 1934 that R. Gane proved that ethylene is synthesized by plants and is responsible for faster ripening.

Another historical practice implicating still a different role for ethylene was the building of bonfires by Puerto Rican pineapple growers and Philippine mango growers near their crops. These farmers apparently believed that the smoke helped to initiate and synchronize flowering. Ethylene causes these effects in both species, so it is almost surely the most active smoke component. Stimulation of fruit ripening is a widespread phenomenon, while promoted flowering appears restricted to mangos and most bromeliad species, including the pineapple.

Still another effect of gases was reported as early as 1864. Before the use of electric lights, streets were lighted with illuminating gas. Sometimes the gas pipes leaked, and in certain German cities this caused the leaves to fall off the shade trees. Ethylene stimulates senescence and abscission of leaves, so again it was presumably responsible.

A Russian physiologist named Dimitry N. Neljubow (1876–1926) first established that ethylene affects plant growth. He identified ethylene in illuminating gas and showed that it causes a triple response on pea seedlings: It inhibits stem elongation, increases stem thickening, and causes a horizontal growth habit. Furthermore, it inhibits leaf expansion and retards normal opening of the epicotyl hook. We shall discuss these and other effects of ethylene (Abeles, 1973) in more detail.

Ethylene Synthesis Ethylene production by various organisms can be readily detected by gas chromatography, since the molecule escapes readily into the atmosphere. Only a few bacteria reportedly produce ethylene, but perhaps all higher plants do. No algae are known to synthesize it, and it generally has little influence on their growth. In seedlings, the shoot apex is an important site of production. This might result from the high amounts of IAA there, because auxins stimulate ethylene formation. Nodes of dicot seedling stems produce much more ethylene than internodes do when equal tissue weights are compared. Roots release relatively small amounts, but again auxin treatment causes the rate to rise. Production in leaves generally rises slowly until the leaves become senescent and abscise. Flowers also synthesize ethylene, and this gas often causes their senescence and abscission (Section 15.7). In many fruits, little ethylene is produced until just before the respiratory climacteric signaling the onset of ripening, when the content of this gas in the intercellular air spaces rises dramatically from almost undetectable amounts to about 0.1 to 1 ppm (μl/l). These concentrations stimulate ripening of many fleshy and nonfleshy fruits if the fruits are sufficiently developed to be susceptible (Coombe, 1976; McGlasson, 1970). Sections of ripe apple or pear fruits or even apple peelings are often used in laboratory demonstrations as an ethylene source.

Strangely, numerous mechanical effects such as gently rubbing a stem or leaf, increased pressure, pathogenic microorganisms, and insects increase ethylene production. As discussed in Chapter 18, this ethylene might contribute to stem shortening and thickening (an example of thigmomorphogenesis) in the affected plants. More ethylene also accumulates in plants grown in waterlogged than in well-aerated soils (Jackson and Campbell, 1976), but this seems to result mainly from a slower diffusion of the gas out of the roots through water (Kawase, 1976). Egyptian civilizations unknowingly took advantage of increased ethylene production resulting from injury by gashing immature sycamore figs to stimulate ripening. When figs only about 16 days old are gashed, they ripen within as little as 4 days.

How is this ethylene synthesized? The sulfur-containing amino acid methionine is the important ethylene source in plants (Yang, 1974; Lieberman, 1975). When methionine* labeled in different carbon atoms with ^{14}C is converted to ethylene, CO_2, NH_3, HCOOH, and methyl thioadenosine are released:

$$\overset{\text{adenosine}}{\underset{|}{}}$$
$$^5CH_3 - S - {}^4CH_2 - {}^3CH_2 - {}^2CHNH_2 - {}^1COOH \rightarrow$$
$$^1CO_2 + NH_3 + H^2COOH + {}^5CH_3 - S\text{-adenosine} +$$
$$^3CH_2 = {}^4CH_2$$

*Further research by D. O. Adams and S. F. Yang (Plant physiology, vol. 60, 1977: 892–896) shows that the actual precursor of ethylene is an activated form of methionine, S-adenosylmethionine, in which the sulfur atom is attached to C-5 of the ribose moiety in adenosine. Thus, R17-1 shows methionine in this form and methylthioadenosine as the sulfur-containing product.

Both the NH_3 and the methylthio group can be metabolized back into methionine, thereby conserving nitrogen and sulfur and allowing ethylene synthesis to continue.

Auxins and physical factors that stimulate ethylene production might enhance formation of an unidentified enzyme needed for ethylene synthesis by R17-1, or they might increase the availability of methionine. Methionine might become more available if permeability of the tonoplast were increased, allowing stored methionine to leak out of the vacuole into the cytoplasm where contact is made with the enzymes. A similar leakage might explain the burst of ethylene production in many fruits just before ripening and in morning glory flower petals during senescence (see Fig. 15-11). A curious autocatalytic ability of ethylene to stimulate its own formation in sensitive cells accompanies senescence of petals and fruits of many species. This autocatalysis may result from increased permeability of the tonoplast to methionine caused by small amounts of ethylene (Kende and Hanson, 1976). Whatever its cause, autocatalysis probably explains the ability of one rotten apple in a barrel to cause overripening of the others.

Ethylene Effects on Flowering The induction of flowering in mangos and bromeliads by ethylene is unusual, because the gas inhibits flowering in most species. Nevertheless, the indirect use of ethylene to promote flowering has been studied in the pineapple industry. Fields have been sprayed with an auxin such as NAA to stimulate ethylene synthesis in the plants. Pineapple fields flower faster and, more importantly, mature fruits appear uniformly, the goal being to allow a one-harvest mechanical operation. Nevertheless, most pineapples in Hawaii are still hand picked.

An ethylene-releasing substance called **Ethrel** (trade name) or **ethephon** (common name) is commercially available.

It is 2-chlorethylphosphonic acid ($Cl—CH_2—CH_2—PO_3H_2$), which rapidly breaks down in water at neutral or alkaline pH values to form ethylene, Cl^- and $H_2PO_4^-$. Since Ethrel can be translocated throughout the plant, this compound also has promise for use with pineapples and is becoming widely used in various aspects of horticulture such as fruit production. It is widely employed in research studies on ethylene.

Ethylene Causes Epinasty of Petioles Normal angles of leaves and branches in dicots and gymnosperms are controlled by the relative elongation rates of the upper and lower sides. If growth is faster on the lower side, the organ bends up, while faster growth above causes downward bending. Exposure to ethylene causes faster growth on the upper than the lower side, resulting in **epinasty** (Greek *epi*, "above"). Epinasty is also caused by ethylene produced when *excess* auxins are applied. Figure 17-6 shows this effect on an NAA-treated cocklebur plant. Herbicidal effects of auxins such as 2,4-D, 2,4,5-T, and MCPA are also accompanied by epinastic effects. Nevertheless, many monocots and some dicots show no epinastic effects from excess auxins or ethylene (Crocker et al., 1932). How ethylene causes epinasty is unknown (Palmer, 1976).

Ethylene Inhibits Elongation of Stems, Roots, and Leaves Although epinasty of petioles generally seems to be caused by ethylene-promoted growth of the upper petiole cells, this gas usually inhibits elongation of stems, roots, and leaves, especially in dicots. Retarded stem elongation in peas forms part of the previously mentioned triple response. Roots swell in the elongating regions much as stems do, and both effects

Figure 17-6 Epinasty in cocklebur. Plant on the left is an untreated control. Plant on the right was dipped into a solution of 1 m*M* NAA about two days before the photograph was taken. Epinasty in response to ethylene is very similar. (Photo by F. B. Salisbury.)

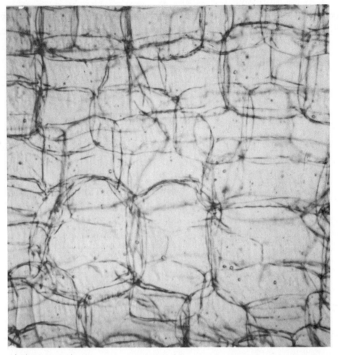

ethylene-treated

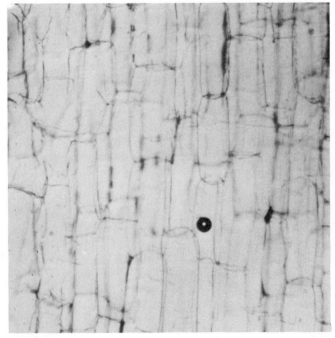

control

Figure 17-7 Effects of ethylene on cell elongation and radial expansion in the upper internode of pea seedlings. Plants were grown 4 days in darkness, then gassed with 0.5 μl/l ethylene or left as controls. Cell sections were made 24 hours after ethylene treatments began. (From R. N. Stewart, M. Lieberman, and A. T. Kunishi, 1974, Plant Physiology 54:1–5.)

result from inhibited elongation and increased radial growth of the individual cells (Fig. 17-7). Altered cell shapes are apparently caused by a more longitudinal orientation of cellulose microfibrils being deposited in the walls, preventing expansion parallel to these microfibrils but allowing expansion perpendicular to them (Apelbaum and Burg, 1971). Epidermal cells of roots and stems also enlarge radially, and since they are not restrained on the outside, they bulge and form a profusion of hairlike structures. Lateral root production is simultaneously inhibited.

Root and stem thickening in dicots caused by ethylene are of survival value for seedlings emerging from the soil. In such species, a hook in the epicotyl or the hypocotyl is formed shortly after germination in response to endogenous ethylene; this hook pushes up through the soil, making a hole through which the cotyledons or young leaves can be safely drawn. If the soil is excessively compact, the hook and primary root become unusually thick, probably because ethylene is synthesized faster when the compacted cells are subjected to increased mechanical pressure, and because the ethylene escapes less rapidly in compacted than in loose soils. This thickness increases the strengths of both stem and root, allowing them to push through the compacted soil. Growth is slow, however, because of retarded elongation.

Another situation in which retarded ethylene diffusion from the cells causes abnormal growth occurs in waterlogged soils. In dicots, the lower leaves become chlorotic, leaf epinasty occurs, stem elongation is slowed, adventitious roots form on the stem, and eventually death of the plant occurs. These symptoms occur in part because ethylene cannot escape from the waterlogged roots as it normally would, because of the water barrier. Some ethylene then apparently moves through the intercellular air space system into the shoot, producing the symptoms. As mentioned previously, delivery of cytokinins and gibberellins from root tips to the shoot is also retarded in waterlogged plants, and cytokinins added to the shoot overcome most of the symptoms, at least in tomato.

Relation of Ethylene to Auxin Responses The ability of IAA and all synthetic auxins to stimulate ethylene production raises the question of whether many auxin effects are really caused by ethylene. Growth stimulation by auxins and enhanced activity of the vascular cambium are responses apparently independent of ethylene. Several other auxin responses already discussed that are caused mainly or partly by ethylene include epinasty, inhibition of stem, root, and leaf elongation, flower induction in bromeliads and mangos, inhibition of epicotyl or hypocotyl hook opening in dicot seedlings, increased percentage of female flowers in dioecious plants, and perhaps apical dominance (Abeles, 1973; Burg, 1973). Also, release of auxin by germinating pollen grains stimulates ethylene production in the stigma, which contributes to senescence of the flower. As described in Section 17.4, abscission of leaves, flowers, and fruits involves interactions between auxins, ethylene, cytokinins, and abscisic acid. Nevertheless, growth promotion, initial stages of adventitious root production, and many other effects of auxins appear to be independent of ethylene production. Only when the auxin concentration becomes relatively high is ethylene production great enough to duplicate certain auxin effects.

How Does Ethylene Act? Much evidence indicates that ethylene binds temporarily to a metal (probably Cu) present in some unidentified enzyme or other protein, perhaps a membrane protein. This binding does not change ethylene, only the plant. The unknown metal also appears to recognize both CO and CO_2. Carbon monoxide causes some of the same effects as ethylene, provided the concentrations are too low to inhibit cytochrome oxidase, but CO_2 is a competitive inhibitor of ethylene action in many responses.

Because of this competition, CO_2 is often used to prevent overripening of picked fruits. Such fruits are stored in an airtight room or container in which the gas composition is controlled. An ideal atmosphere for many fruits contains 5 to 10 percent CO_2, 1 to 3 percent O_2, and no ethylene. Removal of some oxygen is important, because this slows ethylene synthesis. However, if too much O_2 is removed, glycolysis is stimulated by the Pasteur effect, causing excess sugar breakdown. Another technique useful in fruit storage is to evacuate the container partially, thereby removing O_2 and ethylene from the tissues into the atmosphere. Such techniques have also been used to preserve certain valuable flowers.

Many ethylene effects are accompanied by increased synthesis of enzymes, the kind of enzyme depending on the target tissue (Varner and Ho, 1976). When ethylene stimulates leaf abscission, cellulase and other cell wall degrading enzymes appear in the abscission layer (see Section 17.4). When fruit ripening occurs, necessary enzymes are produced in the fruit cells. When cells are injured, phenylalanine ammonia lyase (see Section 14.5) appears, which is an important enzyme in formation of phenolic compounds thought to be involved in wound healing. Nevertheless, we still do not know how ethylene stimulates enzyme synthesis.

17.3 Growth Inhibitors and Miscellaneous Regulating Compounds

It might seem that plants could regulate all of their essential processes and coordinate growth of various organs by synthesizing and transporting promotive hormones, then degrading or altering them when the response is sufficient. Nevertheless, most processes are influenced by another regulatory mechanism that involves growth inhibitors. Plants have many different types of inhibitors. They include abscisic acid (ABA), which seems to be the most important for internal regulation (Milborrow, 1974b), plus numerous phenolic compounds and phenolic derivatives such as coumarins, alkaloids, and miscellaneous compounds that have additional roles in a plant's interaction with other organisms, as discussed in Chapter 14.

Evidence that bud dormancy might be due to growth inhibitors was first obtained by Torsten Hemberg of Sweden, in 1949. He found that dormant buds of potato tubers and ash trees contained potent growth inhibitors, but the levels of these declined when dormancy was broken. In the 1950s, investigation of growth activity in compounds separated from numerous plant extracts showed a pronounced inhibitory zone on paper chromatograms. The substances in this zone were named the β-**inhibitor.** We now know that a principal inhibitory compound in the β-inhibitor zone is ABA, yet in dormant potato buds and certain other buds some phenolic compound is also present, and there is evidence for at least one other kind of inhibitor in this zone.

17.4 Abscisic Acid

Abscisic acid (ABA) was first identified in 1965 by P. F. Wareing and his colleagues in Wales, who were investigating bud dormancy in *Acer pseudoplatanus*, and by F. T. Addicott and coworkers in California, who were studying compounds responsible for abscission of cotton fruits. They agreed to name the compound abscisic acid, but this seems to have been a somewhat unfortunate name, because ABA is apparently much more important in causing dormancy of both buds and seeds and in stomatal control than in abscission.

ABA (Fig. 17-8) is a 15-carbon sesquiterpene synthesized from farnesyl pyrophosphate, which is derived from carbons in three isopentenyl pyrophosphate units (see Fig. 16-11). Thus, early reactions in the synthesis of ABA are identical to those of gibberellins, sterols, carotenoids, and other isoprenoid compounds mentioned in Chapter 14. Synthesis of ABA occurs in chloroplasts, so leaves, stems, and green fruits are important sites of ABA formation (Milborrow, 1974b). Wheat endosperm and roots also synthesize ABA, so this hormone might arise in many tissues and organs, perhaps always in plastids or proplastids. Certain conditions of stress stimulate ABA synthesis in leaves, including drought, mineral starvation, and flooding injury (Vaadia, 1976). There is evidence that ABA increases the plant's resistance to such stresses. The well-documented role of ABA accumulation in causing stomatal closure under drought conditions was discussed in Chapter 3. More recent studies indicate that ABA also accumulates in water-deficient roots and decreases the resistance of the roots to water entry, further protecting the leaves against desiccation (Walton et al., 1976).

Transport of ABA occurs in both xylem and phloem and probably also in living cells outside vascular bundles. Mechanisms for complete degradation of ABA have not been studied, but it can be removed from the cells by conversion to three principal derivatives that are essentially inactive, dihydro-**phaseic acid** and **phaseic acid** (oxidation products), and an ABA **glucose ester** (Fig. 17-8).

All angiosperms and gymnosperms investigated contain ABA, and so do at least one fern, a horsetail, and a moss. Concentrations are in the range of 0.001 to 1 μM. Algae and

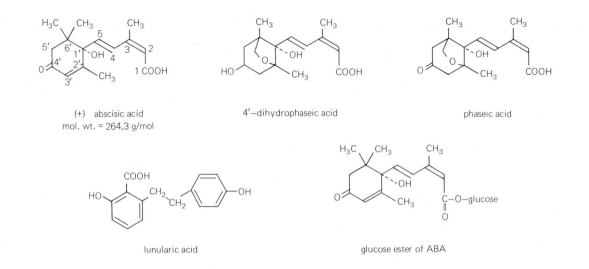

(+) abscisic acid
mol. wt. = 264.3 g/mol

4'–dihydrophaseic acid

phaseic acid

lunularic acid

glucose ester of ABA

Figure 17-8 Structure of abscisic acid (ABA) and some related compounds. ABA (upper left) has one asymmetric carbon atom (1' in the ring). The form synthesized by plants is the dextrorotatory (+) product shown, but commercial ABA is a racemic (±) mixture. Both forms are biologically active.

liverworts seem to replace ABA with a somewhat different growth inhibitor, **lunularic acid** (Fig. 17-8) (Pryce, 1972; Milborrow, 1974b). Fungi and bacteria seem to lack both ABA and lunularic acid.

ABA and Dormancy Although rare instances in which ABA stimulates growth are known, the almost universal response of cells is growth inhibition. Wareing's early results leading to discovery of ABA indicated that levels of the β-inhibitor complex, in which ABA is the most active inhibitor, increased considerably in leaves and buds when dormancy appeared during the shorter daylengths of late summer. Wareing and his colleagues later found that when ABA was added to leaves and growing buds of various species, it caused the buds to stop elongating and simulate normal dormancy (Fig. 17-9). The normal dormant condition involves more than growth inhibition; it is preceded by development of bud scales and other characteristics of winter buds. They reported that ABA caused all of these characteristics on birch (*Betula pubescens*), sycamore maple (*Acer pseudoplatanus*), and black currant (*Ribes nigrum*). No other compound is known to have these effects.

These results suggested that ABA represents the primary bud dormancy hormone, synthesized in leaves that detect daylength and transported through phloem to buds that respond. Several studies also indicate that ABA levels decrease in the buds before spring growth. Numerous other investigations in which ABA levels rather than total β-inhibitor activity were measured support a major role for ABA in bud dormancy. Results of one such experiment with beech (*Fagus sylvatica*), in which only the buds were analyzed, is shown in Fig. 17-10. Note that the buildup of free ABA beginning in early September correlates well with the time of apparent deep winter dormancy, and the subsequent decrease in free ABA occurred before the spring growth. The decrease was accompanied by an increase in the bound form.

Contrary results show that ABA added to leaves of birch

and alder does not induce bud dormancy and that extremely little ^{14}C-ABA moves into the buds (Hocking and Hillman, 1975). In a northern race of red maple from Massachusetts, ABA stops stem elongation, but the terminal buds do not resemble those normally caused by short days and cold temperatures (Perry and Hellmers, 1973). Furthermore, short-day treatments that induce dormancy cause no rise in bud ABA levels in red maple, birch, or sycamore maple. It is now clear that ABA is not the only hormone influencing bud dormancy in midsummer and bud growth in early spring, but that a buildup of other inhibitors and disappearance of growth promoters such as gibberellins and cytokinins are also involved (Alvim et al., 1976; see also Fig. 21-10).

Figure 17-9 Formation of a dormant terminal bud of black currant after application of ABA to the shoot apex. (From H. M. M. ElAntably, P. F. Wareing, and J. Hillman, 1967, Planta 73:74–90.)

To understand seed dormancy may prove equally difficult. ABA is a potent inhibitor of seed germination. Furthermore, many dormant seeds contain ABA. In some species, the ABA levels decrease when exposed to cold temperatures that break dormancy, but in others, no decrease occurs. In studies with both seeds and buds, it is probably essential to analyze the tissues in which growth is restricted (elongating regions of the radicle in seeds and of the subapical meristem in buds), for it is here where changes in ABA, other growth inhibitors, and growth-promoting hormones have their dormancy effects. This more difficult task has not yet been accomplished.

Further discussion of dormancy-regulating compounds in relation to seeds and to daylength effects on buds is found in Chapter 21; for a probable role of ABA in geotropism of roots, see Chapter 18.

How Does ABA Act? ABA seems to have three major effects: one is on the plasma membrane, another is inhibition of RNA synthesis, and the third is inhibition of protein synthesis. The influence on barley root membranes is to make them more positively charged, increasing the tendency with which excised root tips stick to negatively charged glass surfaces. This effect is no doubt involved in the rapid loss of K^+ from guard cells, and perhaps in the ability of ABA to inhibit within 5 minutes or less auxin-induced growth. Interference with synthesis of proteins and other enzymes could explain long-term effects on growth and development, including dormancy of seeds and buds and the inhibition of gibberellin-promoted hydrolase activity in cereal grain seeds. The mechanism of both effects is unknown.

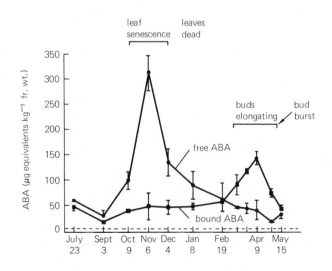

Figure 17-10 Seasonal changes in free and bound ABA in beech buds in England. Dotted line represents free ABA level in leaves measured in July. The bound form is presumably the glucose ester. (From S. T. C. Wright, 1975, Journal of Experimental Botany 26:161–174.)

17.5 Hormones in Senescence and Abscission

Processes of deterioration that accompany aging and that lead to death of an organ or organism are called **senescence** (Beevers, 1976). Although meristems do not senesce and might be potentially immortal, all differentiated cells produced from meristems have restricted lives. Senescence therefore occurs in all nonmeristematic cells, but at different times. In conifers, many needles remain active for several years before they die and abscise. In deciduous trees and shrubs, the leaves die each year, but again, the stem and root systems remain alive for several years. In perennial grasses and herbs such as alfalfa, the aboveground system dies each year, but the crown and roots remain largely viable. In herbaceous annuals, leaf senescence progresses from old to young leaves, followed by death of stem and roots after flowering. Only the seed survives.

What causes senescence? Chemical analyses show that leaf senescence is accompanied by early losses in chlorophyll, RNA, and proteins, including many enzymes. Since these and other cellular constituents are constantly being synthesized and degraded, loss could result from inhibited synthesis or stimulated breakdown, or both. Inhibited synthesis is expected when nutrients normally arriving in an organ are diverted elsewhere, as, for example, when flowering and fruit formation occur. One theory for leaf senescence is, therefore, that flower and fruit development cause a **competition for nutrients.** In Section 15.9, we discussed competition between vegetative and reproductive organs for nutrients essential for growth, and showed in Fig. 15-14 how removal of all flowers postpones leaf senescence in annual soybeans.

Nutrient competition is not the entire explanation of senescence, even for soybeans, because the young flowers could not possibly divert enough nutrients to cause death of the leaves. In cocklebur plants, short-day and long-night conditions induce flowering and leaf senescence, but even if all flower buds are removed, leaf senescence still occurs. Furthermore, development of male flowers on staminate spinach plants induces senescence as effectively as development of both flowers and fruits on female plants, yet staminate flowers divert much less nutrients than do fruits and seeds (Leopold and Kriedemann, 1975). Another example of senescence already discussed is the degradation of food reserves and loss of integrity in food storage cells of seeds. We mentioned that endosperm degradation in cereal grains is stimulated by gibberellins moving from the embryo. Taken together, these results indicate that *senescence is also hormonally controlled.* It appears as though organs or tissues acting as nutrient sinks send hormones to other organs, causing them to release some of their nutrients.

If hormones are involved, we should eventually be able to identify them and explain how they act. For gibberellins and cereal grain seeds, we are approaching both answers. We also mentioned that cytokinins can delay leaf senescence in many angiosperms, and that gibberellins are also effective in some.

For cytokinins, the main effect is to delay breakdown of RNA and proteins, in part by inhibiting synthesis of ribonucleases and proteases. How gibberellins retard leaf senescence is not known. Auxins also delay leaf senescence in some species, but this is less easy to demonstrate than for gibberellins and cytokinins arriving from roots, because leaves normally seem to synthesize enough IAA. Therefore, adding more auxin, even to an excised leaf, usually does not delay senescence.

Contrary to effects of the other hormones, ethylene and ABA stimulate senescence. In fruits, the effect of ethylene is manifested by rapid ripening followed by abscission; in flowers, the result is fading, transport of nutrients, withering, and then abscission; and in leaves, it is loss of chlorophyll, RNA, and protein, and transport of nutrients, followed by abscission. The effect of ethylene is much more dramatic than that of ABA. In fact, the ABA influence might actually be caused indirectly by ethylene, because relatively large amounts of ABA must usually be applied. Such amounts usually stimulate ethylene production, just as do excesses of auxins and cytokinins. To what extent natural rises in the ABA levels contribute to senescence and abscission is still uncertain.

What advantage is there to abscission of senescent leaves, flowers, and fruits? For fruits, the importance in perpetuation of the species is obvious, because fruits contain seeds. For flowers, we suspect that the reasons involve removal of a useless organ that might act as a potential infection source and which, in some species, would shade new leaves the next growing season. We mentioned in Section 15.7 that senescence of flowers is usually accompanied by hydrolysis of RNA and protein. The products are converted into mobile amides and amino acids. Many other large molecules (except for those in cell walls) are also degraded into smaller, more readily translocatable forms in which nutrients are conserved by storage in other parts of the plant. For leaves that abscise, a similar extensive salvage of nutrients by mobilization also precedes abscission. This nutrient economy allows forest trees to grow for several years after a single fertilization with nitrogen or phosphorus and explains their ability to survive on infertile soils. Such senescent leaves could not withstand cold winters and would shade new leaves the next spring, so their loss, preceded by nutrient salvage, apparently increases overall survival and productivity of individual perennial plants.

In most species, abscission of leaves, flowers, or fruits is preceded by formation of an **abscission layer** or **zone** at the base of the organ involved (Kozlowski, 1973). In leaves, this zone is formed across the petiole near its junction with the stem (Fig. 17-11). In many compound leaves, each leaflet also forms an abscission zone. The abscission zone consists of one or more layers of thin-walled parenchyma cells resulting from anticlinal divisions across the petiole (except in the vascular bundle). In some species, these cells are formed even before the leaf is mature. Just before abscission, the middle lamella between certain cells in the *distal region* (that farthest from the stem) of the abscission layer is often digested. This digestion involves synthesis of polysaccharide-hydrolyzing enzymes, most importantly cellulase and pectinases, and their secretion

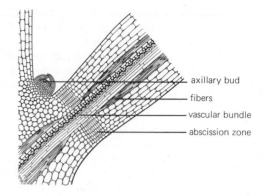

Figure 17-11 The abscission layer. (From F. T. Addicott, 1965, in W. Ruhland, ed., Encyclopedia of plant physiology, vol. 15, part 2, Springer-Verlag, Berlin.)

axillary bud
fibers
vascular bundle
abscission zone

from the cytoplasm into the wall. Formation of these enzymes is accompanied by a rapid rise in respiration in cells of the *proximal* part of the abscission zone (those cells of the zone close to the stem). This rise is similar to that occurring in climacteric fruits and also involves increases in polyribosomes characteristic of cells actively synthesizing proteins. Furthermore, one or more layers of these proximal cells increases in size (both length and diameter), while the cells of the abscission zone distal to the breaking point do not. These wall digestion processes accompanied by pressures resulting from unequal growth in expanding proximal and senescent distal cells of the zone apparently cause a break between them.

Wound healing in cells proximal to the break point involves formation of a corky layer that protects the plant from pathogen invasion and excess water loss. Compounds synthesized during healing include suberin and lignin.

Several environmental factors stimulate abscission, especially drought and nitrogen deficiency, but these first stimulate senescence. This is important, because even though abscission is an active process in cells proximal to the break, the organ being shed first becomes senescent. What role do hormones play in abscission? Our present ideas about leaves are summarized as follows: IAA is synthesized in growing leaves and retards senescence and abscission until its level and rate of transport into the petiole decrease too much when the leaf matures. Cytokinins and perhaps also gibberellins arriving from the roots also delay senescence and abscission for a while. Finally, however, degradation processes overcome synthetic processes in the leaf, partly due to competition with growing organs for nutrients, and senescence occurs. Now another unidentified compound, which is called **senescence factor** (Osborne et al., 1972; Chang and Jacobs, 1973) and is not ABA, moves from the senescent leaf blade or petiole into the abscission zone. Senescence factor stimulates ethylene production in the abscission zone, and ethylene stimulates growth of the proximal cells and formation of cell wall degrading enzymes in them. Abscission then occurs. No clear-cut role for ABA in this has been established, although increases in ABA may contribute to the senescence preceding abscission.

17.6 Hormone Interactions

You should now realize how far we have departed from Julius Sachs's early idea that plants might have specific hormones responsible for growth or formation of a particular organ. Clearly, the rule is that most physiological processes require interactions among several hormones, and a single hormone has several functions. Furthermore, each process generally depends on the kind of cells involved (target organ) and on the species. Apparently, different species sometimes use different hormones or rely on different interactions among them to accomplish their various functions. Khan (1975) reviewed several such interactions.

Besides these problems, other hormones await discovery. There is evidence for a hormone that moves from leaves to underground stems, where it stimulates tuber and bulb formation (Chapter 21), for a flower-inducing hormone (Chapter 22), for a hormone that guides the pollen tube into the embryo sac where fertilization is accomplished, and for such others as senescence factor and rhizocalines. It would obviously be simpler for us to interpret the results if such hormones do not prove to be varying mixtures of known hormones.

18

Differentiation and Differential Growth

We have come a long way in building a conceptual image of a growing plant. In Chapters 8 through 14, we emphasized the many biochemical reactions occurring in plants, each generally catalyzed by a different enzyme. Then, in Chapters 15, 16, and 17, we examined growth and development, emphasizing anatomical and hormonal aspects. We explained how hormones can control growth by enhancing or inhibiting production or action of certain enzymes, even though most enzymes respond little or only quite indirectly to changes in hormone levels. In this and subsequent chapters, we shall describe how the environment can influence growth and development, frequently by modifying hormone levels. Often, a hormone may act as an environmental **transducer,** converting some signal such as light energy or a temperature change into a metabolic response. In other examples, enzymes themselves or other proteins act as transducers, as when light controls the activity of certain Calvin cycle enzymes (see Section 10.4) or when the phytochrome system (see Chapter 19) is activated by red light.

Regardless of its cause, any metabolic effect involves a change in the amount of some enzyme or in its ability to function. Enzyme synthesis depends upon genes, according to the following scheme:

$$DNA \xrightarrow{\text{transcription}} \text{messenger RNA} \xrightarrow{\text{translation}} \text{enzymes}$$

Our first question is how enzymes can control the internal structures and overall shapes of cells, tissues, organs, and organisms. Much of our answer involves speculation, but there are some helpful facts.

18.1 Some Principles of Differentiation

Spontaneous Self-Assembly One principle of differentiation is that *once synthesized by enzymes, many large molecules are arranged into fairly stable three-dimensional structures by spontaneous self-assembly* (Bouck and Brown, 1976). For example, after a polypeptide is synthesized during translation, it spontaneously folds into a structure in which it and the water molecules with which it associates have the lowest free energy

under conditions existing in the cell. No other energy or direction from the cell is needed for this structure to form. If two or more polypeptides are present in a completed protein, these also combine spontaneously to form a stable structure. Ribosomes and certain viruses containing both RNA and numerous protein molecules can be totally dissociated *in vitro,* but if these subunits are again mixed under favorable conditions, each apparently finds its original site, so that active ribosomes and infectious viruses again result. Isolated microtubules can be dissociated into their subunit tubulin proteins by being chilled on ice, then reformed spontaneously in the proper medium by simply being rewarmed to a physiological temperature. If the lipids, ions, and proteins in certain membranes are totally separated, then remixed under favorable conditions, a membrane with properties similar to the original results.

Different cell types in tissues and organs of several animals (e.g., sponges, snails, sea urchins, insects, amphibians, birds, and mammals) can be separated from each other by various techniques, after which they can be brought together to reaggregate, often forming tissues and even organs with complex architectures remarkably like the originals. Such experiments with plant cells have so far shown only minimal success. The plasmodesmata that normally connect the cytoplasms of adjacent plant cells must be partially responsible; when these are broken, plant cells apparently will not reaggregate. Furthermore, membrane surface characteristics must be responsible for reaggregation in animal cells, and the plant cell wall may prevent membranes of separate cells from contacting and thus recognizing each other.

In all these examples, no covalent bonds are formed; the message for arrangements is in the individual component molecules or membranes. The message depends upon the kinds of atoms and their three-dimensional arrangements in each component, as shown, for example, by the specific hydrogen-bonding-dependent association of cellulose "molecules" with each other (see Fig. 10-15). When lignins are synthesized from aromatic alcohol-free radicals, the arrangements of these alcohols depend on the position of the free radical in each, and this is partly controlled by enzymes (see Section 14.6). Therefore, although the kind of lignin product is extremely variable, the association of subunits (here covalently bonded) is again spontaneous.

The implication of self-assembly for production of specific cell architectures is that the cell need only control the kinds of enzymes (i.e., their amino acid sequences) it produces and their activities, as affected by inhibitory and promotive substances. While we believe this is true, we are still a long way from understanding all the complex relations between enzyme products and cell structure.

Higher levels of complexity are the associations of cells that form tissues, organs, and the organism. Let us ask why an organ has a particular shape, why stems are usually long and cylindrical and leaves are often wide and thin. Two explanations come to mind: *the plane of cell division and the direction of cell expansion.* We don't know what causes the cell plate to form in the position it does, except that microtubules are involved. We do know something about why some cells become long and narrow rather than more isodiametric. This depends on the orientation of cellulose microfibrils in the walls, but again microtubule orientation is involved, and we know nothing about what controls it (see Section 15.10). Nevertheless, shapes of organs must be functions of enzyme products.

Gene Activation and Repression Now another important biological question arises: If genes determine the kinds of enzymes a cell produces and if mitosis causes all cells developing from a zygote to have the same kinds of genes, why are all cells in the organism not identical? That is, how can differentiation occur? Clearly, not only are the kinds of enzymes important, but so are their relative rates of synthesis and degradation and how effectively they function. *Although genes control the kinds of enzymes a cell can make, the environment determines whether or not they function effectively.* Two extreme examples of control over enzyme functions may be mentioned: Nitrate reductase in roots will not reduce nitrate if no NO_3^- is available, and amylases cannot function in leaves if the plants have been kept dark long enough to deplete totally their starch reserves. The environment also directly influences the speed of reactions controlled by enzymes. Temperature provides a convenient example. Each enzyme has a different temperature optimum for maximum activity, so even if substrates are available for all enzymes, we predict a different output of enzyme products at different temperatures, even though this difference may be small. Some enzymes are also affected directly or indirectly by light, and we shall discuss the photomorphogenetic effects of this in Chapter 19.

Not only does the environment influence the rates at which enzymes function, it also influences transcription and translation to control the kinds of enzymes that are formed. The environment influences transcription by controlling the kinds of genes that are active at any given time. A corollary of this statement is that not all genes are always active. In fact, at any time most genes are not functioning in RNA synthesis. That is, only a fraction of the enzymes which could be synthesized are actually being formed in a cell. Realizing this, it is much easier to understand how cells differentiate. In a sense, genes are analogous to the keys of a piano. If individual keys are played only at the proper times, a beautiful piece of music may result; if genes function only at the proper times in the correct cells, metabolism is properly coordinated, and the normal cell structure is produced.

How could the environment determine which genes are active? It could act as an off-switch or an on-switch, in which the affected genes would either become totally inactive or fully active. Alternatively, it could just modulate gene action. Most environmental influences are apparently modulatory, but a few seem to approach all-or-nothing responses. An example of gene inactivation occurs in pea buds (shoot apices). James Bonner and his colleagues showed in the early 1960s that these organs do not synthesize detectable amounts of globulin storage protein that is normally abundant in pea cotyledons. That the control for this lies in the chromosomes and does not result from an inhibitory effect upon enzymes that synthesize the globulin (translation effect) was proved by experiments with isolated chromatin. Chromatin from shoot apices could not synthesize the globulin *in vitro* unless histone proteins were removed from the DNA, but chromatin from cotyledons synthesized some globulin whether histones were present or not. This and many similar examples show that there is something about the internal environment of cells that can cause genes to be switched off in places and at times when they are not needed. It also shows that histones act as part of the switch mechanism. This is an example of gene repression. Considered more specifically, when an enzyme or other protein fails to be synthesized, we call the situation **enzyme repression.** Although the phenomenon is complex, the ability of the positively charged histones to bind to the negatively charged DNA and prevent RNA polymerase from transcribing that gene's information is apparently a common part of enzyme repression.

Gene activation leading to enzyme production is called **enzyme induction.** One good example of enzyme induction controlled by the external environment is synthesis of nitrate reductase only when NO_3^- is available to the cells (see Section 13.3). Of course, the external environment must always act by controlling the cell's internal environment. Another example of enzyme induction, in which the control might be on translation rather than on transcription, is gibberellin-stimulated synthesis of certain endosperm-degrading enzymes in cereal grain aleurone layers (see Section 16.2).

If histones prevent genes from functioning, how does enzyme induction occur? We know that interactions between histones and nonhistone proteins can temporarily uncover genes so that information in the DNA molecules can be transcribed by RNA polymerases. Just what subtle yet crucial changes the environment causes to alter these interactions among chromosomal proteins is still a mystery. Nevertheless, the environment does control some differentiation processes in an all-or-nothing way. Examples include control of flowering by temperature or daylength, depending on the species, control of bud and seed dormancy, leaf senescence, and leaf abscission by daylength, drought, and temperature, and control of seed germination by the ratio of red to far-red light or by an extended cold period (Chapters 19, 21, and 22). In conclusion, the cell's heredity determines the kinds of enzymes this cell can produce, but the environment strongly influences those that are produced and the rate of production of each. Continued changes in the internal environment of a developing plant caused by fluctuating external conditions and by cellular interactions dependent on a cell's position in the tissue modify differentiation. The most dramatic effects on differentiation are probably caused by transcriptional effects in which whole blocks of genes are activated or deactivated, while less dramatic effects could result from switching on or off only a few genes or from modulation of the activity of a few genes. Most translational control processes and processes that control the activity of already existing enzymes are seldom involved in dramatic examples of differentiation.

Positional Effects and the Importance of Cellular Interactions in Differentiation Many differentiated tissues exhibit a capacity for self-perpetuation, indicated by their ability to induce adjacent relatively undifferentiated tissues to become like them. A few examples illustrate this phenomenon.

When secondary growth in a young dicot stem begins, an interfasicular cambium develops within the pith ray parenchyma cells between the strands of primary vascular tissue. This cambium arises directly adjacent to the fasicular cambium within each vascular strand, forming a continuous vascular cambium around the stem (Fig. 18-1). A similar ring is formed in roots that undergo secondary growth, except that new cambial initials are formed from the pericycle. These new meristems are produced in both shoots and roots by alteration of existing differentiated pith or pericycle cells, a process called **redifferentiation*** (Sinnott, 1960). The important point is that the new cambial cells begin to form adjacent to the existing procambium. This appears to be an induction process involving release of something special from the procambium into the parenchyma cells that commits the latter to the same function of the former. This is an example of **homeogenetic induction** (Lang, 1974).

Other similar examples are known. If a transverse wedge is cut out of a *Coleus* or geranium (*Pelargonium* sp.) stem, severing a vascular bundle, the gap is closed by redifferentiation of cortex or pith tissues into xylem and phloem immediately below the wounded vascular cells. If a root tip is cut off near the apex, a new tip often regenerates. Xylem and phloem strands are then formed in the new tip in direct continuity with those in the main root.

A different type of induction is called **heterogenetic induction.** Here one cell type seems to induce others to become

---

*Redifferentiation is used in developmental biology to indicate direct transfer of one cell type to another (Spratt, 1971). Although many animal cells have this capability, plant cells generally *dedifferentiate* to the meristematic condition, divide, and then a daughter cell differentiates into the final cell type.

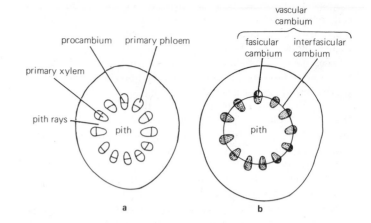

Figure 18-1 Development of the interfasicular cambium in the shoot apex of a stem that undergoes secondary growth. (**a**) Numerous provascular strands (primary tissues). (**b**) Secondary growth is about to begin. The procambium has remained meristematic in each strand and becomes the fasicular cambium. An interfasicular cambium has developed from pith ray parenchyma between each vascular strand. The resulting continuous ring of cambial cells makes up the vascular cambium responsible for production of secondary xylem and secondary phloem.

specialized in such a way that their structures and functions are different from those of the inducing cells. In young roots of *Potamogeton natans*, the immature epidermal cells differentiate into two kinds of cells, root hairs and ordinary epidermal cells. After this differentiation, the hypodermal layer just beneath the epidermis is stimulated to divide, but only under the root hair cells. A number of short cells are produced in the hypodermis internal to the root hairs, while a series of long cells arise between them internal to the ordinary epidermal cells. Similarly, in some dicots, leaf hairs are located only over vascular bundles, suggesting a heterogenetic influence from the vascular cells.

Another universal phenomenon in angiosperms that might be caused by heterogenetic influence occurs in the zygote. Before it divides to form the first daughter cell of the embryo, one end of the zygote becomes different from the other. This difference is presumably caused by something provided from the surrounding cells of the mother plant. For example, at the upper end of the cotton zygote lies the nucleus and most of the plastids and other organelles, while a vacuole predominates at the basal end (Jensen, 1964). We say that the cell is **polarized.** When it divides, the daughter cells are different from each other, the apical cell being smaller and more densely staining than the basal. Functions of the apical and basal cells are also quite different. From the apical cell arise all of the cells of the mature embryo, while the basal cell forms only a row of suspensor cells through which nutrients apparently move into the embryo. We conclude that the original establishment of polarity in the zygote causes changes in the first daughter cells that for a while are perpetuated during embryo formation.

Other short-range influences different from the contagious homeogenetic and heterogenetic effects are frequently referred to by animal physiologists as **field effects.** A plant example of field effects seems to be involved in stomatal patterns. Inspection of Fig. 3-5 shows that such patterns are not random. A minimum although not constant distance is maintained between stomates, yet no excessive stomate-free spaces are left in the epidermis. As the stomatal cells begin to differentiate in various regions, they apparently send out inhibitors into surrounding cells that prevent too close differentiation of other young epidermal cells into guard cells. Another probable example of a field effect occurs in the determination of leaf phyllotaxis at the shoot apical meristem (Section 15.6). Several other examples are also known.

All examples of homeogenetic, heterogenetic, and field effects might involve hormone release from the controlling cells. In homeogenetic induction of new xylem and phloem in severed stems, there is good evidence that IAA is the primary determining factor (Roberts, 1969; Torrey et al., 1971), although cytokinins and sucrose also appear necessary. But in most cases, it is not necessary to postulate specific hormonal involvement. Many heterogenetic and field effects might result simply from local depletion of certain nutrients required in large amounts by the controlling cells.

Differentiation of tissues frequently involves commitment of particular developmental fates in one or a few embryonic cells, often involving little apparent specialization in them. Without previous experience, we could not predict which of the two cells arising from the first division of a zygote would form only suspensor cells and which would give rise to the embryo. Yet once these initial commitment events are established, determination can sometimes appear in numerous progeny cells in the absence of whatever stimulation originally caused commitment.

Tissue Differentiation Usually Requires an Initial Act of Cell Division In both homeogenetic and heterogenetic induction, the first step in redifferentiation normally involves **dedifferentiation** of parenchyma cells to the meristematic condition, followed by mitosis and cytokinesis in them (Meins, 1975). For example, *direct* conversion of a parenchyma cell of the cortex or pith into a xylem or phloem cell during vascular tissue regeneration does not occur. Neither does a vascular cambium cell develop directly into a xylem or phloem cell. Only the first product of division of a cambial cell becomes a xylem or phloem cell. Since xylem differentiates toward the inside and phloem toward the outside of the stem, position effects controlling the internal environment of each cell must determine whether it becomes xylem or phloem. The point relevant to our present discussion is that the cambial cells themselves do not become xylem or phloem. Consider what would happen to the continuity of cambium, phloem, and

xylem if cambial cells did occasionally differentiate directly into vascular cells.

It is commonly true that only after polarity establishment followed by division can the daughter cells differentiate into specific tissue types. The division is often unequal, as when the zygote first divides or when a young root epidermal cell (Fig. 18-2) divides into a **trichoblast** (from which a root hair will develop) and an **atrichoblast** (which becomes an ordinary epidermal cell). Unequal cell divisions of the immature epidermis in young leaves also precede formation of specialized mature epidermal cells, guard cells, and subsidiary cells. Finally, formation of a sieve tube element and companion cell involves unequal partition of a vascular cambium daughter cell.

In gamma plantlets, in which cell division cannot occur because of radiation treatment (Section 15.1), no root hairs, leaf hairs, or stomates are formed, further emphasizing the importance of cytokinesis in differentiation. This requirement for cytokinesis is widespread, but why is it usually essential? Considering that commitment to differentiation first appears in one or both daughter cells, what happened in the mother cell to cause this? If some inducing substance caused a change in the mother cell, what process could it affect? Substantial changes in gene expression are involved, so the substance could act on the genes in any of the cell cycle phases leading to mitosis (G_1, S, or G_2), during mitosis, or during cytokinesis. We are a long way from understanding this phenomenon, but evidence suggests that DNA replication during S phase

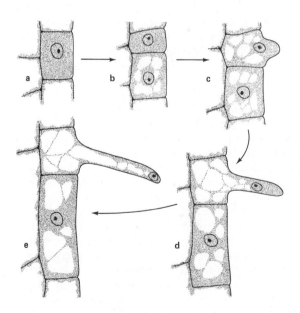

Figure 18-2 (**a**) An unequal cell division of a young epidermal cell precedes formation of a root hair and an ordinary epidermal cell. This division forms (**b**) a trichoblast (upper cell) and an atrichoblast (lower cell). (From Jensen and Salisbury, 1972.)

is the essential part of the cell division requirement for differentiation (Shininger, 1975). Perhaps only when DNA becomes uncovered during its replication can inducing agents combine with this DNA to inactivate some genes and activate others.

18.2 Totipotency and Tissue Cultures

We assumed earlier that cells in a plant become different, even though their genes are identical. How do we know they all have the same kinds of genes? First, our understanding of chromosomal duplication and separation during mitosis strongly suggests this. Second, many plant cells are **totipotent.** By this we mean that a nonembryonic cell has the potential to dedifferentiate into an embryonic cell and then to develop into an entire new plant, if the environment is correct. A root parenchyma cell, for example, may begin to divide and produce an adventitious bud and finally a mature flowering shoot. All the genes for production of the whole plant still exist in such root cells, even though they are somewhat differentiated. This could not happen if their genes had been altered during root development. Totipotency is illustrated by formation of buds on certain kinds of roots that develop into flowering shoots and by development of pith callus tissues into new plants. Partial totipotency occurs when adventitious roots develop from stem cells and when xylem and phloem are regenerated from wounded cortex cells. In fact, totipotency might be advantageous to plants mainly because it provides them with a mechanism for healing many wounds.

In each of these examples of totipotency, several cells cooperate to form primordia from which the whole plant arises. A few examples in which even a single differentiated cell apparently develops into a whole plant are known. Obviously, cell division is essential for this, but we assume from our previous discussion that commitment first appears in the daughter cells. Experiments in which plants develop from single cells were pioneered by F. C. Steward and his coworkers at Cornell University in the 1950s, associated with his work on cytokinins (see Section 17.1). He found that cells broke away from pieces of callus derived from carrot root phloem, and in the proper solution of hormones and other solutes, single cells dedifferentiated. When conditions were changed, one such cell would occasionally divide to form a multicellular embryoid. If these embryoids were cultured on an agar medium containing the proper IAA to cytokinin ratio, callus tissues that formed both roots and buds developed. From these, new plants capable of forming seeds were formed. Figure 18-3a illustrates young tomato plants growing on agar but derived from free cell cultures grown in liquid media by Murray W. Nabors at Colorado State University (see his personal essay on pages 282–283). Even haploid pollen grains develop into callus tissues and then whole plants, as shown in Fig. 18-3b (Sunderland, 1970; Sangwan and Norreel, 1975). In this example, the plants contain predominantly triploid and diploid

Figure 18-3 Development of (**a**) tomato and (**b**) petunia plants from a callus, illustrating totipotency. (Photographs courtesy of Murray Nabors and R. S. Sangwan.)

chromosome numbers, although some of the cells are haploid. Apparently, the diploid and triploid cells result from **endoreduplication** (doubling of chromosomes in mitosis, with lack of subsequent cytokinesis) or nuclear fusion.

18.3 Juvenility and Phase Changes

Although most living cells seem to be totipotent, the life cycles of many perennial species consist of two phases in which certain morphological and physiological characteristics are rather distinct. After germination, most annual and perennial seedlings enter a rapidly growing phase in which flowering usually cannot be induced. A characteristic morphology, especially evident in leaf shapes, is sometimes produced during this time. Plants having these characteristics are said to be in the **juvenile phase,** as opposed to a **mature** or **adult phase.**

The juvenile phase with respect to flowering varies in perennials from only 1 year in certain shrubs up to 40 years in beech (*Fagus sylvatica*), with values of 5 to 20 years common in trees. In Chapter 22, this reaching of maturity after which flowering can occur is described as attaining a ripeness to respond. Such long juvenile phases in conifers and other trees pose serious obstacles to genetic programs designed to improve their quality. Another common physiological difference between perennials in the juvenile and adult phases is the ability of stem cuttings to form adventitious roots. In the adult phase, the rooting ability is usually diminished and sometimes lost.

The juvenile and adult morphologies of leaves are examples of **heterophylly**. Heterophylly of an annual dicot is well illustrated by the bean, which always forms simple primary leaves at first and compound trifoliate leaves later (see Figs. 16-10 and 17-2), and by the pea, which has quite reduced, scalelike juvenile leaves. Among perennials, many junipers form needlelike juvenile leaves and scalelike adult leaves. The many species of *Acacia* and *Eucalyptus* often have strikingly different juvenile leaf forms. English ivy (*Hedera helix*), another perennial, has been studied extensively (Fig. 18-4). Its juvenile growth habit is that of a creeping vine, but later it becomes shrublike and forms flowers. Juvenile leaves are palmate with three or five lobes, while adult leaves are entire and ovate. Although attainment of the adult phase is usually fairly permanent, juvenility in ivy can be induced in shoots that develop from lateral buds of mature stems by treating the leaf just above this lateral bud with GA_3. ABA prevents this reversion caused by GA_3, suggesting that a balance of gibberellins and ABA might normally be involved in the transition from one state to another. On the other hand, GA_3 terminates juvenility and induces flowering in many gymnosperm species in the Cupressaceae and Taxodiaceae families.

18.4 Differential Growth and Turgor Movements

You are probably familiar with a plant's abilities to respond to the environment by displaying unequal or differential growth. You know that roots grow downward and stems upward in response to gravity (**geotropism**) and that stems and leaves frequently grow toward the light (**phototropism**). Less apparent is the phenomenon of **thigmotropism** (Greek *thigma*, "touch"), a response to contact with a solid object exhibited by many climbing plants (see Fig. 16-10) growing around a pole or the stem of another plant (Jaffe and Galston, 1968). These tropisms are responses to environmental stimuli that act with greater intensity from one side than another and are caused by a faster rate of cell elongation on one side than the other.

Other examples of unequal growth (**nastic movements**) occur when an organ is exposed to environmental stimuli equally from all directions or when the stimulus is presented primarily from one side but when growth on the two sides is independent of the direction of the stimulus. For example, leaf primordia on vertically oriented stems grow upward and enclose the apical meristem (see Fig. 15-8) because of faster expansion of cells on the lower than on the upper side

JUVENILE　　GA₃　　GA₃　　MATURE
ABA

Figure 18-4 Reversion of mature *Hedera helix* to the juvenile form with GA_3 and stabilization of the mature (adult) phase with ABA. The GA_3-treated shoot arose from a mature plant treated with 0.005 μmol of GA_3. The shoot labeled GA_3, ABA arose from a mature plant treated with 0.005 μmol of GA_3 and 5 μmol of ABA. (From C. E. Rogler and W. P. Hackett, 1975, Physiologia Plantarum 34:148–152.)

(**hyponasty**). This response is thought to be independent of environmental stimuli such as gravity, because it occurs on horizontal as well as vertical branches. Opening and closing of flower petals of tulips, crocuses, the night-blooming jasmine, and certain other plants in response to changes in temperature and light intensity are also classified by Ball (1969) as nastic movements. They depend on unequal growth on the two sides of the petal. Furthermore, some tendrils are **thigmonastic** instead of thigmotropic, because they bend the same direction regardless of which side is rubbed.

Nastic movements that are reversible and do not involve true growth are also important in affecting the appearance of plants and in their apparent well-being. These result from turgor changes due to water uptake, causing temporary expansion in certain cells. Since no permanent increase in size occurs, we sometimes refer to these as **turgor movements.** Examples include folding of *Mimosa pudica* leaves and leaflets in response to touch, closing the Venus's-flytrap after a trigger hair is touched (both examples of thigmonasty), sleep movements (see Chapter 20) of leaves and flowers (**nyctinasty;** Greek *nyx, nyktos,* "night"), folding of leaves of certain grass species in response to a water deficit, and the forcible discharge of pollen onto insects by the anthers of *Catasecum sacchatum,* an orchid.

Still another growth movement is shown by stems and roots, as studied extensively by Charles Darwin (1880). As these grow from the time of seed germination, the rate of elongation on various sides constantly changes, causing the tip to rotate continuously around the stem axis. This phenomenon Darwin called **circumnutation.** If the environment causes circumnutation, we do not know which factor is involved or how it acts. We shall now discuss several of these effects in more detail.

18.5 Phototropism

Curvature of stems or leaves toward light (**positive phototropism**) is usually caused by a more rapid elongation of cells on the shaded than the illuminated side of the stem or petiole. Usually, the petioles curve so that the leaves hardly overlap, causing **leaf mosaics.** Such mosaics are also shown by many trees, easily observed by standing under them and looking upward, and by ivy climbing on building walls. Maximum absorption of light for photosynthesis thereby occurs. Curvature of stems toward light also results in greater light absorption by the shoot system. Many roots also respond phototropically, but they usually curve away from the light (**negative phototropism**). Advantages of this seem minimal, yet adventitious roots formed aboveground or nongeotropic roots emerging from the soil might grow toward the soil because of it. Roots of numerous other species do not exhibit phototropism, while a few are even positively phototropic.

It was in coleoptiles of canary grass and oat that Charles Darwin (1880) studied phototropism extensively, leading to Frits Went's discovery, published in 1928, of an auxin in oat coleoptiles (see his personal essay in Chapter 16). Since most of what we know about phototropism comes from coleoptile studies, let us begin there. Darwin found that it was the coleoptile tip that absorbed phototropically active light, even though curvature occurred in the elongating cells below. We now believe that curvature of coleoptiles results primarily from an increased IAA content in the elongating cells on the shaded side that is caused by polar transport from the tip. How can light cause more IAA to appear in the shaded than in the illuminated side? If we ask a photochemical question, perhaps we should first find out what color of light is effective, for then we might learn the color of the pigment that absorbs this light and something about the resulting chemical processes. *To identify the pigment responsible for any photochemical process, an essential step is to compare the action spectrum for the process* (see Section 9.3) *with the absorption spectra of pigments suspected to be involved.* This was done as early as 1909 by A. H. Blaauw in Holland, who found that blue light was most effective in causing phototropic curvature. In Figure 18-5, a more modern action spectrum for phototropism is compared to the absorption spectra for two yellow pigments common in plants, β-carotene and riboflavin. The action spectrum shows a major peak in the blue-violet region, so it is logical that the pigment absorbing effective light should be yellow, the complementary color. Near-ultraviolet wavelengths also cause phototropism, but green and longer wavelengths are ineffective. The same appears to be true of phototropism in all plant parts investigated.

Careful investigation of the curves in Fig. 18-5 shows that the absorption spectrum for β-carotene more closely matches the action spectrum in the blue region than does that of riboflavin, but the curve for riboflavin matches the action spectrum much better in the ultraviolet than does that for β-carotene. It is impossible to identify the effective pigment from such experiments, but other data favor riboflavin or a similar flavin

compound as the active pigment. For example, in certain phototropically active fungi where action spectra for various responses are almost identical to that for phototropism, a flavin attached to a protein (**flavoprotein**) appears to be the only pigment involved (Muñoz and Butler, 1975). Upon light absorption, the flavoprotein becomes oxidized by reducing a b-type cytochrome in the plasma membrane (Brain et al., 1977).

How light absorption by the photoreceptor pigment leads to auxin migration toward the shaded side of a coleoptile tip is unknown. That unilateral auxin migration does occur in

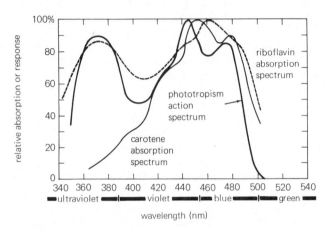

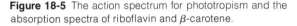

Figure 18-5 The action spectrum for phototropism and the absorption spectra of riboflavin and β-carotene.

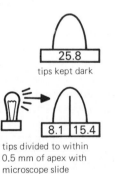
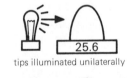

Figure 18-6 Experiments showing that unilateral illumination of corn coleoptile tips leads to a transport of auxin from the illuminated to the shaded side of the tips and does not cause a destruction of auxin. Numbers on the agar blocks represent the degrees of curvature caused by application of the blocks to decapitated oat coleoptile stumps. In the partially split tips, part of the auxin was transported laterally above the dividing barrier, but in the completely split tips, this was not possible. (After W. R. Briggs, 1963, Plant Physiology 38:241.)

coleoptiles was theorized by N. Cholodny and demonstrated by Went in 1926. The proposed migration of auxin in both phototropism and geotropism became known as the **Cholodny-Went theory.** Went's work was confirmed and extended by Winslow Briggs and his colleagues in the early 1960s. Figure 18-6 illustrates their experimental design and the amount of auxin transported basipetally into the agar block (measured by the coleoptile bending bioassay; see Figs. 16-1 and 16-4). Tips exposed to light transported as much auxin as those kept dark did. This shows that auxin destruction by light did not occur. When the agar and part of the coleoptile were divided with a thin piece of mica and the tip exposed to light from one side, the amount of auxin delivered from the shaded side was twice as great as from the illuminated side. Division of the entire tip with mica prevented auxin transport in the tip (lower right); such division also prevents phototropic curvature.

Another important question is how much light energy (or how many photons or einsteins) is required to cause phototropism. Since the process is shown by plants growing near windows on the shaded side of a house or under objects that cause continuous shade, full sunlight is clearly unnecessary. In fact, exposure to as little as 10^{-11} einsteins (6.6×10^{-6} cal) of blue light per cm^2 of tip area causes marked bending of coleoptiles. (Full sunlight provides at sea level a maximum of about 0.6 cal/cm^2/min of visible radiation, but over half the wavelengths are phototropically inactive.) Figure 18-7a illustrates the extent of curvature of oat coleoptiles caused by different exposures to blue light and shows a surprising result. When the exposure exceeds about 10^{-11} einsteins/cm^2 (maximum of **first positive response**), curvature becomes less marked and near 10^{-9} einsteins/cm^2 is even slightly reversed. Still greater exposures again cause curvature toward the light (**second positive response**). In nature, phototropism must almost always be due to the second positive response. Both first and second positive responses are apparently due to the same photoreceptor pigment and, in coleoptiles, to auxin migration.

Much less is known about phototropism in organs other than coleoptiles. All dicot seedlings investigated show both first and second positive responses to blue light, but at light energies between them there is simply a plateau in the response. Furthermore, if the seedlings are grown in the light, little first positive effect occurs. Responses in light-grown and dark-grown radish seedlings are illustrated in Fig. 18-7b. These seedlings are approximately as sensitive and respond as rapidly (beginning within about 30 minutes) as coleoptiles.

Do dicot seedlings also curve because of auxin migration toward the shaded side? For the rapidly growing third internode of week-old pea seedlings grown in darkness (see Fig. 16-5a), the answer appears to be yes (Kang and Burg, 1974). When the very young leaves that synthesize both auxin and gibberellins are removed, bending in unilateral blue light is slightly reduced. Replacement of the young leaves by lanolin pastes containing IAA again stimulates curvature, but so does addition of GA_3 to the apex of intact seedlings. Since both IAA and GA_3 promote growth of cells below the apex, in-

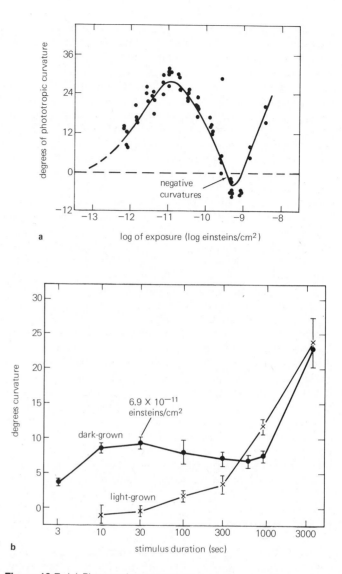

Figure 18-7 (**a**) Phototropic response (first positive, negative, and second positive curvatures) of oat coleoptiles as caused by increasing energies of blue light at 436 nm. Intensity was 1.4×10^{-12} einsteins/cm^2/sec, and exposure times varied to give the total exposures indicated along the abscissa. (From B. K. Zimmermann and W. R. Briggs, 1963, Plant Physiology 38:248.) (**b**) Dose-response curves for radish seedling phototropism. Stimulus energy was 6 erg/cm^{-2}/sec^{-1} with 460 nm light. Stimulus duration was plotted on a log scale. Error bars represent standard errors. (From M. Everett, 1974, Plant Physiology 54:222–225.)

creased curvature might result only from the ability of cells to elongate faster in the bending region, not from light-stimulated lateral migration of either hormone. However, blue light from one side causes a significant preferential movement of radioactive ^{3}H-IAA but not GA_3 into the shaded side.

This work suggests that in dicot stems as well as coleoptiles, auxin migration in response to blue light is the primary cause of phototropism. It also implicates young leaves as the source of this auxin. In light-grown sunflower seedlings, even mature leaves and cotyledons influence phototropism of the stem below. Curvature of the hypocotyl is largely controlled by light absorbed by the cotyledons and much less so by the young leaves or other cells of the epicotyl. This curvature is not accompanied by an increase in IAA in the shaded side of the entire hypocotyl (Bruinsma et al., 1975), although the amounts in opposing sides of the curving zone itself were not determined. In mungbean (*Phaseolus aureus*) seedlings, curvature of the stem is largely controlled by light absorbed by the stem itself, not by the primary leaves above the curving region (Brennan et al., 1976). Little information is available for stems of other dicot species or for leaves.

If one side of a cocklebur leaf blade is covered with aluminum foil to simulate natural shading, two responses can be seen. One response is elongation of the corresponding side of the petiole, so that it bends and displaces the leaf blade toward the opposite direction. This phenomenon presumably also occurs when leaves move away from one another to form leaf mosaics. The assumption is that when one part of a leaf blade is shaded by another, the shaded part transports more auxin to the same side of the petiole than the brightly illuminated side transports to its corresponding petiole part. The second observation is that the shaded half of the leaf blade bends upwardly. These phenomena are important to expose leaves to maximum sunlight, and it is unfortunate that we don't understand them well. Actual analyses of IAA and other hormones in opposing sides of curved petioles or leaf blades are needed and would not be difficult.

18.6 Geotropism

Growth movements toward or away from the earth's gravitational pull are examples of geotropism. Roots are positively geotropic, primary roots generally more so than secondary roots. Tertiary roots and roots of higher orders are hardly geotropic and therefore grow almost horizontally. This difference between the various root types is important in allowing much greater exploration of the soil than would occur if all types grew straight down, side by side. Stems and flower stalks are generally negatively geotropic, although the extent to which this occurs is quite variable. The main stem or tree trunk usually grows 180 deg away from the gravitational stimulus, but branches, petioles, rhizomes, and stolons are usually more horizontal. These differences again allow a plant to best fill space and thus absorb CO_2 and light. The growth of several organs at intermediate angles is given the special term **plagiotropism,** while the vertically oriented stem and root axes exhibit **orthogeotropism.**

Two major questions arise. How does the gravitational stimulus cause curvature, and what causes roots and shoots to generally behave oppositely? Gravity can affect an organ only by causing movement (acceleration of mass) in one of its component parts. Hypothetical particles inside cells that respond to gravity by moving toward the lower side and whose movement causes geotropism are called **statoliths.** Inspection of gravity-detecting cells of various organs in many species indicates that the only subcellular organelle present in and consistently moving to the bottom of such cells is the amyloplast, a colorless, starch-containing plastid made heavy by the starch it contains. Sometimes, however, geotropism occurs slowly in organs from which all or nearly all the starch has been digested. Here the amyloplasts are much less dense but are probably still capable of sedimenting slowly. Smaller bodies such as mitochondria and ribosomes are too light to respond consistently to gravity, while the large nucleus does not have a great enough density to respond. Dictyosomes and endoplasmic reticuli do not show consistent movements expected of statoliths either (Shen-Miller and Hinchman, 1974).

The part of a plant that detects gravity is sometimes several cell layers away from the part that responds, as is also true for phototropism. In horizontally oriented coleoptiles, the tip is primarily responsible for gravity detection, while the cells below respond by differential growth, causing upward curvature. Only in the tip and in a group of cells adjacent to each of the two vascular bundles do the abundant amyloplasts almost totally settle to the bottoms of the cells. In roots, the root cap seems to be solely responsible for detecting gravity, while the elongating cells of the cortex or epidermis behind the cap cause curvature. The role of the cap was shown by failure of the roots to curve when the cap was removed, even though they continued to elongate, then restoration of curvature after a day or two, when a new cap was regenerated from the quiescent center (Wilkins, 1975, 1976). Most root cap cells contain numerous amyloplasts, while other cells in growing roots have little if any starch, although a few leucoplasts without starch are present.

In dicot stems, gravity is apparently detected near the apex and in certain cells of the elongation region that are rich in mobile amyloplasts. These cells, often called a **starch sheath,** frequently form a continuous layer around the vascular bundles similar to the endodermis formed in roots. This sheath extends from a few millimeters to a few centimeters below the shoot apex. Elongating parenchyma cells of the cortex external to the starch sheath have been assumed to be responsible for curvature, but the elongation of epidermal cells may be even more critical. The young leaves and perhaps also the meristematic cells of shoot tips are essential for geotropism, apparently because they supply IAA and gibberellins that stimulate elongation, without which no curvature can occur.

In stems of the grass family, there are two kinds of tissues responsible for curvature, the kind depending on the species (Gould, 1968). In the Panicodeae subfamily (corn, sugarcane, crabgrass, and others) and the Eragrostoideae (bermudagrass, buffalograss, blue grama, and others), curvature is restricted to the growing region of the stem itself at the base of the hollow internode. In the Festucoideae subfamily (oats, barley, bromegrass, ryegrasses, and others), both collenchyma cells (long, narrow cells with thick but nonlignified walls) and parenchyma cells rich in mobile amyloplasts at the base of the cylindrical leaf sheath surrounding the stem retain this growth capacity, so curvature occurs because of the elongation of such cells, especially the collenchyma cells, on the lower side of a horizontal stem (Dayanadan et al., 1976). These geotropically sensitive cells make up part of what is sometimes called a pulvinus, or by others, a **false pulvinus,** to distinguish it from the true pulvinus that controls nyctinastic leaf movements in certain dicots by *temporary* changes in cell sizes (see Section 18.8).

If amyloplasts are statoliths, how could their displacement within individual cells cause curvature of a root or stem? This crucial question is still unanswered (Juniper, 1976). The Cholodny-Went theory, accepted for the last 50 years, is that geotropism is caused by auxin accumulation on the lower side of the organ, promoting growth in stems but inhibiting it in roots. Statolith displacement in cells is assumed somehow to cause polarized auxin transport from the base of each cell toward the lower side of the organ. If other hormones are involved, their migration or metabolism would also have to be controlled by statolith displacement. Let us investigate the evidence for hormone participation in geotropism, beginning with coleoptiles and stems.

For coleoptiles, geotropism is caused by an inhibition of elongation on the upper side and a promotion on the lower side; the two effects are of roughly equal magnitude. Auxin migration toward the lower side of a horizontally oriented tip was reported by Herman Dolk in the 1930s. He cut off coleoptile tips and placed them horizontally with the cut end in contact with an agar block. The block was divided in half by a thin piece of mica (Fig. 18-8a) to prevent mixing of auxin that diffuses into the agar from the upper and lower coleoptile portions. Auxin in each half of the agar was bioassayed with the coleoptile curvature test. Almost twice as much auxin was collected from the lower than from the upper half. This differ-

ence results from auxin migration, not from its destruction or synthesis, and has been verified many times. Similar techniques showed comparable accumulations of auxin on the lower sides of various dicot stems, supporting the Cholodny-Went theory.

In 1972, I. D. J. Phillips discovered that gibberellins might participate in geotropism of stems. The techniques he used for collection of gibberellins (Fig. 18-8b) were similar to those of Dolk. About eight times as much gibberellin activity was consistently recovered from the lower than the upper side of a horizontally placed sunflower stem after a 20-h incubation period. The total amount of gibberellin activity released from horizontal stem sections was about three times as great as from normal upright sections, suggesting that considerable gibberellin synthesis occurs in the lower side and that lateral transport across the stem might not occur. This was confirmed by Phillips and Hartung (1976), who found no unequal distribution of radioactive GA_1 across a horizontal sunflower epicotyl section after adding it to the cut apex in an agar block. In their experiments, ^{14}C-IAA applied similarly in agar blocks also failed to accumulate on the lower side, suggesting that the greater concentrations of endogenous auxin found by earlier investigators on the lower side of dicot stems may have resulted from faster synthesis there. They questioned whether a lateral transport of either IAA or gibberellins caused geotropism in sunflower stems.

In evaluating these evidences for hormonal roles in geotropism, an important question must be asked. Does the hormone accumulate on the proper side of a horizontal organ *rapidly* enough and *in sufficient amounts* to account for the growth promotion occurring there? In 1976, J. Digby and R. D. Firn reviewed previous results and concluded that neither criterion seemed to be satisfied. In oat coleoptiles and sunflower stems, upward curvature is readily detectable within 30 min and sometimes even within 10 min after horizontal placement. As discussed in Chapter 16 (see Fig. 16-8), externally supplied IAA does not begin to promote elongation of stem or coleoptile sections until after a lag period of 10 to 15 min, although part of this lag must involve absorption and accumulation to active levels. Even before curvature begins, 2 to 10 min are required for the tissues to sense the geostimulus (i.e., for statoliths to settle), and at least 10 min seem necessary for detectable IAA migration to the lower side. Digby and Firn concluded that gravity detection, IAA transport, and promotion of elongation were not likely to occur before the observed geotropic responses, although their conclusions are admittedly based on several unproven assumptions. They also concluded that too little IAA accumulates on the lower side to cause the unequal growth observed, but this is still controversial. Since the lag times preceding stimulation of stem growth by gibberellins are sometimes as long as an hour (Adams et al., 1975), they appear even more unlikely to cause geotropism.

In grass nodes, insertion of a mica barrier parallel with the axis of the stem so that flow of any hormone from one side of the cylindrical false pulvinus to the other side would be

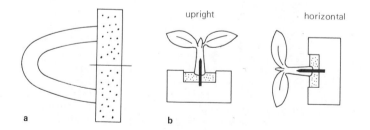

Figure 18-8 (**a**) Herman Dolk's technique for collecting auxin in two halves of an agar block after the hormone is redistributed in a coleoptile tip by gravity. Line dividing the agar block represents a thin mica barrier. (**b**) I. D. J. Phillips's technique for collecting gibberellins in two halves of an agar block after unequal distribution in a horizontal sunflower stem. (From I. D. J. Phillips, 1972, Planta 105:234–244.)

A Role for Tissue Culture in Plant Breeding

Murray W. Nabors

F.B.S. 1976

We are apparently on the threshold of being able to control or adjust the morphogenesis of organisms. Murray W. Nabors, a colleague of Cleon Ross at Colorado State University, sees a great potential for good in some of the recent developments with plants. He first explains to us a bit about the techniques and possibilities of plant tissue culture, cloning, cell fusion, and so on, then relates his personal involvement on this exciting frontier.

The knowledge that many plant cells are totipotent—meaning they contain all the genetic information about the plant—has led to considerable thought about possible useful applications of tissue cultures. Such cultures are produced by removing a small portion of the plant and placing it on a defined nutrient medium. For instance, a small section of stem, cotyledon, or root might be used. All nutrient media contain the major and minor minerals necessary for plant cells, several vitamins normally produced in other portions of the plant, and sucrose as an energy source. Hormones can also be added to the medium, and, depending on the hormone types and concentrations, various developmental patterns can be encouraged in the tissue section.

One might attempt, for example, to maintain the tissue in its already defined developmental pattern. Roots are frequently cultured for many years in this manner. Alternatively, cultured cells can be encouraged to grow in an undifferentiated manner, forming a large cell mass, or callus. By proper manipulation of the culture medium, callus cells can be made to differentiate into buds and roots. In dicot callus, bud formation is encouraged in media with a high ratio of cytokinin to auxin. In monocot callus, bud formation is promoted on a medium containing little or no auxin and no cytokinin. By causing cell proliferation into a callus, then instituting bud formation on it, an investigator can produce large numbers of identical plants by the tissue culture method. A group of genetically identical plants is known as a **clone.** Cloning by the tissue culture method has a potential usefulness in the rapid propagation of rare, valuable plants with either agricultural or horticultural utility (Nabors, 1976).

Methods for regenerating plants from callus tissues are available for some species; for others, further experimentation is needed to find a successful regeneration procedure. For instance, few legumes have been regenerated. In particular, a method is currently lacking for soybean, despite extensive effort in several laboratories, including my own. Recently we have succeeded in regenerating corn plantlets from root callus. Before this discovery, regeneration had been obtained only from a particular portion of very young embryos by C. E. Green and R. L. Phillips at the University of Minnesota. Our strategy for obtaining corn buds was to assume that corn callus is producing its own auxin. Since bud regeneration from monocot callus is promoted by low levels of auxin, we added antiauxins to the medium for successful regeneration.

Callus cultures are usually grown on a medium solidified with agar, but they can be grown in a liquid medium, in which case the cell masses frequently break apart into single cells or cell clumps. Such cultures are known as **cell suspensions.** Either callus cultures or cell suspensions can be utilized to obtain mutant cells and plants with new and potentially useful characteristics. The power of the tissue culture method is that large numbers of cells—and thus plants—can be grown in very small volumes. For instance, a 100-ml cell suspension of tobacco can contain over 10 million cells. Mutations occur spontaneously in such cultures, but the mutation rate can be increased, if desired, by utilizing a mutagenic chemical or ionizing radiation. Useful mutants are obtained by a selection procedure, and plants can be regenerated from mutant cell lines. Such mutant plants must be tested for persistence and inheritability of the mutant trait. Both characteristics have already been reported in the literature for a few mutant types.

The potential usefulness of tissue culture in plant breeding can be markedly enhanced in many cases by techniques for producing haploid plants. Such plants are produced by culturing anthers at a proper stage of development on the correct nutrient medium. If the parent plant is polyploid, anther-derived plants are referred to as "polyhaploid." Haploid plants and cell lines are useful in mutation selection because the possibility of obtaining recessive mutations is markedly enhanced. Haploid mutant cells can be returned to full ploidy by colchicine treatment. Alternatively, the cell walls can be enzymatically removed, and naked protoplasts can be fused together to obtain a fertile product. Potentially, protoplast fusion can also be utilized in attempts to combine useful traits contained in two separate varieties. Peter Carlson and his coworkers have succeeded in regenerating plants from the interspecific hybridization of tobacco protoplasts.

My interest in using tissue culture to produce new, agriculturally useful varieties grew originally from a senior honors project carried out at Yale University under Dr. Ian Sussex. I investigated the tissue culture of the rhizophore of *Selaginella*. Although this plant is certainly not useful as a food crop, the techniques and potentials of tissue culture seemed fascinating, and the experiments were fun to do. After graduate school in 1970, I spent a year at the University of Oregon. Carlson had just published his ingenious studies on auxotrophic mutant selection in cultures of fern spores and of tobacco cells. Basically, he was able to grow large suspension cultures of cells, then to select cells carrying specific mutant traits. Carlson also got his start in tissue culture from Sussex. Some of us at Oregon, including Howard Bonnett, Don Hague, Ken Brooks, and Curt Peterson, spent a good deal of time discussing the potential inherent in Carlson's work and speculating on what sorts of experiments the future would bring. I learned a lot from these discussions, and before long it seemed to me that a most interesting series of experiments might be performed if tissue culture techniques were utilized to solve practical problems. At this time, most people in the field were concentrating on the genetic and biochemical as opposed to the agricultural implications of tissue culture. Granting agencies seemed quite disinterested in providing money to support this kind of research. Nevertheless, I began some preliminary work to select NaCl-resistant mutants in *E. coli*, and later in tobacco cells.

The selection procedure consists of gradually increasing the salt level of the medium. This process encourages the growth and division of tolerant mutants and discourages that of nontolerant, "normal" cells. After several months of selection, an entire culture of salt-resistant mutants is obtained. At present, we have produced cell suspensions of tobacco with over 13 times normal tolerance of NaCl. These cells are growing in a NaCl concentration over one third that of sea water. Regenerated plants are now in the testing stage to determine mutation expression and inheritability. The production of such mutant plants could make it possible to produce more food on brackish soils; in addition, we might avoid the gradual reduction in yields that frequently occurs as irrigation systems add salts to the land. Currently, we are shifting a tobacco-based technology to more useful plants such as corn, oats, and wheat. We are also investigating the selection of other types of useful mutants.

The development of clever selection procedures should make it possible to use tissue culture breeding to produce many different types of crop plants with useful characteristics. Resistance to salts, herbicides, high or low temperatures, drought, or disease are characteristics currently being sought by this method. New varieties more responsive to fertilizers or exhibiting less photorespiration might also arise. The aim of most such studies is to produce plants that grow suitably in existing environments, thus avoiding the expense and energy involved in changing the environment.

prevented does not retard upward curvature (Bridges and Wilkins, 1973). Furthermore, IAA does not accumulate on the lower side of such pulvini in the absence of a barrier. These results indicate that lateral transport of a hormone does not cause geotropism here and that the elongating cells themselves both detect and respond to gravity. To what extent this occurs in coleoptile and stem cells is still worthy of investigation.

For roots, the Cholodny-Went theory suggested that auxin accumulation on the lower side would inhibit growth there, consistent with its general inhibition of root elongation. Supporting evidence for this came in the 1930s, but with few exceptions modern investigations show no such accumulation of either endogenous auxin or ^{14}C-IAA on the lower side. Nevertheless, there is strong evidence that some growth inhibitor accumulates in the lower half of a horizontal root cap, migrates basipetally out of these cells, and reduces growth in the elongation zone (Audus, 1975; Wilkins, 1975, 1976; Juniper, 1976). It is agreed that geotropism in roots is accompanied by growth inhibition, slight on the upper side and pronounced on the lower side.

Figure 18-9 illustrates results obtained by Paul-Emile Pilet in Switzerland, which show that corn root caps produce such an inhibitor. Pilet removed the entire caps and part of the meristem from some roots and half of the caps plus part of the meristem from others. Those without caps hardly curved, indicating that the apical meristem is geotropically inactive. Those with half-caps consistently curved in the direction of the remaining cap cells. This result is easily explained if an inhibitor is transported from the cap cells to the elongating zone.

The inhibitor provided by the root cap is probably ABA. Thus, ABA becomes about three times as concentrated on the lower side of horizontal broad bean roots within 30 minutes, and similar results occur in corn roots. Even more evidence was provided by Henry Wilkins and R. L. Wain (1975). They realized that certain varieties of corn are not positively geotropic in darkness but are in light. (A similar light requirement exists in roots of oat and wheat, but not of pea and cress, *Lepidium sativum*). They showed that light promotes produc-

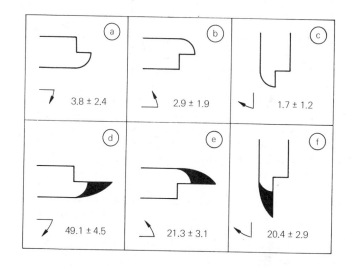

Figure 18-9 Geotropic curvature in corn roots as affected by removal of root cap and meristematic cells of the root apex. Direction of curvature is indicated by arrows, while numbers represent degrees of curvature. In the three roots at the top, the entire cap and part of the meristem was removed, and almost no curvature occurred. (Redrawn from P. E. Pilet, 1973, Planta 111:275–278.)

tion of ABA in the root caps of a light-requiring variety and that in darkness such roots will curve downwardly if ABA is provided to the tips. These facts suggest that the light requirement is only to allow synthesis of ABA in the root cap. Furthermore, ABA will not cause downward curvature if it is added to decapped roots, apparently because it is transported downward only in the cap. Evidence supporting downward transport in the cap is provided by the failure of curvature when a thin impermeable barrier is inserted horizontally into a horizontal cap. Whether enough ABA is then transported to the lower side of the elongating zone, and whether it does so rapidly enough to explain geotropic responses occurring within 30 min or so, has not been shown conclusively.

For plagiotropic responses, even less is known. Plagiotropic secondary roots have hardly been investigated, and similarly responding petioles and branches have been studied only for auxin gradients. Auxin distributions are consistent with leaf and branch angles in several cases, but other hormones should also be investigated.

18.7 Reaction Wood

As a plagiotropic tree branch elongates, it might be expected to bend downward because of its increased weight and distance from the trunk. This is sometimes observed, but it is resisted by formation of **reaction wood** (Scurfield, 1973). Reaction wood is the increased xylem produced on either the upper or lower side of a branch by more rapid division of the vascular cambium on that side. In conifer (softwood) limbs, reaction wood forms on the lower side and by expansion *pushes* such limbs more upright, maintaining a more constant angle. Tracheid walls become abnormally thick and contain more lignin and less cellulose than usual. In angiosperm (hardwood) trees, reaction wood forms on top and *pulls* the branch toward the trunk by tension. Angiosperm reaction wood contains more fibers and fewer vessel elements than normal wood. The fibers and fiber tracheids attain secondary walls that are thicker than in normal wood, because they form an additional cellulose-rich layer in the secondary wall. In contrast to conifer reaction wood, reaction wood in hardwoods does not become unusually lignin rich.

What is reaction wood a reaction to? It could be either a geotropic response or a response to tensions and pressures resulting from bending. If a cable is tied to a pine limb, causing the end to bend down (Fig. 18-10), reaction wood forms on the lower side. Either explanation might account for this effect. If the cable causes an upward bend, reaction wood forms on the upper side, showing that geotropism is not the only explanation. Pressures (compression) on cells of the concave side or tensions on the convex side seem more likely to account for the positions of reaction wood. But if a young leader of a pine tree is wrapped into a complete loop, reaction wood forms on the lower side of both horizontal parts of the loop. Since the lower side of the upper part is under pressure and the lower side of the bottom part is under tension, the second hypothesis also proves inadequate. Redistribution of IAA or other hormones might explain these results, but the cause of hormone movement may prove complex.

18.8 Nyctinasty

Movements of leaves of some species into a more nearly vertical position at night than during the daytime has been recognized for more than 2,000 years. As discussed in Chapter 20, such nyctinastic movements are rhythmic processes controlled by an unknown time-keeping mechanism (biological clock). Light absorbed by a pigment named phytochrome and perhaps also by another unidentified pigment (see Chapter 19) influences nyctinasty. Here we shall emphasize the anatomical and physiological aspects of such movements.

One of the species most carefully investigated is the silk tree, *Albizzia julibrissin* (Satter and Galston, 1973). It has doubly compound leaves, each leaf bearing several pinnae and each pinna bearing several pairs of opposite pinnules (leaflets) attached to a single rachilla. Such doubly compound leaves often exhibit striking sleep movements, as illustrated in Fig. 18-11. At night, the tips of opposite leaflets press closer together, rise upward, and become pointed toward the distal end of the rachilla. This movement is caused by relative changes in cell sizes on opposite sides of the base of the leaflet in a short zone called the **pulvinus** (Fig. 18-12). At night, water moves out of subepidermal cortical cells in the upper (ventral) side of the pulvinus and into cortical cells on the lower (dorsal) side. Swelling of dorsal cells and compression of ventral cells allows closing to occur, while these changes are reversed the next morning. Since changes in size of these motor cells are not permanent, we do not refer to them as growth movements.

What causes water to flow from one side of the pulvinus to another? In 1955, Hideo Toriyama observed for the comparable process in the sensitive plant (*Mimosa pudica*) that K^+ moves out of the cells that lose water. In the early 1970s, Ruth L. Satter, Arthur W. Galston, and others showed that the K^+ concentration of *Albizzia* pulvini is unusually high (almost $0.5\ M$) and that leaflet closing is accompanied by loss of K^+ from ventral cells and absorption of K^+ by dorsal cells, but that the dorsal cells do not receive the K^+ which the ventral cells lose. More recently, they have shown that Cl^- also shows changes similar to those of K^+ (Schrempf et al., 1976). Apparently, other cells or cell walls of the pulvinus act as reservoirs for these ions. Nevertheless, water movement is probably in response to the osmotic effect of ion transport, just as occurs in stomate opening (see Section 3.5). The same is true for nyctinastic movements in other species investigated. The principal problem remaining is to learn how light, temperature, or other aspects of the biological clock affect transport of these ions.

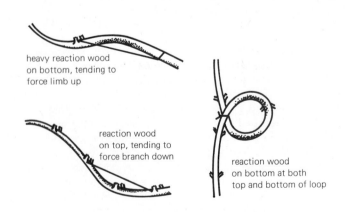

heavy reaction wood on bottom, tending to force limb up

reaction wood on top, tending to force branch down

reaction wood on bottom at both top and bottom of loop

Figure 18-10 Summary of some experiments causing reaction wood.

day

night

Figure 18-11 *Albizzia julibrissin* leaves in normal daytime position and in typical sleep (night) position. (Courtesy Beatrice M. Sweeney.)

18.9 Thigmonasty

Nastic movements resulting from touch were long thought to occur in only a few species, especially members of the Mimosoideae subfamily in the Fabaceae (formerly Leguminosae) family (Ball, 1969). The most notable of these species is *Mimosa pudica,* which has doubly compound leaves with leaflet and pulvinar structures similar to *Albizzia.* Upon being touched, heated, or treated with an electrical stimulus, its leaflets and leaves rapidly fold up (Fig. 18-13). Only one leaflet needs to be stimulated; some stimulus then moves throughout the plant. The advantages of this to the plant are uncertain, but one idea is that the collapse of the leaflets is useful because it startles away insects that alight upon the leaflets before they begin to eat the foliage. Leaflet movement is caused by water transport out of certain motor cells of the pulvinus. Again, an efflux of K^+ precedes water loss, at least in the main pulvinus at the base of each leaf.

How a stimulus can be transported in *Mimosa* has been investigated for many years (Pickard, 1973; Sibaoka, 1969). There is evidence for two distinct mechanisms, one electrical and the other chemical. The electrical response was first studied extensively by J. C. Bose, in India, between 1907 and 1914, then in better experiments by A. L. Houwink, in Holland, during the early 1930s. The electrical fluctuation proved to be an action potential similar to those occurring in animal nerve cells, and it apparently travels through parenchyma cells of the xylem and phloem at velocities of about 2 cm/sec. This action potential will not pass through a pulvinus from one leaflet to another unless the chemical response is also elicited, in which case several leaflets may fold up. The chemical response was first reported by Ubaldo Ricca, an Italian, in 1916. It is caused by a substance that moves through the xylem vessels along with the transpiration stream. Ricca cut through

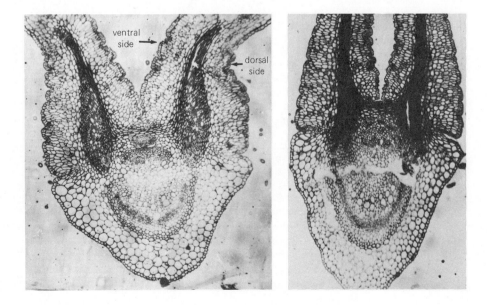

ventral side

dorsal side

Figure 18-12 Longitudinal sections through dorsiventral sections of *Albizzia julibrissin* pulvinules and adjacent leaflet tissues in the open (left) and closed (right) positions. The rachilla to which leaflets are attached is shown in transverse section at left and in oblique section at right. (From R. L. Satter, D. D. Sabnis, and A. W. Galston, 1970, American Journal of Botany 57:374-381.)

The Sense of Touch in Plants

Mark J. Jaffe

F.B.S. 1976

Are there still major discoveries left for you to make in biology or plant physiology? It should be evident from the many unknowns discussed in this text that major breakthroughs will almost certainly have to be made before some of the problems are solved. The slow, bit-by-bit accretion of data may not be enough in some fields. Mordecai (Mark) J. Jaffe (Department of Botany, Ohio University, Athens, Ohio) had the experience of realizing the importance of a phenomenon that had probably been casually noticed by many plant physiologists. His realization that plant responses to mechanical stimuli may well be as important as responses to other environmental factors is perhaps a small thing, but why hadn't most other botanists arrived at the same conclusion? There is certainly a premium on those persons who can see things in a different way. Mark Jaffe tells us his story:

As with many things that are fun, my initial experience with the sense of touch in plants came spontaneously and quite by accident. It was during the early months of my postdoctoral work with Arthur Galston at Yale. I had come, fresh with a Ph.D. in photobiology from Cornell, to study the phytochrome system. At that time, Galston's laboratory was exploring the possible involvement of flavonoids in the photo-morphogenesis of peas. Since they had previously shown that the tendrils of the pea plant produced prodigious quantities of the mole-cule of interest, I set out to use these tendrils as a convenient extraction source. I guess I had always been vaguely aware that tendrils coiled around supports, but I was caught completely off guard by the vigor and sensitivity of their response. Beguiled, I looked into the nineteenth-century literature on the subject (there were no recent studies), including books by Charles Darwin and Hans Fitting. "Well," I thought, "I'll just play with them on the side," and the "side" has grown to encompass more than half of my research. Both I and whatever I have been able to contribute in this field owe much to Arthur Galston, who recognized that I had taken the lure of the coiling tendril and had to follow where it led. Without his enthusiastic support and encouragement for an "offbeat" problem, my interests would undoubtedly be very different today.

Since starting with tendrils, I have branched out to many touch-sensitive systems, working on such plants as the sensitive *Mimosa,* the Venus's-flytrap, and the thigmotropic stamens of *Portulaca grandiflora.* One thing that fascinates me about all of these touch-sensitive organs is that to work with them, you must touch them, but when you touch them, they change. In other words, the very act of asking a question of them changes the answer. This sort of "botanical uncertainty principle" contributes enormously to the challenge of the problem and makes what success is achieved all the more satisfying.

However, to me at least, the thrill of the hunt is at least as exciting as its denouement, and the time came when I thought to pursue the quarry on a broader field. I had worked for 7 or 8 years with specialized organs that were specifically adapted to respond rapidly to touch. But now I asked, "What would happen if an ordinary plant, with no special adaptation, were subjected to mechanical stimulation?" After all, in the state of nature in which they evolved, plants are constantly being mechanically perturbed by wind, raindrops, soil particles and animals or other plants. Might they not have developed some kind of response to this stimulation? When tested, the answer turned out to be an emphatic "yes." All plants, although some more than others, respond to mechanical stimulation by elongating much more slowly and having their radial growth increased. This growth-retarding effect, now shown to be ubiquitous for vascular plants, I have given the name "thigmomorphogenesis." Just as the nastic and tropic responses to touch occur very quickly, so, too, does thigmomorphogenesis, with the earliest response (an electropotential change) being detectable in fractions of a second. If we think of wind as a stress, the plants respond by growing stockier and sturdier to withstand it better. This is mainly why greenhouse-grown plants often appear leggier and more spindly than those growing just outside, exposed to the wind. We have grown bean plants in the greenhouse or growth chamber, rubbing or bending them just once each day. When this is done, they look just like those growing in the garden, whereas unstimulated indoor plants or those growing outside but sheltered from the wind are taller and thinner. Similarly, trees growing in areas of frequent strong winds, such as at mountain timberlines or on cliffs overhanging large bodies of water, always present a gnarled and stunted appearance when compared to their more protected fellows.

I have recently begun to study the effect that mechanical stimulation has on other responses to the environment. The discovery that wind can interact powerfully with gravity or light to modulate geotropism and phototropism underlines the precautions that should be taken when studying those responses. It also helps to bring to the fore yet another environmental factor, the wind, as an influence on plant growth and development.

Until recently, there has been essentially no systematic study of the phenomenon of thigmomorphogenesis. There have, however, been many anecdotal mentions of the response in studies designed for other purposes. For example, in a 1908 U.S. Government bulletin describing field research on cotton boll weevil control, attempts were made to get rid of the insects by lightly hitting the plants with sticks. There are beautiful pictures in the publication of the dwarfing effect on the plants, but there was no further interest in the response itself.

It seems that the sense of touch in plants is quite a widespread phenomenon. There is a general effect on almost all vascular plants in which growth is retarded; and the tropic and especially the nastic responses to touch may be modifications of the morphogenetic effect. As far as the sensory function itself is concerned, it seems to have attributes that are common to all the systems that have been studied: The sensory reception is very rapid, and bioelectrical activity is quickly generated by it. For the future, it will be necessary to unravel the hidden mechanisms that translate the perception of a mechanical perturbation to the profound movements or changes in growth and development that can be observed. But in the present, it is beguiling to contemplate a stand of pine trees, a field of beans, or indeed any floral landscape, and perceive in the mind's eye the changes both subtle and dramatic that are wrought by every errant breeze.

Figure 18-13 Response of the sensitive plant. The tip of one leaf is stimulated in such a manner (in this case, by using a flame) that the entire plant is not jarred. After 14 sec, the petiole of the leaf has collapsed, and many of the leaflets have folded up. As the stimulus is transmitted along the stem, other leaves collapse and their leaflets fold up. Some of this activity can be observed even after 1½ min.

a stem, then connected each cut end with a narrow, water-filled tube. When a leaf on one side of the tube was wounded, a second leaf on the other side folded. The active substance can be extracted from wounded cells, applied to a cut branch, and its folding effects then measured. Although still unidentified, it is known as **Ricca's factor**. Its movement elicits electrical responses that travel ahead of it from one leaflet to another.

Action potentials in other plants have been studied by numerous investigators but usually with relatively crude methods. Besides *Mimosa*, one of the few well-studied examples in which an action potential is obviously useful to a plant involves excitation by an insect of one or more sensory hairs of the Venus's-flytrap (*Dionea muscipula*). Action potentials move into the bilobed leaf and cause the lobes to snap shut within a half second or so. This usually traps the insect, which is then digested by enzymes secreted from the leaf. Amino acids and other nitrogenous digestion products are used by the plant as sources of nitrogen. About 400 other angiosperm species are also carnivorous, and the capture mechanisms are diverse and often independent of action potentials (Heslop-Harrison, 1976).

In the early 1970s, Barbara G. Pickard began to study action potentials in various species at Washington University in St. Louis. She and her colleagues found that these are widespread among plants and can be caused by several stimuli, including heating, chilling, wounding, or treating with salt solutions (Pickard, 1973; van Sambeek and Pickard, 1976). They showed that Ricca's factor exists in several species, although it is still not identified, and that the long-distance spread of an electrical wave depends on its movement through the xylem. Furthermore, the arrival of the electrical potential in a leaf is accompanied by an increased CO_2 release and decreased photosynthesis and transpiration rates; the last two effects probably result from stomatal closure. The overall response to such stimuli is therefore probably an inhibition of growth, also indicated from thigmomorphogenesis studies described in Section 18.10.

18.10 Thigmomorphogenesis

In the early 1970s, M. J. Jaffe began to investigate effects of mechanical stimulation, especially rubbing, on plants not thought to display dramatic responses to such a stimulus (see his personal essay appearing on the facing page). He discovered that most vascular plants investigated respond by elongating more slowly and increasing in diameter slightly more rapidly, causing short, stocky plants (Fig. 18-14), sometimes

Figure 18-14 The effect of the number of rubs (once up and once down between thumb and forefinger, with moderate pressure) on the growth of young bean plants. From left to right, the number of stimuli were 0, 2, 5, 10, 20, or 30. (From M. J. Jaffe, 1976, Zeitschrift für Pflanzenphysiologie 77:437–453.)

| 0 | 2 | 5 | 10 | 20 | 30 |

only 40 to 60 percent as tall as controls. He called these and similar developmental responses to mechanical stress **thigmo-morphogenesis** (Jaffe, 1973). Bending the stems also causes these responses, and in nature the bending effects of wind influence plant development in this way. As a result of thig-momorphogenetic responses, the shorter, stronger plant is less easily lodged by subsequent winds than is a taller plant with a more slender stem. Rubbing by farm machinery and by animals probably has the same effect, but this may be unno-ticeable unless it is repeated frequently. Indeed, removal of the stimulus allows stem growth to resume as fast or faster than before.

The stems are not the only organs to exhibit inhibited growth after mechanical stimulation. F. B. Salisbury dis-covered in the early 1960s that simply measuring the lengths of cocklebur leaves with a ruler at daily intervals slows their growth and causes premature senescence. Others found that young sweetgum (*Liquidamber styraciflua*) trees stop elongating as rapidly and set winter terminal buds when their trunks are vibrated or shaken for only a 30-second period each day. Inhibitory effects on flowering and increased epinasty of a few species have also been observed, and it is likely that several additional thigmomorphogenetic responses will be discov-ered. Such responses may prove to be as common and thus important to plants as response to light, temperature, or gravity. Certainly, they might confuse the outcome of any experiment if controls and treated plants do not receive comparable mechanical stimulation.

What causes thigmomorphogenesis? Since the plants at first show no injury symptoms except an altered growth, a change in hormonal patterns is suspected. The decrease in stem elongation suggests a deficiency in IAA or a gibberellin, while this response, the increased stem thickening, the epinastic effects, and the known stimulation of ethylene pro-duction by mechanically stimulated plants (see Section 17.2) suggest an overabundance of ethylene in the affected epi-dermal and cortical cells (note effects of ethylene on such cells in Fig. 17-7). Since ABA can inhibit growth with little or no accompanying necrosis, an enhanced production or avail-ability of this hormone might also be involved. Nevertheless, the decreased photosynthetic rates accompanying arrival of the electrical wave in leaves, as discussed in the preceding section, probably also caused decreased growth rates.

How a change in hormone balance might come about is unknown, but Jaffe found that the electrical resistance (mem-brane permeability to solutes) of bean stems decreases greatly within a few seconds after rubbing, followed by a slower rise back toward the normal level. This is probably a manifesta-tion of the electrical responses reported by van Sambeek and Pickard (1976). A change in membrane permeability might rapidly affect availability of hormones at the normal sub-cellular sites where they are effective and might also affect subsequent production of hormones by altering the availability of hormone-precursor molecules to hormone-synthesizing enzymes.

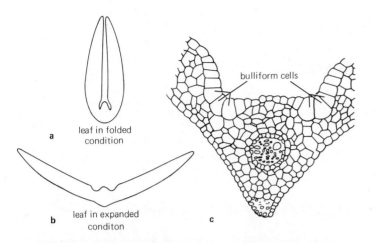

bulliform cells

leaf in folded
condition

a

b

leaf in expanded
conditon

c

Figure 18-15 (**a**) Outline drawing of blue grass (*Poa pratensis*) leaf in folded condition, (**b**) in expanded condition, (**c**) detailed drawing of midportion of the leaf showing the bulliform cells, changes in the turgor of which control the folding and opening of the leaf. (From Meyer and Anderson, 1952.)

18.11 Leaf-Folding and Leaf-Rolling Movements in Grasses

Many grass leaves either fold or roll up when subjected to water stress, and these processes supplement stomatal closure, further minimizing transpiration. Folding and rolling movements are caused by loss of turgor in thin-walled motor cells called **bulliform cells,** as illustrated for blue grass (*Poa pratensis*) in Fig. 18-15. The bulliform cells have little or no cuticle and so lose water faster than other epidermal cells. As their pressure potential decreases, turgor in cells on the lower side of the leaf causes the folding shown. This represents but one of several mechanisms by which plants resist drought, as discussed in Chapter 24.

19

Photomorphogenesis

Light is an important environmental factor controlling plant growth and development. A principal reason for this, of course, is that light causes photosynthesis. Furthermore, as emphasized in the last chapter, light has other effects on development by causing phototropism and by allowing geotropism in roots of some species. Numerous other effects of light independent of photosynthesis or phototropism also occur. Most of these effects control the appearance of the plant, that is, its development or morphogenesis (origin of form). The control of morphogenesis by light is commonly referred to as **photomorphogenesis.** One pigment absorbing such light has been identified and named **phytochrome,** but at least one other apparently exists.

The importance of light in photomorphogenetic responses can be noted easily by comparing seedlings grown in light with those grown in darkness (Fig. 19-1). The latter are **etiolated** (French *etioler,* "to grow pale or weak"). Several differences due to light are apparent:

1. Chlorophyll production is *promoted* by light (as discussed in Section 9.2).

2. Leaf expansion is *promoted* by light, less so in the monocot (corn) than in the dicot (bean).

3. Stem elongation is *inhibited* by light, less so in the monocot than in the dicot.

4. Root development is *promoted* by light in both species.

All these differences seem to be related to the necessity for a young seedling to extend its stem through the soil if its leaves are to reach the light. More of the food reserves in the endosperm (corn) or cotyledons (bean) are apparently used to extend the stem upward in darkness than in light, while less food is used for development of leaves and roots and for formation of chlorophyll in darkness, all of which are less important for a dark-grown plant. Besides these light effects, many others are essential to monocots, dicots, gymnosperms, and some lower plants. Such effects sometimes begin with seed or spore germination and often culminate in control over flowering. We describe several of these phenomena in the present chapter, but mechanisms of action are usually deferred to Special Topic 19.1.

19.1 Discovery of Phytochrome

The discovery and isolation of phytochrome and the demonstration of its importance as a pigment controlling many photomorphogenetic responses represents one of the most brilliant and important plant physiological accomplishments. Most of the research leading to phytochrome isolation was accomplished at the United States Department of Agriculture Research Station in Beltsville, Maryland, mostly in the 1940s, 1950s, and early 1960s. The history of phytochrome discovery has been summarized by Harry A. Borthwick (1972), who was one of its pioneers, and by Briggs (1976a).

An important observation had been made at Beltsville by W. W. Garner and H. A. Allard about 1920. They found that the relative durations of light and dark periods control flowering in certain plants (see Sections 20.7 and 22.2). Then, in 1938, Karl Hammer and James Bonner showed that the cocklebur, which requires nights *longer* than some critical minimum

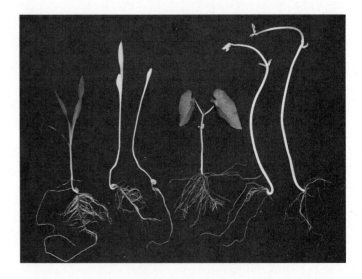

Figure 19-1 Effects of light on seedling development in a monocot (corn) and a dicot (bean). The plant at the left of each group was grown in a greenhouse, while the other representatives of each were grown in continuous darkness for 8 days.

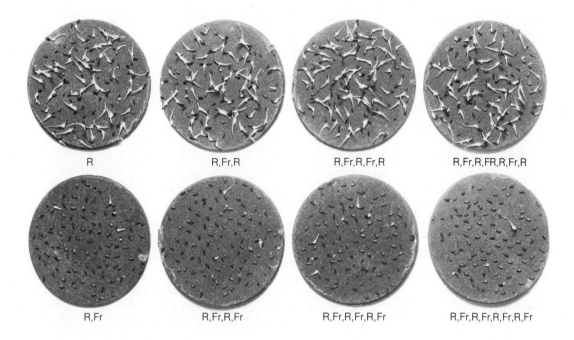

R   R,Fr,R   R,Fr,R,Fr,R   R,Fr,R,FR,R,Fr,R

R,Fr   R,Fr,R,Fr   R,Fr,R,Fr,R,Fr   R,Fr,R,Fr,R,Fr,R,Fr

Figure 19-2 Reversal of lettuce seed germination with red and far-red light. Red exposures were for 1 min and far-red for 4 min. If the last exposure is to red light, seeds germinate; if to far-red, they remain dormant. Temperature during the half hour required to complete the treatments was 7 C; at all other times, it was 19 C. (From Jensen and Salisbury, 1972; courtesy Harry Borthwick.

length to flower (a short-day plant), is prevented from flowering by a brief interruption of the dark period with light. Robert and Alice Withrow, in Indiana, and then Borthwick, Sterling B. Hendricks, and others at Beltsville began in the 1940s to find out what colors of light were most effective in long-night interruption, thereby obtaining clues about the pigment absorbing this light. The Withrows used colored filters, but the workers at Beltsville constructed a large spectrograph with which plants could be irradiated with different colors of light to obtain an action spectrum. Red light proved much more effective than the other wavelengths for interrupting the long nights that would otherwise induce cockleburs and Biloxi soybeans to flower and for promoting expansion of pea leaves. Red light interrupting a dark period was also most effective in stimulating flowering of Wintex barley and other long-day plants that require nights *shorter* than some critical maximum length.

Borthwick and Hendricks then collaborated with E. H. Toole and Vivian K. Toole, who were familiar with seed dormancy in many species. They obtained an action spectrum with a peak in the red for promotion of germination of Grand Rapids lettuce seeds, only 5 to 20 percent of which will sprout in darkness. It had been shown in the 1930s that red light stimulated germination of such seeds, but that after exposure to blue or far-red wavelengths, the percentage of seeds germinating was even less than in darkness. *Far-red includes those wavelengths just longer than the red, covering approximately the range 700 to 800 nm.* (Those longer than about 760 nm are invisible to humans and technically are infrared; as shown in Special Topic 3.1, Fig. 3.1-2. Visible far-red wavelengths appear dark-red to us.) The Beltsville group then made a remarkable discovery. When far-red was added just after a promotive red treatment, promotion was nullified, but if red

was given after far-red, germination was enhanced. By repeatedly alternating brief red and far-red treatments, they found that the light applied last determined whether the seeds germinated or not, red promoting and far-red nullifying that promotion (Fig. 19-2). Even the inhibition of flowering in short-day plants by red could be largely overcome by immediately following red with far-red.

By that point, they realized that a blue pigment was present that absorbed red light but that its concentration was too low to give color to etiolated grass seedlings in which it was present. (Its concentration was later shown to be approximately 0.1 μM in many plant parts, the approximate concentration of most hormones.) They also decided that the pigment could be converted by red light to a different form that absorbed far-red (a form that eventually proved light green in color), and that the blue pigment could be regenerated with far-red. The light-green form produced by red light was deduced to be the active form, while the blue form seemed inactive. These ideas, based only on physiological studies with whole plants or seeds, needed to be verified by extracting the pigment and studying it *in vitro*.

The problem was that when a beam of red light was passed through a sample to measure the absorption of the blue pigment form, the measuring beam itself converted the pigment to the other form, which absorbs far-red and not red. This was a frustrating problem for several years, but a special spectrophotometer was constructed that measured the difference in absorption between alternating beams of red and far-red light. By means of this instrument, the pigment was finally detected in 1959, first in intact etiolated turnip leaves (in which chlorophyll was not present to interfere with spectrophotometric measurements), and then in homogenized etiolated corn shoots (Butler et al., 1959). Red light caused an easily detectable increase in the absorbance of far-red by etiolated leaves or homogenates of such leaves, and exposure to far-red increased the absorbance of red. The results were repeatedly reversible, just as for the physiological processes of seed germination, flowering, and so on. The Beltsville workers brought their spectrophotometer and etiolated corn shoots to the International Botanical Congress in Montreal in 1959, demonstrating the presence of the pigment to a large audience.

Boiling the homogenate prevented such changes, indicating that the pigment might be a denaturable protein. The Beltsville scientists named the pigment phytochrome. In the early 1960s, H. W. Siegelman and others at Beltsville purified phytochrome from such homogenates by column chromatography and other techniques routinely used to purify proteins, demonstrated its color reversibility, and proved that it is a protein. By now, most of the early deductions based upon physiological experiments with whole plants have been verified.

19.2 Physical and Chemical Properties of Phytochrome

Absorption spectra of highly purified phytochrome molecules from angiosperms show maxima in the red at about 667 nm for the red-absorbing blue (P_r) and at about 724 nm for the far-red-absorbing, greenish form (P_{fr}) (Pratt, 1976). Figure 19-3 (top) shows an early example of these absorption spectra, while the middle and lower graphs in that figure illustrate action spectra in the red and far-red for various physiological responses studied at Beltsville. *The similarities of the absorption spectra of phytochrome and the action spectra give important evidence that phytochrome is indeed the pigment causing such responses. A second evidence is that the responses caused by red are almost always nullified by an immediate subsequent exposure to far-red. A third evidence is that very low irradiance levels of either red or far-red capable of interconverting phytochrome from one form to another can also cause these responses.*

Both P_r and P_{fr} absorb violet and blue light, but low irradiance levels of these wavelengths are usually much less effective than red or far-red for the physiological processes we have described so far. Since neither form absorbs green light effectively and our eyes are especially sensitive to green, safelights used in numerous physiological experiments in which phytochrome participates use filters that emit only green wavelengths.

Although phytochrome is a protein, the chromophore responsible for absorbing light is a nonprotein prosthetic group having an open-chain tetrapyrrole structure similar to that in phycobilins, molecules that act as accessory pigments for photosynthesis in red and blue-green algae. The probable structure of the chromophore is shown in Fig. 19-4. The chromophore is attached to the protein by an ester bond involving the carboxyl group of a propionic acid side chain on ring III in the chromophore and an unknown group in the

protein, and probably also by an unidentified bond between ring I and the protein. When converted to P_{fr} by red light, the conjugation of double bonds in ring I is changed, and the bond between ring I and the protein is also apparently altered through unknown subtle changes in protein structure. The change in protein is presumably responsible for the physiological activity of P_{fr} and inactivity of P_r.

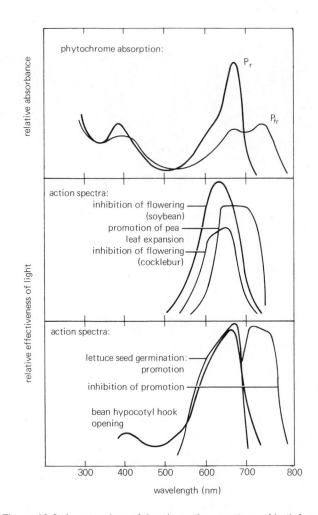

Figure 19-3 A comparison of the absorption spectrum of both forms of phytochrome with action spectra for various physiological processes. (Absorption spectra for P_r and P_{fr} purified from oats are data of W. L. Butler et al., 1965, in T. W. Goodwin, ed., Chemistry and biochemistry of plant pigments, Academic Press, Inc., New York, p. 203. Action spectra shown in the middle graph were redrawn from data of M. W. Parker et al., 1949, American Journal of Botany 36:194. Action spectra for promotion of bean hypocotyl hook opening shown in the lower graph were redrawn from R. B. Withrow et al., 1957, Plant Physiology 32:453. Action spectra for stimulation and subsequent inhibition of lettuce seed [Grand Rapids variety] were redrawn from data of Hendricks, 1960, p. 307.)

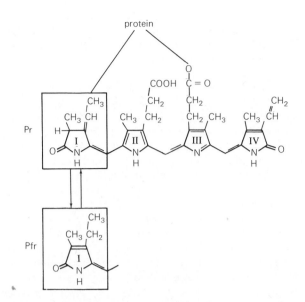

Figure 19-4 Proposed structures for the tetrapyrrole chromophore in P_r and P_{fr}. Only ring I of P_{fr} is shown, to emphasize that all known differences between the chromophores of P_{fr} and P_r exist in this ring. Electron rearrangements occur when the H attached at the upper left in P_r is lost and another is added to the ethylidene group. These changes cause the difference in color of P_r and P_{fr}. (Drawn from structures proposed by J. M. Lhoste and by W. Rüdiger, 1972, in K. Mitrakos and W. Shropshire, eds., Phytochrome, Academic Press, Inc., New York.)

19.3 Distribution of Phytochrome Among Species, Tissues, and Cells

What we have said so far applies to phytochrome of angiosperms. Does the pigment exist in other kinds of plants? It does so in gymnosperms, liverworts, mosses, ferns, and some green and red algae, suggesting that it might be present in all photosynthetic organisms with the exception of photosynthetic bacteria. Little is known about the properties of phytochrome in these species. In a green alga, in several pine species, and in the ancient gymnosperm *Ginko biloba*, the absorption peaks for P_r and P_{fr} occur at slightly shorter wavelengths than in angiosperms. However, even angiosperms display some variability in these peaks, probably because the kinds of neighboring molecules influence pigment absorption spectra, so there is no compelling reason to assume that algal and gymnosperm phytochrome is different from angiosperm phytochrome.

Phytochrome is present in most organs of all plants investigated, including roots. Quantitative measurements have been performed only with etiolated tissues, in which chlorophyll does not interfere with spectrophotometry. In etiolated plants grown in total darkness, phytochrome is present entirely as P_r, apparently because no P_{fr} can be synthesized in darkness. For many years, the only way phytochrome could be measured was by the difference in absorbance of the tissues after one form was produced from the other by exposure of the tissues to red or far-red light, as originally done at Beltsville. Physiological results show that phytochrome certainly occurs in green plants, but these amounts are thought to be lower than in dark-grown plants. In general, measurements in etiolated plants show high concentrations in meristematic zones and rapidly growing zones.

In 1974, Lee H. Pratt and Richard A. Coleman published another technique for phytochrome determination that even allows its identification in specific cells and subcellular organelles. Their technique is close to a thousand times more sensitive than absorbance methods and is reportedly applicable to green tissues (Pratt et al., 1976). They inject rabbits with highly purified phytochrome, allow the rabbit to form a specific antibody against the phytochrome antigen, extract the antibody, and add it to the plant tissues where the antibody and phytochrome then react. Detection of the reaction product is too detailed to discuss here, but both light microscopy and electron microscopy can be used. Their results show that root cap cells of grass seedlings contain high amounts of phytochrome, consistent with absorbance by the cap of light that stimulates geotropic sensitivity in certain grasses. Phytochrome distribution in grass shoots is variable, but oat, rye, rice, and barley seedlings all have high concentrations in the apical regions of the coleoptile, near the shoot apex, and (except for oat) in the growing leaf bases. Subcellularly, they identified phytochrome in etioplasts, in mitochondria, and in the cytoplasm outside any organelle, but not in nuclei or vacuoles. In a green alga, phytochrome also apparently occurs in the plasma membrane, so it is probably widely distributed among the cell's membrane systems. We need to understand its location to understand eventually how it controls photomorphogenesis. Biochemical aspects of phytochrome action are discussed in Special Topic 19.1.

19.4 Phototransformations of Phytochrome and Their Relation to Photomorphogenesis

The absorption spectra of P_r and P_{fr} cross at about 700 nm (Fig. 19-3, top). At this wavelength, the absorbance of each form is the same. Furthermore, at wavelengths less than about 730 nm, both forms absorb, so both will be present. This is an important fact that influences nearly all of our ideas about phytochrome action. In light from any source, a certain ratio of P_r and P_{fr} that depends upon the wavelength is established, red at 667 nm strongly favoring P_{fr} formation and far-red at 724 nm favoring P_r formation. Even with a narrow wavelength band near 667 nm, it is impossible to convert more than three fourths of the P_r to P_{fr}, since P_{fr} absorbs some of this light and is converted back to P_r. In sunlight or under incandescent lights used in growth chambers, the irradiance at

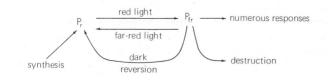

Figure 19-5 A summary of some transformations of phytochrome.

724 nm is close to that at 667 nm (see Fig. 3.1-3 or Fig. 19-6). Nevertheless, P_r absorbs red light more effectively than P_{fr} absorbs far-red, because the absorptivity coefficient for P_r is about 1.5 times as great as for P_{fr}. Therefore, sunlight acts primarily as a red source, forming more P_{fr} than P_r.

In most species, some of the P_{fr} gradually disappears even in darkness. Two processes seem to account for this. The first process is an apparent **destruction,** because after an interval of time in darkness it is no longer possible to regenerate as much P_r in the tissues by a far-red exposure, and the total amount of detectable phytochrome is less. This destruction might involve denaturation, because it has a high Q_{10} value of 3, typical of protein denaturation processes. It might also involve hydrolysis by a protease or simply an irreversible binding to something else in the cell. The second process is an apparent dark **reversion** back to P_r, usually requiring several hours. (It should be emphasized that destruction and reversion also occur in the light but are not caused directly by light.) Reversion seems to occur in most dicots and gymnosperms, but not in monocots or in any of the 10 dicot families often classified as part of the order Centrospermae. Because of these processes, we must modify our idea that light simply sets up a photostationary state between P_{fr} and P_r. Reversion and destruction (where either is applicable) must be added, as shown in Fig. 19-5.

Even Fig. 19-5 may not be sufficient to explain how light controls certain developmental processes via phytochrome. Interconversions of P_r and P_{fr} are not simple, immediate, one-step processes. At least six intermediate forms have been detected by their absorbance properties during P_r conversion to P_{fr}, and at least two different ones for the reverse process. Their functions are unknown. In nature, where both red and far-red light are provided by the sun, low levels of these intermediates are continuously present, and some of them might contribute to the responses caused by P_{fr}. We shall mention a few such responses later.

With a knowledge of some properties of phytochrome and some of the reactions it undergoes, let us now investigate in more detail some of the physiological processes it controls, beginning with seed germination.

19.5 The Role of Light in Seed Germination

Some Examples of Light-Dependent Germination The importance of light for germination of certain seeds has probably been recognized for hundreds of years, but the first comprehensive study was described by Kinzel in 1907 (Rollin, 1972). Kinzel reported that among 964 species, 672 showed enhanced germination in light. Most species that respond to light are small, nondomesticated seeds, rich in fat. Most of our cultivated seeds do not require light, no doubt partly because of man's selection against a light requirement. Seeds of many wild species even show inhibited germination in light, especially because of the blue and far-red present (Black, 1969). Kinzel found 258 of the species he studied to be light inhibited. The far-red and blue wavelengths are thought to be inhibitory because they decrease the amount of P_{fr} in the seed to a level below that needed for germination, either by converting P_{fr} to P_r as proposed by Smith (1973a) or to another form of P_{fr} that is inactive. Evidence supporting the latter idea will be mentioned later. A few of those studied most intensively include *Phacelia tanacetifolia, Amaranthus caudatus,* and *Nemophila insignis.* Interestingly, some even respond to the relative length of light and dark periods by acting as long-day seeds (*Eragrostis ferruginea, Cyperus inflexus,* and *Begonia evansiana*) or as short-day seeds (*Veronica persica*). Certain forest trees also produce seeds that respond to daylength (Evenari, 1965). Long-day seeds germinate only when the days are longer than some critical time period, while short-day seeds respond only to days shorter than some critical time period.

Seeds that require light for germination are said to be **photodormant.** As discussed in Chapter 21, we use **dormancy** as a general description of seeds or buds that fail to grow when exposed to adequate moisture, air, and a favorable temperature for growth. Seeds that normally germinate in darkness but that are inhibited by light also become dormant after the light exposure.

Interactions of Light and Temperature in Photodormant Cells A further complication in interpreting light effects on germination is an interaction with temperature. The temperature both before and after light treatment is much more important than that of the irradiation period itself, because the photochemical interconversions of P_r and P_{fr} are relatively independent of temperature, while the chemical reactions controlled by these pigments and those influencing their destruction are much more temperature sensitive. An example of crucial temperature control occurs in seeds of Grand Rapids lettuce and peppergrass (*Lepidium virginianum*). Light usually promotes their germination, but extended exposures to 35 C temperatures after a single light treatment or in continuous light keeps them dormant. Similarly, seeds of the Great Lakes variety of lettuce usually do not require light to germinate, but if they are soaked at 35 C, they become photodormant and then respond to red light. Still another example is provided by Kentucky bluegrass (*Poa pratensis*), in which alternating 15 and 25 C temperatures substitute for light in causing germination. It is usually assumed that temperature responses such as these are caused by effects on the amounts of P_{fr} in the seeds. Temperature could influence P_{fr} levels through effects on its rate of reversion to P_r, its rate of metabolic destruction, and perhaps also its rate of formation and the extent to which it combines with other compounds necessary to cause its final effect.

Ecological Aspects of Light in Germination Both photodormant and nondormant seeds will usually imbibe water and swell, but only the nondormant ones grow after the imbibition is complete (after the colloids are fully hydrated). Dormancy is normally broken by light only when the seeds are partially or fully imbibed. The time required for imbibition varies from as little as an hour to almost two weeks (Toole, 1973). Apparently, only then is P_r sufficiently hydrated to be transformed to P_{fr}. In seeds that survive many years in the soil, P_r is stable and only awaits the combination of moisture and light to cause germination. When Grand Rapids lettuce seeds are imbibed and exposed to light to form P_{fr}, they can be immediately dehydrated for as long as a year and then will germinate in darkness upon remoistening. This suggests that P_{fr} is also stable in dry seeds for long periods. An implication of stable P_r and P_{fr} is that whether or not a seed requires light to germinate can depend on how much P_{fr} was produced in it during ripening on the mother plant.

What possible ecological benefit is light or daylength to seeds lying in litter near the soil surface? Answers to such questions frequently involve speculation, and this is no exception. For long-day seeds, a clue that spring rather than winter is approaching might be provided by the long-day requirement. This appears to be true for the annual sedge *Cyperus retroflexus*, which also requires a chilling period normally satisfied by winter before long days are effective (Baskin and Baskin, 1976). For buried seeds that are promoted by light essentially irrespective of its length, germination only when they are uncovered more nearly assures that the seedlings will be able to photosynthesize. A light requirement for buried seeds might distribute germination over several years and thus help perpetuate the species, since only a fraction of the seeds present in soil might be disturbed and exposed to light in a given season. An unfavorable growing year might otherwise destroy all or most of the plants. Seeds that are inhibited by light, it was suggested, are prevented from germination until well covered by litter and then would be more likely to have sufficient water to grow. Koller (1969) described two light-inhibited species that inhabit coarse, sandy soils of the Negev desert, in which germination might be prevented unless the seeds are well buried where moisture is more plentiful.*

The idea that phytochrome provides seeds with a clue about whether or not they are likely to be covered by a canopy of other plants or exist in a more open area is appealing to physiological ecologists. The idea develops from two facts.

*These suggestions indicate the difficulty in speculation about ecological advantages of physiological or biochemical processes. We can usually provide an adaptive explanation for almost any natural process, even opposite responses such as a requirement for light or darkness. Such a situation has been called a *tautology*.

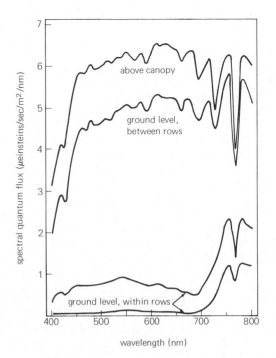

Figure 19-6 Influence of shading upon wavelengths of sunlight present in various regions of a sugarbeet field. Within the rows (bottom two curves) there is much less attenuation of far-red than of the other wavelengths, so shaded plants contain a higher proportion of phytochrome in the P_r form than do unshaded plants. (From M. G. Holmes and H. Smith, 1975, Nature 254:512–514.)

First, extended periods of far-red light usually inhibit germination of light-requiring seeds, and second, leaves in a canopy transmit considerably more far-red than red light. Most of the violet, blue, red, and much of the green wavelengths are removed by leaves through photosynthesis and reflectance, but more of the far-red passes through to seeds below. Figure 19-6 illustrates the spectral distribution of radiation above, between rows, and under the canopy within rows of a sugarbeet field. The lower curves show a small peak of transmission in the green region and much larger ones in the far-red. Under such a canopy, no more than 10 percent of the phytochrome would exist as P_{fr}. If seeds ripen on a plant under a canopy transmitting such a high ratio of far-red to red, they are likely to require natural light to form additional P_{fr} before they will germinate. In one study, 14 of 19 wild species that had no light requirement showed retarded germination when exposed to the far-red rich radiation transmitted by a large rhubarb or other kind of leaf (Górski, 1975). In an evergreen forest, many seeds requiring relatively high amounts of P_{fr} might never sprout until a fire, death of old trees, or timber removal eliminates the canopy.

Contrary to the inhibition of germination in many species by far-red and blue light, certain other species are promoted by exposures to these wavelengths. These seeds apparently are sufficiently sensitive to P_{fr} to respond to the small amounts of it produced by far-red or blue, especially when the exposure is continued for several minutes or hours to maintain the constant though small supply of P_{fr}. Blue light causes 20 to 40 percent of the total phytochrome to exist as P_{fr}, depending on the wavelength.

Is Phytochrome the Only Pigment Active in Germination? Although both promotive and inhibitory effects of blue and far-red radiation have often been attributed to phytochrome, several results are difficult to interpret this way. In tansy (*Phacelia tanacetifolia*), germination is suppressed not only by blue and far-red, but also by red light. If only P_{fr} is important to allow germination, why should wavelengths that cause widely different $P_{fr} : P_r$ ratios all be inhibitory? Furthermore, the amount of light required for inhibition of *Phacelia* is much higher than for typical phytochrome responses. The same is true for far-red and blue inhibition of Grand Rapids lettuce germination and for the inhibition in many other species by the far-red transmitted through a leaf canopy. *Whereas most clearcut phytochrome responses are saturated by energies of red light equal to as little as 10^5 ergs/cm² (less than one hundredth the intensity of visible wavelengths provided by sunlight during a 1-minute period), these far-red and blue effects frequently require at least 100 times more energy.*

Perhaps red, far-red, and blue cause different forms of P_{fr} to be produced. If so, the unknown forms supposedly produced by far-red and blue can be converted into ordinary P_{fr} by red light, because red reverses the inhibitory effects of blue and far-red on germination. If such other forms of P_{fr} neither promote nor retard germination, the inhibited germination of *Phacelia* and other seeds by light could be explained if synthesis of such forms occurs at the expense of ordinary P_{fr}.

Regardless of the explanation, effects of extended blue and far-red exposures on seed germination that are often different from those of classical phytochrome effects and require considerably more energy are also observed in other photomorphogenetic phenomena. Because of the frequent requirement for either a long exposure at a low or moderate irradiance or a shorter exposure at a higher irradiance, they are referred to as **high energy reactions (HER)** or, more commonly now, **high irradiance reactions (HIR)** (Shropshire, 1972; Schäfer, 1976). Whether they all result only from action of some form of phytochrome or from absorption by a quite different pigment remains uncertain, but most HIR responses to red or far-red are apparently caused by phytochrome through P_{fr} formation. The same is true for certain HIR effects of blue light, such as inhibition of elongation in pea stem segments, because these effects are overcome with a brief far-red treatment. Nevertheless, absorption of blue light by one or more flavins can, as in phototropism, produce other photomorphogenetic effects. We shall return to these HIR responses in section 19.6, where more extensive studies help clarify their importance.

The Nature of Photodormancy If P_{fr} is the pigment that causes photodormant seeds such as lettuce to germinate, why do they not sprout in its absence? To answer this, it is essential to identify that part of the seed in which P_{fr} must be formed. This problem was approached in lettuce seeds by separately irradiating the cotelydons and the hypocotyl-radicle tissues with red light. In *Cucurbita pepo*, a laser beam only 1 mm wide allowed even better resolution of the tissues exposed. In both species, germination results only when the hypocotyl-radicle tissues (hereafter called "radicle" in this section) are exposed. P_{fr} is also formed in the cotyledons when they absorb light, but this does not cause germination.

If we excise the lettuce embryo from the surrounding endosperm, seed coat, and fruit coat, the radicle itself now elongates in darkness or after a few minutes' exposure to either red or far-red light, but radicles exposed to red begin to elongate sooner and at a slightly faster rate than those kept dark or given far-red (Fig. 19-7). This growth-promoting effect of red is nullified by a short far-red treatment after the red. Furthermore, if naked embryos are prepared under a green safelight and placed in solutions containing sucrose, mannitol, or polyethylene glycol, those subsequently given red light will absorb water and grow in a solution having a more negative osmotic potential than those kept dark or given far-red (Nabors and Lang, 1971). *Our conclusion is that P_{fr} increases the growth potential of the radicle cells, presumably those in the elongating region, by decreasing their water potential.*

These facts suggest that germination fails in darkness because the radicle cannot grow with sufficient force to break

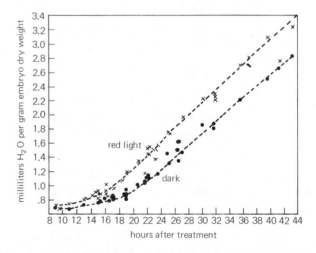

Figure 19-7 Stimulation of growth by red light in naked embryos of lettuce seeds. Seeds were soaked in distilled water for 3.5 hr, then some were given a 10-min treatment with a red light source. Endosperm, seed coat, and fruit coat were then removed under a green safelight, and growth (monitored by measuring water uptake) was measured at the time periods indicated. Each point represents the response of one embryo. (From M. W. Nabors and A. Lang, 1971, Planta 101:1–25.)

through the layers that surround it. Of these layers, the lettuce radicle is restricted almost entirely by the tough endosperm, even though it is only two or three cell layers thick. The endosperm is also the restrictive layer in *Phacelia* seeds; if it is removed, the radicles are no longer light inhibited. In seeds of certain other species, the seed coat or the fruit coat is a more important barrier. To understand photodormancy, we apparently need only to learn how P_{fr} causes elongation of radicle cells. As we explained in Section 15.10, the driving force for cell elongation is water uptake. Therefore, we must consider three not necessarily mutually exclusive alternatives in photodormancy:

1. P_{fr} causes a more negative water potential in the radicle cells by increasing the concentration of osmotically active solutes, making the osmotic potential more negative.
2. P_{fr} decreases the pressure potential (turgor pressure) in the radicle cells by loosening their walls.
3. P_{fr} weakens the endosperm, momentarily decreasing the pressure potential and causing water absorption by the radicle cells, allowing them to break through the endosperm.

Regarding the last alternative, P_{fr} apparently either causes the synthesis or the activation of a mannanase enzyme that degrades **mannans** (mannose polymers), which are unusually abundant in the endosperm walls. This degradation might weaken the endosperm enough to allow radicle protrusion, yet it occurs only after germination is half complete (Halmer et al., 1976). Presumably, the function of this mannanase is only to provide mannose sugars for the growing embryo.

To distinguish the first and second alternatives listed, Nicholas Carpita (1977) studied the influence of short red or far-red light treatments on the osmotic potential and the concentrations of specific solutes in the radicles of naked Grand Rapids lettuce embryos. To eliminate dilution effects resulting from faster water uptake in the red-treated radicles, these radicles had to be placed in a polyethylene glycol solution having a water potential about 3 bars more negative than that in which the embryos exposed to far-red were placed. Thus, radicles of both groups grow at the same rate, but, to equalize the growth rate, those exposed to red must have a water potential 3 bars more negative than those given far-red. About half of this water potential difference results from a more negative osmotic potential, while the other half must then result from a lower pressure potential caused by wall loosening. These results support both the first and second alternatives listed. If such results can be extrapolated to other organs where growth is stimulated by red light, they will help us understand how P_{fr} formation generally promotes growth.

Effects of Hormones upon Photodormancy In most photodormant seeds, applied gibberellins substitute for the light requirement. With a few species such as lettuce, cytokinins also substitute for light, and in several lettuce varieties, these hormones are sometimes much more effective than gibberellins in overcoming dormancy caused by high temperatures. In several fat-rich dicot seeds, including lettuce, cytokinins promote cotyledon expansion (see Section 17.1), suggesting that this expansion might cause germination by forcing the radicle through the endosperm and seed coat. Furthermore, cytokinin-induced swelling of lettuce cotyledons occasionally causes an abnormal germination in which the cotyledons break through the covering layers before the radicle does. Since cytokinins also promote radial expansion of the radicle cells, this force probably also helps break the endosperm barrier.

Auxins do not promote germination of photodormant or nondormant seeds and are instead either innocuous or inhibitory. The role of ethylene is less clear. It cannot break photodormancy, but it can break other kinds of seed dormancy in cocklebur and in certain peanut and clover varieties, and can partially overcome high temperature dormancy in lettuce. Abscisic acid almost always retards germination because of its growth inhibitory effects.

Collectively, these results suggest that P_{fr} might break photodormancy by causing synthesis of a gibberellin or a cytokinin, or by destroying an inhibitor such as ABA. The evidence about this is presently controversial, but no one has yet measured hormone changes only in the radicle cells that are responsible for germination. This seems essential to understand relations among light, growth promoters, and growth inhibitors in photodormancy and other kinds of seed dormancy discussed in Chapter 21.

19.6 The Role of Light in Seedling Establishment and Vegetative Growth

Once germination is accomplished, further development of the plant remains subject to control by light. We introduced some of these controls in Section 19.1 and Fig. 19-1. We shall now evaluate these and other effects and ask whether phytochrome is the only pigment involved.

Development of Grass Seedlings If a seed of a member of the grass family is planted deeply, the coleoptile elongates until its tip breaks through the soil. Just below the base of the coleoptile near the base of the scutellum (see Fig. 16-12), many grasses have an internode called the mesocotyl (first internode) that also elongates considerably in deeply planted seeds. (Barley, wheat, and some other members of the Festucoideae subfamily have no detectable mesocotyl.) The importance of elongation of the mesocotyl, coleoptile, and leaves enclosed by the coleoptile is apparently to carry the leaves into the light and to establish near the soil surface the adventitious (prop) roots produced at the node just above the mesocotyl (Fig. 19-8). The elongation rate of the coleoptile must equal or exceed that of enclosed leaves as they grow upwardly through the soil; otherwise, the leaves would grow out of the coleoptile and be broken off. How the growth rates

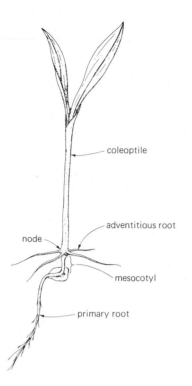

Figure 19-8 Some morphological characteristics of a week-old corn seedling grown in light. The coleoptile has stopped elongating, and two leaves have broken through it and have largely unrolled. The shoot apex is at the node where adventitious (prop) roots originate. The mesocotyl is the first internode formed above the seed storage tissues and the scutellum (cotyledon) in the seed.

of these two organs are coordinated is an interesting question.

Figure 19-8 illustrates a corn seedling grown in light from a shallowly planted seed in which the mesocotyl elongated very little and the first two leaves have emerged from the coleoptile. Each of these leaves was rolled up inside the coleoptile, but when exposed to light they began to unroll (flatten out). In Fig. 19-8, rolling is still evident only at the point of departure from the broken coleoptile.

From Figs. 19-8 and 19-1 (in which the coleoptiles are barely visible where the leaves have protruded through), we may conclude that light has numerous effects on development of grass seedlings. What role does phytochrome play in these? In the mesocotyl, red light inhibits elongation by half at energies as low as 5 μwatt sec/cm^2 (approximately 0.5 erg/cm^2). This energy is approximately equivalent to that in red wavelengths provided by bright moonlight during a 3-minute period. Mesocotyl elongation is apparently the most sensitive photomorphogenetic process known in plants. This red effect is largely overcome by a subsequent far-red treatment of low energy, showing that phytochrome is primarily responsible (Blaauw et al., 1968). A further inhibition decreasing the mesocotyl elongation rate to only 5 percent of the rate in darkness

is caused by much higher doses of red light (about 10^6 μwatt sec/cm^2). This inhibition is not reversible by far-red, as is common to many HIR responses. Both the failure of reversibility and the high irradiance levels indicate that this is a typical HIR effect. Since far-red converts ordinary P_{f_r} to P_r, its failure to overcome the HIR red effect may mean that high energies of red cause formation of a different form of P_{f_r} than do low energies of red, a form which cannot be removed by far-red.

The influence of light on coleoptile extension also involves phytochrome but is complicated by both promotive and inhibitory aspects. If oat seedlings are irradiated with white light when they are only a day or two old, the young coleoptile cells enter the elongation phase more rapidly than in those kept dark, so elongation is at first promoted. But if seedlings are 4 or 5 days old before they are irradiated, the cells are already elongating, and light is then inhibitory. Apparently, light hastens maturation of cells regardless of the phase they are in (Thomson, 1951). In meristematic cells and cells just beginning to elongate, this results in growth promotion, because the cells begin to elongate sooner. The same phenomenon apparently occurs in coleoptiles of wheat and barley (Lawson and Weintraub, 1975). In older cells in which growth is already well underway, the main effect is inhibition in all three species. If day-old seedlings are kept in continuous light, the overall effect always is retarded coleoptile length, even though at first this organ grows more rapidly (Fig. 19-9).

Unrolling of grass leaves is also controlled by a typical phytochrome response, low irradiance levels of red promoting and subsequent far-red nullifying the red effect (Fig. 19-10). Low energies of far-red are without effect, and low-energy blue is only slightly promotive. The process is caused by more rapid growth of mesophyll cells on the concave than the convex side. Whether this growth is caused by wall loosening, solute production, or both is not known, but there is a distinct red-light-promoted degradation of starch to sugars. These sugars not only provide energy and wall materials for the growing cells but are also solutes that cause osmotic water

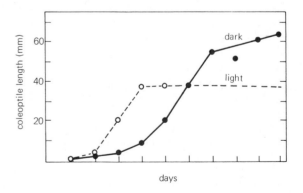

Figure 19-9 Elongation of oat coleoptiles in darkness and in continuous white light. Light at first promotes growth but later is inhibitory. (From B. Thomson, 1954, American Journal of Botany 41:326–332.)

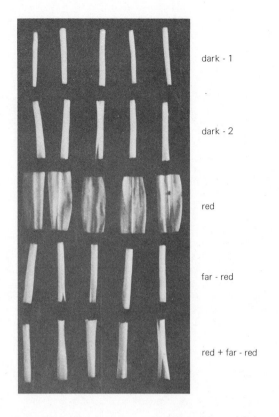

dark - 1

dark - 2

red

far - red

red + far - red

Figure 19-10 Effect of pretreatment with red and far-red light on unrolling of leaf sections from etiolated corn seedlings. Red promotes opening, while a subsequent far-red treatment nullifies the red effect. (From W. H. Klein, L. Price, and K. Mitrakos, 1963, Photochemistry and photobiology 2:233–240.)

uptake. Furthermore, red light stimulates the rates of chloroplast development and chlorophyll formation, thereby hastening photosynthesis in the expanding leaf.

In nature, of course, unfolding of grass leaves is promoted by much higher irradiances than are required to effectively convert P_r to P_{fr}, so the HIR system might again be involved. Investigations by Bente Deutch and Bernhard I. Deutch in Denmark (Deutch, 1976) show that it is involved and, furthermore, that at least one other form of P_{fr} probably participates. Using barley seedlings, they first measured the influence of red light energy levels varying more than 10,000 fold, covering both typical phytochrome and HIR response ranges. Such a wide range had not been studied previously for phytochrome responses. As the time of irradiance increased, leaf unfolding at first rose as expected, but with still longer exposures, the response surprisingly dropped and then rose again when the irradiances reached the HIR range. The resulting response versus irradiance curves are similar to the comparable phototropism curves shown in Fig. 18-7a, where first positive, first negative, and second positive responses are observed at increasing energy levels. Later, they found that although blue

is barely effective at low irradiance levels, increasing exposure generates a curve much like that for red, and in the HIR region the effects of the two colors are almost identical (Deutch, 1976). Low energies of far-red alone are not promotive, and they nullify responses of either low-energy blue or red, but higher energies of far-red do promote unfolding. Thus, in nature, blue, red, and far-red provided at high irradiance levels in sunlight all must contribute to the unfolding process.

To explain their results, Deutch and Deutch developed a computerized model that assumes formation of another active pigment, presumably a modified P_{fr}, by high energies of either red or blue light, but little formation of ordinary P_{fr} by blue. If this model is valid for other HIR responses, it should clarify our understanding of the rather complex effects of various wavelengths at different irradiance levels. The ineffectiveness of far-red at low energies but promotion in the HIR range was not explained by them, yet others have accounted for similar opposite effects of far-red as follows. Low energies simply convert nearly all the P_{fr} to P_r, the rest presumably reverting to P_r in darkness or being destroyed, while continued irradiance constantly reconverts a small amount of P_r back to active P_{fr} because of the overlapping absorption spectra of the two forms. The total supply of phytochrome is presumably maintained by continuous synthesis of P_r. In this way, far-red light acts as an effective red source during experiments of several hours involving low irradiance levels.

Development of Dicot Seedlings In dicots, the cotyledons either remain underground by hypogeal development, as in pea, or emerge above ground epigeally, as in beans, radish, and lettuce. In either case, a hook is formed at the stem apex that pushes up through the soil and pulls with it the fragile young leaves or cotyledons. As mentioned in Section 17.2, this hook forms as a result of unequal growth on the two sides of the hypocotyl or epicotyl in response to ethylene soon after germination. As the hook emerges from the soil, light acting through P_{fr} promotes opening of the hook. In pea and bean seedlings, this opening results from inhibition by light of ethylene synthesis in the hook. Differential growth resulting from faster elongation of cells on the lower (concave) side than on the upper (convex) side represents the driving force for hook opening. Accompanying this, P_{fr} promotes leaf expansion, chlorophyll formation, and chloroplast development, as in grass leaves (Fig. 19-1), and P_{fr} also promotes petiole elongation. If the cotyledons emerge first, essentially the same processes occur in them. In fat-rich cotyledons of **white mustard and radish, light also enhances conversion of** reserve fats and starch into sugars, but more of these sugars are used by the growing cotyledons and less are translocated to the root and shoot systems in light-grown than in dark-grown seedlings.

As the photosynthesizing leaves and the cotyledons begin to produce sugars in light, stem elongation is inhibited for a while, especially in dicots (Fig. 19-1). Of course, the plant will no longer elongate after its food supplies are exhausted, but while sugars are plentiful, light is inhibitory. This inhibition of stem elongation was apparently first recorded by Julius Sachs in 1852. He observed that stems of many species do not grow as fast during daylight as they do at night. We now realize that blue, red, and far-red all contribute to this phenomenon as does the biological clock (Chapter 20).

Much of our information about red light and HIR effects on elongation of dicot seedlings was obtained by the Beltsville group and by Hans Mohr and others in Germany. The Germans have worked extensively with etiolated seedlings of white mustard (*Sinapis alba*) and have measured many responses in these, summarized in Table 19-1. They also measured the action spectrum for inhibition of hypocotyl elongation in etiolated lettuce seedlings. These results are shown in Fig. 19-11. A fairly typical HIR action spectrum was obtained, with a peak in the violet, three peaks in the blue, and one far-red peak at about 720 nm. Note that red light is almost without effect here.

Certain other etiolated dicot seedlings show similar HIR responses to far-red and blue and lack of response to red light, but hypocotyls of etiolated cucumber seedlings are inhibited by HIR effects of red, far-red and blue wavelengths (Fig. 19-12), blue being most inhibitory (Black and Shuttleworth, 1976). In the cucumber, all these inhibitory effects are caused by light absorbed directly by the hypocotyl. The far-red inhibition requires special attention, because it is only observed when these and seedlings of other species are young. As

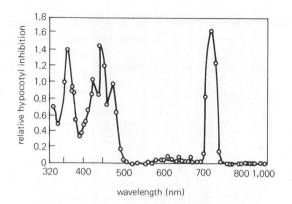

Figure 19-11 Action spectrum for inhibition of hypocotyl elongation in etiolated lettuce seedlings. Data are expressed relative to inhibition by blue light at 447 nm taken as 1.0. Light was applied continuously to the entire seedling for 18 hr, starting 54 hr after planting the seeds. Hypocotyl elongation was measured at the end of the light treatment. (From K. M. Hartmann, 1967, Zeitschrift für Naturforschung 22B:1172–1175.)

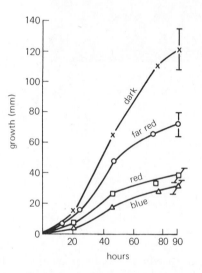

Figure 19-12 Elongation of etiolated cucumber hypocotyls (apical 1 cm section of 5-day-old seedlings) affected by continuous light of various wavelengths. Irradiance levels: blue, 300 μwatt/cm²; red, 350 μwatt/cm², far-red, 500 μwatt/cm². (From M. Black and J. E. Shuttleworth, 1974, Planta 117:57–66.)

Table 19-1 Some Effects of Light on Etiolated White Mustard Seedlings[a]

Inhibition of hypocotyl lengthening
Enlargement of cotyledons
Opening of the hypocotylar ("plumular") hook
Formation of leaf primordia
Development of primary leaves
Synthesis of anthocyanin
Inhibition of translocation from the cotyledons
Increase of the rate of chlorophyll accumulation (in white light)
Unfolding of the lamina of the cotyledons
Elimination of the lag phase of chlorophyll formation (in white light)
Changes in the rate of cell respiration
Increase in the rate of ascorbic acid synthesis
Changes in the rate of degradation of storage protein
Changes in the rate of degradation of storage fat
Increase of negative geotropic reactivity of the hypocotyl
Increase in the rate of long-term protochlorophyll regeneration
Increase of protein synthesis in the cotyledons
Hair formation along the hypocotyl
Formation of tracheary elements
Decrease of RNA contents in the hypocotyl
Increase of RNA contents in the cotyledons
Differentiation of stomata in the epidermis of the cotyledons
Increase in the rate of carotenoid synthesis
Formation of plastids in the mesophyll of the cotyledons
Differentiation of mitochondria in the cotyledons

[a]For references to these effects, see Mohr, 1974.

seedlings become older, far-red progressively loses its effect. The reason for this is unknown, but it certainly affects our conclusions about which wavelengths in natural daylight are responsible for suppressed elongation of stems of older plants, as recorded by Sachs and others. Although the blue and red wavelengths are inhibitory to elongation of more mature plants, far-red becomes stimulatory. We shall return to the ecological aspects of this problem in the next section.

Other important discoveries were made by Black and Shuttleworth after the etiolated cucumber seedlings were exposed to white fluorescent light for 30 hrs to deetiolate them. Upon deetiolation, the hypocotyl hook opens and the cotyledons become green. In such seedlings, far-red is no longer inhibitory to hypocotyl elongation, but red light absorbed by the cotyledons becomes strongly inhibitory. This red effect is reversible by a brief subsequent exposure to far-red, showing that P_{fr} causes the inhibition. When the cotyledons of deetiolated seedlings are covered, red light absorbed by the hypocotyl itself is less inhibitory to elongation than in etiolated seedlings, that is, the red effect partially shifts from the hypocotyl to the cotyledons during deetiolation. Why this shift occurs is puzzling, yet some effect of red light must be transmitted from the cotyledons which absorb such light to the hypocotyl, in which elongation is retarded. This implies control over hormone production. Whether red light inhibits transport of a growth promoter such as IAA, a gibberellin, or perhaps simply sucrose, or promotes transport of a growth inhibitor such as ABA is unknown. Regardless of the mechanism, the expansion of cotyledons in light probably acts as a correlative signal to reduce elongation of the hypocotyl still in the soil. Presumably, further elongation of this hypocotyl is no longer beneficial to the plant.

Let us now examine the effect of blue wavelengths on deetiolated cucumber seedlings. The HIR is clearly involved, because some P_{fr} formation by short exposures of such seedlings to blue does not inhibit hypocotyl elongation. Furthermore, even high irradiance treatments with blue are effective only if the hypocotyl is exposed. Why, if blue acts only through phytochrome, should it not be active in the cotyledons of deetiolated plants where red is effective? This difference and the fine structure of the HIR action spectrum for lettuce hypocotyl elongation (Fig. 19-11) suggest that the blue light causes HIR responses because it is absorbed by a pigment different from phytochrome. The close similarity between the action spectrum in the violet and blue regions shown in Fig. 19-11 to that of phototropism (see Fig. 18-5) is remarkable. The action spectra for various fungal and algal responses shown by Muñoz and Butler (1975) in which a riboflavin-containing protein seems to be responsible for light reception are also very similar to these. *Therefore, we tentatively conclude that although both blue and far-red can act by influencing production of some form of P_{fr}, the major effect of blue in many HIR responses is via a yellow pigment, probably a flavoprotein.* Since far-red is not absorbed by flavoproteins, the simplest explanation is that it acts in the HIR by causing small but continuous levels of P_{fr} to be present. This interpretation is substantiated with several complex experiments performed by the German workers, but we shall not discuss these. In conclusion, plants apparently employ two photosystems to cause many identical responses. If we learn how one pigment acts in these, we should obtain strong clues as to how the other does.

So far, we have only described the effects of light on stem elongation in dicot seedlings that develop epigeally, and we have emphasized only the hypocotyl. Is elongation also suppressed by light in epicotyls? Figure 19-1 indicates that for bean seedlings elongation of both the epicotyl and hypocotyl is inhibited by light, yet in the youngest internodes light usually at first increases the growth rate, even though definite inhibition occurs as the cells become older. In fact, this promotion followed by inhibition also occurs in the stems of oat seedlings and probably in many other grasses. The same explanation as was provided for light effects on coleoptiles (Fig. 19-9) probably accounts for these positive, then negative responses. Light apparently promotes the rate of cell maturation regardless of the growth stage, so young cells begin elongating faster and stop elongating sooner in light than in darkness. For cells already growing, enhanced maturation by light can only reduce elongation.

Photomorphogenetic Effects Later in Vegetative Growth In well-established but still-growing plants, other photomorphogenetic processes occur. If a dicot or a conifer grows under a leaf canopy where the light it receives is primarily far-red, the stems become considerably elongated (Fig. 19-13). This effect

Figure 19-13 Growth of *Chenopodium album* after 21 days under two different red/far-red ratios. Both plants were grown to the three-leaf stage under identical conditions; the one on the right was then provided light enriched in far-red. The estimated ratios of $P_{fr}:P_r$ in the two plants were 0.71 (left) and 0.38 (right). Each plant received the same amount of photosynthetically active radiation (400–700 nm). (From D. C. Morgan and H. Smith, 1976, Nature 262:210–212.)

is therefore opposite to the retarding effect on elongation of *etiolated* seedlings by far-red acting through HIR. Branching of stems is simultaneously retarded in many species under a canopy, so the plant uses a greater fraction of its energy in raising the stem apex toward the top of the canopy than it does when unshaded. In row crops, those plants in the more exposed outer rows are often shorter and more highly branched than those in rows within the field because of this effect. A similar phenomenon is often seen with plants on greenhouse benches. In thick stands of lodgepole pine, such as are abundant in Yellowstone National Park in Wyoming and many other mountainous areas of the northwestern United States, the result of retarded branching is a forest of plants having long, straight trunks that provide excellent timber because they are relatively knot free. The principle is used in selection of distances between transplanted seedlings in reforestation work. All of these effects of far-red light are apparently caused by decreasing the level of P_{fr} compared to that obtained in unshaded plants. *They suggest that a major function of phytochrome in nature is to detect mutual shading and to modify growth accordingly.*

19.7 Photoperiodic Effects of Light

In many species, responses to light absorbed by phytochrome are influenced by the time of day in which light is given. Such effects of light in interrupting the normal dark period or prolonging the normal period of daylight are referred to as **photoperiodic effects** (Vince-Prue, 1975). We have already mentioned short-day seeds and long-day seeds in this chapter, and we shall discuss photoperiodic flowering responses of various species in Chapter 22. In both short-day seeds and short-day plants, interruption of the night with a brief period of red or white light causing P_{fr} formation nullifies the otherwise inductive effect of darkness. In long-day seeds or long-day plants, extension of daylight with light induces germination or flowering.

Light also controls dormancy of buds and cold hardiness in woody plants through photoperiodic effects. As the days become shorter and the nights longer in late summer, buds of temperate zone species usually become dormant, their stems gradually stop elongating (see Fig. 15-4), and their resistance to frost damage rises. Short days thus act as a much more reliable early warning of winter than temperature decreases. Figure 19-14 illustrates the greater growth of young Douglas fir plants after one year on a long, 20-hour photoperiod (or on a shorter 12-hour photoperiod in which the dark period was interrupted with 1 hour of light) than on a 12-hour photoperiod without interruption of darkness. Such growth promotion is caused by P_{fr} formation during an otherwise dark period.

The buds of trees and shrubs remain dormant through most of the winter and are released from dormancy in late winter by increasing daylengths and/or cold temperatures they perceive, allowing them to grow during the warmer tempera-

Figure 19-14 Growth of Douglas fir (*Pseudotsuga menziesii* [Mirb.] Franco) after 12 months on photoperiods of 12 hr (left), 12 hr plus a 1-hr interruption near the middle of the dark period (middle), and 20 hr (right). (From R. J. Downs, 1962, in T. T. Kozlowski, ed., Tree growth, The Ronald Press, New York, p. 133.)

tures of spring. This interaction between photoperiod and temperature causing and breaking bud dormancy is discussed more fully in Chapter 21.

19.8 Light-Enhanced Flavonoid Synthesis

Most plants form anthocyanin pigments and other flavonoids in certain cells of one or more of their organs, and this process is frequently stimulated by light. A simple example is the faster development of the red color resulting from an anthocyanin in apple fruits on the south than on the north side of a tree. Production of flavonoids requires sugar as a source of the phosphoenolpyruvate and erythrose-4-phosphate that provide carbon atoms needed for the B ring and as a source of acetate units needed for the A ring. This sugar can arise from degradation of starch or fat in storage organs during seedling development, or from photosynthesis in chlorophyll-containing cells. It is no surprise, therefore, that anthocyanin synthesis can be stimulated by light acting photosynthetically in leaves or green apple fruit skins, yet light-promoted synthesis of these pigments in organs that photosynthesize little or not at all, including autumn leaves, flower petals, and etiolated seedlings, shows that at least one other pigment participates.

The action spectra for anthocyanin production in several species are shown in Fig. 19-15. In general, maximum responses occur in the red, far-red, and blue regions, while green (approximately 550 nm) is almost without effect. The peaks in the yellow, orange, red, and far-red regions vary considerably both in wavelength and height with the species. Blue is effective in all, and in sorghum seedlings red and far-red are ineffective. A detailed action spectrum in the blue region for synthesis of the aromatic acid precursors of the B ring in gherkin (*Cucumus sativus*) hypocotyls (Smith, 1972) is similar to that of phototropism shown in Fig. 18-10 and to the blue inhibition of hypocotyl elongation in lettuce seedlings shown in Fig. 19-12, suggesting that effective blue wavelengths are absorbed primarily by a flavoprotein. Red and far-red wavelengths act independently of photosynthesis in etiolated seedlings (Mancinelli et al., 1976; Duke et al., 1976), but in green apple skins photosynthesis also contributes. High irradiance levels characteristic of the HIR system are required for these red and far-red effects, and phytochrome is the only pigment known to be involved. The complexity of the action spectra suggests that additional forms of P_{fr} may contribute to these effects, as described for unfolding of grass leaves in Section 19.4.

Numerous attempts to determine the site or sites of light action in the biochemical pathways leading to both the A and B rings of flavonoids have been made. The accumulation of flavonoids in many autumn leaves during senescence suggests a relation between protein hydrolysis, phenylalanine appearance, and the use of phenylalanine in ring B formation. Since phenylalanine can be used in various metabolic pathways, control by light of the first step in its conversion to ring B was suspected. This step requires the enzyme phenylalanine ammonia lyase (see R14-4), and light definitely promotes its activity in various organs of many plants (Smith, 1972, 1973b; Wong, 1976). Nevertheless, several other flavonoid-synthesizing enzymes not mentioned in our book also exhibit increased activity after light treatment (Hahlbrock and Grisebach, 1975), indicating that production of both rings occurs more rapidly in light. As a result, no specific influence of light on a universal rate-controlling reaction of flavonoid synthesis can be identified.

Like the flavonoids, lignins are also formed from the shikimic acid pathway with the participation of phenylalanine ammonia lyase. In seedlings or in immature parts of older plants undergoing xylem differentiation or formation of xylem from the vascular cambium, lignin synthesis and incorporation into the xylem cell walls are promoted by light. This is partly responsible for the greater stiffness of seedlings grown in light than in darkness.

19.9 Effects of Light on Chloroplast Arrangements

When the light intensity is high, chloroplasts are usually aligned along the radial walls of the cells, becoming shaded by each other against light damage. In weak light and often in darkness, they are separated into two groups distributed along the walls nearest to and farthest from the light source, thereby maximizing light absorption. This movement of plastids depends upon the direction of light as well as its intensity and is an example of **phototaxis** (movement of an entire organism or organelle in response to light; Haupt, 1966). In mosses and angiosperms, the phototactic responses to both low- and high-intensity light are maximal under blue wavelengths, and phytochrome apparently does not participate (Inoue and Shibata, 1973). Action spectra suggest that a flavoprotein is again involved. In certain algae, however, phytochrome does absorb low-intensity light responsible for movement of the chloroplast or chloroplasts to the regions of the cells where light absorption is increased. In the green alga *Mougeotia*, the effective phytochrome molecules are probably located in the plasma membrane. Here again there is evidence for participation of an additional form of P_{fr} (Haupt and Bretz, 1976). In general, it appears that the chloroplast itself is not responsible for absorption of light causing phototaxis and that influences on cytoplasmic streaming cause the ATP-dependent process (Haupt, 1973).

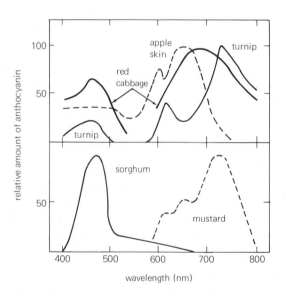

Figure 19-15 Action spectra for anthocyanin formation in various species after prolonged irradiance. The apple fruits contained chlorophyll, but the turnip, cabbage, sorghum, and mustard seedlings were probably chlorophyll free when irradiance began. (Data redrawn from various sources. Apple skin data are from H. W. Siegelman and S. B. Hendricks, 1958, Plant Physiology 33:185–189; red cabbage and turnip, H. W. Siegelman and S. B. Hendricks, 1957, Plant Physiology 32:393–398; sorghum, R. J. Downs and H. W. Siegelman, 1963, Plant Physiology 38:25–30; mustard, H. Mohr, 1957, Planta 49:389–405.)

20

The Biological Clock

Change is the only thing that an organism can count on in its environment. Almost nothing is really constant. In a study of the alpine tundra on the north end of Rocky Mountain National Park in Colorado (Salisbury et al., 1968), several kinds of more or less cyclical changes in environmental factors became evident: Wind velocities changed significantly in less than a second. Temperatures, light intensities, and humidities sometimes changed radically in time intervals from 10 minutes to perhaps 5 or 6 hours. Of course, there was a daily (**diurnal**) change in all these as well as other factors. Weather cycles typically lasted several days. In one summer, for example, heavy storms were separated by intervals of from 10 to 22 days, with an average of about 13. Exceptionally good days were separated by intervals of 5 to 14 days, with an average of about 10. The annual seasonal cycle was extremely evident.

Furthermore, weather trends may be related to the 11-year sun spot cycle, and long-time climatic changes may occur over periods of centuries to millennia. Tidal cycles control the tidal zone environment and may well be important in other habitats as well.

It would be to an organism's advantage to anticipate and adjust to these environmental changes. At least three of them—those related to the mechanics of the solar system— are regular enough that this should be possible: the diurnal, lunar (tidal), and annual cycles. For an organism to anticipate and prepare for these regular changes in its environment, it needs a clock and various associated mechanisms. The clock's system should have at least two broad sets of characteristics.

First, it should be accurate. It should not be strongly influenced by capricious elements of an organism's environment, those that cannot accurately be predicted: temperature, light intensity during the day (which varies due to clouds and shading), wind velocity, moisture, and so on. Even if a biological clock were not highly sensitive to these factors, which would be surprising and impressive, it would be even more impressive if such a clock could run with the accuracy achieved by our own mechanical clock systems. Yet without such accuracy, it might soon get out of phase with the environment and, therefore, be of no benefit to the organism.

There is an alternative to inherent accuracy: The biological clock might frequently and regularly be reset by some dependable feature of the organism's environment. It wouldn't matter much, for example, if the clock were to gain or lose an hour or two a day if it were reset each day at sunrise and/or sunset. Presumably, the clock might have one status appropriate for daytime and another appropriate for night; resetting at dawn and dusk might keep the clock's status in appropriate synchronization with the environment. Incidentally, it is conceivable that a "clock" might simply be driven by changing features in the environment. In this case, the organism would only track environmental changes without anticipating future environmental changes.

Second, there must be mechanisms that allow the clock to be used. It is reasonable, for example, that a plant might conserve energy if it could direct and concentrate its available resources to the photosynthetic mechanism during the day and to other metabolic mechanisms during the night. This requires a coupling system so that the clock can control the available resources within the cell, depending upon the time of day. Probaby the clock must control metabolism in general, plant or animal activities, and possibly many other features that might function more efficiently if time is measured and events are anticipated.

With these things in mind, we might go to nature to observe phenomena that could be manifestations of biological time measurement. Finding such, we could study their features and the roles that they might play in the organism's existence. Indeed, many such phenomena have been observed, and much is known about their manifestations. It appears that virtually all eucaryotic organisms (those with true nuclei) do have biological clocks. Sometimes it is easy to see clock control of metabolism and activity. There are also examples of highly sophisticated clock responses that we might not have expected.

Confronted with these observations, we immediately ask: What is the mechanism of the clock? How does it work? The observed phenomena imply a few things about the nature of the clock, but at the moment little is known about actual clock mechanisms.

20.1 Endogenous or Exogenous?

As illustrated for two plants in Fig. 20-1, the leaves of many species exhibit one position during the daytime (typically nearly horizontal) and another position in the middle of the

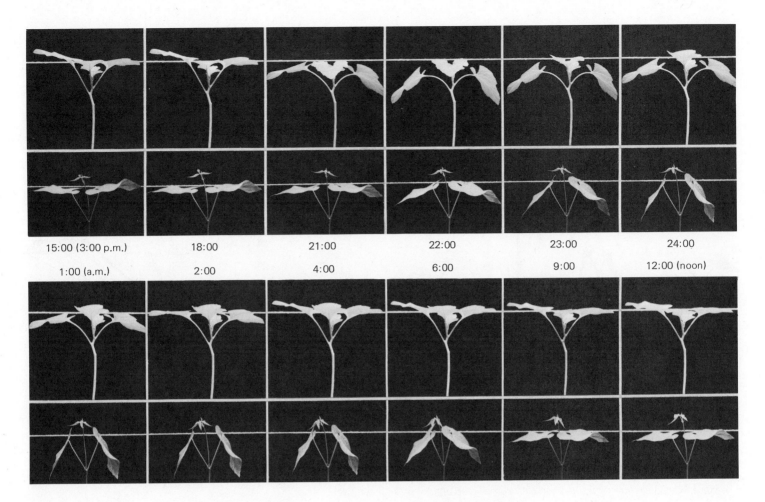

Figure 20-1 Leaf movements in cocklebur (top row) and bean (second row). Plants were photographed at hourly intervals from noon to noon. Twelve photographs were selected for the figure. The bean leaves drop more sharply and somewhat later than the cocklebur leaves.

night (typically nearly vertical). This observation was made at least as early as 400 B.C. by Androsthenes, who was the historian of Alexander the Great.

In 1729, the French astronomer DeMairan was perceptive enough to recognize a fundamental problem in relation to this diurnal cycle of leaf movements in plants. He wondered if the movement were driven by changes in the environment (the daily light-dark cycle), or if it might be controlled by some time-measuring system within the plant. If leaves moved only in response to *external* changes, we could say that timing was **exogenous;** if in response to an *internal* clock, we would say timing was **endogenous.** Using the sensitive plant (*Mimosa* sp.), DeMairan observed the movements even after the plants had been placed in deep shade. Since these motions did not require intense sunlight during part of the 24-hour cycle, he suggested that the movements were endogenously controlled.

A few other early workers, including Charles Darwin and Julius Sachs, were also interested in these rhythms and pub-

lished preliminary studies relating to them. But the early investigator who probably devoted the most time and effort to this topic was Wilhelm Pfeffer, who, from 1875 to 1915, **wrote many papers about the leaf movements of the common bean plant** (*Phaseolus vulgaris*). Much of his extensive work is still of interest. When he began, he was skeptical of an endogenous clock, but by the end of his researches he became convinced that such a clock must exist. Ironically, he was unable to provide experimental data sufficiently convincing to convert other scientists of his time. During Pfeffer's time, scientists were also observing and reporting rhythms in animals, especially diurnal rhythms of activity.

The real breakthrough came in the 1920s. Rose Stoppel in Hamburg, Germany, had continued the researches of Pfeffer on leaf movements in bean plants. Using the method developed by Pfeffer (Fig. 20-2), she attached the bean leaf blade (after the stem and the petiole had been fastened to bamboo sticks) with a thread to a lever that contacted a moving

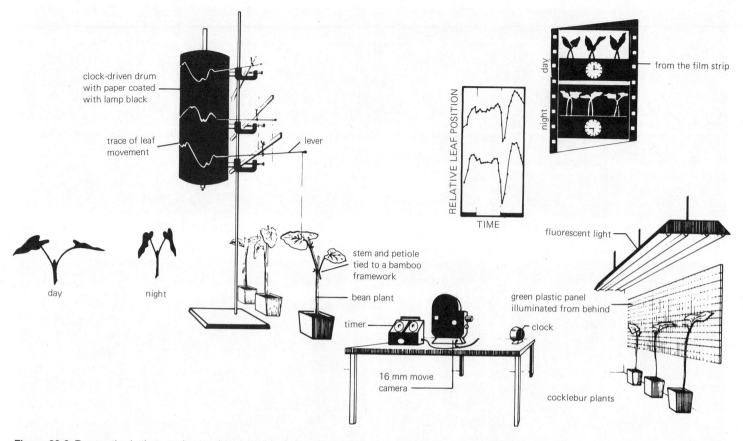

Figure 20-2 Two methods that can be used to record leaf movements. Left, the classical kymograph (clock-driven drum) method of Pfeffer and Bünning; right, a time-lapse photography method. Night photographs in the time-lapse method are time exposures with plants silhouetted against a dim green background.

drum that had been coated with lamp black. The lever traced a record of the leaf movements. (Note that when the leaf is *up,* the end of the lever tracing the record is *down.*)

Stoppel observed that when leaf movements were measured in a dark room at constant temperatures, the maximum vertical position was observed at the same time each day. She reasoned (as we have in the introduction above) that no biological clock could be this accurate; some factor in the environment must be resetting the clock on a daily schedule. Since the plants were in the dark and at constant temperature, this factor could not be daylight or temperature. Stoppel suggested that it was "factor x"!

Two young botanists in Frankfurt, Erwin Bünning and Kurt Stern, were looking for a research problem involving subtle physical factors of the environment, such as the ionic content of the atmosphere. They thought that such a factor might be responsible for timing the leaf movements in Stoppel's experiments. They found that ions have no effect on the rhythms, but they identified Stoppel's factor x. When this factor was eliminated, then Stoppel's prediction held true: The maximum vertical position of the leaves came about an hour and a half later each day, so that the leaf movement cycle was soon out of phase with day and night outside the dark

room. Bünning tells this story in his personal essay on page 312, where he also identifies factor x for you.

Since the rhythms had a so-called **free-running period** that was greater than 24 hours, there seemed to be no other alternative but an endogenous clock. The rhythms were not simply tracking the normal day and night cycle; they would continue in the absence of day and night and at constant temperature. Normally, they were reset or entrained to the natural cycle of day and night, probably by dawn and/or dusk. In the absence of these features of the natural evironment, the clock would "run free," betraying its inaccuracy and, hence, its endogenous nature.*

In the 1950s, Franz Halberg at the University of Min-

*Pfeffer and even A. P. DeCandolla in 1832 had observed circadian free-running periods, but they had not attached proper significance to their observations. Yet Antonia Kleinhoonte, working independently in Delft, Holland, arrived at exactly the same conclusions as those of Bünning, Stern, and Stoppel. Results of her and their experiments were published between 1928 and 1932. Stern later emigrated to America and lost interest in the biological clock; Kleinhoonte also lost interest. Bünning, however, continued his studies until his recent retirement.

nesota suggested that rhythms with a free-running period of approximately but not exactly 24 hours should be called **circadian.** This term is coined from Latin *circa,* which means "approximately," and from *diem,* or "day." The term is now widely applied.

20.2 Circadian and Other Rhythms

There are several modern areas of knowledge that clearly belong in the field of *biology* rather than *botany* or *zoology.* Examples are the principles of inheritance, the pathways of cellular respiration, and the cell theory. Biological clocks and circadian rhythms also occur in both plants and animals, possibly in all eucaryotic cells.

Several rhythms have been studied in single-celled organisms. Phototaxis in the green alga *Euglena* and a mating reaction in *Paramecium* are good examples. Among single-celled organisms, the biological clock has probably been most thoroughly studied in *Gonyaulax polyedra,* a marine dinoflagellate. This organism is mobile, having two flagellae. Beatrice Sweeney (see the personal essay, page 314) and J. Woodland Hastings, then located at the Scripps Institute of Oceanography at La Jolla, California, were the first to work intensively with this organism. They documented three separate rhythms. Most spectacular is a rhythm of bioluminescence observable when a suspension of *Gonyaulax* cells is tapped or otherwise jarred, causing the organisms to emit light. The quantity of light they emit follows a circadian rhythm, with the peak of the rhythm normally occurring near midnight. There is also a rhythm in cell division, with the maximum occurring near dawn. The third rhythm is in photosynthesis. The quantity of CO_2 fixed with a given illumination varies according to a circadian rhythm, with the maximum usually occurring near noon; that is, the photosynthetic mechanism is adjusted by the clock to anticipate the environment, just as we suggested in the introduction that it might be. It is interesting to note that we are dealing here with a population of organisms, rather than with individuals as in Bünning's original studies.

Among the fungi, there is a circadian rhythm in formation of conidiospores in *Neurospora crassa* exhibited as a series of dark bands in the mycelia growing in a growth medium in a long culture tube. This rhythm is being studied intensively at present. Another example is the rhythm of spore discharge in *Pilobolus.*

Many rhythms besides leaf movements have been observed in higher plants. These include petal movements, rates of growth of various organs, concentrations of pigments, stomatal opening and closing, discharge of perfume from flowers, times of cell division, metabolic activity (e.g., photosynthesis and respiration), and even the volume of the nucleus, which was observed to fluctuate according to a circadian rhythm. Of special significance could be the observation that many plant species exhibit a diurnal rhythm in sensitivity of response to certain environmental factors, notably temperature. Many species flower or grow well only when tempera-

tures during the part of the cycle that normally comes at night (**subjective night**) are lower than temperatures during subjective day. Or light given during subjective night may actually inhibit some plant responses. Thus, the clock adjusts a higher plant's metabolism to coincide with its cycling environment.

Several insects exhibit rhythms convenient for study. The time of day at which adult *Drosophila* emerge from the pupae follows a circadian rhythm and provides an intensively studied example, as does the activity of the cockroach. **Activity** or **running cycles** have also been studied in birds and rodents. These rhythms are particularly valuable because they often continue for long intervals (months or even years), and they are relatively easy to study. Various automatic devices may be installed in the cage, allowing for continuous recording of an organism's activity. For example, a microswitch attached to the perch can indicate bird activity, and electrically monitored running cages are often used for rodents.

It is important to realize that the rhythms with obvious outward manifestations (leaf and petal movements, activities, and others) may be less important than the internal metabolic changes controlled by the clock. This is true for plants, as mentioned earlier, and many metabolic cycles (e.g., potassium levels in the blood, urine excretion, and body temperature) have also been documented in animals. These are undoubtedly important in attaining an adjustment between an organism and its environment. A striking observation that may help us realize how subtle these adjustments can be is that an organism may be far more sensitive to toxic chemicals or even to ionizing radiation (particularly X rays) during a part of its circadian cycle. Perhaps animals conserve energy during the inactive part of their cycle by lowering resistance to factors not likely to be encountered. *In any case, anyone doing experiments with plants or animals should be aware of the profound effects of the biological clock on virtually all aspects of an organism's functions.* The *time* when a treatment is given is often decisive.

Many circadian cycles have also been studied in man, although in some ways man is a difficult object because his cycles may be timed by such factors as his wristwatch! Nevertheless, detailed studies in special bunkers near Münich, Germany, for example, or on long flights from Oklahoma to Japan indicate that the circadian rhythms of man (e.g., sleep versus activity, urinary excretion, temperature, and pulse frequency) follow the same general principles that apply to the rhythms of other organisms. A particular feature of one study was that certain of the cycles could be rather easily adjusted to a new cycle, while others were somewhat more resistant.

Although much study remains to be done, many biologists are intrigued with noncircadian rhythms. Short cycles (minutes to hours) are now called **ultradian.** Changes in metabolic components provide examples. Color, activity, and metabolism of fiddler crabs and other organisms have shown cycles in a laboratory closely matched to the tides of the bay where the organisms were collected. Lunar rhythms (called **circalunar**) are closely related to tidal rhythms. The grunion (*Leuresthes tenuis*) is a small fish living off the coast of Southern

Biorhythms and Other Pseudoscience*

There is talk currently about biorhythms. The notion is that human behavior is controlled by three cycles, each of which is initiated at the moment of birth: a physical cycle of 23 days, a sensitivity (emotional) cycle of 28 days, and an intellectual cycle of 33 days. The first half of each cycle is supposed to be the time when one is most positive in the attribute of that cycle; during the second half, one is supposed to be negative. The crossover days from plus to minus or minus to plus are critical days, and if the critical points for two or three cycles fall on the same day (which happens about six times a year for two cycles and once for three), you had better watch out!

The concept was developed from about 1897 to 1932 by certain medical doctors and others in Vienna, Berlin, Innsbruck, Philadelphia, and other locations. It is usually presented as a "scientific" doctrine. Some companies such as the Ohmi Railway Company in Japan calculate the cycles for their employees, warning them of critical days. The accident rate is reported to have dropped by more than 50 percent!

Yet there is little objective evidence to support the hypothesis. Most evidence is anecdotal: Such and such movie star is reported to have had a bad accident on a triple critical day! Many adherents to the doctrine swear by it, but of course, their results might well be examples of the self-fulfilling prophecy: After you have plotted your charts for several months in advance, you will be expecting good and bad days and subconsciously or otherwise you may adjust your life to meet these expectations. If you keep a careful diary, you might test the theory by plotting your charts for a *past* interval covered by the diary and looking to see if anything special happened on the good or bad days—but to be objective, you must also note special things that happened on other days as well.

It should be clear from the discussions in this chapter that the basic premise upon which the concept of biorhythms rests has no foundation in scientific observation. Rhythms clearly exist in or-

ganisms, including man, but they have three features at total variance with those of the biorhythm hypothesis: They are typically *circa*, approximating but almost always varying from an exact period length unless they are continually entrained to a cycling environment; they often vary from individual to individual within a species; and they are relatively easy to shift by various environmental factors. Their periods are *plastic* and not *rigid*. The rhythms discussed in this chapter could never maintain exact periods of 23, 28, or 33 days from the time of birth throughout the proverbial four score years and ten of an individual's life.

Speaking of matters pseudoscientific, the botanical sciences have spawned their share. There are those who suggest that talking to your plants—or praying over them—makes them grow better (and perhaps it does, if you thereby increase the CO_2 concentration or take better care of them). There have been several papers purporting that music makes plants grow better. There are even special records on the market, claiming to provide the best music for plants. (This needs much work, but it is vaguely possible that sound waves might shake up the cellular organelles and influence plant growth—but classical music and not rock music?!)

Probably the most notice has been paid to the "experiments" in which a polygraph (lie detector) was attached to a plant, showing wild responses when bad things happen, such as another plant being "murdered" in the same room or brine shrimp being dunked in boiling water. Do plants really have feelings and emotions?

If so, it surely remains to be demonstrated. The "experiments" apparently worked once, but no one has been able to repeat them consistently. And consistent, objective verification is what we must demand when we encounter outlandish claims that apparently imply that our most fundamental concepts are in need of revision. Truly, progress often depends on startling and unexpected discoveries, but it is as common for such claimed discoveries to be mistaken interpretations or due to poorly designed experiments as for them to be real advances in knowledge. We are entitled, even obligated, to test new claims by insisting upon verification by objective observers who thoroughly understand the role of controls in an experiment, and who understand all the factors that might influence the outcome. For example, the polygraph responses observed when "murder" was perpetrated near the plant attached to the machine were at about the same level as the "noise" to be expected if the polygraph were attached to an inanimate object. Could the results have been due to coincidence? Of course, and that is likely.

*For further information, see:

Gauquelin, M. 1969. The cosmic clocks. Avon Books, New York.
Galston, Arthur W. 1974. The unscientific method. Natural History, March. p. 18–24.
Mackenzie, Jean. 1973. How biorhythms affect your life. Science Digest 74(2): 18–22.
Rodgers, C.W., R. L. Sprinkle, and F. H. Lindbert. 1974. Biorhythms: Three tests of the "critical days" hypothesis. International Journal of Chronobiology 2: 215–310.
Thommen, G. 1973. Biorhythms: Is this your day? Avon Books, New York.

California that spawns from late February to early September during three to four nights at the new and the full moon and during the descending tidal series. Rhythms related only to the moon have also been observed, and because of the 28-day period, it has been suggested that the human menstrual cycle might be a circalunar rhythm. Present evidence is against this. The length of the cycle depends on several factors (e.g., it shortens as menopause approaches) that do not match the lunar cycle.

In certain ground squirrels held under constant conditions, entrance into and termination of hibernation have been shown to follow an annual rhythm of about a year (**circannual rhythm**). Amount of daily wheel-running activity of these animals also follows such a cycle. Germination of certain seeds appears to be best at certain times during the year, even though the seeds have been stored under conditions of constant temperature, light, and moisture. Much more work must be done before all the implications of these apparent circannual rhythms are understood.

20.3 Basic Concepts and Terminology

It is helpful to use the terminology applied to physical oscillating systems, although this terminology is sometimes used in a rather special sense in relation to the rhythms. The oscillations may be thought of as having three characteristics (Fig. 20-3): *First*, the **period** is the time between comparable points

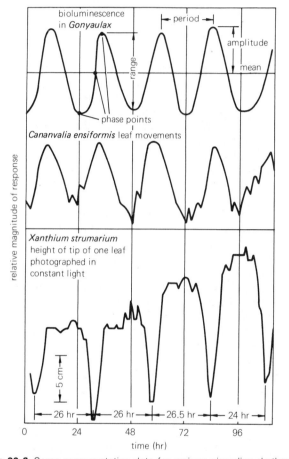

Figure 20-3 Some representative data for various circadian rhythms. Top, bioluminescence intensity in *Gonyaluax* measured for plants kept under constant conditions of dim light. Characteristics of circadian cycles are indicated. Middle, leaf movements of *Cananvalia ensiformis* recorded on a kymograph, so high points on the graph indicate low leaf positions. Light and dark conditions are indicated by the bar. Note gradual shift of the peak during darkness as the cycles progress. Bottom, leaf movement of cocklebur (*Xanthium strumarium*) recorded by the time-lapse photography method illustrated in Fig. 20-2. High points indicate high leaf positions. Period lengths between the troughs are indicated. Note increase in absolute height of the leaves, particularly at the peaks, but also at the troughs. This is largely due to growth of the stem during the course of the experiment, but the increase in range of leaf movement is also apparent. Light was entirely from fluorescent lamps.

on the repeating cycles. Typically, the maxima of the curves are observed and measured, because they show the sharpest changes in slope. Sometimes minima provide a more accurate measurement, or, for that matter, some other point on the cycle might be considered. The term **phase** is used in a specialized sense as any point on a cycle, recognizable by its relationship to the rest of the cycle. The most obvious phase point on a cycle (e.g., a maximum) is called the **acrophase.** Hence, the period is the time between acrophases. In a more general sense, the term phase may mean a recognizable *portion* of a cycle; for example, the part that normally falls during the light period, the so-called photophil phase.

Second, the **amplitude** is the degree to which the observed response varies from the **mean** (Fig. 19-3). The **range** is the difference between the maximum and the minimum values. *Third,* one might consider the **pattern** of the cycle. Usually, the common sine wave comes to mind (as in the bioluminescence rhythms of Fig. 20-3), but there are many variations. A short maximum might be accompanied by a broad minimum, for example, or the slope of the curve approaching the maximum might be steep, while that approaching the minimum might be less so.

When plants or animals are exposed to an environment that fluctuates according to some period, and the rhythms exhibit the same period, they are said to be **entrained** to the environment rather than free running. As we shall discuss later, this entrainment to the environment can be brought about by several factors, particularly an oscillating light environment, with its **dawn** and **dusk**. Such an entraining environmental cycle is called a **synchronizer** or **Zeitgeber,** a German word meaning "timegiver." The term "entrainment" is used when the *Zeitgeber* is a fluctuating environment with several regular cycles. If an environmental stimulus is given only once (e.g., a single flash of light), and the period of the rhythms is shifted in response to it, the rhythm is said to have been **phase-shifted** or **rephased.**

20.4 Rhythm Characteristics: Light

Many investigators have expended much effort in obtaining data relating to the biological clock, especially as it is exhibited by circadian rhythms. A vast amount of detail has accumulated, far too much for extensive discussion here. We shall consider a few effects of light, temperature, and applied chemicals.

With the discovery that the rhythms had free-running periods not exactly equal to 24 hr, it became apparent that they must be entrained by the external environment to account for the normal 24-hr periodicity. The work of Bünning and Stern indicated that entraining factors might be as subtle as a rather weak red light; hence, light was of obvious interest as a possible *Zeitgeber* or synchronizer.

One approach was to see if the rhythms could be entrained to some light-dark cycle other than a 24-hr one. It was readily apparent that this could be done. The rhythms could be entrained by shorter cycles of 20 to 22 hr (in rare cases, even 10 to 16 hr) or longer cycles of 28 to 38 hr.

Another approach was to allow a rhythm to become strongly established by a cycling environment and then to let it run free under constant environmental conditions. In constant darkness, a brief interruption of light may be given at various times during the free-running rhythm. With plants and many animals (especially nocturnal ones), when the flash of light is given during subjective day, there is virtually no effect upon the rhythm. That is, if light comes during the phases typical of day in a natural cycling environment, the following phases of the cycle are not influenced much. When the light interruption comes during early subjective night, however, the rhythm is typically *delayed* (i.e., an acrophase comes later than would have been expected). It is as though the flash of light were acting as *dusk*, but by coming later, a delay resulted. As the light flash is given later and later during subjective night, the extent of the delay increases until a certain point is reached where the flash of light suddenly results in an *advance* of the rhythm rather than in a delay (i.e., an acrophase comes earlier than expected). The flash of light is acting as *dawn* rather than as dusk (Fig. 20-4).

Often, the delay or advance occurs gradually over several cycles, called **transients,** before the new phase relationships become firmly established. The transient cycles are more commonly observed with animal than with plant rhythms. Colin S. Pittendrigh and Victor Bruce suggested that the transients could be understood if the rhythms are actually under the control of two clocks. The first may be readily phase shifted by the light flash; it in turn then gradually entrains the other, which actually controls the rhythms. Entrainment of the second may require several days, accounting for the transients.

By carefully studying curves such as those in Fig. 20-4, and knowing the amount of delay or advance caused by a light flash given during otherwise constant conditions at various times during a free-running cycle, it is possible to account for the phenomenon of entrainment. That is, considering the advancing effects of dawn (lights on) and the retarding effects of dusk (lights out), allows us to predict the phases of a rhythm in relation both to the normal 24-hr cycles and to cycles other than 24 hr in length. Pittendrigh, a zoologist, and his coworkers, first at Princeton and then at Stanford University, have pioneered in this approach.

Entrainment by light is clearly a photobiological process. Certainly some photoreceptor pigment is absorbing the light, and by this absorption it is changed in such a way that it can lead to an advance or a delay of the clock. It would be extremely interesting to understand the photobiochemical mechanism of this response. A first step would be identification of the photoreceptor pigment, and this is usually initially undertaken by a determination of the action spectrum for the response. Flashes of light of carefully controlled spectral qualities are given at various times to test their effectiveness on entrainment.

For a long time, zoologists simply assumed that the receptor was the eye of the animal with which they were working. But it was shown in the 1950s that gonads of ducks would develop in response to long days, even though the eyes were removed. In the 1960s, Michael Menaker at the University of Texas in Austin showed that blind sparrows (both eyes removed) could be entrained in their activity rhythm by light signals. He also confirmed with blind sparrows the development of testes in response to long days, as observed in blind ducks. Furthermore, he has shown that a weak green light has an effect upon the activity rhythm of his birds with normal eyes, but that it does not influence the daylength response. It is the pineal gland of the brain that actually responds to the light. Enough light, especially red light, penetrates the skull to be effective. This gland has been known since antiquity as the "third eye"! It is especially prominent in birds. Menaker and his coworkers (see Zimmermann and Menaker, 1975) have been studying the pineal responses to light and their effects on the clock. The activity and gonadal clocks, at least, seem to be located in the pineal gland; effects are transmitted via hormones rather than nervous impulses.

Action spectra have been determined for the various rhythms of *Gonyaulax*, for a mating rhythm of *Paramecium*, for the conidiation rhythm of *Neurospora* (Fig. 20-5), and for *Drosophila* (fruit fly) and *Pectinophora* (a moth). The action spectra are different in the several organisms, but all do have strong responses in the blue part of the spectrum. The response of *Gonyaulax* to red may be due to chlorophyll, but the reason for the even stronger red response of *Paramecium* is not clear. Victor Muñoz and Warren L. Butler (1975) in La Jolla, California, have isolated a flavo-protein-cytochrome *b* complex from *Neurospora* with an absorption spectrum that

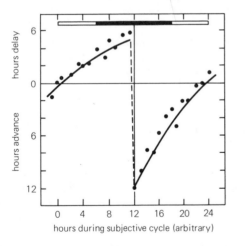

Figure 20-4 Phase shift in petal movements of *Kalanchoe blossfeldiana* following 2-hour exposures to orange light given at various times during an extended period of continuous darkness. Bar at the top indicates subjective status of the rhythm; that is, dark part of the bar indicates petal closure or subjective night. As light interruption approaches the middle of subjective night, there is an increasing delay; following the middle of subjective night, there is an advance, which decreases as subjective day is approached. (Data after Zimmer, 1962.)

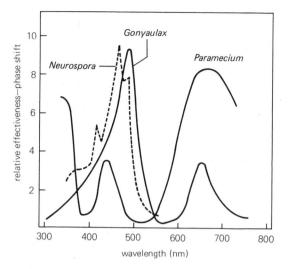

Figure 20-5 Approximate action spectra for phase shifting in the rhythms of *Neurospora*, *Gonyaluax*, and *Paramecium*. (From Ehret, 1960, and Muñoz and Butler, 1975.)

closely matches the action spectrum of Fig. 20-5. They suggest that this or a similar pigment may couple light and the clock in the many organisms, both plant and animal, that respond to blue light.

Lars Lörcher (1957), working in Bünning's laboratory in Tübingen, Germany, determined rough action spectra for the leaf movements of the bean plant. Using dark-grown plants, he found that the rhythms were most effectively established by red light and that this establishment was reversed by an immediate exposure to far-red light. This implicates the phytochrome system. Lörcher found, however, that other wavelengths were also effective in entrainment, providing the plants had been grown in the light rather than in the dark. There is one other suggestive observation relating to phytochrome and the rhythms in flowering plants. The leaf movement rhythms often continue for several days when the plants are maintained at a constant temperature and under continuous light, providing the light is rich in red wavelengths but contains none of the far-red part of the spectrum. When far-red light is present, the rhythms tend to damp out. Therefore, the phytochrome system could be involved in some plants, although the blue-absorbing pigment may be even more important. Phytochrome is probably not present in fungi or animals.

20.5 Rhythm Characteristics: Temperature

Pittendrigh realized that the clock could be of little value to an organism if the rate at which it ran were strongly dependent upon temperature, as are most metabolic functions. He had heard of the eclosion rhythm of *Drosophila* pupae, discovered by H. Kalmus in Germany. Pittendrigh (1954) studied this rhythm at several temperatures and found the period to be nearly constant over a wide temperature range. Thus temperature independence of the biological clock was discovered. Bünning had investigated the question in 1931, but in bean leaf movements temperature independence was not as clear as in the animal examples. Bunning did indicate that the temperature response was unexpectedly low.

Frank Brown and Marguerite Webb at Northwestern University in Chicago also discovered temperature independence in 1948. A color change observable in the fiddler crab exhibited a period length that was almost completely independent of temperature, but Brown and Webb did not deduce from their data a temperature-independent clock. Rather, they considered their results to be evidence *against* an endogenous clock.

Indeed, they returned to Stoppel's concept of a factor x in the environment that was responsible for actual time measurement. Brown has championed this idea until the present. He and his coworkers have observed numerous biological responses to such subtle environmental factors as geomagnetic fields. These are of considerable interest in themselves, but no one can see how they might account for biological time measurement. As Pfeffer nearly a century ago was unable to convince his colleagues that the clock was endogenous (it took the experiments of Bünning, Stern, Stoppel, and Kleinhoonte to do that), so Brown has been unable to convince most of us that the clock is exogenous. Most workers now believe that the experiments of Brown and Webb with fiddler crabs were just an especially outstanding demonstration of the clock's temperature independence.

Yet we are faced with somewhat of a paradox in our discussion of temperature effects on the biological clock. Changes in temperature of only 2.5 C or less may synchronize the rhythms (act as a *Zeitgeber*) in *Neurospora* and other organisms, and temperature may also strongly influence the amplitude of the response. Such effects must surely be important in nature. Still, the period of free-running rhythm is relatively temperature insensitive (it remains nearly constant over a wide range of temperatures). So some aspects of the clocks are sensitive and others are insensitive to temperature.

The temperature insensitivity is especially interesting. In *Gonyaulax* the Q_{10} for the effect of temperature on period length is slightly *less* than 1 (as temperature increases, the free-running period becomes slightly longer), whereas it is equivalent to about 1.3 for the leaf movements of the bean. Various workers have observed Q_{10} values that approach 1.00 very closely. For example, the value is approximately 1.02 for biological time measurement in the flowering of cocklebur (Salisbury, 1963).

In spite of its ecological value, temperature independence of the clock is somewhat surprising, since the Q_{10} for most biochemical reactions is appreciably greater than one. Those involving hydrolysis of ATP, for example, are often about 2.0. How, one might ask, if the living organism is fundamentally a biochemical system, can we account for temperature independence?

Potato Cellars, Trains, and Dreams:
Discovering the Biological Clock

Erwin Bünning

F.B.S. 1963

As we've seen in the chapter, it was Erwin Bünning as a postdoctoral fellow, along with several colleagues, who discovered the free-running, circadian nature of the biological clock in plants, a discovery that made endogenous timing seem to be the only acceptable explanation. In 1963–64, one of us [F.B.S.] spent a sabbatical year with Bünning in Tübingen, where he was Director of the Botanical Institute, and had the opportunity to hear him tell of his early work. This is Salisbury's translation of a letter from Bünning, sent in response to a request that he record some of his experiences. The letter is dated June 2, 1970.

The story went something like this. At the Institute for the Physical Basis of Medicine in Frankfurt, the biophysicist Professor Dessauer (an X-ray specialist) became interested in the effects of the ionic content of the air upon humans. Those were the years when people began to be interested in atmospheric electricity, cosmic rays, and so on. Naturally, humans could not be used as experimental objects, and so in 1928, Dessauer searched for botanists to work on plants. One whom he found was Kurt Stern, who lived in Frankfurt; the other was me, who had just finished my doctoral work in Berlin. So we began in August of 1928 to contemplate the problem. In the process we came upon the work of Rose Stoppel, who had been studying the diurnally periodic movements of *Phaseolus* (common bean) leaves. In the process she had found, as had several other authors, that under "constant" conditions in the darkroom, most leaves reached the maximum extent of their sinking (their maximum night position) at the same time; namely, between 3:00 and 4:00 A.M. Her conclusion: Some unknown factor synchronized the movement. Could this be atmospheric ions? We had Rose Stoppel visit us from Hamburg for two or three weeks so that we could become familiar with her techniques. In our group we always called her *die Stoppelrose* ("stubble rose"), and the name was most appropriate. She was energetic and persistent, so persistent that she just died this January in her 96th year. Her results also appeared in our experiments: The night position usually occurred at the indicated time. Then we investigated the effects of air that had been enriched with ions or air from which all the ions had been removed. The result: Nothing changed—atmospheric ions are not "factor x."

We then decided that our research facilities at the Institute were insufficient. Hence, after *die Stoppelrose* had left, Stern and I moved to his potato cellar, where with the help of a thermostat, we obtained rather constant temperature. Contrary to the practice of Stoppel, who turned on a red "safe" light to water her plants, we went into the cellar just once a day with a very weak flashlight and felt around with our fingers for the pots and recording apparatus so that we could water the plants and so that we could see if everything was in order with the recorders. The flashlight was weakened with a dark red filter so that one could see only for a few centimeters' distance. In those days it was the dogma of all botanical textbooks that red light had absolutely no influence upon plant movements or upon photomorphogenesis. We did one other thing differently from Stoppel. Since Kurt Stern's house was a long way from our laboratory, we didn't make our daily control visit in the morning, rather only in the afternoon. The result: Most of the maximal night positions no longer appeared between

3:00 and 4:00 A.M., but rather between 10:00 and 12:00 A.M. Hence we concluded: The dogma is false. Red light must synchronize the movements so that a night position always appears about 16 hr after the light's action. That was "factor x." When we eliminated this hardly visible red light, we found that the leaf movement period was no longer exactly 24 hr but 25.4 hr. [This circadian feature of the clock was the key to understanding its endogenous nature—Ed.]

So that was about how that story went. Naturally, I could also tell you the story about how I came upon the significance of the endogenous rhythms for photoperiodism. That was about in 1934. Of course, I had already often asked myself how such an endogenous rhythm might ever have any selection value (in evolution), and I had already expressed the opinion in 1932 in a publication (Jahrbuch der Wissenschaftlichen Botanik 77: 283-320) that some interaction between the internal plant rhythms and the external environmental rhythms must be of significance for plant development. But there were coincidences in the story. As a young scientist, one naturally had to allow himself to be seen by the power-wielding people of his field, so that he might receive invitations for promotion. Hence, I traveled in 1934 from Jena to Könisberg in Berlin to introduce myself to the great Professor Kurt Noack. We discussed this and that. He mentioned that discoveries were being made that were so remarkable one simply couldn't believe in them. Such a one, for example, was photoperiodism. He, as a specialist in the field of photosynthesis, must certainly know that it makes no difference whatsoever what program is followed in giving the plant the necessary quantities of light. Then, as I was riding back on the train, the idea came to me—aha, for the plant it does make a difference at which time light is applied, if not exactly in photosynthesis, nevertheless for its development!

I could present a third story, one from very recent years. I have long felt that the daily leaf movement rhythms had no selection value in themselves. As you know, I have recently changed my mind. The movements could indeed be important in avoiding a disturbance of photoperiodic time measurement by moonlight. Before I had begun to test the idea experimentally or even to think about it, it came very simply one night in a dream. The dream (apparently as a memory of one of my visits to the tropics): tropical midnight, the full moon high above on the zenith, in front of me a field of soybeans, the leaves, however, not sunken in the night position, rather broadly horizontal in the day position. My thoughts (in the dream): How shall these plants know that this is not a long day? They had better hide themselves from the moon if they want to flower.

We can envision certain feedback systems. The products of one reaction may inhibit the velocity of an earlier one. Such feedback inhibition systems are well known in living organisms. As temperature increases, the velocity of the first reaction (the time-measuring reaction) might increase, but the inhibitory product would then increase at a proportional rate, and consequently the reaction would maintain a near-constant rate over a wide range of temperatures. The Q_{10} value of less than 1 observed in *Gonyaulax* could be accounted for by such a scheme, assuming that the inhibitory product were produced somewhat faster with increasing temperature than the time-measuring reaction is accelerated. It is difficult to imagine any other way to account for this observation. So such a feedback system (possibly overcompensating as in *Gonyaulax*) could be invoked in a **temperature-compensated system** of time measurement based upon biochemical reactions. ("Temperature compensation" is thus a preferable term to "temperature independence.") Furthermore, some simple enzyme systems are now known that have temperature coefficients close to 1. Or could the clock consist of some undreamed of physical (rather than biochemical) system?

20.6 Rhythm Characteristics: Applied Chemicals

If we could find some chemical that would clearly inhibit or promote biological time measurement, then by our speculations about its mode of action we might gain some insight into the operation of the biological clock. Many chemicals have been studied, but we must be careful in interpreting the results. A given chemical might, and often will, influence the *amplitude* of the rhythm without influencing its *period*, the actual indication of time measurement. That is, the physiological manifestation of time measurement (e.g., leaf movements or animal activities) may be rather easily influenced by some substance such as a powerful respiration inhibitor, although the clock is more resistant (Fig. 20-6). There are some instances in which the amplitude may be completely damped out, yet when the inhibitor is removed, the cycle proves to be in phase with control organisms having the same period and to which no inhibitor was applied. This is one aspect of the general problem of **masking.**

In the early 1960s, Bünning and his coworkers reported that several chemical inhibitors applied to bean plants appeared to act directly on the clock itself, rather than on its manifestation. In these cases, the period was lengthened somewhat, and the amplitude was unaffected or only somewhat decreased. These substances include colchicine, ether, ethyl alcohol, urethane (ethyl carbamate), and some others. They are sometimes but not always effective with other organisms. It has been suggested that the feature they have in common is the ability to influence membranes, and that therefore the clock is associated with cellular membranes. Another clear-cut effect of a chemical upon time measurement is that of heavy water, deuterium oxide. Treatment with this sub-

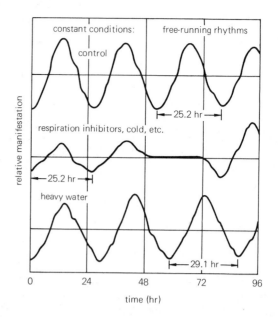

Figure 20-6 Effects of various chemicals and other factors upon free-running rhythms. Most compounds that inhibit metabolism reduce the amplitude but do not affect the period (compare center record with control above). Heavy water increases the length of the period (29.1 hr compared with 25.2 hr for eclosion in *Drosophila*). Curves are schematic and do not represent actual data.

stance caused a considerable increase in the phototaxis period of *Euglena* and also of bean leaf movements, and it slowed the measurement of daylength in flowering. We are left with little in the way of concrete proposals to account for the mechanism of the clock, since heavy water might influence many processes within the plant, but membranes are again good candidates. Valinomycin, a compound that alters membrane transport of K^+, also causes phase shifts in beans and in *Gonyaulax*, further implicating membranes. This was shown by Bünning and Ilse Moser in 1972.

Other recent work is with inhibitors of protein or RNA synthesis that influence the clock. Several such inhibitors do not influence the clock (e.g., actinomycin D), but others clearly do (e.g., cycloheximide and puromycin). Protein synthesis may well be a part of the clock mechanism.

Jerry F. Feldman and others have studied the genetics of the clock in *Neurospora* and other organisms. Period length may be controlled by single genes in these systems. In the past, genetic analysis has led to deeper understanding of underlying mechanisms (e.g., in the metabolic pathways of *Neurospora* or the operon concept of procaryotes and viruses).

20.7 Photoperiodism

Even before the work of Bünning, Stern, Stoppel, and Kleinhoonte, the biological clock had clearly been demonstrated as a component of living plants, although the discovery was not

Women in Science

Beatrice M. Sweeney

There is much concern about the role of women in our society. It has often seemed to me [F. B. S.], whether fact or only an illusion, that an unusually high percentage of women has contributed to our understanding about the biological clock. I was discussing this idea with Beatrice M. Sweeney during a symposium on biological clocks several years ago. She agreed to write a few personal thoughts about these matters for use in our textbook. She is located in the Department of Biological Sciences, University of California at Santa Barbara.

Dear Frank:

Until you remarked upon it at the Christmas meeting at Sacramento, I hadn't noticed the unusual number of women in the field of biological clocks. I am not accustomed to thinking about the sex of my scientific colleagues. In this respect, I suppose I am a truly liberated woman. I think I owe the fact that I am not conscious of whether or not scientists are women to my good fortune in receiving my training in Dr. Kenneth Thimann's laboratory at Harvard, where even long ago all graduate students were on an equal footing. We all regarded ourselves as superior, and so our chances of turning out that way were much increased. This brings me back to the subject of women in science, since I often think women have a difficult time believing that they really can do first-class research in a field usurped exclusively for so long by men, at least in the top prestige bracket. In rhythm work, women have, I believe, found relief from their feelings of inferiority, because they recognize in this field a comfortable familiarity. To get up in the middle of the night, perhaps several times, is not strange to them. What matter that it is a series of tubes full of *Gonyaulax* and not a baby that demands their attention? I notice that the predeliction of women for chronobiology does not seem to be declining. The traditions set by such women as Rose Stoppel, Anthonia Kleinhoonte, Marguerite Webb, and Janet Harker are being ably perpetuated by Audrey Barnett, Ruth Halaban, Marlene Karakashian, Laura Murray, Ruth Satter, Therese van den Driessche, and I'm sure you can name others. Of course, I still regard myself as active and expect any day to crack the problem of the basic circadian oscillator.

I would like to say something personal to the young ladies who are perhaps considering a future as scientists, now that the sole production of children is no longer in the interest of world well-being. My work in science has been the very stuff of life to me, endlessly frustrating and rewarding. I should like to relate to these young women what fun it is to work, rather than be a tourist, in strange parts of the world. Imagine the delight of traveling northward inside the Great Barrier Reef on a 60-foot boat usually devoted to the hunting of alligators, and to have available a microscope with which to see the details of the strange animals and plants, the familiar yet unfamiliar phytoplankton, the algae growing within the giant clams and coral. I could mention also the jungles of New Guinea at night, ringing with the stridulations of locusts at multiple high frequencies, flashing with fireflies. Or the beaches of Jamaica, white and shining and fringed with the bending fronds of the coconut palms like long eyelashes. The knowledge of flora and fauna acquired in scientific study immeasurably increases the pleasure of viewing an unfamiliar part of the world.

But most of all I'd like to say that the day-to-day research and teaching of an academic profession provides an endlessly varying and interesting way of life, to my tastes infinitely more satisfying than cooking, dusting and shopping, even than bringing up children. Perhaps my four children have benefited rather than suffered from the fact that I have other interests and satisfactions than themselves. At least it is interesting to see that three of the four are in some way pursuing science as a career.

widely recognized by those working on rhythms until perhaps the 1950s. Wightman Wells Garner and Henry Ardell Allard were two scientists working in the United States Department of Agriculture research laboratories at Beltsville, Maryland. They had made two observations that they could not explain. Maryland Mammoth tobacco grew at that latitude to a height of 3 to 5 m during the summer months but never came into flower, although it flowered profusely when only about 1 m tall after transplanting to the winter greenhouse. They had also noticed that all the individuals of a given variety of soybean would flower at the same time in the summer regardless of when they had been planted in the spring. That is, the large plants sown in early spring came into flower at the same time as the smaller plants sown only in early summer. Garner and

Allard wondered if some factor of the environment might be responsible for flowering of these two species.

They considered various environmental factors that might differ between the summer fields and the winter greenhouses. They tested light intensity, temperature, soil moisture, and soil nutrient conditions, but no combination of these factors resulted in flowering of the tobacco plants. Garner and Allard realized that daylength varied throughout the season, as well as being a function of latitude, so they tested what seemed like the remote possibility that this could control flower formation. It did. When the days were shorter than some maximum length, the tobacco plants began to bloom. When the days were shorter than some other maximum length, the soybean plants also began to bloom. Results were published in 1920. Garner and Allard named the discovery of flowering in response to photoperiod **photoperiodism,** after a suggestion from A. O. Cook, a colleague at Beltsville.

Subsequent work seemed to indicate that the length of the dark period was in some ways even more important than the photoperiod (see Chapter 22). The point here is that these *plants were capable of measuring the length of the light and/or the dark period.* Here, then, was a clear-cut demonstration of biological time measurement. Completely comparable responses were subsequently discovered in animals, including effects on insect life cycles, fur color (arctic hare), breeding times (gonad size), and migration of many birds and other animals.

Some organisms respond when the daylength *exceeds* some minimum. Garner and Allard called such plants (e.g., Wintex barley and spinach), which flower in response to days longer than some critical length, **long-day plants,** and those such as tobacco and soybean, which flower when the daylength is *less* than some maximum, **short-day plants.** Some of the species they studied showed no response to daylength; these they called **day-neutral plants.**

20.8 Time Memory

While Bünning and Stern were discovering the endogenous nature of the biological clock as it occurs in plants, Ingeborg Behling in Germany (1929) was making an important discovery about the clock in honeybees. She found that it was possible to train honeybees to feed at a certain time during the day. It is as though the clock in the honeybee can have a "rider" attached to it, indicating the time of day and informing the honeybee 24 hours later that it is time to eat. It remains to be seen whether this is the same clock that controls circadian rhythms.

Man has a comparable time-measuring system. Time memory in man is most frequently manifested by an ability to wake up at a predetermined time. This is particularly impressive since he or she must translate an abstract idea into some form that will "adjust the rider" on his or her biological clock. Human time memory can often be most impressively demonstrated under hypnosis, sometimes with posthypnotic suggestion.

20.9 Celestial Navigation

In spite of the extremely intriguing discoveries of the 1920s and 1930s, only a small minority of biologists showed any great interest in this phenomenon until the early 1950s. At that time American botanists began to become interested in the work of Bünning, because it seemed to bear a direct relationship to photoperiodism, a topic in which there was considerable interest. Indeed, Bünning had proposed a theory to correlate photoperiodism and the circadian rhythms (see his personal essay). Zoologists, particularly Pittendrigh at Princeton Uni-

versity, also began to take notice of Bünning's discovery and of other work going on in Europe. The discovery of temperature compensation particularly stimulated interest.

Then Gustav Kramer, K. von Frisch, and others working primarily in Germany found that certain birds and other animals could tell direction upon the earth's surface by the position of the sun in the sky. Since this position changes, the organism must be able to correct for the time of day, apparently by the use of some kind of clock. Up until then, some of the manifestations of the clock (e.g., the leaf movements) did not seem to be of much value to the organism. In the case of **celestial navigation,** however, the clock clearly is used, so this spectacular discovery (along with temperature compensation) caught the attention of biologists all over the world and led to a surge of interest in the biological clock.

20.10 The Biological Clock in Nature

In a cycling environment, it is not difficult to imagine advantages conferred upon organisms by possession of a biological clock. Rhythms of activity in animals, for example, allow a species to occupy a niche not only in space but also in time: A nocturnal animal and a diurnal animal might use the same space but at different times. Plant rhythms in metabolism adjust the plant to the light and temperature environment. Rhythms in flower opening may confer an adaptation to the time factor of natural environments, and the time memory of the honeybees must be an adaptation to this plant feature. The phenomenon of photoperiodism confers upon the organism possessing it the ability to occupy a particular niche in *seasonal* as contrasted to *diurnal* time. The several species that flower in response to photoperiodism may do so at different times and in sequence throughout the season, providing a rather constant source of nectar for insect pollinators throughout an entire growing season. Given a time during the season with minimal flowers but high availability of pollinators, there would be a selective advantage to any species able to flower during that time.

What are the ecological advantages of the leaf movements? Pfeffer asked this question and settled it in his mind by tying the leaves fast to a bamboo framework so that the movements could not occur. Since his plants exhibited no apparent ill effects, he concluded that the movements were some byproduct of the evolutionary process and of no selective value to the plant. Charles Darwin suggested that leaf positions might play a role in heat transfer between a plant and its environment (see Chapter 3). That is, a horizontal leaf is in a good position for the reception of sunlight during the day, but at night heat might radiate better from a horizontal leaf into space. *Vertical* leaves in a plant community, however, might radiate more to each other. Darwin and his son (1888) performed experiments to test this idea and obtained positive although not very striking results. Bünning suggested (see his personal essay) that a horizontal leaf would be in a better position to absorb the light of the full moon (at its zenith at midnight). Since such absorption might upset the photoperiodism response, a vertical leaf position at night could be a protective device to insure successful time measurement in photoperiodism.

Celestial navigation is certainly a spectacular and clear-cut application of the biological clock among animals. But the circadian rhythms of man and his other time-measuring abilities are also important in certain obvious ways, and they may clearly increase in importance (in a negative way) as our modern life becomes ever more complex. Jet air travel across time zones, for example, has a strong effect upon the internal timing systems of the passengers. It may take several days for the clock in a tourist—or a diplomat—to adjust to a new time zone.

All these and other observations suggest that the biological clock offers an exciting and increasingly important field for scientific investigation.

21

Plant Responses to Temperature and Related Phenomena

Plant growth is notoriously sensitive to temperature. Often a difference of a few degrees leads to a noticeable change in growth rate. Each species or variety has, at any given stage in its life cycle and any given set of study conditions, a **minimum temperature** below which it will not grow, an **optimum temperature** (or range of temperatures) at which it grows at a maximum rate, and a **maximum temperature** above which it will not grow and may even die. Some curves for growth rate as a function of temperature are shown in Figure 21-1. The growth of various species is typically adapted to their natural temperature environment. Alpine and arctic species have low minima, optima, and maxima; tropical species have much higher **cardinal temperatures.**

But temperatures may influence much more than tissue growth. Often, critical steps in the life cycle may be initiated in response to certain temperature treatments: Seeds may germinate, flowers may be initiated, and perennial plants may become dormant or break dormancy. And these developmental responses may be influenced by environmental factors in addition to temperature, such as light, photoperiod, and moisture. These interactions are diverse and complex, so the topics of this chapter sometimes stray from its central theme, which is plant response to temperature, especially low temperature.

21.1 The Temperature-Enzyme Dilemma

We usually think of the plant growth response to temperature in terms of enzyme reactions in which two opposing factors seem to operate. With increases in temperature, the increased kinetic energy of the reacting molecules results in an increased rate of reaction, but increasing temperature also results in an increased rate of enzyme denaturation. Subtracting the destruction curve from the reaction curve produces an asymmetrical curve (Fig. 21-2), with its minimum, optimum, and maximum cardinal temperatures. The curve applies to respiration, photosynthesis, and many other plant responses besides growth.

After becoming familiar with positive responses as temperature increases from the minimum to the optimum, it may

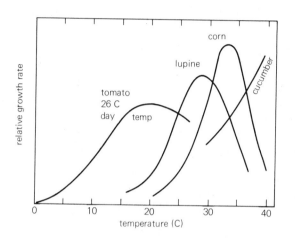

Figure 21-1 Plant growth as a function of temperature for several species. With tomato, day temperature was constant and night temperature varied.

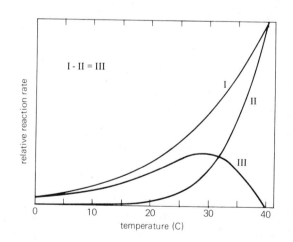

Figure 21-2 Enzyme activity and temperature. I. Rate of a reaction with a Q_{10} of 2. II. A reaction with a Q_{10} of 6, such as denaturation of protein. III. The expected curve for enzyme destruction subtracted from the curve for enzymatically controlled reaction rate.

come as a surprise to learn that certain processes are promoted as temperature decreases toward zero. In **vernalization,** exposing plants to low temperatures for a few weeks results in the formation of flowers, usually after plants are returned to normal temperatures. Low temperatures in the autumn often cause or contribute to the development of dormancy in many seeds or the buds of woody plants, and then the low temperatures of winter may be responsible for the breaking of dormancy in these same seeds and buds. If we oversimplify dormancy induction by thinking of it as a simple *negative* slowing down of plant processes at decreased temperatures, then we must surely consider the breaking of dormancy in an opposite, *positive* sense. This is the interesting dilemma of the low temperature response.

We shall consider five different positive responses to low temperatures: The *first,* vernalization, has been studied extensively, so it provides a good starting place. The *second* is the breaking of seed dormancy by exposure of moist seeds to low temperatures. This treatment is referred to as **stratification.*** We shall have much more to say about seed dormancy and germination, however, than just their responses to temperature. The *third* process is closely related: the breaking of winter dormancy in the buds of perennial woody plants. The *fourth* process has been studied less: the low temperature induction of underground storage organs such as tubers, corms, and bulbs. The *fifth* process has been studied still less: the effect of low temperatures upon the vegetative form and growth of certain plants.

In each of these five processes, we are concerned mostly with **inductive** (delayed) effects upon some developmental plant process. Such effects, in which the response appears some time after completion of the stimulus, are also observed in response to other environmental factors, such as daylength. In fact, low temperature and daylength effects are frequently interrelated in the five plant responses.

Ultimately, the low temperature response must involve activation of specific genes—a switching of the morphogenetic program in response to low temperatures. Could the genes respond directly to low temperature? If not, then what in the cell actually responds, transducing the low temperature signal to a physiological change? Are the transducers located in the cytoplasm or the nuclei of the cells in which the program is readjusted? Or in other cells? There are few answers to these questions, but they could guide future research.

21.2 Vernalization

Vernalization was described in at least 11 publications in the United States during the mid-nineteenth and early twentieth centuries (e.g., in the *New American Farm Book* in 1849), but it was completely overlooked by "establishment" science until

*It is easy to confuse vernalization (an effect on flowering) and stratification (an effect on germination), since both processes can occur when moist seeds are exposed to low temperature.

1910 and 1918, when J. Gustav Gassner in Germany described the vernalization of cereals. Much of the early work on plant development took place in Europe and Great Britain; the United States and Canada were apparently preoccupied with subduing the frontier.

In the 1920s, the term "vernalization" was coined by Trofim Denisovich Lysenko, who, during the reign of Stalin, was allowed to exercise absolute political control over Russian genetics, decreeing that geneticists there should accept the dogma of the inheritance of acquired characteristics (see Caspari and Marshak, 1965). Vernalization, from Latin, would be translated into English as "springization," the implication being that winter varieties were converted to the spring or summer varieties by cold treatment. We realize, although apparently Lysenko did not, that the genetic makeup is not changed by the low temperature treatment. The cold period supplied artificially by the experimenter simply substitutes for the natural cold of winter.

The term "vernalization" has been widely misused. Any plant response to cold has sometimes been referred to as vernalization, as has any promotion of flowering by any treatment (even daylength). We will restrict the term **vernalization** to *a low temperature promotion of flowering.*

The Response Types There are numerous vernalization responses, depending not only upon species but frequently upon varieties within species. In classifying the response types, there are several factors to consider (indicated by italics). To begin with, we may differentiate *inductive* from *noninductive* responses. Most plants that have been studied respond inductively, although a few (e.g., Brussels sprouts) form flowers during the cold treatment instead of later and are thus noninductive.

A profitable way to classify the response types is according to the *age* at which the plant is sensitive to cold. The **winter annuals,** especially cereal grasses, were studied during the 1930s and the 1940s, particularly in Russia and by Frederick G. Gregory and O. Nora Purvis (see Purvis, 1961) at Imperial College in London. They respond to the low temperatures as seedlings or even as seeds, providing that sufficient oxygen and moisture are present. Petkus rye (*Secale cereale*) seeds are normally planted in the fall of the year, when they usually germinate, spending the winter as small seedlings. Or moist seeds may be exposed to low temperatures in a cold chamber for a few weeks. Plants then form flowers at normal temperatures in approximately seven weeks. Without the cold treatment, 14 to 18 weeks are required to form flowers, but ultimately flowers do form. Since the cold requirement is a **quantitative** or **facultative** one (low temperatures result in *faster* flowering) but not a **qualitative** or **absolute** one (in which flowering *absolutely depends* upon cold), we have another basis for classification. Most winter annuals are inductive and quantitative in their response, although some (e.g., Lancer wheat) have an absolute cold requirement.

With Petkus rye there are two interesting complications:

Short-day treatment will substitute to a certain extent for low temperature, and flowering of previously vernalized, growing plants is strongly *promoted by long days.* All winter annuals so far studied are not only promoted by cold but also by the subsequent long days of late spring and early summer.

The **biennials** live two growing seasons, then flower and die. Examples include several varieties of beets, cabbages, kales, Brussels sprouts, carrots, celery, and foxglove. They germinate in the spring, forming vegetative plants that are typically a rosette (see Fig. 21-4, lower right). The leaves often die back in the autumn, but their dead bases protect the crown with its apical meristem. With the coming of the second spring, new leaves form, and there is a rapid elongation of a flowering shoot, a process called **bolting.** Exposure to the winter cold between the two growing seasons induces flowering. Most biennials must experience several days to several weeks of temperatures slightly above the freezing point to subsequently flower; they have an absolute cold requirement, as contrasted to the facultative winter annuals. Sugar beet plants may be kept vegetative for several years by not being exposed to cold (Fig. 21-3). Flowering of many biennials is also promoted by long days following the cold, and some may absolutely require this treatment (e.g., the European henbane *Hyoscyamus niger,* Fig. 21-4). Other biennials are day-neutral following vernalization.

There are many species of cold-requiring plants that do not fall readily into the categories of winter annuals or biennials. Flowering of several perennial grasses, for example, is promoted by cold. Some of these have a subsequent short-day requirement for flowering. The chrysanthemum is a short-day perennial that has been studied extensively because of its photoperiodism response. Its cold requirement, which must be met once before it can respond to the short days, was overlooked because plants are propagated vegetatively, and cuttings carry the vernalization effect with them. Even woody perennials have a low-temperature requirement for flowering (Chouard, 1960), and several annual garden vegetables will flower somewhat earlier in the season if they are exposed to a short vernalization treatment (Thompson, 1953).

To summarize: Many different species are promoted to flower by cold periods, in some cases there is a quantitative and in others a qualitative effect, and flowering of many species also requires or is promoted by suitable daylength. These responses to environment allow a plant to anticipate the annual cycle of climate. We are not dealing with an endogenous timer as in the previous chapter but with a complex system in which a plant responds to one season by becoming prepared for the next.

Location of the Low Temperature Response It is the bud, presumably the meristem, that normally responds to cold by becoming vernalized. Only if buds are cooled will the plants subsequently flower. Embryos or even isolated meristems from rye seeds have also been vernalized. In another approach, various parts of a vernalized plant have been grafted

Figure 21-3 A 41-month sugar beet plant kept vegetative by never being exposed to low temperatures. The technician at the Earhart Plant Research Laboratory at the California Institute of Technology is Helene Fox. (Courtesy Albert Ulrich; see Ulrich, 1955.)

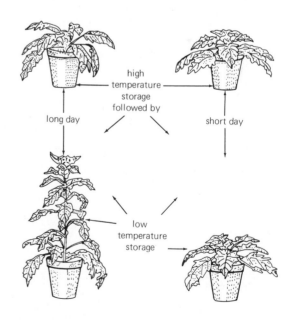

Figure 21-4 Bolting (flowering) response of henbane (*Hyoscyamus niger*), a typical rosette species, to storage at high or low temperature followed by long or short day. Only cold followed by long day induces flowering.

onto an unvernalized plant. If a vernalized meristem is so transplanted, it will ultimately flower, but if a meristem from an unvernalized plant is grafted onto a vernalized plant after removal of the vernalized meristem, the growing transplanted meristem remains vegetative (see Lang, 1965b).

S. J. Wellensiek (1964) in the Netherlands has suggested that vernalization requires dividing cells. Several studies support his conclusion, although some seeds respond even at temperatures a few degrees below the freezing point, where cell division seems unlikely and microscopic investigation has failed to reveal it. If cell division or DNA replication in non-dividing cells proves to be necessary, it could be significant. We noted in Section 18.1 that DNA must replicate before cellular differentiation. Perhaps only when the DNA is temporarily separated from the chromosomal proteins can gene activation or inactivation occur.

Physiological Experiments Sometimes in plant physiology we are unable to study biochemical or biophysical events directly but must take a more indirect approach. Whole plants are manipulated in various ways. Results are observed and deductions made, based on our understanding of plant function at the molecular level. These are **physiological experiments.**

A physiological investigation on vernalization might determine optimum temperatures (Fig. 21-5). Vernalization proceeds at a maximum rate over a fairly wide range of cool temperatures, depending upon the species, and vernalization occurs even at a few degrees below freezing. Usually the lower limit is set by the formation of ice crystals within the tissues. Another physiological study determines the most effective vernalization times. Minimum lengths for any observable effects vary from 4 days to 8 weeks, depending on species. Saturation times vary from 3 weeks for winter wheat to 3 months for henbane.

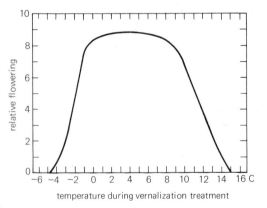

Figure 21-5 Final relative flowering response as a function of temperature during vernalization. The data represent response of Petkus rye to a 6-week period of treatment. (From Salisbury, 1963; see Purvis, 1961.)

If a plant, immediately following vernalization treatment, is exposed to high temperatures, it usually does not flower. This reversal is referred to as **devernalization.** To be strikingly effective, devernalizing temperatures must be about 30 C or higher with winter rye, and they must be applied within 4 or 5 days after the low temperature (somewhat longer in other species). Actually, some devernalization can be observed when plants are exposed to any temperature higher than that which will cause vernalization. In winter rye, 15 C is the neutral temperature; any temperature below this speeds flowering, and any higher temperature delays it. Anaerobic conditions given just after vernalization also cause devernalization, even at neutral temperatures. After devernalization, most species can be revernalized with another cold treatment.

Vernalin and Gibberellins If the apical meristem itself responds to low temperatures, a translocated flowering stimulus or hormone does not sound likely. And in most cases, the vernalization effect is not translocated from one meristem to another, either within the same plant or when a vernalized plant is grafted to a nonvernalized plant. There are exceptions, however, as reported as early as 1937 by Georg Melchers in Germany (see Lang, 1965b). The work has since been extended somewhat in Russia. A vernalized henbane plant was grafted to a vegetative receptor plant that had never experienced low temperature, inducing it to flower. Dissimilar response types will also transmit the flowering stimulus across a graft union; cold-requiring plants can be induced without the cold period by being grafted to a noncold-requiring variety, for example. The reverse, though less clear-cut, will also occur. It must be emphasized, however, that transmission is limited to a few species.

In these experiments, a living graft union must be formed between the two plants, and conditions favoring transport of carbohydrates also favor transport of the stimulus. If the receptor is defoliated or darkened, for example, while photosynthesizing leaves are left on the donor, the receptor must obtain its nutrients from the donor, which favors movement of the vernalization stimulus across the graft union.

Melchers postulated existence of a hypothetical vernalization stimulus called **vernalin.** The logical thing to do would be to isolate and identify it. Many futile attempts have been made, but results with gibberellins show that their properties are similar to those expected for vernalin. Anton Lang found in 1957 (see Lang, 1956b) that gibberellins applied to certain biennials induced them to flower without a low temperature treatment (Fig. 21-6). Others (e.g., Purvis, 1961) induced winter annuals by treating seeds with gibberellins. It was then shown that natural gibberellins build up within several cold-requiring species during exposure to low temperature. Gibberellins clearly seem to be involved in vernalization.

But is gibberellin equivalent to vernalin? For various reasons, plant physiologists have been reluctant to accept this conclusion. When gibberellins are applied to a cold-requiring rosette plant, for example, the first observable

Figure 21-6 Anton Lang's induction of flowering in the carrot by application of GA$_3$ (center) or by vernalization (right). The control plant (left) and the center plant were held above 17 C. The plant at the right was given 8 weeks of cold treatment (see Lang, 1957).

response is the elongation of a vegetative shoot, followed by developing flower buds on this shoot. When plants are induced to flower by exposure to cold, however, flower buds are apparent as soon as the shoot begins to elongate. Could gibberellins induce changes within the plant that in turn lead to flowering or even to production of vernalin? Could several molecules influence the morphogenetic program in the same way?

Mikhail Chailakhyan in Russia suggested that there are two substances involved in flower formation, one a gibberellin or gibberellin-like material, the other a substance called **anthesin.** Low-temperature- and long-day-requiring plants might lack sufficient gibberellins until they have been exposed to the inducing environment, while short-day plants might contain sufficient gibberellins but lack anthesin. Chailakhyan suggested that cold-requiring plants produce vernalin in response to cold, and the vernalin is then converted to gibberellins in response to long days, at least in those plants requiring long days following the cold.

An elegant experiment performed many years ago by Melchers (1937) supports the two-substance point of view. A noninduced short-day plant (Maryland Mammoth tobacco) was grafted to a noninduced cold-requiring plant (henbane),

causing the latter to flower. Apparently, each contained one of the essential substances for the flowering process but had to obtain the other from the plant to which it was grafted, which succeeded for the henbane but not for the tobacco. (In Chapter 22 we shall present evidence that gibberellins are less important in flowering than might be implied here.)

Vernalization and the Induced State Flower development typically follows the cold treatment by days or weeks. How permanent is this **induced state,** the vernalized condition of the plant before flowering? One henbane variety requires low temperatures followed by long days for flowering. After vernalization, flowering can be postponed by providing only short days. No loss of the vernalization stimulus appeared in such plants after 190 days, even though all the original leaves exposed to the cold had died. Only after 300 days was there any loss of the vernalized condition. In many other species the induced state appears to be highly stable. Certain cereal seeds, for example, can be moistened (40 percent water—too little for germination), vernalized, and then dried out and maintained for months to years without loss of the vernalized condition. Yet there are several other species in which the induced state is far less permanent.

21.3 Dormancy

Only a few plants are able to function actively close to the freezing point. How, then, do plants live where temperatures remain close to freezing for several weeks or months each year? Most commonly, plants become dormant or quiescent, in which condition they remain alive but exhibit little metabolic activity. Leaves and buds of many evergreens show this reduced activity during the winter, and deciduous perennials lose their leaves and form special inactive buds. Seeds of most species in cold regions are dormant or quiescent during the winter. Certain changes occur in the protoplasts of dormant tissues that allow them to resist subfreezing temperatures (see Chapters 23 and 24).

It seems appropriate that temperature itself might play a regulatory role in the survival of plants in cold regions. The dormant or quiescent condition of evergreen leaves and buds often develops in response to low temperatures, these being typically combined with such other environmental parameters as short days. Furthermore, subsequent growth in the spring is often dependent upon prolonged exposure of dormant buds and seeds to low temperatures during the winter. The buds or seeds measure time by accumulating or summing the periods of exposure to cold. Thus, they become ecologically adjusted to the time and place.

As we've seen so often in our discussions, the situation becomes complex. Plants commonly respond to multiple environmental cues. Germination of seeds, for example, is influenced not only by temperatures, but also (depending always upon species) by light, breaking of the seed coat to permit penetration of the radicle and perhaps entry of oxygen and/or water, removal of chemical inhibitors, and maturity of the embryo.

Concepts and Terminology A seed may remain **viable** (alive) but unable to germinate or grow for several reasons. These can be roughly classified into *external* or *internal* conditions. An internal situation easy to understand is an embryo that has not reached a morphological maturity capable of germination (e.g., in certain members of the Orchidaceae, Orobancheae, or the genus *Ranunculus*). Only time will allow this maturity to develop. Germination of seeds of wild plants is often limited in this or some other internal way, but seeds of many domestic plants may be limited only by lack of moisture and/or proper temperature.

To distinguish between these two different situations, seed physiologists have utilized two terms: **quiescence** is the condition of a seed when it is unable to germinate only because external conditions *normally required for growth* are not present, and **dormancy** is the condition of a seed when it fails to germinate due to internal conditions, even though external conditions (e.g., temperature, moisture, and atmosphere) are suitable.

There are problems with this terminology. Dormant seeds are frequently induced to germinate by some specific change in the environment such as light or a period of low temperature. Where do we draw the line on conditions "normally required for growth"? Furthermore, in one sense it is *always* the internal conditions that are limiting. (If water is limiting, it is lack of water in the cells of the embryo inside the seed.) In another sense, external conditions always allow germination by influencing the internal ones. We can at least be more precise, stating conditions rather than depending upon the word "normally": **Dormancy*** is the condition of a seed when it fails to germinate, even though ample external *moisture* is available, the seed is exposed to *atmospheric conditions* typical of those found in well-aerated soil or at the earth's surface, and *temperature* is within the range usually associated with physiological activity (say, 10 to 30 C). (A seed physiologist will define conditions still more precisely.) **Quiescence** is the condition of a seed when it fails to germinate *unless the foregoing conditions are available.*

One other term has been widely used in studies in this area. **Afterripening** refers to the changes that go on within the seed (or the bud) during the breaking of dormancy. Some authors have used the term in a more restricted sense, limiting it to the changes that occur during dry storage or to maturation of the embryo (e.g., Leopold and Kriedemann, 1975).

*Researchers who study fruit trees (**pomologists**) employ a different terminology (Samish, 1954). The concept of dormancy as defined in this text is called **rest**, while the term "dormancy" is used in exactly the same sense as "quiescence." You should be on guard as you study the literature, especially for the term "dormancy," which may be used in either sense. In this text, dormancy is never used in the sense of quiescence.

21.4 Seed Longevity and Germination

It is an impressive idea that a living organism can go into a sort of suspended animation, remain alive, but not grow for a long period of time, only to begin active growth when conditions are finally suitable. Stories have circulated that seeds found in the ancient tombs of Egypt or the pueblos of the southwestern United States were capable of germination when hundreds to thousands of years old. Careful investigation has failed to support these stories, but the life span of some seeds is indeed great, often far exceeding the human life span.

Table 21-1 lists the longevities of several seeds. Some were determined in the 1930s by T. Becquerel, who studied dated seeds stored in the National Museum of Paris. *Mimosa glomerata* seeds remained viable for 221 years, but a rather typical life span of seeds is from 10 to 50 years. The documented record for seed survival appears to be that of a lupine (*Lupinus arcticus*). Viable seeds of this species thought to be at least 10,000 years old were found in lemming burrows deeply buried in permanently frozen silt of the Pleistocene Age in the central Yukon (Porsild et al., 1967). If these seeds were indeed that old, the cold temperatures must have protected them immensely.

Storage conditions always influence seed viability. Increased moisture may result in a more rapid loss of viability, but some seeds live longest submerged in water. Oxygen is generally detrimental to seed life spans. Viability is usually lost most rapidly when seeds are stored in humid air at temperatures of 35 C or warmer. Some seeds remain alive longer when buried in the soil than when stored in jars on a laboratory shelf, perhaps because of differences in light, O_2, CO_2, and ethylene. A few seeds have an unusually short life span. Seeds of *Acer saccharinum*, *Zizana aquatica*, *Salix japonica*, and *S. pierotti* lose their viability within a week if kept in air. Seeds of several species may remain viable anywhere from a few months to less than a year.

How can long-lived seeds remain viable so long? While a seed remains alive, it retains its stored foodstuffs within its cells; as soon as it dies, they begin to leak out. Dormant but viable seeds remain intact on wet filter paper for months; as soon as they die, they are overgrown by bacteria and fungal hyphae, which live on the food that leaks out. What maintains the integrity of the membranes? Frits Went (1974) points out that "if the maintenance of the living membranes of 3,000-year-old buried *Lotus* seeds required any metabolic processes, then no more than one sugar molecule would have been available each minute per hundred million protein molecules, truly negligible for any maintenance work." Respiration and, presumably, all metabolic activities must stop completely! Do we even begin to understand life function? What resists the decay to maximum entropy?

And what happens during germination? Although it is an oversimplification, seed physiologists speak of four stages: (1) hydration or imbibition, during which water penetrates into the embryo and hydrates proteins and other colloids, (2) the formation or activation of enzymes, leading to increased

Table 21-1 Some Representative Life Spans for Seeds[a]

Species	Viability (%) Initial	Viability (%) Final	Age at test	Notes
1. Sugar maple (*Acer saccharinum*)	—	—	<1 week	
2. English elm (*Ulmus campestris*)	—	—	ca 6 months	
3. American elm (*Ulmus americana*)	70	28	10 months	dry storage
4. *Heavea, Boehea, Thea,* sugar cane, etc.	—	—	<1 yr	
5. Wild oats (*Avena fatua*)	70	9	1 yr	buried 8″ in soil
6. Alfalfa (*Medicago sativa*)	85	1	6 yr	buried 8″ in soil
7. Yellow foxtail (*Setaria lutescens*)	56	4	10 yr	buried 8″ in soil
8. Cocklebur (*Xanthium strumarium*)	50	15	16 yr	buried 8″ in soil
9. Canada thistle (*Cirsium arvense*)	57	1	21 yr	buried 8″ in soil
10. Kentucky bluegrass (*Poa pratensis*)	91	1	30 yr	buried 8″ in soil
11. Red clover (*Trifolium pratense*)	90	1	30 yr	buried 8″ in soil
12. Tobacco (*Nicotiana tabacum*)	89	13	30 yr	buried 8″ in soil
13. Button clover (*Medicago orbicularis*)	—	—	78 yr	herbarium
14. Clover (*Trifolium striatum*)	—	—	90 yr	herbarium
15. Big trifoil (*Lotus uliginosus*)	—	1	100 yr	dry storage
16. Red clover (*Trifolium pratense*)	—	1	100 yr	dry storage
17. Locoweed (*Astragalus massiliensis*)	—	—	100–150 yr	herbarium
18. Sensitive plant (*Mimosa glomerata*)	—	—	221 yr	herbarium
19. Indian lotus (*Nelumbo nucifera*)	—	—	1,040 yr	peat bog
20. Arctic lupine (*Lupinus arcticus*)	—	—	10,000 yr	frozen silt, lemming burrows

[a]From various sources. See summaries in Altman and Dittmer, 1962, and Mayer and Poljakoff-Mayber, 1963. Reprinted with permission of Pergamon Press, Inc., New York.

metabolic activity, (3) elongation of radicle cells followed by emergence of the radicle (embryonic root) from the seed coat (which is germination proper), and (4) subsequent growth of the seedling. Dormancy can be due to interference with any of the first three stages, and covering layers around the embryo—the endosperm, the seed coat, and the fruit coat—play a decisive role in this interference. In some, these layers prevent the entry of water and/or oxygen; in others, they prevent emergence of the radicle by acting as a mechanical barrier (Section 19.5); while in others, they apparently prevent leaching of inhibitors out of the embryos or contain inhibitors themselves. What are the causes of dormancy, what ecological advantages do dormancy mechanisms confer, and how are various forms of dormancy broken to allow germination?

21.5 Seed Dormancy

Impaction and Scarification One of the easiest examples of dormancy to understand is the presence of a hard seed coat that prevents absorption of oxygen or water. Such a hard seed coat is common in members of the Leguminosae family, although it does not occur in beans or peas, exemplifying that dormancy mechanisms are not common in domesticated species. In a few species, water and oxygen are unable to penetrate certain seeds because entry is blocked by a cork-like filling (the **strophiolar plug**) in a small opening (the

strophiolar cleft) in the seed coat. Vigorous shaking of the seeds sometimes dislodges this plug, allowing germination. The treatment is called **impaction,** and it has been applied to seeds of *Melilotus alba* (sweetclover), *Trigonella arabica,* and *Crotallaria egyptica.*

Breaking the seed coat barrier is called **scarification.** Knives, files, and sandpaper have been used. In nature, the abrasion may be by microbial action, passage of the seed through the digestive tract of a bird or other animal, exposure to alternating temperatures, or movement by water across sand or rocks. In the laboratory and in agriculture (when needed), alcohol or other fat solvents (which dissolve away the waxy materials that sometimes block water entry) or concentrated acids may be used. Cotton seeds, for example, may be soaked for a few minutes to an hour in concentrated sulfuric acid and then washed to remove the acid, after which germination is greatly improved.

Scarification is of considerable ecological importance. The time required for scarification to be completed by some natural means may protect against premature germination in the autumn or during unseasonal warm periods in winter. Scarification in the digestive tracts of birds or other animals leads to germination after the seeds are more widely dispersed. Dean Vest (1952) demonstrated an interesting symbiotic relationship between a fungus and the seeds of shadscale (*Atriplex confertifolia*) growing in the deserts of the Great Basin. The fungus grew on the seed coats, scarifying them so germination could occur. Fungal growth occurred only when temperatures and moisture conditions were suitable during early spring, the most likely time for survival of the seedlings.

Fire is another important natural means of scarification. Several seeds, particularly in situations such as the chaparral vegetation of Mediterranean climates (e.g., Southern California), are effectively scarified by the fires so common in these situations. The result is a relatively rapid recovery of the area following the fires.

Chemical Inhibitors What prevents the seeds in a ripe tomato from germinating inside the fruit? Temperature is usually ideal, and there is ample moisture and oxygen. If the seeds are removed from the fruit, dried, and planted, they germinate rapidly, indicating that they are mature enough for germination. They apparently fail to germinate in the fruit because of chemical germination inhibitors (Mayer and Poljakoff-Mayber, 1963). Tomato juice can be diluted 10 to 20 times and still inhibit germination of many seeds. In many fleshy fruits, the osmotic potential of the fruit juice is simply too negative to permit germination.

There are also chemical inhibitors present in the seeds, and often these must be leached out before germination can occur. In nature, when enough rain falls to leach the inhibitor from the seed, the ground will be adequately wet for survival of the new seedling. This is especially important in the desert where moisture limits more than other factors such as temperature. Vest (1952) found that shadscale seeds contained enough sodium chloride to inhibit them osmotically (see also Koller, 1957). Usually the material is more complex (Evenari, 1957; Ketring, 1973), and inhibitors include representatives from a wide variety of organic classes. Some are cyanide-releasing complexes (especially in rosaceous seeds), while others are ammonia-releasing substances. Mustard oils are common in the Cruciferae. Other important organic compounds include organic acids, unsaturated lactones (especially coumarin, parasorbic acid, and protoanemonin), aldehydes, essential oils, alkaloids, and phenolic compounds. The most widespread and perhaps the most potent inhibitor is ABA.

These substances occur not only in seeds but also in leaves, roots, and other parts of plants. When leached out or released during decay of litter, they may inhibit the germination of seeds or root development in the vicinity of the parent plant (Chapter 14). Substances produced by one organism that negatively influence another are called **allelopathics** (Section 23.2.10). Effects of the allelopathics are not examples of dormancy in the usual sense. Furthermore, some compounds produced by other organisms may act as germination promoters. Nitrate is a commonly used germination promoter in seed physiology laboratories and is produced by decay of virtually any plant or animal residue.

Before leaving the topic, we should note that many known compounds that are not natural products may strongly influ-

ence germination one way or the other. These include many of the growth regulators presently of commercial importance (e.g., 2,4-D, Dalapon, and others). Thiourea is often used in the laboratory as a germination promoter.

Stratification Many seeds, particularly those of rosaceous species such as the stone fruits (peach, plum, cherry), many other deciduous trees, several conifers, and several herbaceous *Polygonum* species will not germinate until they have been exposed to low temperatures in the moist condition in the presence of oxygen for weeks to months (Fig. 21-7). Crocker and Barton (1953) listed 62 such species, and numerous others have been found since. Rarely, moist seeds respond to high temperatures, and several seeds respond best when daily temperatures alternate between high and low levels. The practice of layering the seeds during winter in flats containing moist sand and peat is called **stratification.** In nature, the low temperature requirement protects the seed from precocious germination in the fall or during an unseasonal warm period in winter.

What chemical changes go on within the seed during stratification, allowing it subsequently to germinate when conditions are right? Most cold-requiring seeds are rich in fats and proteins but have little starch (Nikolaeva, 1969; Lang,

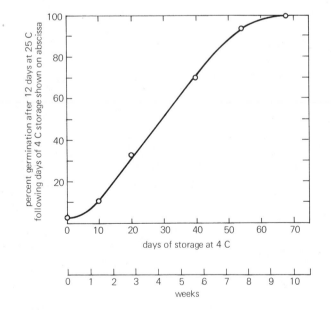

Figure 21-7 Germination of apple seeds as a function of storage time at 4 C. (Data from Villiers, 1972.)

1965a), and during the cold treatment the embryo usually grows extensively by transfer of carbon and nitrogen compounds from food storage cells. Sugars accumulate, which might be required as sources of energy and to attract water osmotically, later causing germination. Even in cold-requiring seeds such as European ash (*Fraxinus excelsior*) where the embryo is already fully developed before stratification, there is a massive degradation of fat in the embryo itself. The protein content rises, and starch then appears.

Perhaps inhibitors disappear during cold treatment, and/or growth promoters such as gibberellins or cytokinins accumulate (Villiers, 1972). Auxins have little effect on germination, but in many cases gibberellins will substitute for all or part of the cold treatment, just as they often do for vernalization. Perhaps they accumulate during stratification in amounts that overcome dormancy, but some cold-requiring seeds such as apple fail to respond to gibberellins. Cytokinin effects are usually less dramatic and are much less widespread.

Both inhibitor disappearance and hormone accumulation in whole seeds have been observed, but there are numerous contradictions. In Section 19.3, we discussed similar measurements in light-requiring seeds. We concluded that the radicles themselves should be analyzed, because changes in the rest of the seed could mask the important changes in the relatively small radicles. Recently, I. Arias and coworkers (1976) measured gibberellins in the embryonic axis and in the food storage cotyledon cells of the hazel tree (*Corylus avellana*), a species in which gibberellins fully overcome the stratification requirement. During chilling, there is little accumulation of gibberellin in either part, yet chilling allows the embryonic axis to synthesize much gibberellin when returned to a germination temperature of 20 C. The much larger cotyledons synthesized relatively little gibberellin, so the GA concentration became 300 times as great in the axis as in the cotyledons. Similar studies are needed with other seeds.

Where does the dormancy mechanism lie? Does the seed coat contain a chemical that inhibits elongation of the radicle? Does the seed coat or endosperm act as a mechanical barrier to elongation? Or does the radicle itself lack the ability to grow until chilled? Isolated, stratified embryos from many seeds will subsequently grow when placed at warmer temperatures, but nonstratified naked embryos will not. Hence, cold temperatures must act directly on the embryo. Consistent with this, growing embryos in stratified walnut seeds can exert mechanical pressure at least 10 bars greater than nonchilled embryos incapable of breaking the shells. In various species of lilac, including *Syringa vulgaris*, stratification has no effect on the strength or the inhibitor content of the endosperm, but radicles of chilled embryos will elongate in a solution having a water potential about 5 bars more negative than that in which nonchilled radicles will grow (Junttila, 1973). Although inhibitors are often present in seed coats, direct evidence that they inhibit radicle growth in such seeds seems to be absent. Whether changes in inhibitors and growth promoters in the radicle account for stratification remains to be seen.

Light In Section 19.5, we saw that light may control germination of many seeds, and we discussed some of the many complications in this response. Clearly there are many environmental cues, often interacting in intricate ways, that control the germination process.

21.6 Bud Dormancy

Seed and bud dormancy have much in common, but with buds we are nearly as concerned with the induction of dormancy as we are with breaking dormancy. Bud dormancy almost always develops before fall color and the senescence of leaves. Buds of many trees stop growing in midsummer, sometimes exhibiting a little growth again in late summer before going into deep dormancy in autumn. Flower buds that will grow the next season may form on fruit trees in midsummer. The leaves remain green and photosynthetically active until early autumn, when leaf senescence occurs in response to short, bright, cool days. As chlorophyll is lost, the yellow and orange carotenoid pigments become apparent, and anthocyanidins (primarily cyanidin glyoside) are synthesized. Fruits such as apples may develop during this time. Frost hardiness also develops in response to the low temperatures and short days of autumn.

Bud dormancy is induced in many species by low temperatures, but there is also a response to daylength, especially if temperatures remain high (see Fig. 19-14). With several deciduous trees studied at Beltsville (Downs and Borthwick, 1956), short-day treatment resulted in formation of a dormant terminal bud and cessation of internode elongation and leaf expansion but often a retention of the leaves. Long nights, each interrupted by an interval of light, acted the same as long days. The buds of birch (*Betula pubescens*) detect the daylength directly, but leaves usually detect the photoperiod although dormancy occurs in the buds (Wareing, 1956). Perhaps this correlative phenomenon, like others, is caused by a hormone, which could be abscisic acid (Section 17.4).

There are always a number of interactions. In the Beltsville study with deciduous trees, the daylength response was observed at temperatures of 21 to 27 C, but at temperatures of 15 C and 21 C, there was little stem growth, so daylength effects were not readily apparent.

There is considerable genetic variability in dormancy responses within a species. For example, Thomas O. Perry and Henry Hellmers (1973) found that a northern (Massachusetts) race of maple (*Acer rubrum*) developed winter dormancy in response to short days and cold temperatures in growth chambers, but a southern race from Florida did not. O. M. Heide (1974) studied Norway spruce (*Picea abies*). Austrian trees (47 deg latitude) stopped elongating at daylengths of 15 hours or less, but trees from northern Norway (64 deg latitude) stopped when daylengths were 21 hours or less (Heide, 1974). Temperature had little influence, but trees from high elevations stopped elongating at daylengths longer than those required to halt growth of trees at the same latitude but at low elevations. Heide also found that the roots did not respond to photoperiods applied to the tops. With few exceptions, roots continue to grow as long as nutrients and water are available, until soil temperatures become too cold (Kozlowski, 1971). Clearly, such trees are well adapted to the environments where they naturally occur. And the Florida maples, because they could not enter dormancy soon enough in the fall, are restricted to warm, southern climates.

Withholding water frequently accelerates development of dormancy, as does the restriction of mineral nutrients, particularly nitrogen. This is probably important for species that go into dormancy to resist the high temperatures and drought occurring in dry climates or in the tropics. Situations are also known in which dormancy develops in response to *changing* daylength (and even to changing soil temperature).

As dormancy first begins to develop, it can easily be reversed by moderate temperatures and long days (or continuous light). The plant is then only in the initial stages of quiescence and is not truly dormant. Gradually, however, attempts to induce active growth fail, and at this time the plant is said to have reached true dormancy (Fig. 21-8).

Morphology is important in dormancy phenomena. A dormant bud typically has greatly shortened internodes and specially modified leaves called **bud scales.** These scales prevent desiccation, insulate against heat loss, and restrict movement of oxygen to the meristems below. They may also respond to the light environment (see later) and/or perform other functions. In a sense, bud scales are analogous to the seed coat.

Overcoming dormancy is also a function of temperature or daylength or both. The temperature effect was studied as early as 1880 to 1914 (see Leopold and Kriedemann, 1975), but the daylength effect has been recognized only within the past two decades. Since leaves respond to daylength in the *induction* of dormancy and in flowering, it seemed reasonable that leaves are the only organs that respond to daylength. But it is now known that dormancy is broken by long days in several leafless trees: beech, birch, larch, yellow poplar, sweetgum, and red oak, for example. Except for beech, these species also respond to cold periods. In other cases, cold must be followed by long days. Even in midwinter, certain deciduous species will respond to long-day treatment (particularly continuous light). A midsummer dormancy occurs in some species (especially evergreens), during which the stems cease to elongate for a period of time. This is typically broken by exposure to more long days.

What organ responds to the long days that overcome dormancy? Apparently the bud scales themselves respond, or enough light penetrates to bring about the response within the primordial leaf tissues inside the bud. Probably both the short-day induction and the long-day breaking of dormancy are phytochrome responses, but the case is not clear cut. In some studies of the short-day induction, red light is most

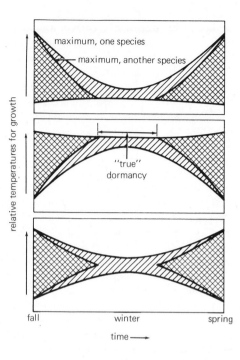

Figure 21-8 An illustration of Vegis's data and theory on dormancy and the temperature range allowing growth (see Vegis, 1964). The three graphs show three patterns of developing and breaking dormancy, each graph indicating responses for two species, one that develops "true" dormancy and another that does not. As the season progresses, the temperature range over which seeds will germinate or buds will grow becomes progressively narrower; finally the range widens again, allowing plant activity. If the range narrows to the point where germination or growth will not occur at any temperature, "true" dormancy is achieved; if not, plants are quiescent over most of the temperature range but will germinate or grow if the suitable narrow temperature range is used. The upper graph represents a situation common in warm climates: The maximum temperature allowing germination or growth decreases with progression of the season, protecting the seed or plant against a hot period of drought. The middle graph represents plants in cold climates: The minimum temperature for activity increases, protecting the plant against a cold winter. The bottom graph represents the most common situation in temperate climates, providing protection against both a hot and dry summer and a cold winter.

effective in the night interruption, and its effect is reversed somewhat by a subsequent exposure to far-red light, but this reversal has failed in other studies.

Dormancy in many buds can be broken by exposure to low temperatures (see the boxed essay and Fig. 21-9). Days to months may be required at temperatures below 10 C. With fruit trees, 5 to 7 C is more effective than 0 C. Considerable work has been done with fruit trees to determine the minimum cold period required to break dormancy. Apples,

The Annual Cycle in Fruit Trees
Schuyler D. Seeley

Schuyler Seeley is an Associate Professor of Pomology at Utah State University. He received the Ph.D. degree from Cornell University in 1971.

The annual cycle of a perennial deciduous tree comes to our attention in the spring when a dormant, seemingly lifeless skeleton buds into a beautiful, green, ecological machine. Early spring cambial growth begins at the base of expanding buds and spreads downward through the branches of the tree. The hormonal stimulus for this is thought to be auxin. Although some development probably occurs at any temperature above freezing, the threshold for significant growth in fruit trees is about 4 C. Bud swell and phenological stages of flower bud opening can be predicted based upon the temperature experience after the end of dormancy. Roots, which have no well-defined dormant period, release metabolites and hormones to the newly activated aerial portion of the tree. The yearly **grand phase of growth** (maximum growth rate) follows bud swell and flowering as shoot elongation and leaf development extend over 4 to 12 weeks.

Growth slows as temperatures exceed the optimum for growth, daylengths change, and water supplies diminish. Terminal buds become dormant during midsummer, and leaves mature and begin to senesce. Growth promoters, present in the growing buds at high concentration during the grand phase of growth, decrease, while inhibitors of plant growth accumulate. Winter scales form around the terminal buds in mid and late summer, and growth virtually stops,

although limited growth may occur in tree trunks and winter buds during the fall. Much root growth occurs after bud dormancy. During midsummer and early fall, signals from the environment cause changes in the plant's physiology, and a dormant or resting phase of growth begins in vegetative parts of the tree, although fruits grow and mature. After leaf fall, chilling of buds and perhaps cambium must occur to complete the dormant phase of growth and development before the next cycle begins.

Phenological data for orchard trees indicate that this chilling response has a temperature optimum (Fig. 21-9). A chill unit is defined as one hour of chilling at 7 C. As temperatures drop to the freezing point or increase to about 15 C, chilling efficiency decreases so that more hours are required. Above 15 C, a negative process occurs in that some of the chilling effects are reversed. Peaches and apples require 800 and 1,300 chill units respectively to completely break dormancy. This must occur before the plant can enter the bud swell, anthesis, and grand phases of growth.

Once the ability to grow has developed during the dormant phase, growth of deciduous orchard trees can be correlated to growing **degree hours with a threshold of 5 C and an optimum somewhere just below 25C.** (For example, 10 hours at 15C = 10 (15−5) = 100 growing deg hr. These orchard trees reach the first noticeable bud swell between 2,000 and 4,500 growing deg hr, depending upon the species. Full bloom in Chinese apricot and red delicious apple required 7,400 and 12,500 deg hr respectively. Cooling by evaporation of water from the buds reduces the rate of accumulation of growing degree hours, so sprinkling of fruit trees can reduce the rate of accumulation of growing degree hours and delay bloom for over two weeks, avoiding late spring frosts 80 percent of the time in some areas.

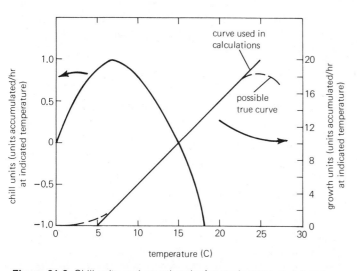

Figure 21-9 Chill units and growth units for apple trees as a function of temperature. Units are calculated by multiplying the number of hours at a given temperature by the factor for that temperature as indicated by the appropriate curve, and then by summing units for the tree's entire temperature experience during the period under consideration. For example, to calculate units for 10 hr at 10 C: 10 × 0.8 = 8 chill units; or, 10 × −0.5 = −5 growth units.

for example, may require 1,000 to 1,400 hr at about 7 C. Some headway has been made in selecting peach varieties with a shorter chilling requirement, allowing them to be cultivated farther south, where the winters are warmer. High temperatures following cold will reinduce dormancy in apple trees, a situation closely analogous to devernalization.

The effects of chilling upon the breaking of dormancy are not translocated within the plant but are localized within the individual buds. A dormant lilac bush, for example, may be placed with one branch protruding outside through a small hole in the greenhouse wall. The branch exposed to the low temperatures of winter will leaf out in early spring, but the rest of the bush inside the greenhouse remains dormant.

Several chemical treatments will break dormancy. 2-Chloroethanol (ClCH$_2$CH$_2$OH), often called ethylene chlorohydrin, has been used with success for many years. Applied in vapor form, it breaks dormancy of fruit trees. Another simple but often effective treatment is immersion of the plant part in a warm water bath (40 to 55 C). Often a short exposure (15 sec) is effective. Such a procedure was developed as early as 1909 by Hans Molisch in Austria.

Applied gibberellins break bud dormancy in many deciduous plants, just as they break dormancy of many cold-requiring seeds and induce flowering of many cold-requiring

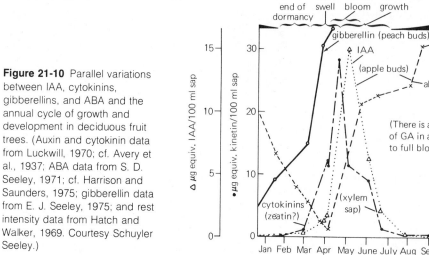

Figure 21-10 Parallel variations between IAA, cytokinins, gibberellins, and ABA and the annual cycle of growth and development in deciduous fruit trees. (Auxin and cytokinin data from Luckwill, 1970; cf. Avery et al., 1937; ABA data from S. D. Seeley, 1971; cf. Harrison and Saunders, 1975; gibberellin data from E. J. Seeley, 1975; and rest intensity data from Hatch and Walker, 1969. Courtesy Schuyler Seeley.)

plants. Furthermore, a number of growth-inhibiting compounds such as abscisic acid have been extracted from dormant seeds and buds, and gibberellins increase in these organs as dormancy is broken. Figure 21-10 summarizes some of these findings.

21.7 Underground Storage Organs

In many cases, temperature conditions will induce the formation of such underground storage organs as bulbs, corms, and tubers. Dormancy may also be broken or subsequent growth influenced by storage temperatures. Daylength may also influence formation of the organs.

The Potato Under usual greenhouse conditions, potato tubers form in response to short days. Such tubers develop from swellings at the tips of underground stems called **stolons,*** which are derived from nodes at the base of the stem in the soil. All the expected features of photoperiodism are present (including a critical night and inhibitory effect of a light interruption during the dark period—see Chapter 22). Tuber formation does not *require* short days, however, but proceeds at any daylength (a day-neutral response) if the night temperature is below 20 C. Tuberization is optimal at night temperatures of about 12 C. This interaction between photoperiodism and temperature is common, as we explained in relation to vernalization and dormancy.

Since the underground stolons are in darkness, some "tuber-inducing principle" must be produced in the leaves in response to short-day treatment and translocated to the stolons.

*Stolons are usually defined as aboveground stems, as in the strawberry. Potato stolons are usually underground, but they can be aerial; in darkness, even aboveground buds develop into stolons.

Low temperatures stimulating tuberization are also detected by the *tops* of the plants, rather than the underground parts. On long days, no tubers will form at any soil temperatures unless the shoots are exposed to low temperatures. Stolons form over a wide range of soil and air temperatures and daylengths, but only the conditions just described cause these to develop into tubers.

Among the known hormones, auxins are relatively inactive in tuber formation, while gibberellins inhibit formation and promote stolon elongation. Ethylene causes increased radial growth and decreased elongation of the stolon tip cells, as it does other tissues. This stolon swelling is morphologically somewhat similar to that of normal tubers, but these swellings contain little starch. In normal tuber formation, starch formation precedes visual morphological changes. Thus ethylene may not trigger tuber formation, but it may contribute to continued radial expansion. Low concentrations of kinetin stimulate formation of starch-rich tubers, but it is not known whether endogenous cytokinins normally control this process.

Because it is an underground stem, the potato tuber exhibits stem characteristics. Its eyes are the axillary buds, and they remain inactive in response to the presence of the apical bud. When the potato is cut up to produce seed pieces, this apical dominance is lost, and the axillary buds grow if dormancy has been broken. There are practical reasons both to prolong and to break tuber dormancy. The longer tubers can be stored during winter and spring in the dormant condition, the higher their price when sold. In potato "seed" certification, however, it is desirable to break dormancy prematurely to test for pathogens in sample tubers. The time normally required to break dormancy is somewhat shorter when the tubers are stored at about 20 C than at lower temperatures, but there is no clear-cut temperature effect. Certainly there is no cold requirement.

It is possible to break dormancy in potato tubers by the chemical treatments that are effective in breaking bud dormancy of aboveground stems (2-chloroethanol, gibberellins, hot water, and so on). Thiourea also causes sprouting but may result in as many as eight sprouts from a single eye rather than the usual single sprout. Dormancy may also be induced or prolonged by spraying such growth regulators as maleic hydrazide or chloropropham on the foliage before harvest or on the tubers after harvest. Storage temperature is also important. Tubers sprout somewhat prematurely at high temperatures. The ideal compromise seems to be 10 C.

Bulbs and Corms There has been little investigation of the induction of bulbs, corms, and rhizomes, but much work has been done in Holland, supported largely by the Dutch bulb

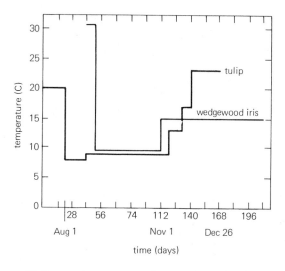

Figure 21-11 Temperature treatment for early flowering of *Tulipa gesneriana* W. Copland and of *Iris xiphium* Imperator. With the tulip, flower initiation begins and is well under way during the 20 C treatment. Moving to storage rooms at 8 and 9 C provides an acceleration in blooming, so that flowers are produced at Christmas. Continuous 9 C treatment gives equal earliness, but quality is poor unless the 20 C treatment is given first. The bulbs are planted in a controlled temperature greenhouse about midway during the low temperature treatment. The temperature is first raised when the leaf tips are visible, then again when they are 3 cm long, and finally when they are 6 cm long. With iris, the short period at high temperature is completely essential to flowering, although actual initiation of flower primordia does not occur until the bulbs have been moved from low temperature to 15 C, at which time the sprouts are about 6 cm long. Again the 9 C treatment is to ensure earliness. At temperatures much above 15 C during the last part of the treatment, abnormal flowers may be produced. Low light intensities will also result in "blasted" flowers at this time, especially if the temperatures are not right. If extremely high temperatures (38 C) are used during the first flower induction period, flower parts are increased or decreased, or tetramerous, pentamerous, or dimerous flowers result. (Data from Annie M. Hartsema, 1961; figure from Salisbury, 1963.)

industry, to determine the optimum storage conditions (primarily storage temperature as a function of time) that will result in the formation of leaves, flowers, and stems at desirable times and with the desirable properties. The approach was to observe the morphology of the bulb carefully in the field during a normal season, and then to repeat this using bulbs stored under accurately controlled temperature conditions. This work has been going on since the 1920s (see Hartsema, 1961; Rees, 1972).

We shall make a few generalizations: Bulbs must reach a critical size, which may require 2 or 3 years, before they begin to respond to temperature storage conditions by the formation of flower primordia. In some cases (e.g., tulip), leaf primordia are formed before the flowers, but sometimes leaf and flower formation are nearly simultaneous. Specific temperatures are often required for flower initiation or subsequent stem elongation. The course of change and the optimum temperatures usually match the climate at the location native to the bulbs.

There are several patterns: In some species, flower primordia form before the bulbs can be harvested. This allows little control during storage, so these have not been studied very much. In others, flower primordia form during the storage period after harvest in the summer but before replanting in the fall, making control easier. Figure 21-11 shows a temperature storage regimen designed to cause rapid flowering of tulips in time for Christmas. Note that temperatures inducing flowers are relatively high compared to those effective in vernalization of seeds and whole plants. Nevertheless, there is a parallel response.

In most bulbous irises (Fig. 21-11) the actual flower primordia appear during the low temperatures of winter (9 to 13 C optimum), but a high temperature (20 to 30 C) pretreatment is essential if flower formation is to occur at all. This is a true inductive effect similar to vernalization, but the response is to high rather than low temperatures.

21.8 Thermoperiodism

Growth rates of the vegetative parts of a plant are always strongly influenced by temperatures, as we have seen (see Fig. 21-1). Sometimes plant form is changed in rather specific and even inductive ways. This is often most noticeable in the vegetative structures associated with flowering, as in the bolting stalks of rosette plants formed in response to vernalization and/or long-day treatment. Stratification of seeds sometimes has a strong inductive effect on growth, in addition to its dormancy-breaking action. If the embryos of peach seedlings are excised from their cotyledons, they germinate without stratification, but the seedlings are frequently stunted and abnormal. When excised embryos are given low temperature, they grow into normal seedlings, so it is stratification and not the presence of the cotyledons that insures their normality. The accumulation of gibberellins or other hormones *during stratification* could account for these results, or stratification could increase the potential to synthesize gibberellins.

Frits Went (1957) described **thermoperiodism,** a phenomenon in which growth and/or development is promoted by alternating day and night temperatures. We noted before that potato tubers form in response to low night temperatures; fruit set on tomato plants is also strongly influenced by the night temperature. Stem elongation and flower initiation are also thermoperiodic responses in some species. An original implication of the thermoperiodism concept was that plant productivity was higher under a thermoperiodic environment. For some species, including certain tomato varieties, this may be true, but fluctuating day and night temperatures are not essential for optimum growth of numerous other species. Cocklebur, sugar beet, wheat, oats, bean, and pea grow as well at an *optimum* constant temperature as they do when day and night temperatures vary. An experimenter must be careful to compare various thermoperiodic regimens with the optimum constant temperature rather than some other temperature (Friend and Helson, 1976).

Generally, plants do better when the environment fluctuates on a 24-hr cycle, presumably to coincide with the phases of their circadian clock. Thus, some species grow poorly when both light and temperature are constant. Varying temperature on a 24-hr cycle prevents the injury caused on tomato plants by continuous light and temperature, if light intensities are high enough. Nevertheless, many thermoperiodic responses interact with the light environment, typically via photoperiodism and balances in the phytochrome system.

One of the most spectacular examples of thermoperiodism reported by Went (1957) involved *Baeria charysostoma,* a small annual composite commonly seen during spring in California. This plant normally occurs in mountain valleys and foothills and occasionally in the west Mojave Desert. It is extremely sensitive to night temperature. Grown under short-day conditions, plants survive only two months when the night temperature is 20 C. At lower temperatures, they grow for at least 100 days. They die rapidly at night temperatures of 26 C. The concept of 26 C as a lethal temperature is unexpected. Many species do not grow particularly well at night temperatures this high, but how can we account for *death* at this temperature? The *Baeria* plants flourish when the day temperature is well above 26 C, providing only that the night temperature is low enough. Other plants native to California acted similarly in Went's experiments.

21.9 Mechanisms of the Low Temperature Response

How can we understand positive plant responses to low temperature? We might be dealing with some kind of hormonal or metabolic block. Such a block could be a chemical inhibitor or the *lack* of some necessary substance within the plant, or both. An inhibitor could disappear, or a growth regulator could arise at low temperatures, influencing flowering, germination, subsequent seedling growth, and so on. Gibberellins and ABA often seem to play a role. Are the mechanisms the same in the several responses we have described? Surely the diversity is great enough that we wouldn't expect a common mechanism, but in many cases there are striking similarities.

Remember the paradox introduced at the beginning of the chapter. If low temperatures reduce the rate of chemical reactions, how can we account for *increased production* of some growth promoter or *increased destruction* of an inhibitor at low temperatures compared to high? Melchers and Lang, and Purvis and Gregory, simultaneously and independently suggested a model (Fig. 21-12), not unlike that of Fig. 21-2. There might be two hypothetical interacting reactions, one (**I**) with a fairly low temperature coefficient or Q_{10}, the other (**II**) with a higher Q_{10}. Products of reaction I are acted upon by reaction II. If the rate of reaction I exceeds that of II, then the product of reaction I (*B*) will accumulate; if the reverse is true, the product of reaction II (*C*) will accumulate. Even if the Q_{10} of reaction I is relatively low, but the reaction progresses at low temperatures more rapidly than reaction II, then we can explain the accumulation of *B* at low temperatures. With increasing temperature, the rate of reaction II increases much more rapidly than the rate of reaction I, so at some critical temperature, *B* will be utilized as fast as it is produced and hence will not accumulate. Reaction II would be devernalization, and the fact that devernalization fails after two or three days at the neutral temperature might indicate a third reaction (III), which converts *B* to *D*, a stable end product. Of course the model is naive, because many other factors could play a role:

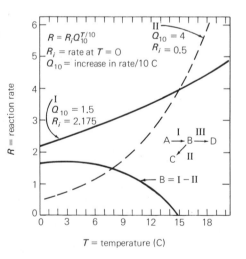

Figure 21-12 Sample curves showing hypothetical reaction rates as a function of temperatures for reactions with Q_{10} values of 1.5 or 4.0. If the reaction with $Q_{10} = 1.5$ is considered to be reaction I in the formula in the text, and the reaction with $Q_{10} = 4.0$ to be reaction II, then the hypothetical product *B* will be proportional to curve 2 minus curve 1, as shown (curve *B*). Compare the shape of curve *B* with the curves in Figs. 21-1, 21-2, and 21-5. (From Salisbury, 1963.)

enzyme synthesis, membrane permeability changes, phase changes, and so on. But the principle could apply in various ways.

For example, the compensated feedback systems discussed in relation to temperature independence in the biological clock (Section 20.5) could provide for an overall reaction with a negative temperature coefficient. The product of one reaction might inhibit the *rate* of another. Or, at low temperatures, a substance might accumulate because another compound inhibiting its production might not. Again, different temperature coefficients would be required. Since gibberellins increase in some seeds and buds as dormancy is broken, they may be equivalent to B or D in Fig. 21-12. Or gibberellins might leak out of a storage compartment when membranes become much more permeable at low temperatures (Arias et al., 1976). With a few species, cytokinins or ethylene could play this role.

What if we are dealing with *destruction of an inhibitor* at low temperatures rather than the synthesis of a promoter? We have only to reverse the roles of the two hypothetical reactions in the model. The destruction (or conversion) reaction must have the fairly rapid rate at low temperatures and the low Q_{10}. The synthesis of the inhibitor, on the other hand, must be low at low temperatures but have the high Q_{10}.

Photoperiodism

The synchronization of organisms with seasonal time is a truly spectacular manifestation. Often, this synchronization is concerned with reproduction: It is appropriate and adaptive for young animals to be born at specified times of year, for all members of a given angiosperm species to flower at the same time (insuring an opportunity for cross-pollination), and for mosses, ferns, conifers, and even some algae to form reproductive structures in a given season. Many other plant responses, such as stem elongation, leaf growth, dormancy, formation of storage organs, leaf fall, and development of frost resistance, also occur during certain seasons. Frequently, these seasonal responses are synchronized by photoperiodism. Much of what we see happening in the natural world is happening because plants and animals are able to detect the relative lengths of day and night.

22.1 Detecting Seasonal Time by Measuring Daylength

In a nonmountainous region on the equator, sunrise and sunset occur at the same time each day, so the relative lengths of day and night remain constant throughout the year. Exactly at the poles, the sun remains above the horizon for six months each year and below for the other six months. Again, the day and the night are about equal; each is six months long! Traveling from the equator toward the poles, the days become longer in summer and shorter in winter (Fig. 22-1). This is because the equator is tipped 23.5 deg to the plane of the ecliptic (the earth's orbit around the sun), so that during winter the north pole is tipped 23.5 deg away from the sun and during summer 23.5 deg toward the sun.

The rate at which daylength changes varies during the year (Fig. 22-1). Near the times of the summer and winter solstices, when days are longest and shortest, there is little change from day to day; during spring and autumn, the rate of change is much more rapid, as days become longer during spring, shorter during fall. Thus, an organism might detect the season by measuring day and night lengths and how they change, but because the absolute daylengths at any time of year depend so strongly upon latitude, organisms must be "calibrated" to take this into account.

Study of the photoperiodic responses of organisms could contribute important information to our understanding of natural ecosystems, but of the approximately 300,000 species of plants, only a few hundred have been grown with different artificial photoperiods. Not many surprising facts have come to light through this work. As one would expect, plants that grow at latitudes far from the equator respond to longer days than plants growing closer to the equator. It was surprising, however, to learn that many tropical plants respond to daylength, detecting the slight changes that occur 5 to 15 deg from the equator.

It was also interesting to learn that different ecotypes (see Section 23.2.9) within a single species often have different responses to daylength. In three representative studies, specimens of two short-day plants, lambsquarters (*Chenopodium album*; Cumming, 1969) and cocklebur (*Xanthium strumarium*; McMillan, 1974a), and one long-day plant, alpine sorrel (*Oxyria digyna*; Mooney and Billings, 1961) were collected at various latitudes throughout North America. Alpine sorrel plants were also collected in the Arctic and cocklebur plants from all over the world. In each case, the daylength that induced flowering was longer for individuals collected farther north. Often, no morphological differences could be detected between individuals having a greatly different flowering response. Charles Olmsted (1944) even found that sideoats grama (*Bouteloua curtipendula*) had short-day strains (ecotypes) at the southern end of its range and long-day strains at the northern end. Varieties of many other species (e.g., cotton, soybean, rice, wheat, chrysanthemum, and other native grasses) have also been compared, although not always from a wide latitudinal range, and the same diversity has become apparent. Some soybean varieties grown in northern latitudes are so sensitive to daylength that they will grow best only within a 50-mile range of latitude. So varieties are genetically "calibrated" to a habitat. Sometimes one gene will make the difference between a daylength-sensitive variety and one that is not sensitive to daylength. In other species, several genes are involved. These results indicate the importance of using a carefully standardized variety in a continuing scientific study, and they also indicate that classifications of response types, such as in Table 22-1 (see later), are at best only approximations.

Why hasn't the role of photoperiodism in ecology been studied more intensively? Partially, perhaps, because the

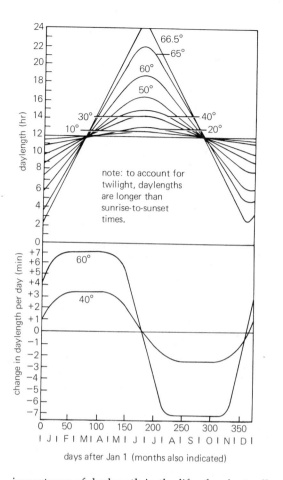

Figure 22-1 Top, daylength at various latitudes as a function of time during the year. The latitudes shown pass near the following geographic locations:

  0 deg latitude: Borneo, Ecuador, Nairobi;
10 deg latitude: Saigon, Trinidad, Caracas, Port Guinea;
20 deg latitude: Hawaii, Hong Kong, Hanoi, Mexico City, Puerto Rico, Mecca, Bombay;
30 deg latitude: New Orleans, Cairo, Shanghai.
40 deg latitude: Denver, Philadelphia, Madrid, the "heel" of Italy, Peking.
50 deg latitude: Vancouver, Winnipeg, the southern tip of England, Frankfurt.
60 deg latitude: Anchorage, southern tip of Greenland, Oslo, Helsinki, Leningrad;
70 deg latitude: Point Barrow, Northern Norway, Northern Siberia;
80 deg latitude: Northern Greenland, Spitsbergen.

Bottom, rate of change in daylength at two latitudes as a function of time during the year.

importance of daylength in the life of a plant calls no attention to itself until the plant is moved to another daylength or to artificial conditions of light and temperature. Plants must be well adapted to the daylengths at the latitudes where they exist or they could not exist there, so ecologists are not likely to notice the daylength response. Plant physiologists are concerned with such things, but so far they have faced the challenges of understanding the mechanism of photoperiodism rather than its ecological significance.

22.2 Some General Principles of Flowering Physiology

Since the days of Tournois, Klebs, and Garner and Allard (see essay, p. 338), well over a thousand papers have been published describing studies on photoperiodism. The most striking initial impression to be gained from this vast body of facts might be that there are no generalities, no sweeping laws to help us understand the photoperiodism response. Each species and often each variety within a species seems to have its own features of response; probably no two respond exactly alike.

Such a situation certainly poses a challenge for a plant physiology student and for the authors of a plant physiology

text! But things are not quite as hopeless as they may sound. Though every rule seems to have its exception, some generalizations can nonetheless be made: principles of photoperiodism that apply whether or not the process being controlled is the initiation of a soybean flower or a female pine cone, or the development of a potato tuber. In the next seven sections, we shall present seven generalizations. We shall present some experimental data based upon several plant species, and we shall note a few exceptions. In your reading, don't worry about remembering all the details—experts in the field have difficulty doing that—but let those details help you to understand the generalization presented at the end of each section.

We shall document only a few points with specific references to the literature. Several recent reviews include detailed references (e.g., Evans, 1969 and 1975; Schwabe, 1971; Vince-Prue, 1975; and Zeevaart, 1976a and 1976b).

22.3 The Response Types

In Figure 22-2, the relative number of days from the beginning of treatment to the appearance of flowers is plotted as a function of the hours of light in each 24-hr cycle. In a truly day-neutral plant, which is probably relatively rare, the time to the appearance of flowers is independent of daylength, so a

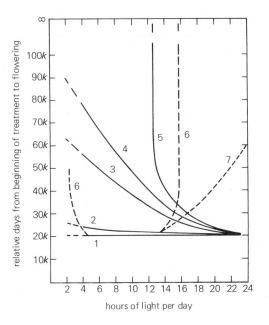

Figure 22-2 Representative flowering responses to different daylengths, plotted as time required for appearance of flowers. *K* following the numbers on the ordinate indicates that these numbers are arbitrary. Curve 1 is for a true day-neutral plant. Other curves belong to two families of curves, those for LDPs* and those for SDPs.* Curve 2 would also be considered a day-neutral response, although flowering is slightly promoted by long days. Curves 3 and 4 represent quantitative LDPs, but 4 is more sensitive to daylength than 3. Curve 5 is a true (absolute) LDP. Curve 6 is a true (absolute) SDP, with its minimum day (about 2.5 hr) indicated on the left and its minimum night (8.5 hr; a 15.5-hr day) on the right. Curve 7 is a quantitative SDP. (Modified from Salisbury, 1963b.)

horizontal line appears in the figure. Long-day plants* flower sooner on LD, so their curves (solid lines) slope downward to the right. SDPs take longer to flower on LD, so their curves (dotted lines) slope upward to the right. As in vernalization (Section 21.2), there is both a facultative and an absolute response to photoperiodism. Curves representing plants with an absolute daylength requirement become vertical at the daylength called the **critical day.** It is also possible to speak of the **critical night,** that nightlength that must be exceeded for flowering of SDPs or inhibition of flowering of LDPs. One other complication is apparent. SDPs will often not flower when only a few hours of light are received each day. This brief daylength requirement is more than photosynthesis. Plants that can be maintained indefinitely in the dark by being

*We will use the following abbreviations: LD(s) = long day(s), SD(s) = short day(s), LDP(s) = long-day plant(s), and SDP(s) = short-day plant(s).

fed sucrose nevertheless show the minimum light requirement for the photoperiodism response. Note that this requirement is represented by curves that slope downward to the right. Such curves express the long-day response, so short-day plants become long-day plants when the days are extremely short!

Note that SDPs and LDPs respond in a completely opposite way. Line 6 in Fig. 22-2 could represent cocklebur, a classic SDP. Line 5 could represent henbane (*Hyoscyamus niger*), a classic LDP. Both flower when days are about 12.5 to 15.7 hr long (nights 11.5 to 8.3 hr).

Sometimes the critical day or night can be determined for a population of plants within rather narrow limits (e.g., 5 to 10 min). In other cases, the limits are far less exact and may spread out over 1 or more hours. With cocklebur, the greater the number of inductive cycles, the more exact is the critical day or night.

Still other ways in which species differ is in their ripeness to respond (see discussion following) and in the number of SD or LD cycles required to induce flowering. Species that require only a single inductive cycle for flowering (Table 22-1) have been widely used, because they permit experimental treatments (e.g., application of a chemical) to be performed at various times in relation to the single cycle and without complications due to effects on previous or subsequent cycles.

Representative LD, SD, and day-neutral plants with some critical daylengths are shown in Fig. 22-3 and Table 22-1. The three response types observed by Garner and Allard form the foundation for any classification of photoperiod response types. We have already encountered a few embellishments on this simple scheme, however, and there are more, including several valid categories nearly equal in importance to those of Garner and Allard. To produce flowers, a few species (item 5, Table 22-1) require LDs followed by SDs, as occurs in late summer and fall. When maintained either under continuous LDs or continuous SDs, they remain vegetative. The counterparts to these **long-short-day plants,** the **short-long-day plants** (item 6), require SDs followed by LDs, as in spring. There are one or two species at least (item 7) that flower only on **intermediate daylength** and remain vegetative when days are either too short or too long. They have counterparts, which remain vegetative on intermediate daylength, flowering only on longer or shorter days (item 8).

There are several interesting interactions between photoperiod and temperature. We have seen that a vernalization requirement is often followed by a requirement for LDs, consistent with the LDs of late spring following winter. In other cases, a plant may exhibit a given response type at one temperature but not at another. It may, for example, have a qualitative or quantitative SD response at temperatures above, say, 20 C but be essentially day-neutral at cooler temperatures. Two examples of plants that are absolute SDPs at high temperature and absolute LDPs at low temperature have been reported: poinsettia (*Euphorbia pulcherrima*), and morning glory (*Impomea purpurea*). They are day-neutral only at the intermediate temperature. *Silene armeria,* a LDP, is induced on

SD   LD
Tomato, ca 110 days

LD→SD   LD
Lambsquarters, ca 110 days

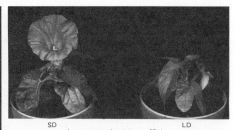

SD   LD
Japanese morning glory, ca 35 days

SD   LD
Cocklebur, ca 60 days

SD   SD→LD
Henbane, ca 110 days

SD   LD
Radish, ca 35 days

SD   LD
Muskmelon, ca 46 days

SD SD→LD LD→SD LD
Petunia, 54 days (27 SD, 27 LD)

Figure 22-3 Some representative day-neutral (tomato), SD (lambsquarters, Japanese morning glory, and cocklebur), and LD (henbane, radish, muskmelon, petunia, barley, and spinach) plants. In tomato, note strong effects of daylength on vegetative form, although both plants are flowering.

SD   LD
Barley, ca 35 days

SD   LD
Spinach, ca 35 days

Table 22-1 Representative Species Having Various Flowering Responses to Photoperiodic Treatment[a]

I. Known species that flower in response to a single inductive cycle

SHORT-DAY PLANTS	Approximate Critical Night[b]
Chenopodium polyspermum Goosefoot	
Chenopodium rubrum Red goosefoot	
Lemna paucicostata Duckweed	
Lemna perpusilla Strain 6746 Duckweed	12
Oryza sativa cv Zuiho Rice	12
Pharbitis nil cv Violet Japanese morning glory	9–10
Wolffia microscopia Duckweed	
Xanthium strumarium Cocklebur	8.3

LONG-DAY PLANTS	Approximate Critical Day[b]
Anagallis arvensis Scarlet pimpernel	12–12.5
Anethum graveolens Dill	11
Anthriscus cerefolium Salidcherril	
Brassica campestris Bird rape	
Lemna gibba Swollen duckweed	
Lolium temulentum Darnel ryegrass	14–16
Sinapis alba White mustard	ca 14
Spinacia oleracea Spinach	13

II. Some species that require several cycles for induction[c]

SHORT-DAY PLANTS

1. SDP (qualitative or absolute)

  Cattleya trianae   Orchid
  Chrysanthemum morifolium   Chrysanthemum
    variety
  Cosmos sulphureus   cv Yellow   Cosmos
  Glycine max   Soybean
  Kalanchoe blossfeldiana   Kalanchoe
  Perilla crispa   Purple common perilla
  Zea mays   Maize or corn
  SDP at high temperature; quantitative SDP at low temperature
    Fragaria × ananassa   Pine strawberry
  SDP at high temperature; day-neutral at low
    temperature
    Pharbitis nil E   Japanese morning glory
    Nicotiana tabacum   Maryland mammoth
      tobacco
  SDP at low temperature; day-neutral at high
    temperature
    Cosmos sulphureus   cv Orange flare   Cosmos
  SDP at high temperature; LDP at low temperature
    Euphorbia pulcherrima   Poinsettia
    Ipomoea purpurea   cv Heavenly Blue   Morning
      glory

2. Quantitative SDP

  Cannabis sativa   cv Kentucky   Hemp or marijuana
  Chrysanthemum morifolium   Chrysanthemum
    variety
  Datura stramonium H (older plants are day-neutral)
    Jimsonweed datura

  Gossypium hirsutum   Upland cotton
  Helianthus annuus   Sunflower
  Saccharum spontaneum   Sugar cane
  Quantitative SDP; require or accelerated by low-
    temperature vernalization
    Allium cepa   Onion
    Chrysanthemum morifolium   Chrysanthemum
      variety

LONG-DAY PLANTS

3. LDP (qualitative or absolute)

  Agropyron smithii   Bluestem wheatgrass
  Arabidopsis thaliana   Mouse ear cress
  Avena sativa, spring strains   Oats
  Chrysanthemum maximum   Pyrenees
    chrysanthemum
  Dianthus superbus   Lilac pink carnation
  Fuchsia hybrida   cv Lord Byron   Fuchsia
  Hibiscus syriacus   Hibiscus
  Hyoscyamus niger, annual strain   Black henbane
  Nicotiana sylvestris   Tobacco
  Raphanus sativus   Radish
  Rudbeckia hirta   Black-eyed Susan
  Sedum spectabile   Showy stonecrop
  LDP; require or accelerated by low-temperature
    vernalization
    Arabidopsis thaliana, biennial strains   Mouse ear
      cress
    Avena sativa, winter strains   Oats
    Beta saccharifera   Sugar beet
    Bromus inermis   Smooth bromegrass
    Hordeum vulgare   Winter barley
    Hyoscyamus niger, biennial strain   Black henbane
    Lolium temulentum   Darnel ryegrass
    Triticum aestivum   Winter wheat
  LDP at low temperature; quantitative LDP at high
    temperature
    Beta vulgaris   Common beet
  LDP at high temperature, day-neutral at low
    temperature
    Cichorium intybus   Chicory
  LDP at low temperature; day-neutral at high
    temperature
    Delphinium cultorum   Florists larkspur
    Rudbeckia bicolor   Pinewoods coneflower
  LDP; low temperature vernalization will substitute
    (at least partly) for the LD requirement
    Spinacia oleracea   cv Nobel   Spinach
    Silene armeria   Sweetwilliam silene

4. Quantitative LDP

  Hordeum vulgare   Spring barley
  Lolium temulentum   cv Ba 3081 Darnel ryegrass
  Nicotiana tabacum   cv Havana *A*   Tobacco
  Secale cereale   Winter rye
  Triticum aestivum   Spring wheat

Table 22-1 (continued)

Quantitative LDP; require or accelerated by
low-temperature vernalization
Digitalis purpurea   Foxglove
Pisum sativum   Late flowering garden pea
Secale cereale   Winter rye
Quantitative LDP at high temperature; day-neutral
at low temperature
Lactuca sativa   Lettuce
Petunia hybrida   Petunia

DUAL-DAYLENGTH PLANTS

5. Long-short-day plants

Aloe bulbilifera   Aloe
Kalanchoe laxiflora   Kalanchoe
Cestrum nocturnum (at 23 C, day-neutral at >24 C),
night-blooming jasmine

6. Short-long-day plants

Trifolium repens   White clover
Short-long-day plants; require or accelerated by
low-temperature vernalization
Dactylis glomerata   Orchardgrass
Poa pratensis   Kentucky bluegrass
(in these plants, the SD is required for induction
and LD for development of the inflorescence)
Short-long-day plants; low temperature substitutes
for the SD effect and, after low temperature,
plants respond as LDP
Campanula medium   Canterbury bells

INTERMEDIATE-DAY PLANTS

7. Plants flower when days are neither too short nor too long

Chenopodium album   Lambsquarters goosefoot
Coleus hybrida   cv Autumn coleus
Saccharum spontaneum   Sugar cane

AMBIPHOTOPERIODIC PLANTS

8. Plants quantitatively inhibited by intermediate
daylengths

Chenopodium rubrum ecotype 62° 46′ N at 25°C E
(responds as quantitative intermediate-day
plant at 15 to 20 C and as a quantitative LDP at
30 C)   Goosefoot
Madia elegans   Tarweed
Setaria verticillata   Hooked bristlegrass

DAY-NEUTRAL PLANTS

9. Day-neutral plants: These are the plants with least
response to daylength for flowering. They flower at
about the same time under all daylengths but
may be promoted by high or low temperature, or by
a temperature alternation

Cucumis sativus   Cucumber
Fragaria-vesca semperflorens   European alpine
strawberry
Gomphrina globosa   Globe amaranth
Gossypium hirsutum   Upland cotton
Helianthus annuus   Sunflower
Helianthus tuberosus   Jerusalem artichoke
Lunaria annua   Dollar plant
Nicotiana tabacum   Tobacco
Oryza sativa   Rice
Phaseolus vulgaris   Kidneybean
Pisum sativum   Garden pea
Zea mays   Maize or corn
Day-neutral plants; require or accelerated by
low-temperature vernalization
Allium cepa   Onion
Daucus carota   Wild carrot
Geum sp   Avens
Lunaria annua   Dollar plant

[a]Mostly from Vince-Prue, 1975 and Salisbury, 1963b.

[b]Critical night or day often depends on conditions (e.g., temperature), age of the plant, number of inductive cycles, and cultivar;
hence, some are not shown, and those that are shown are only representative.

[c]Note that single species often appear in several categories, indicating variabilities of varieties within species. To conserve space,
the lists have been greatly abbreviated.

SDs by either high (i.e., 32 C) or low (5 C) temperatures. Specific daylength requirements only at certain temperatures prove to be quite common.

Vernalization and a given photoperiodic treatment are sometimes interchangeable. For example, in a variety of Canterbury bells (*Campanula medium*), vernalization is fully replaced by exposure to SDs, but LDs are required following either treatment. Still other complications and interactions are known. For example, Japanese morning glory has become a prototype SDP. Yet it can be induced under LDs by low temperatures, high-intensity light, treatment with growth retardants, removal of roots, and low nutrient levels. Surely such a diversity of response types should be of ecological importance.

The generalization that concludes this section is simply stated: **1. There is a wide diversity of response types.**

If daylength plays such a decisive role, why wasn't photoperiodism discovered sooner? To be sure, A. Henfrey had suggested in 1852 that daylength might influence plant distribution, but the measurement of time by plants must have seemed unlikely to nineteenth-century botanists. Even the discoverers seemed to resist their own ideas generated by their data. Probably the first to realize the role of daylength was Julien Tournois, who began studying the flowering and sexuality of hops *(Houblon japonais)* and hemp *(Cannabis sativa)* in Paris in 1910. He noticed the extremely early flowering of his plants in the winter greenhouses, but at first he convinced himself that they were flowering in response to the decreased *quantity* of light rather than its duration. In his third paper, published in 1914, he finally grasped the point: "Precocious flowering in young plants of hemp and hops occurs when, from germination, they are exposed to very short periods of daily illumination." And: "Precocious flowering is not so much caused by shortening of the days as by lengthening of the nights." He had planned further experiments but was killed at the front in World War I.

Across the lines in Heidelberg, Germany, Georg Klebs (1918) had probably also discovered the role of daylength. He made plants of houseleek *(Sempervivum funkii)* flower by exposing them to several days of continuous illumination. Most scientists of that time thought that nutrition controlled reproduction in plants, but Klebs felt that the additional light, which caused his plants to flower, was acting catalytically and not as a nutritional factor. Since *Sempervivum* was a long-day plant, perhaps flowering required the additional photosynthesis provided by the added light, it could be argued. Tournois's hops and hemp, however, flowered with *less* light, so he tried lower intensities extended over a longer time to see if they would provide the same response as short durations; they did not, so the time factor seemed to be controlling. Garner and Allard (Section 20.7) followed the same line of reasoning, separating in their experiments effects of light quantity from those of light duration.

Incidentally, nutritional factors often do play at least a quantitative role in plant reproduction. The acceleration of flowering in many species by high carbohydrates and low nitrogen was also being documented during World War I (Fischer, 1916; Klebs, 1918; and Kraus and Kraybill, 1918). The effects of this knowledge on agriculture have been at least as profound as our understanding of photoperiodism. Tomato and apple yields, for example, are increased by withholding nitrogen at appropriate times.

22.4 Ripeness to Respond

Only a few plants respond to photoperiod when they are small seedlings. The Japanese morning glory (*Pharbitis nil*) responds to SDs in the cotyledonary stage, and some species of goosefoot or lambsquarters (*Chenopodium* sp.; Cumming, 1959) respond and flower as minute seedlings. In laboratory studies, they are grown on filter paper in a Petri dish. Most species, such as the cocklebur, must attain a somewhat larger size; cotyledons do not respond. Henbane must be 10 to 30 days old before it will respond to LDs. Certain monocarpic bamboo species and several polycarpic trees will not flower until they are 5 to 40 or more years old. Klebs called the condition a plant must achieve before it will flower in response to the environment **ripeness to flower** (German *Blühreife*), but a more descriptive term might be **ripeness to respond.** In many species, the number of required photoperiodic cycles decreases as the plant gets older; that is, the level of ripeness to respond increases with age. Often, the plant finally flowers independently of the photoperiod; that is, it becomes day-neutral.

Individual leaves must also reach a ripeness to respond. In several species, the leaf is maximally sensitive when it is first mature (fully expanded). Cocklebur leaves less than 1 cm long will not respond, but the half-expanded leaf, the one growing most rapidly, is most sensitive. On the other hand, leaves of the scarlet pimpernel (*Anagallis arvensis*) are most sensitive to LDs when the plant is a seedling; young leaves on older plants are less sensitive.

The concept of ripeness to respond is almost identical to that of juvenility defined as the condition of a plant before it is mature enough to flower (Section 18.3). The term "juvenility" is often used, however, to emphasize the special morphological features (particularly leaf shape) characteristic of juvenile plants. Another related concept is that of **minimum leaf number,** which is the minimum number of leaves from seedling to earliest flower under the most ideal conditions for flowering.

The conclusion of this section: **2. Before a plant can flower in response to its environment (particularly daylength and temperature) the organs that detect the environmental change, usually leaves or meristems, must reach a condition called ripeness to respond. There is a great diversity among species and plant organs in the age at which they achieve this condition.**

22.5 Phytochrome and the Role of the Dark Period

In 1938, James Bonner went to the University of Chicago to spend the summer working with Karl C. Hamner on photoperiodism in cocklebur. In the brief time available, a number of pioneering experiments were completed. Among other things, Hamner and Bonner wondered about the relative importance of day and night in photoperiodic induction. They took two experimental approaches. In one series of experiments, days and nights were varied to give cycles that did not

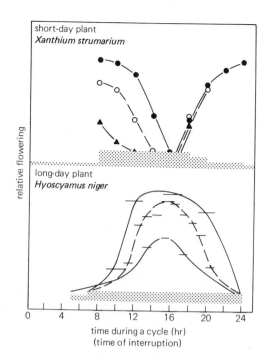

Figure 22-4 Effects of a light interruption given at various times during dark periods of various lengths (shaded bars) on subsequent flowering of an SDP and an LDP. Interruptions with *Xanthium* were 60 sec; with *Hyoscyamus*, times are indicated by length of data lines. (Data for *Xanthium* from Salisbury and Bonner, 1956; data for *Hyoscyamus* from Claes and Lang, 1947.)

equal 24 hours. The *critical night*, but not the critical day remained constant, indicating the importance of the dark period. In the other approach, days were interrupted with darkness or nights with light. Dark interruption of the day had little or no effect, but night interruption by light inhibited flowering in SDPs and (in later experiments) promoted flowering in LDPs (Fig. 22-4). This was the discovery of the **night interruption phenomenon.**

Once it was known that a light break during the dark period would nullify the effect of darkness, several possibilities for experimentation became immediately apparent. Researchers could ask: Which is more important, the intensity of light used or the total quantity of light energy as calculated by multiplying its intensity times the time interval over which it is applied? Within rough limits, the total quantity of energy proves to be the determining factor. It's like when you expose the film in photography: If you have a longer exposure time (slower shutter speed), you must compensate by letting less light strike the film (smaller lens aperture = larger f-stop number). Researchers could then ask: How dark is dark? Light applied during the entire dark period is effective (especially in inhibiting flowering of SDPs) at very low intensities. In

some species, for example, it is effective at 3 to 10 times the light from the full moon or 0.02 ft-c.

When is the light most effective? Usually at some constant time after the beginning of an inductive dark period using SDPs or an inhibitory dark period using LDPs (Fig. 22-4). This time of most effectiveness is often equivalent to the critical night.

Which wavelengths of light are most effective? In the early 1940s, it became apparent that *red light* was considerably more effective than other wavelengths. Action spectra for inhibition in SDPs and promotion in LDPs are typical of those for other phytochrome responses (see Fig. 19-3). Thus in the early 1950s, immediately after far-red reversibility was discovered in lettuce seed germination, cocklebur plants were illuminated in the middle of a long inductive dark period with red light followed by far-red. If the far-red followed the red illumination immediately, plants flowered; if about 30 minutes were allowed to elapse between the red and the far-red exposures, the far-red no longer reversed the effects of the red. Apparently P_{fr} completes its inhibitory act within 30 minutes in cocklebur.

Let's state a preliminary conclusion: **3A. The dark period plays an important role in the photoperiodic response, since a light break inhibits flowering of short-day plants and promotes flowering of long-day plants. Phytochrome apparently detects the light break, and its effectiveness depends upon the time of illumination.**

Now we'll examine some complications. LDPs are often less sensitive and somewhat more quantitative in their response to a night break than are SDPs. Using four photoflood lamps, for example, flowering is completely inhibited in cocklebur plants with a few seconds of light given about 8 hr after the beginning of an inductive dark period. With many LDPs, however, flowering continues to be promoted as the duration of a night break (using comparable high intensities) increases from seconds to hours. Furthermore, whereas red light is most effective in a night break with SDPs, a mixture of red and far-red wavelengths is almost always more effective with LDPs, whether applied as a night break or during the continuous light that best induces flowering in most LDPs. With SDPs, far-red given at the beginning of a dark period may inhibit if the dark period is exceptionally long but have no effect if it is short!

It would help considerably if we could measure the various forms of phytochrome in leaves at different times during photoperiodic induction. So far, it has not been possible to make these measurements in green tissues. In one study, no phytochrome at all could be found in some of the most sensitive SDPs, including cocklebur, perilla (*Perilla* sp), and soybean (*Glycine max*), although phytochrome was easily detected in extracts of leaves of the LDPs henbane, spinach (*Spinacia* sp.), beet (*Beta maritima*), and barley (*Hordeum vulgare*), and some other SDPs, including Maryland mammoth tobacco (*Nicotiana*) and sorghum (Lane et al., 1963). No one doubts that phytochrome exists in the leaves where it could not be detected; **our current methods limit us.**

In recent years, an ingenious approach has been developed to provide an indirect measure of the ratio of the two forms of phytochrome at any given time. This is called the **null response technique.** Plants are illuminated with mixtures of red and far-red light. The mixture that provides *no response* must be a measure of P_{fr}/P_r at the time of illumination. That is, a given ratio of wavelengths is known from experiments with etiolated tissues to produce a known ratio of P_{fr}/P_r. If the illumination has no effect, that ratio must have already been present at the time of illumination. There is an important limitation to the null point technique, however: Because of the overlapping absorption spectra of the two forms of phytochrome, no wavelength or mixture of wavelengths can produce more than about 80 percent P_{fr} or less than 1 or 2 percent P_{fr}. Since the determining quantities in the photoperiodic response could involve these percentages, the null point technique cannot solve all our problems. It has provided some rather interesting data, however.

In the SDPs *Chenopodium* and Japanese morning glory, P_{fr} is relatively high during the first 3 to 6 hr of an inductive dark period, and then there is an abrupt drop. In LDPs (e.g., perennial ryegrass, *Lolium temulentum*; henbane; and a long-day duckweed, *Lemna gibba*), there is also a drop in P_{fr} during the first 5 hr of a dark period. But perhaps the results of light experiments are not due only to P_{fr} and P_r. The HIR (Section 19.5) could be operating, phytochrome intermediates (Section 19.4) could play a role, or some other pigment could be involved. At the moment, we have no satisfactory answer to the problem of phytochrome in photoperiodism, but we can state an operational conclusion anyway: **3B. A night interruption inhibits flowering of short-day plants and promotes flowering of long-day plants. Red light is more effective with short-day plants and a mixture of red and far-red with long-day plants.**

22.6 Time Measurement in Photoperiodism

The central characteristic of photoperiodism is measurement of seasonal time by detecting relative lengths of day and night. But how is time measured in photoperiodism? It seems logical to place photoperiodism in the context of other examples of biological time measurement: circadian rhythms, celestial navigation, and so on (Chapter 20). During the past two decades, plant physiologists have taken this approach to the problem. Why wasn't it taken sooner? Because it seemed equally logical to imagine that time measurement was simply the time required for the completion of some yet-to-be-discovered metabolic reactions.

If time is measured by the interval required for some metabolite to be converted to another form, this is analogous to an hourglass that measures the interval of time required for the sand to fall from top to bottom through a narrow opening. In such a system, only one interval of time can be measured, and then some outside influence must restart the system

(invert the hourglass). Circadian rhythms that function over long intervals of time under constant conditions of light, temperature, and other factors are not analogous to a simple hourglass. Rather, an oscillator such as a pendulum comes more readily to mind. Is time measurement in photoperiodism analogous to an hourglass or to a pendulum?

Some features certainly seem to have the hourglass mode of operation. Let's consider a simple experiment in which we expose SDPs to dark periods of various lengths and then observe the level of flowering several days later. Such an experiment has been done with many SDPs, but cocklebur is convenient because it will respond to a single inductive night, and the degree of flowering can be conveniently observed by examining the apical buds under a dissecting microscope and classifying them according to a series of floral stages (Fig. 22-5). Two or three days after the single long dark period, the buds begin to develop through the stages shown in the figure, but the rates at which they develop will depend upon the degree to which they were induced by the long dark period; 9 days after the dark period, plants receiving 10 hr of darkness may have only reached stages 1 or 2, whereas plants receiving 16 hr of darkness may have reached stages 6 or 7 (Fig. 22-6). It is clear that the flowering stimulus increases from about the 9th hour (the critical night) to the 11th to 16th

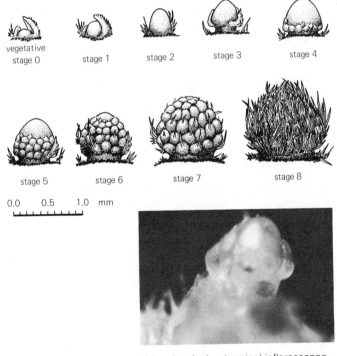

Figure 22-5 Top, drawings of the developing terminal inflorescence primordium (staminate) of cocklebur, illustrating the system of floral stages devised by Salisbury (1955). Below, a photograph through a dissecting microscope of a cocklebur inflorescence primordium at stage 3.

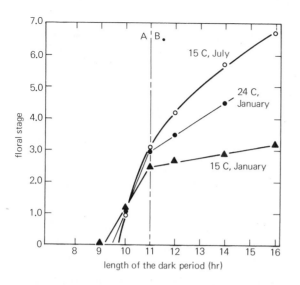

Figure 22-6 Some examples of cocklebur flowering response to different nightlengths. (From Salisbury, 1963.)

hour, depending upon conditions. If we are dealing with the synthesis of a flower-inducing hormone during this time (see later), then this might be an hourglass kind of reaction.

During the day, plants are normally exposed to light that converts most of the phytochrome pigment to the far-red absorbing form (P_{fr}). Yet when a night break is given, red light is the most effective, indicating that much of the pigment is in the red-absorbing form (P_r). The inevitable conclusion is that plants detect darkness as their phytochrome shifts from P_{fr} to P_r. When this was discovered in the early 1950s, it was immediately suggested that the shift from P_{fr} to P_r might account for time measurement. Perhaps the critical night is the time it takes for P_{fr} to drop below some critical level.

As the data continued to come in during the 1950s and 1960s, it became apparent that this explanation was too simple. Phytochrome does shift from P_{fr} to P_r when plants are placed in darkness, and that is how the plant "knows" it is in darkness, but the shift occurs within 1 or 2 hours in most species, far too rapidly to account for time measurement. It seems unlikely, anyway, that simple chemical reactions could account for time measurement in photoperiodism, since such reactions are notoriously sensitive to temperature, whereas photoperiodic time measurement is not. Figure 22-7 presents data indicating that the initial pigment shift is sensitive to temperature, whereas the subsequent time measuring mechanism is not.

Let's formulate the first part of our conclusion to this section: **4A. Part of timing in photoperiodism has the characteristics of an hourglass, particularly pigment shift and synthesis of a flowering hormone.**

Bünning suggested in the early 1930s that plants might use their oscillating-circadian clocks in photoperiodic time measurement (see bibliography for Chapter 20). There are two rather simple ways to look for rhythmical phenomena in

photoperiodism. Both are illustrated for soybeans in Fig. 22-8. In one, plants were given seven cycles, each including an extended, 64-hr dark period, which was interrupted at various times with a 4-hr night break. With soybean and many other SDPs and LDPs, there is a rhythmical response to the night breaks: At one time, the light interruption inhibits flowering, and this is repeated about 24 and 48 hr later; between the times of inhibition, there are periods of promotion. This experiment is certainly suggestive of an oscillating timer. A few species do not respond this way, particularly those such as cocklebur that flower in response to a single inductive cycle.

In the second experiment of Fig. 22-8, plants were given various combinations of light and dark periods, and their subsequent flowering is plotted as a function of total cycle length. It is evident with soybean (and a few other species) that maximum flowering occurs when the light and the dark periods total 24 hr, 48 hr, and 72 hr. Hence, the dark period is not all controlling, but the combination of the light and the dark periods plays the decisive role. Again, some species do not show such a response. Cocklebur plants can be kept under continuous light for several weeks, exposed to a single dark period longer than the critical night, and returned to continuous light until flowers have developed.

Yet even with cocklebur, it was possible to observe features similar to those observed in the circadian rhythms. In one set of experiments, cocklebur plants were given a 7-hr **phasing dark period,** too short to induce flowering. Then they were given an **intervening light period** that varied in duration,

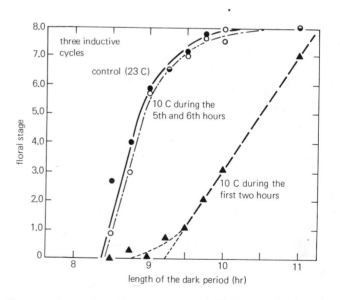

Figure 22-7 Flowering response of cocklebur to various night lengths as influenced by 10 C treatments applied during the first 2 hr of the dark period or between the beginning of the fifth and the end of the sixth hours, as compared to controls with no 10 C treatments. Plants were treated with three dark periods, beginning July 18, 1962. Temperature during the dark period other than treatment times was 23 C. (From Salisbury, 1963b.)

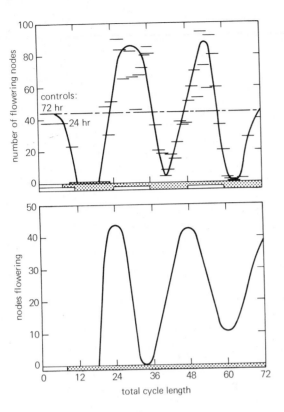

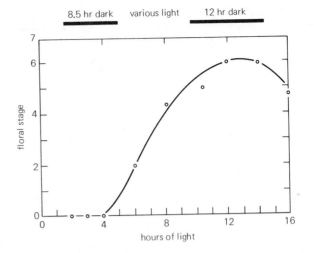

Figure 22-9 Flowering response as a function of length of the intervening light period in an experiment symbolized by the bars above the figure. (Data from Salisbury, 1965.)

Figure 22-8 Rhythmical responses in flowering of Biloxi soybean. Top, flowering response of soybean to 4-hr interruptions (indicated as scattered horizontal lines) applied at various times during the 64-hr dark period. Plants received 7 cycles of 8 hr of light followed by 64 hr of darkness, as indicated by the top bar at the bottom. Bottom bar shows postulated subjective days and subjective nights. Bottom, flowering response of soybeans to 7 cycles, including 8 hr of light and different dark periods to provide the total cycle lengths as indicated. (Data for both figures from Hamner, 1963, and other publications; figures from Salisbury, 1963b.)

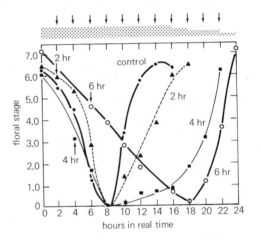

Figure 22-10 Clock resetting in flowering of *Xanthium*. Effects of a light interruption given at various times during long test dark periods that followed phasing dark periods of various lengths. Arrows above the bars at the top indicate times when light interruptions were given and correspond to the data points. Of course, a given set of plants receiving a given test dark period received only one light interruption during the test dark period and following the interruptions a 0, 2, 4, or 6 hr. (From Papenfuss and Salisbury, 1967.)

after which they were given a 12- to 16-hr **test dark period** (depending upon the experiment) that would normally induce flowering. Floral stage after 9 days is shown as a function of length of the intervening light period in Fig. 22-9. With brief intervening light periods there was no flowering, even though the subsequent test dark period was several hours longer than the critical night. When the intervening light period was about 5 hr, plants began to flower; flowering increased until it was 12 hr—a typical long-day behavior. An optimal dark period for flowering in cocklebur is about 12 hr (see Fig. 22-7); combining this with the 12-hr optimum light period again gives the magic 24-hr value.

An action spectrum was determined for the intervening light period, and red light, even at low intensity, was by far the most effective in *promoting* flowering; far-red inhibited. This is another typical LD response. Thus something closely

analogous to the night break experiment with soybeans of Fig. 22-8 became apparent with cocklebur: Red light (P_{fr}) promotes flowering at one time and inhibits it about 12 hr later.

In another rather involved set of experiments, cocklebur plants were given a long dark period that was interrupted twice. The first interruption was given 2, 4, or 6 hr after beginning of the dark period, and the second interruption was given at various times after the first. Floral stages after 9 days were plotted as a function of the time of the second interruption (Fig. 22-10). Control plants given only one interruption at various times were most sensitive about 8 hr after beginning of the dark period, as were plants given their first interruption at 2 or 4 hr. Plants given their first interruption 6 hr after the beginning of the dark period, however, were most sensitive to the second interruption 18 hr after the beginning of the dark period. The time of maximum sensitivity was delayed 10 hr when the first interruption was given at 6 hr. Such phase shifting is highly reminiscent of circadian rhythms (compare Fig. 20-4).

Do circadian manifestations such as leaf movements compare with time measurement in photoperiodism? Carbon dioxide evolution by a SD variety of duckweed (*Lemna perpusilla*) exhibits a circadian periodicity closely related to photoperiodic timing in this plant (Hillman, 1976). On the other hand, it was possible to separate the leaf movement rhythms in cocklebur from photoperiodic timing (Salisbury and Denney, 1974). Our conclusion to this section: **4B. The measurement of time in photoperiodism has some elements of an hourglass timer, but an oscillating timer is also clearly involved.**

22.7 The Florigen Concept: Flowering Hormones and Inhibitors

Not long after photoperiodism was discovered, workers around the world wondered which part of the plant detected daylength. It was soon apparent that the leaf responded. Using a SDP, for example, an experimenter could enclose the leaf for 16 hr in a black paper envelope, while leaving the rest of the plant under LDs or continuous light. Such a treatment soon induced flowering. A similar experiment with a LDP prevented flowering. Covering the bud but leaving the leaves under LDs did not lead to flowering in SDPs but did in LDPs. (Actually, the green stems of some species will respond to photoperiod, if they are given a sufficient number of inductive cycles.)

If the leaf detects the photoperiod, but the bud becomes the flower, there must be some stimulus transmitted from the leaf to the bud. In the 1930s, Michail Chailakhyan in Russia (see his 1968 review) grafted induced plants to noninduced plants held under noninducing daylengths, observing that the flowering stimulus would cross a graft union. He suggested that the stimulus was a chemical substance, a hormone, as opposed to some electrical or nervous stimulus. Chailakhyan coined the term **florigen** (Latin *flora*, "flower," and Greek *genno*, "to beget").

The grafting studies have provided two important bits of information: Florigen moves only through a living tissue union between the two graft partners, and probably only through phloem tissue. Furthermore, florigen frequently seems to move with the assimilate stream. If the receptor part-

ner is defoliated or held under low light intensity, movement of assimilates into the receptor is promoted and so is movement of florigen. On the other hand, in some species hormone is exported from extremely young leaves that would be expected to be importing assimilates, so florigen might move by other mechanisms as well as in the assimilate stream.

Many different varieties or species representing different response types have been grafted to each other (see Lang, 1961; Vince-Prue, 1975). Florigen produced by one response type will often induce flowering in another type; an induced SDP grafted to a LDP in SDs will induce the LDP to flower, for example. In the most extreme example, an induced SDP (*Xanthium strumarium*) caused a vegetative LDP (*Silene armeria*) to flower, although the two species belong to unrelated dicot families. The conclusion is that, although florigen is produced in response to widely different environmental conditions, it is the same compound or physiologically equivalent in virtually all angiosperms.

A third experiment that is difficult to interpret in any way but by the florigen concept was first performed in the early 1950s. Plants from several species that require only a single inductive cycle were defoliated at various times after that cycle, and the level of flowering sometime later was plotted as a function of the time of defoliation. Cocklebur results are shown in Fig. 22-11; the SD Japanese morning glory and the

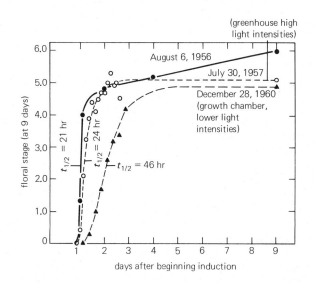

Figure 22-11 Three translocation curves obtained by defoliation of plants at various times following a 16-hr inductive dark period. Times of defoliation are shown, with floral stages of all plants being determined on the ninth day. Numbers on the abscissa represent noon of the indicated day, while the bar on the abscissa indicates the inductive dark period. Approximate times after beginning of the dark period when half the stimulus was out of the leaf are indicated by $t_{1/2}$ Dates refer to the day plants were subjected to the dark treatment. Plants represented by the $t_{1/2} = 46$-hr curve were kept under 2,000 ft-c of fluorescent light in growth chambers (23 C). (From Salisbury, 1963b.)

LD perennial ryegrass have also been studied. Apparently, when the leaf is cut off immediately following the inductive long dark period (SDPs) or long light period (LDPs), plants remain vegetative because the hormone has not yet been exported from the leaf. Export is complete some hours later, however, because defoliated plants flower almost as well as when the leaf is not removed at all. This experiment contributes some information about the rate at which florigen moves out of the leaf. In several species, florigen apparently moves much more slowly than do assimilates produced by photosynthesis, as indicated by $^{14}CO_2$ or ^{14}C-sugars. Proving the rule that no generalization in the field of photoperiodism is without its exceptions, however, florigen apparently moves as rapidly from induced Japanese morning glory cotyledons as do photosynthetic assimilates. In any case, florigen's rate of movement is influenced by temperature and other factors, as implied in Fig. 22-11.

In spite of the strong evidence for flowering promoters, there is equally strong evidence for inhibitory substances or processes in plant reproduction. Indeed, both promoters and inhibitors must influence flowering. If a plant is induced by exposing only one leaf to suitable photoperiodic conditions, flowering is frequently inhibited by the noninduced leaves. The presence of leaves on a receptor plant in a graft experiment also reduces receptor flowering. These inhibitory effects are often due to influences on the translocation of assimilates. For example, a long-day leaf growing between an induced short-day leaf and the bud may be exporting assimilates directly to the bud and thereby effectively blocking movement of assimilates—and florigen—from the induced leaf to the bud. Many inhibitor experiments can be understood in this way, and often the explanations are supported by tracer experiments.

Some effects cannot be explained on the basis of translocation, however. If the whole explanation were photosynthesis and transport of assimilates, then light intensities should be important. Yet a brief night interruption at low intensity will often produce a typical long-day inhibitory effect in a single leaf on a short-day plant (Gibby et al., 1971; King and Zeevaart, 1973). In some species, flowering seems to be repressed under noninductive conditions strictly by inhibitors coming from the leaves. Such plants will flower when defoliated. Examples are the LD henbane and various SD varieties of strawberry (*Fragaria* sp.). In some species (e.g., cocklebur, Japanese morning glory) promoters may be dominant although modified by inhibitors, in others there is a true balance (darnel ryegrass), while in still others (henbane, strawberry varieties) inhibitors may be dominant but modified by promoters.

We need to isolate and identify florigen(s) and flower inhibitor(s). So far, this has not been completely successful, primarily due to the lack of a reliable bioassay. Numerous solvents and application techniques have been tried in attempts to extract induced plants and apply the extracts or fractions to noninduced plants. Sporadic successes have been reported. The best-known method (see Lincoln et al., 1966) ex-

tracts quick-frozen, induced cocklebur leaves with organic solvents. Extracts made from vegetative plants have no effect, but extracts from a day-neutral sunflower and even from a fungus (*Calonectria*) have sometimes been effective. Only a low level of flowering is induced in test plants, and GA applied with the extracts sometimes promotes flower development. Activity in the extracts follows the acid fraction and has been called **florogenic acid.**

The experiment has succeeded in Hamner's laboratory (Hodson and Hamner, 1970) but not always in other laboratories. Ilabanta Mukherjee and Salisbury visited Hamner's laboratory and went through the procedure. Salisbury returned to Utah and, with Alice Denney, was sometimes able to perform the experiment successfully (unpublished). Mukherjee returned to the laboratory of Jan A. D. Zeevaart in Michigan but was never able to obtain an effective extract (a different variety of cocklebur, perhaps?)—although some years previously he had been successful in the laboratory of Dennis J. Carr (1967) in Ireland! Such have been the problems of florigen extraction.

Since florigen apparently moves only in phloem tissue, Charles F. Cleland and A. Ajami (1974) collected honeydew from aphids feeding on induced cocklebur and other species. Using a LD duckweed as a test plant, strong florigenic activity was found. Two active substances proved to be salicylic acid and acetylsalicylic acid—which is common aspirin! Unfortunately, salicylic acid and aspirin are completely ineffective when applied to vegetative cocklebur or other plants. Furthermore, salicylic acid concentrations are equal in the honeydew of both vegetative and induced cocklebur plants. Duckweed flowering is influenced in many complex ways by the nutrient medium, and apparently salicylic acid is one of these. A few other successes have also been reported (e.g., Biswas et al., 1967), but these have not been confirmed.

Reports from William L. Wardell (1976) may or may not relate to this problem. DNA was extracted from stem tissue close to the flowers of a day-neutral tobacco species that flowers only in response to high light intensities. A test plant was decapitated near an axillary bud, and the stem was hollowed out by removing some of the pith. Profuse flowering of the axillary bud under low light intensity was induced when the DNA solution was put in the hole. Extracts from vegetative plants had no flower-inducing effect, but DNA that had been heated and cooled rapidly to separate the two chains in the double helix was more effective than unheated DNA. DNA treated with DNase was completely ineffective.

When DNA was extracted from a strain with one type of flower and applied to a strain with a different flower type, floral characteristics of the type from which the DNA was extracted appeared occasionally in the flowers induced by the DNA but more frequently in the flowers of plants obtained from the seeds of flowers induced by DNA. Thus, the experiments imply a transfer of genetic information from one strain to another by transferring DNA. Such a phenomenon is well known with bacteria and has been reported for higher plants and animals.

So far, few attempts have been made to extract and isolate floral inhibitors, although gallic acid appears to be a specific flower inhibiting substance extracted from vegetative plants of the SD *Kalanchoe blossfeldiana* growing in LDs (Pryce, 1972). We must remember that an inhibitory effect might be a process rather than a substance. For example, LD leaves on a SDP might in some way absorb and destroy promoting substances produced in SD leaves.

Because attempts to extract promoters and inhibitors have been so disappointing, other more indirect approaches have been taken. Experimenters have added various antimetabolites to LDPs and SDPs, seeking those that apparently inhibit the flowering process in a specific way. For example, a compound may be effective only when applied during florigen synthesis. Again, results have not been promising. It appears that respiration and protein and nucleic acid synthesis are involved in flowering, but this is not surprising, since these processes are involved in virtually all of the life of a plant. Before we are carried further afield, let us state the conclusion for this section: **5. There is much circumstantial evidence that flower initiation is controlled by hormones: one or more positive acting florigens and one or more negatively acting inhibitors. These substances remain to be identified.**

22.8 Responses to Applied Plant Hormones and Growth Regulators

Since plant hormones and growth regulators influence virtually every aspect of plant growth and development, it is logical to investigate their effects on flowering. Many compounds are now known that will induce or inhibit flowering in one species or another. There is an important potential for practical application of this knowledge, since the induction of flowers (e.g., in ornamentals, fruits, pineapple) or the inhibition of flowering (e.g., in sugar cane, sugar beets, various root crops) plays such an important role in agriculture. Work with hormones and growth regulators could also lead to better understanding of the flowering process. Again, however, there are about as many exceptions as rules. Let's summarize a few tentative observations.

Auxins and Ethylene More often than not, auxins inhibit flowering. In SDPs, this occurs before translocation of florigen from the leaf is complete, after which there may be marginal promotive effects. Promotions have also been observed in LDPs held under days just too short for induction, and auxin clearly promotes flowering in some bromeliads, including the pineapple. In pineapple and cocklebur, applied auxins cause a production of ethylene, which itself influences flowering the same as auxins (inhibition in SDPs, promotion in bromeliads). In the bromeliads, IAA is relatively ineffective, since it is apparently broken down by the plant's enzymes. Thus, synthetic auxins such as NAA or 2,4-D must be used. Auxin concentrations required to inhibit flowering usually produce severe epinasties and other responses, and measured plant auxin levels seldom correlate with flowering in any meaningful way. Hence, although endogenous auxins may influence flowering to some extent, they probably do not control it.

Gibberellins Gibberellins (GA) will substitute for the cold requirement of several species that require vernalization and also for the long-day requirement of several LDPs. There are some important exceptions among LDs (e.g., *Scrofularia hyecium* and *Melandrium* sp.), and gibberellins normally fail to replace SDs in SDPs—although again there are a few exceptions (e.g., cosmos and rice). In many species, gibberellins increase under LDs, so it seems reasonable to assume that gibberellins might account for at least part of the flowering requirement (a florigen complex?) in long-day plants. But the situation is not simple. In LDPs, for example, LDs increase both the rate of synthesis and the rate of destruction of GA, and many studies now show a lack of correlation between extractable GAs and flowering. A recent review (Zeevaart, 1976) states that there is now "conclusive evidence" that flower formation and stem elongation are separate processes, with GA promoting stem elongation only. Growth retardants inhibit stem elongation but not flower formation in *Silene*, for example. Although GA was reduced to levels below those that could be detected, flowering occurred. It has also been shown (Wellensiek, 1973) that flower formation and stem elongation are under control of two separate but closely linked genes. Thus, flowering seems to be quite independent of GA in some species—but, of course, the situation could be different in other species.

GAs seem to be particularly important in the flowering of conifers. Y. Kato et al. (1958) in Japan, Richard Pharis et al. (1976) in Calgary, Canada, and others have pioneered in this work. Most conifers require several years before they attain ripeness to respond, but Pharis was able to induce the formation of staminate strobili on Arizona cyprus (*Cupressus arizonica*) when plants were only 55 days old by spraying with GA_3. He and his coworkers were unable to induce cone formation in members of the Pinaceae with GAs, but they now achieve this by applying the less polar GAs (GA_4, GA_5, GA_7, GA_9). Because breeding of conifers is important to the lumber industry, these observations could be significant. Breeding times might be reduced from years to months.

Cytokinins Cytokinins have been observed to promote flower formation in several plants. A combination of a cytokinin (benzyladenine) and GA_5 induced flowering in one SD variety of chrysanthemum. In another variety, benzyladenine could substitute for the latter part of photoinduction. In most LDPs and SDPs, cytokinins do not affect flowering.

Abscisic Acid Abscisic acid (ABA) was reported to induce flowering in SD blackcurrants, but it is now clear that there is no clear-cut promotion under noninductive conditions. If plants were already slightly induced, ABA will often promote the level of flowering. This is also true in the SDPs, Japanese morning glory and *Chenopodium,* but ABA inhibits flowering in some other species and is completely innocuous in others. ABA is generally but not always higher in plants under LDs.

Sterols A substance called TDEAP (tris-[2-diethylaminoethyl]-phosphate trihydrochloride), which inhibits synthesis of cholesterol in animal systems, inhibits flowering of cocklebur and Japanese morning glory. Does inhibition by TDEAP imply that florigen is related to cholesterol; that is, that it is a sterol of some kind? Not according to results so far. Application of various sterols does not overcome the effect of TDEAP, nor can any significant changes in sterol fractions be detected during induction. Sterol biosynthesis can be inhibited by compounds other than TDEAP without influencing flowering one way or the other. Thus, we have no idea what TDEAP is doing in the plant, and evidence that florigen is a sterol is at best flimsy. But flowering can be induced in two SDPs by application of sitosterol and lanosterol (in *Chrysanthemum morifolium* cv Princess Anne) and estradiol (in *Callistephus chinensis;* see Vince-Prue, 1975).

To state our growth regulator conclusion: **6. The flowering response is often influenced by applied growth regulators and hormones, but the few patterns that can be tentatively discerned all have several exceptions. Although several compounds will cause flower formation, there is no convincing evidence that florigen is one or more of the well-known plant hormones.**

22.9 The Induced State and Floral Development

To illustrate the uniqueness of studies in flowering physiology, let us briefly consider how the flowering hormone must act in different species. In cocklebur, young leaves that are allowed to grow out on an induced plant can be grafted to vegetative receptors, causing them to flower. The young leaves apparently become induced by the older leaves even under LDs. As a matter of fact, as many as five vegetative cocklebur plants have been grafted in series to an induced one at the end of the chain, with flowers forming on all plants.

Perilla, on the other hand, acts quite differently. An excised leaf may be induced by SDs and can then be grafted to a series of vegetative receptors for several months, inducing each to flower. But none of the leaves on the receptor plants become induced. Thus, the flowering condition in perilla is not as "contagious" as it is in cocklebur.

Upon arrival of florigen at the apical meristem, the course of meristematic development changes from the vegetative to the reproductive mode. In many species, arrival of the stimulus leads to an immediate increase in mitotic activity, and nuclear size often increases, as does the size of the nucleolus. Frequently, there is a buildup in the number of ribo-

somes, mitochondria, and RNA in the apical cells. The buildup in RNA may occur *before* the stimulus arrives. Does this mean that some other stimulus precedes the main one? Or that florigen at a concentration too low to cause flowering arrives sooner, causing the observed changes in RNA and other factors?

Our final conclusion: **7. The induced state has unique properties in different species, and the changes that take place at the flowering apex could help us understand development.**

Where Do We Go from Here? Since the first edition of this book was published, much has been written about photoperiodism and the flowering process, yet no major breakthroughs have appeared, and activity in this research field has decreased noticeably. Most workers apparently feel that we have reached a temporary dead end. There are obvious things to do, but none is really promising. We have realized with increasing vividness that the problems are exceedingly complex. Differences between species are extensive and significant, much more so than we might have imagined one or two decades ago. Studies on reproduction biochemistry have told us virtually nothing. Except within broad limits, there is no way to predict how a given growth regulator will influence flowering of a species not yet tested.

Yet the unanswered questions are central to understanding development, which remains one of the most significant unsolved problems in modern biology. What is time measurement in photoperiodism? What is induction? What is the floral stimulus? How does it act? Hopefully, in the not too distant future, someone will think of new approaches, so this potentially fascinating and important field will again become a hub of active research.

22.10 Other Responses to Photoperiod

Photoperiodism is an extremely widespread phenomenon in nature. Garner and Allard in their early papers suggested that bird migrations might be controlled by photoperiod, and soon photoperiodism in birds was demonstrated (Rowan, 1925). Since then, animal responses to photoperiod have been documented, including several developmental changes in insects, promotion of reproduction in insects, reptiles, birds, and mammals and fur (pelage) changes in mammals. There are also many plant responses besides the initiation of flowers. Virtually every aspect of plant growth and development is influenced by photoperiod (Table 22-2; Vince-Prue, 1975).

The Developing Flower Some plants such as *Chrysanthemum* abort their flowers when transferred to noninductive daylengths. These plants are not capable of induction. Flower development, even in plants that can be induced with a single cycle, is often promoted by further inductive cycles. In a few cases, infertile flowers may develop under noninductive day-

Table 22-2 Plant Responses to Photoperiod Other than Flower Initiation[a]

Response	Short Day	Long Day	Nightbreak Is Effective (Clock)	Red–Far-Red Reversibility (Phytochrome)	Leaf Responds (Stimulus)	Effects of Applied Growth Regulators
The developing flower						
Flower development	+ −	− +	+	+	+	GA promotes many SDP but inhibits some species; auxin promotes or inhibits various species
Flower fertility	− (+)	+ (−)	+			Some evidence for control by auxins, GAs, and cytokinins
Sex expression	− +	+ −	+			Auxin almost always increases ♀ ethylene almost always increases ♀ GA: ♀ in some plants, ♂ in others ABA: ♀ in *Cannabis,* cucurbits Cytokinins: ♀ in *Vitis* clones Steroids: estradiol, ♀ testosterone, ♂
Vegetative reproduction	−	+				Cytokinins promote in *Bryophyllum*
Dormancy						
Induction of bud dormancy	+	−	+	+	+	Natural promoters and inhibitors apparently play a role: ABA and narringenin; applied ABA is sometimes effective
Breaking of dormancy	−	+			Bud	GA breaks dormancy of some species; cytokinins may play a role
Seed germination	+ −	− +	+	+		
Development of cold resistance	+	−				
Leaf fall	+	−				
Stem elongation	−	+	+	+		GA also promotes
Storage organs						
Tubers	+	−	+	+	+	Cytokinins promote; ABA inhibits; ethylene may promote; GA and auxin may promote
Bulbs (onions)	−	+	+	+	+	Cytokinins promote
The vegetative plant						
Leaf growth	−	+	+			
Branching	+	−				
Chlorosis of tomato, etc.	−	+	+			
Chlorophyll production	+	−				
Anthocyanin production	+ −	− +				
Rooting of cuttings	−	+				
Biochemical responses	+ −	− +				

[a]The information in this table is from many sources, summarized in Vince-Prue, 1975.

lengths. Typically, pollen fertility is more sensitive to day-length than ovule fertility. In many monoecious and dioecious species, the relative number of staminate to pistillate flowers depends upon photoperiod. Either femaleness or maleness can be promoted by either SDs or LDs, depending upon species. Often, growth regulators influence these processes of flower development, although few generalizations can be made except in the case of sex expression; auxin and ethylene almost always increase femaleness. In cucurbits, ethylene is apparently an endogenous regulator that promotes female-ness. Under reduced pressure, which allows more ethylene to escape, maleness is increased; ethylene added under reduced pressure restores femaleness. Carbon dioxide, a competitive inhibitor of ethylene action, also promotes maleness. One of the most interesting studies in this field dates back to 1945 (Löve and Löve). Femaleness in *Melandrium rubrum* was increased by estrogens, which are female sex hormones in animals; maleness was increased by testosterone and other male hormones. Yet both estrogens and androgens increased femaleness in *Cucumis sativus,* and neither had any effect in several species of *Rumex.*

In a few cases, researchers have shown that a night break is equivalent to LD in some of these flower development phenomena. In at least one case, red/far-red reversibility was also shown, indicating the participation of phytochrome.

Some plants increase their rate of vegetative reproduction in response to photoperiod. Strawberries produce many more runners under LDs, for example, and *Bryophyllum* also forms plantlets at the edges of its leaves in response to LDs.

Dormancy We have already discussed dormancy of buds and seeds in Sections 21.3 and 21.5. As a rule, SDs induce bud dormancy. The response is closely analogous to flower initiation: There are absolute and facultative responses, some species are day-neutral (often becoming dormant in response only to low temperatures), night breaks give a LD effect, and red/far-red reversibility has sometimes been demonstrated. The leaves respond to the SDs that induce dormancy, so a stimulus must be involved, but if LDs are effective in breaking dormancy, it is the buds themselves or the scales surrounding them that detect the daylength. As we saw, ABA will sometimes but not always mimic SDs, causing plants to become dormant. GA breaks the dormancy of many but not all species. The germination of a few seeds is a photoperiodic response. There are SD and LD seeds, a night break produces a LD response, and phytochrome plays a role. In a few species of woody plants (e.g., apple), the induction of dormancy may be relatively independent of daylength, but development of cold resistance and sometimes abscission of leaves may be promoted by SDs. Often, these responses are much more strongly influenced by temperature, however.

Stem Elongation Stems of many species tend to elongate more under LDs than under SDs. There is an interesting anomaly, however, in that far-red given at the beginning of the night also increases elongation. Removal of P_{fr} just before darkness might be expected to promote a SD response, but this is clearly not the case with stem elongation. Actually, mixtures of red and far-red light act in ways reminiscent of those observed in flowering of LD species. We do not understand the situation in either case.

Storage Organs Formation of all storage organs except onion (*Allium*) bulbs is favored by SD. As usual, there are interactions with temperature, and all the usual features of photoperiodism have been demonstrated: Night length is important, a night break is effective, phytochrome is involved, there is a stimulus produced in the leaves, and the response is inductive. Indeed, the response is closely similar to flowering of many SDPs. Only onion formation is different, being similar to flowering of LDPs. Though there are confusing results with gibberellins, auxin, and ethylene, cytokinins are capable of inducing the development of a range of storage organs (stem tubers in potato, bulbs in onions, root tubers in *Ipomoea*) on plants with different photoperiodic requirements (SD in potatoes, LD in onions, and day-neutral in *Ipomoea*). So far, however, the transmissible, tuber-inducing stimulus formed in the leaves in response to SDs has not been identified as a cytokinin.

The Vegetative Plant Leaf growth of many species is often strongly influenced by photoperiod. LDs lead to the growth of longer, larger, thinner, yellower leaves. In succulent plants, succulence is often reduced under LD. The leaf response is independent of the flowering response. Other aspects of the vegetative plant may also be influenced by photoperiod. Branching of many species is promoted by SDs. Tomato and a few other species develop a severe interveinal chlorosis when plants are given light for more than 18 hr each day, although many other species do not respond this way. In any case, it is not unusual to observe that chlorophyll production is often decreased by SDs. Anthocyanin synthesis is also sometimes affected, parallel to or independently of the flowering response. Rooting of cuttings seems to be most often promoted by LDs, but few species have been studied. In several cases, the photoperiod applied to the plants before the cuttings were taken has an important effect; in other cases, the cutting itself responds.

Biochemical Responses Many biochemical constituents of leaves have been shown to vary widely among plants growing in different photoperiods. This includes several enzymes, total organic acids, various hormones, and miscellaneous responses such as changes in cytoplasmic viscosity. LDs or SDs promote a given reaction, depending upon species and the material being considered.

In all these cases, it is possible to consider the adaptive role played by plant response to photoperiod. As suggested at the beginning of the chapter, this is a fertile field for ecological research that so far remains almost completely undisturbed.

Section Four

Environmental Physiology

The desert east of Phoenix, Arizona.

23

Principles of Environmental Physiology

It's high summer in the Rockies. You are on your way to the west coast when you and your traveling companions decide that you will take an extra day to detour through Rocky Mountain National Park over the Continental Divide. You are afraid that the subcompact might begin to boil, but to your great relief, you make it without mishap to the high point on the Trail Ridge Road—12,183 feet above sea level! You pull into a parking stall, turn off the engine, and set out for a short tramp on the tundra.

Those little alpine plants are really something, you think to yourself. How can they be so healthy looking when they are covered with snow 9 or 10 months of the year? They must have to photosynthesize at peak efficiency and at low temperatures to succeed in a place like this. How do their genes differ from those of the prairie plants that you saw yesterday?

Just last spring, you completed the undergraduate course in plant physiology. Your copy of Salisbury and Ross (which you prudently decided not to turn in at the student book exchange!) lies on your pile of gear back in the car. Remembering the book's general outline, you begin to formulate questions and observations in your mind: Is there something about the water potential of these plants that allows them to withstand frost? The cells in their leaves must be turgid because of osmosis. The soil seems to have ample moisture; would transpiration cool these plants even below the rather chilly and breezy air temperature? How do they respond to the low temperatures up here? Is there anything special about the way they absorb minerals from this cold soil, transporting them across their membranes to the cytoplasm inside? Is there anything unusual about translocation of assimilates in these plants? Probably not. Do their enzymes have special configurations that allow them to function optimally at low temperatures? Photosynthesis must be highly efficient in these plants, and the low temperatures at night probably reduce dark respiration. Do they by any chance utilize C-4 photosynthesis? Do the ones growing almost on the bare rock have some kind of nitrogen fixation—mycorrhiza, associated bacteria or something? The flowers seem brightly colored. Are their pigments produced through some sort of special biochemistry? Could the colors be intensified some way by the bright light and high ultraviolet intensities? (Are the plants being "sunburned"?)

Being dormant so much of the year must mean that these plants have special timing mechanisms. Do they actually grow under the snow, so that they are ready to go when summer finally comes? Do they go dormant in the fall in response to the shorter days? What hormonal mechanisms mediated these responses to environment? Do the leaves fold up at night to resist radiant loss of heat to the cold sky above? Surely auxins, gibberellins, cytokinins, ethylene, and inhibitors play important roles in making these plants what they are, where they are. How do they do it?

With questions like these, you are beginning to think like an environmental physiologist. And, of course, you don't have to be standing on the Trail Ridge to have such thoughts. Perhaps your walk was in the Sonora Desert near Tucson, Arizona, or the relic grasslands of Nebraska. Were you lucky enough to visit the steaming jungles of Brazil? Or if not, then the near-tropics of the Florida Everglades? Maybe your questions developed as you walked along the Appalachian Trail, or in the Blue Ridge Mountains of Kentucky. Similar questions could develop in Central Park in New York City, or as you putter in your backyard garden.

In all these situations and thousands more, you can ask related questions about plants in their environments. Indeed, you can develop these questions into a science. Though the questions may be similar, the sciences that develop from them may have different names, depending upon your interest. If you are a dyed-in-the-wool physiologist, you might be quite happy with the term **environmental physiology.** If your field is agriculture and you are interested in the response of crop plants to their environments, you may call your science **crop ecology** or even **crop physiology.** Traditionally, if you work with vegetables or ornamentals, you may call yourself a **horticulturist,** but you may still be asking the questions of environmental physiology. If your interest is in field crops (cereals, forage crops, root crops, and others), you may call your science **agronomy.** If you are interested in how plants grow in their natural environments, asking questions similar to those just outlined, then your field is **physiological ecology,** which also has its applied aspects such as **forestry** and **range management.** Indeed, most students in agriculture and in forestry and range management are required to take a course in plant physiology, mainly for the environmental aspects.

23.1 The Problems of Environmental Physiology

Specific guiding questions for research will depend upon the specific field of endeavor. In agriculture, most research will be guided by the economics of obtaining the highest maximum yields or the highest possible quality for the lowest energy inputs. Environmental physiology always plays a role in such research.

Studies in physiological ecology apply the methods of physiology to the problems of ecology. Traditionally, these problems have centered on the question of plant and animal distribution. The ever-present assumption is that organisms occur where they do in nature because they are well adapted to their environments. Plants that grow in deserts can withstand drought and high temperatures, for example, while deciduous trees only grow where there is ample moisture and moderate temperature. Hence, studies in physiological ecology may attempt to measure the microenvironments of plants or animals in the field, and then duplicate these environments to study the same organisms in the laboratory. But ecology has many interesting problems besides those of distribution. Much interest during the past two decades has centered around the transfer of energy through ecosystems. Clearly, studies of energy flow require the application of many physiological methods, including careful measurements of photosynthesis and respiration.

Ecology is also concerned with interactions besides those involved in the food chain. There are numerous examples in which one organism influences another organism in either positive or negative ways. The ecologist speaks of parasitism, commensalism, symbiosis, and so on. Again, an understanding of these relationships typically requires the application of physiological methods. Competition is one of the most important and commonly mentioned interactions between organisms in an ecosystem—yet it remains to be understood in really fundamental ways.

Perhaps most ecosystem interactions are involved in the phenomenon of **community succession.** As members of one biotic community grow and develop, they often change the environment, and other organisms may then be better suited. A successional sequence may occur from bare ground until a **climax community** is achieved. For many years, it was felt by most ecologists that such a community was highly ordered and specific, resembling in certain fundamental respects a kind of superorganism. Individual plants would be the "cells," for example. It is now apparent that environments typically vary continuously rather than sharply with changing position on the earth's surface, and that biotic communities vary in a similar manner, so that organisms are distributed according to a **continuum.** They do not occur in **discrete communities,** although the dominant vegetation allows us to recognize general types. This is true even if succession has gone to completion over all the region being studied—if succession can ever really go to completion, considering long-term climatic and geological changes. Environmental physiology has helped to create the intellectual climate in which ideas of the continuum could develop and can now help us to understand the concepts better.

Unfortunately, ecosystem studies soon become extremely complex. To understand the distribution, energy exchange, and succession occurring in a given biotic community (i.e., to understand its **synecology,** to use a somewhat outdated term), one should understand the physiological ecology of each individual in the community (i.e., its **autecology**). At present, detailed environmental physiological data are available only for a few plant and animal species. And if we had such data for all the individuals in a community, how could we combine them to understand the total ecology of the system? This would be an extremely complex undertaking, but perhaps computers with their ability to handle vast quantities of data might make it possible.

We have been summarizing the problems of ecology: distribution, energy flow, and ecosystem interactions. At one time James Bonner (who was then mostly a plant physiologist) summarized these problems in a much more zestful way with the following three questions: *Who lives where and why? Who eats whom? And what is the physiology of togetherness?*

Clearly, environmental physiology is a large field. Virtually anything already discussed in this book might apply to it. If each point were illustrated with examples, volumes could be written. Thus, it becomes essential to limit our approach in this chapter. We will begin by summarizing several concepts as they might apply in the various fields of environmental physiology. Then we shall have a word to say about the philosophy and approach to research in environmental physiology. Finally, we shall examine the various elements of a plant's environment, thinking of how plants may respond. This will provide an opportunity to summarize and review some of the concepts discussed in this book.

23.2 Some Principles of Plant Response to Environment

Most of the following principles of environmental physiology have already been encountered in this text, but let's review 10 ideas in the context of physiological ecology or agriculture.

1. Saturation, the Cardinal Points, and Limiting Factors Perhaps the most fundamental principle of plant response to environment—and the one most frequently encountered in this text—is that of **saturation.** Organisms respond to virtually any environmental parameter according to a common pattern: As the parameter increases in concentration or intensity, it reaches a **threshold** above which it begins to have an effect, after which response increases until the system becomes saturated by the parameter. Then, as the parameter intensity

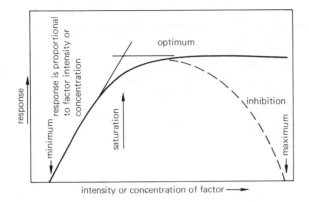

Figure 23-1 A generalized curve showing organism response to an environmental parameter. Minimum, optimum, and maximum are called cardinal points.

or concentration continues to increase, response may remain constant or may begin to decrease if the parameter at its high levels becomes toxic or inhibitory. Figure 23-1 shows the expected pattern. Looking back through the text, there are figures that illustrate the phenomenon for temperature (Fig. 21-1), photosynthesis (Figs. 11-3 and 11-12), mineral nutrition (Fig. 6-5), enzyme action (Fig. 8-10), transport of ions across membranes (Fig. 5-11), response to various plant hormones (Figs. 16-4 and 16-5), and so on.

In many of these cases, it is possible to speak of the **cardinal points**: the **minimum,** the **optimum,** and the **maximum.** It is easy to understand these curves between the minimum and the optimum and to understand the concept of saturation. The organism simply utilizes the factor being considered until its capacity for this utilization is used up or saturated. But what about the descending right-hand features present in so many of these examples? Explanations for this part of the curve differ, depending upon the phenomenon considered. When growth is inhibited by high temperature, we have suggested enzyme denaturation as the explanation (see Fig. 21-2), but this explanation is not always satisfying. Remember *Baeria chrysostoma* (Section 21.8) that dies within 30 days when night temperatures are 26 C? Surely enzymes are not being denatured at these temperatures, since day temperatures of 26 C or above are not harmful. The effects are no doubt complex, involving perhaps the production of an inhibitor during warm nights (implying a phytochrome interaction?). Superoptimal concentrations of mineral nutrients might become toxic because they begin to interact with systems in the organism other than those that were responding on the ascending part of the curves.

In 1840, Justus von Liebig formulated his **Law of the Minimum,** which in retrospect can be derived from and understood on the basis of saturation curves. The law states that: "The growth of a plant is dependent upon the amount of foodstuff that is presented to it in minimum quantities." Applied to mineral nutrition or photosynthesis, for example,

we might expect the ascending part of the saturation curves to be identical at two levels of one factor when another factor is presented to the organism in relatively small (limiting) amounts. The two curves will have different saturation levels, however, depending upon the different quantities of the second **limiting factor.** The principle was discussed and the term "limiting factor" proposed in a paper by F. F. Blackman in 1905. It was later illustrated by his experiments on photosynthesis of algae. Such curves are often referred to as **Blackman curves.**

Figure 23-2 illustrates this with a simple experiment in mineral nutrition involving two levels of phosphorus given to plants over a wide range of nitrogen concentrations. The threshold for nitrogen is extremely low, but below that level plants do not grow at all. As nitrogen increases, plants respond the same at both phosphorus levels until phosphorus becomes limiting at the nitrogen saturation level. Higher phosphorus leads to a higher saturation level for nitrogen.

The practical implications of Liebig's law were and continue to be obvious and important. In agriculture (Liebig was

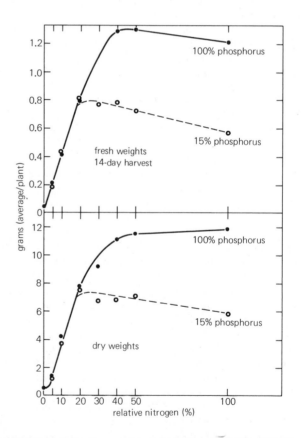

Figure 23-2 Results of an experiment in mineral nutrition. Tomato plants were grown in vermiculite and watered with nutrient solutions containing various concentrations of nitrate (NO_3^-) and one of two concentrations of dihydrogen phosphate ($H_2PO_4^-$), as shown. Curves represent fresh and dry weights of 14-day-old plants. Data are averages of six plants. (From Salisbury, 1975.)

probably the first agricultural chemist), the challenge is to discover the limiting factor and to remedy it. If plants are limited in their growth by insufficient nitrogen fertilizer, then more nitrogen is applied to increase the yield. When enough nitrogen has been applied, then perhaps phosphorus becomes limiting and needs to be applied. This approach has had spectacular success since 1840, so that much more food can be produced by a single farmer now than was possible then. In physiological ecology studies, a plant's distribution might be limited at its boundary by some single environmental factor presented to it in the "least" amount (this always being a relative matter, since highly disparate quantities of the different elements and environmental parameters are required by plants—see Tables 6-1 and 6-3, for example).

2. Interaction of Factors Unfortunately, things are not as simple as Liebig's law might imply. Under carefully controlled conditions such as those of Fig. 23-2, everything may work out as the law predicts. In the real world, things seldom work out so well; the curves are not identical in the ascending parts where only one factor is supposed to be limiting. Figure 23-3 shows a somewhat more typical example, although the experiment was performed under conditions similar to those of Fig. 23-2.

There are several ways to explain a failure of Liebig's law. Most probably boil down to a single idea, however: The extent to which Liebig's law might function (i.e., Blackman curves might be obtained in multiple factor studies) will depend upon the *extent* to which the factors under consideration are able to enter into reactions within the organism. Inability of a factor to enter into such a reaction may depend upon several things, such as restrictions upon diffusion of CO_2 through stomates or movement of ions in the free space, or ultimately the chemical rate constants for the appropriate reactions.

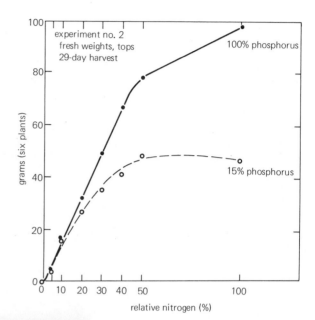

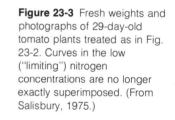

Figure 23-3 Fresh weights and photographs of 29-day-old tomato plants treated as in Fig. 23-2. Curves in the low ("limiting") nitrogen concentrations are no longer exactly superimposed. (From Salisbury, 1975.)

The idea can be understood by thinking of a single reaction that involves two precursors. If the reaction has a large equilibrium constant so that the precursors are almost completely used up (they enter into the reaction to the greatest extent possible), then one can obtain Blackman curves by plotting the product as a function of precursor concentration (Fig. 23-4). If, however, the reaction does not go to completion, for whatever reason (the equilibrium constant is small), then Liebig's law does not work out as well (Fig. 23-4).

Can Liebig's law guide our speculations in environmental physiology? It is not a bad starting place. We are entitled to begin our studies of agricultural yields or plant distribution by looking for "limiting factors," but we will probably have to progress beyond these initial steps to understand the real world.

Some powerful mathematical tools have been developed to help us understand factor interactions in nature or in controlled experiments. One of these is **regression analysis,** which is a highly valuable tool in such situations as field observations where data must be taken as they come. When an

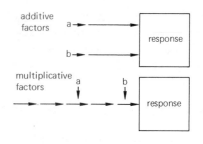

Figure 23-5 An illustration of additive and multiplicative factors according to the analysis of Mohr (1972).

experiment can be carefully designed in advance (i.e., where treatments can be set up according to a table of random numbers, and so on), **Fisher's analysis of variance** is used widely and appropriately (see statistics textbooks for descriptions of these methods).

Such studies indicate whether or not two environmental factors interact. If they both influence a given response but do *not* interact, they may be additive in their effects or they may be multiplicative (Fig. 23-5). When they are additive, they act upon different causal sequences that lead to the response. Say that a compound is made in two different compartments in the cell; one factor may influence one compartment and another factor the other compartment. Stem growth in the white mustard plant, for example, can be influenced oppositely by gibberellic acid and by red light (phytochrome: P_{fr}). The two responses are perfectly additive, as shown in Fig. 23-6. Multiplicative responses are more common. The two factors act on different steps in the same causal sequence (Fig. 23-5), so that the effect of one is always some fraction of the effect of the other. For example, rate of stem growth in the white mustard plant as influenced by red light is determined by the concentration of ions or sucrose in the growth medium (Fig. 23-7). Analysis of variance shows when responses are additive or multiplicative; that is, when they are *not* interactions. Any other result in an analysis of variance indicates an interaction of factors, and there are many kinds of interaction (see Lockhart, 1965 and Mohr, 1972).

In addition to the quantitative manner in which an organism responds to environment (as discussed earlier), there are other ways of classifying responses. The next six topics were part of a classification scheme for environmental responses suggested by Anton Lang at the Annual Meetings of the American Institute of Biological Sciences held at Purdue University in 1961. We have modified them somewhat.

3. Direct (Noninductive) Environmental Effects Sometimes, as an environmental factor changes, the plant response changes directly and immediately (or in other cases, *almost* immediately). Photosynthesis is an excellent example. When light intensity is limiting, the rate of photosynthesis changes within

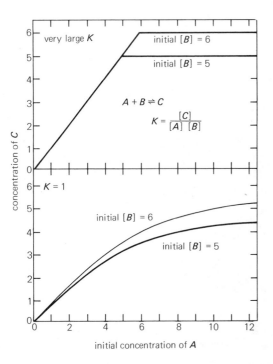

Figure 23-4 An illustration of the principle of limiting factors as it might be observed in a chemical equilibrium reaction. Concentration of a hypothetical product (C) is plotted as a function of initial concentration of one reactant (A) in the presence of two initial concentrations of the other reactant (B). Top, with a very small equilibrium constant, virtually all available A enters into the formation of C until B becomes limiting at its two concentrations. This is the ideal limiting factor response. If K is as large as 1 (bottom), then even at low concentrations, both A and B limit the amount of product formed.

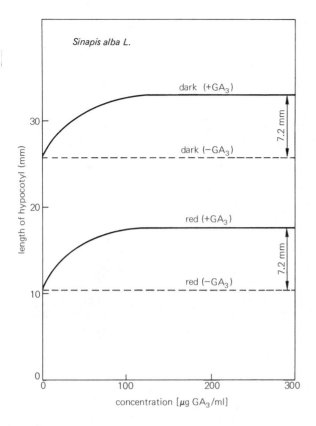

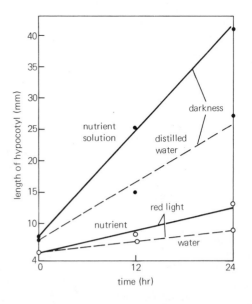

Figure 23-7 An illustration of multiplicative effects. Growth of the hypocotyl of white mustard (*Sinapis alba*) is plotted as a function of time. Red light inhibits compared to darkness, and distilled water inhibits compared to a complete nutrient solution (Knop's solution). The percentage of inhibition is constant at all times. (From Schopfer, 1969.)

Figure 23-6 An empirical example of "numerically additive behavior" of two factors. Hypocotyl lengthening in the mustard seedling was investigated under the control of light (continuous standard red light) and exogenously applied gibberellic acid (GA₃). It is apparent that the dose-response curve for exogenously applied GA₃ is the same with and without light. Note that GA₃ promotes lengthening whereas red light inhibits lengthening of the hypocotyl. Hypocotyl length was measured 72 hr after sowing. (After Mohr, 1972.)

seconds as light intensity changes. Transpiration is a comparable situation, but this is complicated somewhat by the stomatal response, which may be a little less immediate and direct. Many enzyme-controlled reactions respond directly to temperature or, in some cases, to the presence of specific molecules in their environment. This is true of allosteric enzymes. As the promoting or inhibiting molecule increases in concentration, enzyme activity immediately becomes proportional.

4. Triggered, or On-Off, Inductive Responses There are a few plant responses to environment (initiated by some environmental change) that may then proceed even if the environment returns to its original state. When this happens, the response is said to be **inductive.** Frequently, there is a **delay** between the triggering action and the response itself. Often,

there is also an **amplification;** that is, the energy supplied to the plant by the environmental change is well below the energy required to bring about the response. The energy for such a response is provided by the plant's metabolism. Seed germination in response to light is an excellent example of such a triggered response. For a given seed, when the absorbed light crosses a threshold, germination begins and will continue even if the triggering light is removed. Actually, examples of triggered responses are not easy to find. Most responses that initially appear to be of the triggered type turn out to belong to the next category.

5. Modulated, or Quantitative, Inductive Responses In this category, the response is not clearly on or off, but the *level* of the response is determined by the level of the environmental factor. Again, there is frequently an amplification, but this amplification is still proportional to the environmental input. There is also a delay in such responses, as in the triggered responses. Phototropism is an excellent example of a modulated response. As we discussed earlier (see Fig. 18-7), the degree of bending of the stem is a function of the number of quanta of light absorbed by the stem or coleoptile, but the light does not supply the energy for bending.

Modulated, amplified responses are inductive. When the delay is long, we are especially likely to speak of induction. Vernalization is an excellent example. Flowering typically occurs long after the plants have been exposed to the low temperature that caused it. Numerous responses in this category are also time related. In photoperiodism (flowering, dormancy, tuber formation, and so on), for example, we saw that the *time* of the light interruption during the dark period is extremely important (see Fig. 22-5). Time of light application is also a controlling factor in circadian rhythms.

6. Homeostasis and Feedback In the middle of the nineteenth century, the great animal physiologist Claude Bernard stated the principle that body mechanisms strive to maintain a constant inner environment through self-regulation. This phenomenon is called **homeostasis.** It was studied extensively in animals (where body temperature, blood chemistry, and numerous other parameters tend to remain constant), but the principle also applies to plants. We have already studied several mechanisms that illustrate the point, such as stomatal regulation of transpiration, maintenance of various hormone levels, and so on. Often, homeostasis is known to be achieved by feedback mechanisms both in plants and in animals; indeed, feedback may be the only manner in which such regulation can be achieved. As a thermometer detects the temperature, switching heating or cooling systems on or off in a thermostatically controlled system, so carbon dioxide levels control stomatal apertures, and products often affect reaction rates by acting allosterically on the enzymes that catalyze their production.

7. Conditioning Effects In this situation, there is not much delay. The change is typically a gradual one occurring in response to a continual application of the environmental factor. Development of frost and drought resistance are good examples. The more the plant is exposed to low temperatures and short days, the more frost- or drought-hardy it tends to become. Sometimes (e.g., in peach seeds) the germination temperature determines the nature and extent of subsequent seedling growth. This may also be an example of a conditioning effect.

8. Carryover Effects In the early 1950s, Harry Highkin, at the California Institute of Technology, discovered that genetically pure, inbred pea plants grew poorly when the day and night temperatures were equal and held constant (10, 17, or 20 C) (see Highkin and Lang, 1966). When the pea plants were grown for several generations under these adverse conditions, each generation (up to about the fifth) grew more poorly than the previous one. When he reversed the situation (germinated seeds from stunted plants under optimum conditions), it required at least three generations to reach the maximum level of growth. Such a carryover of environmental effects from one generation to the next seems quite contrary to most of our concepts of genetics. (The environment does not change the genes.) Nevertheless, the phenomenon is real, and it has since been confirmed by other workers. Highkin and the

others were careful to demonstrate that the effects were not due to genetic selection. Apparently, the developing embryo (or perhaps the stored food material in the cotyledons) is in some way conditioned by the environment so that the effect carries over through a number of succeeding generations.

9. Ecotypes: The Role of Genetics We assume that the distribution of a given species is determined by its genetics. *But what if there is genetic diversity within a species?* Could the dwarf *Potentilla glandulosa* plants of the high Sierra Nevada, for example, have a genetic composition that allows them to get along at relatively low temperatures, while the larger *Potentilla glandulosa* plants found at lower elevations have a genetic composition that allows them to do well only at higher temperatures? When we think about the principles of evolutionary gene-pool change, we might certainly expect such situations. The reproductive processes are relatively slow, which makes the rate of gene flow within the gene pool relatively slow, but climatic pressures on the population differ depending upon location. Hence, we might expect the genetic composition of a population to vary throughout the range of that population.

We can imagine two possible explanations for the dwarfed versus large *Potentillas*: First, their genetic compositions might be alike, but their different appearances might be due to the different climates under which they occur; and second, the differences could be due to an actual genetic diversity. How do we distinguish between these two possible explanations? Obviously, the thing to do is to bring the different plants together, growing them either in a **uniform garden** or in a controlled environment facility. In the 1920s, Göte Turesson in Sweden developed the uniform-garden approach to a high level of finesse. Jens Clausen, William Hiesey, and David Keck (1940) of the Carnegie Laboratory at Stanford University in California and others followed suit. As it turns out, both environment and genetics are important.

Effects of environment upon plant morphology (i.e., appearance) and physiology are common. Turesson called plants with similar genetics that exhibit differences due to varied natural environments **ecophenes.** This is usually not emphasized in discussions of this type, because the genetic differences we are about to discuss are so obviously important. Nevertheless, we should realize that the environment can and does produce many different ecophenes from any uniform genetic stock. Numerous effects of temperature, light, nutrients, and other factors on plant growth and development have been emphasized in this and several other chapters.

Turesson and others also found genetic differences in representatives taken from the different areas of a species's distribution. These different genetic representatives of the population were called **ecotypes.** When the *Potentillas* from the Sierra Nevada, the Coast Range, and other locations were brought together in a uniform garden, they continued to exhibit striking and significant morphological differences (Fig. 23-8). Many species have now been studied, and it seems obvious that different environments will exert different selection pressures, resulting in different genetic compositions that are directly correlated with geography.

As might be expected, selection also works for the physiological responses to environment (Billings, 1970). For example, photoperiodic ecotypes have been demonstrated in several species (Section 22.1). Alpine sorrel plants collected from several locations in the Arctic flowered in response to longer days (20 hr or longer) than those collected from the southern Rockies (15 hr), for example. The arctic plants also reached peak photosynthesis rates at lower temperatures than their southern counterparts, but the alpine plants that grow at high elevations and relatively lower carbon-dioxide pressures were more efficient in utilizing carbon dioxide. Any competent taxonomist would classify all the alpine sorrel specimens as the same species, but careful observation revealed a number of morphological differences between the northern and the southern representatives, as well as the physiological differences.

The carryover effects discussed in the last subsection could be an important complication in uniform garden studies. Apparent differences in the first few generations could be due to carryover rather than genetics. Regrettably, this possibility has been largely ignored by most workers (but see Clements et al., 1950).

10. Allelochemics and Allelopathy An **allelochemic** is a substance produced by one organism that influences another organism (Whittaker, 1975). Production by plants of allelochemics that are harmful to other plants is called **allelopathy.** Fungi produce allelochemics, called **antibiotics,** that are effective against bacteria, and higher plants also produce allelochemics that are effective against other species. The walnut (*Juglans* sp.) produces *juglone,* a quinone, and soft chaparral produces terpenes that inhibit the germination or root growth of other plants. Rarely do they act directly against a mature plant. The compounds may be produced in leaves and released when they decay, leached from the leaves by rain, or secreted by roots. In any case, when the plant releases these compounds, it assures for itself a sort of "territory" in which it will not be bothered by competition from other species. Study of allelochemics is becoming an important part of environmental physiology.

Many plants produce **narcotics** that exert profound effects upon the physiology or even psychology of animals that consume them. Most are alkaloids. When these allelochemics result in sickness or death, we call them poisons. They provide defense mechanisms for the plants that produce them, since grazing animals avoid eating them. Only man is perverse enough to cultivate tobacco, tea, coffee, and other narcotic producers, although, properly used, some narcotics can be beneficial to man; we are also aware of their harmful effects. Alcohol (produced by yeasts) is not an alkaloid or an allelochemic, but its effects are similar to many narcotics.

23.3 How It's Done: Field and Laboratory

It seems appropriate to say a few words about the methods of environmental physiology. As stated, any physiological methods might be used in such studies, but the ones that are used will depend upon the viewpoint and goals of the research at hand. Some purely theoretical studies that may qualify as environmental physiology will be carried out strictly in the laboratory. A researcher is interested, for example, only in the response kinetics of his particular experimental organism grown under his particular set of highly artificial conditions. The response may be to temperature, light intensity, photoperiod, nutrient levels, humidity, or other factors. The goal is to obtain data that can be analyzed mathematically, and there may be little or no interest in what the results might imply about plants in more natural environments.

Most environmental physiology, however, is concerned with plants in cultivated or natural ecosystems. In either case, it is imperative to establish a proper balance between field and laboratory studies. In the laboratory, work is motivated by a desire to understand responses in the field, and

Coastal   Mid-Sierran   Alpine

Figure 23-8 Photograph of three *Potentilla glandulosa* specimens grown in a uniform garden at Mather, California, and collected on June 5 to 18, 1935. Plants were dug at three locations in the Coastal, Mid-Sierran, and Alpine stations in California 5 to 13 years previously. (From Clausen, Keck, and Hiesey, 1940.)

studies in the field are guided by relationships established in the laboratory. It is easy to find oneself becoming more and more involved in either one of these two phases, but one should resist this, since final answers will almost certainly be more satisfying when a suitable balance between field and laboratory studies has been established.

In spite of the desirability of maintaining proper balance, some problems seem to have been studied more in the field while others have been emphasized in the laboratory. Many workers, for example, have studied plant-water relations in the field, and others have devised field methods for measuring photosynthesis. Such studies are usually backed by appropriate investigations in the laboratory, where techniques must be developed. On the other hand, responses of plants to temperature and photoperiod have been studied extensively in the laboratory. Temperature studies are often extended to the field, but application of photoperiod investigations to natural or agricultural environments needs more emphasis (Section 22.1).

The guiding concept of field research is to measure and deduce. Many environmental parameters may be measured, and then statistical methods such as those mentioned may be applied to deduce which factors are influencing plant growth and in what ways. Laboratory research, on the other hand, may emphasize the control of environment. An ideal would be to control all environmental parameters, holding all constant but one, which varies and therefore must be responsible for the observed plant response. This is the principle of analysis. Extensive, controlled environment laboratories for such studies have been established at several dozen locations throughout the world. If they emphasize the study of plants, they may be called **phytotrons.** At least one such facility includes both plants and animals and is called a **biotron.** It is located at the University of Wisconsin. In addition to the large-scale phytotrons, smaller plant growth chambers are commercially available and widely used (Fig. 23-9).

Much technology is involved in the construction of such analytical tools. Temperature is relatively simple to control, although humidity is somewhat more difficult. Length of the photoperiod is easy to control, but the production of artificial light matching sunlight in quality and intensity is extremely difficult and costly. Xenon arc lamps with water filters or other advanced technology can nearly simulate sunlight, but to do so is expensive. Most growth chambers and phytotrons utilize fluorescent tubes combined with incandescent bulbs. The incandescent bulbs are much less efficient (supply less light for a given power input) than the fluorescent lamps, but they supply red and far-red wavelengths not supplied, or not supplied in sufficient quantity, by the fluorescent tubes. Tubes with special phosphors producing light that is rich in blue and red wavelengths have been designed for plant growth and are rather widely used (Special Topic 3.1). Some studies indicate that they are advantageous; others suggest that they are no better than ordinary tubes. Some chambers use a trans-

Figure 23-9 A small walk-in chamber used in controlled environment studies. (Courtesy Environmental Growth Chambers, Chagrin Falls, Ohio.)

parent barrier between the lights and the plants. Such a barrier allows the temperature around the lamps to be controlled independently of that around the plants. Since the light output of fluorescent tubes is strongly influenced by temperature, this is important, but the barrier absorbs a portion of the light energy and collects dirt.

In addition to environmental control, studies in environmental physiology might apply any of the methods of physiology. Some exciting studies have shown, for example, that the proportion of unsaturated fatty acids in cell membranes and in storage fats increases noticeably for plants or animals grown at lower temperatures (Chapter 24). Or the enzymes of bacteria adapted to saline conditions may differ considerably from comparable enzymes in comparable bacteria that do not require high salt concentrations in their environment.

23.4 The Environment and Plant Response

To summarize our thoughts on environmental physiology, let's list the factors of the environment, consider how they vary, and examine a few of the known plant responses to these factors. To have a true understanding of environmental physiology, we need to understand the responses of individual species and of ecotypes within species. Such a task may be hopeless, so we may be restricted to a few broad generalizations and occasional intensive work with a few ecotypes of special interest. Nevertheless, remember the importance of differences among species.

Light Light has been the subject of discussion in much of this book (Chapters 3, 9, 10, 11, 19, 20, and 22). We have seen that light may vary in three ways: (1) in intensity, (2) in quality, and (3) these parameters as a function of time. All three ways are extremely important to plants growing in the open. We have discussed photosynthesis, chlorophyll synthesis, phototropism, numerous phytochrome-controlled responses, phototaxis, and photoperiodism. Clearly, the light environment (more properly, the radiation environment, including short and long wavelengths beyond the visible part of the spectrum) varies over a wide range in natural environments. Daylengths vary with latitude and time of year. Intensities vary with time of day, shading, cloudiness, elevation, and height of the sun above the horizon. Intensities are especially high in deserts (low cloudiness) and in the alpine tundra (high elevations). (See Chapter 24.) Qualities (colors) also vary with time of day, elevation, height of the sun above the horizon, atmospheric conditions (e.g., atmospheric pollutants), leaves through which light passes, depth in the water, snow depth, and so on. As we have seen, all these things might influence plant response. Ultraviolet light is a special feature discussed by Martyn Caldwell in Special Topic 23.1.

Temperature There is hardly a chapter in this book in which temperature has not been mentioned as a controlling factor

in the response under consideration. Temperature influences water potential, transpiration, transport across membranes, the translocation of assimilates, metabolism and virtually all enzyme action, growth and development, photomorphogenesis, the biological clock, dormancy mechanisms, and photoperiodism. We have seen that temperature is especially important in some of these, such as transpiration and heat transfer, cellular respiration, and the low-temperature responses of plants. Situations in which plants grow and respond under what might be unexpected conditions of temperature or other factors are especially interesting. For example, some plants grow actively under the snow; others (mostly algae) grow in hot springs.

Wind and Other Mechanical Stimuli Air movement is an important part of virtually all terrestrial environments. Yet, in contrast to temperature and light, wind has seldom been mentioned in this text (mostly in Chapter 3 on transpiration and heat transfer). We can imagine that wind might have at least two kinds of effects on plants: It will certainly influence the exchanges that take place between a plant and its surrounding atmosphere, exchanges of heat, carbon dioxide, oxygen, water vapor, and other gases; it will also move the plant mechanically, causing thigmomorphogenetic effects (Section 18.10).

Precipitation Water and water potential formed the topic of the first part of this book, and enzymes function in a water milieu, hormones are in water solution, and so on. In agriculture, yields can often be increased by irrigation. The vegetation of natural ecosystems is probably more strongly influenced by water than by any other factor (see Table 11-1). Dry areas, whether hot or cold, are deserts (Chapter 24), and regions of abundant precipitation are jungles when they are warm or boreal forests or tundras when cold. Too much water, on the other hand, leads to ethylene accumulation in plants, retarding growth and causing chlorosis and epinastic effects characteristic of waterlogged plants.

Humidity Atmospheric humidity has scarcely been mentioned in this text except in relation to transpiration. Indeed, it has been the consensus of most plant physiologists that humidity has little effect upon plant growth, as long as it is above, perhaps, 50 percent. Several studies seemed to indicate this. Recent studies indicate that humidity does indeed influence plant growth, however (see boxed essay, p. 360).

The Soil Clearly, the soil is extremely important to plant growth, although we have considered it only in the early chapters. Soil water is especially important, as implied by the discussion of precipitation. The effect of soil temperature on plant growth is undoubtedly also of considerable importance, although it has seldom been studied as an independent variable. Effects of root temperatures on growth of alpine plants have been studied (Spomer and Salisbury, 1968; Higgins and Spomer, 1976).

Of course, soil nutrients are extremely important in both agricultural and natural situations. When plants grow well on soil that is highly impoverished in one or more of the mineral nutrient elements, it is because their physiology in some way enables them to do so. On highly acid, yellow material (the exposed cross sections of ancient hot springs) in the western United States, ponderosa pine and other trees, shrubs, and herbs are able to grow, while other plants are unable to do so (Salisbury, 1964). Similar situations occur on serpentine soils and over vast areas in Australia and New Zealand where certain micronutrient elements are present only in very low concentrations.

These examples vividly recall the important lesson of this chapter: that plant species and ecotypes differ widely in their responses to environmental parameters, that these differences are genetically controlled, that they are little understood at this time, and that they must, nevertheless, account for the differing abilities of agricultural species to grow in various regions and for the natural distribution of plants in undisturbed habitats.

Other Organisms An important part of an organism's environment consists of other organisms. This was implied by the discussions of community succession and allelopathy, but **plant pathogens** play another important role in both physiological ecology and agriculture. A few of these are bacteria, but most are fungi, viruses, and nemotodes. Because these parasites compete with us for food, they are of great importance in agricultural science. **Plant pathology** for many years was only descriptive and empirical, but now pathologists are beginning to understand the physiological interactions between a parasite and its host. These interactions are largely at the molecular level, involving compounds produced by both host and parasite. Penelope Hanchey-Bauer from Colorado State University has prepared an essay describing this work (Special Topic 23.2). In a sense, it is environmental physiology carried to the biochemical level.

So there are many opportunities in plant physiology, opportunities to understand an important part of our universe: how the plants that surround us and provide us with food for life function. Our studies and researches may consider whole plants growing in their environments, substances moving into and out of cells, transport of materials from leaves to roots, enzymes catalyzing a host of reactions, genes transferring information to messenger-RNA molecules, phytochrome molecules changing as quanta of red light are absorbed, or many other processes and substances. So keep your copy of Salisbury and Ross handy during your travels; don't be afraid to cross things out and write in the margins as you encounter new discoveries. And drop us a line if you get the urge!

24

Stress Physiology

What are the environmental limits for the existence of life on earth? What are the mechanisms within plants and animals that establish these limits? How do these mechanisms break down under the stresses caused by environmental extremes? These questions are important, because the stresses occurring at a plant population's boundary often control the population's distribution, and stresses often limit agricultural productivity.

24.1 Drought

Considering world vegetation types, plant productivity is more closely related to available water than to any other environmental factor (see Table 11-1). Think of hot jungle rain forests, or temperate rain forests, as in Washington and Oregon, compared to hot or cold deserts. So lack of water has long been studied as a stress factor. Ecologists classify plants according to their response to water: **Hydrophytes** grow where water is superabundant; **mesophytes,** where water availability is intermediate; and **xerophytes,** where water is scarce. Solutes strongly influence water potential, so ecologists further classify plants that are sensitive to relatively high salt concentrations as **glycophytes** and those that are able to grow in the presence of high salts as **halophytes.**

The Desert Drought is characteristic of deserts. A desert is an area of low rainfall—less than about 20 to 40 cm per year, depending upon temperatures, season of precipitation, and other factors. Global air movement is one of the several climatic conditions that can be responsible for a low annual precipitation. The most extensive deserts occur in the so-called horse latitudes, where air that has ascended in other latitudes descends and is thereby compressed and warmed. Warm air can hold more moisture than cold air, so precipitation does not occur. This effect is responsible for the Sahara Desert, the deserts of Mexico, and the western coastal deserts of South America.

Another cause of deserts is the **rain shadow effect.** North of the horse latitudes, in the temperate zone, global air movement is predominantly from west to east. Storm systems moving this way rotate in a counterclockwise direction, so a storm center is preceded by south winds and followed by north or west winds. As these storm systems approach a mountain range such as the Sierra Nevada, the rising air expands, cools, and can hold less moisture, resulting in precipitation on the western slopes. On the eastern slopes, the air descends, compresses, warms, and can hold more moisture. Such descending warm and dry winds on the eastern slopes of the Rockies are called **chinooks** or "snow eaters." The deserts of the Great Basin occur in the rain shadow east of the Sierras, but the plains east of the Rockies are not deserts because they receive moisture moving north from the Gulf of Mexico and south from the Arctic.

Deserts in the horse latitudes are typically hot and dry all year, while rain shadow deserts may be cold during winter. Because air above deserts is typically dry, it absorbs relatively little incoming sunlight or outgoing, long-wave thermal radiation. Thus, deserts are often hot during the daytime and relatively cold at night. Because of these temperature extremes, wind is often common. Air cools at higher elevations and flows down canyons and washes during the night. This can lead to dune formation, although the great sand dunes associated with deserts by moviemakers and others are not as common as we are led to believe. Perhaps so-called **desert pavement** is more common. This consists of a surface layer of small stones, the finer material having been eroded away, mostly by wind.

Desert soils are typically salty, because the low rainfall does not leach away the salts as they form by weathering. The actual status of a desert soil depends considerably upon the time during the year when rain does fall. In Mediterranean climates, rain may fall only during winter with 6 to 9 months of summer drought. Yet when the rain does fall, precipitation is high, so these soils tend to be less salty than other desert soils. The summer showers in the Arizona desert lead to a unique and relatively lush desert vegetation. Cold deserts of the Great Basin experience winter storms and occasional summer thunderstorms.

The Xerophytes We need to consider two sets of terms: Jacob Levitt (1972), who has contributed much to our understanding of stress, suggests that we should distinguish between **resistance** and **hardiness** to any given stress factor. H. L. Shantz (1927) classified xerophytes as those plants that *escape, resist,* *avoid,* or *endure* drought. Those that escape, resist, or avoid drought have Levitt's resistance; only those that endure have hardiness. Let's examine Shantz's four groups.

Annual plants in the desert **escape** the drought by existing as dormant seeds during the dry season. When enough rain falls to wet the soil to a considerable depth, these seeds may germinate, perhaps in response to the leaching away of germination inhibitors. Many of these plants are able to grow to maturity and set at least one seed per plant before all soil moisture has been exhausted. They are eminently well adapted to dry regions and thus are xerophytes in the true sense of the word, yet their active and metabolizing protoplasm is never exposed to a high water stress (highly negative water potential) and is not drought hardy. They are resistant to drought only by escaping it.

Shantz called the **succulent species** in dry regions, such as the cacti, century plant *(Agave americana),* and various other crassulacean acid metabolism (CAM) plants (Section 10.5), **drought resisters.** They resist the drought by storing water in their succulent tissue. Enough water is stored, and its rate of loss is so extremely low (due to an exceptionally thick cuticle and stomatal closure during the daytime), that they can exist for long periods without added moisture. D. T. MacDougal and E. S. Spaulding reported in 1910 that a stem of *Maximowiczia sonorae* (a desert succulent in the cucumber family) stored "dry" in a museum formed new growth every summer for eight consecutive summers, decreasing in weight only from 7.5 to 3.5 kg! Since their protoplasm is not subjected to extremely negative water potentials, succulents also are drought resistant but not hardy.

Many plants of dry regions are **nonsucculent perennials.** Some of these **avoid** the drought by various anatomical modifications. The most effective is a deep root system extending to the water table. Such deep rooted desert plants as palms, alfalfa, and mesquite always have an ample water supply, so they are also resistant but not drought hardy. Similar modifications, none of which is as effective as roots that reach the water table, include extensive shallow root systems capable of absorbing surface moisture after a storm (common in cacti), reduction in size of leaf blades (which increases heat transfer by convection, lowers leaf temperature, and thus reduces transpiration; see Chapter 3), sunken stomates, shedding of leaves during dry periods, and such other factors as heavy pubescence on leaf surfaces (Ehleringer et al., 1976). Although these modifications may indeed reduce the loss of water, they never completely prevent it and are by themselves insufficient protection against extreme drought.

Among nonsucculent perennials are those that simply **endure** drought. That is, they lose exceptionally large quantities of water so that their protoplasm is subjected to extremely negative water potentials, yet they are not killed. Such **euxerophytes** (true xerophytes) exhibit hardiness rather than merely resistance. Plants that only resist drought are of interest to ecologists, but they do not challenge our understanding as the euxerophytes do. Incidentally, many of the characteristics of drought avoiders, such as small leaves and sunken sto-

mates, also occur in the drought endurers. Yet in the euxerophytes the ultimate weapon against drought is the ability to endure it, to be drought hardy.

The ability of some euxerophytes to endure drought is phenomenal. Water content of creosote bush *(Larrea divaricata),* a desert shrub of both North and South America, drops to only 30 percent of the final fresh weight before leaves die. With most plants, levels of 50 to 75 percent are lethal. Some of the most spectacular euxerophytes are mosses and ferns— plants that we normally associate with wet situations. Their ability to dry out and then become metabolically active immediately upon rehydration apparently depends upon special features not common to other plants.

Much recent work on desert xerophytes has been done in the Negev Desert of Israel. Researchers (Evenari et al., 1975) there use a somewhat more complex classification scheme than that developed by Shantz, but the same principles apply. They have studied algae, lichens, and mosses that can tolerate extreme and prolonged desiccation, as well as extreme cold and heat when they are dry. They can take up water directly and instantaneously from dew, rain, or even a moist atmosphere (some when relative humidity is 80 percent or above), and such water absorption leads to an instantaneous switching on of metabolic activity.

Among higher plants, Israeli workers have found most of the features just discussed, plus a few others. An important adaptation of desert plants, for example, is that of **heteroblasty,** or the property of producing seeds from the same plant that have several different germination requirements. Thus, only a few seeds in a given crop germinate at any given time; others germinate at different times. These workers have also studied stomatal regulatory mechanisms in desert plants. In the most interesting examples, when water stress was low in the leaf, stomates opened when the temperature was raised; when the same was done at higher water stress, stomates closed.

In the Negev Desert, plants often utilize dew. It is never as spectacular, however, as in a case reported by Frits Went (1975; studied by Fusa Sudzuki and others) from the Pampa del Tamarugal, a vast desert plain in the northern, tropical part of Chile. This is a perfectly flat, rainless, and fogless area that supports forests of *Prosopis tamarugo,* a mesquite. The soil surrounding the roots of these trees is saturated with water during the entire year, although above and below this root zone the soil is dry at all times. Apparently, because the area is perfectly flat, wind is extremely uncommon. Thus, a single air mass will remain in place for days to weeks. Whenever the temperature drops below the dewpoint at night, an event that occurs frequently during at least part of the year, dew forms on the leaves, where it is absorbed and transmitted to the root zone. Water enters the atmosphere by transpiration during the day, but the water lost to the stagnant atmosphere is closely equivalent to the water reabsorbed from that same atmosphere during cool nights. So, the special physiological feature enabling the mesquite trees to grow where there is no rain is their ability to absorb moisture readily from dew and actually secrete it into the soil around their roots, where it is stored.

Incidentally, insect physiologists (Edney, 1975) have been interested for several years in the ability of certain insects to absorb water from an atmosphere with a relative humidity as low as 50 percent—and thus a water potential of almost –1,000 bars! Liquid within the insect's body might have a water potential of only –10 to –20 bars and would be in equilibrium with an atmosphere of about 99 percent relative humidity. How absorption occurs remains a mystery. We need to know much more about the fine structure and function of insect cuticles and about the properties of insect epidermal cell membranes.

There are numerous other adaptations of desert plants that cannot be discussed for lack of space. The allelochemics (Section 23.2.10) are often produced by desert plants, restricting the germination or growth of competing plants, which reduces competition for water. Furthermore, as in other environments, desert plants sometimes profoundly influence the soil upon which they grow. M. Fireman and H. E. Hayward (1952) found, for example, that greasewood (*Sarcobatus vermiculatus*) in the Escalante Desert of Utah brought salts from depths, depositing them on the surface (Fig. 24-1). Probably the salts were contained in the leaves and released upon their fall and decay. The result was high salt concentrations beneath the greasewood plants, especially when compared with soil beneath big sagebrush (*Artemesia tridentata*) plants that did not redistribute salt in this manner. Clearly, these two species have different physiologies that are of ecological significance.

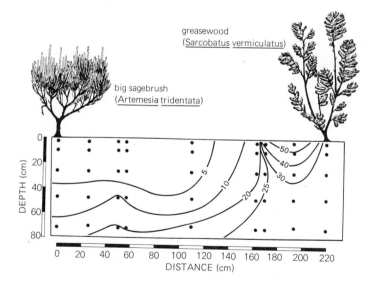

Figure 24-1 Salt concentrations in the soil (numbers in lines = percent exchangeable sodium) as a function of position under a sagebrush plant and under a greasewood plant in the Escalante Desert of southern Utah. Note high sodium under greasewood. Dots are sampling points. (From Fireman and Hayward, 1952.)

Plant Responses to Water Stress Desert xerophytes are certainly of interest to physiological ecologists, but the most work on water stress has been done by agriculturists. Water is virtually everywhere a limiting factor in agriculture. This may be due to unexpected dry periods or to normally low rainfall, necessitating regular irrigation. If one must irrigate to obtain a crop, how much and how often should water be applied to the field? The answer, which could influence the expenditure of many millions of dollars in a given state and a given year, may be strongly influenced by research on stress physiology. Here we are seldom concerned with the severe stresses endured by desert plants; rather, we are interested in the extent to which withholding relatively small amounts of water might influence crop yield.

Research has been extensive, continuing for well over a century. Thousands of papers on plant responses to drought have now been published. Theodore Hsiao (1973) and others have reviewed work in this field, evaluating the conclusions of older work in light of work in the 1960s and early 1970s that utilized a number of new important techniques. In Table 24-1, Hsiao summarizes the sensitivity to water stress of several plant processes. The table outlines the sequence of events that occurs when water stress develops rather gradually in a plant, as when water is withheld from a plant growing in a substantial volume of soil. It is important to realize that the later events are almost undoubtedly indirect responses to one or more of the early events rather than to water stress *per se*.

Cellular growth appears to be the most sensitive to water stress (Fig. 24-2). Decreasing the external water potential by only a bar or less results in a perceptible decrease in cellular growth (which is irreversible cell enlargement). Hsiao suggests that this sensitivity is responsible for the common observation that many plants grow only at night when water stress is lowest. (But endogenous rhythms could also be involved.) In any case, the response of cellular growth to water stress is manifest as a slowing of shoot and root growth. This is usually followed closely by a reduction in cell wall synthesis. Protein synthesis in the cell may be almost equally sensitive to water stress. These responses are observed only in tissues that are normally growing rapidly (synthesizing cell wall polysaccharides and protein as well as expanding). It has long been observed that cell wall synthesis depends upon cell growth (Section 15.10). The effects on protein synthesis are apparently controlled at the translational level, the level of ribosome activity.

At slightly more negative water potentials, protochlorophyll formation is inhibited, although this observation is based on only a few studies. Many studies indicate that activities of certain enzymes, especially nitrate reductase, phenylalanine ammonialyase, and a few others, decrease quite sharply as water stress increases. A few enzymes, such as α-amylase and ribonuclease, show increased activities. It was thought that such hydrolytic enzymes might break down starches and other materials to make the osmotic potential more negative, thereby resisting the drought, but careful studies negate this idea. Nitrogen fixation and reduction also drop with water stress, consistent with the observed drop in nitrate reductase activity. At levels of stress that cause observable changes in enzyme activities, cell division is also inhibited. And stomates begin to close, leading to a reduction in transpiration and photosynthesis.

At about this level of stress, abscisic acid (ABA) begins to increase markedly in leaf tissues. Increases of as much as 40-fold have been observed. Of course, this could be responsible for stomatal closure, since applied ABA causes stomates to close rapidly. Perhaps ABA closes stomates only in mesophytes. Limited studies show little accumulation of ABA in moisture-stressed aquatic plants or in plants adapted to arid environments. Furthermore, such plants also have unusually low ABA levels (Kriedemann and Loveys, 1974).

An increased production of ethylene may be an even more sensitive indicator of water stress than increasing ABA levels. Ethylene is produced in response to various kinds of stress besides drought, including excess water, plant pathogens, air pollution, root pruning, transplanting, and handling (Section 17.2).

Although stresses are relatively mild at $\psi = -3$ to -8 bars, interactions and indirect responses begin to be the rule. Cytokinins increase in at least a few species at about these levels. At slightly higher stress levels, the amino acid proline begins to increase sharply, sometimes building up to levels of 1 percent of the tissue dry weight! Other amino acids and amides

also accumulate when the stress is prolonged. Proline arises *de novo* from glutamic acid and ultimately probably from carbohydrates. What is the proline doing? It might act only as a storage pool for reduced carbon and nitrogen during stress (Hsiao, 1973), but its presence as a solute would significantly lower the osmotic potential, and this could be important in the osmotic regulation (osmoregulation) of halophytes and other plants, helping to match the internal water potential more closely to that of the surrounding medium (see Section 24.5).

At higher levels of stress ($\psi = -10$ to -20 bars), respiration, translocation of assimilates, and CO_2 assimilation drop to levels near zero. Hydrolytic enzyme activity increases considerably, and ion transport can be slowed. Parts of the transpiration stream in the xylem eventually cavitate, so the xylem becomes blocked by vapor space. This was observed by placing sensitive microphones in contact with stems under stress and listening to the sounds of water cavitating.

Plants usually recover if watered at this point, although growth and photosynthesis in young leaves may not reach the original rate for several days, and old leaves may be shed. Clearly, since growth is especially sensitive to water stress, yields may be noticeably decreased even with moderate drought. Cells are smaller and leaves develop less during water stress, resulting in reduced area for photosynthesis. Furthermore, plants may be especially sensitive to drought during certain stages, such as tassel formation in maize (corn).

Table 24-1 Generalized Sensitivity to Water Stress of Plant Processes or Parameters[a,b]

	Sensitivity to Stress →			
	Very Sensitive		Relatively Insensitive	
	Reduction in Tissue ψ Required to Affect Process[c] →			
Process or Parameter Affected	0 bar	10 bars	20 bars	Remarks
Cell growth	——— - - -			Fast-growing tissue
Wall synthesis	———			Fast-growing tissue
Protein synthesis	———			Etiolated leaves
Protochlorophyll formation	——			
Nitrate reductase level	——			
ABA accumulation	- - - ——			
Cytokinin level	——			
Stomatal opening	- - - - —————— - - - -			Depends on species
CO₂ assimilation	- - - - —————— - - - -			Depends on species
Respiration	- - - ——			
Proline accumulation	- - - ——			
Sugar accumulation	——			

[a]From Hsiao, 1973.

[b]Length of the horizontal lines represents the range of stress levels within which a process first becomes affected. Dashed lines signify deductions based on more tenuous data.

[c]With ψ of well-watered plants under mild evaporative demand as the reference point.

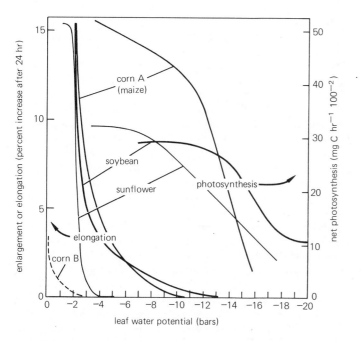

Figure 24-2 Leaf enlargement (soybean and sunflower) or elongation (corn) and net photosynthesis, both as a function of leaf water potential (water stress). Data for corn A, soybean, and sunflower after Boyer (1970). The dotted curve (corn B) indicates relative elongation rate for leaves (originally measured in micrometers per minute, but actual figures are not shown) as a function of *soil* water potential (data from Acevedo et al., 1971); elongation data plotted as a function of *leaf* water potential produce a curve similar to that of Boyer (corn A).

Mechanism of Plant Response to Water Stress How does a euxerophyte differ from ordinary mesophytes? Protoplasmic viscosity usually increases with high water stress, often to a point where the protoplasm becomes brittle. Euxerophytes maintain protoplasmic plasticity much better at a given water stress than do mesophytes. The hydrolytic activity (breakdown of starch, protein, and so on) is also less noticeable in euxerophytes.

But what about the responses of mesophytes to mild water stress as previously outlined? At least five possibilities have been discussed. First, it is known that water activity (indicating its ability to enter into chemical reactions) is a function of water potential and is thus lowered by water stress. Yet at the maximum stress levels of interest to agriculturists ($\psi = -10$ to -20 bars), water activity is lowered only slightly, probably not enough to be of any real consequence in chemical reactions. Second, solutes increase in concentration as water is lost. This could be important but is probably not very important under mild water stress, simply because the concentration changes would amount to only a few percent. Third, water stress might result in special changes in membranes. Such

effects have indeed been demonstrated, but comparable effects can be caused by other factors without noticeable plant response, so it does not seem likely that this is an important aspect of plant response to water stress. Fourth, water stress might upset the hydration of macromolecules, the "ice structure" of the water molecules surrounding enzymes, nucleic acids, and so on. If this water of hydration (see Special Topic 2.1) is upset, function would also be influenced. Levitt (1962) has suggested that dehydration of key enzymes would cause disulfide bonds within proteins to break and reform, sometimes between adjacent molecules, leading to enzyme denaturation when molecules are rehydrated. But again, it is easy to calculate that mild water stress would not have much influence on the structure of water of hydration, which involves only a few percent of the water in a cell anyway. Amazingly enough, studies have shown that considerable water can be lost from a cell before enzyme function is noticeably influenced. But there could be exceptional enzymes that have not been studied.

Fifth, even the mildest water stress may profoundly change the turgor pressure within plant cells. Pressure changes of this magnitude ($\psi_p = 1$ to 10 bars) probably have little effect upon enzyme activities (judging by observed responses as well as thermodynamic principles), but such changes could be the stimulus to which some special response mechanism in the cell reacts in transducing water stress to the observed cellular responses. In the large-celled marine green algae, *Valonia*, ion uptake decreases with slight increases in cellular turgor pressure. Such responses have not been observed in most higher plant cells, although red beet tissue responds to decreased turgor. The observation may serve as a model for what might be taking place (see discussion in Hellebust, 1976).

Mild water stress might be transduced into a plant metabolic response by special mechanisms existing in the plant just for this purpose rather than by the more general mechanisms that we have mentioned. Thus, response to mild water stress would be a plant regulatory response rather than environmental damage to the plant. The regulatory pressure probably involves changes in cellular turgor pressure. The coupling of wall synthesis and protein synthesis to growth could be important but has yet to be explored. Furthermore, the slowing of growth (6 percent or more within 10 minutes) must result in a rapid accumulation of many metabolites, and this could produce many secondary effects.

Plants exposed to low water levels, high light intensities, and such other factors as high phosphorus and low nitrogen fertilization become drought hardy compared to plants of the same species not treated in this way. That is, they become **acclimated** to drought, a process of considerable importance to agriculture. This is a good example of a conditioning effect. Russian scientists (Henckel, 1964) have reported that plants are drought hardy when grown from seeds that were soaked in water for two days and then air dried. Other scientists have had only limited success in duplicating this **presowing drought hardening**.

24.2 High Temperatures

Elevated temperatures typically accompany drought conditions and are an important environmental stress factor in themselves. This is especially true for euxerophytes that are hardly cooled by transpiration.

The upper temperature limits permitting survival have long been of interest to biologists. Plants typically die when exposed to temperatures of 44 to 50 C, but some can tolerate higher temperatures. Stem tissue near the soil line of plants in the desert, for example, may reach levels considerably above this, and *Tidestromia oblongifolia* does well in Death Valley, California, at these temperatures (Section 11.2).

In thermal springs of Yellowstone National Park, bacteria grow at the boiling point (above 90 C at that elevation) and blue-green algae in waters of 73 to 75 C, about the "melting" point of DNA (the temperature at which the strands in the double helix disassociate). Microorganisms with nuclei (e.g., green algae) have not been found at temperatures above 56 to 60 C (Brock and Darland, 1970).

The upper limit for animals seems to be about 45 to 51 C. Interestingly enough, many enzymes are denatured at temperatures below these, but apparently this doesn't occur in the few organisms that can survive such temperatures. Bacteria have been found growing at the bottoms of oil wells at temperatures somewhat over 100 C but at greatly elevated pressures. These organisms will repeat this performance in the laboratory but will not grow at such high temperature levels when pressure is reduced. Several dry spores and seeds of higher plants will *survive* temperatures above 100 C. These do not actively *grow* at these temperatures, however. In general, dry and dormant structures withstand stress well.

Death of a plant because of exposure to high temperatures results from several interacting factors. Two are especially important. Rapid warming causes a coagulation of protein and consequent disruption of protoplasmic structure. More gradual warming results in a breakdown of protein with a release of ammonia, which is toxic (Henckel, 1964). Of course, plants may be damaged and yields reduced at temperatures not high enough to cause death.

Plants that are hardy to high temperatures exhibit high levels of **bound water** (water of hydration) and high protoplasmic viscosity, characteristics that are also exhibited by euxerophytes. High-temperature plants also are able to synthesize at high rates when temperatures become elevated, allowing synthetic rates to equal breakdown rates, thus avoiding ammonia poisoning. Plants can be acclimated somewhat to high temperatures, but this is minimal compared to acclimation to drought or to freezing temperatures.

24.3 Low Temperatures

Although productivity of world ecosystems is probably limited more by water than any other environmental factor, low temperature is probably most limiting to plant distribution (Parker,

1963). To grow even in subtropical regions subject occasionally to near-freezing temperatures, plants must be capable of some acclimation to low temperatures. Plants that grow in polar regions must be capable of extreme low temperature hardiness, and only a few species can achieve this. Frost and low temperature damage in crop plants is an important hazard in most agricultural regions of the world. For example, tropical or subtropical plants normally grown in southern regions of North America are sometimes damaged by frost or even by temperatures slightly above freezing (as in the winter of 1976–77). Such crops include citrus, cotton, maize, rice, sorghum, soybean, sugarcane, and sweet potato. Certain tropical fruits such as bananas are damaged even by a few hours below 13 C. (Never put bananas in the refrigerator!) Timing is crucial: Rice plants exposed to temperatures below 16 C at the time of pollen-mother cell division will not produce a crop. It has been estimated that the world rice production would decrease 40 percent if world mean temperature dropped only 0.5 to 1.0 C.

Actively growing plants, especially herbaceous species, are damaged or killed by temperatures of only –1 to –5 C, but many of these plants can be acclimated to survive winter temperatures of –25 C or lower. In regions where air temperatures drop below this, many plants have underground meristems protected from extreme air temperatures by soil or snow. They resist or escape cold rather than being cold hardy. Most of the species that survive freezing temperatures do so by tolerating some ice formation in their tissues. Generally, hardier plants can survive with more of their water frozen than less hardy plants. But there are apparently several mechanisms of hardiness (see excellent discussion in Burke et al., 1976).

Most deciduous forest species and fruit trees avoid freezing in some of their tissues by **deep supercooling** to temperatures as low as about –40 C (lowest on record is –47 C). Ice forms in the bark and buds of such acclimated plants when temperatures are only a few degrees below the freezing point of pure water, but the ice crystals form extracellularly (in the spaces between cells), as in herbaceous and other drought-hardy plants; tissues are not damaged by this. Xylem tissue in hardwoods is too compact to permit the formation of such ice crystals, however. When freezing occurs, xylem ray parenchyma cells are killed, the wood becomes dark and discolored, and vessels become filled with gummy occlusions. Wood-rotting organisms often invade such injured trees. But the cell sap in xylem rays of most temperate woody species is capable of being supercooled without freezing. Pure water can be supercooled to –38 C if ice nucleation is prevented, and the solute-containing water in these acclimated xylem cells is apparently protected by the plasmalemma from such nucleation; the cells can often be deep supercooled to –40 C or below. This is observed by measuring stem temperatures with a small thermocouple as temperature is lowered (Fig. 24-3). When the xylem ray cells finally freeze, the released heat of fusion causes a sharp temperature rise, as when osmotic potentials are determined cryoscopically (see Fig. 2-7).

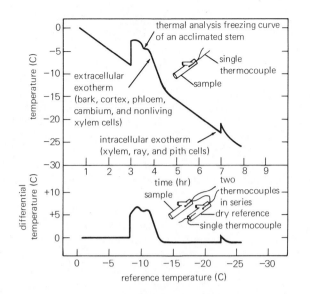

Figure 24-3 Thermal analysis and differential thermal analysis as methods to observe deep supercooling in stem samples. In both experiments, the sample is cooled over a period of time, and the sample temperature (top) or the differential temperature between the sample and a dry reference (bottom) is monitored. The peaks or exotherms show release of the heat of fusion and thus indicate freezing points. The first peak (left—warmest temperature) indicates freezing of extracellular water; the second peak (right—coldest temperature) presumably indicates freezing of water in cells. Cell freezing typically leads to the death of the cells. (From Burke et al., 1976.)

Apparently, deep supercooling is a survival mechanism for some plant tissues. When freezing does occur, ice probably forms within the living cells, and they are killed. Such species do not grow where winter temperatures drop below about −40 C.

Extremely hardy woody plants native to the boreal forests of North America and Asia do not deep supercool. The extracellular freezing process is similar to that found in less hardy herbaceous plants, but ultimate hardiness is much greater. Large ice crystals form, drawing water from the cells until all but the water of hydration (bound water) is removed. Studies extended over many years have failed to show any relationship between the extent of hardiness and the amount of bound water in such plants, but their hardiness is apparently a function of their cell's ability to withstand extreme dehydration. In any case, such dormant, winter-hardy, woody plants (typically softwoods) can readily survive the liquid nitrogen temperature of −196 C! These same plants, when actively growing, may be killed by −3 C! Thus cold acclimation is much more spectacular than drought or heat hardening.

There is apparently no lower temperature limit for survival of spores, seeds, and even lichens and certain mosses in the dry condition. Such test objects have been held within a fraction of a degree of absolute zero for several hours with no apparent damage. Even active tissue may survive these low temperatures if it is experimentally cooled so rapidly that water **vitrifies** into a solid-liquid state, like glass, rather than freezing into a crystal lattice. Certain bacteria actually grow at temperatures of −22 C. Molds have exhibited active growth in cold storage lockers at temperatures of −38 C, and spores have been formed by these organisms at temperatures of −47 C! These unusual examples are observed only rarely, and one would like to know more about how temperatures were measured and maintained. (If such examples were common, cold-storage lockers could not be used.) Several higher plants, such as winter rye, will grow actively at the freezing point and perhaps a few degrees below. Such responses are especially striking when compared with the extreme temperature sensitivity of tropical plants such as the banana.

In practical terms, minor increases in hardiness could have a major impact on world food production. Winter wheats and winter rye yield 25 to 40 percent more than comparable spring varieties, for example, because they make better use of spring rains. If the winter wheats and rye could be made 2 C more cold hardy, they could replace much of the large acreages of spring wheats and rye in North America and the U.S.S.R.

Measuring Frost Hardiness Hardiness is measured in the laboratory by placing pieces of tissue at successively lower temperatures (e.g., 5 C intervals) with intervals of one or two hours at each temperature. As each lower temperature is reached, samples of tissue are removed, thawed out, and tested for viability. Several viability tests may be used. A common one is the application of colorless tetrazolium hydrochloride, which is reduced by active dehydrogenase in living cells to a bright red form.

Frost hardiness typically develops during exposure to relatively low temperatures (e.g., 5 C) for several days. Temperatures down to −3 C are sometimes required for maximum acclimation (Weiser, 1970). Short days also promote acclimation in several species, and there are indications that a stimulus may move from leaf tissue to the stems. The development of frost hardiness is a metabolic process requiring an energy source. Apparently this can be provided by light and photosynthesis. Factors that promote more rapid growth inhibit acclimation: high nitrogen in the soil, pruning, irrigation, and so on. In general, nongrowing or slowly growing plants are more resistant to several environmental extremes, including air pollution. Water-stressed plants are more resistant to air pollution partially because their stomates are closed.

Frost Killing It is "logical" that death of the plant results from its expansion upon freezing and subsequent disruption of cell walls and other anatomical features. Careful examination during the early decades of the nineteenth century showed, however, that plants actually contract rather than expand upon freezing. This is because ice crystals grow into the extracellular air spaces. Furthermore, ice is almost never observed within the living cells of tissues that have frozen naturally. (It is observed in the dead xylem cells of trees in winter.) Nor is damage to cells other than collapse observed. There is no rupture of cell walls or even cell membranes. With very rapid cooling of tissue in the laboratory (e.g., 20 C in 1 hour), ice can be seen within the cells, and damage to cellular components can be observed.

Such rapid cooling does occur in nature. Acclimated American arborvitae (*Thuja occidentalis*) tissues were capable of withstanding temperatures to –85 C when cooled slowly, but southwest facing foliage was injured when the temperature dropped 10 C per minute from 2 C to –8 C at sunset. Such changes duplicated in the laboratory also injured plants. Injury symptoms could not be duplicated by any form of desiccation, and it was concluded that the winter burn was caused by the rapid temperature drop (White and Weiser, 1964).

Typically, ice crystals begin to form in the extracellular spaces, and water from within the cells diffuses out and condenses on the growing ice crystal, which may become several thousand times as large as an individual cell. In frost-hardy plants, when these ice crystals melt, the water goes back into the cells, and they resume their metabolism. In nonacclimated plants, metabolism cannot be resumed, and the water does not reenter the cells completely. Again, there are significant differences in protoplasm between hardy and sensitive plants. Hardy plants frequently contain higher concentrations of solutes; their protoplasm is more elastic and remains more elastic during freezing. But the molecular basis of hardiness is not understood.

Charles R. Olien (1967, 1974) studied redistribution of water during freezing. Among other things, Olien examined the interference of freezing caused by large water-soluble polymers extracted from the cell walls of hardened plants. These substances interfered with freezing by competing with water molecules for sites in the ice lattice at the liquid interface. The result was that they tended to stop crystal growth, causing an imperfect ice mass to form. Polymers extracted from nonhardy winter cereals had little effect upon ice formation, but polymers from hardy varieties resulted in small, highly imperfect crystals within the plant. The polymers were mainly polysaccharides, containing large amounts of xylose and arabinose. Olien has developed a technique for screening for hardiness in seedling populations of winter cereals based on these differences. He freezes aqueous plant extracts in rotating drums and identifies hardy plant extracts by visual observation of the type of ice that is formed.

Chilling Damage As in our discussion of water stress effects, we are faced with the question of mild chilling effects as contrasted to the death caused by severe freezing. Many mechanisms have been proposed to account for chilling injury. Since chilling disrupts the entire metabolic and physiological processes in plants, it seemed almost futile to look for a single

key reaction that might be responsible. Nevertheless, it now appears that just such a key response might have been identified (Lyons, 1973). As the temperature is lowered in chilling-sensitive plants, lipids in cellular membranes solidify (crystallize) at a critical temperature that is determined by the ratio of saturated to unsaturated fatty acids. And this critical temperature proves to be equivalent to the temperature that causes chilling damage. As indicated, this can be as high as 10 to 13 C in sensitive species of tropical origin. Development of frost hardiness in chilling and frost-sensitive plants apparently involves changes in this ratio. An increase in the portion of unsaturated fatty acids results in the membrane remaining functional at lower temperatures.

The hypothesis suggests that the membrane normally exists in a liquid-crystalline condition. In this state, its enzymes have their optimal activity, and its permeability is thus under control. Below the critical temperature, it exists in a solid gel state. This change in state should bring about a contraction resulting in cracks or channels that lead to increased permeability. This would lead to the upset ion balances (ion leakage from chill-damaged cells or mitochondria) that are observed. Enzyme activities would also be upset, leading to imbalances with nonmembrane-bound enzyme systems. Thus, metabolites such as those produced in glycolysis would be expected to accumulate, because they could not be acted upon by the mitochondrial enzyme systems. Such accumulation had indeed been observed. Little ATP would be formed, because of the importance of membranes in its formation and because of these imbalances between the mitochondria and the glycolytic systems. Similar events would probably take place between the chloroplast and the cytoplasm around it.

If the temperature is raised soon enough, the membranes return to the liquid-crystalline state (since this phase transition is completely reversible), and the cell recovers. If metabolite buildup and solute leakage are allowed to occur to any great extent, however, then cells are injured or killed.

It has been observed that some cultivars are more sensitive to chilling than others, although their fatty acid ratios appear to be the same. These differences could be due to different sensitivities to the accumulated metabolites, rather than to the initial effects on the membranes. Chilling effects can often be avoided if tissues are exposed to high temperatures for brief intervals between chilling periods, provided that initial chilling was not too prolonged. In terms of the hypothesis given, this would allow the metabolism of accumulating metabolites, so that toxic levels are not allowed to develop.

24.4 The Alpine Tundra

Plants growing in the Arctic and above timberline in high mountains (Fig. 24-4) are exposed to cold stress even during the growing season (Billings, 1973; Billings and Mooney, 1968; Bliss, 1971; Tranquillini, 1964). Two factors in the alpine (mountain) tundra result from the increased elevations and are particularly important in controlling plant response: Light

Figure 24-4 The north end of the Neversummer Range in Colorado, showing treeline with eternal snows and alpine tundra above.

intensity increases due to a thinner layer of scattering atmosphere, and temperature decreases. Even when the sky is overcast, the diffused light measurable within clouds shrouding the tundra is much brighter than below the clouds in the valleys. Highest intensities are observed when the sky is partly cloudy, and direct rays of the sun and reflected rays from the clouds both fall on the same area. Intensities may be as high as 2.2 cal/cm²/min (solar constant = 1.94 cal/cm²/min), or 16,000 ft-c in one measurement when clear sky intensities equalled 13,000 ft-c (Salisbury and Spomer, 1964).

Warm air rises, expanding and cooling as it does so. The **adiabatic lapse rate,** or rate of cooling as dry air rises, is about −1.0 C/100 m. If the temperature gradient in the atmosphere is less than this, dry air will not rise by convection (Fig. 24-5). For example, given air at one elevation at 25 C, atmospheric stability will be achieved if air 1,000 m higher is 15 C or warmer. Hence, air on high mountains can remain cooler than air in the valleys. Because the layer of air between contains moisture and carbon dioxide, both of which absorb infrared radiation, plant and soil surfaces at low elevations will cool less rapidly by radiation into space, especially at night. As a result, alpine temperatures are virtually always lower than temperatures at lower elevations. Since plants in the Arctic tundra closely resemble those in the alpine tundra, and low temperature is the factor most common to the two habitats, this low temperature may well be the factor most responsible for the tundra type of vegetation.

Frequent high winds are also characteristic of most alpine areas. They are typically gusty rather than constant, and such

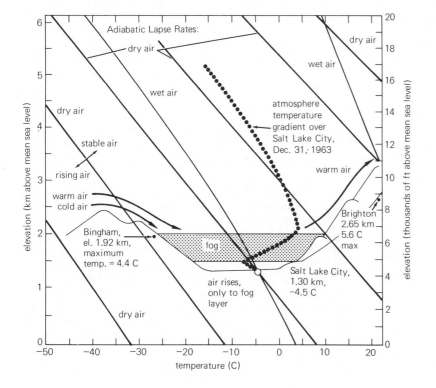

Figure 24-5 The adiabatic lapse rates. Unbroken lines indicate stable temperatures at various elevations. Note different slopes for lapse rates of wet and dry air. For wet air, the lines are slightly curved; for dry air, they are straight. Actual temperatures as a function of elevation are shown by the dotted line. Data were taken above Salt Lake City, Utah, on December 31, 1963. Gray part of figure schematically shows conditions in the Salt Lake Valley and vicinity on that date. As indicated by the temperatures, there was a strong inversion and much of the valley was full of fog, although air movement occurred upward in the bottom part of the valley (to the bottom of the inversion layer). Winds were moving from west to east (left to right on the figure). Cold air could penetrate the fog layer, but warm air moved above it. Nature of the inversion is strongly indicated by higher temperatures at Bingham and Brighton than in the Salt Lake Valley at a much lower elevation. (Curves from U. S. Department of Commerce, Weather Bureau chart, supplied by Arlo Richardson, who also furnished the data. See also Lois M. Cox, 1967, Utah science 28(2):53.)

Stress Physiology   **369**

a variability in level of a given environmental factor is also highly characteristic of alpine regions. Temperatures often fluctuate over a wide range during each daily cycle and over considerable ranges within relatively short intervals during the day or night. Light intensity fluctuates extremely on partially cloudy days. We might well expect that response of alpine plants is strongly adjusted to this fluctuating environment.

In many alpine regions, water is seldom limiting, and frequently humidities are relatively high. In ranges far inland such as the Rocky Mountains, however, periods of drought lasting several weeks may occur. Winter snow accumulation has a profound effect upon alpine plant distribution, since snowbanks may last late into the summer.

Transpiration and Water Balance Depending upon specific habitat in the tundra, some plants transpire much larger quantities of water than others. In some plants, there is a decrease in transpiration rate around noon. The actual amounts of water transpired per unit leaf area are comparable to those transpired by lowland plants. In the Austrian tundra, species that transpire the most still utilize only about one quarter of the water available throughout the year, and much precipitation falls during summer, so water is never limiting during the growing season. In the Colorado Rockies, however, periods of drought may limit plant growth. In late winter, water may become severely limiting to some alpine plants because the ground is frozen. Plants particularly sensitive to desiccation may require snow protection to survive (e.g., *Rhododendron* spp.), while others can be exposed during the entire winter. Many plants in the Colorado Rockies can survive winter exposure, particularly cushion plants such as *Silene acaulis* and even *Geum rossii* (a perennial rosette plant), which also grows in snow accumulation areas. Much of the Colorado tundra is blown free of snow during most of the winter and is snow covered only after spring storms, during which wet and heavy snow falls without much wind. The cushion plants tend to have low (negative) water potentials and thus low transpiration rates. This allows them to reach temperatures well above ambient air temperatures.

Frost Resistance It is difficult to separate the effects of winter drought from those of freezing. Frost hardiness is relatively low in tundra plants during the summer. Although summer frosts are common, temperatures drop at most to only a few degrees below freezing. Frost hardiness increases from midsummer on, reaching a peak at midwinter. Sugar concentrations and osmotic potentials parallel this, osmotic potential becoming more negative with the development of frost hardiness. There are exceptions, however, indicating that frost hardiness is probably not caused by this increase in solute concentration.

Pinus cembra may withstand winter temperatures of −42 C. Several other Austrian plants could withstand temperatures from −24 to −39 C, but others were severely damaged when temperatures dropped below −20 C. This latter group requires snow protection for survival.

Photosynthesis Photosynthesis of alpine plants is efficient compared to that of plants of lower elevations, although alpine plants are C-3 plants. Light intensities that bring about saturation are unusually high, and they become higher with increasing elevations where temperatures are lower. The optimum temperature for photosynthesis of alpine plants is lower than that for lowland species. Values for tundra plants often range around 12 to 15 C, whereas plants of warmer climates photosynthesize optimally at 25 to 30 C. Minimum photosynthesis temperatures are usually around −5 to −2 C, but photosynthesis has been detected at −14 C. Alpine plants can also utilize carbon dioxide more efficiently at lower partial pressures than can lowland plants.

The net photosynthesis (photosynthesis compared to respiration) varies as might be expected, but Tranquillini (1964) found that photosynthetic levels are greatly reduced on days following exceptionally cold nights, particularly if frost occurred during the night. This response could establish the upper tree limit (timberline). If trees cannot produce more carbohydrate by photosynthesis than is used up in respiration, they cannot build new wood and therefore cannot survive. In the spring, respiration capacity is restored much more rapidly than photosynthetic capacity.

Many alpine plants are evergreens, with leaves ready to grow in spring. Even deciduous species have unusually rapid leaf growth rates at low temperatures. Respiration is also efficient at low temperatures, producing the ATP and carbon skeletons essential for rapid growth. Translocation of assimilates is also unusually active at low temperatures. The overriding adaptation of tundra plants is their ability to utilize even short periods above freezing for growth and development—just as xerophytes utilize brief periods when moisture is available.

Growth and Development W. Dwight Billings at Duke University and his students collected examples of *Oxyria digyna* (mountain sorrel) from several latitudinal locations (Sections 22.3 and 23.2.9). All were long-day plants, but those collected from more northern latitudes had much longer critical daylengths. George G. Spomer and Salisbury (1968) studied the physiological ecology of *Geum rossii*. They demonstrated the importance of root temperature in growth of the plant and studied effects of light and temperature on dormancy and flowering. In growth chambers, plants went into dormancy only when daily average temperatures were 8 C or less, yet in the field the same dormancy symptoms became apparent during the middle of August when temperatures were at or near the maximum for the season. There was no daylength effect on dormancy or flowering. Flower primordia are formed in the field during July, and mature flowers appear immediately after plants begin to grow the following spring. Yet no combinations of temperature or daylength would result in flower primordia formation in growth chambers. Several intriguing problems await study in this area.

24.5 Salt

A widespread factor restricting growth in many temperate regions is soil salinity. Millions of acres have gone out of production as salt from irrigation water accumulates in the soil. The plant faces two problems in such areas, one of obtaining water from a soil of negative osmotic potential and another of dealing with the high concentrations of potentially toxic sodium carbonate and chloride ions. Some crop plants (e.g., beets, tomatoes, rye) are much more salt hardy than others (e.g., onions, peas), and salt hardiness can be increased somewhat by exposure to saline conditions. Salt hardening is minimal, however, compared to drought or cold acclimation.

In the study of salt hardiness, the **obligate halophytes** are particularly interesting. Several such species will grow only where salt levels in the soil are high, as in deserts or near-brackish waters on the sea coasts or close to the shores of

extremely salty waters such as the Great Salt Lake, where the salt content may be saturated at levels as high as 27 percent by weight. The following genera are good examples: *Allenrolfea* (iodinebush), *Salicornia* (the pickleweed or samphire), and *Limonium* (sea lavender, marsh rosemary). *Atriplex* (shadscale) and *Sarcobatus* (black greasewood) also grow in somewhat less salty soils, and certain bacteria and blue-green algae live in the waters of the Great Salt Lake. In the case of the terrestrial halophytes, the osmotic potential of leaf cell sap is invariably highly negative. Tissues from actively growing *Atriplex* species, having no special cold hardiness, for example, freeze only when temperatures drop below −14 C, implying that their osmotic potentials are as low as about −170 bars. This contrasts with a normal −20 to −30 bars in most plants. In some cases, the xylem sap does not have a highly negative osmotic potential but may be almost pure water. To obtain water from the surrounding soil, water potential within the xylem sap must then be greatly lowered by tension. This was demonstrated by Scholander and his coworkers (1965) for mangrove trees (Section 4.6).

Some halophytes are referred to as **salt accumulators.** In these plants the osmotic potential continues to become more

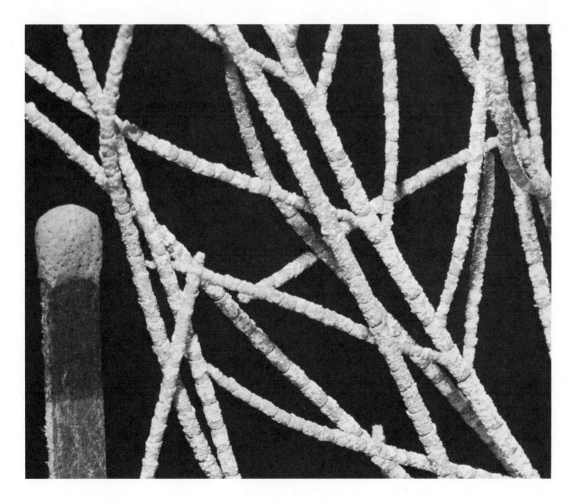

Figure 24-6 Tamarisk leaves (*Tamarix pentandra*) collected at Barstow in the Mojave Desert, California, showing heavy incrustations of salt. The paper match indicates the scale.

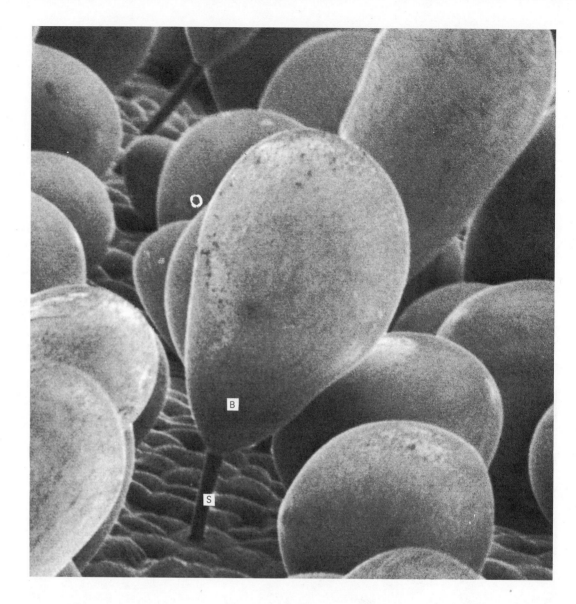

Figure 24-7 A scanning electron micrograph of salt bladders on the leaf of saltbush (*Atriplex spongiosa*). Saltbush is indigenous to Australia and is a salt-tolerant species because it has developed a mechanism to control the Na^+ and Cl^- concentrations of its tissues. The epidermal bladders on the surface of the aerial parts of the plant are specialized cells that accumulate salt. Salt from the leaf tissues is transferred through the small stalk cell (S) and into the balloonlike bladder cell (B). As the leaf ages, the salt concentration in the cell increases, and eventually the cell bursts or falls off the leaf, releasing the salt outside the leaf. ×450. (From John Troughton and Lesley A. Donaldson, 1972, Probing plant structure, McGraw-Hill Book Company, New York.)

negative throughout the growing season as salt is absorbed. Even in these plants, however, the soil solution is not taken directly into the plant. It is easy to calculate, based upon quantities of water transpired by the plant, that if the complete soil solution were absorbed, the plant would contain 10 to 100 times as much salt as is actually observed. Instead, water moves into the plant osmotically and not simply in bulk flow. The endodermal layer in the roots probably provides the osmotic barrier.

Halophytes in which the salt concentration within the plant does not increase during the growing season are known as **salt regulators.** Often salt does enter the plant, but the leaves swell by absorbing water, so concentrations do not increase. This leads to the development of **succulence** (a high volume/surface ratio), a common morphological feature of halophytes. Sometimes excess salt is exuded on the surface of the leaves, helping to maintain a constant salt concentration within the tissue (Fig. 24-6). In certain halophytes there are readily observable salt glands on the leaves, sometimes consisting of only two cells (Fig. 24-7). Although Na^+ ions are essential for some salt-tolerant species, it is probable that sodium pumps in cell membranes actively transport much of the ion out of the cytoplasm of both root and leaf cells, *inwardly* to the central vacuoles and *outwardly* to the extracellular spaces.

Actually, large quantities of both organic and inorganic materials may be leached from the leaves of many plants, both halophytes and glycophytes. Some of the leaching brought about by washing the leaves may be due to removal of materials from within the tissue as well as washing off materials that have been exuded at the surface. In any case, these materials may be washed from the leaves to the soil (see Fig. 24-1) and thus recycled within an individual plant and to other plants. This certainly has important ecological implications, particularly in relation to mineral nutrition.

Frequently, halophytes synthesize large quantities of the amino acid proline (Fig. 24-8), which accumulates in the vacuoles and perhaps also in the cytoplasm of the cells. This results in a highly negative osmotic potential without the toxic effects of the sodium ion. Proline accumulation is widespread in both angiosperms and gymnosperms subjected to water deficiency, and many algae that grow in saline waters also synthesize high concentrations of proline. Yet the halophyte *Plantago maritima* does not accumulate proline during salt stress, and other compounds such as galactosyl glycerol and organic acids sometimes function in osmotic adjustment (Hellebust, 1976). Proline accumulation seems to occur strictly as a response to salt or water stress, and there is presently little evidence that species normally having high proline levels are any more tolerant to low water potentials than others.

Another potential problem for plants growing on saline soils is obtaining enough potassium. This is because sodium ions compete with the uptake of K^+ by a low affinity mechanism (Chapter 5), and K^+ is commonly present in such soils in much lower concentrations than is Na^+. In this respect, the presence of Ca^{2+} appears to be crucial. If sufficient calcium is present, a high affinity uptake system having preference for transport of K^+ can operate well, and the plants can then obtain

sufficient potassium and restrict sodium (LaHaye and Epstein, 1969). It is possible that fertilization of some saline soils with Ca^{2+} might increase their agricultural productivity. A favorable effect of Ca^{2+} on soil structure could also be important. Gypsum ($CaSO_4$) is sometimes used, providing both Ca^{2+} and some acidity, which helps in leaching out the Na^+. Elemental sulfur is also sometimes applied. It oxidizes to produce sulfuric acid, which aids in leaching.

24.6 Soil *pH*

Plants are found growing on soils over a *pH* range of at least 3 to 9, and the extremes provide another stress to which some species are adapted. Cranberries, for example, grow on acid bogs, while certain desert species normally grow only on high *pH* soils. In general, we know far too little about why some plants are native to low *pH* soils and others to soils with higher *pH* values. Certainly one of the reasons is competition. If we use hydroponic techniques to study the growth of various species apparently preferring different *pH* values, we usually find that they do reasonably well over a fairly wide *pH* range. But in nature even a slight advantage of one species over another can eventually lead to elimination of the less well-adapted one.

Soil factors closely correlated with *pH* are probably more important than the concentration of H^+ ions *per se*. For example, high rainfall leads to leaching of calcium and formation of acidic soils, so calcium is usually low in acidic soils and abundant in soils of high *pH* (calcareous soils). Rather high concentrations of this element favor development of root nodules on many legumes (Chapter 13), so nitrogen-fixing legumes will grow better on soils rich in calcium than on most acidic soils. The less abundant calcium on acidic soils may also limit plant growth simply because H^+ is much more toxic to roots in the absence of calcium. One of the beneficial effects of liming acid soils no doubt derives from this fact.

The *pH* also strongly influences the solubility of certain elements in the soil and the rate at which they are absorbed by plants. Iron, zinc, copper, and manganese are less soluble in alkaline than acidic soils because they precipitate as hydroxides at high *pH*. Iron deficiency chlorosis is thus common on soils in the western United States, which are often alkaline. Phosphate, absorbed largely as the monovalent $H_2PO_4^-$ ion, is more readily absorbed from nutrient solutions having *pH* values of 5.5 to 6.5 than at lower or higher *pH* values. In soils of high *pH*, more of the phosphate is present as the less readily absorbed divalent $HPO_4^=$ ion. Furthermore, much of this is usually present as insoluble calcium phosphates. In soils of low *pH* where $H_2PO_4^-$ should predominate, the frequent high concentrations of aluminum ions cause its precipitation as aluminum phosphate.

The relatively high concentrations of available aluminum in many acidic soils (less than about *pH* 4.7) can inhibit growth of some species not only because of detrimental effects on phosphate availability but apparently also by inhibiting ab-

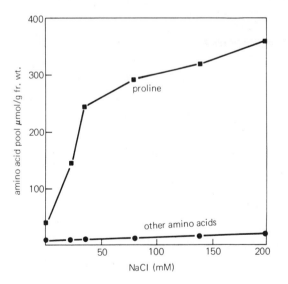

Figure 24-8 Amino acid and proline accumulation in *Triglochin maritima* grown at different salinities. Cuttings of *T. maritima* were grown in nonsaline media for two weeks before being transferred to saline medium. Shoot tissue was harvested for analysis after 10 days of saline treatment. (From Stewart and Lee, 1974.)

sorption of iron and by direct toxic effects on plant metabolism. Some species (e.g., azaleas) not only tolerate these high aluminum concentrations but thrive on such soils. Still other species tolerate amounts of various heavy metals that are toxic to most plants. An example is bentgrass (*Agrostis tenuis*) that grows in Wales and Scotland on mine tailings having unusually large amounts of lead, zinc, copper, and nickel. This grass does not exclude such toxic metals but somehow accumulates them without being injured appreciably. We don't understand the tolerance mechanism, although it was suggested that specific chelating agents (e.g., in root cell walls) form strong complexes with the metal ions and prevent their reaction with sensitive protoplasmic constituents such as enzymes. Secretion of these metals into the vacuoles would also decrease their toxic effects.

24.7 A Final Note

We began this text by discussing water; now it is apparent that in several stress situations, negative water potentials are involved. In drought, this is obvious. In frost, water is removed from the cells, resulting in more negative water potentials within the protoplast. High salt concentrations have the same result. Yet high-temperature damage is probably not closely related to negative water potentials, although it frequently accompanies drought. Effects of pH are not obviously related to water, and the earliest effects of drought, chilling, and salt seem to differ from each other, as we have seen. But water always seems to be important, and plants that have become acclimated to one stress are often hardy to other stresses as well.

Section Five

Special Topics

A cluster of *Pelargonium* (geranium) flowers.

Special Topic 1.1
The Water Milieu

Plant physiology is, to a surprising degree, the study of water. Since water is fluid and virtually incompressible, and since much of a plant is water, a plant is a hydraulic system. When the water pressure drops within soft tissues such as leaves, the tissues collapse; we say they wilt. A dried and shrunken mummy testifies that even animals depend upon internal water for their normal appearance. Plants grow as the water pressure in their cells increases, causing the cells to expand. Many plant parts such as petals and leaves move as water moves in or out of special cells. For example, the leaves of the sensitive plant (*Mimosa* sp.) fold as water moves out of special cells at their base. Stomates on leaf surfaces open as their guard cells take up water and close when water moves out of these cells. The plant's hydraulic system also includes transport mechanisms: Dissolved substances move passively as the water flows through the conducting elements in the wood and the bark. Dissolved materials also diffuse in the water within the plant and through the many membranes around and within each water-filled cell.

Furthermore, a plant or animal in its environment is a strong expression of certain more subtle properties of water than its fluidity and incompressibility. Water has a high heat capacity and high heats of fusion and vaporization. Thus, a plant or animal (made mostly of water) can absorb relatively large inputs of heat energy with relatively minor changes in temperature, especially if melting or evaporation are involved. Indeed, much of the heat absorbed by a plant in the sun is returned to the environment via the transpiration of water from the plant. Some species have an amazingly efficient system of evaporative cooling.

But water in the plant is much more than a hydraulic, transport, and temperature-regulating medium. Protoplasm itself is an expression of the properties of water. Its protein and nucleic acid components owe their molecular structures, and hence their biological activities, to their close association with water molecules. Indeed, virtually all the molecules of protoplasm owe their specific chemical activities to the water milieu in which they exist. Exceptions are the molecules contained in cellular oil droplets or the liquid (fatty) portions of membranes—but the droplets and the membranes are themselves strongly influenced by their water milieu. Last but not least, water molecules enter actively into the chemistry that is life. Along with carbon dioxide molecules, they constitute the raw materials of photosynthesis, for example. Little metabolism would be possible without the utilization or production of water molecules.

The Hydrogen Bond, Key to Water's Properties

Most of the unique properties of water can be ascribed to the interesting fact that the lines connecting the centers of the two hydrogen atoms with that of the oxygen atom do not form a straight line. Instead, they form an angle of about 105 deg—closer to a right angle than to a straight line (Fig. 1.1-1). The angle is exact in ice but only an average in liquid water. The two electrons that fill the first shell of the hydrogen atom (the hydrogen's own and the other borrowed from the oxygen) are not equally distributed around the hydrogen nucleus but are usually closer to the oxygen nucleus. Thus, the hydrogen atoms approximate naked protons on the surface of the oxygen atom. While the net charge for the molecule as a whole is, of course, neutral, the protons distributed 105 deg apart

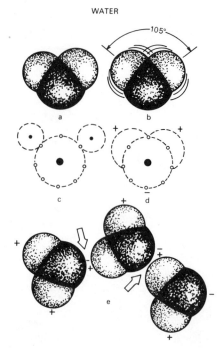

WATER

Figure 1.1-1 Water molecules, indicating (a and b) the angle between the two hydrogen atoms (lighter) attached to the oxygen atom (darker). As shown, the angle is not absolutely stable (b) but represents an average sharing of electrons and distribution of charge (c and d). Attraction of the negative side of one water molecule to the positive side of another produces the hydrogen bond (e). Arrows indicate hydrogen bonds.

on the surface of the oxygen atom provide a slight positive charge on one side of the molecule. This is balanced by an equal negative charge on the other side of the molecule. Thus, the molecule is said to be **polar.** The result is that the positive side of one molecule is attracted to the negative side of another (Fig. 1.1-1). The bond formed this way between the two molecules is called a **hydrogen bond.**

The strongest chemical bonds are **ionic bonds,** in which an electron moves from one atom and becomes attached to another. This is the situation in a crystal of sodium chloride, for example, where sodium atoms each lose an electron (to become a sodium **ion**) and chlorine atoms each gain an electron (making a chloride **ion**). The energy in the ionic bonds of NaCl is equivalent to 183 kcal/mol. Representative ionic bond strengths are 139 kcal/mol for CsI to 240 kcal/mol for LiF. The strengths of **covalent bonds,** in which electrons are shared by two atoms, overlap with these but are usually weaker. Representative values (all given in kcal/mol) are 33 for the O—O bond, 70 for C—N, 83 for C—C, 84 for C—O, 99 for C—H, 110 for O—H (as in water), 145 for C=C, and 198 for C≡C. **Hydrogen bond** strengths usually lie in the range of 2 to 10 kcal/mol, depending upon the atom that bonds with the hydrogen atom. The bonds of hydrogen with various atoms have decreasing strengths in the order shown: F, O, N, Cl, and S. Hydrogen bonds with O and N are especially important in plants.

Van der Waals attractive forces are the weakest of all, having bond strengths of about 1.0 kcal/mol. In neutral, nonpolar molecules, they result from the fact that electrons are continually in motion so that the molecule's center of negative charges does not always correspond with its center of positive charges. Thus, as two like molecules approach each other very closely, they may induce slight polarizations in each other with the ends of unlike charge attracting each other. Such forces hold the molecules together in the liquid hydrocarbons, for example, but also in membranes and the internal parts of proteins. All these molecular bonds are electrical in nature, although ionic are strongest and van der Waals bonds are weakest.

Some Properties of Water Important to Life

Liquid at Room Temperature As a general rule, the higher the molecular weight of an element or a compound, the greater the likelihood that this material will be a liquid or a solid at a given temperature—say, room temperature. The lower the molecular weight, the greater the likelihood of its being a liquid or a gas. The larger the molecule, the more energy (heat) is required to cause it to break its binding forces with surrounding molecules and become a liquid or a gas. For example, the low-molecular-weight hydrocarbons are gases or liquids at room temperature, while the higher ones are liquids or solids (Fig. 1.1-2). But compare water, having a molecular weight of 18, with ammonia (mol. wt. = 17) or methane (mol. wt. = 16), or even carbon dioxide (mol. wt. = 44). Only water is a liquid at room temperature, while the others are gases and must be cooled to very low temperatures before they become liquids or solids. The explanation is the hydrogen bond, which provides a disproportionately high attractive force among water molecules, preventing them from separating and escaping as vapor. The hydrocarbons have only van der Waals forces among molecules in the liquid state.

It is interesting to note that other liquids with low molecular weights are also polar molecules with hydrogen bonding between them. Good examples are the lower alcohols (methyl = CH_3OH) or the lower organic acids (formic = $CHOOH$, or acetic = CH_3COOH). The presence of the oxygen atom makes hydrogen bonding possible in these compounds.

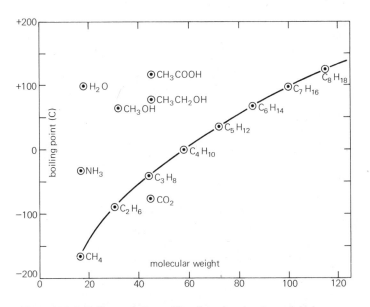

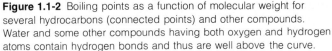

Figure 1.1-2 Boiling points as a function of molecular weight for several hydrocarbons (connected points) and other compounds. Water and some other compounds having both oxygen and hydrogen atoms contain hydrogen bonds and thus are well above the curve.

Specific Heat At 14.5 C, exactly 1 small **calorie** (cal) is required to raise 1 g of pure water 1 C (by definition; note further that a kilocalorie or large Calorie = 1,000 small calories). The amount of energy required to raise the temperature of a substance 1 C is called its **specific heat.** This value of 1 cal varies only slightly for water over the entire range of temperatures at which water is a liquid, and it is higher than that measured for any other substance except liquid ammonia. The high value comes about because the arrangement of molecules in liquid water is such that the hydrogen and oxygen atoms are allowed to vibrate very freely, almost as if they were free ions. Thus, they can absorb large quantities of energy without much temperature increase. Plants (think of a large succulent cactus) and animals consist largely of water and thus have relative temperature stability.

Latent Heats of Vaporization and Fusion Some 586 calories are required to convert 1 g of water to 1 g of water vapor at 20 C. This unusually high **latent heat of vaporization** can again be ascribed to the tenacity of the hydrogen bond. It is important in cooling leaves by transpiration.

To melt 1 g of ice at 0 C, 80 cal must be supplied. This is also a high **latent heat of fusion,** due again to the hydrogen bonds, although ice has fewer per molecule than water. Each H_2O molecule in ice is surrounded by *four* others, forming a tetrahedral structure (each oxygen atom attracts two extra hydrogen atoms). The tetrahedrons are arranged in such a manner that the ice crystal is basically hexagonal, as demonstrated in the pattern of snowflakes. As is the usual situation during conversion from the solid to the liquid state, molecules of water move farther apart during melting (from an average distance of 2.72 A to 2.90 A). Yet water is extremely unusual because its total volume decreases at melting. This is because the molecules are packed more efficiently in the liquid than in the solid. Each molecule in the liquid is surrounded by *five* or more others (each oxygen attracts three or more extra hydrogens).

The result of this packing difference is that water expands as it turns to ice, so that ice has a lower density than water. This keeps ice floating on top of lakes in the winter rather than going to the bottom where it might remain without thawing through the coming summer. The expansion may also result in damage to plant or animal tissue during ice crystal formation (but usually doesn't; see p. 368). Because water expands upon freezing, increased pressure will make ice melt at a *lower* temperature (pressure lowers the melting point). With other substances, increased pressure usually results in a raised melting point.

Viscosity Since hydrogen bonds must be broken for water to flow, one might expect a viscosity or resistance to flow considerably higher than that actually encountered. But in liquid water each hydrogen bond is shared on the average by two other molecules, and thus the bonds are somewhat weakened and fairly easily broken. Water can flow readily through plants. In ice there are fewer bonds per oxygen; hence, each is stronger. Viscosity of water is strongly influenced by temperature (Table 1.1-1).

Adhesive and Cohesive Forces of Water Because of its polar nature, water is attracted to many other substances; that is, it wets them. The cellulose, starch, and protein of living tissues constitute excellent examples. This attraction between unlike molecules (water and other molecules) is called **adhesion.** It often involves hydrogen bonding between the water and the other molecules. The attraction of water molecules for each other (because of hydrogen bonding) is called **cohesion.** It bestows upon water an unusually high tensile strength. In a thin, con-

Table 1.1-1 Viscosities of Fluids

Fluid	Temperature	Coefficient of Viscosity[a] (Centipoises)	Percent of H_2O Viscosity at 20 C
Water	0 C	1.787	177
	10 C	1.307	130
	20 C	1.002	100
	30 C	.7975	80
	40 C	.6529	65
	60 C	.4665	47
	80 C	.3547	35
	100 C	.2818	28
Ethyl alcohol	20 C	1.20	120
Benzene	20 C	.65	65
Glycerine	20 C	830.0	83,000
Mercury	20 C	1.60	160
Machine oil	19 C	120.0	12,000

[a]In centipoise = poise multiplied by 100. One poise represents a force of 1 dyne per cm² required to displace a large plane surface in contact with the upper surface of a layer of liquid 1 cm thick over a distance of 1 cm in 1 sec. A more convenient method of measurement notes the time required for a given volume of liquid to flow by gravity through a tube of given dimensions and then applies a suitable equation.

fined column of water such as that in the xylem elements of a stem, this tensile strength may reach high values, allowing water to be pulled to the tops of tall trees.

Cohesion between water molecules also accounts for **surface tension.** The molecules at the surface are continually being pulled into the liquid by the cohesive (hydrogen bond) forces. The result is that a drop of water acts as if it were covered by a tight elastic skin. Indeed, it is the surface tension that makes a falling drop round. The surface tension of water is higher than that of virtually all other liquids. Surface tension often plays a role in the physiology of plants. For example, the passage of air bubbles through openings in cell walls may be prevented because surface tension forces are too great to allow deformation of the bubble.

Water as a Solvent Water will dissolve more substances than any other common liquid. This is partially because it has one of the highest known **dielectric constants,** which is a measure of the capacity to neutralize the attraction between electrical charges. Because of this property, water is an especially powerful solvent for electrolytes. The positive side of the water molecule is attracted to the negative ion and the negative side to the positive ion. Water molecules thus form a "cage" around the ions so that they are unable to unite with each other and precipitate. Actually, water proves to be a surprisingly good solvent for certain nonpolar molecules as well. Methane gas is a good example; though only slightly soluble, it is much more soluble than might have been expected. Its molecules are exactly the right size to fit into certain "holes" or "cages" in the molecular structure of water.

If water contains dissolved electrolytes, then these will carry a charge, and water becomes a good conductor of electricity. If water is absolutely pure, however (and pure water is extremely difficult to obtain), then it is a very poor conductor. Hydrogen bonding makes it too rigid to carry a charge readily.

Although water is the best solvent known, it is still rather inert and innocuous, entering readily into only a relatively few chemical reactions. This allows it to function even better as the milieu for protoplasmic activities.

Ionization of Water and the pH Scale Some of the molecules in water separate into hydrogen and hydroxyl ions. The tendency for these ions to recombine is a function of the chances for collisions between the ions; that is, recombination depends upon the relative number of ions present in the solution. This **mass law relationship** may be expressed mathematically by saying that the product of the molal concentrations equals a constant: $[H^+] \cdot [OH^-] = K$. Near room temperature, $K = 10^{-14}$, so in pure water both $[OH^-]$ and $[H^+] = 10^{-7} M$. (To *multiply* 10^{-7} by 10^{-7}, exponents are *added* to give 10^{-14}.) Water is seldom pure enough to allow such an exact distribution between hydrogen and hydroxyl ions. Presence of dissolved carbon dioxide, the usual case, may raise the hydrogen ion concentration as high as $10^{-4} M$. The hydroxyl ion concentration is then determined at $10^{-10} M$, since the product of these two concentrations must equal 10^{-14}.

The hydrogen ion concentration is indicated by the **pH scale**, on which $pH = -\log [H^+]$. Another way to state this is that pH equals the absolute value of the hydrogen ion concentration expressed as a negative exponent of 10. For example, when $[H^+] = 10^{-4} M$, then $pH = 4$. Neutrality is expressed by $pH = 7$, and values below 7 indicate increasing acidity; above 7, increasing alkalinity. We should remind ourselves that the pH units are multiples of 10 on a logarithmic scale and should therefore not be added together or averaged. Only a tenth as many H^+ need to be added to a solution to change the pH from 7 to 6 as from 6 to 5.

Light Absorption by Water Water is essentially transparent to visible light, although the slight absorption of red makes large bodies of it appear bluish green. The hydrogen bond absorbs very effectively at infrared wavelengths of about 3 micrometers (μm), and long wavelength thermal radiation (10 to 30 μm) is also strongly absorbed by water molecules (and by most other substances). This is important in the absorption of radiant heat energy by water vapor in the atmosphere or by water in the plant itself. The near-ultraviolet portion of the spectrum penetrates pure water, but far ultraviolet is absorbed.

Water and Life

Would life on another planet also be based on water? Or could we imagine a life form based upon liquid ammonia, for example? Perhaps, but considering the many special properties of water and the widespread occurrence of hydrogen and oxygen in the universe, it is probably safe to predict that life as a general rule depends upon water. Clearly, water is uniquely well suited to the functions of living things.

Special Topic 2.1
Colloids

Protoplasm—life, that is—is characterized by two special features: highly complex and unique molecules and a predominance of surfaces or interfaces where numerous reactions and interactions take place.

The physical nature of the surfaces in plants and the soils in which they grow is the main subject of this essay. Such interfacial systems are not restricted to soils and protoplasm, however, although they reach their highest levels of complexity there. Even modern technology takes advantage of such systems in its water softeners, catalytic converters, and numerous other applications.

Colloids: Minute Particles with Vast Surfaces

The special physical nature of protoplasm is due largely to the presence of particles too small to settle out by gravity, but larger than the atoms, small molecules, and ions that are true solute particles. These larger particles suspended in water sometimes form a glue, so they have been termed **colloids** from the Greek word *kolla,* or "glue."

Why don't colloids settle out? They are surrounded in water by much smaller water molecules, each of which is in motion, as discussed in Chapter 1. The result is that a colloidal particle is never at rest. It is small enough that the random velocities of the impacting molecules do not average out on its surface, so there is always a high probability that such a particle will be bombarded more strongly on one side than on the other. Thus, when such particles are observed in a light microscope by illumination with a strong beam of light from the side so that they appear as points of light (the **Tyndall effect,** first noticed by John Tyndall, 1820–1893), they appear to dance around with many random jerks per second. The largest (brightest) particles dance less than the smaller (dimmer) particles.

This **Brownian movement,** discovered by the Scottish botanist Robert Brown in 1827, is a beautiful and even spectacular confirmation of kinetic theory. It is this erratic and continuous motion that keeps colloidal particles from settling. Indeed, we might define such a particle as one that is small enough to remain in suspension due to its Brownian movement. Slightly larger particles are less influenced by the random bombardments of the molecules making up their milieu, so gravity wins out and they settle.

The upper size limit (diameter) for particles exhibiting Brownian movement proves to be about 100 to 2000 nanometers (nm), depending on their shape and density. Since light waves in the visible part of the spectrum are 385 to 776 nm long, only the larger colloidal particles are "large enough to cast a shadow." The smaller ones cause light waves to be refracted, accounting for the Tyndall effect, but they are not actually visible in the light microscope. The electron microscope, on the other hand, which operates with electron beams having wavelengths less than 0.1 nm, easily resolves most colloidal particles, the smallest of which have diameters of about 10 nm.

Actually, the size of the smallest objects visible with an electron microscope is limited by our ability to make magnetic lenses and by the power of the beam rather than its wavelength. An electron beam destroys most small organic molecules, but by coating these molecules with heavy metals and by using other special preparative techniques, it has been possible to resolve particles less than about 2 nm in diameter; some workers claim resolutions down to 0.2 to 0.5 nm. In any case, it is important to realize that many of the particles in a cell, including the ribosomes and all the single protein molecules that act as enzymes, are in the colloidal size range.

Most colloidal particles will pass through filter paper but not through the pores of cellophane, as will true solute particles. Suspension particles are too large to pass through filter paper.

Although colloidal particles are small, they are large enough to present a *surface* (a layer of atoms) to their surrounding water molecules and the solute particles dissolved in them. And, almost paradoxically, the consequence of their small size is that the amount of surface they

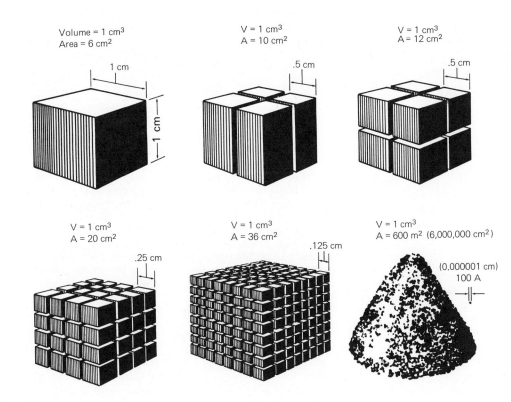

Volume = 1 cm³
Area = 6 cm²

1 cm
1 cm

V = 1 cm³
A = 10 cm²

.5 cm

V = 1 cm³
A = 12 cm²

.5 cm

V = 1 cm³
A = 20 cm²

.25 cm

V = 1 cm³
A = 36 cm²

.125 cm

V = 1 cm³
A = 600 m² (6,000,000 cm²)

(0.000001 cm)
100 A

Figure 2.1-1 The relationship between surface area and volume as particle size decreases.

present is relatively huge. Imagine a solid cube of material 1 cm long on each of its edges (as in Fig. 2.1-1). There are six faces, so it has a surface area of 6 cm². Cut it once. You expose 2 cm² more of surface. Continue slicing until you have reduced each particle to a cube 10 nm long on each of its edges. Now the total surface area exposed is 6,000,000 cm² (600 m²). A single cube with the same surface area would be 10 m high with a volume of 1,000 m³! Protein molecules are not cubes, but they are of comparable size. The reactions of life occur on surfaces. It is easy to see how relatively large surface areas can exist in a single cell.

Colloidal Surface Charge and Its Consequences

The surfaces of most colloidal particles are electrically charged. Proteins have ionic groups (carboxyl, amino, and others) on their surfaces as a result of the molecular structures of the amino acids that make up protein (see Chapter 8). Clay particles also have surface charges, and the microfibrils in cell walls (Chapter 10) are also associated with charged groups (especially pectins). Consider a protein molecule or a clay particle (Fig. 2.1-2) with its predominantly negative surface charges. which is the typical situation in nature. Since other similar particles are also negatively charged, all repel each other. Thus, they will not contact and stick to each other but will remain in suspension.

The negative charges on the surface of a colloidal particle will attract positive ions in solution and also the slightly positive sides of water molecules by hydrogen bonding. The positive ions in turn attract negative ions, and the exposed negative sides of the water molecules attract the positive sides of other water molecules. Thus, layers of water molecules with intermingled solute ions are built up around a colloidal particle. This layer of adsorbed water cushions the particles from each other and helps maintain them in suspension.

Ions are not all attracted equally to the surfaces of colloidal particles or to other charged surfaces. The common ions have been arranged according to the **Hofmeister** or **lyotropic series,** with the most tenaciously adsorbed cations listed first as follows:

$$Al^{+3} > H^+ > Ba^{+2} > Mg^{+2} > Ca^{+2} > Mg^{+2} > K^+ = NH_4^+ > Na^+ > Li^+$$

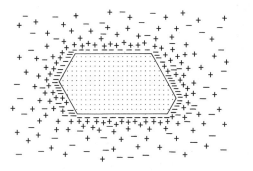

Figure 2.1-2 The electrical double layer of ionic distribution around a charged colloidal particle. Charges next to the surface of the particle are meant to indicate surface charge. This is mostly negative but is interrupted at three points with positive charge. Distribution of ionic charge is statistically opposite to the distribution of surface charge, but a few irregularities are apparent.

When one is dealing with the minerals readily available in soils, the list is shortened to:

$$H^+ > Ca^{+2} > Mg^{+2} > K^+ = NH_4^+ > Na^+$$

The attraction of anions to positively charged surfaces has also been studied. The situation is somewhat more complex, however, and it is of less importance for our applications.

The lyotropic series has two important consequences: First, if enough ions become adsorbed on colloidal particles, they will neutralize the surface charge, thereby removing the protective layers of water and neutralizing the particles so that they no longer repel each other. The particles then contact each other, sticking together by van der Waals forces until aggregates are large enough to settle in response to gravity; they **precipitate out** or **flocculate.** The ions that are most tenaciously adsorbed on colloidal surfaces (e.g., Al^{+3}, H^+, Ca^{+2}) are most efficient in causing flocculation. That is, when a colloid is treated with **equivalent concentrations** (equal numbers of charges per unit volume of water: $6H^+ = 3Ca^{+2} = 2Al^{+3}$), the most efficient ones in the lyotropic series cause the most flocculation.

A second consequence is **cation exchange.** The most efficient ions in the lyotropic series will replace less efficient ions when both are present at equivalent concentrations. Consider, for example, a cation exchange reaction occurring on a clay particle in a soil:

$$2NH_4^+ + clay \cdot Ca \rightleftharpoons Ca^{+2} + clay \cdot (NH_4)_2$$

Calcium is more tenaciously adsorbed than ammonium at equivalent concentrations, so the reaction tends to move strongly to the left. If the ammonium concentration is increased far above the calcium concentration, however, the reaction may be driven toward the right.

The negative surfaces of clay particles, cell walls, proteins, and so on are all capable of this cation exchange, and this is often of considerable importance. Much of soil science is based upon the cation exchange properties of clay colloids. In areas of high rainfall, for example, the incoming moisture contains dissolved CO_2, which reacts with the water to produce carbonic acid (H_2CO_3). The hydrogen ions (H^+) of carbonic acid are more effective in the lyotropic series than any other ion except Al^{+3}, so they gradually tend to replace all the other ions in the soil as these are produced by breakdown of minerals. The result is that in areas of high rainfall, such as the tropical rainforest, the soil becomes highly acidic, and most nutrient ions are leached away to the water table. Many of the ions that remain are held in the vegetation. Cation exchange also occurs in cell walls and within the protoplasm of plants.

A home water softener is based on the cation exchange principle. A highly effective, colloidal-like material (usually zeolite) is first charged with a high concentration of sodium ions from common salt, which displace other cations. As water is then run through the system, "hard" calcium ions replace the "soft" sodium ions on the exchange material. Calcium ions are "hard" because they form an insoluble calcium soap (the ring around the bathtub), while sodium ions form a water-soluble soap that runs down the drain. Ion exchange purifiers are also used in the laboratory to produce highly pure water and to separate various organic constituents.

In systems where relatively large, nonmobile, charged surfaces exist, such as in cells (especially plant cells with their walls), the concentrations of electrolytes become influenced by these surfaces. If the surface charges are negative, cations tend to build up close to the surface, while anions are decreased in concentration. This has a number of important consequences, including the establishment of pH gra-

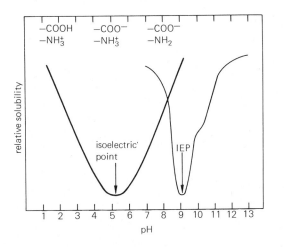

Figure 2.1-3 The relative solubility of amphoteric colloids as a function of pH. The two curves represent two proteins. Note that specific shape of a curve will depend upon the specific protein, as will the exact position of the isoelectric point. Conditions of free carboxyl and amino groups of protein are indicated at the top of the graph.

dients and gradients of anions and cations. Gradients of water potential result from these solute gradients.

Often the charges on the surface of a colloid are strongly influenced by pH. This is especially true with proteins, which have free carboxyl, amino, and other ionizing groups on their surfaces. If the pH is low (acidic), the excess hydrogen ions tend to neutralize the carboxyl groups and to become attached to the amino groups, giving them a positive charge. Thus a protein in a solution with a pH below its neutral point has a positively charged surface rather than negative. At higher pH, the hydroxyl (OH^-) groups react with the hydrogen ions of both carboxyl and amino groups, leaving the amino groups neutral and the carboxyl groups negatively charged. Thus, in either highly acidic or highly basic solutions, proteins are charged, and they remain in suspension because of their mutual repulsion for each other and because of the associated water layers. There will be a pH value between these extremes where the protein molecule is electrically neutral; if there are positive and negative charges on its surface, they are equal to each other. At this pH, called the **isoelectric pH** or isoelectric point (pI), the mutually repelling charges are absent from the protein molecules, and the water layers are minimal; hence flocculation readily occurs. That is, the solubility of a protein is lowest at its isoelectric point (Fig. 2.1-3).

Colloids that can have either a positive or a negative charge depending upon pH are called **amphoteric colloids**. The clay particles of soil as well as proteins are amphoteric. As a rule, however, both proteins and clay particles have a net negative charge at the pH levels found in protoplasm and in soil. Noncolloidal surfaces (as in a cell wall) can also be amphoteric.

Sols and Gels

Colloidal particles form a variety of systems such as the haze of the Blue Ridge Mountains (minute hydrocarbon particles suspended in air) or a stained glass window (colloidal particles of mineral suspended in glass). Two of these systems are especially interesting and important

to an understanding of protoplasm: the sol-gel system and the emulsion.

You are probably familiar with how gelatin dessert is prepared. Gelatin, a protein obtained from the hooves of cattle, combined with flavoring and sugar (which not only tastes good but lowers the water potential when it is dissolved), is added to hot water and stirred. The colloidal protein molecules go into suspension, forming a **sol.** The system remains highly fluid because of the high kinetic activities of the water molecules in which the gelatin is suspended. The Brownian movement of the elongate gelatin molecules is so high that contacts between molecules are extremely temporary. As the sol cools, the gelatin particles make less energetic impacts with each other and begin to adhere by van der Waals forces instead of bouncing apart. When enough of these contacts have been made, the sol has become a **gel.** The semisolid, tenuously rigid ("nervous pudding") condition of the gel is familiar to us all.

Dyes and other solutes diffuse through a gel nearly as rapidly as they diffuse through pure water (e.g., in a capillary tube too small to permit convection currents). Thus, a gel must not consist of compartments of protein enclosing droplets of water; rather, it has a structure more similar to a pile of brush (the **brush pile theory** of gels). As a mouse runs rapidly through a brush pile, so solute particles diffuse rapidly through a gel.

A sol-gel system has a number of interesting characteristics. Often it is **reversible,** as in the gelatin dessert. Adding kinetic energy by heating causes the contact points to break. Sometimes, however, it is irreversible, as in a blood clot. Protoplasm often has the characteristics of a reversible sol-gel system. An amoeba moves by reversibly changing its protoplasm from a gel to a sol, which allows its pseudopod to flow forward, after which it becomes a gel again. Some gels shrink with time, forcing out droplets of liquid (**syneresis**). **Thixotropic gels** become sols when they are disturbed. (A thixotropic paint is on the market.) In **hysteresis,** a gel exhibits a kind of "memory." One gel is prepared with a large quantity of water and another with less water. Both are dried to complete dryness. When they are placed in water, the one originally made up with the most water will reabsorb the most. A dry gel absorbs water by **hydration,** as discussed later.

An **emulsion** is a suspension of droplets of one liquid in another liquid with which the first is immiscible (won't mix). If the droplets are oil and the continuous phase in which they are suspended is water, we have a **cream-type emulsion.** If the droplets are water and the continuous phase oil, the emulsion is a **butter type.** Often the two types are interchangeable, depending upon what solutes are present. A calcium soap (the calcium salt of a long-chain fatty acid) is more soluble in fat than water, lowering the surface tension of the fat more than the water, so that a butter-type emulsion results. Sodium soap dissolves more in the water, producing a cream-type emulsion. Membranes in living cells have some of the properties of a butter-type emulsion (see Chapter 5).

Hydration

Strange things happen to water in the plant and in the soil. Ordinarily, water under tension vacuum boils, but the water in the xylem of a stem is nearly always under tension, yet it doesn't vacuum boil (see Chapter 4). Furthermore, the idea of "ice" at room temperatures and pressures certainly seems contradictory to our everyday experience, yet in the process of hydration, water apparently forms a sort of ice. **Hydration** or **imbibition** occurs when colloidal or other water-attracting surfaces adsorb water molecules. Most modern research has been done with the clay colloids found in soils, but the principle probably applies as well to proteins and other colloids and surfaces in organisms. Hydration is a spontaneous process, indicating that the free energy of the system decreases. Often heat is given off, as when water is added to dry starch or cellulose. Apparently, as polar water molecules come within the electrical fields on the surface of the colloids, they are "held in place," so that their velocities are actually reduced. The kinetic energy lost by these molecules may appear in the rest of the system as heat.

Water releases heat when it freezes as well as during hydration on clay or protein. And in hydration, there is evidence that the water molecules form the same crystalline structure found in ice—although temperatures could approach the boiling point! One evidence for the ice structure is that the density of adsorbed water is, like ice, below that of ordinary water. On a clay particle, a majority of the water forms the pseudoice structure; the remaining adsorbed water is influenced by exchangeable cations so that the structure is distorted. Yet the water influenced by the ions is most tightly adsorbed.

Certainly, something drastic happens during hydration. Not only is heat given off, but enormous pressures may be developed by the swelling material. To estimate pressures that could develop, the tenacity with which the water molecules are held on the hydrating surface is measured. For example, one might measure the vapor pressure of air in equilibrium with the dry material. It is possible to calculate the water potential of the air (see equation 8 on page 25), and at equilibrium this is equivalent to the water potential of the water molecules on the surface (the matric potential—see page 29). Another method is to find a solution with a water potential so low that no water can be adsorbed by the hydrating material, implying that the system was at equilibrium to begin with. Such methods have yielded matric potential values as negative as $-1,000$ to $-3,000$ bars.

Such measurements express the free energy decrease of the system during hydration and thus imply the maximum useful work that could be done. Such work can be expressed as real pressure when the hydrating system is confined. A swelling or expansion occurs as the layers of water molecules build up on the hydrating surface. Thus seeds swell as they imbibe water, as do doors in wet weather! Matric potentials imply that pressures of 1,000 to 3,000 bars might develop.

The pressures produced during growth of some plant parts are the pressures of hydration. Such hydrating materials as starch and protein are synthesized as roots grow, and these then swell upon imbibing water, causing the powerful expansion forces often observed in nature. Roots growing in a crack of a rock may split the rock. Grass seeds trapped under an asphalt pavement often develop enough hydrational force to push the developing seedlings up through the asphalt. Walter P. Cottam, a botanist at the University of Utah, observed a group of mushrooms that caused a concrete driveway to shatter. Enough stress was developed in the concrete before breaking took place that, upon release of this stress, a noise like a rifle shot was produced, and pieces of concrete were actually displaced around the small "crater" containing the mushrooms. Hydration is a powerful force in nature!

Several factors influence the rate and extent of hydration. With increasing temperature, the rate of hydration increases, because the more rapidly moving water molecules "find their place" on the hydrating surface more rapidly. Yet the total quantity of water imbibed is decreased, because water molecules with more kinetic energy are harder to hold on the hydrating surface. Hence, in the making of oatmeal or wheat mush, a softer product can be obtained if the material is allowed to imbibe water overnight at low temperatures, but the mush can be prepared faster if it is boiled gently. Ions influence the hydration process, because they decrease the water potential and in specific ways that depend upon their adsorption on the hydrating surface (i.e., according to the lyotropic series). The pH influences the net charge on a hydrating surface, as we have seen, and hence influences the hydration process.

Protein Purification

Although the physical effects of colloids that we have been discussing are extremely important to living things, the chemical effects are even more so. Life is an aggregation of active protein molecules: the **enzymes.** These molecules catalyze in highly specific ways the thousands of chemical reactions going on in cells (see Chapter 8). Each enzyme does this by adsorbing a substrate molecule of such a reaction at a highly specific point on its surface, the **active site,** thereby catalyzing the production of product molecule(s). Because of this, reaction rates are greatly speeded up, and life becomes possible.

Thus, most biochemists are enzyme chemists, and as such, they will often want to purify a single species of protein (a single enzyme) from a cellular extract that may contain dozens to thousands of others. The procedures are based upon the properties of colloids that we have been discussing in this essay. By reviewing some of these procedures, we can review some of the principles.

The initial procedures will often involve separation of proteins by differential precipitation or flocculation. A common method is called **salting out.** Some electrolyte such as ammonium sulfate, which is highly soluble in cold water (760 g/l at 0 C), is added to the extract. This electrolyte will neutralize the surface charges on the proteins, doing so in ways that are highly specific and dependent upon the nature of these charges and the concentrations of salts employed. Some proteins become insoluble when treated with a solution 30 percent saturated with $(NH_4)_2SO_4$, for example, while others will not precipitate unless more salt is added. Those that are precipitated at a given salt concentration can be separated from the others by filtration and/or centrifugation.

Another method of causing flocculation is **isoelectric precipitation.** As the pH is adjusted, some proteins will flocculate out when their isoelectric point is reached, while others remain in solution because they have different isoelectric points. After the proteins have been precipitated, either isoelectrically or by salting out, they can be resuspended in a buffer of suitable pH to form a sol.

Another means of precipitating some proteins while others are left in solution involves heat. Heating changes the protein's structure so that hydrophobic groups are exposed; hence the protective water layer is removed. Still other methods involve addition of organic solvents such as acetone or heavy metal ions such as tungsten. In these treatments, structures may be changed, charges may be neutralized, and/or water layers may be removed, depending upon the nature of the protein.

Often it is desirable to remove low molecular weight molecules that contaminate a protein preparation. This can be done by the process of **dialysis,** which is based upon the principle that solute particles will pass through the pores of cellophane while the larger colloidal proteins will not. The mixture is poured into a cellophane dialysis bag, which is then placed in circulating water (Fig. 2.1-4), usually in a refrigerator or cold room.

Another way to separate small molecules from proteins is to use a three-dimensional polysaccharide network called **Sephadex,** or certain polyacrylamide gels in the form of tiny beads. These materials act as **molecular sieves** and are also used to separate mixtures of proteins differing in size. The materials are packed into a glass column, and the crude protein mixture is then added at the top. As the proteins pass through, the smaller molecules penetrate and diffuse into and out of the pores in the beads of the gel and are thereby slowed in their passage through the column. The larger molecules, however, are too large to enter the beads and thus move more rapidly through the column. Fractions collected at various times then contain different-sized molecules.

Protein molecules have surface charges depending upon their

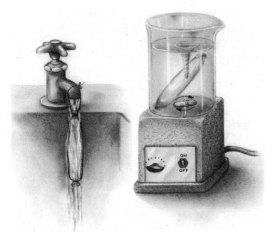

Figure 2.1-4 A laboratory method of dialysis.

species and upon the *p*H. Hence, they may be separated by **electrophoresis** as described in Chapter 8.

Since proteins are large molecules in the colloidal size range, they can settle out in a centrifuge, providing the accelerational forces due to spinning are great enough. Such a high-speed centrifuge is called an **ultracentrifuge.** The rate of settling will be a function of the size and shape of the protein molecules, and arbitrary size designations (daltons*) may be assigned to different proteins based upon their rates of settling in an ultracentrifuge. A beam of light may be passed through the spinning tubes on the centrifuge and then through an optical system that indicates index of refraction to show the position of the migrating protein molecules. A centrifuge equipped with such an optical system is called an **analytical ultracentrifuge.** Using all these procedures, it has been possible to prepare several crystalline enzymes essentially free of all contaminants.

Special Topic 3.1
Radiant Energy: Some Definitions

Plants are strongly influenced by the radiant energy in their environment. Hence, a plant physiologist should understand the nature of radiant energy and how it might interact with a plant. The principles of radiation and how radiation interacts with matter are taught in physics and physical chemistry classes. Thus, the following is presented primarily with two purposes in mind: to provide a brief review of the topic, and to act as an easily accessible reference source for ideas or terms once understood but no longer remembered clearly. The format is a series of definitions. For reference purposes, it would be convenient to list these alphabetically, but, assuming that you will want to review the entire topic in a logical fashion, the definitions have been arranged so that one builds upon another. Individual terms can be found by scanning the list.

---

*One dalton = the mass of one hydrogen atom (one proton plus one electron). Thus, a dalton weight is often nearly equivalent to a molecular weight, but the dalton weight is used when a worker is not certain that he is measuring the mass of a single molecular species (e.g., when the preparation might consist of a mixture of similar proteins).

Radiant energy A form of energy that is emitted or propagated through space or some material medium. It is said to be *electromagnetic,* and it is propagated in the form of pulsations or waves. Certain concepts and equations appropriately describe the wave nature of radiant energy, but this energy might also be thought of as consisting of "particles." These "particles" without rest mass can also be described by certain equations and by reference to certain manifestations. Clearly, we do not yet fully understand radiant energy, but the term is sometimes extended to include streams of subatomic or atomic particles that do have mass, such as electrons, alpha or beta particles, or certain cosmic rays.

The wave nature of radiant energy Several phenomena, including diffraction, interference, and polarization (mentioned later) suggest that radiant energy is propagated in the form of waves. Since familiar waves (e.g., water or sound waves) are propagated through a medium, it was postulated that radiant energy was propagated through the *ether.* Careful experiments around the turn of the century designed to prove the existence of the ether failed, and the concept has been rejected. Nevertheless, the wave nature of light continues to be apparent, even though no medium for its propagation is involved.

Frequency (ν) The number of wave crests (peaks in energy) passing a given point in a given interval of time. Frequency is usually given in terms of energy crests (vibrations or waves) per second. Green light has a frequency of about 6×10^{14} pulsations sec^{-1}; radio waves about 10^4 to 10^{11} sec^{-1}.

Velocity (c) The distance traveled by a peak of radiant energy in some specified interval of time. Velocity of all forms of radiant energy is constant in a vacuum and is equal to 3.00×10^{10} cm sec^{-1}* (300,000 km sec^{-1} or 186,000 miles sec^{-1}). It is virtually identical in air but slower in media such as water (2.25×10^{10} cm sec^{-1}) or crown glass (1.98×10^{10} cm sec^{-1}).

Wavelength (λ) The distance between waves or crests of energy in electromagnetic radiation. The wavelength is equal to the velocity divided by the frequency: $\lambda = \frac{c}{\nu}$. Likewise, the frequency is equal to the velocity divided by the wavelength: $\nu = \frac{c}{\lambda}$. Wavelengths of radiant energy vary from much shorter than the diameter of an atom to several kilometers in length (see *Electromagnetic spectrum,* p. 384). Green light has a wavelength of about 500 nm or 5×10^{-5} cm; radio waves about 10^{-1} to 10^6 cm.

Wave number ($\bar{\nu}$) A term convenient in some applications and equal to the reciprocal of the wavelength measured in centimeters: $\bar{\nu} = \frac{1}{\lambda}$ From the foregoing equations, it is apparent that the wave number is equal to the frequency divided by the velocity (in cm sec^{-1}) of radiant energy: $\bar{\nu} = \frac{\nu}{c}$ Thus, wave number is simply another way of expressing frequency. Green light has a wave number of about 2×10^4 cm^{-1}; radio waves about 10 to 10^{-6} cm^{-1}.

Refraction The change in direction ("bending") that takes place when a ray of radiant energy passes from one medium into another in which its velocity is different. Refraction at the surface between glass and air makes the construction of lenses possible. Light is refracted within leaves as it passes from air into a cell wall or the cytoplasm; it may be refracted several times within a leaf. Since different wavelengths

---

*There are three ways of writing units when some occupy the position of a denominator; all three are equivalent: cm per sec. cm/sec, and cm sec^{-1}. The last is coming into wide use, partially because it makes it easier to cancel units in equations such as those that follow (see *Einstein* and *Photosynthetically or photochemically active irradiance*).

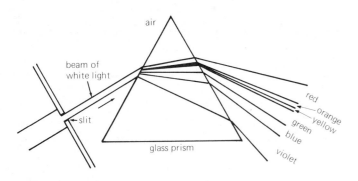

Figure 3.1-1 White light is dispersed into its component colors by refraction when passed through a glass prism.

are refracted to different degrees, wavelengths are separated when they pass through a prism (Fig. 3.1-1).

Diffraction and interference Diffraction includes those phenomena produced by the spreading of waves around and past obstacles that are comparable in size to the wavelength. Interference phenomena are caused by reinforcement when energy crests (waves) are superimposed upon each other (are in phase), or by the opposite effect, which occurs when waves are out of phase, canceling or dampening each other. Thus, as waves are diffracted, they may reinforce or cancel each other by interference, producing a rainbow effect (separating the various wavelengths). Two devices often used by plant physiologists operate according to these principles: a **diffraction grating**, which consists of fine lines ruled close together on a transparent surface and which separates a mixture of wavelengths with the end result similar to that produced by a prism; and *interference filters*, which have a thin layer of a reflective medium on the surface of glass, the layer being of such thickness that one wavelength is strongly reinforced by passing through the filter while other wavelengths are canceled.

Polarization Light waves normally vibrate in many planes parallel to their direction of propagation. When light is polarized, the energy is made to vibrate in more or less one plane. Light becomes polarized when it is passed through certain substances or is reflected. Many molecules important in plants and other living things will, in solution, rotate the plane of polarization of a beam of polarized light. These **optically active molecules** typically contain at least one asymmetric carbon atom (an atom with four different groups attached to it).

The particulate nature of radiant energy Radiant energy exists in units that cannot be further subdivided. In the photoelectric effect, for example, one electron may be ejected from a surface upon the absorption of one "particle" or "packet" of radiant energy. The particulate nature of radiant energy is described by certain equations, many of which include terms for the frequency or the wavelength—terms derived from the wave concepts of radiant energy.

Quantum or photon Terms used interchangeably for the "particles" of energy in radiation.

Quantum energy (E) The energy (E) of a quantum or photon is equivalent to the frequency (ν) times Planck's constant (h): $E = h\nu$. Thus, the energy of a photon is directly proportional to the frequency of the radiation; higher frequencies have *more* energetic photons. Since the frequency is equal to the velocity divided by the wavelength, the energy of a photon will be inversely proportional to the wavelength:

$E = \dfrac{hc}{\lambda}$. Longer wavelengths (lower frequencies) have *less* energetic photons. These equations are useful in calculating the energy relations of photosynthesis and other plant processes that depend on light energy.

Planck's constant (h) A universal constant of nature relating the energy of a photon to the frequency of the oscillator that emitted it (i.e., the frequency of the radiant energy). Its dimensions are energy times time. It is equal to 1.58×10^{-34} cal sec or 6.624×10^{-27} erg sec.

Quantum yield (ϕ) An expression of efficiency when absorption of a photon by a molecule results in some photochemical reaction. The quantum yield (ϕ) is equal to the ratio of the number of molecules reacted (M) to the number of photons absorbed (Q): $\phi = M/Q$. Quantum yields for photosynthesis and for interconversion of the two forms of phytochrome (see Chapters 9 and 19) are widely studied values.

Einstein A number of quanta or photons equal to Avogadro's number: 6.02×10^{23} atoms mol^{-1}; hence, the same as a "mole of quanta." The energy in an einstein of red light ($\lambda = 660$ nm or 6.6×10^{-5} cm, $\nu = 4.545 \times 10^{14}$ sec^{-1}) can be calculated as follows:

$$E = \frac{hc}{\lambda} = h\nu$$

$$E = \frac{(1.58 \times 10^{-34} \text{ cal sec})(3.0 \times 10^{10} \text{ cm sec}^{-1})}{6.6 \times 10^{-5} \text{ cm}} =$$

$(1.58 \times 10^{-34} \text{ cal sec})(4.545 \times 10^{14} \text{ sec}^{-1}) = 7.18 \times 10^{-20}$ cal photon^{-1}

$$E = (7.18 \times 10^{-20} \text{ cal photon}^{-1})(6.02 \times 10^{23} \text{ photon einstein}^{-1})$$
$$E = 43,200 \text{ cal einstein}^{-1}$$

Blue light ($\lambda = 4.50 \times 10^{-5}$ cm, $\nu = 6.67 \times 10^{14}$ sec^{-1}) has a frequency 6.67/4.545 times that of red, so its energy per photon will be 1.466 times that of red $= 10.5 \times 10^{-16}$ cal photon^{-1} or 63,400 cal einstein^{-1}

The electromagnetic spectrum The known distribution of electromagnetic energies arranged according to wavelengths, frequencies, or photon energies (Fig. 3.1-2). At one end of the spectrum is radiant energy of extremely short wavelengths and consequently extremely high frequencies and energetic photons. At this end of the spectrum are cosmic rays. Slightly longer wavelengths (lower frequencies, less energetic photons) are gamma rays, which overlap broadly with the X-ray part of the spectrum. Ultraviolet radiation is expressed by wavelengths slightly shorter than those in the visible or light part of the spectrum, and infrared radiation has wavelengths longer than those of visible light. Radiowaves are still longer. The entire spectrum extends over at least 20 orders of magnitude with the visible portion being a part of only one order of magnitude.

Light In the correct sense, the visible portion of the spectrum; sometimes extended to the ultraviolet and the infrared portions as well.

Color The appearance of objects as determined by the response of the eye to the wavelengths of light coming from these objects. Short wavelengths of light produce a sensation of violet or blue; the longest wavelengths produce a sensation of red. Colors of things are due to *pigments* (see later).

Light sources Since plant physiologists deal continually with responses of plants to light, it is important to know something about the possible sources of light to which the plants are exposed; for example, in special plant growth chambers. Spectral distributions for several light sources are shown in Fig. 3.1-3. Note that large portions of the solar or xenon spectra are in the infrared part of the spectra, this being even more the case for incandescent light. Light from fluorescent tubes

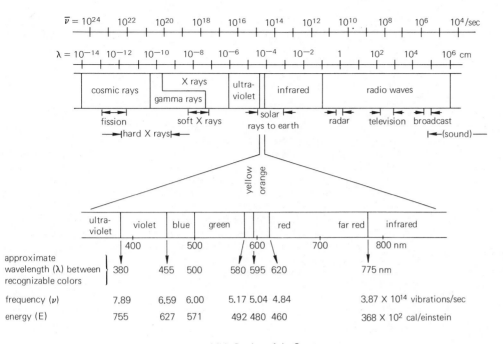

Figure 3.1-2 The electromagnetic spectrum, using both wave number ($\bar{\nu}$) and wavelength (λ) in cm. Various portions of the spectrum are shown, and the visible portion is expanded to indicate the region that appears to the human eye to have various colors.

Visible Portion of the Spectrum

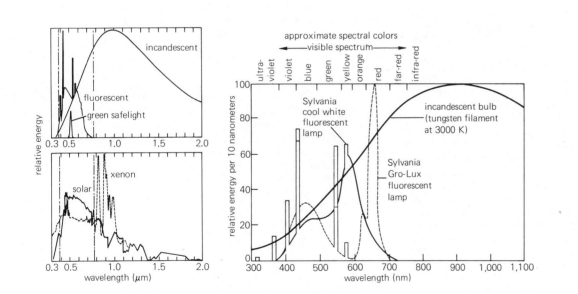

Figure 3.1-3 Emission spectra for several sources. Note incandescent lamp peak at about 1.0 μm; mercury emission lines in fluorescent lamp spectra; infrared peaks (0.85 to 1.05 μm) from the xenon lamp; and solar peak in the midpart of the visible (indicated by dashed vertical lines).

is mostly in the visible portion of the spectrum with minor amounts of ultraviolet and virtually no near-infrared radiation.

Light intensity **Radiant flux** is the radiant energy falling upon a surface in an interval of time (e.g., joules sec^{-1}). **Irradiance** is the radiant flux density received on a unit surface (e.g., joules sec^{-1} cm^{-2}). Since 1 watt = 1 joule sec^{-1}, joules sec^{-1} cm^{-2} = watts cm^{-2}. If irradiance levels are extremely low, it is convenient to use units of ergs cm^{-2} sec^{-1}; when intensity is higher, joules (joule = 10^7 erg) or calories (cal = 4.185 joules) are often used. The total energy in sunlight or other light sources for plant growth is often measured in cal cm^{-2} min^{-1}. One cal cm^{-2} is called

a **langley** (**ly**). In the literature of plant physiology, light intensity is sometimes given in units of illumination rather than total energy. Such units (e.g., 10.76 lux = 1.0 foot candle, ft-c) are defined in terms of the sensitivity of the human eye. Plants, however, respond to the spectrum in ways quite different from the human eye, so such a measurement is virtually meaningless. Indeed, it has no value at all unless some information is given about the nature of the light source (e.g., sunlight, light from an incandescent or fluorescent lamp). Because plants respond to some wavelengths of light but do not absorb others, even a measurement given in energy units has little if any value when the spectral distribution is not also given. Nevertheless, we have used lux and ft-c in this text because these are often the only data available to us. We have tried to specify sources.

Photosynthetically or photochemically active irradiance In a photochemical process such as photosynthesis, the end product depends upon the number of quanta absorbed rather than the total light energy absorbed. A single red photon has the same effect in photosynthesis as a single blue photon, for example, although the blue photon has more energy. Hence, in the recent literature it has become common to refer to the number of photons per unit area per unit time. Einsteins (for photosynthesis) or microeinsteins (for low light responses) are used. A worker might use an instrument that responds only to light between the wavelengths of 400 to 700 nm (photosynthetically active), for example. Say that he measures a light intensity value of 0.5 cal cm^{-2} min^{-1}. If the energy of photons in the green ($\lambda = 5.5 \times 10^{-5}$ cm, $\nu = 5.454 \times 10^{14}$ sec^{-1}) part of the spectrum is representative of the light source, then the light has $h\nu$ times Avogadro's number of calories per einstein:

$$(1.58 \times 10^{-34}\text{ cal sec photon}^{-1})(5.454 \times 10^{14}\text{ sec}^{-1})(6.02 \times 10^{23}$$
$$\text{photons einstein}^{-1}) = 5.188 \times 10^4\text{ cal einstein}^{-1}$$

The light source then has

$$\frac{0.5\text{ cal cm}^{-2}\text{ min}^{-1}}{5.188 \times 10^4\text{ cal einstein}^{-1}} = 9.64 \times 10^{-6}\text{ einsteins cm}^{-2}\text{ min}^{-1}$$

$$= 5.78\text{ einsteins m}^{-2}\text{ hr}^{-1}$$

Pigment Any substance that absorbs light energy. If all the visible spectrum is absorbed, the substance appears black to the human eye; if all but the wavelengths in the green part of the spectrum are absorbed, the substance appears green.

Ground state and excited state (energy level) When atoms, ions, or molecules absorb radiant energy, they are raised to a higher **energy level**. Before absorbing the energy, they are said to be in the **ground state;** after, in the **excited state.** The actual change in energy content may be achieved by changes in the vibration or rotation of the atom, ion, or molecule, or by changes in the electronic configuration of the atoms involved. Typically, an electron is moved a greater distance from the nucleus as energy is absorbed; it is said to be moved to a higher energy level. Due to the wave motions of the electrons orbiting the nuclei of atoms, there is not a continuous series of energy levels; rather, electrons can exist only at certain discrete distances (energy levels) from the nucleus. To move an electron to the next higher level requires, then, a discrete amount of energy. A pigment absorbs only those photons that have exactly the required quantity of energy to bring about the change in electronic configuration. Chlorophyll, for example, absorbs photons in the blue and the red spectral regions preferentially. Actually, there is a range of photon energies that can be absorbed in any given case because some translational, vibrational, or rotational energy changes can occur in addition to the electronic changes.

Fluorescence and phosphorescence An atom, ion, or molecule in an excited state may lose its excitation energy in any of three ways: First, it may be immediately lost as heat; that is, totally converted to translational, vibrational, or rotational energy. Second, it may be partially lost as heat with the remainder emitted as visible light of a wavelength longer (a photon of lower energy) than that absorbed. If this occurs within 10^{-9} to 10^{-5} sec after absorption of the original photon, it is called **fluorescence**. If the delay is longer than that (10^{-4} to 10 sec or more), it is called **phosphorescence**. Third, the energy may be used to cause a chemical reaction such as in photosynthesis.

Black body A surface that absorbs all the radiation falling upon it. Usually the term is used in reference to some portion of the spectrum under consideration. One may speak of a black body with reference to visible light and/or infrared radiation, for example. Carbon black or black velvet provide surfaces that approximate a black body. A still more perfect approach to true black body conditions would be a small opening in the surface of a large sphere lined with carbon black. Obviously, only a minute portion of the radiation entering this opening would ever leave through the opening.

Absorptivity coefficient A decimal fraction expressing the portion of impinging radiation that is absorbed. A leaf, for example, has an absorptivity coefficient of about 0.98 in the infrared portion of the spectrum.

Transmission and reflection Radiant energy that is not absorbed is either transmitted or reflected. Transmission or reflection are usually expressed as decimal fractions or as percentages (see Section 11.4).

Transmittance (T) or percent transmission The fraction of light transmitted by a substance expressed as a decimal fraction, $T = I/I_0$, or a percentage ($I/I_0 \times 100$), where: I_0 = intensity of *incident* radiant energy. I = intensity of *transmitted* radiant energy.

Absorbance (A) Formally, *optical density*. The logarithm of the reciprocal of transmittance (T):

$$A = \log \frac{1}{T} = -\log T = \log \frac{I_0}{I}$$

Absorbance is often proportional to concentration of a pigment in a transparent solution, according to the following laws:

Beer's law Each molecule of a dissolved pigment absorbs the same fraction of light incident upon it. Thus, in a nonabsorbing medium, the light absorbed should be proportional to dissolved pigment concentration. The law often holds for dilute solutions but fails as the light-absorbing properties of pigment molecules change at higher concentrations.

Lambert's law Each layer of equal thickness absorbs an equal *fraction* of the light that traverses it. This idea can be combined with Beer's law as the **Beer-Lambert law:** *The fraction of incident radiation absorbed is proportional to the number of absorbing molecules in its path.*

Extinction coefficient (ϵ) The Beer-Lambert law can be stated mathematically as follows:

$$A = \log \frac{I_0}{I} = \epsilon c l$$

where ϵ = extinction coefficient
c = concentration of the pigment solute
l = length of the path of light (e.g., through a special quartz cell) in centimeters
(A, I_0, and I are defined earlier)

The extinction coefficient is a constant for a given pigment in dilute solution and can be determined by solving the equation just given:

$$\epsilon = \frac{A}{cl}$$

If the concentration of solute is in moles liter^{-1} (molarity), ϵ is called the **molar extinction coefficient** (with units of liters mole^{-1} cm^{-1}). If concentration is known only in grams liter^{-1}, ϵ is the **specific extinction coefficient** (usually with the symbol a_s). The extinction coefficient is a characteristic of a given absorbing molecule in a given solvent with light of a specified wavelength. It is independent of concentration only when the Beer-Lambert law holds. The more intensely colored a pigment is at a given concentration, the larger its extinction coefficient.

Stefan-Boltzmann law All objects above the absolute zero emit radiant energy. The quantity (Q) emitted is a function of the fourth power of the absolute temperature (T) of the emitting surface, according to the Stefan-Boltzmann law:

$$Q = e\delta T^4$$

where: Q = the quantity of energy radiated (in calories, using δ as below)

e = the emissivity (about 0.98 for leaves at growing temperatures)

δ = the Stefan-Boltzmann constant (8.132×10^{-11} cal cm^{-2} min^{-1} deg K^{-4})

T = the absolute temperature in deg K (deg C + 273)

The fourth power of absolute temperature in the expression means that the emission of radiant energy will increase greatly as temperature increases. Although the normal range of temperatures encountered by plants is narrow on the absolute temperature scale, the fourth power function means that energy radiated by bodies in this narrow range nevertheless varies considerably (Table 3.1-1).

Table 3.1-1 Radiation Emitted from Black Body Surfaces at Various Temperatures

deg C	deg K	T^4	Q = cal cm^{-2} min^{-1}	% of Q at 0 C
0	273	5.55×10^9	0.431	100
20	293	7.38×10^9	0.572	133
30	303	8.42×10^9	0.651	151
5,477[a]	5,750[a]	1.10×10^{15}	84,900.	19,700,000

[a]Average surface temperature of the sun.

Net radiation The difference between the radiation absorbed by an object and that emitted is the net radiation. A leaf, for example, emits radiation according to the Stefan-Boltzmann law. At normal temperatures, most of this is in the far-infrared portion of the spectrum. Such emission leads to cooling. If the leaf is being illuminated by sunlight, however, it absorbs a portion of that (according to its absorptivity coefficient), which warms the leaf. Whether the leaf increases or decreases in temperature will depend upon whether more or less radiation is being absorbed or emitted—and also upon other mechanisms (i.e., convection, transpiration, and so on) that add heat to or remove heat from the leaf.

Wien's law Not only is the quantity of radiant energy emitted by an object a function of its temperature, but the quality is also influenced by temperature. With increasing temperature, the peak of emitted radiant energy (λ_{max}) shifts toward the shorter wavelengths. This peak

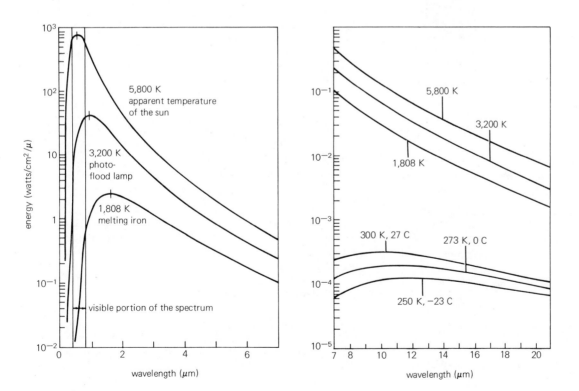

Figure 3.1-4 Black body emission spectra compared over a wide range of energy emission and wavelength. The spectra would apply for any perfect black body radiator. Note shift in the peaks toward longer wavelengths (Wien's law), flattening of the curves, and decrease in total energy (Stefan-Boltzmann law) as temperatures decrease.

times the absolute temperature (*T*) is equal to a constant, Wien's displacement constant (*w*): $\lambda_{max} T = w$. This is Wien's law, and it is illustrated for a wide range of temperatures in Fig. 3.1-4. At room temperatures, objects are emitting maximally in the far-infrared part of the spectrum (λ_{max} = approximately 10μm or 10,000 nm). At temperatures approximating those of an incandescent filament, the peak of emission is in the near infrared (λ_{max} = about 1 μm), and at the temperature of the sun or other stars the emission peak is in the visible part of the spectrum (λ_{max} = about 0.50 μm).

Color temperature As expressed by Wien's law, the emission peak of objects is a function of absolute temperature. Thus, the spectral distribution of the light emitted by incandescent sources (such as an incandescent filament or the surface of a star) can be indicated in terms of absolute temperature. Epsilon Orionis and Sirius, with surface temperatures of 28,000 K and 13,600 K, are blue-white stars; the sun (5,800 K) has an emission peak in the green-yellow part of the spectrum; and Betelgeuse (3,600 K) is a red star. Color temperatures are widely used in photography; sensitivity of film must be balanced to the color temperature of the light source, for example.

Emissivity The curves shown in Fig. 3.1-4 are for perfect black body radiators. Actually, such an ideal is seldom achieved, especially in objects near room temperature. In practice, as objects fail to absorb some wavelengths, they also fail to emit some wavelengths, the curve for emission being the same as the curve for absorption. Thus, as a leaf has an absorptivity coefficient of about 0.98 in the far-infrared part of the spectrum, it also has an emissivity of about 0.98, with most of the radiant energy being emitted in that part of the spectrum. Incidentally, the atmosphere is far from a perfect black body, even in the far infrared, as indicated by the solar spectrum in Fig. 3.1-3. This must be taken into account in calculating the thermal radiation coming from the atmosphere and the thermal radiation absorbed by the atmosphere after being emitted by objects on the ground.

Special Topic 4.1
Soil Water—Jerome J. Jurinak

Jerome J. Jurinak is a Professor of Soil Chemistry at Utah State University. He has done research on the chemistry of soil water systems and is currently devoting much of his time to a study of problems of salinity in the Upper Colorado River Basin.

Water is an essential component of the soil system. The soil must supply large quantities of water to meet the transpirational demands of growing plants, and water with its dissolved salts constitutes the soil solution that provides nutrients for plant metabolism. Soil water is also intimately involved in controlling soil temperature and soil aeration. Our interest in soil water is obvious.

The behavior of water in soil and its availability to plants have been unified into a single energy concept that considers the soil-plant-atmosphere chain as a continuum. The energy concept of soil water is used to explain why water enters and moves in soil, is absorbed and transported through plants, and then is transpired into the atmosphere. The phenomena are energy related.

Energy of Soil Water

The capacity of the soil to store water is the result of the attraction of the soil matrix (solids) for water. Water interacting with the soil loses energy,

so free energy or water potential of the water retained by the soil is less than that of pure free water. The interaction of ions and other solutes with water (osmotic forces) further reduces the free energy of water in the soil, and gravitational forces also affect the energy of soil water, particularly under saturated moisture conditions. An important consequence of the low free energy of soil water is that the removal of water from the soil matrix—say, by a plant root—requires an expenditure of energy, ultimately that supplied by sunlight.

The energy required to extract a quantity of water from the soil is expressed by the water potential, ψ, the value of which is the sum of the contribution of all the various forces acting on water in the soil. This is given by the water potential equation already presented (see equation 13 on p. 31), except that a factor for the effect of gravity (ϕ_g) replaces the pressure term:

$$\psi = \phi_g + \psi_c + \psi_\pi + \cdots$$

For our purpose, the water potential is considered to be the sum of ϕ_g, the gravitational potential; ψ_c, the matric potential; and ψ_π, the osmotic potential. Under normal field conditions, after free drainage has occurred, ϕ_g is considered insignificant relative to ψ_c and ψ_π. Although ψ is usually expressed in pressure units (bars, atmospheres, or cm of water), it indicates the *energy* at which the soil retains water; it does not give any information as to the amount of water in the soil at that potential. Remember that ψ is a negative term representing the free energy difference between soil water and pure free water.

The Matric and Osmotic Potentials

The matric potential, ψ_c, results from the combined attraction of capillary and adsorption forces. The matric potential not only gives the soil its water-storing ability, it also functions to determine soil water movement. The matric potential is always negative, and water moves by diffusion from regions of high matric potential (high free energy) to regions of low (more negative) matric potentials (low free energy). But the rate is much slower (soil conductivity is much lower) when soil water potential is lower (Fig. 4.1-1).

The osmotic potential, ψ_π, which is negative, originates from the presence of solutes, both ionic and nonionic, in the soil solution. Unlike the matric potential, the osmotic potential does not have any effect on the movement of soil water, nor does it affect the retention of water by the soil matrix. Its principal effect is an influence on water uptake by plant roots. Root tissue acts as a differentially permeable membrane that transmits water more readily than solutes. A soil solution with a high solute content effectively restricts the movement of water through the root membrane if, due to the presence of salt, it has a lower free energy than water in the plant root.

Incidentally, negative potentials have in the past commonly been given positive units and have been called *suctions* or *tensions* in discussions of soil water. For example, matric potential and matric suction are identical except for the sign.

Soil Moisture Characteristic Curve

The water content of the soil and the soil water potential are related. As water is removed from a soil, it necessarily follows that the remaining water is held at a lower or more negative potential. Figure 4.1-2 shows the experimentally determined relation between matric potential and moisture content for several soils of different textures. This is called a soil moisture characteristic curve, and it is similar to the curves for plant matric potentials in Fig. 2-11, p. 30. Because soil water potentials were measured with a pressure membrane, osmotic potentials are not included as part of soil water potential in Fig. 4.1-2.

Under normal nonsaline conditions, the matric potential is dominant in controlling the availability and movement of soil water. Note that the curves are smooth, indicating that the potential-moisture content relation is a continuous function. At any given potential value, the heavier-textured soils with more colloidal material (clays) hold more water than do such lighter-textured soils as loams and sands.

After free drainage has occurred in a saturated soil ($\phi_g = 0$), the soil water is held principally in the soil pores as well as in a continuous film around the soil particles. The matric potential, for most soils, at this point is about $-\frac{1}{3}$ bar. The moisture content of such a soil is commonly referred to as the **field capacity** of the soil.

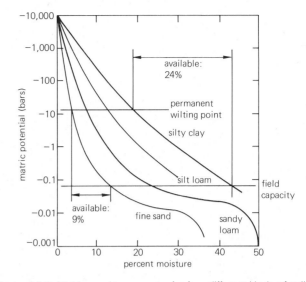

Figure 4.1-2 Moisture release curves for four different kinds of soils. Note that much more water is available to plants from clay soil than from sand. (Redrawn from M. B. Russell, 1939, Soil science society of America proceedings 4:51–54.)

As water is removed from the soil, the matric potential decreases until it becomes equal to that of the plant so that the plant can no longer obtain water; that is, turgor pressure in the leaf cells drops to zero so that the plant permanently wilts and will not recover if put in an atmosphere having 100 percent relative humidity. Since different plants can be reduced to different water potentials before permanent wilting occurs, the point of permanent wilting on the soil moisture characteristic curve will depend on the species, but it is commonly reached at a water potential of about -15 bars. Hence, the amount of moisture retained in a given soil at -15 bars is referred to by agronomists as the **permanent wilting point** or **percentage.** The water held at water potentials more negative than -15 bars is usually considered unavailable to plants, and most of the gravitational water above field capacity drains from a soil before plants can absorb it, so the amount of water held by a soil between $-\frac{1}{3}$ and -15 bars matric potential is considered as the **available water.** Since a few plants can obtain water from soils with a water potential below -15 bars (e.g., creosotebush, *Larrea divaricata,* can absorb water to a water potential of -60 bars), there can be more available water in a given soil for some species than for others. Yet at such negative soil water potentials, very little water is present, so the difference between the actual quantity of water held by a soil at -15 bars and that held at -60 bars may be almost negligible.

Special Topic 7.1
What Is a Plant? The Significance of
Vacuoles and Cell Walls
—Herman H. Wiebe

Herman H. Wiebe is a plant physiologist at Utah State University, where he teaches and conducts research on the water relations of plants. He studied this topic as a graduate student under Paul Kramer at Duke University, finishing his doctorate in 1953. As a colleague of Salisbury and the major professor of Ross (1959–1961), Prof. Wiebe has taken a

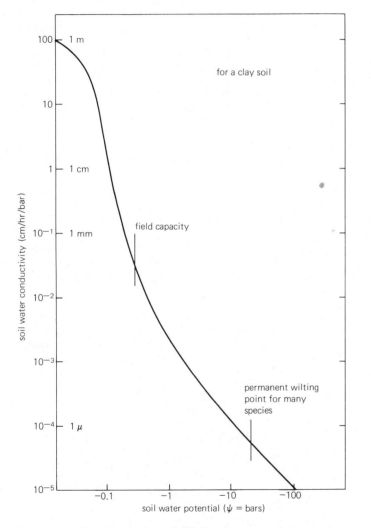

Figure 4.1-1 Conductivity (rate of diffusional movement) of water in a clay soil as a function of the tension with which the water is held by the soil (that is, soil water potential). (From W. R. Gardner, 1960, UNESCO arid zone research 15:37–61.)

close personal interest in the preparation of this text, both first and second editions. His total concern with a student's ability to grasp a given topic has resulted in several extensive revisions—as have his constant efforts to emphasize the whole plant and to suggest applications of knowledge that put ideas in their proper context. The following essay is of that genre.

Vacuoles and cellulose cell walls are characteristic features of plant cells and are always present in the vegetative tissues of green algae, bryophytes, and vascular plants. They are just as constantly absent from animal cells. In fact, large central vacuoles and cell walls constitute a characteristic and universal difference between plants and animals; even more basic, perhaps, than the presence of chlorophyll. Albino mutants and parasitic plants are common, but none lacking vacuoles is known.

The large central vacuoles are quantitatively and perhaps also structurally different from the small vesicles found in most, perhaps all, eucaryotic cells. These vesicles, which unfortunately are often called vacuoles, typically occupy far less than 10 percent of the cell volume, whereas the large central vacuole may occupy over 90 percent of the mature plant cell volume.

The Role of Vacuoles

It would be surprising if features that are so universally present as vacuoles and cell walls did not have profound significance. We can only assume that if they were not of great importance to plants, they would have been eliminated by rigorous selection pressure, at least from occasional plant species. Yet there are no exceptions. In the leaves, stems, roots, and flowers of all plant species, all mature cells contain vacuoles. Only in seeds and meristems are the vacuoles lacking or not yet developed.

The clue to the significance of vacuoles is the manner by which plants receive raw materials from their environment. Except for water and oxygen, these raw materials are all present in dilute form. Making up only three parts per 10,000, carbon dioxide is extremely dilute in the air, even more dilute than argon and almost as dilute as some pollutants such as sulfur dioxide. In water also, at least in the neutral and acid range, dissolved carbon dioxide and carbonate are very dilute. Energetically, sunlight is also diffuse and subject to shading. It never contains more than about 2 cal cm^{-2} min^{-1}, only part of which can be used in photosynthesis. A man positioned perpendicular to the sun would intercept some 6×10^6 cal per day, while 1.5×10^6 cal are required just to maintain him. Plants with a large surface area have a much better chance to intercept light and carbon dioxide. This is even truer of submerged aquatics growing where light intensities are lower than on land. It is the vacuoles that help plants achieve this area.

In soil also, the plant encounters raw materials in dilute or "hard-to-get" form. Water movement, except for a short period after rain, is extremely slow in soil (see Fig. 4.1-1, p. 389). Essential mineral ions move even more slowly in soil than water does and are often very dilute. Only plants with a large root system are able to penetrate and explore a large enough volume of soil to supply the water lost in transpiration, especially when rains are infrequent and insufficient, and to obtain the required minerals. That the vacuole aids in root growth is nowhere more clearly seen than in root hair elongation, which involves almost exclusively vacuolar enlargement and which brings the plant into intimate contact with a larger soil volume.

In submerged aquatics as well, the increased surface provided by large vacuoles aids in absorbing essential ions, which may be extremely dilute in the surrounding water. This is apparent in fresh water but may be equally true of critical essential elements such as potassium in brackish water and in oceans.

Animals get their food in concentrated form. Most are mobile and forage about in search of food. This places a premium on mobility and thus reduction in weight, so animals are compact organisms with a minimum of external surface. Internally, they are also compact; there are no water-filled vacuoles and no cell walls. Animals absorb material through an internal tube (alimentary canal) or invaginations (lungs), usually aided by muscular action (eating, swallowing, peristalsis, breathing, heartbeat).

Dendritic Form

Plants never use motion to obtain the highly dilute raw materials they require. Instead, they assume a branched filamentous or dendritic growth form that places them in intimate contact with a much larger volume of their environment than would ever be possible with a compact form. It is as if the stationary plant were reaching out with tentacles. Although branching increases surface area, the benefits accruing from form and area are somewhat different. Area is beneficial when the permeability of membranes is rate limiting. Increased area separate from dendritic form is seen in the transfer cells (see Chapter 7) and glands of some plants, and in villae of the gut. Large surface is also important where reactions are associated with membranes, as in the endoplasmic reticulum, cristae and microvillae of mitochondria, and chloroplast thylakoids.

Dendritic form is advantageous when movement of essential materials from the environment to the organism is the rate-limiting factor. This will most likely happen when materials move to the organism by the slow process of diffusion. A dendritic form will bring an organism into contact with a larger volume of its environment than any other form, thereby enabling maximum material uptake by diffusion. The dendritic plant form is best developed in roots in soil, and generally also in aquatics. Leaves in air are often laminar, having added strength. The laminar form is not a serious disadvantage, because diffusion is 10,000 times faster in the air than in water and often is also facilitated by wind. The overall shoot is typically dendritic—in fact providing the root of the term.

Several systems in animals also have a dendritic or nearly dendritic form. Sometimes the advantages bestowed by this kind of structure are similar to those obtained by plant roots and shoots, and in other cases there seems to be a subtle difference. The capillary network in animal tissues explores the volume of those tissues much as the roots explore a volume of soil. On the other hand, the lungs (or trachea in insects), which exhibit a striking dendritic form compacted in a small volume, provide extensive surface area for gas exchange between inhaled air and circulating blood. Roots penetrate a large volume of their environment with vast surfaces arranged dendritically; lungs and alimentary canals (with their microvilli) with the help of breathing and peristalsis, consume a part of the environment, as it were, and process it, thanks to the surface provided by their dendritic structure.

Economics of Water in Structures

In plants, the dendritic form, with maximum environmental contact, is achieved with a minimum of protoplasm, using water that is relatively abundant and energetically cheap to fill the vacuoles and provide most of the plant volume. To achieve the same surface area with protoplasm alone—proteins, membranes, and so on—would require far more food (photosynthate), which in energy terms is far more expensive. It is conceivable that leaf cells, stem cells, and root hair cells could consist entirely of protoplasm as many animal organs do. It is obvious, however, that selection in plants has favored cells in which the protoplasm often occupies less than 10 percent of the cell volume and the vacuole up to 90 percent. This is clearly an economy measure. It may also reflect the frequent shortage of fixed nitrogen. A plant that utilizes cheap water for a large fraction of its structure is able to devote more of its resources to storage and to its offspring, insuring survival of the species.

Position of Cytoplasm Aids Diffusion

The position of the cytoplasm—as a thin bag surrounding the vacuole, actually a thin layer between the vacuole and the wall—may be significant. In this way the diffusion distance, and especially the liquid phase distance, between the environment and the mitochondria and chloroplasts is reduced to the minimum. A structure favoring gaseous diffusion is nowhere more apparent than in leaves. The ramifying air spaces are in contact with up to 90 percent of the surface area of nearly all mesophyll cells, leaving as little as 10 percent of surface for structural contact with adjacent cells. When correction is made for the very low concentration of CO_2, leaves are more efficient at absorbing CO_2 from air than are lungs in absorbing O_2. Furthermore, leaves are completely unaided by any breathing action.

An alternative internal cell structure might conceivably consist of a more dilute protoplasm with chloroplasts and mitochondria scattered throughout the cell volume. Vacuole and cytoplasm would lose separate identity and be blended. Selection has obviously not favored such a dilute cytoplasm in plants. Perhaps the important factors were high concentrations of reactants and enzymes and an arrangement of all organelles as near the cell surface as possible, thus minimizing liquid diffusion resistance for carbon dioxide, oxygen, and reactants.

Cell Walls and Stressed Composite Structures

To achieve a "rigid" structure, the vacuole requires a relatively inelastic cell wall with high tensile strength. This cell wall consists partly of cellulose, which has remarkably high tensile strength comparable to steel piano wire. Water alone is generally considered far too fluid to be a useful structural material. When enclosed by a membrane and cell wall, however, its incompressibility becomes very useful. (Even air can be a "structural" material, as in pneumatic buildings.) The cellulose walls provide tensile strength, the water, compression strength; together they form a rigid structure. A football pumped full of water has the impact of a rock. This structural rigidity of turgid cells alone accounts for the rigidity of leaves, young stems, and indeed of many entire herbaceous plants. Witness the loss of structure or "rigidity" as the cells lose water—turgor pressure—and the plant wilts. Cellulose fibrils, for all their tensile strength alone, are incapable of holding the plant upright; they need the compression strength of water (or lignin in secondary tissues) to keep the plant upright.

The opposing of tensile and compressive forces leads to a "prestressed" type of structure, discovered by plants eons ago, and recently rediscovered by engineers in prestressed concrete and similar structures. Wood is also an example of prestressed structure. While the cells destined to become wood are extended by turgor pressure, lignin—a compound with compression strength—is gradually deposited between and among the stretched cellulose fibrils. Finally, the cell dies, leaving behind the cell wall consisting of stressed cellulose fibrils and compressed lignin: wood, a material with structural properties that have long been appreciated by man, and which is still unsurpassed for many uses—skis, tools, buildings, and so on.

Cellulose, although thermodynamically much more expensive than water, requires only slightly more energy in its synthesis than the simple sugar glucose and far less than lipids, proteins, or the chitin found in insect exoskeletons. Besides being the most abundant organic compound on earth, it is also one of the cheapest in terms of energy required for its construction, and one of the strongest.

While it seems clear that the primary role of vacuoles and cell walls is to provide a fairly rigid structure with maximum contact with its environment, it would not be surprising if these structures played other roles as well. Could the vacuoles serve as a reserve water supply, as has been suggested? A rapidly transpiring nonsucculent plant would lose all its vacuolar water in only one day, and it is ridiculous to consider the vacuoles of submerged plants as a water reservoir. Yet the succulent growth habit seems to be an adaptation for water storage in vacuoles, and this is often aided by closed stomates during the day.

Plants lack the complex excretory system of higher animals, but materials of no apparent function are often found in vacuoles. It is reasonable to suppose that such products are isolated from contact with the metabolic machinery of the cytoplasm by secretion into vacuoles, but since only a few vacuoles contain such materials, this waste storage is apparently one of their secondary functions. It would seem that short-lived root hairs could secrete wastes directly to the soil. Another such vacuolar function may be storage of essential ions. This may well be important in plants, but it is relevant to note that animals, including aquatic animals of various degrees of complexity, manage to survive quite nicely without vacuoles. They apparently depend upon plants to perform the first step of concentrating essential electrolytes from the dilute environment, just as they depend upon plants for their chemical energy, reduced carbon and nitrogen, vitamins, and miscellaneous other materials.

Mobility and Food Reserves

Plants have completely forfeited mobility, except in seeds and pollen (or spores and sperm in lower forms). The large dendritic form, built from compressed vacuolar water and cell walls (sometimes lignified), makes possible efficient absorption of dilute raw materials from the environment, but also results in structures that are too cumbersome to move. Neither is there much advantage in movement. Carbon dioxide and light come to the plant, as does rain, although less dependably. Adverse seasons are survived in the dormant state, or as seeds.

In plants, foods are typically stored as starch, although other polysaccharides and sucrose are also found. These compounds contain less than half as much energy per unit weight than fats and oils. But they are thermodynamically more efficient, because they are formed directly from the glucose phosphate produced in photosynthesis with little energy loss. Photosynthate may first be converted to fats and then back to acetyl CoA for oxidation, but at a cost of 12 percent or more of the potential energy, which is lost in the fat conversions. This "inefficiency" is obviously "tolerated" in animals, where mobility is at a premium and fat is the main reserve food. In plants, where mobility is not a factor and weight only a minor consideration, selection has favored the more thermodynamically efficient starch as the reserve food energy. Virtually all the energy resulting from photosynthesis is stored, available for later use, instead of only the 88 percent that would be stored if the carbohydrate were first converted to fat before storage.

It is only in seed (and spores) that any significant fat storage is found. Seeds are faced with two conflicting demands: seed dispersal, which is favored by light weight, and seedling establishment, which is favored by a large food or energy supply. It is certainly no accident that seeds of over 80 percent of plant species contain food reserves as fat rather than as starches. This is especially true of smaller seeds, such as orchids. The most notable exceptions are found in the agricultural cereal grasses, where man insures the dispersal.

Special Topic 9.1
A Theory of ATP Synthesis

For years scientists have asked how light causes ATP synthesis. No universally accepted answer to that question is yet available, but the most popular theory now is called the **chemiosmotic theory.** It was first suggested in 1961 by Peter Mitchell in England and applies also to ATP synthesis in mitochondria and procaryotes. In an abbreviated form it is really quite simple.

ATP synthesis from ADP and H_2PO_4 involves the formation of H_2O. It is this frequently overlooked H_2O that usually causes ATP to be broken down to ADP and $H_2PO_4^-$ again, and the equilibrium constant for the reaction strongly favors breakdown. Mitchell pointed out that if organisms had a way to remove most of this H_2O, ATP could not be broken down (hydrolyzed) so readily, and the laws of mass action would make its synthesis favorable. Adenosine triphosphatase enzymes (**ATPases**) catalyze this hydrolysis of ATP, and because enzymes simply speed up the rate at which a reversible reaction comes to equilibrium (see Section 8.2), ATPases must also speed the rate of ATP synthesis when thermodynamic conditions allow this.

Let us assume that ATP formation from ADP and $H_2PO_4^-$ occurs in hydrophobic membranes from which H_2O is being removed, as in Fig. 9.1-1. This figure indicates that H_2O is removed by transport of OH^- to one side of the membrane and H^+ to the other side. What would pull these ions away from each other? One answer is a *preformed* pH *gradient* in which the OH^- being removed goes to the side of the membrane previously loaded with more H^+ than OH^-, while the H^+ being removed goes to the other side, where OH^- predominates. Clearly, on both sides this meeting of OH^- and H^+ will cause formation of stable water molecules. Another answer is a *preformed electropotential gradient* caused by preferential buildup of cations (other than H^+) relative

to anions on one side of the membrane, because OH^- from H_2O will then be attracted to the electropositive side, and H^+ will be drawn toward the electronegative side. We shall emphasize the pH gradient and leave the discussion of electropotential gradients to more advanced treatment.

The problem in ATP formation, then, seems to be how to establish a pH gradient across an ATPase-containing membrane so that the elements of H_2O will be attracted out of the membrane. Mitchell's theory predicted that light-driven electron flow across chloroplast lamellae, oxygen-pulled electron flow in mitochondrial and certain procaryotic membranes, and even nitrate- or sulfate-pulled electron flow in anaerobic bacterial membranes, could cause the buildup of H^+ across these membranes. Figure 9-11 shows this buildup for chloroplasts. A second important aspect of Mitchell's chemiosmotic theory is that such membranes must be impermeable to both OH^- and H^+, and indeed they are, or nearly so.

Evidence for the chemiosmotic theory for ATP synthesis in chloroplasts was first provided by André Jagendorf and G. Hind at Cornell University in the early 1960s. First, they demonstrated that the pH of the stroma increased as much as 1 unit during light-driven electron flow. Later work of others showed that the pH decreases in the thylakoid channels. Second, if ADP and $H_2PO_4^-$ were withheld from the thylakoids while the lights were on, subsequent addition of those ions (with Mg^{2+} as a coenzyme) allowed ATP synthesis even in darkness. This meant that light caused the formation of something rich in energy that could be used to form ATP in the dark. Third, ATP was also synthesized in darkness when an H^+ gradient was imposed by an acid-to-base bath treatment. This treatment consisted of first exposing the chloroplasts to a weak organic acid (pH 4) that would penetrate into the thylakoid channels, ionize, and build up the H^+ concentration there. When the plastids were injected into a solution of pH 8 containing ADP, $H_2PO_4^-$, and Mg^{2+}, ATP was formed. This indicated that an H^+ gradient could replace the light requirement and that the previously mentioned energy-rich product of light is an H^+ gradient.

Subsequent results of Jagendorf and others showed that certain chemicals that destroy the pH gradient across a membrane will prevent ATP synthesis. These compounds are called **uncouplers,** because they remove the normal interdependence (**coupling**) between ATP synthesis and electron flow. Two kinds of uncouplers are known: the "ice-pick" type and the "ferryboat" type. The ice-pick type includes certain detergents that effectively disrupt the lipid-protein associations of the membranes so that H^+ and OH^- leak through. This nullifies the pH gradient and, of course, ATP synthesis, but it may actually speed up electron transport via the model of Fig. 9-11. This speed-up would

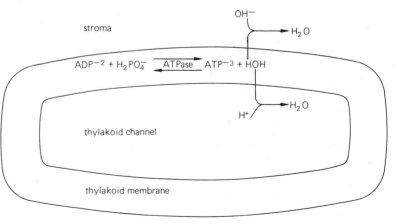

Figure 9.1-1 Proposed mechanism of ATP formation in hydrophobic membranes.

occur because then no H^+ or OH^- accumulation is occurring to create a backpressure against H^+ removal from the stroma or H^+ secretion into thylakoid channels.

The ferryboat uncouplers are weak bases that can pick up an H^+ on the acidic side of the membrane, carry it across as a neutral complex, then release it to OH^- on the other side, again destroying the pH gradient. Ammonium salts act as ferryboat uncouplers because of the free base ammonia with which they are in equilibrium, and this is apparently how NH_4^+ inhibits ATP synthesis in both chloroplasts and mitochondria. Dinitrophenol is another ferryboat uncoupler, and the resulting block of ATP formation in root mitochondria probably accounts for its inhibition of ion uptake (see Section 5.10).

One of the problems with the chemiosmotic theory has been to understand how electron transport in membranes could ever cause formation of a pH gradient. Critics have correctly mentioned that only if the carriers of H^+ and electrons were located asymmetrically in the membranes could such pickup of H^+ on one side and release on the other occur. Evidence is now accumulating that such asymmetry does exist.

Special Topic 16.1
Herbicides—John O. Evans

F.B.S. 1977

John O. Evans is an Associate Professor of Plant Science at Utah State University. His first love is the rational use of herbicides and other methods to control weeds in crops, a science that he learned at the University of Minnesota (Ph.D. in 1970). His special topic is one of the most important applied fields of modern plant physiology.

In addition to chemical fertilizers, modern crop production is increasingly dependent upon the numerous synthetic agricultural chemicals that have been developed over the past three or four decades, especially those used for weed and insect control. Improving production efficiency and reducing the human burden of raising crops involves the use of larger and more powerful tractors to farm larger acreages, chemical fertilizers to improve production, and insecticides and especially **herbicides** (weed killers) to replace hand labor. Herbicides also offer an alternative to the tremendous expenditure of fuel otherwise required to control weeds by mechanical means. Within one or two decades, most acreages of one or more of the major agronomic crops in the United States will be farmed under minimum or no-tillage concepts to save fuel and intensify production. Such programs rely heavily on herbicides and insecticides.

Agricultural chemicals will, then, be an integral part of our food production in the foreseeable future. In this brief discussion of the kinds of materials and their behaviors in plants and soils, only major classes will be mentioned. Although all members in a class may behave alike, the phenomenal results obtained with herbicides have resulted because the slightest modification of a basic chemical structure may drastically alter its effect on plants. For example, the methyl ester of **dichlorprop,** a phenoxypropionic herbicide, is active against dicots but relatively inactive against monocots. When the dichlorophenoxy moiety is separated from the alkyl portion of the molecule by having a second phenoxy ring placed between them, the plant response is totally reversed. The second herbicide is safe for such dicots as sugarbeets, soybeans, alfalfa, and many others, while it eliminates such monocots as wild oats, foxtail, and barnyardgrass!

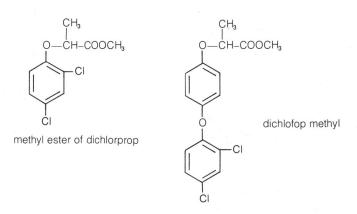

methyl ester of dichlorprop

dichlofop methyl

Equally fascinating is the herbicide **glyphosate** (trade name Roundup), which is highly phytotoxic to nearly all plant species although its chemical structure is composed of two parts, both common substances in plants. Glyphosate (N-phosphonomethyl glycine) is simply the amino acid glycine attached through a methylene group to phosphoric acid:

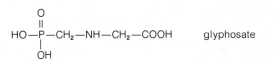

glyphosate

It is extremely reactive in plants and moves freely throughout the plant system to kill thoroughly the underground parts of such creeping perennials as quackgrass (*Agropyron repens* L.), johnsongrass (*Sorghum halepense* L.), Canada thistle (*Cirsium arvense* L.), and many others. On the other hand, it is extremely short lived in soils and plant residues, so treated acreages can be replanted with desirable crops after a few days. Production of nearly every crop will benefit from the discovery and registration of this valuable substance. (For other equally interesting examples, see Ashton and Crafts, 1973; Anderson, 1977.)

Aliphatic Acid Herbicides

These herbicides are aliphatic acids such as acetic, propionic, or butyric, with chlorine, bromine, or other activating groups substituted for hydrogen atom(s) to impart greater biological activity (usually making the compound more acid) and less susceptibility to plant or microorganism degradation. Glyphosate may be considered as phosphonomethyl amine, replacing one hydrogen on carbon number two of acetic acid. Dalapon (2,2-dichloropropionic acid) and TCA (2,2,2,-trichloroacetic acid) have been in use for over 20 years. They limit growth of grasses by preventing the development of leaf waxes. Both monocots and dicots exhibit reduced cuticles when treated with dalapon or TCA, but the lubricative function of the waxes in the expansion of grass leaves makes them much more sensitive to these herbicides.

How do they limit wax formation? We're not quite sure, but TCA as a strong protein precipitant may denature the enzymes required for lipid biosynthesis. TCA is not as selective as dalapon, probably because it also indiscriminately precipitates other critical enzymes. Dalapon's structure closely resembles pantoic acid, a precursor of coenzyme A, which is intricately involved in lipid and hence cuticle biosynthesis. Dalapon probably competes with pantoic acid. In either case, concentration is critical; higher concentrations kill both dicots and monocots.

Phenoxy Herbicides

These materials probably mimic the natural auxins. They upset the hormone balance in plants and cause callus growth on roots and stems, undifferentiated growth of phloem tissue, epinasty, and abnormal tropic responses. The 20 or so phenoxy herbicides may alter such processes as DNA transcription and protein synthesis. This places the most sensitive species in a competitive disadvantage in a mixed plant population, as when 2,4-D controls mustard or redroot pigweeds in cereals.

Benzoic Acid Herbicides

Like the phenoxy herbicides, this class of a half dozen or so registered materials consists of auxin-like compounds. They disrupt the normal balance of growth-regulating systems, causing epinasty of stems and leaves, and calluslike growths . . . on roots. They disrupt the vital transport of sugars in the symplast by inducing abnormal differentiation or possibly by causing some differentiated tissue cells to resume stages of active division.

Some compounds in this class are used as foliar-applied herbicides to reduce growth of older, well-established weeds, dicamba being an excellent example. Chloramben (3-amino-2,5-dichlorobenzoic acid) is applied to the soils of freshly planted soybean fields to selectively control many annual weeds in this crop. Their mechanism of action on plants is probably not greatly different from that of 2,4-D.

s-Triazine Herbicides

These herbicides inhibit the Hill reaction (splitting of water) in photosynthesis. Plants from a broad spectrum of types are susceptible to s-triazine toxicity; only those plants that escape by one means or another are spared. For example, corn degrades **atrazine** to an inactive product with the help of **benzyoxazinone,** a natural part of the sap in corn tissue:

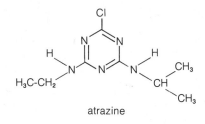

atrazine

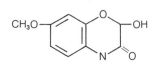

benzyoxazinone

When the chlorine atom in the s-triazine molecule is replaced with a hydroxyl group, the resulting compound is nonphytotoxic. Other plants besides corn can inactivate the chloro-s-triazine herbicides, some in the manner just explained, while other species remove the alkyl sub-

stituents from the amino side groups. Other factors influence the tolerance of a species to the chloro-s-triazines. The herbicide can be stored harmlessly in special tissues such as the lysigenous glands of cotton. Other species (e.g., fruit trees) avoid its toxic action with deep roots below the soil where it is applied.

Atrazine is one of the most important herbicides today. It is absorbed by the roots and moved in the xylem to the leaves, causing chlorosis, necrosis, and death. Atrazine disrupts the chloroplast structure as it inhibits photosynthesis. Each of the approximately 30 herbicides in this class owe their phytotoxicity to this reaction, but each has its own unique selective action and usefulness.

Dinitroaniline Herbicides

The dinitroaniline herbicides are some of the most popular in use today. They possess extremely low water solubility, so they are degraded slowly in plants and soils and are taken up by roots only in extremely small amounts. They enter primarily as vapors, being absorbed into the coleoptile or hypocotyl about the time seedlings emerge from the soil. Herbicidal effectiveness disappears soon after the plants emerge.

These herbicides are poorly translocated within plants. Their action results from the extremely powerful inhibiting action on plant cell division. They stop lateral root formation, and they can stop the growth of the primary root tip, especially when it is very young. This group is often referred to as "root pruners." They are selective because plant entry, metabolism, and movement are often greater for the weeds than for the crops in which they are used.

Most dinitroaniline herbicides readily vaporize and disappear rapidly when applied to the soil surface. Hence, it is advisable to combine herbicide application and soil mixing into a single operation where possible.

Carbamate Herbicides

Chemical compounds that are derived by substitution of alkyl or aryl groups onto carbamate, thiocarbamate, or dithiocarbamate are important herbicides, insecticides, and fungicides. The important herbicides are derived primarily from carbamate and thiocarbamate, with approximately 10 widely used herbicides in each of these groups. The formulated herbicides have stable shelf lives and are convenient to use. When applied to moist soil, they are degraded in 1 week to 2 or 3 months. In addition, they volatilize readily and, like the dinitroaniline herbicides, should be immediately mixed with the soil. For the most part, these compounds enter the plant through the roots and move upward in the transpiration stream to stems and leaves, although a few are taken up by leaves.

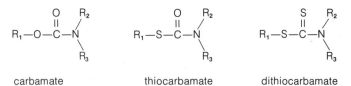

carbamate   thiocarbamate   dithiocarbamate

There are interesting differences in the herbicidal actions of the carbamates and the thiocarbamates. Both inhibit stem and leaf growth, but only the carbamate group inhibits root growth. Thiocarbamate herbicides inhibit grass shoot elongation in about the same way as the aliphatic herbicides, producing closely similar symptoms. The carbamate herbicides, on the other hand, act much like the dinitroanilines; they cause root tip swelling and stunting and inhibit mitosis of roots and shoots. Finally, the carbamate herbicides but not the thiocarbamates inhibit photosynthesis. The most plausible basis for these differences rests in the basic structures of the two groups: Carbamates have a peptide-type linkage ($R\!-\!O\!-\!\overset{\overset{O}{\|}}{C}\!-\!\overset{\overset{H}{|}}{N}\!-\!R$) capable of strong hydrogen bonding, whereas thiocarbamates, to be herbicidally active, have both amine hydrogens replaced by other groups ($R\!-\!S\!-\!\overset{\overset{O}{\|}}{C}\!-\!\overset{\overset{R}{|}}{N}\!-\!R$), which greatly restricts their bonding potential.

Urea Herbicides

Nearly every student of plant physiology has witnessed the results of an overdose of urea fertilizer on lawns or crops. Large amounts of this substance are damaging to annual plants or weeds. About 30 years ago, agronomists discovered that certain substitutions for the hydrogens of urea molecules made them much more toxic and, of even greater importance, capable of killing specific plants in mixed populations. Today there are about two dozen commercial substituted urea herbicides, and dozens more are being tested. They form one of the largest classes of herbicides, both in numbers and in economic importance.

The herbicide analogues of urea are usually much more stable in soils and plants than urea, remaining active from a few weeks to several months. They are absorbed by plant roots and carried to the aerial parts where they inhibit the Hill reaction of photosynthesis. In fact, DCMU (**diuron**, Karmex®) and CMU (**monuron**) have contributed significantly to our present understanding of photosynthesis. These two herbicides were among the first materials discovered in this class, are largely nonselective (tolerated primarily by plants that have deep tap roots when herbicides are applied to soil surface), and they are quite persistent in soils. Monuron is commonly considered a soil sterilant, since its inhibitory action may last from one to three years in soils. Numerous recent additions to the urea herbicides have improved selectivity and crop safety. Some of the more recent ones have shown increased potential for foliar penetration of plants, allowing an additional degree of freedom in how they are applied.

Uracil Herbicides

Although the uracil herbicides structurally resemble uracil, substituting for it (e.g., in nucleic acids) or inhibition of uracil metabolism apparently does not account for their toxic action. Like the *s*-triazine and the urea herbicides, they inhibit the Hill reaction of photosynthesis. All other metabolic systems studied have either proven to be unaffected, or they are at least one thousand times more tolerant to uracil-like herbicides than the chloroplastic processes. Both monocots and dicots are susceptible to **bromocil** (Hyvar®) and **terbacil** (Sinbar®), presently the two most prominent members of the class. Yet there is a practical selectivity. For example, deep-rooted perennial plants like citrus, fruit trees, alfalfa and mint, escape the toxic action of substituted uracil herbicides, while shallow-rooted annual weeds cannot escape. Physiological differences that allow the roots of one species to take up more herbicide or to translocate it more efficiently to its site of action also contribute substantially to differential susceptibility.

These herbicides often remain in the soil for one year and sometimes longer, depending on soil type and microbiological activity.

Several classes of soil microorganisms provide the primary means of degrading the uracil-like herbicides in the environment. Generally speaking, herbicides have little or no effect upon native soil microorganisms.

That all the effective agricultural chemicals have not been discovered yet has been substantiated recently with the discovery of herbicides such as glyphosate and many others. In fact, most of the major discoveries in chemical technology in agriculture probably are yet to be made. Teams of chemists, biochemists, physiologists, and agronomists, now a common phenomenon, are more likely to be successful than individuals working independently, since their approach is more methodical and has a broader base. Empirical screening under a narrow set of circumstances often fails to recognize the true potential of a given substance; unquestionably, many worthwhile substances have been set aside. With the appropriate interests and ambitions, honed to a razor's edge by training in plant physiology and agriculture, you could become a member of one of those teams destined to revolutionize agriculture still further!

Special Topic 19.1
Possible Mechanisms of Light Action in Photomorphogenesis

Responses to Blue Light

We discussed certain responses to blue light in Chapters 18 and 19, including phototropism, phototaxis, and various HIR effects such as inhibition of stem elongation, promotion of grass leaf unrolling, inhibition of seed germination, and promotion of flavonoid synthesis. Our question now is how light causes such responses. For those in which phytochrome plays little or no role (phototropism and phototaxis) and for others in which a yellow flavoprotein participates in addition to phytochrome, a major task is to identify this flavoprotein and learn how and where it acts. Present evidence suggests that it is in the plasma membrane and that, upon absorbing blue or violet light, it reduces a cytochrome also present there. The remaining biochemical processes await discovery (Briggs, 1976).

Responses to Phytochrome

When we consider the numerous morphogenetic effects caused by P_{fr}, it becomes apparent that many chemical reactions must be affected by this pigment. Even for etiolated mustard seedlings, at least two dozen effects have been found by Hans Mohr and others in Germany (see Table 19-1). Their research in the 1960s was guided by a hypothetical model in which P_{fr} somehow activates previously repressed genes, so that messenger RNA and enzyme molecules not previously present could be formed. These enzymes were logically assumed to cause the responses observed. In fact, they measured changes (usually increases) in activities of several enzymes in these seedlings (Mohr, 1974). In different plants, other investigators have also observed changes in activity of certain enzymes consistent with that model. For enzymes exhibiting increased activity, some are synthesized *de novo* (anew) in response to P_{fr}. Inhibitors of protein synthesis such as cycloheximide or puromycin usually prevent these increases, and sometimes actinomycin D, an antibiotic that blocks RNA synthesis, is also inhibitory.

The incorporation into these enzymes of radioactive amino acids or amino acids labeled with deuterium (D = ^{2}H) or ^{18}O by allowing the tissues to absorb heavy water (D$_2$O) or H$_2^{18}$O substantiates de novo synthesis. There is usually a lag of at least 3 hours after light treatment before increases are detectable. This would allow sufficient time for production of new messenger RNA molecules that code for the enzymes (transcription) and for enzyme synthesis itself (translation). Yet it is possible that P_{fr} enhances enzyme synthesis only by effects on translation.

A principal reason for suspecting that P_{fr} stimulates translation is that red light causes conversion of monoribosomes (free ribosomes) to polyribosomes in certain tissues prior to a morphogenetic response. This is true for photodormant lettuce seeds, Pinus thunbergii seeds, and etiolated bean and corn leaves. In corn leaves, even the free cytoplasmic ribosomes are somehow activated by P_{fr}, because they become about 70 percent more effective in protein synthesis when used in an in vitro system to which is added an artificial messenger RNA, transfer RNAs, amino acids, ATP, GTP, K$^+$, and so on (Travis et al., 1974). The effect is detectable within 30 minutes. Since it occurs with an artificial messenger RNA, any effect of P_{fr} upon messenger RNA synthesis in the leaves is separate from this response. Furthermore, a brief exposure to red light (nullifiable by immediate subsequent far-red) or to continuous far-red (which maintains P_{fr} at a low but constant level) enhances the conversion of monoribosomes to polyribosomes in etiolated bean leaves when messenger RNA synthesis is almost totally blocked by cordycepin (Smith, 1976b). The response involves both cytoplasmic and chloroplastic polyribosomes and occurs within 1 hour. These results indicate that enzyme synthesis can be stimulated by P_{fr} in the absence of transcription, strongly arguing against Mohr's earlier interpretations. There is also much evidence that P_{fr} rapidly becomes attached to some unidentified membrane after it is formed by red light (Chapters 8, 9, and 11 in Smith, 1976a). Since this attachment occurs in many species and takes place in less than a minute, it is probably an important primary effect that precedes other photomorphogenetic responses, but the biochemical relationships are still unknown.

The effects of P_{fr} on enzyme synthesis are probably indirect; that is, not caused by P_{fr} itself but by some metabolic change it produces. The first evidence for a more rapid morphologically observable response to P_{fr} came from studies of sleep movement in Mimosa pudica and Albizzia julibrissin (see Section 18.8, Fig. 18-11). An HIR response to blue and far-red is responsible for return of the leaflets to their normal day position at sunrise, but low-intensity light absorbed by phytochrome controls whether or not the leaflets close when placed in darkness. If the leaflets are exposed briefly to red light just before dark, they almost immediately begin to close slowly in darkness. If a short far-red treatment is given instead, the leaflets remain open in the day position. Both red and far-red effects are reversible by light of the other color if the first treatment is followed immediately by the second. The rapid beginning of closure in red light cannot result from effects on either transcription or translation but is caused by movement of K$^+$ salts and water out of cells in the ventral side of the pulvinus and into cells in the dorsal side (see Fig. 18-12). A quick change in permeability of the membranes to K$^+$ seems to be responsible (Galston and Satter, 1976).

Another rapidly observable response to P_{fr} was discovered in 1968 by Takuma Tanada, at Beltsville. In the presence of IAA, ATP, ascorbic acid, K$^+$, Cl$^-$, and Mn^{2+}, barley root tips stick within seconds to a negatively charged glass beaker (made negative by washing with phosphate) if phytochrome is in the P_{fr} form. If P_{fr} is then converted to P_r, the tips are rapidly released, especially if ABA is present. This is known as the **Tanada effect.** The phytochrome responsible for this effect is localized in the root caps. A similar response occurs with mung beans,

except that ABA promotes attachment and facilitates release, and IAA has opposite effects. The mung bean response is accompanied by development of a positive electrical potential at the tip surface when P_{fr} is formed and a negative potential when P_{fr} is converted to P_r. Similar electrical potential changes have also been found in oat coleoptiles and pulvini of Samanea saman and are probably caused by movement of H$^+$ and/or K$^+$ across the membranes, again indicating a membrane permeability change. Since transport of both K$^+$ and H$^+$ is apparently driven by a plasma membrane-bound ATPase (see Section 5.10), P_{fr} might influence such an enzyme rather directly. Consistent with this, phytochrome exists in the plasma membrane of the alga Mougeotia (Haupt, 1973), and P_{fr} becomes bound to some cellular membranes of higher plants, as mentioned earlier.

At least two enzymes that exhibit rapid rises in activity after P_{fr} formation are exceptions to those that are synthesized de novo only after a lag of about 3 hours. These exceptions are phenylalanine ammonia lyase (Attridge and Johnson, 1976) and nitrate reductase (Johnson, 1976) in white mustard seedlings. For nitrate reductase, the rise in activity immediately follows light treatment, showing that activation of pre-existing enzymes occurs and that de novo synthesis is not involved. Thus, we have mentioned three ways in which P_{fr} can stimulate enzyme activity: through relatively slow effects on transcription and on translation requiring 30 minutes to about 3 hours, and an effect involving activation, not synthesis, that is sometimes immediate.

Another rapid response to P_{fr} occurs during unrolling of etiolated wheat and barley leaves. In wheat leaf sections, the unrolling process itself requires about 18 hours for completion, and no stimulation by red light can be observed until about 12 hours after exposure. GA$_3$ is as effective as red light in promoting unrolling, but its effects also require hours to be detected. The hypothesis that P_{fr} acts by promoting gibberellin synthesis was tested by measuring amounts of gibberellins in leaf sections kept dark or treated with red light. Several reports show that a far-red reversible promotion of gibberellin levels can be observed within 10 min after red light is given (see Evans, 1976). These processes occur mainly or entirely in etioplasts and can be caused even when the etioplasts are removed from the cells. Furthermore, much of the rise is caused by P_{fr}-enhanced efflux of gibberellins from the etioplasts into the cytoplasm or, in the case of isolated etioplasts, into the medium (Evans, 1976; Cooke and Kendrick, 1976). Two significant conclusions can be made: Phytochrome probably exists in the envelopes of etio-

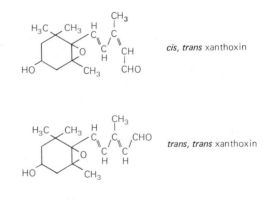

Figure 19.1-1 Two isomeric forms of a growth inhibitor, xanthoxin. The only difference between the two is the arrangement of the methyl group, H atom, and aldehyde group at the terminal end of the side chain. Both contain an oxygen atom attached to the ring in an epoxide form.

plasts (and therefore probably also in envelopes of mature chloroplasts), and P_{fr} promotes gibberellin release from them independent of any effect on the cell's plasma membrane. Since gibberellins also break photodormancy of seeds, P_{fr} might promote germination by enhancing gibberellin production in the radicles or release of gibberellins from proplastids in the radicles.

Now let us deal with the inhibitory effects of light on stem elongation. Light might either cause formation of an inhibitor such as ABA or stimulate destruction of a growth promoter, most likely a gibberellin. Whatever the answer, the responses are relatively fast in hypocotyls of etiolated gherkin seedlings. The inhibition by blue light occurs dramatically within a minute, red within 30 min, and far-red within an hour (Meijer, 1968). For etiolated pea seedlings, however, stem elongation is not inhibited by red light until after a 6-hour lag period and is not extensive unless brief red treatments are repeated at frequent intervals. (This need for repeated exposures is not unusual in phytochrome responses and is apparently necessitated by destruction of P_{fr} and its dark reversion to P_r.)

Part of the P_{fr}-retarded elongation of pea stems seems to result from enhanced synthesis of a potent aldehyde inhibitor similar to absicisic acid called **xanthoxin** (Wain, 1975). Xanthoxin exists in plants in two isomeric forms, the *trans, trans* and the much more inhibitory *cis, trans* form (Fig. 19.1-1). Xanthoxin is synthesized by degradation of the carotenoid pigment **violaxanthin,** a xanthophyll. In etiolated pea seedlings exposed to 1 min of red light every 30 min, growth inhibition is accompanied by a large rise in the amount of violaxanthin and other carotenoids and of both xanthoxin isomers (Anstis et al., 1975). The rise in carotenoids might result from P_{fr}-stimulated development of etioplasts into chloroplasts, and the xanthoxin then arises from an additional light-promoted degradation of violaxanthin. These effects might be rapid enough to explain growth inhibition by red light in etiolated stems, but blue effects were not tested. Perhaps blue light absorbed in the etioplasts by a flavoprotein or by violaxanthin itself could cause xanthoxin synthesis soon enough to explain its immediate inhibition of gherkin hypocotyl elongation. Nevertheless, there is also evidence that P_{fr} inhibits gibberellin synthesis in whole *Phaseolus coccineus* seedlings in which stem elongation is inhibited by light (Bown et al., 1975). To understand photomorphogenesis, we shall apparently have to learn more about changes in both inhibitory and stimulatory hormones, and analyses of these hormones will probably have to be made in those cells that actually respond to light.

(For references, see the bibliography for Chapter 19.)

Special Topic 23.1
Ultraviolet Radiation and Plants
—Martyn M. Caldwell

Martyn Caldwell is a Professor of Ecology at Utah State University. Since his doctoral work at Duke University (Ph.D., 1967) he has been concerned with solar ultraviolet effects on plants, serving with various evaluation committees for NASA, the Department of Transportation, and the National Academy of Sciences.

Ultraviolet radiation is nonionizing electromagnetic radiation of wavelengths just shorter than those normally perceived by the human eye. Because photons of radiation have higher energy levels at shorter wavelengths, UV radiation is particularly effective in causing photo-

chemical reactions, and this also applies to biological systems. Interest in the possible effects of UV radiation on plant growth and development was initiated in the latter part of the last century following the discovery that it was UV radiation in sunlight that caused sunburn of human skin and that UV radiation could also kill bacteria.

By convention, UV radiation is normally considered in three basic wavebands: UV-A, 315–400 nm; UV-B, 280–315 nm; and UV-C, shorter than 280 nm. Ultraviolet radiation in sunlight striking the earth's surface is composed almost exclusively of wavelengths longer than 295 nm because of the efficient absorption of shorter wavelengths by the thin ozone layer in the upper atmosphere. This is fortunate, since the most energetic photons occur in the UV-C part of the UV spectrum. These wavelengths are highly efficient in killing most microorganisms with relatively short exposures. This effectiveness has been used extensively in sterilization procedures with low-energy germicidal lamps, which emit radiation principally at 254 nm and are not only effective in killing microorganisms but, with sufficient exposure, can kill or severely injure most higher plants as well.

The principal compounds in plants that absorb the short-wavelength radiation are proteins and nucleic acids. The latter group, particularly DNA, can be effectively incapacitated from performing its normal functions in the plant cell when even rather low doses of UV-C radiation are absorbed. It is not surprising that a variety of physiological manifestations of altered DNA molecules has been noted when either higher or lower plants are exposed to UV-C radiation: increased mutation rates, chromosomal aberrations, disruption of membranes, and so on. If the exposure lasts very long, the plants are normally killed. Undoubtedly, when plant life on earth was still evolving and the earth's atmosphere still had only trace amounts of oxygen (and therefore almost no ozone, since ozone is formed in the stratosphere from oxygen), the shortwave UV radiation must have been a major constraint in evolution of plant life, particularly in terrestrial environments. We shall discuss later the remarkably elaborate mechanisms that occur in all organisms to repair the UV damage to nucleic acids.

The UV-A portion of normal sunlight also causes a variety of photochemical reactions and consequent physiological manifestations in plants. These include such phenomena as repression of growth, changes in secondary metabolites such as flavonoids and alkaloids, and induction of chloroplast movement in certain algae. These various effects do not seem to be related to a common type of absorbing pigment. Several chromophores for UV-A effects have been proposed. These include carotenoids and flavins as responsible for phototropic bending in *Avena* coleoptiles, and cytochromes or quinones have been implicated in the UV-A-induced impairment of respiration. Also, both forms of phytochome absorb in the UV-A portion of the spectrum and thus could be responsible for many of the phenomena resulting from exposure to UV-A radiation. Most of the physiological phenomena brought about by UV-A radiation can be induced even more effectively by visible radiation. Certainly, most responses to UV-A radiation are not nearly so pronounced as are responses to UV-C radiation. Therefore, both plant physiologists and ecologists have shown comparatively little interest in specific photobiological reactions caused by UV-A.

In many ways the UV-B is the most interesting portion of the spectrum, because many of the plant responses to UV-C radiation also occur in the UV-B. In an ecological vein, it is the UV-B portion of sunlight that is also of greatest interest. This is the most variable portion of the solar spectrum reaching the earth's surface, and it is also the most **actinic;** that is, for radiation reaching the earth's surface, these are the most photochemically potent wavelengths. It is of more recent concern that manmade agents, which might partially destroy the earth's thin ozone layer, would affect the UV-B portion of the terrestrial solar spectrum.

Solar Ultraviolet Radiation
as an Ecological Factor

During the Precambrian age, about 700 million years ago, oxygen concentrations of our atmosphere may have been only 1/1000 of present-day values, so there was no effective ozone in the atmosphere, and UV radiation of wavelengths as short as 210 nm reached the surface of the earth and probably penetrated as much as 5 m into the clear waters of oceans and lakes. Until the oxygen (and consequently ozone) concentrations increased due to the evolution and proliferation of green plants, solar UV-C radiation was undoubtedly a severe limitation for terrestrial plant and animal life. Under present-day conditions, there is an equilibrium amount of ozone that is continually formed and broken down by short-wavelength UV-C radiation in the upper reaches of the stratosphere. If this ozone layer were condensed to standard temperature and pressure, it would be only about 3 mm thick. Nevertheless, this very thin layer is sufficient to absorb completely all UV-C radiation and almost all UV-B radiation of wavelengths shorter than approximately 295 nm. Thus the amount of UV-B radiation penetrating the atmosphere constitutes a fraction of 1 percent of the total solar radiation incident on the earth's surface. Radiation in the UV-A band is more abundant and is not appreciably reduced by ozone absorption, but it is not of great interest anyway, as we have seen.

UV-B radiation is actinic because tails of the absorption spectra of proteins and nucleic acids slip into the UV-B spectrum—in fact, to wavelengths as long as 315 nm. Protein and nucleic acids do not absorb nearly as strongly in the UV-B as they do in the UV-C, but what they do absorb is enough to be of substantial biological consequence. In man, sunburning and most skin cancers can be attributed to the minute amount of UV-B radiation absorbed by these sensitive biological molecules.

Photosynthesizing plants have, of course, evolved to be efficient in collecting solar radiation to harvest the sun's energy. A necessary consequence is the absorption of the actinic UV-B radiation. It turns out that for most plants, very little of this UV-B radiation is reflected from the leaves. None is transmitted, so it is absorbed.

The consequences of solar UV-B absorption by plants in nature are not well understood. Although a wide range of physiological and photobiological studies have been conducted in laboratory situations with artificial sources of UV radiation (particularly with 254-nm radiation, which is so effectively absorbed by nucleic acids) the effectiveness of UV-B radiation at the intensities received on the earth's surface has not been well studied. There are several reasons for this. First, UV-B is a difficult type of radiation to produce by lamps, particularly if a realistic simulation of terrestrial UV-B radiation is sought. Second, most of the UV-B responses to the intensities received on the earth's surface are rather subtle. Undoubtedly, plants have evolved rather well to cope with the amount of UV-B radiation to which they are normally exposed.

Coping with Solar UV-B Radiation

There are two basic mechanisms by which plants can cope with this actinic radiation. The first is avoidance. Since green plants require the solar radiation containing UV-B for photosynthesis, avoidance of UV-B must take place by screening or filtering of the radiation that impinges on the leaves or other plant parts. Cuticular waxes may in some cases absorb some of this UV-B radiation, but the compounds that appear to be of greatest benefit are the flavonoids and related compounds. These compounds, which can either be colorless (and hence transmit the visible wavelengths effective in photosynthesis) or compounds such as the red anthocyanins, are likely candidates as important screening compounds: (1) They have high absorption coefficients in the UV-B portion of the spectrum, (2) they do not themselves appear to be chromophores (that is, pigments that absorb photons and mediate subsequent photochemical reactions), and (3) they do not appear to have any other known metabolic functions. In work I have conducted with alpine plants in high altitudes, where the solar ultraviolet radiation is somewhat more intense, I concluded that these compounds are important components in reducing the transmittance of the epidermis of these plants to UV-B radiation. Also, I found that the degree of damage that could be effected by intense UV-B radiation under laboratory conditions was inversely correlated in some of these plants with the amount of these flavonoid pigments.

Nevertheless, a certain amount of this UV-B irradiance is still probably absorbed by physiological targets, especially nucleic acids. Thus a second means of coping is to repair the damage done by UV radiation. For nucleic acids, the types of damage that can occur include breaking of the DNA chain, formation of linkages between DNA and protein molecules, or formation of linkages between different strands of DNA, all of which can impair DNA function. By far the most frequent lesions, however, are **thymine dimers** formed between adjacent thymine bases in the DNA chain. These dimers also impair normal DNA function in the cell by preventing normal replication and also by interfering with control of RNA and hence protein synthesis. Such thymine dimers are thought to account for many of the physiological lesions induced by UV-C and, to a certain extent, by UV-B radiation.

A process knows as **photoreactivation** was first discovered in certain soil microorganisms by Albert Kelner at the Carnegie Laboratories at Cold Spring Harbor, New York, in 1948. He found that if these organisms were exposed to the daylight coming through a north window following UV irradiation, a substantial amount of the damage induced by UV radiation was repaired. It has since been learned that photoreactivation results in splitting of the thymine dimers. Such repair has been found in a wide range of species throughout the plant and animal kingdoms. Photoreactivation is accomplished by an enzyme system that must be driven by radiation in the UV-A and the blue part of the visible spectrum. Since this radiation is normally abundant in sunlight, either concomitant or subsequent exposure of plants to these longer wavelengths can be effective in repairing a large percentage of the thymine dimer lesions.

There are also some very intricate repair mechanisms that do not require light and are referred to as dark repair systems. Until recently, it was thought that these did not occur in green photosynthesizing plants, but the recent experiments of Gary Howland at the Oak Ridge National Laboratory have demonstrated that dark repair systems do operate in green plants. **Excision repair** involves a complement of enzymes that first survey and recognize lesions in the DNA, excise the damaged segment of the DNA strand, synthesize a replacement segment using the complementary DNA strand as a template, reinsert this replacement segment, and couple the strand back together. Such an elaborate sequence of enzyme-mediated steps is a very effective means of repairing any type of DNA lesion. Although it is quite effective in repairing UV-induced lesions, it may also be an important system for repairing other mistakes or lesions in DNA caused by other environmental factors.

Although it might be argued that these repair systems evolved when the earth's atmosphere was lacking in ozone to absorb UV-C radiation and is now only a vestigial element in plants and animals, this is rather unlikely. The apparent ubiquity of these repair systems throughout the plant and animal kingdoms among species that are normally exposed to solar radiation makes it unlikely that these systems would be a vestigial element held over for the last 700 million years in so many diverse groups of organisms. This is indicated by several lines of evidence. In humans, for example, there is a hereditary skin condition known as *xeroderma pigmentosum* that has a syndrome of severe sunburn and skin cancers developing at an early age in life (in fact, individuals rarely live beyond the mid-teen years). This condition has been traced to a single deficient enzyme in the dark radiation repair process. Such individuals must take extreme precaution not to be exposed even to indirect sunlight for a matter of minutes. Repair-deficient mutants of certain bacteria have also been developed. These bacteria are killed by even the briefest exposure to sunlight. Analogous cases among higher plants in nature are not known, largely because if mutant individuals did appear in the natural populations, they would be eliminated before they might be recognized. A few years ago, I performed an experiment in which alpine plants were exposed to UV-B radiation carefully simulated in quantity and exposure rate to be equal to that received in a single day in the alpine environment. This UV-B radiation was, however, presented to the plants without concomitant or subsequent exposure to UV-A or visible radiation for a period of twenty-four hours. Several of these plants developed severe UV radiation lesions, while control plants that were exposed to the same UV-B treatment but with a subsequent exposure to visible light did not develop the lesions. These lines of evidence suggest that repair systems are important for survival under present-day UV-B radiation conditions.

Implications of a Reduced Atmospheric Ozone Layer

Plants have apparently evolved to cope with solar UV-B irradiance by screening and repair mechanisms; in fact, experiments performed in the field where solar UV radiation was removed by special filters have generally indicated that this removal does not really enhance plant growth or survival. I performed such an experiment with alpine plants in a natural alpine community at 3700 m elevation in the Colorado Rockies and found little difference in plant growth or success. This refuted some of the earlier experiments performed in Europe that had suggested there were substantial differences due to solar UV radiation removal. These older experiments were, unfortunately, confounded with a number of experimental problems. More recently, similar UV removal experiments with agronomic crops in Florida also suggested that removal of natural solar UV radiation did little to enhance or change the growth and yield of agronomic plants. Thus, plants seem to have evolved to cope with present-day intensities of this irradiance.

In the last few years, a strong interest has developed in UV-B radiation because of suggestions that man's activities may lead to a partial depletion of the high-altitude atmospheric ozone layer. Oxides of nitrogen, which are contained in the exhaust emissions of supersonic aircraft, or chlorine resulting from the breakdown of the freon compounds that are stable until they reach the upper atmosphere, will catalyze a partial destruction of the atmospheric ozone layer. The freon compounds come from refrigerants and the propellants used in aerosol spray cans. The exact amount of destruction is still in debate, yet it would probably be less than 15 percent. It is possible to calculate the amount of increased solar UV-B radiation that would penetrate the atmosphere. There would be a very small increase of shorter-wavelength UV-B radiation. Yet because the extra short wavelengths that would be transmitted to the earth's surface by a reduced ozone layer are so biologically effective, there is certainly valid reason for concern. Although some predictions have been made concerning the increased incidence in human skin cancer, it is not yet possible to anticipate effects on plant life. The key question that needs to be resolved is whether or not the avoidance and repair capacities of plants would be sufficient to cope with an added load of UV-B radiation. Some of the most recent experiments my graduate students have performed indicate that exposure of plants to carefully simulated UV-B irradiance that would be expected with reduced atmospheric ozone may lead to decreased photosynthesis and leaf expansion in some sensitive species. This might in turn lead to a change in the competitive balance among plants. Nevertheless, this is presently difficult to extrapolate to plants growing in nature.

It is apparent from these studies and also from the earlier studies of Cline and Salisbury and others that plant species vary considerably in their sensitivity to ultraviolet radiation. This undoubtedly would be reflected in a differential effect on these species if solar UV-B intensities were increased. Some of this variation in sensitivity may be due to the ability of plants to screen out UV radiation before it reaches physiologically sensitive targets. Perhaps part of the differential sensitivity is due to differences in molecular radiation repair capacity, such as photoreactivation of excision repair. Certainly, much remains to be learned, not only about how plants cope with current-day intensities of this irradiance but how they might be affected by increases of this irradiance that could result from man's activities.

Special Topic 23.2
Physiological Plant Pathology
—Penelope Hanchey-Bauer

Penelope Hanchey-Bauer is an Associate Professor of Plant Pathology at Colorado State University, where she studies the physiology and ultrastructure of host-parasite relations in plants. Her doctoral degree was earned at the University of Kentucky in 1968.

It is not surprising that plant parasites affect the metabolism and chemical compositions of their hosts. Alterations have been recorded in cell walls, membrane permeability, photosynthesis, respiration, growth regulator metabolism, and secondary products of metabolism. Obviously, many of these changes are secondary consequences of infection. Yet the physiological plant pathologist has examined even these in attempting to understand the entire disease syndrome.

Plants are exposed to incalculable types of bacteria, fungi, viruses, insects, and nematodes, *yet susceptibility to these agents of disease is the exception, not the rule.* And the nature of the damage caused by these pathogenic agents on their respective "host" plants is dependent not only upon the type of pathogen but also upon the physiological condition of the plant and the environment.

The symptoms of a typical diseased plant encountered in nature frequently include stunted growth and some chlorosis or even death. In short, the "host" plant appears starved, so much so, in fact, that it seems more appropriate to term the plant a victim rather than a host. As the pathogen becomes active within the plant, it derives nutrients for its continued existence and/or reproduction from the plant. Just this inability to control its own resources will eventually doom the plant.

Carbohydrate Metabolism of the Diseased Plant

Effects on photosynthesis are a common result of pathogenesis, yet generalizations are difficult to make. Virus infections usually lead to decreased photosynthesis. This decrease is not simply a result of less chlorophyll, since CO_2 fixation is lower even when expressed per unit of chlorophyll or per chloroplast. Viruses that invade phloem may interfere with its function and thus disrupt normal translocation. Indeed, an accumulation of carbohydrates in leaves infected with such viruses has been found. These carbohydrates could inhibit photosynthesis by feedback effects or interfere with chlorophyll synthesis.

An enhanced dark fixation of $^{14}CO_2$ has been reported for several fungal diseases. Some of this increase may be due to the ability of the fungus to fix CO_2, but stimulations of photosynthesis have also been observed in the absence of the living pathogen following application of toxic products of the pathogen known to cause fully all symptoms of the disease (**pathotoxins**). Enhanced CO_2 fixation may occur via a C_3 acid (malic enzyme or phosphoenolpyruvic carboxylase) or by increased turnover of ribulose-5-phosphate and ribulose-1, 5-diphosphate. Photosynthesis may also be stimulated in noninfected leaves of diseased plants when photosynthesis of infected leaves is inhibited. Associated with this stimulation is an increased translocation from noninfected to infected leaves. This activity may benefit the parasite, since the nutrients can be used to further its growth.

In later stages of infection, certain parasites, termed **obligate parasites** because in nature they require a living plant on which to complete their life cycles, induce the appearance of regions termed **green islands** around infection sites. Although the rest of the infected tissue may have yellowed, the tissues around the parasite act as a metabolic sink and remain green and photosynthetically active. It is possible that a parasite product, acting much like a cytokinin, may be responsible for green islands. Nevertheless, the infected tissue remains functional long enough for the obligate parasite to extract substances necessary for its own growth.

Infection of plants is also characterized by an increased respiration. The rate of respiration is influenced by several factors, one of the most important being the availability of ADP. The levels of ADP are higher when there is an increased turnover of ATP such as will occur during rapid growth. Indeed, in many diseases caused by obligate parasites, there is an increased size and dry weight of infected tissue. A shift from a predominantly glycolytic-Krebs cycle sequence of metabolism to the pentose phosphate respiratory pathway has also been described. In some cases, this shift parallels the increased respiration. In other diseases, the shift occurs much later and may be predominantly associated with fungal metabolism.

When a pathotoxin was substituted for a pathogen, a shift to the pentose phosphate pathway paralleled the increased respiration. The fate of ^{14}C-labeled glucose was followed and found to appear in several cell wall components. The significance of this change may be related to the unusual cell wall modifications found early in many diseases. These wall modifications appear to form in part by the accumulation of products of dictyosomes between the plasma membrane and cell wall. Their function is not known but may represent an attempt by the host to repair damaged areas.

Offensive Aspects of Pathogenesis

One of the pioneers in physiological studies of plant diseases, Ernst Gaümann, once said: "Micro-organisms are pathogenic only if they are toxigenic." This rather sweeping generalization was made to emphasize the importance of pathogen-produced products that are harmful to their host. Using the broadest possible definition of the term "toxin," a biologically produced compound that is harmful to another organism, we can include such substances as degradative enzymes, hormones in unusual concentrations, and compounds of small or large molecular weight that interfere with the normal functioning of cells.

For a pathogen to penetrate a plant, it must either enter through natural openings or wounds or it must soften or degrade cell walls. Regardless of the initial mechanism of entry, it will at some point come in contact with cell walls. All fungal and bacterial pathogens studied produce cell-wall-degrading enzymes in culture, but to demonstrate a role in disease, the enzymes must be shown in the diseased plant. Cellulases, hemicellulases, and various types of pectic enzymes have been found in diseased plants. The pectic enzymes are undoubtedly most important, since destruction of pectins in the middle lamella will cause a dissolution of the cell wall and allow the cellulose microfibrils to slide freely past one another. The most important pectic enzymes in disease are the **endopolygalacturonases**. These enzymes cleave polygalacturonic acid in a random fashion as opposed to cleavage of single terminal residues.

A significant question regarding the role of cell wall enzymes is why, when most pathogens produce them easily in culture, are they effective in infection only within the appropriate host? Peter Albersheim, a cell wall biochemist at the University of Colorado, has devoted years to the study of this question. He found that the cell wall compositions of various plants are not sufficiently distinct to cause differential induction of a pathogen's polysaccharide-degrading enzymes. Inhibitors of endopolygalacturonases have been found in some plants, however, and if present in sufficiently high quantities, they could prevent the action of these enzymes. There is a problem, though: Many pathogens are highly specific and will attack only certain varieties of a particular species, but the inhibitors are not sufficiently different among varieties to account for this specificity.

Other cell wall factors may also be involved in resistance. Modifications to cell walls of older tissue, such as the formation of calcium pectates, make them resistant to wall-degrading enzymes, and this may explain changing susceptibilities with age in some diseases. Higher plants also frequently possess several enzymes such as **chitinase**, which degrade walls of pathogens. The latter enzymes may in some cases be inhibited by a product of the pathogen. From such studies we conclude that there may be many offensive and defensive strategies employed by both the pathogen and the plant, and the interaction between the two determines the success or failure of pathogenesis.

Other offensive products produced by pathogens are nonenzymatic toxins. A toxin, when released into a susceptible host, may incite all or part of the symptoms of the disease. To prove a causal role in pathogenicity, several rigorous criteria should be satisfied. Mere isolation of a toxic compound from a culture medium in which the pathogen grew or from the diseased plant is insufficient, since such a compound could very well be an artifact of the isolation procedures or produced too late to be involved in initial interactions. A few instances are known in which pathogenic strains produce the toxin but nonpathogens do not, the toxin affects only host varieties normally attacked by the pathogen, and the toxin reproduces accurately all symptoms of the disease. These toxins have been termed **pathotoxins**. Others that do not satisfy such rigorous criteria as causal agents of disease, yet are believed to contribute to the disease syndrome, are called **phytotoxins**. Examples of known pathotoxins are (1) *Helminthosporium maydis* toxin, named for the fungus that destroyed almost 1 billion dollar's worth of maize in the United States in the severe epidemic (Southern corn leaf blight) of 1970–1971; (2) *Helminthosporium victoriae* toxin (**victorin**), responsible for the symptoms of Victoria blight of oats, a serious disease in the late 1940s; (3) *Helminthosporium saccharri* toxin (**helminthosporoside**), involved in eyespot disease of sugarcane; and (4) **Amylovorin**, a toxin produced in the fireblight disease of apples and pears and caused by the bacterium *Erwinia amylovora*.

Pathotoxins have been employed in breeding programs as rapid screening agents for disease-resistant plants. Harry Wheeler and H. H. Luke, working at Louisiana State University in 1955, treated over 45 million oat seedlings with the pathotoxin victorin. They isolated over 900 seedlings insensitive to the toxin and thus resistant to the pathogen. This technique has since been routinely employed in searches for resistance to other pathotoxin-caused diseases.

The mode of action of such toxins depends upon the compound in question. Some are peptides or glycopeptides, but most are substances of small molecular weight with no particular chemical entity in common. They may affect membrane permeability, mitochondrial respiration, or inhibit various essential enzymatic reactions; for others, the mechanism of action remains unknown. Victorin induces depolarization of membrane electropotentials as early as 2 minutes after treatment of susceptible oat roots. Electrolyte leakage is detected within 5 minutes. A partial recovery of membrane potentials followed by apparently irreversible effects on permeability, respiration, and other processes after a 20-minute lag suggested that the toxin may act on some other, perhaps intracellular, site in addition to the plasma membrane.

Although nonenzymatic toxins have been regarded as nothing more than biological curiosities by many, they have proven a favorite research tool for persons interested in physiological plant pathology. Since the symptoms of the disease are reproduced by the toxin, researchers can avoid the difficulties encountered when two living organisms, the host and the pathogen, contribute to any complex changes observed. Furthermore, many effects can be studied in a short time, since one does not have to wait the hours, or even days, necessary for infection by the pathogen to occur.

Mechanisms of Disease Resistance

A popular concept 40 years ago was that disease-resistant plants contained a toxic compound that inhibited the growth of potential pathogens. Investigations in J. C. Walker's laboratory at the University of Wisconsin in 1929 demonstrated that phenolic compounds were associated with resistance to the fungus *Collectotrichum circinans* in certain onion varieties. These phenolics diffused to the dead scale surfaces and prevented fungal spore germination.

Phenolics include a large group of compounds such as anthocyanins, leucoanthocyanins, coumarins, anthoxanthins, hydroxybenzoic acids, and glycosides. The term "polyphenol" is frequently used and refers to the presence of more than one hydroxyl group on a benzene ring.

The toxicity of o- or p-diphenols occurs by virtue of their oxidation to the corresponding quinones, which can readily oxidize sulfhydryl groups on enzymes, thereby disrupting the normal oxidation-reduction balance in cells. The hypothesis that quinones are inhibitors of fungal development is supported by increased phenolic biosynthesis in many infected tissues and the activity of phenol-oxidizing enzymes. Of course, these phenolic compounds are also phytotoxic, and this may explain the rapid collapse of a limited number of cells in some resistant interactions (the hypersensitive response).

Relatively few cases are known in which preformed inhibitory compounds determine resistance to infection. More often, toxic compounds are sequestered as glycosides and are hydrolyzed after infection; or, increased biosynthesis of toxic substances may contribute to resistance. In some cases, pathogens detoxify the compounds and thus spread and cause a susceptible reaction.

Disease is characterized by many postinfectional metabolic changes, including increased synthesis of preinfectional toxic compounds and the formation of substances known as **phytoalexins**. Phytoalexins (Greek: "plant warding-off compounds") are usually not present in healthy tissues or are present in low concentrations, and are usually synthesized only at the area of infection. Examples include **pisatin** in peas, **orchirol** in orchids, **phaseollin** in beans, **ipomeamorone** in sweet potato, and **medicarpin** in alfalfa.

Several lines of evidence support the conclusion that phytoalexins play a role in disease resistance: (1) Upon induction, they are synthesized more rapidly in resistant than susceptible plants after infection; (2) treatments that induce resistance also stimulate phytoalexin synthesis; (3) they are induced in response to many types of stress; and (4) in some cases pathogens can detoxify the phytoalexin, whereas nonpathogens cannot. Although most studies have dealt with the antifungal activity of phytoalexins, they may also be toxic to the plants that produce them.

Despite the compelling evidence that they play a role in disease resistance, some question still exists about whether phytoalexins are the determinants of specificity. For one thing, phytoalexins accumulate in a very localized area, and it is difficult to determine whether concentrations are sufficient at the point of infection to prevent spread of the pathogen. Second, phytoalexins may not accumulate in large enough concentrations, and fungal growth may not be inhibited until after the localized hypersensitive necrosis associated with resistance has occurred. Professor Kiraly of the Plant Protection Institute in Budapest suggests that hypersensitivity associated with phytoalexin production is a consequence, rather than the cause, of host resistance to infection. His experiments with several fungal diseases show that death or inhibition of the parasite then precedes the hypersensitive death of the host. The damaged parasite then releases an **endotoxin** or **elicitor** that induces death of host cells and phytoalexin production. For these reasons, many researchers currently propose that phytoalexins are not the determinants of specificity but may play an important role in the overall defense mechanisms.

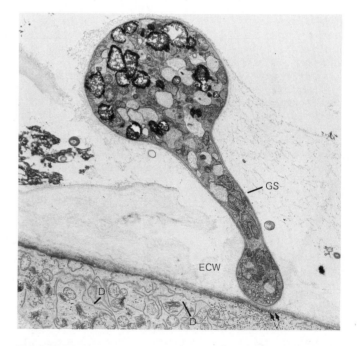

Figure 23.2-1 A germ tube of *Phytophthora parasitica* var. *nicotiana* (T) elicits a host response of numerous vesicles (V) before the fungus has fully penetrated the root epidermal cell wall (W) of a resistant variety of tobacco.

Increasing evidence suggests that some type of recognition reaction occurs very early in infection and may precede actual penetration of the host. In the black shank disease of tobacco caused by the fungus *Phytophthora parasitica* var. *nicotiana,* a host response occurs when the spore germ tube comes in contact with the root epidermal cell wall of resistant varieties (Fig. 23.2-1). The reaction consists of the formation of vesicles originating from the endoplasmic reticulum and the dictyosomes. The vesicle contents are released through the plasma membrane and accumulate against the cell wall. Shortly after penetration, the cell dies and fungal growth is inhibited. Susceptible cells do not react as quickly, and the fungal filaments ramify through other cells.

What determines this very early specificity? James DeVay and his colleagues working at the University of California, Davis have found a correlation between susceptibility and the occurrence of similar proteins in the host and pathogen. In resistant interactions, the host and pathogen proteins are dissimilar. The intracellular function of the proteins has not been established, but data from many host-pathogen combinations suggest that the host may have the capacity to recognize foreign **antigens** (proteins) and thus react against them. Successful pathogens, through coevolution, may have selected for proteins similar to those found in their natural hosts and thus delay detection and associated defense reactions until infection is well established. There is no evidence, however, that plants have an immunological system like that of higher animals. The mechanism of recognition or nonrecognition of proteins remains unclear.

Albersheim has proposed a recognition system based upon the carbohydrate chemistry of the cell walls of both host and pathogen. Such a recognition system is not uncommon. For example, the yeast *Hansenula wingei* will undergo sexual mating only when two different mating types (sexes) come together. The two mating types apparently recognize the carbohydrates of the other's cell wall. Other studies have shown that the *Rhizobia* that nodulate soybeans have a carbohydrate surface different from that of strains that do not form nodules. Even the ability of pollen grains to germinate and penetrate only the appropriate species has been attributed to carbohydrate surface characteristics. Albersheim's work promises interesting findings and will hopefully explain the specificity of pathogens for particular hosts.

The ultimate aim of plant pathologists is to seek better methods for controlling plant diseases. The physiological studies mentioned will hopefully someday lead to new approaches. By studying the physiological basis of susceptibility or biochemical mechanisms of recognition or resistance, identification of the genes determining specificity can be made. The incorporation of specific genes with known functions into plant breeding programs will enable us to select new varieties in a more logical manner than the trial and error programs that we are now too often dependent upon.

The following special topic is out of order because it was added at the last minute.

Special Topic 1.2
An Introduction to the Laws
of Thermodynamics
—J. Clair Batty

F.B.S. 1978

J. Clair Batty is an Associate Professor of Mechanical Engineering at Utah State University, where he applies the principles of thermodynamics to assess and improve the efficiency of energy consumption by twentieth century society. He became enamored with the philosophies underlying thermodynamic principles while he was a student at the Massachusetts Institute of Technology, where he obtained the Doctor of Science Degree in 1969.

In the sense of seeking conformity to political edict, laws of nature are not really laws at all but are simply man's attempts to predict what must always happen under certain prescribed conditions. In order to eliminate exceptions to the predicted behavior, the prescribed conditions tend to be restrictive, while informed regard for the law tends to increase with generality of application.

The purpose of this brief essay is to give the serious student a glimpse of the beautifully simple basis for the often bewildering maze of statements and corollaries found in standard expositions of thermodynamics. If properly stated, the truly elegant laws of thermodynamics seem to apply to all known systems whether that system is a Rolls Royce engine, a petunia, a volcano, a solar system, or a bumblebee.

System, Boundary, Surroundings

We shall consider the word **system** to mean that region of space or quantity of matter upon which we have focused our interest and attention. Thus, as implied above, a system could be an electron, a liter of air, a leaf, the Ural Mountains, or the Milky Way Galaxy. Everything not in the system is referred to as the **surroundings** (environment). The system is separated from the surroundings by the **boundary**. In most cases, the boundary must be imagined. Now, we ask ourselves, in what way can the surroundings communicate or **interact** with the system across the boundary?

The Work and Heat Interactions

It may seem that there are many ways of interaction, but basically an interaction involves only a pushing, pulling, or striking of some kind. If the pushing or pulling across the boundary is in a highly organized fashion, we call that a **work interaction.** A work interaction usually involves many, many particles organized to move together in the same direction and at the same time, such as a spinning shaft, a hammer striking a nail, or a stream of electrons flowing through a wire. In fact, the thermodynamic definition of work might be stated as: **Work is an interaction between a system and its surroundings such as that which happens at the interaction boundary *could be repeated* in such a way that the sole effect would be the raising or lowering of a weight in the surroundings.** Notice that the change in level of a weight necessitates the marshalling of many particles into motion in the same direction and at the same time.

If the pushing, pulling, or striking at the interaction boundary is random, chaotic, and on a microscopic scale, by which we mean the scale of atoms and molecules, we call that a **heat interaction.** A lighted match held against the bottom of a metal table will not lift the table because, even though, on the average, the molecules in the hot combustion gases are striking the molecules in the table surface more often and with greater vigor than the molecules of the table strike back, they lack **organization.** Some interactions are apparently neither all heat nor all work. An example is a beam of light (photons) striking the system. The photons certainly have some degree of organization because of direction and uniform speed, but much of that organization is lost upon striking the surface of the system.

Clearly, systems do not possess work or heat, since work and heat are interactions and recognizable only at a system boundary. Heat and work are, therefore, not properties of a system.

System Properties

Energy: A System Property One might next ask the question, what property does a system possess by virtue of its motion, either microscopic or macroscopic, or its proximity to other systems? We believe systems possess such a property and assign to that property the name of **energy**. While we have difficulty in rigorously defining energy, we do know how to measure it. **Kinetic energy**, or the energy associated with the motion of a system, is evaluated by computing the work required (what weight moving through what distance) in a perfectly organized process (no heat interaction) to give the system that motion relative to some arbitrary reference. The kinetic energies of a jet aircraft, a bumblebee, or an electron in flight are all measured in the same fashion. **Potential energy**, or the energy associated with a system's position, is evaluated by computing the work required in a perfectly organized process to bring the system into that position relative to some arbitrary reference point. The height of a hawk above the earth or the position of a molecule or an electron subjected to the electrical push or pull of its neighbors thus becomes an effective way to store energy.

If a system consists of many individual particles, we can evaluate only the bulk kinetic or potential energy of the system. It is not usually possible to evaluate the kinetic or potential energies of each individual particle, so we refer to these collectively as the **internal energy** of the system.

Entropy: A System Property It should now be apparent that an important system property in addition to energy is the degree to which that energy is organized, or rather disorganized. Attempts to describe this relative disorganization or randomness have resulted in the concept of the property **entropy**. Entropy is defined as: $S = K \ln w$, where K is simply a constant of proportionality (Boltzmann's constant), and $\ln w$ is the natural logarithm of w, which is the **thermodynamic probability**. The thermodynamic probability is simply the number of ways the system can be arranged microscopically and still exhibit the same macroscopic features. (How many ways can the molecules of the petunia be arranged and still have the resulting arrangement be a petunia?)

Imagine a classroom with 10 chairs and 10 students. We ask ourselves in how many ways can those 10 people arrange themselves on those 10 chairs? The first person to sit down has 10 choices, the second has 9 choices, the third has 8 choices, and so on. We quickly see that there are $10 \times 9 \times 8 \times 7 \times 6 \times 5 \times 4 \times 3 \times 2 \times 1 = 3,628,800$ ways of arranging 10 students on 10 chairs. If we had no additional information regarding the seating habits of the students, we must conclude that all arrangements are equally likely, and the probability of finding the students in one particular arrangement is 1 divided by 3,628,800. Obviously, as the number of possible arrangements increases, the probability of observing one certain arrangement decreases. Entropy thus is a measure of our uncertainty or lack of information regarding the microscopic arrangement of a system, and the "entropy" associated with this example could be computed as:

$$S = K \ln 3,628,800 = 15.1 \, K$$

If we had carefully instructed the students where to sit beforehand and had total assurance that those instructions would be followed precisely, then we would have no uncertainty regarding the arrangement we would find, and the entropy would be:

$$S = K \ln 1 = 0$$

On the other hand, if we had not previously instructed the students even to be in the room at all, the number of possible arrangements would include those where some could be at home, at the movies, or anywhere else. The probability of finding those students seated in one particular arrangement on our 10 chairs would be small indeed, and the entropy or uncertainty associated with this situation would be large.

For more complicated systems, such as alligators or petunias, the number of possible molecule arrangements is impossible to evaluate. Yet as large as that number is, it is negligibly small compared to the incomprehensibly large number of arrangements those same molecules could have if they remained disorganized and scattered throughout the swamp, the flower garden, the atmosphere, or elsewhere.

It is an intriguing notion that the value of a system property depends on the knowledge of the observer, but that is surely the case with entropy. Consider a rancher with 20 cows in each of two adjoining pens. Somehow the gate between the pens is pushed open, and the cows are mixed. The rancher's wife discovers the open gate and, watching until there are 20 cows in each pen, closes the gate and reports smugly to her husband that the situation has not deteriorated. If she were the only observer, the entropy did not increase. The husband, however, knows each cow by name and realizes that there has been an increase in disorganization; so, for him, entropy increases. If we replace the cows of this simple example with molecules, we realize that mixing gases, for example, results in an increase in entropy (disorganization) only if someone can tell which molecules are which—and cares!

The Law of Stable Equilibrium

The notion of **equilibrium** is also basic to an understanding of the laws of thermodynamics. A system is said to be in **stable equilibrium** if no changes may take place in observable system properties without a finite, permanent change in the surroundings. We conclude that a mixture of gasoline and oxygen is not in stable equilibrium, because only an infinitesimally small spark introduced from the surroundings is required to bring about rather dramatic changes in system properties. A golf ball resting in the bottom of the hole is considered to be in a state of stable equilibrium, since a finite, permanent change in the surroundings is required to remove it. Again it is only by a lack of knowledge of the observer that systems may be claimed to be in stable equilibrium. Visualize as a system a closed container of air consisting of say 10^{25} molecules. Without interaction from the surroundings, the pressure, temperature, volume, energy, and entropy of the system would not change, and the system could be said to be in a state of stable equilibrium. If the observer were even casually acquainted with a few of the individual molecules, however, the system would never achieve a state of stable equilibrium, because the observer would seldom, if ever, find his molecule friends in the same location moving in the same direction and with the same speed as noted on a previous glance.

In the world of the atom and molecule there is no such thing as equilibrium, because the individual particles remain in constant motion. We do not expect the molecules of air in a carefully isolated room to ever become tired and lie motionless on the floor for the simple reason that they are so small that they can respond to the random hits, kicks, and punches of their similar-sized neighbors in the walls of the container. Macroscopic systems such as Ping-Pong balls do become tired and lie motionless on the floor. They are so massive that they cannot respond to the disorganized hits, kicks, and punches of the vibrating molecules in the surface of the floor.

An interesting excursion in thought is to realize that a perfectly elastic ball that would bounce back to the height from which it was dropped transfers *none* of its macroscopic, organized potential energy into microscopic, disorganized random internal energy. A ball that bounces higher than the height from which it was dropped would have the awesome capability of commanding the randomly vibrating molecules in the rebound surfaces to get themselves organized and all push in the same direction at exactly the right instant.

Hopefully, all this jumping from a macroscopic to a microscopic point of view and back again has not been terribly confusing, and you are now prepared to ponder the **law of stable equilibrium** as propounded by Hatsopoulos and Keenan: **A system having specified allowed states and an upper bound in volume can reach from any given state one and only one stable state and leave no effect on its environment (surroundings).** Again, it may be useful to visualize some specific examples. Imagine a number of Ping-Pong balls in motion in an isolated room. The walls of the room provide an upper bound in volume so that the balls don't fly off forever into space and thus never achieve equilibrium. The fact that no two balls could be expected to occupy the same space at the same time suggests that only certain specified states are allowable (as no two students could sit on the same chair). As the bouncing balls settle down, the bulk kinetic and potential energies associated with the bouncing balls become transformed into the kinetic and potential energies of the molecules that make up the balls and the walls. The greater the initial kinetic and potential energies of the balls, the greater the final internal energies will be, provided no energy and pressure are lost to the environment. The law of stable equilibrium simply suggests that this final equilibrium condition is fully determined by the initial condition. This surely seems reasonable and consistent with our total experience.

The First Law of Thermodynamics

The reader is probably familiar with a statement to the effect that energy is always conserved if we can but properly account for it. Such statements are often referred to as the **first law of thermodynamics.** Actually, the law of stable equilibrium expresses the familiar version of the first law of thermodynamics without encountering the difficulty of expressing a fundamental law in terms of an undefined parameter—namely, energy.

A rigorous demonstration that the conservation-of-energy principle arises from the law of stable equilibrium would take more space than can be tolerated here, but, hopefully, another brief example might further convince you that such an assertion is at least plausible.

Imagine a small container of oxygen located inside a larger evacuated container. Also imagine that the larger container has been carefully isolated from the surroundings. Somehow the small container ruptures. What happens next? The oxygen molecules rush forth to fill the larger container, of course, and the final equilibrium macrostate can be predicted exactly as the law of stable equilibrium suggests, provided we know the initial specified state of the oxygen. The calculational procedure we would use to predict that final condition involves the assumption that the energy of the final state is identical with the energy of the initial state, since no energy was allowed to escape through the walls of the larger container to affect the surroundings.

The Second Law of Thermodynamics

There are many versions of the **second law of thermodynamics** such as:

> **It is impossible to construct a device that will operate in a cycle and produce no effect other than the transfer of heat from a cooler body to a hotter body.** (Operating in a cycle means that at the conclusion of the process the final properties of the device are identical to the initial properties.) Or, **Heat cannot be converted entirely into work in a cyclic process**. Or, **The entropy of an isolated system cannot decrease**.

It is not difficult to demonstrate that all these statements are more or less equivalent and that they arise directly from the law of stable equilibrium. To express the second law of thermodynamics one need only say: **Stable equilibrium states exist**.

If stable equilibrium states did not exist, then we could expect Ping-Pong balls lying quietly on the floor to begin to quiver and then to bounce as the random internal energy in the floor and in the balls organized itself. We might even see balls bounce higher than the height from which they were dropped, leaving slightly cooler spots behind them as they bounded about collecting and organizing random disorganized energy. We would occasionally see all the molecules of air in a room organize themselves and on the count of three all rush over to one corner, leaving us gasping for breath. We would, perhaps, occasionally observe molecules of soil, water, and air of their own volition without interactions with the surroundings, organize themselves into a petunia or a person.

But in fact stable states apparently exist in nature, and we have never observed such behavior as described above. There are vastly many more possible microscopic arrangements resulting in the random disorganized configuration associated with stable equilibrium than in the relatively few possible arrangements resulting in the organized nonequilibrium configuration. Systems left to their own devices, therefore, always tend toward equilibrium. *Thus one eventually comes to regard the laws of thermodynamics as the inevitable consequence of relative size (between microscopic and macroscopic worlds) and simple probability.*

Bibliography

Prologue

Buvat, Roger. 1969. Plant cells: An introduction to plant protoplasm. McGraw-Hill Book Company, New York.

Clarkson, David T., ed. 1973. Ion transport and cell structure in plants. John Wiley & Sons, Inc., New York.

Cutter, Elizabeth. 1969. Experiment and interpretation; Cells and tissues. Addison-Wesley Publishing Co., Inc., Reading, Mass.

Hall, J. L., T. J. Flowers, and R. M. Roberts, eds. 1974. Plant cell structure and metabolism. Longman, London.

Jensen, William A., 1970. The plant cell, 2nd ed. Wadsworth Publishing Company, Belmont, Ca.

Ledbetter, Myron C. and Keith R. Porter. 1970. Introduction to the fine structure of plant cells. Springer-Verlag, New York.

Novikoff, Alex B. and Eric Holtzman. 1976. Cells and organelles, 2nd ed. Holt, Rinehart and Winston, New York.

Robards, A. W., ed. 1974. Dynamic aspects of plant ultrastructure. McGraw-Hill Book Company, New York.

Roberts, Lorin W. 1976. Cytodifferentiation in plants. Cambridge University Press, Cambridge.

Chapter 1

Kramer, P. J. 1969. Plant and soil water relationships. McGraw-Hill Book Company, New York.

Ling, G.n. 1967. effects of temperature on the state of water in the living cell. Pages 5–24 in A. H. Rose, ed., Thermobiology. Academic Press, Inc., New York.

Meyer, B. S. 1938. The water relations of plant cells. Botanical Review 4:531–547. (Formulation of the system long used by plant physiologists; now superseded by the water potential terminology.)

Meyer, B. S. and D. B. Anderson. 1952. Plant physiology. Second edition. D. Van Nostrand Company, Princeton, N. J.

Meyer, B. S., D. B. Anderson, R. H. Bohning, D. G.Fratianne. 1973. Introduction to plant physiology (2nd edition). D. Van Nostrand Company, New York.

Nobel, Park S. 1974. Introduction to biophysical plant physiology. W. H. Freeman and Company Publishers, San Francisco.

Renner, O. 1915. Theoretisches und Experimentelles zur Kohäsionstheorie der Wasserbewegung [Theoretical and experimental considerations of the cohesion theory of water transport]. Jahrbuch für Wissenschaftliche Botanik 56:617–667 (Discovery of the water potential concept).

Slatyer, R. O. 1967. Plant-water relationships. Academic Press, Inc., New York.

Spanner, D. C. 1964. Introduction to thermodynamics. Academic Press, Inc., New York. (Written by a plant physiologist.)

Sutcliffe, J. F. 1968. Plants and water. St. Martin's Press, Inc., New York.

Wyllie, G. 1965. Kinetic theory and transport in ice and water. Symposia of the Society for Experimental Biology 19:33–54.

Chapter 2

Boyer, J. S. 1967. Leaf water potentials measured with a pressure chamber. Plant Physiology 42:133–137.

Boyer, J. S. 1967. Matric potentials of leaves. Plant Physiology 42:213–217.

Boyer, J. S. and E. B. Knipling. 1965. Isopiestic technique for measuring leaf water potentials with a thermocouple psychrometer. Proceedings of the National Academy of Sciences 54(4):1044–1051.

Campbell, Gaylon S. and G. S. Harris. 1975. Effect of soil water potential on soil moisture absorption, transpiration rate, plant water potential and growth for Artemisia tridentata. US/IBP Desert Biome Res. Mem. 75–44. Utah State University, Logan, Utah.

Cary, J. W. and H. D. Fisher. 1971. Plant water potential gradients measured in the field by freezing point. Physiologia Plantarum 24:397–402.

Cline, Richard G. and Gaylon S. Campbell. 1976. Seasonal and diurnal water relations of selected forest species. Ecology 57:367–373.

Gardner, W. R. and C. F. Ehlig. 1965. Physical aspects of the internal water relations of plant leaves. Plant Physiology 40:705–710.

Green, P. B. and R. W. Stanton. 1967. Turgor pressure: Direct manometric measurement in single cells of Nitella. Science 155:1675–1676.

Harris, J. A. 1934. The Physico-chemical properties of plant saps in relation to phytogeography. Data on native vegetation in its natural environment. University of Minnesota Press, Minneapolis.

Hellkvist, J., G. P. Richards, and P. G. Jarvis. 1974. Vertical gradients of water potential and tissue water relations in sitka spruce trees measured with the pressure chamber. Journal of Applied Ecology 11:637–667.

Knipling, E. B. and P. J. Kramer. 1967. Comparison of the dye method with the thermocouple psychrometer for measuring leaf water potentials. Plant Physiology 42:1315–1320.

Kramer, Paul J., Edward B. Knipling, and Lee N. Miller. 1966. Terminology of cell-water relations. Science 153:889–890.

Lang, A. R. G. and H. D. Barrs. 1965. An apparatus for measuring water potentials in the xylem of intact plants. Australian Journal of Biological Science 18:487–497.

Levitt, J. 1953. Further remarks on the thermodynamics of active (non-osmotic) water absorption. Physiologia Plantarum 6:240–252.

Lockhart, James A. 1959. A new method for the determination of osmotic pressure. American Journal of Botany, 46:704–708.

Manohar, Man Singh. 1971. Which water potential? Differences between isopiestic thermocouple psychrometer measurements of intact and excised plant materials. Biologia Plantarum, 13:247–256.

Meyer, B. S. and D. B. Anderson. 1939. Plant physiology. D. Van Nostrand Company. New York.

Nobel, Park S. 1974. Introduction to biophysical plant physiology. W. H. Freeman and Company Publishers, San Francisco.

Pisek, Arthur. 1956. Der Wasserhaushalt der Mesound Hygrophyten [The water relations of mesoand hydrophytes]. Encyclopedia of Plant Physiology 3:825–853.

Ray, P. M. 1960. On the theory of osmotic water movement. Plant Physiology 35:783–795.

Salisbury, F. B. and R. V. Park. 1965. Vascular plants: Form and function. Wadsworth Publishing Company, Inc., Belmont, Calif.

Scholander, P. F., H. T. Hammel, E. D. Bradstreet, and E. A. Hemmingsen. 1965. Sap pressure in vascular plants. Science 148:339–346.

Slatyer, R. O. 1967. Plant-water relationships. Academic Press, Inc., New York.

Slatyer, R. O. and S. A. Taylor. 1960. Terminology in plant-soil-water relations. Nature 187:922–924.

Sutcliffe, J. F. 1968. Plants and water. St. Martin's Press, Inc., New York.

Tyree, M. T. and J. Dainty. 1973. The water relations of hemlock (Tsuga canadensis). II. The kinetics of water exchange between the symplast and apoplast. Canadian Journal of Botany 51:1479–1481.

Tyree, M. T. and H. T. Hammel. 1972. The measurement of the turgor pressure and the water relations of plants by the pressure-bomb technique. Journal of Experimental Botany 23(74):267–282.

Waring, R. H. and B. D. Cleary. 1967. Plant moisture stress: Evaluation by pressure bomb. Science 155:1248–1254.

West, D. W. and D. F. Gaff. 1971. An error in the calibration of xylem-water potential against leaf-water potential. Journal of Experimental Botany 22(71):342–346.

Wiebe, Herman H. 1966. Matric potential of several plant tissues and biocolloids. Plant Physiology 41:1439–1442.

Chapter 3

Aylor, Donald E., Jean-Yves Parlange, and A. D. Krikorian. 1973. Stomatal mechanics. American Journal of Botany 60:163–171.

Bartlett, Peter N. and David M. Gates. 1967. The energy budget of a lizard on a tree trunk. Ecology 48:315–322.

Cowan, I. R. and J. H. Troughton. 1971. The relative role of stomata in transpiration and assimilation. Planta 97:325–336.

Drake, B. G. 1967. Unpublished master's thesis, Colorado State University, Fort Collins.

Drake, B. G. and F. B. Salisbury. 1972. Aftereffects of low and high temperature pretreatment on leaf resistance, transpiration, and leaf temperature in *Xanthium*. Plant Physiology 50:572–575.

Fischer, R. A. 1973. The relationship of stomatal aperture and guardcell turgor pressure in *Vicia faba*. Journal of Experimental Botany 24(79): 387–399.

Fischer, R. A. and Theodore C. Hsiao. 1968. Stomatal opening in isolated epidermal strips of *Vicia faba*. II. Responses to KCl concentration and the role of potassium absorption. Plant Physiology 43:1953–1958.

Fujino, M. 1959. Stomatal movement and active migration of potassium [in Japanese]. Kagaku 29:660–661.

Gates, David M. 1968. Transpiration and leaf temperature. Annual Review of Plant Physiology 19:211–238.

Gates, David M. 1971. The flow of energy in the biosphere. Scientific American 255(3):88–100.

Gates, David M., Harry J. Keegan, John C. Schleter, and Victor R. Wiedner. 1965. Spectral properties of plants. Applied Optics 4:11–20.

Hanks, R. J. and R. W. Shawcroft. 1965. An economical lysimeter for evapotranspiration studies. Agronomy Journal 57:634–636.

Hsiao, Theodore C. 1973. Plant responses to water stress. Annual Review of Plant Physiology 24:519–570.

Hsiao, Theodore C. and W. G. Allaway. 1973. Action spectra for guardcell Rb+ uptake and stomatal opening in *Vicia faba*. Plant Physiology 51:82–88.

Humble, G. D. and K. Raschke. 1971. Stomatal opening quantitatively related to potassium transport: Evidence from electron probe analysis. Plant Physiology 48:447–453.

Idso, S. B. and D. G. Baker. 1967. Relative importance of reradiation, convection, and transpiration in heat transfer from plants. Plant Physiology 42:631–640.

Imamura, S. 1943. Untersuchunger über den Mechanismus der Turgorschwankung der Spaltöffnungsschlieszellen [Investigations of the mechanisms of turgor changes in guard cells]. Japanese Journal of Botany 12:251–346.

Kriedemann, P. E., B. R. Loveys, G. L. Fuller, and A. C. Leopold. 1972. Abscisic acid and stomatal regulation. Plant Physiology 49:842–847.

Lange, O. L. and E. D. Schulze. 1971. Measurement of CO_2 gas exchange and transpiration in the beech (*Fagus silvatica* L.). Ecological Studies, Analysis, and Synthesis 2:16–28.

Lange, O. L., R. Lösch, E. D. Schulze, and L. Kappen. 1971. Responses of stomata to changes in humidity. Planta 100:76–86.

Larque-Saavedra, A. and R. L. Wain. 1974. Abscisic acid levels in relation to drought tolerance in varieties of *Zea mays* L. Nature 251:716–717.

Levitt, J. 1974. The mechanism of stomatal movement—once more. Protoplasma 82:1–17.

Meidner, Hans and T. A. Mansfield. 1968. Physiology of stomata. McGraw-Hill Book Company, New York.

Ogunkanmi, A. B., A. R. Wellbugrn, and T. A Mansfield. 1974. Detection and preliminary identification of endogenous antitranspirants in waterstressed *Sorghum* plants. Planta 117:293–302.

Penny, M. G. and D. J. F. Bowling. 1974. A study of potassium gradients in the epidermis of intact leaves of *Commelina communis* L. in relation to stomatal opening. Planta 119:17–25.

Penny, M. G. and D. J. F. Bowling. 1975. Direct determination of pH in the stomatal complex of *Commelina*. Planta 122:209–212.

Raschke, Klaus. 1975. Stomatal action. Annual Review of Plant Physiology 26:309–340.

Raschke, Klaus. 1976. How stomata resolve the dilemma of opposing priorities. Philosophical Transactions of the Royal Society of London, Biological Sciences 273:551–560.

Raschke, Klaus and M. P. Fellows. 1971. Stomatal movement in *Zea mays*: Shuttle of potassium and chloride between guard cells and subsidiary cells. Planta 101:296–316.

Raschke, Klaus and G. D. Humble. 1973. No uptake of anions required by opening stomata of *Vicia faba*: Guard cells release hydrogen ions. Planta 115:47–57.

Salisbury, Frank B. and George G. Spomer. 1964. Leaf temperatures of alpine plants in the field. Planta 60:497–505.

Selirio, I. S., D. M. Brown, and K. M. King. 1971. Estimation of net and solar radiation. Canadian Journal of Plant Science 51:35–39.

Tranquillini, Walter. 1969. Photosynthese und Transpiration einiger Holzarten bei verschieden starkem Wind [Photosynthesis and transpiration of several woody species at various wind velocities]. Centralblatt für das Gesamte Forstwesen 86:35–48.

Troughton, John and Lesley A. Donaldson. Probing plant structure. McGraw-Hill, New York.

Wellburn, A. R., A. B. Ogumkanmi, R. Fenton, and T. A. Mansfield. 1974. All-trans-farnesol: A naturally occurring antitranspirant? Planta 120:255–263.

Willmer, C. M. and J. E. Pallas, Jr. 1973. A survey of stomatal movements and associated potassium fluxes in the plant kingdom. Canadian Journal of Botany 51:37–42.

Ziegenspeck, H. 1955. Die Farbenmikrophotographie, ein Hilfsmittel zum objektiven Nachweis submikroscopeischer Strukturelemente. Die Radiomicellierung und Filierung der Schleisszellen von *Ophioderma pendulum* [Color photomicrography, an aid to the objective study of submicroscopic structural elements. Radial micellation and filiation of guard cell of *Ophioderma pendulum*]. Photographie und Wissenschaft 4:19–22.

Chapter 4

Apfel, R. E. 1972. The tensile strength of liquids. Scientific American 227(6):58–71.

Barrs, H. D. 1966. Root pressure and leaf water potential. Science 152:1266–1268.

Bowling, D. J. F., 1968. Active and passive ion transport in relation to transpiration in *Helianthus annuus*. Planta 83:53–59.

Boyer, J. S. 1969. Free-energy transfer in plants. Science 163:1219–1220.

————. 1974. Water transport in plants: Mechanism of apparent changes in resistance during absorption. Planta 117:187–207.

Briggs, G. E. and R. N. Robertson. 1957. Apparent free space. Annual Review of Plant Physiology 8:11–29.

Briggs, Lyman J. 1950. Limiting negative pressure of water. Journal of Applied Physics 21:721–722.

Brogardh, T., A. Johnsson, and R. Klockcare. 1974. Oscillatory, transpiration and water uptake of *Avena* plants. V. Influence of the water potential of the root medium. Physiologia Plantarum 32:258–267.

Cooke, R. and I. D. Kuntz. 1974. The properties of water in biological systems. Annual Review of Biophysics and Bioengineering 3:95–126.

Crafts, A. S. and T. C. Broyer. 1938. Migration of salts and water into xylem of the roots of higher plants. American Journal of Botany 25:529–535.

Dixon, Henry H. 1914. Transpiration and the ascent of sap in plants. Macmillan and Company, Ltd., London.

Esau, Katherine. 1965. Plant anatomy. John Wiley & Sons, Inc., New York.

Hammel, H. T. 1967. Freezing of xylem sap without cavitation. Plant Physiology 42:55–66.

Harvis, John R. 1971. water movement in woody stems during freezing. Cryobiology 8:581–585.

Hayward, A. T. J. 1971. Negative pressure in liquids: Can it be harnessed to serve man? American Scientist 59:434–443.

Heine, R. W. and D. J. Farr. 1973. Comparison of heat-pulse and radioisotope tracer method for determining sap-flow in stem setments of poplar. Journal of Experimental Botany 24:649–654.

Hooymans, J. J. M. 1969. The influence of the transpiration rate on uptake and transport of K+ in barley plants. Planta 88:369.

Jarvis, P. and C. R. House. 1970. Evidence for symplastic ion transport in maize roots. Journal of Experimental Botany 21:83–90.

Jarvis, P. G. and R. O. Slayter. 1970. The role of the mesophyll cell wall in transpiration. Planta 90:303–322.

Kaku, S. 1971. A possible role of the endodermis as a barrier for ice propagation in the freezing of pine needles. Plant and Cell Physiology 12:941–948.

Levitt, J. 1957. The significance of "apparent free space" (A.F.S.) in ion absorption. Physiologia Plantarum 10:882–888.

Morohashi, Y. and M. Shimorkoriyama. 1971. Apparent free space of germinating pea seeds. Physiologia Plantarum 25:341–345.

Oertli, J. J. 1971. The stability of water under tension in the xylem. Zeitschrift für Pflanzenphysiologie 65:195–209.

O'Leary, J. W. 1970. Can there be a positive water potential in plants? Bioscience 20:858–859.

O'Leary, J. W. and P. J. Kramer. 1964. Root pressure in conifers. Science 145:284–285.

Philip, J. R. 1966. Plant-water relations: Some physical aspects. Annual Review of Plant Physiology 17:245–368.

Plumb, R. C. and W. B. Bridgman. 1972. Ascent of sap in trees. Science 176:1129–1131.

Redshaw, A. J. and H. Meidner. 1970. A thermal method for estimating continuously the rate of flow of sapthrough an intact plant. Zeitschrift für Pflanzenphysiologie 62:405–416.

Renner, O. 1911. Experimentelle Beiträge zur Kenntnis der Wasserbewegung [Experimental contributions to the knowledge of water movement]. Flora 103:171.

————. 1915. Theoretisches und Experimentelles zur Kohäsionstheorie der Wasserbewegung [Theoretical and experimental contribution to the cohesions theory of water movement]. Jahrbuch für Wissenschaftliche Botanik 56:617–667.

Russell, R. S. and D. A. Barber. 1960. The relationship between salt uptake and the absorption of water by intact plants. Annual Review of Plant Physiology 11:127–140.

Rutter, A. J. and F. H. Whitehead, eds. 1961. The water relations of plants. A symposium of the

British Ecological Society. John Wiley & Sons, Inc., New York.

Scholander, P. F. 1968. How mangroves desalinate sea water. Physiologia Plantarum 21:251–268.

Scholander, P. F., H. T. Hammel, E. D. Bradstreet, and E. A. Hemmingsen. 1965. Sap pressure in vascular plants. Science 148:339–346.

Scholander, P. F., E. Hemmingsen, and W. Garey. 1961. Cohesive lift of sap in the rattan vine. Science 134:1835–1838.

Scholander, P. F. 1968. How mangroves desalinate sea water. Physiologia Plantarum 21:251–268.

Scholander, P. F., H. T. Hammel, E. D. Bradstreet, and E. A. Hemmingsen. 1965. Sap pressure in vascular plants. Science 148:339–346.

Scholander, P. F., E. Hemmingsen, and W. Garey. 1961. Cohesive lift of sap in the rattan vine. Science 134:1835–1838.

Sheriff, D. W. 1974. Magnetohydynamic sap flux meters: An instrument for laboratory use and theory for calibration. Journal of Experimental Botany, 25:675–683.

Sheriff, D. W. and H. Meidner. 1974. Water pathway in leaves of Hedera helix L. and Tradescantia virginiana L. Journal of Experimental Botany 25:1147–1156.

Spanner, D. C. 1973. The components of the water potential in plants and soils. Journal of Experimental Botany 24:816–819.

Spomer, G. G. 1968. Sensors monitor tension in transpiration streams of trees. Science 161: 484–485.

Stone, J. F. and G. A. Shirazi. 1975. On the heat-pulse method for the measurement of apparent sap velocity in stems. Planta 122:166–177.

Sucoff, E. 1968. Freezing of conifer xylem and the cohesion-tension theory. Physiologia Plantarum 22:424–431.

Tobussen, P., P. W. Rundel, and R. E. Stecker. 1971. Water potential gradient in tall Sequoiadendron. Plant Physiology 48:303–304.

Troughton, John and Lesley A. Donaldson, 1972. Probing plant structure. McGraw-Hill, New York.

Tyree, M. T. and M. H. Zimmermann. 1971. The theory and practice of measuring transport coefficients and sap flow in the xylem of red maple stems (Acer rubrum). Journal of Experimental Botany 22:1–18.

van Fleet, D. S. 1961. Histochemistry and function of the endodermis. Botanical Review 27:165–220.

Wiebe, H. H., R. W. Brown, T. W. Daniel, and E. Campbell. 1970. Water potential measurement in trees. Bioscience 20:225–226.

Chapter 5

Anderson, W. P. 1972. Ion transport in the cells of higher plant tissues. Annual Review of Plant Physiology 23:51–72.

———. 1975. Ion transport through roots. Pages 437–463 in Torrey, J. G. and D. T. Clarkson, eds., The development and function of roots. Academic Press, London.

———. 1976. Transport through roots. Pages 129–156 in Lüttge, U. and M. G. Pitman, eds., Transport in plants. Encyclopedia of Plant Physiology, new series, vol. 2, part B. Springer-Verlag, Berlin.

Baker, D. A. and J. L. Hall, eds. 1975. Ion transport in plant cells and tissues. North-Holland Publishing Company, Amsterdam.

Benson, A. A. and A. T. Jokela. 1976. Cell membranes. Pages 65–89 in J. Bonner and J. E. Varner, eds., Plant biochemistry, Academic Press, Inc., New York.

Caldwell, M. and L. B. Camp. 1974. Below-ground productivity of two cool desert communities. Oceologic 17:123–130.

Chung, H. and P. J. Kramer. 1975. Absorption of water and ^{32}P through suberized and unsuberized roots of loblolly pine. Canadian Journal of Forest Research 5:229–235.

Clarkson, D. T. 1974. Ion transport and cell structure in plants. McGraw-Hill Book Company, New York.

Clarkson, D. T. and A. W. Robards. 1975. The endodermis, its structural development and physiological role. Pages 415–436 in Torrey, J. G. and D. T. Clarkson, eds., The development and function of roots. Academic Press, Inc., New York.

Collander, R. 1959. Cell membranes: Their resistance to penetration and their capacity for transport. Pages 3–102 in Steward, F. C., ed., Plant physiology. Vol. 2. Academic Press, Inc., New York.

Copaldi, R. A. 1974. A dynamic model of cell membranes. Scientific American 230(3):26–33.

Cox, G. and P. B. Tinker. 1976. Translocation and transfer of nutrients in vesicular-arbuscular mycorrhizas. New Phytologist 77:371–378.

Davies, D. D. 1973. Control of and by pH. Symposium for the Society of Experimental Biology 27:513–529.

Drew, M. C. 1975. Comparison of the effects of a localized supply of phosphate, nitrate, ammonium and potassium on the growth of the seminal root system and the shoot in barley. New Phytologist 75:479–490.

Eisenberg, M. and S. McLaughlin. 1976. Lipid bilayers as models of biological membranes. BioScience 26:436–443.

Epstein, E. 1972. Mineral nutrition of plants: Principles and perspectives. John Wiley & Sons, Inc., New York.

———. 1973. Mechanisms of ion transport through plant cell membranes. International Review of Cytology 34:123–168.

———. 1973. Roots. Scientific American 228(5): 48–58.

Gerdemann, J. W. 1975. Vesicular-arbuscular mycorrhizae. Pages 575–591 in Torrey, J. G. and D. T. Clarkson, 1975, eds., The development and function of roots. Academic Press, Inc., New York.

Glass, A. D. M. 1976. Regulation of potassium absorption in barley roots. An allosteric model. Plant Physiology 58:33–37.

Greenwood, D. J. and D. Goodman. 1971. Studies on the supply of oxygen to the roots of mustard seedlings. New Phytologist 70:85–96.

Grun, Paul. 1963. Ultrastructure of plant plasma and vacuolar membranes. Journal of Ultrastructural Research 9:198–208.

Higinbotham, N. 1973. Electropotentials of plant cells. Annual Review of Plant Physiology 24: 25–46.

———. 1973. The mineral absorption process in plants. Botanical Review 39;15–69.

Higinbotham, N. and W. P. Anderson. 1974. Electrogenic pumps in higher plant cells. Canadian Journal of Botany 52:1011–1021.

Hodges, T. K. 1973. Ion absorption by plant roots. Advances in Agronomy 25:163–207.

———. 1976. ATPase associated with membranes of plant cells. Pages 260–283 in Lüttge, U. and M. G. Pitman, eds., Encyclopedia of Plant Physiology, new series, vol. 2, part A. Springer-Verlag, Berlin.

Hsiao, T. C., E. Acevedo, E. Fereres, and D. W. Henderson. 1976. Stress metabolism. Water stress, growth, and osmotic adjustment. Philosophical Transactions of the Royal Society of London, series B, 273:479–500.

Kormanik, P. P. and C. L. Brown. 1967. Root buds and the development of rootsuckers in sweetgum. Forest Science 13:338–348.

Laties, G. G. 1975. Solute transport in relation to metabolism and membrane permeability in plant tissues. Pages 98–151 in Davies, P. J., ed., Historical and current aspects of plant physiology: A symposium honoring F. C. Steward. New York State College of Agriculture and Life Sciences, Ithaca.

Marks, G. C. and T. T. Kozlowski, eds. 1973. Ectomycorrhizae, their ecology and physiology. Academic Press, Inc., New York.

Mosse, B. 1973. Advances in the study of vesicular-arbuscular mycorrhiza. Annual Review of Phytopathology 11:171–196.

———. 1975. A microbiologist's view of anatomy. Pages 39–65 in Walker, N., ed., Soil microbiology. John Wiley & Sons, Inc., New York.

Nissen, P. 1974. Uptake mechanisms: Inorganic and organic. Annual Review of Plant Physiology 25:53–79.

Noggle, J. C. 1966. Ionic balance and growth of sixteen plant species. Soil Science of America Proceedings 30:763–766.

Pitman, M. G., W. P. Anderson, and U. Lüttge. 1976. Transport processes in roots. Pages 37–69 in Lüttge, U. and M. G. Pitman, eds., Transport in plants. Encyclopedia of plant physiology, new series, vol. 2, part B. Springer-Verlag, Berlin.

Raven, J. A. and F. A. Smith. 1974. Significance of hydrogen ion transport in plant cells. Canadian Journal of Botany 52:1035–1048.

Reisenauer, H. M. 1966. Mineral nutrients in soil solution. Pages 507–508 in Altman, P. L. and D. S. Dittmer, eds., Environmental biology. Federation of American Societies for Experimental Biology, Bethesda, Md.

Reynolds, E. R. C. 1975. Tree rootlets and their distribution. Pages 163–177 in Torrey, J. G. and D. T. Clarkson, eds., The development and function of roots. Academic Press, Inc., New York.

Robards, A. W. 1975. Plasmodesmata. Annual Review of Plant physiology 26:13–29.

———. 1976. Plasmodesmata in higher plants. Pages 16–57 in B. E. S. Gunning and A. W. Robards, eds., Intercellular communication in plants: Studies on plasmodesmata. Springer-Verlag, Berlin.

Robards, A. W. and D. T. Clarkson. 1976. The role of plasmodesmata in the transport of water and nutrients across roots. Pages 181–201 in Gunning, B. E. S. and A. W. Robards, eds., Intercellular communication in plants: Studies on plasmodesmata. Springer-Verlag, Berlin.

Robards, A. W. and S. M. Jackson. 1976. Root structure and function—an integrated approach. Pages 413–422 in Sunderland, N., ed., Perspectives in experimental biology. Vol. 2, Botany. Pergamon Press, Inc., Elmsford, N. Y.

Rovira, A. D. and C. B. Davey. 1974. Biology of the rhizosphere. Pages 153–204 in Carson, E. W., ed., The plant root and its environment. The University Press of Virginia, Charlottesville.

Sanders, F. E., B. Mosse, and P. B. Tinker, eds. 1976. Endomycorrhizas. Academic Press, Inc., New York.

Singer, S. J. 1974. The molecular organization of membranes. Annual Review of Biochemistry 43:805–833.

Singer, S. J. and G. L. Nicolson. 1972. The fluid mosaic model of the structure of cell membranes. Science 175:720–731.

Skou, J. C. 1975. The (Na$^+$ + K$^+$) activated enzyme system and its relationship to transport of sodium and potassium. Quarterly Reviews of Biophysics 7:401–434.

Torrey, J. G. and D. T. Clarkson, eds. 1975. The development and function of roots. Academic Press, Inc., New York.

Troughton, John and Lesley A. Donaldson. 1972. Probing plant structure. McGraw-Hill, New York.

Vlamis, J. 1944. Effects of oxygen tension on certain

physiological responses of rice, barley, and tomato. Plant Physiology 19:33–51.

Wiebe, H. and Paul Kramer. 1954. Translocation of radioactive isotopes from various regions of roots and barley seedlings. Plant Physiology 29:342–348.

Zimmermann, U. and J. Dainty, eds. 1974. Membrane transport in plants. Springer-Verlag, Berlin.

Chapter 6

Broyer, T. C. and P. R. Stout. 1959. The macronutrient elements. Annual Review of Plant Physiology 10:277–300.

Dixon, N. E., C. Gazzola, R. L. Blakeley, and B. Zerner. 1976. metal ions in enzymes using ammonia or amides. Science 191:1144–1150.

Epstein, E. 1972. Mineral nutrition of plants: Principles and perspective. John Wiley & Sons, Inc., New York.

Gauch, H. G. 1972. Inorganic plant nutrition. Dowden, Hutchinson & Ross, Inc., Stroudsberg, Pa.

Hewitt, E. J. and T. A. Smith. 1975. Plant mineral nutrition. John Wiley & Sons, Inc., New York.

Jones, R. G. W. and O. E. Lunt. 1967. The functions of calcium in plants. Botanical Review 33:407–426.

Lewin, J. and B. E. F. Reimann. 1969. Silicon and plant growth. Annual Review of Plant Physiology 20:289–304.

Lindsay, W. L. 1974. Role of chelation in micronutrient availability. Pages 507–524 in Carson, E. W., ed., The plant root and its environment. The University Press of Virginia, Charlottesville.

Miller, L. P. and F. Flemion. 1973. The role of minerals in phytochemistry. Pages 1–40 in L. P. Miller, ed., Phytochemistry. Vol. 3, Inorganic elements and special groups of chemicals. Van Nostrand Reinhold Company, New York.

Oertli, J. J. and E. Gregurevic. 1975. Effect of pH on the absorption of boron by excised barley roots. Agronomy Journal 67:278-280.

Rains, D. W. 1976. Mineral metabolism. Pages 561–597 in J. Bonner and J. E. Varner, eds. Plant biochemistry, Academic Press, Inc., New York.

Reisenauer, H. M. 1966. Mineral nutrients in soil solution. Pages 507–508 in Altman, P. L. and D. S. Dittmer, eds., Environmental biology. Federation of American Societies for Experimental Biology, Bethesda, Md.

Schmidtling, R. C. 1973. Intensive culture increases growth without affecting wood quality of young southern pines. Canadian Journal of Forest Research 3:565–573.

Shrift, A. 1969. Aspects of selenium metabolism in higher plants. Annual Review of Plant Physiology 20:475–494.

———. 1972. Selenium toxicity. Pages 145–162 in Harborne J. B., ed., Phytochemical ecology. Academic Press, Inc., New York.

Sprague, H. B., ed. 1964. Hunger signs in crops. A symposium. David McKay Co., Inc., New York.

Stadtman, T. C. 1974. Selenium biochemistry. Science 183:915–922.

Turner, J. and P. R. Olson. 1976. Nitrogen relations in a Douglas-fir plantation. Annals of Botany 40:1185–1193.

van den Driessche, R. 1974. Prediction of mineral nutrient status of trees by foliar analysis. Botanical Review 40:347-394.

Chapter 7

Arnold, W. N. 1968. The selection of sucrose as the translocate of higher plants. Journal of Theoretical Biology 21:13–20.

Behnke, H. D. 1971. The contents of the sieve-plate pores in Aristolchia. Journal of Ultrastructural Research 36:493–498.

Biddulph, O. and R. Cory. 1960. Demonstration of two translocation mechanisms in studies of bidirectional movement. Plant Physiology 35: 689–695.

———. 1965. Translocation of ^{14}C-metabolites in the phloem of the bean plant. Plant Physiology 40:119–129.

Bollard, E. G. 1960. Transport in the xylem. Annual Review of Plant Physiology 11:141–166.

Botha, C. E. J., R. F. Evert, and R. D. Walmsley. 1975. Observations of the penetration of the phloem in leaves of Nerium oleander (Linn.) by stylets of the aphid, Aphis nerii (B. de F.). Protoplasma 86: 309–319.

Bowling, D. J. F. 1969. Evidence for the electroosmosis theory of transport in the phloem. Biochimica et Biophysica Acta 183:230–232.

Buttery, B. R. and S. G. Boatman. 1966. Manometric measurement of turgor pressures in laticeferous phloem tissues. Journal of Experimental Botany 17:283–296.

Canny, M. J. P. 1971. Translocation: mechanisms and kinetics. Annual Review of Plant Physiology 22:237–260.

Crafts, A. S. and C. E. Crisp. 1971. Phloem transport in plants. W. H. Freeman and Company Publishers, San Francisco.

Crafts, A. S. and O. Lorenz. 1944. Fruit growth and food transport in cucurbits. Plant Physiology 19:131–138.

Cronshaw, J. and K. Esau. 1968. P-protein in the phloem of Cucurbita. II. The P-protein of mature sieve elements. Journal of Cell Biology 38:293–303.

De Vries, H. 1885. Über die Bedeutung der Circulation und der Rotation des Protoplasma für das Stofftransport in der Pflanze [On the significance of the circulation and rotation of protoplasm for translocation in plants]. Botanische Zeitung 43:1–26.

Esau, Katherine. 1965. Plant Anatomy. John Wiley & Sons, Inc., New York.

Esau, Katherine and V. I. Cheadle. 1965. Cytologic studies on phloem. University of California Publications in Botany 36:253–344.

Eschrich, W. 1967. Bidirektionelle Translokation in Siebröhren [Bidirectional translocation in sieve tubes]. Planta 73:37–49.

Fensom, D. S. 1972. A theory of translocation in phloem of Heracleum by contractile protein microfibrillar material. Canadian Journal of Botany 50:479–497.

Ferrier, Jack M. and Melvin T. Tyree. 1976. Further analysis of the moving strand model of translocation using a numerical calculation. Canadian Journal of Botany 54:1271–1282.

Field, R. J. and A. J. Peel. 1971. The movement of growth regulators and herbicides into the sieve elements of willow. New Phytologist 70:997–1003.

Fisher, Donald B. 1975. Structure of functional soybean sieve elements. Plant Physiology 56:555–569.

Gardner, D. C. J. and A. J. Peel. 1972. Some observations on the role of ATP in sieve tube translocation. Planta 107:217–226.

Geiger, D. R. and A. L. Christy. 1971. Effect of sink region anoxia on translocation rate. Plant Physiology 47:172–174.

Geiger, D. R., R. T. Giaquinta, S. A. Sovonick, and R. J. Fellows. 1973. Solute distribution in sugar beet leaves in relation to phloem loading and translocation. Plant Physiology 52:585–589.

Geiger, D. R., S. A. Sovonick, T. L. Shock, and R. J. Fellows. 1974. Role of free space in translocation in sugar beet. Plant Physiology 54:892–898.

Giaquinta, R. T. and D. R. Geiger. 1972. Mechanisms of inhibition of translocation by localized chilling. Plant Physiology 51:372–377.

Gunning, Brian E. S. and John S. Pate. 1974. Transfer cells. Pages 441–480 in A. W. Robards, ed., Dynamic aspects of plant ultrastructure. McGraw-Hill Book Company, New York.

Hammel, H. T. 1968. Measurement of turgor pressure and its gradient in the phloem of oak. Plant Physiology 43:1042–1048.

Ho, L. C. and D. C. Mortimer. 1971. The site of cyanide inhibition of sugar translocation in sugar beet leaf. Canadian Journal of Botany 49:1769–1775.

Ho, L. C. and A. J. Peel. 1969. Investigation of bidirectional movement of tracers in sieve tubes of Salix viminalis L. Annals of Botany 33:833–844.

Jarvis, P. and R. Thaine. 1971. Strands in sections of sieve elements cut in a cryostat. Nature 232: 236–237.

Kursanov, A. L. 1963. Metabolism and the transport of organic substances in the phloem. Advances in Botanical Research 1:209–274.

Läuchli, Andre. 1972. Translocation of inorganic solutes. Annual Review of Plant Physiology 23:197–218.

MacRobbie, E. A. C. 1971. Phloem translocation. Facts and mechanisms: A comparative survey. Biological Review 46:429–481.

McCready, C. C. 1966. Translocation of growth regulators. Annual Review of Plant Physiology 17:283–294.

Münch, E. 1930. Die Stoffbewegungen in der Pflanze [Translocation in plants]. Fischer, Jena.

Palevitz, Barry A. and Peter K. Hepler. 1975. Is P-protein actin-like?—Not yet. Planta 125:261–271.

Pate, J. S. and B. E. S. Gunning. 1972. Transfer cells. Annual Review of Plant Physiology 23:173–196.

Peel, A. J. 1974. Transport of nutrients in plants. John Wiley & Sons, Inc., New York.

Peterson, Carol A. and H. B. Currier. 1969. An investigation of bidirectional translocation in the phloem. Physiologia Plantarum, 22:1238-1250.

Phillis, E. and T. G. Mason. 1933. Studies on the transport of carbohydrates in the cotton plant. III. The polar distribution of sugar in the foliage leaves. Annals of Botany 47:585–634. An early study that suggested phloem loading.

Roeckle, B. 1949. Nachweise eines Konzentrationshubs zwischen Palisadenzellen und Siebröhren [Evidence for a concentration buildup between palisade cells and sieve tubes]. Planta 36:530–550.

Rogers, S. and A. J. Peel. 1975. Some evidence for the existence of turgor pressure gradients in the sieve tubes of willow. Planta 126:259–267.

Salisbury, F. B. 1966. Translocation: the movement of dissolved substances in plants. Pages 70–91 in W. A. Jensen and L. G. Kavaljian, eds., Plant biology today: Advances and challenges. Wadsworth Publishing Company, Inc., Belmont, Ca.

Sauter, J. J., W. Iten, and M. H. Zimmermann. 1973. Studies on the release of sugar into the vessels of sugar maple (Acer saccharum). Canadian Journal of Botany 51(1):1–8.

Sovonick, S., D. R. Geiger, and R. J. Fellows. 1974. Evidence for active phloem loading in the minor veins of sugar beet. Plant Physiology 54:886–891.

Spanner, D. C. 1970. The electro-osmosis theory of phloem transport in the light of recent measurements on Heracleum phloem. Journal of Experimental Botany 21:325–334.

Thaine, R. 1964. The protoplasmic-streaming theory of

phloem transport. Journal of Experimental Botany 15:470–484.

Thaine, R. and M. E. De Maria. 1973. Transcellular strands of cytoplasm in sieve tubes of squash. Nature 245:161–163.

Thaine, R., M. C. Probine, and P. Y. Dyer. 1967. The existence of transcellular strands in sieve elements. Journal of Experimental Botany 18:110–127.

Trip, P. and P. R. Gorham. 1968. Bidirectional translocation of sugars in sieve tubes of squash plants. Plant Physiology 43:877–882.

Wiebe, Herman H. 1962. Physiological response of plants to drought. Utah Science 23(3):70–71.

Wiersum, L. K. 1967. The mass-flow theory of phloem transport: A supporting calculation. Journal of Experimental Botany 18:160–162.

Zimmermann, Martin H. 1960. Transport in the phloem. Annual Review of Plant Physiology 11:167–190.

———. 1963. How sap moves in trees. Scientific American 208(3):132–142.

Zimmermann, M. H. and J. A. Milburn, eds. 1975. Transportation in plants I. Phloem transport. In Lüttge, U. and M. G. Pitman, eds., Encyclopedia of plant physiology, new series, vol. 1. Springer-Verlag, Berlin.

Chapter 8

Anfinson, C. B. 1959. The molecular basis of evolution. John Wiley & Sons, Inc., New York.

———. 1972. The formation and stabilization of protein structure. Biochemical Journal 128: 737–749.

Bell, R. M. and D. E. Koshland, Jr. 1971. Covalent enzyme-substrate intermediates. Science 172: 1253–1256.

Boulter, D., J. A. M. Ramshaw, E. W. Thompson, M. Richardson, and R. H. Brown. 1972. A phylogeny of higher plants based on the amino acid sequences of cytochrome c and its biological implications. Proceedings of the Royal Society of London, series B, 181:441–455.

Cammack, R., D. Hall, and K. Rao. 1971. Ferredoxins: Are they living fossils? New Scientist and Science Journal 23:696–698.

Changeux, J. P. 1965. The control of biochemical reactions. Scientific American 212(4):36–45.

Chisolm, J. J., Jr. 1971. Lead poisoning. Scientific American 224(2):15–23.

Dixon, M. and E. C. Webb. 1964. Enzymes. Academic Press, Inc., New York.

Goldwater, L. J. 1971. Mercury in the environment. Scientific American 224(5):15–21.

Gottlieb, L. D. 1971. Gel electrophoresis: New approach to the study of evolution. Bioscience 21:939–944.

Harborne, J. B. and C. F. van Sumere, eds. 1975. The chemistry and biochemistry of plant proteins. Academic Press, Inc., New York.

Harpstead, D. D. 1971. High-lysine corn. Scientific American 225(2):34–42.

Johnson, V. A. and C. L. Lay. 1974. Genetic improvement of plant protein. Journal of Agricultural and Food Chemistry 22:558–566.

Kirschner, K. and H. Bisswanger. 1976. Multifunctional proteins. Annual Review of Biochemistry 45: 143–166.

Koshland, D. E., Jr. 1973. Protein shape and biological control. Scientific American 229(4):52–64.

Liener, I. E. 1976. Phytohemagglutinins (phytolectins). Annual Review of Plant Physiology 27:291–319.

Markert, C. L., ed. 1975. Isozymes II, Physiological Function. Academic Press, Inc., New York.

Millerd, A. 1975. Biochemistry of legume seed proteins. Annual Review of Plant Physiology 26:53–72.

Orr, M. L. and B. K. Watt. 1957. Amino acid content of foods. United States Department of Agriculture Home Economics Research Report 4:1–82.

Preiss, J. and T. Kosuge. 1976. Regulation of enzyme activity in metabolic pathways. Pages 277–336 in Bonner J. and J. E. Varner, eds., Plant biochemistry. Academic Press, Inc., New York.

Scientific American. 1976. 235(3):30–205. (The entire issue of 12 articles is devoted to food and agriculture.)

Shannon, L. M. 1968. Plant isoenzymes. Annual Review of Plant Physiology 19:187–210.

Sharon, N. 1974. Glycoproteins of higher plants. Pages 235–252 in Pridham, J. B., ed., Plant carbohydrate biochemistry. Academic Press, Inc., New York.

———. 1977. Lectins. Scientific American 236(6): 108–119.

Spiker, S. 1975. An evolutionary comparison of plant histones. Biochimica et Biophysica Acta 400: 461–467.

Stadtman, E. R. 1966. Allosteric regulation of enzyme activity. Pages 41–154 in Nord, F. F., ed., Advances in enzymology and related subjects of biochemistry, vol. 28. Interscience Publishers, Inc., New York.

Sylvester-Bradley, R. and B. F. Folkes. 1976. Cereal grains: Their protein components and nutritional quality. Scientific Progress 63:241–263.

Trewavas, A. 1976. Post-transitional modification of proteins by phosphorylation. Annual Review of Plant Physiology 27:349–374.

Wolfe, S. L. 1972. Biology of the cell. Wadsworth Publishing Company, Belmont, Calif.

Chapter 9

Anderson, J. M. 1975. The molecular organization of chloroplast thylakoids. Biochimica et Biophysica Acta 416:191–235.

Avron, M. 1975. The electron transport chain in chloroplasts. Pages 374–386 in Govindjee, ed., Bioenergetics of photosynthesis. Academic Press, Inc., New York.

Beale, S. I., S. P. Gough, and S. Granick. 1975. Biosynthesis of delta-aminolevulinic acid from the intact carbon skeleton of glutamic acid in greening barley. Proceedings of the National Academy of Sciences 72:2719–2723.

Bearden, A. J. and R. Malkin. 1975. Primary photochemical reactions in chloroplast photosynthesis. Quarterly Reviews of Biophysics 7:131–177.

Black, C. C. 1976. How herbicides work. Weeds Today 8:13–15.

Brown, J. S. 1972. Forms of chlorophyll in vivo. Annual Review of Plant Physiology 23:73–86.

Egneus, H., U. Heber, U. Matthiesen, and M. Kirk. 1975. Reduction of oxygen by the electron transport chain of chloroplasts during assimilation of carbon dioxide. Biochimica et Biophysica Acta 408:252–268.

Goodenough, U. and R. P. Levine. 1970. The genetic activity of mitochondria and chloroplasts. Scientific American 223(5):22–29.

Goodwin, T. W. 1973. Carotenoids. Pages 112–142 in Miller, L. P., ed., Phytochemistry. Vol. I, The process and products of photosynthesis. Van Nostrand Reinhold Company, New York.

Govindjee, ed. 1975. Bioenergetics of photosynthesis. Academic Press, Inc., New York.

Govindjee and R. Govindjee. 1974. The primary events of photosynthesis. Scientific American 231(6): 68–82.

Inada, K. 1976. Action spectra for photosynthesis in higher plants. Plant and Cell Physiology 17:355–365.

Joliot, P. and B. Kok. 1975. Oxygen evolution in photosynthesis. Pages 388–412 in Govindjee, ed., Bioenergetics of photosynthesis. Academic Press, Inc., New York.

Kirk, J. T. O. 1972. The genetic control of plastid formation: Recent advances and strategies for the future. Sub-Cellular Biochemistry 1:333–361.

Kirk, J. T. O. and R. A. E. Tilney-Bassett. 1967. The plastids. W. H. Freeman and Company Publishers, San Francisco.

Kok, B. 1976. Photosynthesis: The path of energy. Pages 845–885 in Bonner, J. and J. E. Varner, eds., Plant biochemistry. Academic Press, Inc., New York.

Lohr, J. B. and H. C. Friedmann. 1976. New pathway for δ-aminolevulinic acid biosynthesis: formation from α-ketoglutaric acid by two partially purified plant enzymes. Biochemical and Biophysical Research Communications 69:908–913.

Myers, J. 1974. Conceptual developments in photosynthesis, 1924–1974. Plant Physiology 54: 420–426.

Park, R. B. 1976. The chloroplast. Pages 115–145 in Bonner, J. and J. E. Varner, eds., Plant biochemistry. Academic Press, Inc., New York.

Possingham, J. V. 1973. Effect of light quality on chloroplast replication in spinach. Journal of Experimental Botany 24:1247–1260.

Possingham, J. V. and R. J. Rose. 1976. Chloroplast replication and chloroplast DNA synthesis in spinach leaves. Proceedings of the Royal Society of London, series B. 193:295–305.

Radmer, R. J. and B. Kok. 1976. Photoreduction of O_2 primes and replaces CO_2 assimilation. Plant Physiology 58:336–340.

Rebeiz, C. A. and P. A. Castelfranco. 1973. Protochlorophyll and chlorophyll biosynthesis in cell-free systems. Annual Review of Plant Physiology 24:129–172.

Satoh, K., R. Strasser and W. L. Butler. 1976. A demonstration of energy transfer from photosystem II to photosystem I in chloroplasts. Biochimica et Biophysica Acta 440:337–345.

Simonis, W. and W. Urbach. 1973. Photophosphorylation in vivo. Annual Review of Plant Physiology 24:89–114.

Stemler, A. and R. Radmer. 1975. Source of photosynthetic oxygen in bicarbonate-stimulated Hill reaction. Science 190:457–458.

Trebst, A. 1974. Energy conservation in photosynthetic electron transport of chloroplasts. Annual Review of Plant Physiology 25:423–458.

Vernon, L. P. and G. R. Seely, eds. 1966. The chlorophylls. Academic Press, Inc., New York.

Chapter 10

Akazawa, T. 1976. Polysaccharides. Pages 381–403 in J. Bonner and J. E. Varner, eds. Plant biochemistry. Academic Press, Inc., New York.

Albersheim, P. 1976. The primary cell wall. Pages 225–274 in J. Bonner and J. E. Varner, eds. Plant biochemistry. Academic Press, Inc., New York.

Anderson, L. E. and M. Avron. 1976. Light modulation of enzyme activity in chloroplasts. Generation of membrane-bound vicinal-dithiol groups by photosynthetic electron transport. Plant Physiology 57:209–213.

Bamberger, E. S., B. A. Ehrlich, and M. Gibbs. 1975. The glyceraldehyde 3-phosphate and glycerate

3-phosphate shuttle and carbon dioxide assimilation in intact spinach chloroplasts. Plant Physiology 55:1023–1030.

Björkman, O. 1973. Comparative studies on photosynthesis in higher plants. Pages 1–63 in A. C. Giese, ed., Photophysiology, vol. 8. Academic Press, Inc., New York.

Black, C. C. 1971. Ecological implications of dividing plants into groups with distinct photosynthetic production capacities. Advances in Ecological Research 7:87–114.

———. 1973. Photosynthetic carbon fixation in relation to net CO_2 uptake. Annual Review of Plant Physiology 24:253–286.

Black, C. C., W. H. Campbell, T. M. Chen, and P. Dittrich. 1973. The monocotyledons: Their evolution and comparative biology. III. Pathways of carbon metabolism related to net carbon dioxide assimilation by monocotyledons. Quarterly Review of Biology 48:299–313.

Bowes, G., W. L. Ogren, and R. H. Hageman. 1975. pH dependence of the K_m (CO_2) of ribulose-1, 5-diphosphate carboxylase. Plant Physiology 56:630–633.

Burris, R. H. and C. C. Black, eds. 1976. CO_2 metabolism and plant productivity. University Park Press, Baltimore.

Chen, T. M., P. Dittrich, W. H. Campbell, and C. C. Black. 1974. Metabolism of epidermal tissues, mesophyll cells, and bundle sheath strands resolved from mature nutsedge leaves. Archives of Biochemistry and Biophysics 163:246–262.

Chu, D. K. and J. A. Bassham. 1975. Regulation of ribulose-1, 5-diphosphate carboxylase by substrates and other metabolites. Plant Physiology 55:720–726.

Cleland, W. W. 1964. Dithiothreitol, a new protective reagent for SH groups. Biochemistry 3:480–482.

Daly, L. S., T. B. Ray, H. M. Vines, and C. C. Black. 1977. Characterization of phosphoenolpyruvate carboxykinase from pineapple leaves Ananas cosmosus (L.) Merr. Plant Physiology 59:618–622.

Doesburg, J. J. 1973. The pectic substances. Pages 271–296 in L. P. Miller, ed., Phytochemistry. Vol. 1, the process and products of photosynthesis. Van Nostrand Reinhold Company, New York.

Downton, W. J. S. 1975. The occurrence of C-4 photosynthesis among plants. Photosynthetica 9:96–105.

Gander, J. E. 1976. Mono- and oligosaccharides. Pages 337–380 in J. Bonner and J. E. Varner, eds., Plant biochemistry. Academic Press, Inc., New York.

Gibbs, M. 1966. Carbohydrates: Their role in plant metabolism and nutrition. Pages 3–115 in F. C. Steward, ed., Plant physiology. Academic Press, Inc., New York.

Givan, C. V. and J. L. Harwood. 1976. Biosynthesis of small molecules in chloroplasts of higher plants. Biological Reviews 51:365–406.

Gutierrez, M., V. E. Gracen, and G. E. Edwards. 1974a. Biochemical and cytological relationships in C-4 plants. Planta 119:279–300.

Gutierrez, M., R. Kanai, S. C. Huber, S. B. Ku, and G. E. Edwards. 1974b. Photosynthesis in mesophyll protoplasts and bundle sheath cells of various types of C-4 plants. I. Carboxylases and CO_2 fixation studies. Zeitschrift für Pflanzenphysiologie 72:305–309.

Hatch, M. D. 1976. Photosynthesis: the path of carbon. Pages 797–844 in J. Bonner and J. E. Varner, eds. Plant biochemistry. Academic Press, Inc., New York.

Hatch, M. D. and T. Kagawa. 1976. Photosynthetic activities of isolated bundle sheath cells in relation to differing mechanisms of C-4 pathway photosynthesis. Archives of Biochemistry and Biophysics 175:39–53.

Hatch, M. D., C. R. Slack, and T. A. Bull. 1969. Light-induced changes in the content of some enzymes of the C-4-dicarboxylic acid pathway of photosynthesis and its effect on other characteristics of photosynthesis. Phytochemistry 8:697–706.

Hatch, M. D., C. R. Slack, and H. S. Johnson. 1967. Further studies on a new pathway of photosynthetic carbon dioxide fixation in sugar cane and its occurrence in other plant species. Biochemical Journal 102:417–422.

Heber, U. 1974. Metabolite exchange between chloroplasts and cytoplasm. Annual Review of Plant Physiology 25:393–421.

Karr, A. L. 1976. Cell wall biogenesis. Pages 405–426 in J. Bonner and J. E. Varner, eds., Plant biochemistry. Academic Press, Inc., New York.

Kauss, H. 1974. Biosynthesis of pectin and hemicelluloses. Pages 191–205 in J. B. Pridham, ed., Plant carbohydrate biochemistry. Academic Press, Inc., New York.

Kortschak, H. P., C. E. Hartt, and G. O. Burr. 1965. Carbon dioxide fixation in sugarcane leaves. Plant Physiology 40:209–213.

Krenzer, E. G., Jr., D. N. Moss, and R. K. Crookston. 1975. Carbon dioxide compensation points of flowering plants. Plant Physiology 56:194–206.

Ku, S. B. and G. E. Edwards. 1975. Photosynthesis in mesophyll protoplasts and bundle sheath cells of various types of C-4 plants. IV. Enzymes of respiratory metabolism and energy utilizing enzymes of photosynthetic pathways. Zeitschrift für Pflanzenphysiologie 77:16–32.

Laetsch, W. M. 1974. The C-4 syndrome: A structural analysis. Annual Review of Plant Physiology 25:27–52.

Manners, D. J. 1973. Starch and inulin. Pages 176–197 in L. P. Miller, ed., Phytochemistry. Vol. 1, The process and products of photosynthesis. Van Nostrand Reinhold Company, New York.

———. 1974. The structure and metabolism of starch. Pages 37–71 in P. N. Campbell and F. Dickens, eds., Essays in biochemistry, vol. 10. Academic Press, Inc., New York.

Miller, L. P. 1973. Mono- and oligosaccharides. Pages 145–175 in L. P. Miller, ed., Phytochemistry. Vol. 1, The process and products of photosynthesis. Van Nostrand Reinhold Company, New York.

Neales, T. F. 1975. The gas exchange patterns of CAM plants. Pages 299–310 in R. Marcelle, ed., Environmental and biological control of photosynthesis. Dr. W. Junk, The Hague.

Northcote, D. H. 1974. Sites of synthesis of the polysaccharides in the cell wall. Pages 165–181 in J. B. Pridham, ed., Plant carbohydrate biochemistry. Academic Press, Inc., New York.

Preiss, J. and T. Kosuge. 1976. Regulation of enzyme activity in metabolic pathways. Pages 277–336 in J. Bonner and J. E. Varner, eds., Plant biochemistry. Academic Press, Inc., New York.

Preston, R. D. 1974. Plant cell walls. Pages 256–309 in A. W. Robards, ed., Dynamic aspects of plant ultrastructure. McGraw-Hill Book Company, New York.

Rathnam, C. K. M., A. S. Raghavendra, and V. S. Rama Das. 1976. Diversity in the arrangements of mesophyll cells among leaves of certain C-4 cotyledons in relation of C-4 physiology. Zeitschrift für Pflanzenphysiologie 77:283–291.

Ting, I. P. 1970. Nonautotrophic CO_2 fixation and crassulacean acid metabolism. Pages 169–185 in M. D. Hatch, C. B. Osmond, and R. O. Slatyer, eds., Photosynthesis and photorespiration. Wiley-Interscience, New York.

White, L. M. 1973. Carbohydrate reserves of grasses: A review. Journal of Range Management 26: 13–18.

Wolf, F. T. 1972. Effects of light on the enzymatic activities of green plants. Pages 19–96 in Chandra, L., ed., Advancing Frontiers in Plant Science, vol. 29. Impex India, Delhi.

Chapter 11

Anderson, J. M., D. J. Goodchild, and N. K. Boardman. 1973. Composition of the photosystems and chloroplast structure in extreme shade plants. Biochimica et Biophysica Acta 325:573–585.

Berry, J. A. 1975. Adaptation of photosynthetic processes to stress. Science 188:644–650.

Björkman, O. 1973. Comparative studies on photosynthesis in higher plants. Pages 1–63 in Giese, A. C., ed., Photophysiology, vol. 8. Academic Press, Inc., New York.

Björkman, O. and J. A. Berry. 1973. High-efficiency photosynthesis. Scientific American 229(4): 80–93.

Björkman, O., J. Troughton, and M. Nobs. 1974. Photosynthesis in relation to leaf structure. Brookhaven Symposia in Biology 25:206–226.

Black, C. C. 1973. Photosynthetic carbon fixation in relation to net CO_2 uptake. Annual Review of Plant Physiology 24:253–286.

Bolin, B. 1970. The carbon cycle. Scientific American 223(3):124–132.

Bowes, G. and W. L. Ogren. 1972. Oxygen inhibition and other properties of soybean ribulose 1,5 diphosphate carboxylase. Journal of Biological Chemistry 247:2171–2176.

Breidenbach, R. W. 1976. Microbodies. Pages 91–114 in J. Bonner and J. E. Varner, eds., Plant biochemistry. Academic Press, inc., New York.

Burris, R. H. and C. C. Black, eds. 1976. CO_2 Metabolism and Plant Productivity. University Park Press, Baltimore, Md.

Chabot, B. F. and A. R. Lewis. 1976. Thermal acclimation of photosynthesis in northern red oak. Photosynthetica 10:130–135.

Chollet, R. and W. L. Ogren. 1975. Regulation of photorespiration in C-3 and C-4 species. Botanical Review 41:137–179.

Cloud, P. and A. Gibor. 1970. The oxygen cycle. Scientific American 223(3):110–123.

Cooper, J. P., ed. 1975. Photosynthesis and productivity in different environments. Cambridge University Press, Cambridge.

Doley, D. and D. J. Yates. 1976. Gas exchange of Mitchell grass [Astrebla lappacea (Lindl.) Domin] in relation to irradiance, carbon dioxide supply, leaf temperature, and temperature history. Australian Journal of Plant Physiology 3:471–487.

Gibbs, M. 1970. The inhibition of photosynthesis by oxygen. American Scientist 58:634–640.

Hartsock, T. L. and P. S. Nobel. 1976. Watering converts a CAM plant to daytime CO_2 uptake. Nature 262:574–576.

Hatch, M. D., C. B. Osmond, and R. O. Slatyer, eds. 1971. Photosynthesis and photorespiration. John Wiley & Sons, Inc., New York.

Heath, O. V. S. 1969. The physiological aspects of photosynthesis. Stanford University Press, Stanford, Ca.

Jackson, W. A. and R. J. Volk. 1970. Photorespiration.

Annual Review of Plant Physiology 21:385–432.

Jarvis, P. G. and M. S. Jarvis. 1964. Growth rates of woody plants. Physiologia Plantarum 17:654–666.

Kriedemann, P. E. 1971. Crop energetics and horticulture. Hortscience 6:432–438.

Ku, S. and G. E. Edwards, 1977. Oxygen inhibition of photosynthesis. I. Temperature dependence and relation to O_2/CO_2 solubility ratio. Plant Physiology 59:986–990.

Larcher, W. 1969. The effect of environmental and physiological variables on the carbon dioxide gas exchange of trees. Photosynthetica 3:167–198.

Leopold, A. C. and P. E. Kriedemann. 1975. Plant growth and development. McGraw-Hill Book Company, New York.

Loomis, R. S., W. A. Williams, and A. E. Hall. 1971. Agricultural productivity. Annual Review of Plant Physiology 22:431–468.

Lorimer, G. H., T. J. Andrews, and N. E. Tolbert. 1973. Ribulose diphosphate oxygenase. II. Further proof of reaction products and mechanism of action. Biochemistry 12:18–23.

Mooney, H. A. 1972. The carbon balance of plants. Annual Review of Ecology 33:72–86.

Mooney, H. A., J. Ehleringer, and J. A. Berry. 1976. High photosynthetic capacity of a winter annual in Death Valley. Science 194:322–324.

Moss, D. N. and L. H. Smith. 1972. A simple classroom demonstration of differences in photosynthetic capacity among species. Journal of Agronomic Education 1:16–17.

Neales, T. F. and L. D. Incoll. 1968. The control of leaf photosynthetic rate by the level of assimilate concentration in the leaf: A review of the hypothesis. Botanical Review 34:107–125.

Pallas, J. E., Jr. and Y. B. Samish. 1974. Photosynthetic response of peanut. Crop Science 14: 478–482.

Patterson, D. T. 1975. Photosynthetic acclimation to irradiance in *Celastrus orbiculatus* Thunb. Photosynthetica 9:140–144.

Penning de Vries, F. W. T. 1975. The cost of maintenance processes in plant cells. Annals of Botany 39:77–92.

Penning de Vries, F. W. T., A. H. M. Brunsting, and H. H. van Laar. 1974. Products, requirements and efficiency of biosynthesis: A quantitative approach. Journal of Theoretical Biology 45: 339–377.

Tolbert, N. E. 1971. Microbodies-perioxisomes and glyoxysomes. Annual Review of Plant Physiology 22:45–74.

————. 1973. Compartmentation and control in microbodies. Pages 215–239 in D. D. Davies, ed., Symposia of the Society for Experimental Biology, no. 27. Cambridge University Press, Cambridge.

Vigil, E. L. 1973. Structure and function of plant microbodies. Sub-Cellular Biochemistry 2:237–285.

Whittaker, R. H. 1975. Communities and ecosystems. 2nd ed. Macmillan, Inc., New York.

Wildner, G. F. and J. Henkel. 1976. Specific inhibition of the oxygenase activity of ribulose 1,5-bisphosphate carboxylase. Biochemical and Biophysical Research communications 69:268–275.

Wittwer, S. H. 1974. Maximum production capacity of food crops. Bioscience 24:216–224.

Wittwer, S. H. and W. Robb. 1964. Carbon dioxide enrichment of greenhouse atmospheres for food crop production. Economic Botany 18:34–56.

Woodwell, G. M. and E. V. Pecan. 1973. Carbon and the biosphere. Twenty-fourth Brookhaven Symposium in Biology, (Technical Information Center, Office of Information Services, U.S. Atomic Energy Commission.

Zelitch, I. 1971. Photosynthesis, photorespiration, and plant productivity. Academic Press, Inc., New York.

————. 1975a. Improving the efficiency of photosynthesis. Science 188:626–633.

————. 1975b. Pathways of carbon fixation in green plants. Annual Review of Biochemistry 44:123–145.

Chapter 12

ap Rees, T. 1974. Pathways of carbohydrate breakdown in higher plants. Pages 87–127 in D. H. Northcote, ed., MTP International Review of Science, Biochemistry series 1, vol. II. Butterworth & Co., Ltd., London.

Armstrong, W. and T. J. Gaynard. 1976. the critical oxygen pressures for respiration in intact plants. Physiologia Plantarum 37:200–206.

Axelrod, B. and H. Beevers. 1956. Mechanisms of carbohydrate breakdown in plants. Annual Review of Plant Physiology 7:267–298.

Bajrachara, D., H. Falk, and P. Schopfer. 1976. Phytochrome-mediated development of mitochondria in the cotyledons of mustard (*Sinapis alba* L.) seedlings. Planta 131:253–261.

Beevers, H. 1961. Respiratory metabolism in plants. Harper & Row, Publishers, New York.

————. 1974. Conceptual developments in metabolic control, 1924–1974. Plant Physiology 54:437–442.

Bidwell, R. G. S. and G. P. Bebee. 1974. Carbon monoxide fixation by plants. Canadian Journal of Botany 52:1841–1847.

Bonner, W. D., Jr. 1973. Mitochondria and plant respiration. Pages 221–261 in L. P. Miller, ed., Phytochemistry. Vol. 3, Inorganic elements and special groups of chemicals. Van Nostrand Reinhold Company, New York.

Davies, D. D., S. Grego, and P. Kenworthy. 1974. The control of the production of lactate and ethanol by higher plants. Planta 118:297–310.

Downton, W. J. S. and J. S. Hawker. 1975. Response of starch synthesis to temperature in chilling-sensitive plants. Pages 81–88 in R. Marcelle, ed., Environmental and biological control of photosynthesis. Dr. W. Junk, The Hague.

Dunn, G. 1974. A model for starch breakdown in higher plants. Phytochemistry 13:1341–1346.

Gibbs, M. 1959. Metabolism of carbon compounds. Annual Review of Plant Physiology 10:329–369.

Glier, J. H. and J. L. Caruso. 1974. The influence of low temperatures on activities of starch degradative enzymes in a cold-requiring plant. Biochemical and Biophysical Research Communications 58:573–578.

Goddard, D. R. and W. D. Bonner. 1960. Cellular respiration. Pages 209–312 in F. C. Steward, ed. Plant physiology. Vol. 1A, Cellular organization and respiration. Academic Press, Inc., New York.

Henry, M. F. and E. J. Nyns. 1975. Cyanide-insensitive respiration. An alternative mitochondrial pathway. Sub-Cellular Biochemistry 4:1–65.

Hulme, A. C., ed. 1970. The biochemistry of fruits and their products, vol. 1. Academic Press, Inc. New York.

Ikuma, H. 1972. Electron transport in plant respiration. Annual Review of Plant Physiology 23;419–436.

Inman, R. E., R. B. Ingersoll, and E. A. Levy. 1971. Soil: A natural sink for carbon monoxide. Science 172:1229–1231.

James, W. O. 1963. Plant physiology. 6th ed. Oxford University Press, London.

Jensen, W. A. and F. B. Salisbury. 1974. Botany: An ecological approach. Wadsworth Publishing Company, Inc. Belmont, Calif.

Kortschak, H. P. and L. G. Nickell. 1973. Photosynthetic carbon monoxide metabolism by sugarcane leaves. Plant Science Letters 1:211–216.

Laties, G. G. 1975. Solute transport in relation to metabolism and membrane permeability in plant tissues. Pages 98–151 in P. J. Davies, ed., Historical and current aspects of plant physiology: A symposium honoring F. C. Steward. New York State College of Agriculture and Life Sciences, Ithaca, New York.

Lipmann, F. 1975. Reminiscences of Embden's formulation of the Embden-Meyerhof cycle. Molecular and Cellular Biochemistry 6:171–175.

Malone, C., D. E. Koeppe, and R. J. Miller. 1974. Corn mitochondria swelling and contraction—an alternate interpretation. Plant Physiology 53: 918–927.

Manners, D. J. 1974. Some aspects of the enzymic degradation of starch. Pages 109–125 in J. B. Pridham, ed., Plant carbohydrate biochemistry. Academic Press, Inc., New York.

Meeuse, B. J. D. 1966. The voodoo lily. Scientific American 215(1):80–88.

————. 1975. Thermogenic respiration in aroids. Annual Review of Plant Physiology 26:117–126.

O'Brien, T. P. and M. E. McCully. 1969. Plant structure and development. Collier-Macmillan, Limited, London.

Öpik, H. 1974. Mitochondria. Pages 52–83 in A. W. Robards, ed., Dynamic aspects of plant ultrastructure. McGraw-Hill Book Company, New York.

Palmer, J. M. 1976. The organization and regulation of electron transport in plant mitochondria. Annual Review of Plant Physiology 27:133–157.

Reid, E. E., P. Thompson, C. R. Lyttle, and D. T. Dennis, 1970. Pyruvate dehydrogenase complex from higher plant mitochondria and proplastids. Plant Physiology 59:842–848.

Rhodes, M. J. C. 1970. The climacteric and ripening of fruits. Pages 521–533 in A. C. Hulme, ed., The biochemistry of fruits and their products, vol. 1. Academic Press, Inc., New York.

Schnarrenberger, C., A. Oeser, and N. E. Tolbert. 1973. Two isoenzymes each of glucose-6-phosphate dehydrogenase and 6-phosphogluconate dehydrogenase in spinach leaves. Archives of Biochemistry and Biophysics 154: 438–448.

Steward, F. C., ed. 1966. Plant physiology. Vol. IVB, Metabolism: Intermediary metabolism and pathology. Academic Press, Inc., New York.

Stiles, W. and W. Leach. 1960. Respiration in plants. John Wiley & Sons, Inc., New York.

Storey, B. T. 1976. Respiratory chain of plant mitochondria, XVIII. Point of interaction of the alternate oxidase with the respiratory chain. Plant Physiology 58:521–525.

Turner, J. F. and D. H. Turner. 1975. The regulation of carbohydrate metabolism. Annual Review of Plant Physiology 26:159–186.

Vartapetian, B. B., A. I. Maslov, and I. N. Andreeva. 1975. Cytochromes and respiratory activity of mitochondria in anaerobically grown rice coleoptiles. Plant Science Letters 4:1–8.

Chapter 13

Ashton, F. M. 1976. Mobilization of storage proteins of seeds. Annual Review of Plant Physiology 27: 95–117.

Atkins, C. A., J. S. Pate, and P. J. Sharkey. 1975. Asparagine metabolism—key to the nitrogen nutrition of developing legume seeds. Plant Physiology 56:807–812.

Becking, J. H. 1975. Root nodules in non-legumes. Pages 508–566 in J. G. Torrey and D. T. Clarkson, eds., The development and function of roots. Academic Press, Inc., New York.

Beevers, L. 1976. Nitrogen metabolism in plants. American Elsevier Publishing Company, Inc., New York.

Bell, E. A. 1976. "Uncommon" amino acids in plants. FEBS Letters (Review Letter) 64:29–35.

Bond, G. 1976. The results of the IBP survey of root-nodule formation in non-leguminous angiosperms. Pages 443–474 in P. S. Nutman, ed., Symbiotic nitrogen fixation in plants. International Biological Programme, vol. 7. Cambridge University Press, Cambridge.

Bryan, J. K. 1976. Amino acid biosynthesis and its regulation. Pages 525–560 in J. Bonner and J. E. Varner, eds., Plant biochemistry. Academic Press, Inc., New York.

Burns, R. C. and R. W. F. Hardy, eds. 1975. Nitrogen fixation in bacteria and higher plants. Springer-Verlag, Berlin.

Burris, R. H. 1974. Biological nitrogen fixation, 1924–1974. Plant Physiology 54:443–449.

———. 1976. nitrogen fixation. Pages 887–908 in J. Bonner and J. E. Varner, eds., Plant biochemistry. Academic Press, Inc., New York.

Chrispeels, M. J., B. Baumgartner, and N. Harris. 1976. Regulation of reserve protein metabolism in the cotyledons of mung bean seedlings. Proceedings of the National Academy of Sciences 73:3168–3172.

Dart, P. J. 1975. Legume root nodule initiation and development. Pages 467–506 in J. G. Torrey and D. T. Clarkson, eds., The development and function of roots. Academic Press, Inc., New York.

Delwiche, C. C. 1970. The nitrogen cycle. Scientific American 223(3):136-146.

Delwiche, C. C. and B. A. Bryan. 1976. Denitrification. Annual Review of Microbiology 30:241–262.

Filner, P., J. L. Wray, and J. E. Varner. 1969. Enzyme induction in higher plants. Science 165:358–367.

Fowden, L. 1973. The non-protein amino acids of plants: Concepts of biosynthetic control. Pages 323–339 in B. V. Milborrow, ed., Biosynthesis and its control in plants. Academic Press, Inc., New York.

Hardy, R. W. F. and U. D. Havelka. 1975. Nitrogen fixation research: A key to world food? Science 188:633–643.

Hardy, R. W. F., R. D. Holsten, E. K. Jackson, and R. C. Burns. 1968. The acetylene-ethylene assay for N_2 fixation: Laboratory and field evaluation. Plant Physiology 43:1185–1207.

Hewitt, E. J. 1975. Assimilatory nitrate-nitrite reduction. Annual Review of Plant Physiology 26:73–100.

Hewitt, E. J., D. P. Hucklesby, and B. A. Notton. 1976. Nitrate metabolism. Pages 633–681 in J. Bonner and J. E. Varner, eds., plant biochemistry. Academic Press, Inc., New York.

Marcus, A. 1970. Enzyme induction in plants. Annual Review of Plant Physiology 22:313–336.

McKee, H. S. 1962. Nitrogen metabolism in plants. Oxford University Press (Clarendon), London.

Miflin, B. J. 1973. Amino acid biosynthesis and its control in plants. Pages 49–68 in B. V. Milborrow, ed., Biosynthesis and its control in plants. Academic Press, Inc., New York.

Miflin, B. J. and P. J. Lea. 1976. The pathway of nitrogen assimilation in plants. Phytochemistry 15:873–885.

Millerd, A. 1975. Biochemistry of legume seed proteins. Annual Review of Plant Physiology 26:53–72.

Minchin, F. R. and J. S. Pate. 1973. The carbon balance of a legume and the functional economy of its root nodules. Journal of Experimental Botany 24:259–271.

Richmond, D. V. 1973. Sulfur compounds. Pages 41–73 in L. P. Miller, ed., Phytochemistry. Vol. 3, Inorganic elements and special groups of chemicals. Van Nostrand Reinhold Company, New York.

Schmidt, A. 1976. The adenosine-5-phosphosulfate sulfotransferase from spinach (Spinacea oleracea L.). Stabilization, partial purification, and properties. Planta 130:257–263.

Sinclair, T. R. and C. T. deWit. 1975. Photosynthate and nitrogen requirements for seed production by various crops. Science 189:565–567.

Smith, R. L., J. H. Bouton, S. C. Schank, K. H. Quesenberry, M. E. Tyler, J. R. Milam, M. H. Gaskins, and R. C. Littell. 1976. Nitrogen fixation in grasses inoculated with Spirillum lipoferum. Science 193:1003–1005.

Solomonson, L. P. and A. M. Spehar. 1977. Model for the regulation of nitrate assimilation. Nature 265:373–375.

Stewart, W. D. P., ed. 1975. Nitrogen fixation by free-living microorganisms. Cambridge University Press, Cambridge.

von Bülow, J. F. W. and J. Döbereiner. 1975. Potential for nitrogen fixation in maize genotypes in Brazil. Proceedings of the National Academy of Sciences 72:2389–2393.

Wilson, C. M. 1975. Plant nucleases. Annual Review of Plant Physiology 26:187–208.

Wilson, L. G. and Z. Reuveny. 1976. Sulfate reduction. Pages 599–632 in J. Bonner and J. E. Varner, eds., Plant biochemistry. Academic Press, Inc., New York.

Winter, H. C. and R. H. Burris. 1976. Nitrogenase. Annual Review of Biochemistry 45:409–426.

Chapter 14

Archer, B. L. and B. G. Audley. 1973. Rubber, gutta percha, and chicle. Pages 310–343 in L. P. Miller, ed., Phytochemistry. Vol. 2, Organic metabolites. Van Nostrand Reinhold Company, New York.

Arnold, G. W. and J. L. Hill. 1972. Chemical factors affecting selection of food plants by ruminants. Pages 72–102 in J. B. Harborne, ed., Phytochemical ecology. Academic Press, Inc., New York.

Barz, W. and W. Hosel. 1975. Metabolism of flavonoids. Pages 916–969 in J. B. Harborne, T. J. Mabry, and H. Mabry, eds., The flavonoids. Academic Press, Inc., New York.

Bate-Smith, E. C. 1972. Attractants and repellants in higher animals. Pages 45–56 in J. B. Harborne, ed., Phytochemical ecology. Academic Press, Inc., New York.

Beck, S. D. and J. C. Reese. 1976. Insect-plant interactions: Nutrition and metabolism. Pages 41–92 in J. W. Wallace and R. L. Mansell, eds., Recent advances in phytochemistry. Vol. 10, Biochemical interactions between plants and insects. Plenum Press, New York.

Beevers, H. 1975. Organelles from castor bean seedlings: Biochemical roles in gluconeogenesis and phospholipid biosynthesis. Pages 287–299 in T. Galliard and E. I. Mercer, eds., Recent advances in the chemistry and biochemistry of plant lipids. Academic Press, Inc., New York.

Beytia, E. D. and J. W. Porter. 1976. Biochemistry of polyisoprenoid biosynthesis. Annual Review of Biochemistry 45:113–142.

Bowers, W. S. et al. 1976. Science 193:542–547.

Breidenbach, R. W. 1976. Microbodies. Pages 91–114 in J. Bonner and J. E. Varner, eds. Plant Biochemistry. Academic Press, Inc., New York.

Caldicott, A. B. and G. Eglinton. 1973. Surface waxes. Pages 162–194 in L. P. Miller, ed., Phytochemistry. Vol. 3, Inorganic elements and special groups of chemicals. Van Nostrand Reinhold Company, New York.

Deverall, B. J. 1977. Defence mechanisms of plants. Cambridge University Press, Cambridge.

Dodson, C. H., R. L. Dressler, H. G. Hills, R. M. Adams, and N. H. Williams. 1969. Biologically active compounds in orchid fragrances. Science 164:1243–1249.

Farnsworth, N. R. 1973. Importance of secondary plant constituents as drugs. Pages 351–380 in L. P. Miller, ed., Phytochemistry. Vol. 3, Inorganic elements and special groups of chemicals. Van Nostrand Reinhold Company, New York.

Feeny, P. 1976. Plant apparency and chemical defense. Pages 1–40 in J. W. Wallace and R. L. Mansell, eds., Recent advances in phytochemistry. Vol. 10, Biochemical interaction between plants and insects. Plenum Press, New York.

Freudenberg, K. and A. C. Neish. 1968. Constitution and biosynthesis of lignin. Springer-Verlag, Berlin.

Galliard, T. and E. I. Mercer, eds. 1975. Recent advances in the chemistry and biochemistry of plant lipids. Academic Press, Inc., New York.

Goodwin, T. W. 1973. Carotenoids. Pages 112–142 in L. P. Miller, ed., Phytochemistry. Vol. 1, The process and products of photosynthesis. Van Nostrand Reinhold Company, New York.

———. 1976. Chemistry and biochemistry of plant pigments, vol. 1. Academic Press, Inc., New York.

Green, F. B. and M. R. Corcoran. 1975. Inhibitory action of five tannins on growth induced by several gibberellins. Plant Physiology 56:801–806.

Gross, D. 1975. Growth regulating substances of plant origin (a review). Phytochemistry 14:2105–2112.

Grunwald, C. 1975. Plant sterols. Annual Review of Plant Physiology 26:209–236.

Hall, F. K. 1974. Wood pulp. Scientific American 230(4):52–62.

Harborne, J. B. 1975. Biochemical systematics of flavonoids. Pages 1056–1095 in J. B. Harborne, T. J. Mabry, and H. Mabry, eds., The flavonoids. Academic Press, Inc., New York.

———. 1976. Functions of flavonoids in plants. Pages 736–778 in T. W. Goodwin, eds., Chemistry and biochemistry of plant pigments, vol. 1. Academic Press, Inc., New York.

Harborne, J. B., T. J. Mabry, and H. Mabry, eds. 1975. The flavonoids. Academic Press, Inc., New York.

Harwood, J. L. 1975. Fatty acid biosynthesis. Pages 44–93 in T. Galliard and E. I. Mercer, eds., Recent advances in the chemistry and biochemistry of plant lipids. Academic Press, Inc.,New York.

Hedin, P. A., A. C. Thompson, and R. C. Gueldner. 1976. Cotton plant and insect constituents that control boll weevil behavior and development. Pages 271–350 in J. W. Wallace and R. L. Mansell, eds., Recent advances in phytochemistry. Vol. 10, Biochemical interaction between plants and insects. Plenum Press, New York.

Heftmann, E. 1973. Steroids. Pages 171–226 in L. P. Miller, ed., Phytochemistry. Vol. 2, Organic metabolites. Van Nostrand Reinhold Company, New York.

Hendry, L. B., J. G. Kostelc, D. M. Hindenlang, J. K. Wichmann, C. J. Fix, and S. H. Korzeniowski.

1976. Chemical messengers in insects and plants. Pages 351–384 in J. W. Wallace and R. L. Mansell, eds., Recent advances in phytochemistry. Vol. 10, Biochemical interactions between plants and insects. Plenum Press, New York.

Hitchcock, C. 1975. Structure and distribution of plant acyl lipids. Pages 1–19 in T. Galliard and E. I. Mercer, eds., Recent advances in the chemistry and biochemistry of plant lipids. Academic Press, Inc., New York.

Hitchcock, C. and B. W. Nichols. 1971. Plant lipid biochemistry. Academic Press, Inc., New York.

Hughes, D. W. and K. Genest. 1973. Alkaloids. Pages 118–170 in L. P. Miller, ed., Phytochemistry. Vol. 2, Organic metabolites. Van Nostrand Reinhold Company, New York.

Jack, R. C. 1973. Lipids. Pages 227–253 in L. P. Miller, ed., Phytochemistry. Vol. 2, Organic metabolites. Van Nostrand Reinhold Company, New York.

Kates, M. and M. O. Marshall. 1975. Biosynthesis of phosphoglycerides in plants. Pages 115–159 in T. Galliard and E. I. Mercer, eds., Recent advances in the chemistry and biochemistry of plant lipids. Academic Press, Inc., New York.

Keeler, R. F. 1975. Toxins and teratogens of higher plants. Lloydia 38:56–86.

Kollatukudy, P. E. 1975. Biochemistry of cutin, suberin, and waxes—the lipid barriers on plants. Pages 203–246 in T. Galliard and E. I. Mercer, eds., Recent advances in the chemistry and biochemistry of plant lipids. Academic Press, Inc., New York.

Lindstedt, K. J. 1971. Chemical control of feeding behavior. Comparative Biochemistry and Physiology 39A:553–581.

Lodhi, M. A. K. 1976. Role of allelopathy as expressed by dominating trees in a lowland forest in controlling the productivity and patterns of herbaceous growth. American Journal of Botany 63: 1–8.

Martin, J. R. 1973. Cutins and suberins. Pages 154–161 in L. P. Miller, ed., Phytochemistry. Vol. 3, Inorganic elements and special groups of chemicals. Van Nostrand Reinhold Company, New York.

Mazliak, P. 1973. Lipid metabolism in plants. Annual Review of Plant Physiology 24:287–310.

McClure, J. W. 1975. Physiology and functions of flavonoids. Pages 970–1055 in J. B. Harborne, T. J. Mabry, and H. Mabry, eds., The flavonoids. Academic Press, Inc., New York.

Miller, L. P. 1973. Plants as organic laboratories. Pages 1–14 in L. P. Miller, ed., Phytochemistry. Vol. 1, The process and products of photosynthesis. Van Nostrand Reinhold Company, New York.

Mudd, J. B. and R. Garcia. 1975. Biosynthesis of glycolipids. Pages 162–201 in T. Galliard and E. I. Mercer, eds., Recent advances in the chemistry and biochemistry of plant lipids. Academic Press, Inc., New York.

Muller, C. H. and C. h. Chou. 1972. Phytotoxins: An ecological phase of phytochemistry. Pages 201–216 in J. B. Harborne, ed., Phytochemical ecology. Academic Press, Inc., New York.

Nicholas, H. J. 1973a. Miscellaneous volatile plant products. Pages 381–399 in L. P. Miller, ed., Phytochemistry. Vol. 2, Organic metabolites. Van Nostrand Reinhold Company, New York.

———. 1973b. Terpenes. Pages 254–309 in L. P. Miller, ed., Phytochemistry. Vol. 2, Organic metabolites. Van Nostrand Reinhold Company, New York.

Nichols, B. W. 1974. The structure and function of plant glycolipids. Pages 97–108 in J. B. Pridham, ed., Plant carbohydrate biochemistry. Academic Press, Inc., New York.

Nurnsten, H. E. 1970. Volatile compounds: The aroma of fruits. Pages 239–269 in A. C. Hulme, ed., The biochemistry of fruits and their products, vol. 1. Academic Press, London.

Piatelli, M. 1976. Betalains. Pages 560–596 in T. W. Goodwin, ed., Chemistry and biochemistry of plant pigments, vol. 1. Academic Press, Inc., New York.

Rice, E. L. 1974. Allelopathy. Academic Press, Inc., New York.

Robinson, T. 1967. The organic constituents of higher plants: Their chemistry and interrelationships. Burgess Publishing Company, Minneapolis, Minn.

Roeske, C. N., J. N. Seiber, L. P. Brower, and C. M. Moffitt. 1976. Milkweed cardenolides and their comparative processing by monarch butterflies. Pages 93–167 in J. W. Wallace and R. L. Mansell, eds., Recent advances in phytochemistry. Vol. 10, Biochemical interaction between plants and insects. Plenum Press, New York.

Rothschild, M. 1972. Some observations on the relationship between plants, toxic insects, and birds. Pages 1–12 in J. B. Harborne, ed., Phytochemical ecology. Academic Press Inc., New York.

Rudinsky, J. A. 1970. Sequence of Douglas-fir beetle attraction and its ecological significance. Contributions from Boyce Thompson Institute 24: 311–314.

Sarkanen, K. V. and C. H. Ludwig, eds. 1971. Lignins. John Wiley & Sons, Inc., New York.

Schubert, W. J. 1973. Lignin. Pages 132–153 in L. P. Miller, ed., Phytochemistry. Vol. 3, Inorganic elements and special groups of chemicals. Van Nostrand Reinhold Company, New York.

Schultes, R. E. 1970. The botanical and chemical distribution of hallucinogens. Annual Review of Plant Physiology 21:571–598.

Sedgwick, B. 1973. The control of fatty acid biosynthesis in plants. Pages 179–217 in B. V. Milborrow, ed., Biosynthesis and its control in plants. Academic Press, Inc., New York.

Shutt, D. A. 1976. The effects of plant estrogens on animal reproduction. Endeavour 35(126):110–113.

Sinclair, T. R. and C. T. DeWit. 1975. Photosynthate and nitrogen requirements for seed production by various crops. Science 189:565–567.

Smith, H. 1972. The photocontrol of flavonoid biosynthesis. Pages 433–481 in K. Mitrakos and W. Shropshire, eds., Phytochrome. Academic Press, Inc., New York.

Sondheimer, E. and J. B. Simeone, eds. 1970. Chemical ecology. Academic Press, Inc., New York.

Stafford, H. A. 1974. The metabolism of aromatic compounds. Annual Review of Plant Physiology 25:459–486.

Stout, G. H. and R. E. Schultes. 1973. Importance of plant chemicals in human affairs. Pages 381–399 in L. P. Miller, ed., Phytochemistry. Vol. 3, Inorganic elements and special groups of chemicals. Van Nostrand Reinhold Company, New York.

Street, H. E. and H. Öpik. 1970. The physiology of flowering plants: Their growth and development. American Elsevier Publishing Co., Inc., New York.

Stumpf, P. K. 1976. Lipid metabolism. Pages 427–461 in J. Bonner and J. E. Varner, eds., Plant biochemistry. Academic Press, Inc., New York.

Swain, T. 1976. Nature and properties of flavonoids. Pages 425–463 in T. W. Goodwin, ed., Chemistry and biochemistry of plant pigments, vol. 1. Academic Press, Inc., New York.

Vigil, E. L. 1973. Structure and function of plant micro-bodies. Sub-Cellular Biochemistry 2:237–285.

Wengenmayer, H. J. Ebel, and H. Grisebach. 1976. Enzymic synthesis of lignin precursors. European Journal of Biochemistry 65:529–536.

Went, F. W. 1974. Reflections and speculations. Annual Review of Plant Physiology 25:1–26.

Whittaker, R. H. and P. P. Feeny. 1971. Allelochemicals: Chemical interactions between species. Science 171:757–770.

Wolff, I. A. 1966. Seed lipids. Science 154:1140–1149.

Wong, E. 1976. Biosynthesis of flavonoids. Pages 464–526 in T. W. Goodwin, ed., Chemistry and biochemistry of plant pigments, vol. 1. Academic Press, Inc., New York.

Yatsu, L. Y., T. J. Jacks, and T. P. Hensarling. 1971. Isolation of spherosomes (oleosomes) fromonion, cabbage, and cottonseed tissues. Plant Physiology 48:675–682.

Chapter 15

Barlow, P. W. 1975. The root cap. Pages 21–54 in J. G. Torrey and D. T. Clarkson, eds., The development and function of roots. Academic Press, Inc., New York.

Berlyn, G. P. 1972. Seed germination and morphogenesis. Pages 223–312 in T. T. Kozlowski, ed., Seed biology, vol. 1. Academic Press, Inc., New York.

Beyer, E. M., Jr. 1976. A potent inhibitor of ethylene action in plants. Plant Physiology 58:268–271.

Brown, R. 1972a. Cell growth and cell development. Pages 91–130 in F. C. Steward, ed., Plant physiology. Vol. 6C, Physiology of development: From seeds to sexuality. Academic Press, Inc., New York.

———. 1972b. Germination. Pages 3–48 in F. C. Steward, ed., Plant physiology. Vol. 6C, Physiology of development: From seeds to sexuality. Academic Press, Inc., New York.

Brown, R. and A. F. Dyer. 1972. Cell division in higher plants. Pages 49–90 in F. C. Steward, ed., Plant physiology. Vol. 6C, Physiology of development: From seeds to sexuality. Academic Press, Inc., New York.

Cleland, R. 1971. Cell wall extension. Annual Review of Plant Physiology 22:197–222.

Clowes, F. A. L. 1961. Apical meristems. Oxford University Press, London.

———. 1967. The functioning of meristems. Science Progress 55:529–542.

———. 1975. The quiescent center. Pages 3–19 in J. G. Torrey and D. T. Clarkson, eds., The development and function of roots. Academic Press, Inc., New York.

Coombe, B. G. 1976. The development of fleshy fruits. Annual Review of Plant Physiology 27:207–228.

Cutter, E. G. 1969. Plant anatomy: Experiment and interpretation. Part I, Cells and tissues. Addison-Wesley Publishing Co., Inc., Reading, Mass.

Dure, L. S., III. 1975. Seed formation. Annual Review of Plant Physiology 26:259–278.

Esau, K. 1971. Anatomy of seed plants, 2nd ed. John Wiley & Sons, Inc., New York.

Greulach, V. A. 1973. Plant function and structure. Macmillan, Inc., New York.

Hepler, P. K. 1976. Plant microtubules. Pages 147–187 in J. Bonner and J. E. Varner, eds., Plant biochemistry. Academic press, Inc., New York.

Hepler, P. K. and B. A. Palevitz. 1974. Microtubules and microfilaments. Annual Review of Plant Physiology 25:309–362.

Hsiao, T. C. 1973. Plant responses to water stress. Annual Review of Plant Physiology 24:519–570.

Hulme, A. C. 1970. The biochemistry of fruits and their products, vol. I. Academic Press, Inc., New York.

Jensen, W. H. and R. B. Park. 1967. Cell ultrastructure. Wadsworth Publishing Company, Inc., Belmont, California.

Jensen, W. A. and F. B. Salisbury. Botany: An ecological approach. Wadsworth Publishing Company, Inc., Belmont, Calif.

Kramer, P. J. and T. T. Kozlowski. 1960. Physiology of Trees. McGraw-Hill Book Company, NewYork.

Leopold, A. C. and P. E. Kreidemann. 1975. Plant growth and development. McGraw-Hill Book Company, New York.

McCully, M. E. 1975. The development of lateral roots. Pages 105–124 in J. G. Torrey and D. T. Clarkson, eds., The development and function of roots. Academic Press, Inc., New York.

McNeil, D. L. 1976. The basis of osmotic pressure maintenance during expansion growth in *Helianthus annuus* hypocotyls. Australian Journal of Plant Physiology 3:311–324.

Nitsch, J. P. 1970. Hormonal factors in growth and development. Pages 428–472 in A. C. Hulme, ed., The biochemistry of fruits and their products, vol. I. Academic Press, Inc., New York.

Nurnsten, H. E. 1970. Volatile compounds: The aroma of fruits. Pages 239–268 in A. C. Hulme, ed., The biochemistry of fruits and their products, vol. I. Academic Press, Inc., New York.

Oertli, J. J. 1975. Effects of external solute supply on cell elongation in barley coleoptiles. Zeitschrift für Pflanzenphysiologie 74:440–450.

Perry, T. O. 1971. Dormancy of trees in winter. Science 171:29–36.

Pickett-Heaps, J. D. 1974. Plant microtubules. Pages 219–255 in A. W. Robards, ed., Dynamic aspects of plant ultrastructure. McGraw-Hill Book Company, New York.

Ray, P. M., P. B. Green, and R. Cleland. 1972. Role of turgor in plant cell growth. Nature 239:163–164.

Richards, F. J. 1969. The quantitative analysis of growth. Pages 3–76 in F. C. Steward, ed., Plant physiology. Vol. 5A, Analysis of growth: Behavior of plants and their organs. Academic Press, Inc., New York.

Richards, F. J. and W. W. Schwabe. 1969. Pages 79–116 in F. C. Steward, ed., Plant physiology. Vol. 5A, Analysis of growth: Behavior of plants and their organs. Academic Press, Inc., New York.

Sachs, R. M. 1965. Stem elongation. Annual Review of Plant Physiology 16:73–96.

Sinnott, E. W. 1960. Plant morphogenesis. McGraw-Hill Book Company, New York.

Snyder, J. A. and J. R. McIntosh. 1976. Biochemistry and physiology of microtubules. Annual Review of Biochemistry 45:699–720.

Spratt, N. E., Jr. 1971. Developmental biology. Wadsworth Publishing Company, Inc. Belmont, Calif.

Steeves, T. A. and I. M. Sussex. 1972. Patterns in plant development. Prentice-Hall, Inc., Englewood Cliffs, N.J.

Torrey, J. G. 1967. Development in flowering plants. Macmillan, Inc., New York.

———. 1976. Root hormones and plant growth. Annual Review of Plant Physiology 27:435–459.

Torrey, J. G. and W. D. Wallace. 1975. Further studies on primary vascular tissue pattern formation in roots. Pages 91–103 in J. G. Torrey and D. T. Clarkson, eds., The development and function of roots. Academic Press, Inc., New York.

Troughton, J. and L. A. Donaldson. 1972. Probling plant structure. McGraw-Hill Book Company, New York.

higher plants. Pages 152–165 in Brookhaven Symposia in Biology, no. 25. Brookhaven National Laboratory, Upton, N. Y.

Wardlaw, C. W. 1968. Morphogenesis in plants. Methuen and Co., Ltd., London.

Went, F. W. 1957. The experimental control of plant growth. The Ronald Press, New York.

Whittington, W. J., ed. 1969. Root growth. Butterworth and Co., Ltd., London.

Yeoman, M. M., ed. 1976. Cell division in higher plants. Academic Press, Inc., New York.

Zimmermann, M. H. and C. L. Brown. 1971. Trees—structure and function. Springer-Verlag, Berlin.

Chapter 16

Adams, P. A., M. J. Montague, M. Tepfer, D. L. Rayle, H. Ikuma, and P. B. Kaufman. 1975. Effect of gibberellic acid on the plasticity and elasticity of *Avena* stem segments. Plant Physiology 56: [757–760].

Ashton, F. M. and A. S. Crafts. 1973. Mode of action of herbicides. John Wiley & Sons, Inc., New York.

Audus, L. J. 1972. Plant growth substances, Vol. 1, Chemistry and physiology. Leonard Hill Books, London.

Batra, M. W., K. L. Edwards, and T. K. Scott. 1975. Auxin transport in roots. Pages 300–325 in J. G. Torrey and D. T. Clarkson, eds., The development and function of roots. Academic Press, Inc., New York.

Batt, S. and M. A. Venis. 1976. Separation and localization of two classes of auxin binding sites in corn coleoptile membranes. Planta 130:15–21.

Briggs, D. E. 1973. Hormones and carbohydrate metabolism in germinating cereal grains. Pages 219–277 in B. V. Milborrow, ed., Biosynthesis and its control in plants. Academic Press, Inc., New York.

Broughton, W. J. and A. J. McComb. 1971. Changes in the pattern of enzyme development in gibberellin-treated pea internodes. Annals of Botany 35:213– 228.

Carr, D. J., ed. 1972. Plant growth substances 1970. Springer-Verlag, Berlin.

Carter, J. L., L. A. Garrard, and S. H. West. 1973. Effect of gibberellic acid on starch-degrading enzymes in leaves of *Digitaria decumbens*. Phytochemistry 12:251–254.

Cleland, R. 1971. Cell wall extension. Annual Review of Plant Physiology 22:197–222.

———. 1976. Kinetics of hormone-induced H+ excretion. Plant Physiology 58: 210–213.

Coolbaugh, R. C. and R. Hamilton. 1976. Inhibition of *ent*-kaurene oxidation and growth by α-cyclopropyl-α-(p-methoxyphenyl)-5-pyrimidine methyl alcohol. Plant Physiology 57:245–248.

Davies, E. and O. Özbay. 1975. Comparative effects of indoleacetic acid and gibberellic acid on growth of decapitated etiolated epicotyls of *Pisum sativum* cv. Alaska. Physiologia Plantarum 35: 279–285.

Douglas, T. J. and L. G. Paleg. 1974. Plant growth retardants as inhibitors of sterol biosynthesis in tobacco seedlings. Plant Physiology 54:238–295.

Evans, M. L. 1974. Rapid responses to plant hormones. Annual Review of Plant Physiology 25:195–223.

Evans, M. L. and P. M. Ray. 1969. Timing of the auxin response in coleoptiles and its implications regarding auxin action. Journal of General Physiology 53:1–20.

Frost, R. G. and C. A. West. 1977. Properties of kaurene synthesis from *Marah macrocarpus*. Plant Physiology 59:22–29.

Galston, A. W. 1964. The life of the green plant. 2nd ed. Prentice-Hall, Inc., Englewood Cliffs, N.J.

Glasziou, K. T. 1969. Control of enzyme formation and inactivation in plants. Annual Review of Plant Physiology 20:63–88.

Gove, J. P. and M. C. Hoyle. 1975. The isozymic similarity of indoleacetic acid oxidase to peroxidase in birch and horseradish. Plant Physiology 56:684–687.

Guilfoyle, T. J., C. Y. Lin, Y. M. Chen, R. T. Nagao, and J. L. Key. 1975. Enhancement of soybean RNA polymerase I by auxin. Proceedings of the National Academy of Sciences 72:69–72.

Haissig, B. E. 1974. Origins of adventitious roots. New Zealand Journal of Forestry Science 4:229–310.

Hasson, E. P. and C. A. West. 1976. Properties of the system for the mixed function oxidation of kaurene and kaurene derivatives in microsomes of the immature seed of *Marah macrocarpus*. Electron transfer components. Plant Physiology 58:479–484.

Jones, R. L. 1973. Gibberellins: Their physiological role. Annual Review of Plant Physiology 24:571–598.

Jones, R. L. and Phillips, I. D. J. 1964. Agar diffusion technique for estimating gibberellin production by plant organs. Nature 204:497–499.

Kende, H. and G. Gardner. 1976. Hormone binding in plants. Annual Review of Plant Physiology 27: 267–290.

Key, J. L. 1969. Hormones and nucleic acid metabolism. Annual Review of Plant Physiology 20: 449–474.

Key, J. L., N. M. Barnett, and C. Y. Lin. 1967. RNA and protein biosynthesis and the regulation of cell elongation by auxin. Annals of the New York Academy of Sciences 144: 49–62.

Kormandy, E. J., T. F. Sherman, F. B. Salisbury, N. T. Spratt, Jr., and E. McCann. 1977. Biology. Wadsworth Publishing Company, Inc., Belmont, Calif.

Krishnamoorthy, H. N., ed. 1975. Gibberellins and plant growth. Wiley Eastern, Ltd., New Delhi.

Lang, A. 1970. Gibberellins: Structure and metabolism. Annual Review of Plant Physiology 21: 537–570.

Liu, P. B. W. and J. B. Loy. 1976. Action of gibberellic acid on cell proliferation in the subapical shoot meristem of watermelon seedlings. American Journal of Botany 63:700–704.

MacMillan, J. 1974. Recent aspects of the chemistry and biosynthesis of the gibberellins. Pages 1–19 in V. C. Runeckles, E. Sondheimer, and D. C. Walton, eds., the chemistry and biochemistry of plant hormones. Academic Press, Inc., New York.

MacMillan, J. and R. J. Pryce. 1973. The gibberellins. Pages 283–326 in L. P. Miller, ed., Phytochemistry. Vol. 3, Inorganic elements and special groups of chemicals. Van Nostrand Reinhold Company, New York.

Mann, J. D. 1975. Mechanism of action of gibberellins. Pages 239–287 in H. N. Krishnamoorthy, ed., Gibberellins and plant growth. Wiley Eastern, Ltd., New Delhi.

O'Malley, B. W. and W. T. Schrader. 1976. The receptors of steroid hormones. Scientific American. 234(2):32s–43.

Percival, F. W. and R. S. Bandurski. 1976. Esters of indole-3-acetic acid from *Avena* seeds. Plant Physiology 58:60–67.

Pharis, R. P. and C. C. Kuo. 1977. Physiology of gibberellins in conifers. Canadian Journal of Forest Research 7:299–325.

Phillips, I. D. J. 1971. Introduction to the biochemistry and physiology of plant growth hormones. McGraw-Hill Book Company, New York.

———. 1975. Apical dominance. Annual Review of Plant Physiology 26:341–367.

Raven, J. A. 1975. Transport of indoleacetic acid in

plant cells in relation to pH and electrical potential gradients, and its significance for polar IAA transport. New Phytologist 74:163–172.

Ray, P. M. 1974. The biochemistry of the action of indoleacetic acid on plant growth. Pages 93–122 in V. C. Runeckles, E. Sondheimer, and D. C. Walton, eds., The chemistry and biochemistry of plant hormones. Academic Press, Inc., New York.

———. 1977. Auxin-binding sites of maize coleoptiles are localized on membranes of the endoplasmic reticulum. Plant Physiology 59:594–599.

Rayle, D. and R. Cleland.1977. Control of plant cell enlargement by hydrogen ions. Developmental Biology (in press).

Reeve, D. R. and A. Crozier. 1974. An assessment of gibberellin structure-activity relationships. Journal of Experimental Botany 25:431–445.

Rubery, P. H. and A. R. Sheldrake. 1974. Carrier-mediated auxin transport. Planta 118:101–121.

Sachs, R. M. 1965. Stem elongation. Annual Review of Plant Physiology 16:73–96.

Salisbury, F. B. and R. V. Park, 1964. Vascular plants: Form and function. Wadsworth Publishing Company, Inc., Belmont, Calif.

Schneider, A. and F. Wightman. 1974. Metabolism of auxin in higher plants. Annual Review of Plant Physiology 25:487–513.

Scott, T. K. 1972. Auxins and roots. Annual Review of Plant Physiology 23:235–258.

Sheldrake, A. R. 1973. The production of hormones in higher plants. Biological Reviews 48:509–559.

Steward, F. C. and A. D. Krikorian. 1971. Plants, chemicals, and growth. Academic Press, Inc., New York.

Stuart, D. A. and R. L. Jones. 1977. Roles of extensibility and turgor in gibberellin- and dark-stimulated growth. Plant Physiology 59:61–68.

Stuart, N. W. and H. M. Cathey. 1961. Applied aspects of the gibberellins. Annual Review of Plant Physiology 12:369–394.

Thimann, K. V. 1972. The natural plant hormones. Pages 3–332 in F. C. Steward, ed., Plant Physiology. Vol. VIB, Physiology of development: The hormones. Academic Press, Inc., New York.

———. 1974. Fifty years of plant hormone research. Plant Physiology 54:450–453.

Torrey, J. G. 1976. Root hormones and plant growth. Annual Review of Plant Physiology 27:435–459.

Travis, R. L. and J. L. Key. 1976. Auxin-induced changes in the incorporation of ^{3}H-amino acids into soybean ribosomal proteins. Plant Physiology 57:936–938.

Vanderhoef, L. N., T. S. Lu, and C. A. Williams. 1977. Comparison of auxin-induced and acid-induced elongation in soybean hypocotyl. Plant Physiology 59:1004–1007.

Varner, J. E. and D. T. Ho. 1976. Hormones. Pages 713–770 in J. Bonner and J. E. Varner, eds., Plant biochemistry. Academic Press, Inc., New York.

Vlitos, A. J. and B. H. Most. 1973. Endogenous plant growth regulators. Pages 262–282 in L. P. Miller, ed., Phytochemistry. Vol. 3, Inorganic elements and special groups of chemicals. Van Nostrand Reinhold Company, New York.

Wangermann, E. 1974. The pathway of transport of applied indolyl-acetic acid through internode segments. New Phytologist 73:623–636.

Weaver, R. J. 1972. Plant growth substances in agriculture. W. H. Freeman Company, San Francisco.

Went, F. and K. V. Thimann. 1937. Phytohormones. Macmillan, Inc., New York.

Chapter 17

Abeles, F. B. 1972. Biosynthesis and mechanism of action of ethylene. Annual Review of Plant Physiology 23:259–292.

———. 1973. Ethylene in plant biology. Academic Press, Inc., New York.

Addicott, F. T. 1970. Plant hormones in the control of abscission. Biological Reviews 45:485–524.

Addicott, F. T. and J. L. Lyon. 1969. Physiology of abscisic acid and related substances. Annual Review of Plant Physiology 20:139–164.

Alvim, R., E. W. Hewett, and P. F. Saunders. 1976. Seasonal variation in the hormone content of willow. I. Changes in abscisic acid content and cytokinin activity in the xylem sap. Plant Physiology 57:474–476.

Apelbaum, A. and S. P. Burg. 1971. Altered cell microfilbril orientation in ethylene-treated Pisum sativum stems. Plant Physiology 48:648–652.

Beevers, L. 1976. Senescence. Pages 771–794 in J. Bonner and J. E. Varner, eds., Plant biochemistry. Academic Press, Inc., New York.

Burg, S. P. 1973. Ethylene in plant growth. Proceedings of the National Academy of Sciences 70:591–597.

Burrows, W. J. 1975. Mechanism of action of cytokinins. Current Advances in Plant Science 7:837-847.

Carr, D. J., ed. 1972. Plant growth substances 1970. Springer-Verlag, Berlin.

Chang, Y. and W. P. Jacobs. 1973. The regulation of abscission and IAA by senescence factor and abscisic acid. American Journal of Botany 60:10– 16.

Colquhoun, A. J. and J. R. Hillman. 1975. Endogenous abscisic acid and the senescence of leaves of Phaseolus vulgaris L. Zeitschrift für Pflanzenphysiologie 76:326–332.

Coombe, B. G. 1976. The development of fleshy fruits. Annual Review of Plant Physiology 27:207–228.

Crocker, W., P. W. Zimmerman, and A. E. Hitchcock. 1932. Ethylene-induced epinasty of leaves and the relation of gravity to it. Contributions of the Boyce Thompson Institute 4:177–218.

Fosket, D. E. and K. C. Short. 1973. The role of cytokinin in the regulation of growth, DNA synthesis and cell proliferation in cultured soybean tissues (Glycine max var. Biloxi). Physiologia Plantarum 28:14–23.

Fox, J. E. and J. L. Erion. 1975. A cytokinin binding protein from higher plant ribosomes. Biochemical and Biophysical Research Communications 64:694–700.

Gordon, M. E. and D. S. Letham. 1975. Regulators of cell division in plant tissues. XXII. Physiological aspects of cytokinin-induced radish cotyledon growth. Australian Journal of Plant Physiology 2:129–154.

Hall, R. H. 1973. Cytokinins as a probe of developmental processes. Annual Review of Plant Physiology 24:415–444.

Harvey, B. M. R., B. C. Lu, and R. A. Fletcher. 1974. Benzyladenine accelerates chloroplast differentiation and stimulates photosynthetic enzyme activity in cucumber cotyledons. Canadian Journal of Botany 52:2581–2586.

Hocking, T. J. and J. R. Hillman. 1975. Studies on the role of abscisic acid in the initiation of bud dormancy in Alnus glutinosa and Betula pubescens. Planta 125:235–242.

Horgan, R., E. W. Hewett, J. M. Horgan, J. Purse, and P. F. Wareing. 1975. A new cytokinin from Populus × Robusta. Phytochemistry 14:1005– 1008.

Huff, A. K. and C. W. Ross. 1975. Promotion of radish cotyledon enlargement and reducing sugar content by zeatin and red light. Plant Physiology 56:429–433.

Jackson, M. B. and D. J. Campbell. 1976. Waterlogging and petiole epinasty in tomato: The role of ethylene and low oxygen. New Phytologist 76:21–29.

Kahn, A. A. 1975. Primary, preventive and permissive roles of hormones in plant systems. Botanical Review 41:391–420.

Kawase, M. 1976. Ethylene accumulation in flooded plants. Physiologia Plantarum 36:236–241.

Kende, H. 1971. The cytokinins. International Review of Cytology 31:301–338.

Kende, H. and G. Gardner. 1976. Hormone binding in plants. Annual Review of Plant Physiology 27:267–290.

Kende, H. and A. D. Hanson. 1976. Relationship between ethylene evolution and senescence in morning-glory flower tissue. Plant Physiology 57:523–527.

Kozlowski, T. T., ed. 1973. The shedding of plant parts. Academic Press, Inc., New York.

Leonard, N. J. 1974. Chemistry of the cytokinins. Pages 21–56 in V. C. Runeckles, E. Sonderheimer, and D. C. Walton, eds., The chemistry and biochemistry of plant hormones. Vol. 7, Recent advances in phytochemistry. Academic Press, Inc., New York.

Leopold, A. C. and P. E. Kriedemann. 1975. Plant growth and development. 2nd ed. McGraw-Hill Book Company, New York.

Letham, D. S. 1971. Regulators of cell division in plant tissues. XII. A cytokinin bioassay using excised radish cotyledons. Physiologia Plantarum 25:391–396.

———. 1973. Cytokinins from Zea mays. Phytochemistry, 12:2445–2455.

———. 1974. Regulators of cell division in plant tissues XX. The cytokinins of coconut milk. Physiologia Plantarum 32:66–70.

Lieberman, M. 1975. Biosynthesis and regulatory control of ethylene in fruit ripening. A review. Physiologie vegetale 13:489–499.

McGlasson, W. B. 1970. The ethylene factor. Pages 475–519 in A. C. Hulme, ed., The biochemistry of fruits and their products, vol. I. Academic Press, Inc., New York.

Milborrow, B. V. 1974a. Chemistry and biochemistry of abscisic acid. Pages 57–91 in V. C. Runeckles, E. Sondheimer, and D. C. Walton, eds., The chemistry and biochemistry of plant hormones. Vol. 7, Recent advances in phytochemistry. Academic Press, Inc., New York.

———. 1974b. The chemistry and physiology of abscisic acid. Annual Review of Plant Physiology 25:259–307.

Narain, A. and M. M. Laloraya. 1974. Cucumber cotyledon expansion as a bioassay for cytokinins. Zeitschrift für Pflanzenphysiologie 71: 313–322.

Osborne, D. J., M. B. Jackson, and B. V. Milborrow. 1972. Physiological properties of abscission accelerator from senescent leaves. Nature New Biology 240:98–101.

Palmer, C. E. and W. G. Barker, 1973. Influence of ethylene and kinetin on tuberization and enzyme activity in Solanum tuberosum L. stolons cultured in vitro. Annals of Botany 37:85–93.

Palmer, J. H. 1976. Failure of ethylene to change the distribution of indoleacetic acid in the petiole of Coleus blumei × frederici during epinasty. Plant physiology 58:513–515.

Perry, T. O. and H. Hellmers. 1973. Effects of abscisic acid on growth and dormancy of two races of red maple. Botanical Gazette 134:283–289.

Powell, L. E. and S. D. Seeley. 1974. The metabolism of abscisic acid to a water soluble complex in apple. Journal of the American Society for Horticultural Science 99:439–441.

Pratt, H. K. and J. D. Goeschl. 1969. Physiological roles of ethylene in plants. Annual Review of Plant Physiology 20:5418584.

Pryce, R. J. 1972. The occurrence of lunularic and abscisic acids in plants. Phytochemistry 11: 1759–1761.

Railton, I. D. and D. M. Reid. 1973. Effects of benzyl-adenine on the growth of waterlogged tomato plants. Planta 111:261–266.

Sacher, J. A. 1973. Senescence and postharvest physiology. Annual Review of Plant Physiology 24:197–224.

Seyer, P., D. Marty, A. M. Lescure, and C. Péaud-Lenoël. 1975. Effect of cytokinin on chloroplast cyclic differentiation in cultured tobacco cells. Cell Differentiation 4:187–197.

Skene, K. G. M. 1975. Cytokinin production by roots as a factor in the control of plant growth. Pages 365–396 in J. G. Torrey and D. T. Clarkson, eds., The development and function of roots. Academic Press, Inc., New York.

Skoog, F. and D. J. Armstrong. 1970. Cytokinins. Annual Review of Plant Physiology 21:359–384.

Skoog, F. and R. Y. Schmitz. 1972. Cytokinins. Pages 181–213 in F. C. Steward, ed., Plant physiology. Vol. 6B, Physiology of development: The hormones. Academic Press, Inc., New York.

Stetler, D. A. and W. M. Laetsch. 1965. Kinetin-induced chloroplast maturation in cultures of tobacco tissue. Science 149:1387–1388.

Takegami, T. and K. Yoshida. 1975. Isolation and purification of cytokinin binding protein from tobacco leaves by affinity column chromatography. Biochemical and Biophysical Research Communications 67:782–789.

Torrey, J. G. 1976. Root hormones and plant growth. Annual Review of Plant Physiology 27:435–459.

Vaadia, Y. 1976. Plant hormones and water stress. Philosophical Transactions of the Royal Society of London, series B, pp. 273, 513, 522.

Vanderhoef, L. N. and C. A. Stahl. 1975. Separation of two responses to auxin by means of cytokinin inhibition. Proceedings of the National Academy of Sciences 72:1822–1825.

van Staden, J. 1975. Occurrence of a cytokinin glucoside in the leaves and honeydew of Salix babylonica. Physiologia Plantarum 36:225–228.

van Staden, J. and S. E. Drewes. 1975. Identification of zeatin and zeatin riboside in coconut milk. Physiologia Plantarum 34:106–109.

Varner, J. E. and D. T. Ho. 1976. Hormones. Pages 713–770 in J. Bonner and J. E. Varner, eds., Plant biochemistry. Academic Press, Inc., New York.

Walton, D. C., M. A. Harrison, and P. Coté. 1976. The effects of water stress on abscisic-acid levels and metabolism in roots of Phaseolus vulgaris L. and other plants. Planta 131:141–144.

Wareing, P. F. 1969. The control of bud dormancy in seed plants. Pages 241–262 in H. W. Woolhouse, ed., Dormancy and survival. Symposia for the Society of Experimental Biology no. 23. Academic Press, Inc., New York.

Wareing, P. F. and P. F. Saunders. 1971. Hormones and dormancy. Annual Review of Plant Physiology 22:261–288.

Webster, B. D. 1970. A morphogenetic study of leaf abscission in Phaseolus. American Journal of Botany 57:443–451.

Yang, S. F. 1974. The biochemistry of ethylene: Biogenesis and metabolism. Pages 131–164 in V. C. Runeckles, E. Sondheimer, and D. C. Walton, eds., The chemistry and biochemistry of plant hormones. Vol. 7, Recent advances in phytochemistry. Academic Press, Inc., New York.

Chapter 18

Adams, P. A., M. J. Montague, M. Tepfer, D. L. Rayle, H. Ikuma, and P. B. Kaufman. 1975. Effect of gibberellic acid on the plasticity and elasticity of Avena stem segments. Plant Physiology 56:757–760.

Arslan, N. and T. A. Bennet-Clark. 1960. Geotropic behavior of grass nodes. Journal of Experimental Botany 11:1–12.

Audus, L. J. 1969. Geotropism. Pages 205–242 in M. B. Wilkins, ed., The physiology of plant growth and development. McGraw-Hill Book Company, N. Y.

_____. 1975. Geotropism in roots. Pages 327–363 in J. G. Torrey and D. T. Clarkson, eds., The development and function of roots. Academic Press, Inc.,N. Y.

Ball, N. G. 1969. Nastic responses. Pages 277–300 in M. B. Wilkins, ed., Physiology of plant growth and development. McGraw-Hill Book Company, New York.

Blaauw, A. H. 1909. Die Perzeption des Lichtes [The perception of light]. Rec. Trav. Bot. Neerl. 5:209–272.

Bouck, G. B. and D. L. Brown. 1976. Self-assembly in development. Annual Review of Plant Physiology 27:71–94.

Brain, R. D., J. A. Freeberg, C. V. Weiss, and W. R. Briggs. 1977. Blue light-induced absorbance changes in membrane fractions from corn and Neurospora. Plant Physiology 59:948–952.

Brennan, T., J. E. Gunckel, and C. Frenkel. 1976. Stem sensitivity and ethylene involvement in phototropism of mung bean. Plant Physiology 57:286–289.

Bridges, I. G. and M. B. Wilkins. 1973. Growth initiation in the geotropic response of the wheat node. Planta 112:191–200.

_____. 1974. The role of reducing sugars in the geotropic response of the wheat node. Planta 117:243–250.

Bruinsma, J., C. M. Karssen, M. Benschop, and J. B. Van Dort. 1975. Hormonal regulation of phototropism in the light grown sunflower seedling Helianthus annuus L.: Immobility of endogenous indoleacetic acid and inhibition of hypocotyl growth by illuminated cotyledons. Journal of Experimental Botany 26:411–418.

Carlson, P. S. 1973. The use of protoplasts for genetic research. Proceedings of the National Academy of Sciences 70:598–602.

Carlson, P. S. and J. C. Polacco. 1975. Plant cell cultures: Genetic aspects of crop improvement. Science 188:622–625.

Casjens, S. and J. King. 1975. Virus assembly. Annual Review of Biochemistry 44:555–611.

Darwin, C. 1880. The power of movement in plants. John Murray, London.

Dayanadan, P., F. V. Hebard, and P. B. Kaufman. 1976. Cell elongation in the grass pulvinus in response to geotropic stimulation and auxin application. Planta 131:245–252.

Digby, J. and R. D. Firn. 1976. A critical assessment of the Cholodny-Went theory of shoot geotropism. Current Advances in Plant Science 8:953–960.

El-Antably, H. M. M. 1975. Redistribution of endogenous indoleacetic acid, abscisic acid and gibberellins in geotropically stimulated Ribes nigrum shoots. Physiologia Plantarum 34:167–170.

Firn, R. D., J. Digby, and C. Pinsent. 1977. Evidence against the involvement of gibberellin in the differential growth which causes geotropic curvature. Zeitschrift für Pflanzenphysiologie 82:179–185.

Glasziou, K. T. 1969. Control of enzyme formation and inactivation in plants. Annual Review of Plant Physiology 20:63–88.

Gould, F. W. 1968. Grass systematics. McGraw-Hill Book Company, New York.

Hamner, P. A., C. A. Mitchell, and T. C. Weiler. 1974. Height control in greenhouse chrysanthemum by mechanical stress. Hortscience 9:474–475.

Heslop-Harrison, J. 1967. Differentiation. Annual Review of Plant Physiology 18:325–348.

_____. 1972. Genetics and the development of higher plants: a summary of current concepts. Pages 341–366 in F. C. Steward, ed., Plant physiology. Vol., VIC, Physiology of development: From seeds to sexuality. Academic Press, Inc., New York.

_____. 1976. Carnivorous plants a century after Darwin. Endeavour 35:114–122.

Jaffe, M. J. 1973. Thigmomorphogenesis: The response of plant growth and development to mechanical stimulation. Planta 114:143–157.

_____. 1976. Thigmomorphogenesis: a detailed characterization of the response of beans (Phaseolus vulgaris L.) to mechanical stimulation. Zeitschrift für Pflanzenphysiologie 77:437–453.

Jaffe, M. J. and A. W. Galston. 1968. The physiology of tendrils. Annual Review of Plant Physiology 19:417–434.

Jensen, W. A. 1964. Cell development during plant embryogenesis. Pages 179–202 in Meristems and differentiation. Brookhaven Symposium in Biology, No. 16. Brookhaven National Laboratory, Upton, N. Y.

Jensen, W. A. and F. B. Salisbury. 1972. Botany: An ecological approach. Wadsworth Publishing Company, Inc., Belmont, Calif.

Juniper, B. E. 1976. Geotropism. Annual Review of Plant Physiology 27:385–406.

Kang, B. G. and S. P. Burg. 1974. Red light enhancement of the phototropic response of etiolated pea stems. Plant Physiology 53:445–448.

Lake, J. V. and G. Slack. 1961. Dependence on light of geotropism in plant roots. Nature 191:300–302.

Lang, A. 1974. Inductive phenomena in plant development. Pages 129–144 in Basic mechanisms in plant morphogenesis. Brookhaven Symposium in Biology, no. 25. Brookhaven National Laboratory, N. Y.

Meins, F., Jr. 1975. Cell division and the determination phase of cytodifferentiation in plants. Pages 151–175 in Results and problems in cell differentiation. Vol. 7, Cell cycle and differentiation. Springer-Verlag, Berlin.

Meyer, B. S. and D. B. Anderson. 1952. Plant physiology. 2nd ed. D. Van Nostrand Company, Inc., New York.

Muñoz, V. and W. L. Butler. 1975. Photoreceptor pigments for blue light in Neurospora crassa. Plant Physiology 55:421–426.

Murashige, T. 1974. Plant propagation through tissue cultures. Annual Review of Plant Physiology 25:135–166.

Nabors, M. W. 1976. Using spontaneously occurring and induced mutations to obtain agriculturally useful plants. Bioscience 26:761–767.

Phillips, I. D. J. and W. Hartung. 1976. Longitudinal and lateral transport of [3,4–³H] gibberellin A₁ and 3-indolyl (acetic acid-2-¹⁴C) in upright and geotropically responding green internode segments from *Helianthus annuus*. New Phytologist 76:1–9.

Pickard, B. G. 1973. Action potentials in higher plants. Botanical Review 39:172–201.

Roberts, L. W. 1969. The initiation of xylem differentiation. Botanical Review 35:201–250.

Sangwan, R. S. and B. Norreel. 1975. Induction of plants from pollen grains of *Petunia* cultured *in vitro*. Nature 257:222–224.

Satter, R. L. and A. W. Galston. 1973. Leaf movements: Rosetta stone of plant behavior? Bioscience 23:407–416.

Schrempf, M., R. L. Satter, and A. W. Galston. 1976. Potassium-linked chloride fluxes during rhythmic leaf movement of *Albizzia julibrissin*. Plant Physiology 58:190–192.

Scurfield, G. 1973. Reaction wood: Its structure and function. Science 179:647–655.

Shen-Miller, J. and R. R. Hinchman. 1974. Gravity sensing in plants: A critique of the statolith theory. Bioscience 24:643–651.

Shininger, T. L. 1975. Is DNA synthesis required for induction of differentiation in quiescent root cortical parenchyma? Developmental Biology 45:137–150.

Sibaoka, T. 1969. Physiology of rapid movements in higher plants. Annual Review of Plant Physiology 20:165–184.

Sinnott, E. W. 1960. Plant morphogenesis. McGraw-Hill Book Company, New York.

Spratt, N. T., Jr. 1971. Developmental biology. Wadsworth Publishing Company, Inc., Belmont, Ca.

Street, H. E. 1973a. Plant cell cultures: Their potential for metabolic studies. Pages 93–125 in B. V. Milborrow, ed., Biosynthesis and its control in plants. Academic Press, Inc., New York.

———, ed. 1973b. Plant tissue and cell structure. University of California Press, Berkeley.

———, ed. 1974. Tissue culture and plant science. Academic Press, Inc., New York.

Sunderland, N. 1970. Pollen plants and their significance. New Scientist 47:142–144.

Torrey, J. G., D. E. Fosket, and P. K. Hepler. 1971. Xylem formation: A paradigm of cytodifferentiation in higher plants. American Scientist 59:338–352.

Van Sambeek, J. W. and B. G. Pickard. 1976. Mediation of rapid electrical, metabolic, transpirational, and photosynthetic changes by factors released from wounds. III. Measurements of CO_2 and H_2O flux. Canadian Journal of Botany 54:2662–2671.

Westing, A. H. 1965. Formation and function of compression wood in gymnosperms. Botanical Review 31:381–480.

Wilkins, M. B. 1975. The role of the root cap in root geotropism. Current Advances in Plant Science 8:317–328.

———. 1976. Gravity-sensing guidance systems in plants. Scientific Progress 63:187–217.

Wilkins, H. and R. L. Wain. 1975. Abscisic acid and the response of the roots of *Zea mays* L. seedlings to gravity. Planta 126:19–23.

Chapter 19

Anstis, P. J. P., J. Friend, and D. C. J. Gardner. 1975. The role of xanthoxin in the inhibition of pea seedling growth by red light. Phytochemistry 14:31–35.

Attridge, T. H. and C. B. Johnson. 1976. Photo control of enzyme levels. Pages 185–192 in H. Smith, ed., Light and plant development. Butterworth & Co., Ltd., London.

Bajrachara, D., H. Falk, and P. Schopfer. 1976. Phytochrome-mediated development of mitochondria in the cotyledons of mustard (*Sinapis alba*) seedlings. Planta 131:253–261.

Baskin, J. M. and C. C. Baskin. 1976. Effect of photoperiod on germination of *Cyperus inflexus* seeds. Botanical Gazette 137:269–273.

Bickford, E. D. and S. Dunn. 1972. Lighting for plant growth. Kent State University Press, Kent, Ohio.

Blaauw, O. H., G. Blaauw-Jansen, and W. J. van Leeuwen. 1968. An irreversible red-light-induced growth response in *Avena*. Planta 82:87–104.

Black, M. 1969. Light-controlled germination of seeds. Pages 193–217 in H. H. Woolhouse, ed., Dormancy and survival. Symposia of the Society for Experimental Biology, no. 23. Academic Press, Inc. New York.

Black, M. and J. Shuttleworth. 1976. Inter-organ effects in the control of growth. Pages 317–332 in H. Smith ed., Light and plant development. Butterworth & Co., Ltd., London.

Borthwick, H. 1972. History of phytochrome. Pages 3–23 in K. Mitrakos and W. Shropshire, Jr., eds., Phytochrome. Academic Press, Inc., N. Y.

Bown, A. W., D. R. Reeve, and A. Crozier. 1975. The effect of light on gibberellin metabolism and growth of *Phaseolus coccineus* seedlings. Planta 126:83–91.

Briggs, W. R. 1976a. H. A. Borthwick and S. B. Hendricks—pioneers of photomorphogenesis. Pages 1–6 in H. Smith, ed., Light and plant development. Butterworth & Co., Ltd., London.

———. 1976b. The nature of the blue light photoreceptor in higher plants and fungi. Pages 7–18 in H. Smith, ed., Light and plant development. Butterworth & Co., Ltd., London.

Briggs, W. R. and H. V. Rice. 1972. Phytochrome: Chemical and physical properties and mechanism of action. Annual Review of Plant Physiology 23:293–334.

Butler, W. L., K. H. Norris, H. W. Siegelman, and S. B. Hendricks. 1959. Detection, assay, and preliminary purification of the pigment controlling photoresponsive development of plants. Proceedings of the National Academy of Sciences 45: 1703–1708.

Carpita, N. C. 1977. The mechanism of phytochrome-mediated lettuce seed germination. Ph.D. thesis, Colorado State University, Fort Collins, Co.

Chen, S. S. C. and J. E. Varner. 1973. Hormones and seed dormancy. Seed Science and Technology 1: 325–338.

Cooke, R. J. and R. E. Kendrick. 1976. Phytochrome controlled gibberellin metabolism in etioplast envelopes. Planta 131:303–307.

Deutch, B. 1976. Barley leaf unfolding under mixtures of red and blue light. Physiologia Plantarum 38: 57–60.

Downs, R. J. and H. Hellmers. 1975. Environment and the experimental control of plant growth. Academic Press, Inc., New York.

Duke, S. O., S. B. Fox, and A. W. Naylor. 1976. Photosynthetic independence of light-induced anthocyanin formation in *Zea* seedlings. Plant Physiology 57:192–196.

Esashi, Y., K. Kotaki, and Y. Ohhara. 1976. Induction of cocklebur seed germination by anaerobiosis: A question about the "inhibitor hypothesis" of seed dormancy. Planta 129:109–112.

Evans, A. 1976. Etioplast phytochrome and its *in vitro* control of gibberellin efflux. Pages 129–142 *in* H.

Smith, ed., Light and plant development. Butterworth & Co., Ltd., London.

Evenari, M. 1965. Light and seed dormancy. Pages 804–877 in W. Ruhland, ed., Differentiation and development, Encyclopedia of plant physiology, vol. 15, part 2. Springer-Verlag, Berlin.

Frankland, B. 1976. Phytochrome control of seed germination in relation to the light environment. Pages 477–492 in H. Smith, ed., Light and plant development. Butterworth & Co., Ltd. London.

Galston, A. W. 1974. Plant photobiology in the last half-century. Plant Physiology 54:427–436.

Galston, A. W. and R. L. Satter. 1976. Light, clocks, and ion flux: An analysis of leaf movements (review paper). Pages 159–184 *in* H. Smith, ed., Light and plant development. Butterworth & Co., Ltd., London.

Górski, T. 1975. Germination of seeds in the shadow of plants. Physiologia Plantarum 34:342–346.

Hahlbrock, K. and H. Grisebach. 1975. Biosynthesis of flavonoids. Pages 866–915 in J. B. Harborne, T. J. Mabry, and H. Mabry, eds., The flavonoids. Academic Press, Inc., New York.

Halmer, P., J. D. Bewley, and T. A. Thorpe. 1976. An enzyme to degrade lettuce endosperm cell walls. Appearance of a mannanase following phytochrome- and gibberellin-induced germination. Planta 130:189–196.

Haupt, W. 1966. Phototaxis in plants. International Review of Cytology 19:267–299.

———. 1973. Role of light in chloroplast movement. Bioscience 23:289–296.

Haupt, W. and N. Bretz. 1976. Short-term reactions of phytochrome in Mougeotia? Planta 128:1–3.

Hendricks, S. B. 1960. Comparative biochemistry of photoreactive systems. Academic Press, Inc., New York.

Heydecker, W. 1973. Seed ecology. The Pennsylvania State University Press, University Park, Pa.

Holmes, M. G. 1976. Spectral energy distribution in the natural environment and its implications for phytochrome function. Pages 467–476 *in* H. Smith, ed., Light and plant development, Butterworth & Co., Ltd., London.

Inoue, Y. and K. Shibata. 1973. Light-induced chloroplast rearrangements and their action spectra as measured by absorption spectrophotometry. Planta 114:341–358.

Jensen, W. A. and F. B. Salisbury. 1972. Botany: An ecological approach. Wadsworth Publishing Company, Inc., Belmont, Calif.

Johnson, C. B. 1976. Rapid activation by phytochrome of nitrate reductase in the cotyledons of *Sinapis alba*. Planta 128:127–131.

Kendrick, R. E. and C. J. P. Spruit. 1976. Intermediates in the photoconversion of phytochrome. Pages 31–44 *in* H. Smith, ed., Light and plant development. Butterworth & Co., Ltd., London.

Kendrick, R. E. and H. Smith. 1976. Assay and isolation of phytochrome. Pages 334–364 in T. W. Goodwin, ed. Chemistry and biochemistry of plant pigments, vol. 2. Academic Press, Inc., New York.

Koller, D. 1969. The physiology of dormancy and survival of plants in desert environments. Pages 449–469 *in* H. W. Woolhouse, ed. Dormancy and survival. Symposia for the Society of Experimental Biology, no. 23. Academic Press, Inc., New York.

———. 1972. Environmental control of seed germination. Pages 2–101 *in* T. T. Kozlowski, ed., Seed Biology, vol. 2. Academic Press, Inc., New York.

Lawson, V. R. and R. L. Weintraub. 1975. Effects of red light on the growth of intact wheat and barley coleoptiles. Plant Physiology 56:44–50.

Mancinelli, A. L., C. H. Yang, I. Rabino, and K. M. Ruzmanoff. 1976. Photocontrol of anthocyanin synthesis. V. Further evidence against the involvement of photosynthesis in the high irradiation reaction anthocyanin synthesis of young seedlings. Plant Physiology 58:214–217.

Mayer, A. M. and A. Poljakoff-Mayber. 1975. The germination of seeds, 2nd ed. Pergamon Press, Inc., New York.

McClure, J. W. 1973. Action spectra for phenylalanine ammonia lyase in *Hordeum vulgare*. Phytochemistry 13:1071–1073.

Meijer, G. 1968. Rapid growth inhibition of gherkin hypocotyls in blue light. Acta Botanica Neerlandica 17:9–14.

Mitrakos, K. and W. Shropshire, Jr., eds. 1972. Phytochrome. Academic Press, Inc., New York.

Mohr, H. 1969. Photomorphogenesis. Pages 507–556 in M. B. Wilkins, ed., Physiology of plant growth and development. McGraw-Hill Book Company, New York.

———. 1974. The role of phytochrome in controlling enzyme levels in plants. Pages 37–81 in J. Paul, ed., The biochemistry of cell differentiation. M. T. P. International Review of Science, vol. 9. Butterworth & Co. Ltd., London.

Morgan, C. and H. Smith. 1976. Linear relationship between phytochrome photoequilibrium and growth in plants under simulated natural radiation. Nature 262:210–212.

Muñoz, V. and W. L. Butler. 1975. Photoreceptor pigments for blue light in *Neurospora crassa*. Plant Physiology 55:421–426.

Nabors, M. W. and A. Lang. 1971. The growth physics and water relations of red-light-induced germination in lettuce seeds. I. Embryos germinating in osmoticum. Planta 101:1–25.

Pratt, L. H. 1976. Re-examination of photochemical properties and absorption characteristics of phytochrome using high-molecular-weight preparations. Pages 19–30 in H. Smith, ed., Light and plant development. Butterworth & Co., Ltd., London.

Pratt, L. H., R. A. Coleman, and J. M. MacKenzie, Jr. 1976. Immunological visualization of phytochrome. Pages 75–94 in H. Smith, ed., Light and plant development. Butterworth & Co., Ltd., London.

Quail, P. H. 1976. Phytochrome. Pages 683–711 in J. Bonner and J. E. Varner, eds., Plant biochemistry. Academic Press, Inc., New York.

Racusen, R. H. 1976. Phytochrome control of electrical potentials and intercellular coupling in oat-coleoptile tissue. Planta 132:25–29.

Racusen, R. H. and B. Etherton. 1975. Role of membrane-bound, fixed-charge changes in phytochrome-mediated mung bean root tip adherence phenomenon. Plant Physiology 55:491–495.

Roberts, E. H. 1972. Dormancy: A factor affecting seed survival in the soil. Pages 321–359 in E. H. Roberts, ed., Viability of seeds. Syracuse University Press, Syracuse, N.Y.

Rollin, P. 1972. Phytochrome control of seed germination. Pages 229–254 in K. Mitrakos and W. Shropshire, Jr., eds., Phytochrome. Academic Press, Inc., New York.

Satter, R. L. and A. W. Galston. 1976. The physiological functions of phytochrome. Pages 680–735 in T. W. Goodwin, ed., Chemistry and biochemistry of plant pigments, vol. 1. Academic Press, Inc., New York.

Schäfer, E. 1976. The "high irradiance reaction." Pages 45–62 in H. Smith, ed., Light and plant development. Academic Press, Inc., New York.

Shropshire, W. 1972. Action spectroscopy. Pages 162–181 in K. Mitrakos and W. Shropshire, Jr., eds., Phytochrome. Academic Press, Inc., New York.

Smith, H. 1972. The photocontrol of flavonoid biosynthesis. Pages 433–481 in K. Mitrakos and W. Shropshire, Jr., eds., Phytochrome Academic Press, Inc., New York.

———. 1973a. Light quality and germination: Ecological implications. Pages 219–231 in W. Heydecker, ed., Seed ecology. The Pennsylvania State University Press, University Park, Pa.

———. 1973b. Regulatory mechanisms in the photocontrol of flavonoid biosynthesis. Pages 303–321 in B. V. Milborrow, ed., Biosynthesis and its control in plants. Academic Press, Inc., New York.

———. ed. 1976a. Light and plant development. Butterworth & Co., Ltd., London.

———. 1976b. Phytochrome-mediated assembly of polyribosomes in etiolated bean leaves. Evidence for post-transcriptional regulation of development. European Journal of Biochemistry 65:161–170.

Smith, H. and R. E. Kendrick. 1976. The structure and properties of phytochrome. Pages 377–423 in T. W. Goodwin, ed., Chemistry and biochemistry of plant pigments, vol. 1. Academic Press, Inc., New York.

Tanada, T. 1968. A rapid photoreversible response of barley root tips in the presence of 3-indoleacetic acid. Proceedings of the National Academy of Sciences 59:376–380.

Tezuka, T. and Y. Yamamoto. 1974. Kinetics of activation of nicotinamide adenine dinucleotide kinase by phytochrome–far red absorbing form. Plant Physiology 53:717–722.

Thomson, B. 1951. The relation between age at the time of exposure and response of parts of the *Avena* seedling to light. American Journal of Botany 38:635–638.

Travis, R. L., J. L. Key, and C. W. Ross. 1974. Activation of 80 S ribosomes by red light treatment of dark-grown seedlings. Plant Physiology 53:28–31.

Toole, V. K. 1973. Effects of light, temperature and their interactions on the germination of seeds. Seed Science and Technology 1:339–396.

Villiers, T. A. 1972. Seed dormancy. Pages 220–281 in T. T. Kozlowski, ed., Seed biology. vol. 2. Academic Press, Inc., New York.

Vince-Prue, D. 1975. Photoperiodism in Plants. McGraw-Hill Book Company, New York.

Wain, R. L. 1975. Some developments in research on plant growth regulators. Proceedings of the Royal Society of London, series B, 191:335–352.

Wong, E. 1976. Biosynthesis of flavonoids. Pages 464–526 in T. W. Goodwin, ed., Chemistry and biochemistry of plant pigments, vol. 1. Academic Press, Inc., New York.

Zucker, Milton. 1972. Light and enzymes. Annual Review of Plant Physiology 23:133–156.

Chapter 20

Aschoff, Jürgen, ed. 1965. Circadian clocks. North-Holland Publishing Co., Amsterdam.

——— 1965. Circadian rhythms in man. Science 148:1427–1432.

Behling, Ingeborg. 1929. Über das Zeitgedächtnis der Bienen [On the time memory in honey bees]. Zeitschrift für vergleichende Physiologie 9:259–338.

Brown, Frank A., Jr. 1972. The "clocks" timing biological rhythms. American Scientist 60:756–766.

Brown, Frank A., Jr. and Carol S. Chow. 1973a. Interorganismic and environmental influences through extremely weak electromagnetic fields. Biological Bulletin 144:437–461. (See also 145:265–278.)

Brown, F. A., Jr. and H. M. Webb. 1948. Temperature relations of an endogenous daily rhythmicity in the fiddler crab, *Uca*. Physiologie Zoologie 21:371–381.

Bruce, Victor B. 1974. Recombinants between clock mutants of *Chlamydomonas reinhardi*. Genetics 77:221–230.

Bünning, Erwin. 1932. Über die Erblichkeit der Tagesperiodität bei den *Phaseolus*-Blättern [On the inheritance of diurnal periodicity in *Phaseolus* leaf movements]. Jahrbuch für wissenschaftliche Botanik 77:283–320.

——— 1937. Die endonome Tagersrhythmik als Grundlage der photoperiodischen Reaktion [The endogenous daily rhythm as the basis of the photoperiodic reaction]. Berichte der Deutschen Botanischen Gesellschaft 54:590–607.

——— 1960. Opening Address: Biological clocks. Cold Spring Harbor Symposia on Quantitative Biology 25:1–9.

——— 1969. Die Bedeutung tagesperiodischer Blattbewegungen für die Präzision der Tageslängenmessung [The importance of circadian leaf movements for the precision of daylength measurement—summary in English]. Planta 86:209–217.

——— 1973. The physiological clock. 3rd ed. Academic Press, Inc., New York.

Bünning, E. and J. Baltes. 1961. Wirkung von Äthylakohol auf die physiologische Uhr [Effects of ethyl alcohol on the physiological clock]. Die Naturwissenschaften 49(1):1–2.

Bünning, Erwin and Ilse Moser. 1967. Weitere Versuche zur Lichtwirkung auf die circadiane Rhythmik von *Phaseolus multiflorus* [Further experiments concerning the influence of light on the circadian rhythm of *Phaseolus multiflorus*—summary in English]. Planta 77:99–107.

Bünning, E. and K. Stern. 1930. Über die tagesperiodischen Bewegungen der Primärblätter von *Phaseolus multiflorus*. II. Die Bewegungen bei Thermokonstanz [The diurnal movements of the primary leaves of *Phaseolus multiflorus*. II. Movement at constant temperature]. Berichte der Deutschen Botanischen Gesellschaft 48:227–252.

Bünning E., K. Stern, and R. Stoppel. 1930. Versuche über den Einfluss von Luftionen auf die Schlafbewegungen von *Phaseolus* [Experiments on the influence of atmospheric ions on the sleep movements of *Phaseolus*]. Planta/Archiv für wissenschaftliche Botanik 11(1):67–74.

Darwin, Charles and Francis Darwin. 1881. The power of movement in plants. Appleton, New York.

Ehret, Charles F. 1960. Action spectra and nucleic acid metabolism in circadian rhythms at the cellular level. Pages 149–158 in Cold Spring Harbor Symposia on Quantitative Biology, vol. 25, Biological clocks. The Biological Laboratory, Cold Spring Harbor, New York.

Evans, Hugh L., William B. Ghiselli, and Robert A. Patton. 1973. Diurnal rhythm in behavioral effects of methamphetamine, p-chloromethamphetamine and scopolamine. The Journal of Pharmacology and Experimental Therapeutics 186 (1):10–17.

Feldman, Jerry F. 1975. Circadian periodicity in *Neurospora*: alteration by inhibitors of cyclic AMP phosphodiesterase. Science 190:789–790.

Feldman, Jerry F. and Marian N. Hoyle. 1973. Isolation of circadian clock mutants of *Neurospora crassa*. Genetics 75:605–613.

Garner, W. W. and H. A. Allard. 1920. Effect of the relative length of day and night and other factors of the environment on growth and reproduction in plants. Journal of Agricultural Research 18:553–606.

Griffin, Donald R. 1964. Bird migration. Doubleday & Company, Inc., Garden City, New York.

Halaban, Ruth. 1969. Effects of light quality on the circadian rhythm of leaf movement of a short-day plant. Plant Physiology 44:973–977.

Hillman, William S. 1976. Biological rhythms and physiological timing. Annual Review of Plant Physiology 27:159–179.

Lörcher, L. 1957. Die Wirkung verschiedener Lichtqualitäten auf die endogen Tagersrhythmik von Phaseolus [Effect of various light qualities on the endogenous diurnal rhythms of Phaseolus]. Zeitschrift für Botanik 46:209–241.

Luce, Gay Gaer. 1971. Body time. Pantheon Books, New York.

Menaker, Michael. 1971. Synchronization with the photic environment via extraretinal receptors in the avian brain. Pages 315–333 in Michael Menaker, ed., Biochronometry. National Academy of Sciences, Washington, D. C.

Müller, Dieter. 1962. Über jahres- und lunarperiodische Erscheinungen bei einigen Braunalgen [Manifestations with yearly and lunar periods in certain brown algae]. Botanica Marina 4:140–154.

Muñoz, Victor and Warren L. Butler. 1975. Photoreceptor pigment for blue light in Neurospora crassa. Plant Physiology 55:421–426.

Pittendrigh, Colin S. 1954. On temperature independence in the clock system controlling emergence time in Drosophila. Proceedings of the National Academy of Sciences 40:1018–1029.

Pittendrigh, Colin S. and Victor G. Bruce. 1959. Daily rhythms as coupled oscillator systems and their relation to thermoperiodism and photoperiodism. Pages 475–505 in Robert B. Withrow, ed., Photoperiodism and related phenomena in plants and animals. The American Association for the Advancement of Science, pub. no. 55, Washington, D. C.

Rappe, G. 1966. A yearly rhythm in production capacity of gramineous plants. B: III. Soil cultures in photothermostats. Tests and determinations on soils and plants harvested. Acta oecologica Scandinavica, supplementum 7.

Salisbury, F. B. 1963. Biological timing and hormone synthesis in flowering of Xanthium. Planta 59:518–534.

Salisbury, Frank B., George, G. Spomer, Martha Sobral, and Richard T. Ward. 1968. Analysis of an alpine environment. Botanical Gazette 129(1):16–32.

Satter, Ruth L. and Arthur W. Galston. 1973. Leaf movements: Rosetta stone of plant behavior? Bio-Science 23:407–416.

Siegel, Peter V., Siegfried J. Gerathewhol, and Stanley R. Mohler. 1969. Time-zone effects. Science 164:1249–1255.

Sweeney, Beatrice M. 1969. Rhythmic phenomena in plants. Academic Press, Inc., New York.

———. 1974. A physiological model for circadian rhythms derived from the Acetabularia rhythm-paradoxes. International Journal of Chronobiology 2:25–33.

von Frisch, Karl. 1950. Bees: Their vision, chemical senses, and language. Cornell Paperbacks, Cornell University Press, Ithaca, New York.

Ward, Ritchie R. 1971. The living clocks. Alfred A. Knopf, Inc., New York.

Zimmer, Rose. 1962. Phasenverschiebung und andere Störlichtwirkungen auf die endogen tagesperiodischen Blütenblattbewegungen [Phase shift and other light-interruption effects on endogenous diurnal petal movements]. Planta 58:283–300.

Zimmermann, Natille H. and Michael Menaker. 1975. Neural connections of sparrow pineal: Role in circadian control of activity. Science 190:477–479.

Chapter 21

Amen, R. D. 1967. The effects of gibberellic acid and scarification on the seed dormancy and germination in Luzula spicata. Physiologia Plantarum 20:6–12.

Altman, P. L. and Dorothy S. Dittmer, eds. 1962. Growth, including reproduction and morphological development. Federation of the American Society for Experimental Biology, Washington, D.C.

Arias, I., P. M. Williams, and J. W. Bradbeer. 1976. Studies in seed dormancy. IX. The role of gibberellin biosynthesis and the release of bound gibberellin in the post-chilling accumulation of gibberellin in seeds of Corylus avellana L. Planta 131:135–139.

Avery, G. S., P. R. Burkholder, and H. B. Creighton. 1937. Production and distribution of growth hormone activity in shoots of Aesculus and Malus and its probable role in stimulating cambial activity. American Journal of Botany 24:51–58.

Caspari, E. W. and R. W. Marshak. 1965. The rise and fall of Lysenko. Science 149:275–278.

Chailakhyan, M. 1968. Internal factors of plant flowering. Annual Review of Plant Physiology 19:1–36.

Chouard, P. 1960. Vernalization and its relations to dormancy. Annual Review of Plant Physiology 11:191–238.

Crocker, W. and L. V. Barton. 1953. Physiology of seeds. Chronica Botanica Co., Waltham, Mass.

Doorenbos, J. and S. J. Wellensiek. 1959. Photoperiodic control of floral induction. Annual Review of Plant Physiology 10:147–184.

Downs, R. J. and H. A. Borthwick. 1956. Effects of photoperiod on growth of trees. Botanical Gazette 117:310–326.

Esashi, Y., K. Kotaki, and Y. Ohhara. 1976. Induction of cocklebur seed germination by anaerobiosis: A question about the "inhibitor hypothesis" of seed dormancy. Planta 129:109–112.

Evenari, M. 1957. The physiological action and biological importance of germination inhibitors. Society of Experimental Biology Symposium 11:21–43.

Friend, D. J. C. and V. A. Helson. 1976. Thermoperiodic effects on the growth and photosynthesis of wheat and other crop plants. Botanical Gazette 137(1):75–84.

Gassner, G. 1918. Beiträge zur physiologischen Charakteristik Sommer und Winter annueller Gewächse insbesondere der Getreidepflanzen (Contributions to the physiological characteristics of summer and winter annual growth, particularly with cereals.) Zeitschrift für Botanik 10:417–430.

Gregory, L. E. 1965. Physiology of tuberization in plants. Pages 1328–1354 in W. Ruhland, ed., Encyclopedia of plant physiology, vol. 15(1). Springer-Verlag, Berlin.

Harrison, M. A. and P. F. Saunders. 1975. The abscisic acid content of dormant birch buds. Planta 123:291–298.

Hartsema, Annie M. 1961. Influence of temperatures on flower formation and flowering of bulbous and tuberous plants. Pages 123–167 in W. Ruhland, ed., Encyclopedia of plant physiology, vol. 16. Springer-Verlag, Berlin.

Hatch, A. H. and D. R. Walker. 1969. Rest intensity of dormant peach and apricot leaf buds as influenced by temperature, cold hardiness, and respiration. Journal for the American Society of Horticultural Science 94:304–307.

Heide, O. M. 1974. Growth and dormancy in Norway spruce ecotypes (Picea abies). I. Interaction of photoperiod and temperature. Physiologie Plantarum 30:1–12.

Hurd, R. G. and O. N. Purvis. 1964. The effect of gibberellic acid on the flowering of spring and winter rye. Annals of Botany 38:137–151.

Junttila, O. 1973. The mechanism of low temperature dormancy in mature seeds of Syringa species. Physiologia Plantarum 29:256–263.

Ketring, D. L. 1973. Germination inhibitors. Seed Science Technology 1:305–324.

Kivilaan, A. and R. S. Bandurski. 1973. The ninety-year period for Dr. Beal's seed viability experiment. American Journal of Botany 60:140–145.

Koller, D. 1957. Germination-regulating mechanisms in some desert seeds. IV. Atriplex dimosphostegia. Ecology 38:1–13.

Lang, Anton. 1957. The effect of gibberellin upon flower formation. Proceedings of the National Academy of Sciences 43:709–711.

———. 1965. Effects of some internal and external conditions on seed germination. Pages 849–893 in W. Ruhland, ed., Encyclopedia of plant physiology, vol. 15(2). Springer-Verlag, Berlin.

———. 1965. Physiology of flower initiation. Pages 1380–1536 in W. Ruhland, ed., Encyclopedia of plant physiology, vol. 15(1). Springer-Verlag, Berlin.

Lang, A. and E. Reinhard. 1961. Gibberellins and flower formation. Advances in Chemistry Series 28:71–79.

Leopold, A. C. and Paul E. Kriedemann. 1975. Plant growth and development. 2nd ed. McGraw-Hill Book Company, New York.

Luckwill, L. C. 1970. The control of growth and fruitfulness of apple trees. Pages 237–255 in Physiology of tree crops. L. C. Luckwill and C. V. Cutting, eds. Academic Press, Inc., New York.

Mayer, A. M. and Y. Shain. 1974. Control of seed germination. Annual Review of Plant Physiology 25:167–193.

Mayer, A. M. and A. Poljakoff-Mayber. 1963. The germination of seeds. Pergamon Press, Inc., New York.

Melchers, G. 1937. Die Wirkung von Genen, tiefen Temperaturen und blühenden Pfropfpartnern auf die Blühreife von Hyoscymas niger L. [The effect of genes, low temperatures, and flowering graft partners on ripeness to flower of Hyoscymas niger.]. Biologisches Zentralblatt 57:568–614.

———, ed. 1961. Encyclopedia of plant physiology. Vol. 16, External factors affecting growth and development. Springer-Verlag, Berlin.

Melchers, G. and A. Lang. 1948. Die Physiologie der Blütenbildung (Ein Übersichtsbericht) [The physiology of flowering (a review)]. Biologisches Zentralblatt 67:105–174.

Nikolaeva, M. G. 1969. Physiology of deep dormancy in seeds. Trans., Israel Program for Scientific Translations. National Science Foundation, Washington, D.C.

Perry, T. O. and H. Hellmers. 1973. Effects of abscisic acid on growth and dormancy of two races of red maple. Botanical Gazette 134:283–289.

Porsild, A. E., C. R. Harington, and G. A. Mulligan. 1967. *Lupinus articus* Wats. Grown from seeds of Pleistocene age. Science 158:113–114.

Purvis, O. N. 1961. The physiological analysis of vernalization. Pages 76–122 in W. Ruhland, ed., Encyclopedia of plant physiology, vol. 16. Springer-Verlag, Berlin.

Rees, A. R. 1972. The growth of bulbs. Academic Press, Inc., New York.

Salisbury, F. B. 1963. The flowering process. Pergamon Press, Inc., New York.

Samish, R. M. 1954. Dormancy in woody plants. Annual Review of Plant Physiology 5:183–204. (A pomologist who uses the term "rest.")

Seeley, E. J. 1975. Application of instrumental gibberellic acid analysis techniques: With special reference to gibberellic acids in after-ripening peach seeds and developing flower buds. Ph.D.thesis, Utah State University, Logan, Utah.

Seeley, S. D. 1971. Electron capture gas chromatography of plant hormones with special reference to abscisic acid in apple bud dormancy. Ph.D. thesis, Cornell University, Ithaca, N.Y.

Thompson, H. C. 1953. Vernalization of growing plants. Pages 179–196 in W. E. Loomis, ed., Growth and differentiation in plants. Iowa State College Press, Ames.

Tuan, Dorothy Y. and J. Bonner. 1964. Dormancy associated with repression of genetic activity. Plant Physiology 39:768–772.

Ulrich, Albert. 1955. Influence of night temperature and nitrogen nutrition on the growth, sucrose acccumulation and leaf minerals of sugar beet plants. Plant Physiology 30:250–257.

Vegis, A. 1964. Dormancy in higher plants. Annual Review of Plant Physiology 15:185–224.

Vest, E. D. 1952. A preliminary study of some of the germination characteristics of *Atriplex confertifolia*. Unpublished master's thesis, University of Utah, Salt Lake City.

Villiers, T. A. 1972. Seed dormancy. Pages 219–281 in T. T. Kozlowski, ed., Seed biology, vol. II. Academic Press, Inc., New York.

Wareing, P. F. 1956. Photoperiodism in woody plants. Annual Review of Plant Physiology 7:191–214.

Wellensiek, S. J. 1964. Dividing cells as the prerequisite for vernalization. Plant Physiology 39: 832–835.

Went, F. W. 1957. The experimental control of plant growth. Chronica Botanica Co., Waltham, Mass.

———. 1974. Reflections and speculations. Annual Review of Plant Physiology 25:1–26.

Chapter 22

Bennett, R. D., S. T. Ko, and E. Heftmann. 1966. Effect of photoperiodic floral induction on the metabolism of a gibberellin precursor, (–) kaurene, in *Pharbitis nil*. Plant Physiology 41:1360–1363.

Biswas, P. K., K. B. Paul, and J. H. M. Henderson. 1967. Effects of steroids on chrysanthemum in relation to growth and flowering. Nature 213:917–918.

Bonner, J., E. Heftmann, and J. A. D. Zeevaart. 1963. Suppression of floral induction by inhibitors of steroid synthesis. Plant Physiology 38:81–88.

Carr, D. J. 1967. The relationship between florigen and the flower hormones. Pages 305–312 in J. F. Fredrick and E. M. Weyer, eds., Plant growth regulators. Annals of the New York Academy of Sciences, pub. no. 144.

Chailakhyan, M. Kh. 1975. Forty years of research on the hormonal basis of plant development— some personal reflections. Botanical Review 41:1–29.

Claes, H. and A. Lang. 1947. Die Wirkung von β-Indolylessigsäure und 2,3,5-Trijodbenzoesäure auf die Blüten Bildung von *Hyoscyamus niger* [The action of β-indolacetic acid and 2,3,5-Triiodobenzoic acid on flowering of *Hyoscyamusniger*]. Zeitschrift für Naturforschung 26:56–63.

Cleland, C. F. and A. Ajami. 1974. Identification of flower-inducing factor isolated from aphid honeydew as being salicylic acid. Plant Physiology 54:904–906. (See also 54:899–903.)

Cumming, B. G. 1963. Evidence of a requirement for phytochrome-Pfr in the floral initiation of *Chenopodium rubrum*. Canadian Journal of Botany 41:901–926.

———. 1969. Circadian rhythms of flower induction and their significance in photoperiodic response. Canadian Journal of Botany 47:309–324.

Dass, H. C., G. S. Randhawa, and S. P. Negi. 1975. Flowering in pineapple as influenced by Ethephon and its combinations with urea and calcium carbonate. Scientia Horticuliuria 3:231–238.

de Fossard, R. A. 1974. Flower initiation in tissue and organ cultures. Pages 193–212 in H. E. Street, ed., Tissue culture and plant science. Academic Press, Inc., New York.

Evans, L. T., ed. 1969. The induction of flowering, some case histories. Macmillan Company of Australia Pty. Ltd., Victoria.

———. 1975. Daylength and the flowering of plants. W. A. Benjamin, Inc., Menlo Park, Ca.

Fischer, J. 1916. Zur Frage der Kohlensäureernährung der Pflanze [On the question of carbon dioxide nutrition of plants]. Gartenflora 65:232.

Friend, D. J. C. 1976. Light requirements for photoperiodic sensitivity in cotyledons of dark-grown *Pharbitis nil*. Physiologia Plantarum 35:286–296.

Gibby, David D. and F. B. Salisbury. 1971. Participation of long-day inhibition in flowering of *Xanthium strumarium* L. Plant Physiology 47: 784–789.

Hamner, K. C. 1963. Endogenous rhythms in controlled environments. Pages 215–232 in L. T. Evans, ed., Environmental control of plant growth. Academic Press, Inc., New York.

Hamner, K. C. and J. Bonner. 1938. Photoperiodism in relation to hormones as factors in floral initiation and development. Botanical Gazette 100:388.

Havelange, A., G. Bernier, and A. Jacqmard. 1974. Description and quantitative study of ultrastructural changes in the apical meristem of mustard in transition to flowering. II. The cytoplasm, mitochondria, and proplastids. Journal of Cell Science 16:421–432.

Henfrey, A. 1852. The vegetation of Europe, its condition and causes. J. van Voorst, London.

Hillman, W. S. 1976. Biological rhythms and physiological timing. Annual Review of Plant Physiology 27:159–179.

Hodson, H. K. and K. C. Hamner. 1970. Floral inducing extract from *Xanthium*. Science 167:384–385.

Kato, Y., N. Fukunharu, and R. Kobayashi. 1958. Stimulation of flower bud differentiation of conifers by gibberellin. Pages 67–68 in Abstracts of the 2nd meeting of the Japan gibberellin research association.

King, R. W. and B. C. Cumming. 1972. The role of phytochrome in photoperiodic time measurement and its relation to rhythmic timekeeping in the control of flowering in *Chenopodium rubrum*. Planta 108:39–57.

King, R. W. and J. A. D. Zeevaart. 1973. Floral stimulus movement in *Perilla* and flower inhibition caused by noninduced leaves. Plant Physiology 51: 727–738.

Klebs, G. 1918. Über die Blütenbildüng bei *Sempervivum* [On flower formation in *Sempervivum*]. Flora (Jena) 111/112:128.

Kopcewicz, J. and Z. Porazinski. 1974. Effects of growth regulators, steroids, and estrogen fraction from sage plants on flowering of a long-day plant, *Salvia splendens*, grown under noninductive light conditions. Biologia Plantarum 16:132–135.

Kraus, E. J. and H. R. Kraybill. 1918. Vegetation and reproduction with special reference to the tomato. Oregon Agricultural Experiment Station Bulletin 149:5.

Lane, H. C., H. W. Siegelman, W. L. Butler, and E. N. Firer. 1963. Detection of phytochrome in green plants. Plant Physiology 38:414.

Lang, Anton. 1961. Physiology of flower initiation. Pages 1380–1535 in W. Ruhland, ed., Encyclopedia of plant physiology, Vol. 15. Springer-Verlag, Berlin. (The most complete review up to its time.)

Lincoln, R. G.; A. Cunningham, B. H. Carpenter, J. Alexander, and D. L. Mayfield. 1966. Florigenic acid from fungal cultures. Plant Physiology 41:1079–1080.

Löve, A. and Doris Löve. 1946. Experiments on the effects of animal sex hormones on dioecious plants. Arkiv für Botanik 32A, 13:1–60.

McMillan, C. 1974. Photoperiodic responses in experimental hybrids of cockleburs. Nature 249: 183–186.

———. 1975. Experimental hybridization of *Xanthium strumarium* (Compositae) from Asia and America. I. Response of F¹ hybrids to photoperiod and temperature. American Journal of Botany 62:41–47.

Mooney, H. A. and W. D. Billings. 1961. Comparative physiological ecology of arctic and alpine populations of *Oxyria digyna*. Ecological Monographs 31:1–29.

Olmstead, C. E. 1944. Growth and development in range grasses. IV. Photoperiodic responses in twelve geographic strains of side-oats grama. Botanical Gazette 106:46–74.

Papenfuss, H. D. and F. B. Salisbury. 1967. Aspects of clock resetting in flowering of *Xanthium*. Plant Physiology 42:1562–1568.

Pharis, R. P. and C. G. Kuo. 1977. Physiology of gibberellins in conifers. Canadian Journal of Forestry Research 7(2):299–325.

Pryce, R. J. 1972. Gallic acid as a natural inhibitor of flowering in *Kalanchoe blossfeldiana*. Phytochemistry 11:1911–1918.

Ray, P. M. and W. E. Alexander. 1966. Photoperiodic adaptation to latitude in *Xanthium strumarium*. American Journal of Botany 53:806–816.

Reid, H. B., P. H. Moore, and K. C. Hamner. 1967. Control of flowering of *Xanthium pensylvanicum* by red and far-red light. Plant Physiology 42:532–540.

Rowan, W. 1925. Relation of light to bird migration and developmental changes. Nature 115:494– 495.

Salisbury, F. B. 1955. The dual role of auxin in flowering. Plant Physiology 30:327–334.

———. 1963a. Biological timing and hormone synthesis in flowering of *Xanthium*. Planta 59:518–534.

———. 1963b. The flowering process. Pergamon Press, Inc., New York.

———. 1965. Time measurement and the light period in flowering. Planta 66:1–26.

———. 1969. *Xanthium strumarium* L. Pages 14–61 in L. T. Evans, ed., The induction of flowering. Macmillan Company of Australia Pty. Ltd., Victoria.

———. 1971. The biology of flowering. The Natural History Press, Garden City, New York.

——— and J. Bonner. 1956. The reactions of the photoinductive dark period. Plant Physiology 31:141–147.

——— and Alice Denney. 1974. Noncorrelation of leaf

movements and photoperiodic clocks in *Xanthium strumarium* L. Pages 679–686 in Proceedings of the international society of chronobiology, Igaku Shoin, Ltd., Tokyo.

Schwabe, W. W. 1971. Physiology of vegetative reproduction and flowering. Pages 233–411 in F. C. Steward, ed., Plant physiology—a treatise, vol. 7A. Academic Press, Inc., New York.

Tournois, J. 1914. Etudes sur la sexualite du houblon [Studies on the sexuality of hops]. Annals des Sciences Naturelles (Botanique) 19:49–191.

Vince-Prue, D. 1975. Photoperiodism in plants. McGraw-Hill Book Company, New York. (An excellent modern review.)

Wardell, W. L. 1976. Floral activity in solutions of deoxyribonucleic acid extracted from tobacco stems. Plant Physiology 57:855–861.

Wellensiek, S. J. 1970. The floral hormones in *Silene armeria* L. and *Xanthium strumarium* L. Zeitschrift für Pflanzenphysiologie 63:25–30.

———. 1973. Genetics and flower formation of annual *Lunaria*. Netherland Journal of Agricultural Science 21:163–166.

Zeevaart, J. A. D. 1957. Studies on flowering by means of grafting. II. Photoperiodic treatment of detached *Perilla* and *Xanthium* leaves. Koninklijke Nederlandsche Akademie van Wetenschappen (Series C) 60:332–337.

———. 1976. Physiology of flower formation. Annual Review of Plant Physiology 27:321–348.

Chapter 23

Björn, Lars Olof. 1976. Light and life. Crane, Russack & Co., New York.

Bickford, E. D. and S. Dunn. 1972. Lighting for plant growth. Kent State University Press, Kent, Ohio.

Billings, W. D. 1970. Plants, man, and the ecosystem. 2nd ed. Wadsworth Publishing Company, Inc., Belmont, Ca.

Blackman, F. F. 1905. Optima and lighting factors. Annals of Botany 14(74):281–295.

Clausen, Jens, David D. Keck, and William M. Hiesey. 1940. Experimental studies on the nature of species. I. Effect of varied environments on Western North American plants. Carnegie Institution of Washington, pub. no. 520, Washington, D.C.

Clements, Frederic E., Emmett V. Martin, and Frances L. Long. 1950. Adaptation and origin in the plant world. Chronica Botanica Co., Waltham, Mass.

Downs, R. J. and H. Hellmers. 1975. Environment and the experimental control of plant growth. Academic Press, Inc., New York.

Hiesey, W. M. and H. W. Milner. 1965. Physiology of ecological races and species. Annual Review of Plant Physiology 16:203–216.

Higgins, Paul D. and George G. Spomer. 1976. Soil temperature effects on root respiration and the ecology of alpine and subalpine plants. Botanical Gazette 137:110–120.

Highkin, H. R. and A. Lang. 1966. Residual effect of germination temperature on the growth of peas. Planta 68:94–98.

Kimball, S. L. and F. B. Salisbury. 1974. Plant development under snow. Botanical Gazette 135(2): 147–149.

Levitt, J. 1972. Responses of plants to environmental stresses. Academic Press, Inc., New York.

Lockhart, James A. 1965. The analysis of interactions of physical and chemical factors on plant growth. Annual Review of Plant Physiology 16:37–52.

Mirov, N. T. and R. G. Stanley. 1959. The pine tree. Annual Review of Plant Physiology 10:223–238.

Mohr, H. 1972. Lectures on photomorphogenesis. Springer-Verlag, Berlin.

Riker, A. J. 1966. Plant pathology and human welfare. Science 152:1027–1032.

Salisbury, F. B. 1964. Soil formation and vegetation on hydrothermally altered rock material in Utah. Ecology 45:1–9.

———. 1975. Multiple factor effects on plants. Pages 501–520 in F. John Vernberg, ed., Physiological adaptation to the environment. Intext Publishers Group, New York.

Salisbury, F. B., G. G. Spomer, Martha Sobral, and R. T. Ward. 1968. Analysis of an alpine environment. Botanical Gazette 129(1):16–32.

Schopfer, Peter. 1969. Die Hemmung des Streckungs-Wachstums durch Phytochrom—ein Stoffaufnahme erfordernder Prozess? [Inhibition of elongation growth by phytochrome—a process requiring substrate uptake?] Planta 85:383–388.

Spomer, G. G. and F. B. Salisbury. 1968. Eco-physiology of *Geum turbinatum* and implications concerning alpine environments. Botanical Gazette 129(1):33–49.

Vernberg, F. J., ed. 1975. Physiological adaptation to the environment. Intext Publishers Group, New York.

Vernberg, F. J. and Winona B. Vernberg. 1970. The animal and the environment. Holt, Rinehart and Winston, New York.

Waisel, Yoav. 1972. Biology of halophytes. Academic Press, Inc., New York.

Went, F. W. 1957. The experimental control of plant growth. Chronica Botanica Co., Waltham, Mass.

Whittaker, R. H. 1975. Communities and ecosystems. 2nd ed. Macmillan, Inc., New York.

Chapter 24

Acevedo, Edmundo, Theodore C. Hsiao, and D. W. Henderson. 1971. Immediate and subsequent growth responses of maize leaves to changes in water status. Plant Physiology 48:631–636.

Alden, J. and R. K. Hermann. 1971. Aspects of the cold-hardiness mechanism in plants. Botanical Review 37:37–142.

Allen, T. B. 1964. The quest. Chilton Book Company, Radnor, Pa. (Some interesting chapters on life under extreme conditions.)

Bernstein, L. and H. C. Hayward. 1958. Physiology of salt tolerance. Annual Review of Plant Physiology 9:25–46.

Billings, W. D. 1973. Arctic and alpine vegetations: Similarities, differences, and susceptibility to disturbance. BioScience 23:697–704.

———. 1974. Adaptations and origins of alpine plants. Arctic and Alpine Research 6(2):129–142.

Billings, W. D. and H. A. Mooney. 1968. The ecology of arctic and alpine plants. Biological Review 43:481–529.

Bliss, L. C. 1971. Arctic and alpine plant life cycles. Annual Review of Ecology and Systematics 2:405–438.

Boyer, J. S. 1970. Leaf enlargement and metabolic rates in corn, soybean, and sunflower at various leaf water potentials. Plant Physiology 46:233–235.

Brock, T. D. and G. K. Darland. 1970. Limits of microbial existence: Temperature and pH. Science 169:1316–1318.

Burke, J. J., L. V. Gusta, H. A. Quamme, C. J. Weiser, and P. H. Li. 1976. Freezing and injury in plants. Annual Review of Plant Physiology 27:507–528.

Cram, W. J. 1976. Negative feedback regulation of transport cells. The maintenance of turgor, volume and nutrient supply. Pages 284–316 in U. Lüttge and M. G. Pitman, eds., Cells, Encyclopedia of Plant Physiology, new series, Transport in plants, vol. 2, part A. Springer-Verlag, Berlin.

Edney, E. B. 1975. Absorption of water vapor from unsaturated air. Pages 77–98 in F. John Vernberg, ed., Physiological adaptation to the environment. Intext Publishers Group, New York.

Ehleringer, J., Olle Björkman, and Harold A. Mooney. 1976. Leaf pubescence: Effects on absorptance and photosynthesis in a desert shrub. Science 192:376–377.

Evenari, M., E. D. Schultze, L. Kappen, U. Buschbom, and O. L. Lange. 1975. Adaptive mechanisms in desert plants. Pages 111–130 in F. John Vernberg, ed., Physiological adaptation to the environment. Intext Publishers Group, New York.

Fireman, M. and H. E. Hayward. 1952. Indicator significance of some shrubs in the Escalante Desert, Utah. Botanical Gazette 114:143–154.

Hellebust, Johan A. 1976. Osmoregulation. Annual Review of Plant Physiology 27:485–505.

Henckel, P. A. 1964. Physiology of plants under drought. Annual Review of Plant Physiology 15:363–386.

Hsiao, T. C. 1973. Plant responses to water stress. Annual Review of Plant Physiology 24:519–570.

Kohn, H. and J. Levitt. 1966. Interrelations between photoperiod, frost hardiness and sulfhydryl groups in cabbage. Plant Physiology 41:792–796.

Kriedemann, P. E. and B. R. Loveys. 1974. Hormonal mediation of plant responses to environmental stress. Pages 461–465 in R. L. Bieleski, A. R. Ferguson, and M. M. Creswell, eds., Mechanisms of regulation of plant growth. The Royal Society of New Zealand, Wellington.

LaHaye, P. A. and E. Epstein. 1969. Salt toleration by plants: Enhancement with calcium. Science 166:395–396.

Langridge, J. and J. R. McWilliam. 1967. Heat responses of higher plants. Pages 231–292 in A. H. Rose, ed., Thermobiology. Academic Press, Inc., New York.

Levitt, J. 1962. A sulfhydryl-disulfide hypothesis of frost injury and resistance in plants. Journal of Theoretical Biology 3:355–391.

———. 1972. Responses of plants to environmental stresses. Academic Press, Inc., New York.

Lyons, J. M. 1973. Chilling injury in plants. Annual Review of Plant Physiology 24:445–466.

MacDougal, D. T. and E. S. Spaulding. 1910. The water-balance of succulent plants. Carnegie Institution of Washington, no. 141.

Olien, C. R. 1967. Freezing stresses and survival. Annual Review of Plant Physiology 18:387–408.

———. 1974. Energies of freezing and frost desiccation. Plant Physiology 53:764–767.

Parker, J. 1963. Cold resistance in woody plants. Botanical Review 29:124–201.

———. 1969. Further studies of drought resistance in woody plants. Botanical Review 35:317–371.

Prosser, C. L. 1967. Molecular mechanisms of temperature adaptation. American Association for Advancement of Science, Washington, D.C.

Sakai, A. and S. Yoshida. 1967. Survival of plant tissue at superlow temperatures. IV. Effects of cooling and rewarming rates on survival. Plant Physiology 42:1695–1701.

Salisbury, F. B. and G. G. Spomer. 1964. Leaf temperatures of alpine plants in the field. Planta 60:497–505.

Salisbury, F. B., G. G. Spomer, M. Sobral, and R. T. Ward. 1968. Analysis of an alpine environment. Botanical Gazette. 129:16–32.

Scholander, P. F., H. T. Hammel, Edda D. Bradstreet, and E. A. Hemmingsen. 1965. Sap pressure in vascular plants. Science 148:339–346.

Shantz, H. C. 1927. Drought resistance and soil moisture. Ecology 8:145–157.

Spomer, George G. and F. B. Salisbury. 1968. Ecophysiology of *Geum turbinatum* and implications concerning alpine environments. Botanical Gazette 129:33–49.

Stewart, G. R. and J. A. Lee. 1974. The role of proline accumulation in halophytes. Planta 120:279–289.

Tranquillini, W. 1964. The physiology of plants at high altitudes. Annual Review of Plant Physiology 15:345–362.

Troughton, J. and L. A. Donaldson. 1970. Probing plant structure. McGraw-Hill Book Company, New York.

Weiser, C. J. 1970. Cold resistance and injury in woody plants. Science 169:1269–1278.

Went, F. W. 1955. The ecology of desert plants. Scientific American 4:68–75.

――――. 1975. Water vapor absorption in *Prosopis*. Pages 67–76 in F. John Vernberg, ed., Physiological adaptation to the environment. Intext Publishers Group, New York.

White, W. C. and C. J. Weiser. 1964. The relation of tissue desiccation, extreme cold, and rapid temperature fluctuations to winter injury of American Arborvitae. American Society for Horticultural Science 85:554–563.

Special Topic 1.1

Bernal, J. D. 1965. The structure of water and its biological implications. Symposia of the Society of Experimental Biology 19:17–32.

Buswell, A. M., and Worth H. Rodebush. 1956. Water. Scientific American 194(4):77–89.

Crafts, Aldon S. 1968. Water structure and water in the plant body. Pages 23–47 in Theodore T. Kozlowski, ed., Water deficits and plant growth, vol. 1. Academic Press, Inc., New York.

Frank, Henry S. 1970. The structure of ordinary water. Science 169:635–641.

Franks, Felix, ed. 1972. Water: A comprehensive treatise. Plenum Press, New York.

Kohn, P. G. 1965. Tables of some physical and chemical properties of water. Symposia of the Society for Experimental Biology 19:3–16.

Leopold, Luna B., Kenneth S. Davies, and the Editors of *Life*. 1966. Water. Time-Life Books, A Division of Time Inc., New York.

Nobel, Park S. 1974. Introduction to biophysical plant physiology. W. H. Freeman and Company, Publishers, San Francisco.

Runnels, L. K. 1966. Ice. Scientific American 215(6):118–126.

Special Topic 1.2

Fast, J.D. 1962, Entropy. McGraw Hill Book Company, New York.

Fay, James A. 1965. Molecular thermodynamics. Addison-Wesley Publishing Co., Inc., Reading, Mass.

Hatsopoulos, George N. and Joseph H. Keenan. 1965. Principles of general thermodynamics. John Wiley & Sons, Inc., New York.

Klein, J.J. 1970. Maxwell, his demon and the second law of thermodynamics. American Scientist 58(1); 89.

Miller, G. Tyler, Jr. 1971. Energetics, kinetics, and life. Wadsworth Publishing Company, Inc., Belmont, California.

Morowitz, H.J. 1970. Entropy for biologists. Academic Press, Inc., New York.

Special Topic 3.1

Bickford, Elwood D., and Stuart Dunn. 1972. Light for plant growth. Kent State University Press, Kent, Ohio.

Clayton, Roderick K. 1970. Light and living matter. Vol. 1, The physical part. McGraw Hill Book Company, New York.

――――. 1971. Light and living matter. Vol. 2, The biological part. McGraw Hill Book Company, New York.

Minnaert, M. 1954. The nature of light and color in the open air. Trans., H. M. Kremer-Priest; rev., K. E. Brian Jay. Dover Publications, Inc., New York.

Mousseron-Canet, M. and J.-C. Mani. 1972. Photochemistry and molecular reactions [Photochimie et réactions moleculaires]. Trans., J. Schmorak. Israel Program for Scientific Translations, Jerusalem.

Oster, G. 1968. The chemical effects of light. Scientific American 219(3):158–170.

Weisskopf, V. F. 1968. How light interacts with matter. Scientific American 219(3):60–71.

Special Topic 9.1

Hanstein, W. G. 1976. Uncoupling of oxidative phosphorylation. Biochimica et Biophysica Acta 456:129–148.

Jagendorf, A. T. 1975a. Chloroplast membranes and coupling factor confirmations. Federation Proceedings 34:1718–1722.

――――. 1975b. Mechanisms of photophosphorylation. Pages 414–492 in Govindjee, ed., Bioenergetics of photosynthesis. Academic Press, Inc., New York.

Jagendorf, A. T. and E. Uribe. 1966. Photophosphorylation and the chemiosmotic hypothesis. Pages 215–245 in Energy conversion by the photosynthetic apparatus. Brookhaven Symposium in Biology, no. 19.

Kessler, R. J., C. A. Tyson, and D. E. Green. 1976. Mechanism of uncoupling in mitochondria: Uncouplers as ionophores for cycling cations and protons. Proceedings of the National Academy of Sciences 73:3141–3145.

Mitchell, P. 1961. Coupling of phosphorylation to electron and hydrogen transfer by a chemiosmotic type of mechanism. Nature 191:144–148.

――――. 1966. Chemiosmotic coupling in oxidative and photosynthetic phosphorylation. Biological Review 41:445–502.

――――. 1976. Possible molecular mechanisms of the protonmotive function of cytochrome systems. Journal of Theoretical Biology 62:327–368.

Portis, Archie R., Jr. and Richard E. McCarty. 1976. Quantitative relationships between phosphorylation, electron flow, and internal hydrogen ion concentrations in spinach chloroplasts. Journal of Biological Chemistry 251:1610–1617.

Special Topic 16.1

Anderson, Wood Powell. 1977. Weed science: Principles. West Publishing Company, St. Paul, Mn.

Ashton, Floyd M. and Alden S. Crafts. 1973. Mode of action of herbicides. John Wiley & Sons, Inc., New York.

Audus, L. J., ed. 1976. Herbicides: Physiology, biochemistry, ecology, Vols. 1 and 2. Academic Press, Inc., New York.

Kearney, P. C. and D. D. Kaufman, eds. 1975. Herbicides: Chemistry, degradation, and mode of action, Vols. 1 and 2. Marcel Dekker, Inc., New York.

Parka, S. J. and O. F. Soper. 1977. The physiology and mode of action of the dinitroaniline herbicides. Weed Science 25(1):79–87.

Special Topic 23.1

Caldwell, M. M. 1968. Solar ultraviolet radiation as an ecological factor for alpine plants. Ecological Monographs 28:243–268.

――――. 1971. Solar ultraviolet radiation and the growth and development of higher plants. Pages 131–177 in A. C. Giese, ed. Photophysiology, vol. 6. New York: Academic Press, Inc.

Cline, Morris and F. B. Salisbury. 1966. Effects of ultraviolet radiation on the leaves of higher plants. Radiation Botany 6:151–163.

Howland, G. P. 1975. Dark-repair of ultraviolet-induced pyrimidine dimers in the DNA of wild carrot protoplasts. Nature 254:160–161.

Jagger, J. 1967. Introduction to research in ultraviolet photobiology. Englewood Cliffs, N.J.: Prentice-Hall, Inc.

Johnston, H. 1971. Reduction of stratospheric ozone by nitrogen oxide catalysts from supersonic transport exhaust. Science 173:517–522.

Molina, J. M. and F. S. Rowland. 1974. Stratospheric sink for chlorofluoromethanes: Chlorine atomcatalysed destruction of ozone. Nature 249:810–812.

Nachtwey, D. S., M. M. Caldwell, and R. H. Biggs, eds. 1975. Impacts of climatic change on the biosphere: Part I. Ultraviolet radiation effects. Washington, D.C.: U.S. Dept. of Transportation Climatic Impact Assessment Program (DOT-TST-75-55).

Index

Author Index

Abeles, F.B. 264, 266
Acevedo, E. 365
Adams, D.O. 264
Adams, P.A. 256, 281
Addicott, F.T. 267,270
Ahmed, S. 84
Ajami, A. 344
Akazawa, T. I50
Albersheim, P. 154–157, 400–402
Albert, L.S. 91
Allard, H.A. 290, 314–315, 333–334, 338, 346
Altman, P.L. 323
Alvim, R. 268
Anderson, D.B. 28, 289
Anderson, J.L. 255
Anderson, J.M. 132,164
Anderson, L.E. 145
Anderson, W.P. 76, 248, 393
Androsthenes 305
Anstis, P.J. 397
Apelbaum, A. 262, 266
Arias, I. 325, 331
Armstrong, W. 189
Arnon, D. 124
Ashton, F.M. 203, 245, 393
Askenasy, E. 58
Atkins, C.A. 201
Atkins, W.R.G. 77
Attridge, T.H. 396
Audus, L.J. 243, 283
Avery, G.S. 328
Avron, M. 145
Axelrod, B. 184
Aylor, D.A. 37

Baas Becking, L.G.M. 188, 189
Bacon, F. 250
Bajrachara, D. 180
Baker, D.N. 168
Ball, N.G. 277, 285
Bamberger, E.S. 147
Bandurski, R.S. 242
Barker, W.G. 263
Barnett, A. 314
Barton, L.V. 324
Baskin, C.C. 295
Baskin, J.M. 295
Bassham, J.A. 136–137, 145, 148, 152
Batra, M.W. 244
Batty, J.C. 402
Bauer, W.D. 156
Baumgartner, B. 234
Bebee, G.P. 186
Beck, W.A. 238
Becquerel, T. 322
Beevers, H. 176, 183–184. 209
Beevers, L. 260, 269
Beling, I. 315
Bell, R.M. 118

Benson, A.A. 136, 149
Bergersen, F.J. 195
Berlyn, G.P. 228
Bernard, C. 356
Berry, J.A. 164, 171
Beyer, E. 234
Biale, J. 191
Biddulph, O. 95, 106–107
Bidwell, R.G.S. 186
Billings, W.D. 332, 357, 368, 370
Biswas, P.K. 344
Björkman, O. 166, 171
Blaauw, A.H. 250, 278
Black, C.C. 133, 143, 161, 165, 294, 300–301
Blackman, F.F. 352
Bliss, L. 368
Bohnert, H. 127
Bolin, B. 161
Bollard, E.G. 93
Boltzmann, L.E. 16
Bond, G. 194
Bonner, J. 156, 244, 273, 290, 338–339, 351
Bonnett, H. 282
Borthwick, H.A. 290–291, 325
Bose, J.C. 285
Boulter, D. 115
Bowers, W.S. 218
Bowes, G. 166
Bowling, J.F. 39, 101, 108
Bown, A.W. 397
Boyer, J.S. 77, 365
Brain, R.D. 278
Breidenbach, R.W. 167, 209
Bretz, N. 303
Bridges, I.G. 283
Briggs, D.E. 254
Briggs, L. 59
Briggs, L.F. 62
Briggs, W.R. 278–279, 290, 395
Brock, T.D. 366
Brooks, K. 282
Broughton, W.J. 256
Brown C.L. 226
Brown, F. 311
Brown, J.S. I3I
Brown, R. 229, 379
Brownell, Peter F. 82–83
Broyer, T.C. 57
Bruce, V. 310
Bruinsma, J. 280
Bryan, B.A. 194
Buggeln, R.C. 184,188
Bünning, E. 306, 309, 311–313, 315, 341
Burg, S.P. 262, 268, 279
Burke, E.L. 67
Burke, J.J. 366–367
Burr, G.O. 140
Burrows, W.J. 258

Butler, W.L. 278, 292, 301, 310–311

Caldwell, M.M. 65, 397
Calvin, M. 136, 138, 148, 152
Camp, L.B 65
Campbell, D.J. 264
Campbell, G.S. 28
Cammack, R. 115
Canny, M.J.P. 105
Carlson, P. 282
Carpita, N.C. 128, 202, 243, 297
Carr, D.J. 344
Carter, J.L. 256
Caruso, J.L. 177
Caspari, E.W. 318
Cathey, H.M. 249
Chabot, B.F. 171
Chailakhayan, M. 321, 343
Chang, Y. 270
Chardakov, V.S. 25
Chen, J. 188
Ching, T.M. 210
Chisolm, J.J., Jr. 119
Chollet, R. 166
Cholodny, N. 279
Chou, C.H. 215
Chouard, P. 319
Chrispeels, M.J. 203
Chu, D.K. 145
Chung, H. 69
Claes, H. 339
Clark, J.B. 130
Clarkston, D.T. 68
Clausen, Jens 356–357
Cleland, C.F. 344
Cleland, R. 237, 247, 252–253
Cleland, W.W. 145
Clements, F.E. 357
Cline, M. 399
Cline, R.G. 28
Cloud, P. 160
Clowes, F.A.L. 229
Coleman, R.A. 293
Collander, R. 72
Comar, C. 130
Cook, A.O. 315
Cooke, R.J. 396
Coolbaugh, R.C. 251
Coombs, B.G. 226, 234, 264
Cooper, J.P. 161, 172
Corcoran, M.R. 218
Cort, W.W. 47
Cory, S.R. 95, 106–107
Cottam, W.P. 382
Cox, L.M. 369
Crafts, A.S. 57, 95, 97, 245, 393
Crocker, W. 265, 324
Cronquist, A. 146

Crossland, C.J. 82–83
Crozier, A. 249
Cumming, B.F. 332, 338
Currier, H.B. 106
Cutter, E. 229

Daly, L.S. 146
Darland, G.K. 366
Dart, P.J. 194
Darwin, C. 221, 277–278, 286, 305, 315
Davies, D.D. 76
Dayanadan, P. 281
DeCandolla, A.P. 306
Decker, J.P. 164
Delwiche, C.C. 193
DeMairan 305
DeMaria, M.E. 101, 104
Denny, A. 343–344
de Saussure, N. 123
Dessauer, F. 312
Deutch, B. 299
Deutch, B.I. 299
Devaux, H. 189
DeVay, J. 401
Deverall, B.J. 218
de Vries, H. 100
de Wit, C.T. 196, 203
Dickerson, R.E. 115
Digby, J. 281
Dittmer, D.S. 323
Dixon, H. 51, 58
Dixon, M. 110
Dixon, N.E. 84
Döbereiner, J. 196
Dodson, C.H. 215
Doley, D. 171
Dolk, H. E. 250, 281
Donaldson, L.A. 35–36, 321, 372
Douglas, T.J. 251
Downs, R.J. 302–303, 325
Downton, W.J.S. 140
Drake, B.G. 46
Drew, M.C. 65
Drewes, S.E. 258
Duke, S.O. 303
Dunn, G. 175
Dure, L. 234

Edney, E.B. 363
Edwards, G.E. 170
Egneus, H. 135
Ehleringer, J. 362
ElAntably, H.M.M. 268
Emerson, R. 131
Emmerling, A. 203
Epstein, E. 72, 74, 81, 84, 373
Erion, J.L. 264
Eschrich, W. 106
Evans, A. 396
Evans, H.J. 81, 84, 92
Evans, J.O. 245, 393

Evans, L.T. 333
Evans, M.L. 246–247
Evenari, M. 324, 362
Everett, M. 279

Feeny, P.P. 214
Feldman, J. F. 313
Fensom, D.S. 100, 107
Ferrier, J.M. 108
Filner, P. 198
Fireman, M. 363
Firn, R.D. 281
Fischer, E. 118
Fischer, J. 338
Fischer, R.A. 39
Fiscus, E. 77
Fisher, D. 104–105
Fitting, H. 286
Folkes, B.F. 121
Forrester, M.L. 166
Fox, H. 319
Fox, J.E. 264
Franck, J. 161
Frederick, S.E. 142
Friend, D.J.C. 330
Führ, I. 117
Fujino, M. 39

Gaastra, P. 169
Galston, A.W. 243, 277, 284–286, 396
Gander, J.E. 159
Gane, R. 264
Gardner, G. 246
Gardner, W. 63, 389
Garner, W.W. 290, 314–315, 333–334, 338, 346
Gassner, J.G. 318
Gates, D.M. 47
Gates, F.C. 47
Gauch, H.G. 88, 92
Gaumann, E. 400
Gaynard, T.J. I89
Geiger, D.R. 98
Geis, I. 115
Genest, K. 222
Gerdemann, J.W. 66
Gibbs, J.W. 10–11, 16
Gibbs, M. 184
Gibby, D.D. 344
Gibor, A. 160
Glasziou, K.T. 256
Gleason, H.A. 47
Glier, J.H. 177
Goldwater, L.J. 119
Goodchild, D.J. 195
Goodman, D. 76
Goodwin, T.W. 214, 292
Gordon, M.E. 261
Gorski, T. 295
Gorter, C.J. 250
Gould, F.W. 280
Gove, J.P. 242
Green, C.E. 282

Green, F.B. 218
Green, P.B. 29
Greenwood, D.J. 76
Gregory, F.G. 318, 330
Grgurevic, E. 91
Grisebach, H. 303
Grun, P. 70
Grunwald, C. 214
Guilfoyle, T.J. 246
Gundry, C.S. 92
Gutierrez, M. 142

Haberlandt, G. 258
Hackett, W.P. 277
Hacskaylo, E. 66
Hagan, R. 63
Hagar, A. 252
Hague, D. 282
Hahlbrock, K. 303
Haissig, B.E. 244
Halaban, R. 314
Halberg, F. 306
Hales, S. 8, l6, 32, 93, 123
Hall, F.K. 219
Hall, R.H. 258
Hamilton, R. 25l
Hamner, K.C. 290, 338, 342, 344
Hanchey–Bauer, P.J. 125, 360, 399
Hanks, J. 32
Hanson, A.D. 265
Hanson, H.C. 164
Harborne, J.B. 220–221
Hardy, R.W.F. 193, 196
Harker, J. 314
Harpstead, D.D. 121
Harris, J.A. 28
Harrison, M.A. 328
Hartmann, K. 300
Hartney, V.J. 146
Hartsema, A.M. 329
Hartt, C.E. 140
Hartung, W. 281
Harvey, B.M.R. 262
Harvey, W. 93
Hasson, E.P. 249
Hastings, J.W. 307
Hatch, A.H. 328
Hatch, M.D. 140
Hatsopoulos, G.N. 404
Haupt, W. 303, 396
Havelka, U.V. 193, 196
Hayashi, T. 247
Hayward, H.E. 363
Heber, U. 138, 147
Heftmann, E. 214
Heide, O.M. 325–326
Hellebust, J.E. 365, 373
Hellmers, H. 268, 325
Helson, V.A. 330
Hemberg, T. 267
Henckel, P.A. 365–366
Hendricks, S.B. 156, 291–292, 303
Hendrickson, A.J. 62
Henfrey, A. 338
Henkel, J. 173
Henry, M.F. 184
Hepler, P.K. 104, 236
Herrmann, R. 127
Hesketh, J.D. 169
Heslop–Harrison, J. 287
Hess, C. 188
Hess, D. 36, 71, 237
Hewitt, E.J. 88, 92, 197
Hiesey, W. 356–357
Higgins, P.D. 360
Highkin, H. 356
Higinbotham, N. 76

Hill, G. 161
Hill, R. 124
Hillman, J.R. 268
Hillman, W.S. 343
Hinchman, R.R. 280
Hind, G. 392
Hitchcock, C. 206
Ho, D.T. 254, 267
Ho, L.C. 106
Hoagland, D.R. 81
Hocking, T.J. 268
Hodson, H.K. 344
Höfler, K. 22
Holger, B. 77
Holmes, M.G. 295
Horgan, R. 259
Houwink, A.L. 285
Howland, Gary 398
Hoyle, M.C. 242
Hsiao, T.C. 39, 75, 238, 363
Huff, A.K. 262—263
Hughes, D.W. 222
Hulme, A.C. 191, 234
Humble, G.D. 38–39
Hungerford, H.B. 47

Ikuma, H. 256
Incoll, L.D. 172
Imamura, S. 39
Ingenhousz, J. 123
Inman, R.E. 186
Inoue, Y. 303

Jackson, M.B. 264
Jacobs, W.P. 270
Jaffe, M.J. 277, 286–288
Jagendorf, A. 392
James, W.O. 189
Jeans, J. 11
Jensen, W.A. 175, 180, 230–231, 239, 274–275, 291
Johnson, C.B. 396
Johnson, T. 188
Johnson, V.A. 121
Joley, J. 58
Joliot, P. 132
Jones, D.D. 158
Jones, R.L. 251, 254, 256
Juniper, B.E. 281, 283
Jurinak, J.J. 388
Junttila, O. 325

Kagawa, T. 142
Kalmus, H. 311
Kamen, M.D. 152
Kang, B.G. 279
Karakshian, Marlene 314
Karr, A.L. 159
Kato, Y. 345
Kaufman, P.B. 256
Kawase, M. 260, 264
Keck, D. 356–357
Keegstra, K. 156
Keeler, R.F. 222
Keenan, J.H. 404
Kelner, A. 398
Kende, H. 234, 246, 265
Kendrick, R.E. 396
Ketring, D.L. 324
Key, J.L. 246
Khan, A.A. 271
Kidd, F. 190
King, R.W. 344
Kirk, J.T.O. 127
Klebs, G. 333, 338
Klein, R.M. 244
Klein, W.H. 299
Kleinhoonte, A. 306, 311, 314

Klima, J.R. 184, 188
Kluyver, A.J. 188
Knop, W. 80, 86
Kok, B. 132, 135
Kolattukudy, P.E. 213
Koller, D. 295, 324
Kormandy, E.J. 245
Kortschak, H. P. 140
Koshland, D.E. 118
Kosuge, T. 145
Kowalik, R.V. 127
Kozlowski, T.T. 66, 226, 270, 302, 326
Kramer, G. 315
Kramer, P.J. 9, 16, 62, 68–69, 77, 226–227, 389
Kraus, E.J. 338
Kraybill, H.R. 338
Krebs, H.A. 180
Kreith, F. 47
Krenzer, E.G., Jr. 140
Kriedemann, P.E. 172, 236, 269, 322, 326, 364
Krikorian, A.D. 37
Krotkov, G. 166
Ku, S. 170
Kunishi, A.T. 266
Kuo, C.C. 247
Kurosawa, E. 247
Kursanov, A.L. 98

Laetsch, W.M. 140, 262
Laloraya, M.M. 261
Lang, A. 296, 320, 335–330, 339, 343, 354, 356
Larcher, W. 161–162
Larsen, A. 254
Lavoisier, A. 123
Lawrence, D. 47
Lawrence, E.O. 148, 152–153
Lawson, V.R. 298
Lay, C.L. 121
Lea, P.J. 200
Leavitt, J. 361, 365
Lee, J.A. 373
LaHaye, P.A. 373
Leopold, A.C. 172, 236, 260, 269, 322, 326
Letham, D.S. 259, 261
Lewin, J. 83
Lewis, A.R. 171
Lhoste, J. M. 293
Lieberman, M. 264, 266
Lincoln, R.G. 344
Lindsay, W.L. 87
Lindstedt, K.J. 218
Lipmann, F. 178
Lister, G.R. 130
Liu, P.B.W. 255
Livingston, B.E. 77
Lockhart, J.A. 27, 354
Lodhi, M.A.K. 218
Lohr, J.B. 128
Loomis, W.E. 161, 172, 235
Lörcher, L. 311
Lorenz, O. 95, 97
Löve, A. 348
Löve, D. 348
Loveys, B.R. 364
Loy, J.B. 255
Luckwill, L.C. 328
Luke, H.H. 401
Lysenko, T.D. 318
Lyons, J.M. 368

MacDougal, D.T. 362
McClure, J.W. 221
McComb, A.J. 256

McCree, K.J. 130
McGlasson, W.B. 264
McKee, H.S. 203
McMillan, C. 332
McNeil, D.L. 238
Malone, C. 180
Malphighi, M. 93
Mancinelli, A.L. 303
Mann, J.D. 257
Marcus, A. 198
Markert, C. 117
Marks, G.C. 66
Marr, J. 47
Marré, E. 252–253
Marshak, R.W. 318
Maskell, E.J. 63
Maxwell, J.C. 9, 16
Mayer, A.M. 323–324
Meeuse, B.J.D. 184, 188–189
Meijer, G. 397
Meins, F., Jr. 275
Melchers, G. 320–321, 330
Menaker, M. 310
Meyer, B.S. 16, 28, 289
Mexal, J.G. 67
Miflin, B.J. 200
Milborrow, B.V. 267–268
Miller, C. 258
Milthorpe, F. 63
Minchin, F.R. 196
Mitchell, J.W. 248
Mitchell, P. 392
Mitrakos, K. 293, 299
Mohr, H. 300, 303, 354–355, 395, 396
Molisch, H. 327
Mollenhauer, H.H. 158
Mooney, H. A. 332, 368
Morgan, D.C. 301
Morre, D.J. 158
Moser, I. 313
Moss, D.N. 169
Mosse, B. 194
Mukherjee, I. 344
Muller, C.H. 215
Münch, E. 56–57, 99–100
Muñoz, V. 278, 301, 310–311
Murray, L. 314
Musgrave, R.B. 168

Nabors, M.W. 276, 282, 296
Narain, A. 261
Nason, A. 81
Neales, T.F. 146, 176
Neljubow, D.N. 264
Nelson, C.D. 166
Newcomb, E.H. 142
Nichols, B.W. 206
Nicolson, G.L. 71
Nissen, P. 72
Nitsch, J.P. 235
Noack, K. 312
Nobel, P.S. 9, 16
Noggle, J.C. 75
Noodén, L.D. 246
Norreel, B. 276
Northcote, D.H. 159
Nurnsten, H.E. 234
Nutman, P.S. 194
Nyns, E.J. 184

Oele, J. 360
Oertli, J.J. 91
Ogren, W.L. 166
Olason, D. 188
Olien, C.R. 368
Olmstead, C. 332

Olmstead, G. 156
O'Malley, B.W. 246
Öpik, H. 180
Orionis, E. 388
Orr, M.L. 121
Osborne, D.J. 270
Overton, C.E. 72

Paleg, L.G. 251
Palevitz, B.A. 104, 236
Palmer, C.E. 263
Palmer, J.M. 182
Palzkill, D.A. 360
Papenfuss, H.D. 342
Park, R.B. 239
Parker, J. 366
Parker, M.W. 292
Parlange, J.Y. 37
Pasteur, L. 189
Pate, J.S. 196–197
Patterson, A.A. 146
Peel, A.J. 106
Penman, H.L. 63
Penning de Vries, F.W. 172
Penny, M.B. 39
Percival, F.W. 242
Perry, T.O. 268, 325
Peterson, C.A. 106, 282
Pfeffer, W.F.P. 16, 26, 305–306, 311, 315
Pharis, R.P. 247, 345
Phillips, I.D.J. 245, 251, 281
Phillips, R.L. 282
Piatelli, M. 222
Pickard, B.G. 285, 287–288
Pilet, P.E. 283
Pisek, A. 28
Pittendrigh, C.S. 310–311, 315
Polanyi, M. 152
Polijakoff–Mayber, A. 323–324
Possingham, J.V. 127
Pratt, L.H. 292–293
Preiss, J. 145
Price, L. 299
Priestly, J. 123
Pryce, R.J. 345
Purvis, O.N. 318, 320, 330

Rains, D.W. 88, 92
Radmer, R. 124, 135
Rains, D.W. 88, 92
Raschke, K. 38–41
Raven, J.A. 76
Ray, P.M. 23, 228, 238, 246–247
Rayle, D.L. 247, 252
Reeve, D.R. 249
Reid, C.P.P. 67
Reid, D.M. 262
Reid, E.E. 180
Reimann, B.E.F. 83
Reisenauer, H.M. 81
Reith, W.S. 229
Renner, O.R. 16, 58–59, 77
Reuben, S. 124, 148, 152
Reuveny, Z. 205
Ricca, U. 285
Rice, E. 218
Richards, F.J. 225, 232
Richardson, A. 369
Richmond, D.V. 205
Robards, A.W. 69
Robb, W. 170
Robbins, W.W. 232
Roberts, L.W. 275
Robinson, E. 229
Roeckl, B. 98

Roeske, C.N. 214
Rogler, C.E. 277
Rollin, P. 294
Ronco, F. 165
Rose, R.J. 127
Ross, C.W. 225, 243, 262–263, 282
Rowan, W. 346
Rüdiger, W. 293
Rudinsky, J.A. 215
Ruhland, W. 226, 270
Russel, M.B. 284
Russell, R.S. 77

Sabnis, D.D. 285
Sachs, J. 80, 86, 123, 225, 229, 240, 244, 271, 300, 305
Sachs, R.M. 230
Salisbury, F.B. 48, 100, 103, 175, 180, 230–231, 265, 275, 288, 291, 304, 311, 330, 337, 339–344, 352–353, 360, 369–370, 399
Samish, R.M. 322
Sanders, F.E. 66
Sangwan, R.S. 276
Satoh, K. 131
Satter, R.L. 284–285, 314, 396
Saunders, P.F. 328
Scarisbrick, R. 124
Schäfer, E. 296
Schleiden, M. 2
Schmidt, A. 205
Schmidt, J.M. 127
Schnarrenberger, C. 186
Schneider, A. 241–242
Schofield, R.K. 63
Scholander, P.F. 51, 59–61, 371
Schopfer, F. 355
Schrader, W.T. 246
Schrempf, M. 284
Schwabe, W.W. 232, 333
Schwann, T. 2
Scott, K. 244
Scurfield, G. 284

Seeley, E.J. 328
Seeley, S.D. 328
Senebier, J. 123
Shantz, H.L. 62, 361–362
Sharon, N. 114
Shen-Miller, J. 280
Shibata, K. 303
Shininger, T.L. 276
Shmidtling, R.C. 85
Shropshire, W. 293, 296
Shutt, D.A. 221
Shuttleworth, J.E. 300–301
Sibaoka, T. 285
Siegelman, H.W. 292, 303
Simonis, W. 135
Sinclair, T.R. 196, 203
Singer, S.J. 71–72
Sinnott, E.W. 225, 274
Skaggs, D.P. 248
Skene, K.G.M. 260, 262
Skoog, F. 246, 258–260
Skou, J.C. 74
Slack, C.R. 140
Slatyer, R.O. 9, 16, 62–63
Smith, B. 188
Smith, F.A. 76
Smith, H. 221, 294, 301, 303, 396
Smith, L.H. 169
Smith, R.L. 196
Smith, T.A. 88, 92
Sorger, G.J. 92
Spanner, D.C. 100
Spaulding, E.S. 362
Spiker, S. 115
Spomer, G.G. 48, 360, 369–370
Sprague, H.B. 88
Spratt, N.T., Jr. 274
Stahl, C.A. 262
Stalin, J. 318
Stanton, F.W. 29
Stemler, A. 124
Stern, K. 306, 309, 311–312, 315
Stetler, D.A. 262
Steward, F.C. 258, 276
Stewart, G.R. 373

Stewart, R.N. 266
Stocking, C.R. 232
Stoppel, R. 305–306, 311–312, 314
Stout, P.R. 82
Strasburger, E.A. 50
Stuart, D.A. 256
Stuart, N.W. 249
Stumpf, P.K. 206, 208–209
Sucoff, E. 60
Sudzuke, F. 362
Suelter, C.H. 92
Sunderland, N. 276
Sussex, I. 282
Sutcliffe, J.F. 9
Swain, T. 220
Sweeney, B. 307, 314
Sylvester-Bradley, R. 121

Takegumi, T. 264
Talmadge, K. 156
Tanada, T. 396
Tantraporn, W. 47
Taylor, S. 16, 63
Thaine, R. 101, 104, 108
Thayer, S.S. 188
Thimann, K.V. 156, 240–241, 243–244, 314
Thomas, M. 161
Thompson, H.C. 319
Thomson, B. 298
Tibbitts, T.W. 360
Tilney-Bassett, R.A.E. 127
Tinbergen, N. 188
Ting, I.P. 117, 146
Tolbert, N.E. 167
Toole, E.H. 291
Toole, V.K. 291, 295
Toriyama, H. 284
Torrey, J.G. 244, 275
Tournois, J. 333, 338
Tranquillini, W. 368, 370
Transeau, E.N. 77
Traube, M. 72
Travis, R.L. 396
Trebst, A. 132
Trewavas, A. 117
Tromp, J. 360

Troughton, J. 35–36, 231, 372
Turesson, G. 356
Turner, D.H. 189
Turner, J.F. 189
Tyndall, J. 379
Tyree, M.T. 108

Urbach, W. 135

Vaadia, Y. 267
van den Driessche, R. 85, 314
van den Honert, T.H. 63
Vanderhoef, L.N. 247, 262, 263
van Iterson, G., Jr. 188
van Niel, C.B. 124
van Overbeek, J. 156, 188, 258
van Sambeek, J.W. 287–288
van Staden, J. 258
van't Hoff, J.H. 16, 21
Varner, J.E. 137, 254, 267
Vartapetian, B.B. 179
Vegis, A. 326
Veihmeyer, F.J. 62
Vest, D. 323–324
Vigil, E. 209
Villiers, T.A. 324–325
Vince–Prue, D. 302, 333, 337, 343, 346
von Bülow, J.F.W. 196
von Frisch, K. 315
von Liebig, J. 352

Wain, R.L. 283, 397
Walker, D.R. 328
Walton, D.C. 267
Warburg, O. 164
Wardell, W. L. 344
Wareing, P.F. 267–268, 325
Watt, B.K. 121
Weaver, R.J. 255
Webb, E.C. 110
Webb, M. 311, 314
Weintraub, R.L. 298

Weir, T.E. 126, 232
Weiser, C.J. 367–368
Welch, P.S. 47
Wellensiek, S.J. 320, 345
Wengenmayer, H. 220
Went, F.W. 156, 188, 215, 226, 240, 243–245, 250–252, 278, 322, 330, 362
West, C.A. 249
Whaley, W.G. 226
Wheeler, H. 401
White, W.C. 368
Whittaker, R.H. 160, 214, 357
Wiebe, H.H. 4, 16, 41, 68, 95, 170, 389
Wightman, F. 241–242
Wilbur, K. 77
Wildner, G.F. 173
Wilkins, H. 283
Wilkins, M.B. 280, 283
Williams, P.H. 360
Wilson, A. 153
Wilson, C.E. 235
Wilson, C.M. 91, 203
Wilson, L. 205
Withrow, R.B. 292
Wittwer, S.H. 170
Wolf, F.T. 145
Wong, E. 221
Wright, S.T.C. 269

Yabuta, T. 247
Yang, S.F. 264
Yates, D.J. 171
Yatsu, L.Y. 207
Yoshida, K. 264

Zeevaart, J.A.D. 333, 344–345
Zelitch, I. 173
Zenk, M. 252
Ziegenspeck, H. 37
Zimmerman, M.H. 96, 226
Zimmerman, N.H. 310
Zimmermann, B.K. 279
Zscheile, F. 130
Zschoche, W.C. 117

Index of Species and Subjects

Page numbers shown in **bold type** indicate that the word appears in boldface type on that page (is defined in context); page numbers shown in *italics* indicate that the term is illustrated on that page, appears in a figure or table, or its molecular structure is shown. To save space, words have sometimes been omitted when relatives with closely similar spelling appear on the same page (e.g., 2-phosphoglyceric acid and 3-phosphoglyceric acid; phosphoglyceric acid kinase and phosphoglycerokinase). Common and scientific names are not repeated when they are identical; they are cross referenced for many familiar species (common or scientific name in parentheses).

ABA *see* abscisic acid
abrasion 98
abscisic acid 40–42, 92, 213, **240**, 243, 255, 258, 266–**268**–271, 301, 324–330, 346–*347*–348, *364*; flowering and 345; geotropism and 277, 283; seed dormancy and 297
abscission 233, 246, 264, 266–267, 270–271, 273, 348; layer *268*, 270
absolute 334, *336*
absorbance **386**
absorption spectrum 130, 340; leaf, 43–44; photosynthetic 130, 278; phytochrome 292–293, 299
absorptivity coefficient 130, **386**, 388 of phytochrome, 294
Acacia 277

accessory cells 35, 39
acclimated 365–367, 374
accumulation 73, 76
Acer: rubrum 28, 325; *pseudo-platanus* 155, 267–268; *saccharum 164*, 322–*323*
acetyl coenzyme A 180–181, 187, 208–211, 214, 220, 249
acetic 377
acetylene 196
acetylsalicylic acid 344
achenes 235
Achlya bisexualis 214
acid growth theory 252–253
acidic soils 67
acid rain 90
aconitase 181

acorns 177
acropetal direction **229**
acrophase 309–310
actidione **246**
actin 104
actinic light **397**–398
actinomycetes 194
actinomycin D 246, 313, 395
action potentials 285
action spectrum 130–131, 163, 278, 300, 310–311, 342; for anthocyanin synthesis 303
active: energy requiring process 57; site 118, **382**; theories of phloem transport 99–101, 107; uptake **76**
activity 13–14, 21, 72-73, 308, 315; cycles **307**; of water 365

additive: behavior *355;* factors *354*
adenine 258–259
adenosine 259, 264
adenosine diphosphate *see* ADP
adenosine diphosphoglucose *(see also* ADPG) 147
adenosine-5-phosphosulfate 204–205
adenosine triphosphate *see* ATP
adequate zone **84**
adhesion 50–51, 59, **377**
adiabatic lapse rate *369, 369*
ADP 124, 132, 135, 138–139, **147**–151, 154, 177–179, 181–182, 199–200, 208, 217, 392; availability 400; magnesium chelate 114; role of nitrogen and phosphorus 92

ADPG **147**, 151, 154, 176
ADPG-starch transglucosylase **151**
adsorption 28, 388
adult phase **276**–277
advance the clock 310
adventitious: buds 65; roots, 244, *245*, 253, 258, 202, 266, 276–278, *297–298*
aeration 74
aerosol spray cans 399
afterripening **322**
Agave: americana 145–146, 162, 227, 362, *horrida 146*
Ageratum houstonianum 218
agricultural chemicals 393–395
agriculture 5, 350, 363–365
agronomy 5, **350**
Agropyron: repens 141, 393; *smithii 336*
Agrostis: alba 141; *tenuis* 374
air 58; pollution 215; travel 316
alanine 140–144
Albizzia julibrissin 284–*285*, 396
albuminous cells **102**, 104
alcohol 357, *377*; dehydrogenase 92, 178
aldehydes 324
alder 194, 197, 268
aldolase 139, 178
aleurone: grains **201**; layer **254**, 273
alfalfa *55*, 161–164, 168, 196–197, 218, 269, *323*, 362, 395
algae 1, 73–74, 248, 259, 267, 303, 314, 362, 373; no anthocyanins in 220; phytochrome in 293
alimentary canal 390
alkaloids 187, **222**, 267, 324, 357, 397
allelochemics 214, 216, 221, **357**, 363
allelopathics 215, **324**, **357**
Allenrolfea 371; *occidentalis 28*
allium 348; *cepa 336*–337
allosteric: effectors 72, 116, **122**, 246; enzymes, 121–**122**, 355; inhibition 145; sites **122**
Alnus 194
Alocasia macrorrhiza 163
Aloe: bulbilifera 337; obscura 146
alpha-amylase 92, 175–176, **254**, 257
alpha-glycerophosphate 208–209
alpha-ketoglutaric acid **128**, 181, 187, 199–200; dehydrogenase 181
alpha-oxidation **209**
alpha particles 383
alpha-pinene 215–*215*
alpine: plants 46, 48, 398–399; sorrel 332, 357; tundras 14, 304, **368**–371
aluminum 67–68, 79–80, 83, 87, 90, 373
Amaranthus: albus 141; caudatus 294; retroflexus 141; tricolor 83
ambiphotoperiodic plants *337*
American elm *323*
amides 95–96, 112–113, 203–204, 233–234, 254, 270; translocatable compounds 196–197
amines 184, 188; oxidases 92
amino acids 2, **4**, 92, 96, 99, *112*, 203–204, 233–234, 237, 254–255, 261, 270, 396; aromatic 216–217; general formula 111; photosynthetic products 136; sequences 372; soils 193
amino groups *381*
ammonia 65, 73, 366–377, 380; inhibition of electron transport 199; metabolism of 199–201; nitrogen cycle 192; salt 393; sulfate 382
ammonification 192–**193**
Amo-1618 236, 249, **251**, 253
amoeba 381

AMP 134, 143–144, 204–205, 209
amphoteric colloids **381**, *381*
amplification 246, **355**
amplitude 309, *309*, 311, 313
amylases 151, 256, 273
amylopectin 150–*151*, 175–176
amyloplasts **3**, 150; geotropism and 280–281
amylose 150–*151*, 175–176
amylovorin **400**
anabasine 222
anabolism 174
anaerobic conditions 74–75, 321
Anagallis arvensis 336, 338
analogies 8
analysis 358; of variance 354
analytical ultracentrifuge **383**
anatomy **4**, 53–56, 59
ancymidol 249, **251**
androgens 348
Andropogon scoparius 141
Anethum graveolens 336
angiosperms 55, 66, 102; sieve elements in 98
angstroms *3*
animals 5, 18, 315, 366, 390
anions 67, 68, 75–76, 92, 380–381
annual cycles 304, 327
annual: plants **227**, 362; rhythm 308; rings 226
anoxia 107
antheridiol 214–*214*
anthesin 321
anthesis **233**
anthocyanidins 220–*220*–221
anthocyanins 187, 216, *220*–221, 234, 300–303, 325, 398, 401; in nutrient deficiency 88–89; synthesis 220, 302–303
anthoxanthins 401
Anthriscus cerefolium 336
antiauxins 282
antibiotics **357**
anticlinal divisions **232**, 270
anticodons 258
antigens **401**
antimetabolites 345
antitransparents 41
aphids 95–*96*, *106*–108, 344
apical cells 346
apical dominance **245**, 261, 266
apical meristem **55**, 229–231, 258, 275, 277
apoplast 56–57, 69, 98, *100*, 102, 106, 108
apparent free space **56**, *56*–57
apples 91, 96, 172, 189–190, 226, 235–236, 244, 264, 302–303, *324*–327, 338, 348, 360; leaf scab 217
apricot 96, 226, 235, 327
Arabidopsis thaliana 336
arabinoglactan **155**
arabinose 154–*155*, 368
Arachis hypogea 141
arborvitae 368
arctic 368; hare 316; lupine *323*; tundra 369
Aretemesia tridentata 28, 363
arginine *112*–116
argon 390
aquatics 390
arid regions 62
Aristida: hamulosa 141; *purpurea* 141
Aristolochia brasiliensis 105
Arizona cyprus 345
artifacts **102**
Arum: lilies 184, 188–189; *maculatum* 188
ascorbic acid 300; oxidase 92

ash 267
asparagine 112–*112*–115, 187, *201*, 234; metabolism 201–204; synthetase 201; translocation 196
aspartic acid 112–*112*–116, 122, 140–143; 186, *201*–204, 222; transcarbamylase 122
aspen 65
aspirin 344
assimilates **93**, 105, 343–344
assimilation **364**
Asteraceae(*see also* Compositae) 228
Astragalus massiliensis 323
asymmetric carbon atom 384
atmosphere 21, 47, 49, 57, 63, 388; ions in 312; temperature gradient *369*
atoms 404
ATP **3**, 76, 98–101, 104, 107, 124–125, 132–135, 138–140, 143–151, 154, 165–168, 174–179, 181–185, 188–190, 194–195, 199–201, 204–205, 207–209, 211, 217, 224, 242, 256, 368, 400; magnesium chelate of 114; role of nitrogen and phosphorus in 92; phototaxis and 303; synthesis 392–393
ATPase 132–**133**, 247, **392**, *392*
atrazine 133, 198, **394**
atrichoblast 275
Atriplex: confertifolia 28, 323; hastata 141, 163–164, 168; *patula 163; rosea* 141; *semibaccata* 141; *spongiosa 371*-372; *vesicaria 82*
Atropa belladonna 222
atropine **222**
Australian saltbrush 141
autecology **351**
autoradiography **94**–*95*, 98, *136*–*137*, 149, 229
autumn colors *337*
auxins (*see also* IAA) 187, **229**–230, 235–239, **240**–*241*–247, 258–266, 269–270, 278–280, **284**, 325, 328, *347*–348, **394**; bioassays 242–243; bound **242**; discovery 240; flowering and 345; geotropism and 282, 284, 297; herbicides 245–246; mechanisms of action 247, 252–253; metabolism of 241; protectors **242**, transport 242; zinc and synthesis of 91, 92
available water **389**
avalanche lily 1
Avena (oat): *fatua* 254, *323; sativa* 141, 247, *336*, 397
Avens *337*
avocado 226
Avogadro's number 384
avoid the drought 362
axillary buds 228, 245
azaleas 374
6-azauracil 246
azide 183

bacteria 4, 92, 193–194, 205, 214, 248, 250, 258–259, 268, 322, 344, 357, 360, 366–367, 371, 397, 399
bacteroids 194–*195*–196
Baeria charysostoma 330, 352
Bahiagrass 141
bakanae disease 247
balls bouncing 404–405
bamboos 140, 227, 233, 338
Bambusa 227
banana 150, 177, 226, 235, 264, 366–367
bar 21, 49, 51
bark 57, 93

barley *68*–*69*, *74*, 76, 85, 121, 141, 170, 197, 228, *254*–257, 269, 280, 291, 293, 297–298, *335*–*336*–*339*, 396
barnyardgrass 141
barometer *49*
barometric height *49*
basal meristem 233
basipetal direction 229
beans (Phaseolus) 85, 107, 120–121, 141, 174, 197, 233, *248*, **260**, 277, 286, *288*, *290*, 299, 301, 305, 311–312, 330, 396; phaseollin in 218
beech 242, 268–269, 276, 326
beechwood 219
Beer-Lambert law **386**
beet 91, 141, 200, 227, 319, *336*, 339, 365, 371; nitrate accumulation in 198
Begonia evansiana 294
Benedict's solution 96, 147
bentgrass 374
benzyladenine 259–*259*–262, 345
benzyoxazinone **394**
Bermudagrass 141, 280
Bertholletia excelsa 208
beta amylase 175–176, 254
beta-carotene 126–127, **130**, 214, 278
betacyanins 221–222
betafructofuranosidases 177–178
beta-inhibitor 267–268
betalain pigments, betanidin, betanin, 221–222
Beta: maritima 339; nitrate accumulation in 198; *pubescens* 268, 325; *sacchariferia 336; vulgaris* 141, 227, *336*
beta-oxidation **209**, 211
beta particles 95, 383
beta-pinene 215–*215*
beta-sitosterol *214*, 251
betaxanthins 221–222
Betelgeuse 388
bezoic acid herbicides 394
bicarbonate ion (HCO$_3$⁻) 67, 76
bidirectional movement **99**, 100–101, 105–*106*, 108
biennials 227, 319
big sage 28
big trifoil *323*
bilayer 71
bioassays 242, 258, 344; auxin 243; cytokinin 260–261; gibberellin 249, 251, 254
biochemical responses 347–348
biochemistry 1
biological clock 304–*315*–316
biological time measurement 340
biology 1, 307
bioluminescence 307, *309*
biophysics 42
biorhythms 308
biosphere 47
biotin 92, **208**
biotron **358**
birch 268, 325–326
bird 307, 315, 323; migration 346
bird rape *336*
bisulfite 90
black: body *386*–387–388; currant 268, 345; henbane *336*; locust 98; mustard 141; nightshade 222; shank disease of tobacco 401; velvet 386; walnut 227
black-eyed Susan *336*
Blackman curves **352**–354
bladder cell *372*
bleeding from phloem 104
blood circulation 93; clot 381; pressure 16

blue-absorbing pigment 311, 396
bluegrama grass 141, 280
bluegrass 28, 141, 289
blue-green algae 4, 92, 194–214, 292, 366, 371
blue light 311, 395, 397
blue panicum 141
blue spruce 28
bluestem wheatgrass 336
Blühreife 338
Boehea 323
Boer lovegrass 141
bog myrtle 194
boiling point 26, 377
boldface type 2
bolting 227, 319, 319
Boltzmann's constant 403
Borago 198
bordered pits 54, 54
boric acid 91–92
boron 36, 84, 99; deficiency symptoms 89, 91; metabolic roles 92
botany 4
bottle brush 228
boundary 403; layer 44–45–46, 169
bound: auxins 242; 28, 366–367
Bouteloua: curtipendula 141, 332; gracilis 141
Boyle's law 29
branching 347–348
branch roots 65, 69, 228–230, 244, 261
Brassica: campestris 336; nigra 141
Brazil nut 203
brine shrimp 308
bristlecone pine 227
broad bean (Vicia) 35, 38–39, 197, 203, 228, 283
broccoli 91
bromacil 133
bromegrass 280
Bromeliaceae 145
bromeliads 264–266, 345
bromide ion (Br⁻) 74
bromocil 395
Bromus inermis 336
brown algae 29, 259
Brownian movement 379, 381
brush pile theory 381
brussels sprouts 261, 318–319
Bryophyllum 38, 347–348; tubiflorum 83
bryophytes 245, 390
bubble 57, 59–60
Buchloë dactyloides 141
bud 319; axillary 328; dormancy 253, 267–268, 273, 302, 325–328, 347; of perennials 318; scales 326
buffalograss 141, 280
buffers 92
bulbs 150, 177, 211, 228, 271, 328–329, 347–348
bulk flow 13, 18, 23, 100
bulliform cells 289
bundle sheath 34, 98, 140–144, 165–166
buttercup 55
butter emulsion 381
button clover 323

cabbage 205, 248, 253, 303, 319, 360
Cactaceae 145
cacti 38, 362, 377
caffeic acid 216–217–220, 242
caffeine 222–222
Calcmus 60
calcareous soils 87, 373
calcium 68, 71, 74, 99, 360, 373; compounds 92; deficiency 41,

89–90; metabolic roles 92; role in starch degradation 175; soap 380–381
Callistemon 228
Callistephus chinensis 346
callose 103–104
callus 244, 260–263, 276, 282, 394
Calonectria 344
calorie 377–377, 385
calorigen 189
Caltha leptosepala 41
Calvin cycle 135, 138–139–147, 150, 152, 167, 179, 184, 186, 272
cambium 55–55, 68, 103–104, 275
Campanula medium 337
campesterol 214–214
camphene 215–215
camphor 215
CAM plants (see also crassulacean acid metabolism) 63, 145–147, 161–165, 171
Canada thistle 228, 245, 323, 393
Cananvalia ensiformis 309
canarygrass 141, 278
cancer, skin 398–399
Cannabis 347; sativa 336, 338
canterbury bells 337
capillary: network 390; tube 381
carbamate herbicides 394–395
carbamyl phosphate 122
carbohydrates 96; surfaces of 402
carbon 32; radioactive 94–95, 148; black 386; cycle 161
carbon dioxide (CO₂) 9, 32, 35, 40–41, 63, 67, 343, 348, 357, 390–391; compensation point 165, 169; free air 38–40; radioactive 106, 377–378; stomate response to 38
carbonate 390
carbonic acid (H₂CO₃) 67–68, 247, 380
carbonic anhydrase 92, 124
carbon monoxide 183, 207
carboxydismutase (see also ribulose diphosphate carboxylase) 137
carboxyl groups 75, 381
cardinal: points 352, 352, temperatures 317
carnation 211, 212, 213, 336
carotenes 126–127, 214
carotenoids 126, 129–133, 165, 187, 213–214, 220, 267, 300, 397
carriers 72
carrot 141, 189, 200, 253, 276, 319, 321
carryover effects 356–357
caryophyllales 221
Casaurina equisetifolia 194
Casparian strips 55, 55–57, 70
castor beans 207–208
castor oil 207
catabolism 187
catalase 92, 167, 168
catalysis 6
Catasecum sacchatum 277
cation 67–68, 75–76, 92, 381; exchange 67–68, 75, 380
Cattleya trianae 336
cauliflower 91
cavitation 49, 57–60
CCC 236, 249, 251
Ceanothus 194
celery 91, 255, 319
celestial navigation 315–316
cell 2, 6; division 5, 256, 261–262, 273, 275, 307, 364, 394; expansion and growth 22, 42, 363–364; osmotic system 18; plate 236–237, 273; pressure 29; respiration 3; suspension cultures 282; theory 2; turgid

18, 29; typical plant 2; wall (see below)
cellophane 379, 382
cellulase 157, 267, 270, 400
cellulose 4, 15, 31, 50, 54, 71, 79, 139, 150, 154–155, 157, 159, 161, 194, 218, 237–238, 284, 377, 391; microfibrils 50, 70–71, 155–157–159, 238–239, 262, 266, 272–273, 380, 400
cell wall 2, 4, 8, 21–22, 54–60, 69–71, 92, 98, 101–102, 125, 128, 158, 195, 252, 254, 257, 272, 282, 380, 390–391, 401–402; apparatus 103; constituents 75; in osmosis 18; lignins in 218; modifications 400; polysaccharides in 151, 154–159; synthesis 363
Cenchrus pauciflorus 141
Cenococcum graniforme 67
Centaurea cyanus 221
centimeters 3
centipoises 378
centrifugation 382
Centrospermae 294
century plant 162, 227, 233, 362
cephalins 213
cereal 318, 321, 368, 392, 394
Cestrum nocturnum 337
change 304
chaparral 324, 357
Chara 104
Chardakov's method 24–25
charges 67, 76, 92
check value 58–59
chelates 85–87, 92, 374
chemical: bonds, 12, 376; reactions 10, 12; potential 12–15, 63, 76, 111
chemiosmotic theory 183, 392–393
chemosynthetic bacteria 123
chemotropism 65
Chenopodium 198, 340, 346; album 36, 141, 301, 332, 337–338; polyspermum 336; rubrum 336–337
cherry 226, 228, 234–235, 324
chickweed 28
chicory 336
chilling injury 368
chill units 327
chinook 361
chitinase 400
chloramben 394
chloramphenicol 246
Chlorella 136; pyrenoidosa 137
chlorine (chloride) 68, 73–74, 376, 399; deficiency symptoms 90; in photosynthesis 132; metabolic roles 39, 90, 92
Chloris sp. 141
chloroethanol 327, 329
chlorogenic acid 216–217, 242, 246
chlorophyll 92, 186–187, 205, 213–214, 260, 262, 269–270, 290, 299–300, 386, 390; a and b 126–131, 165; absorption spectra 130; excitation 129; production 127, 347–348
chlorophyllide a 127
chloroplasts 2–3, 37, 70–71, 97, 104, 110, 116, 123–124, 125, 141–143–147, 150–151, 154, 163–167–168, 171, 175–176, 181–187, 198, 203–205, 208, 220–221, 234, 249, 256, 260, 262, 267, 299, 303, 368, 393, 397; development 125–129; movement 397; nitrogen metabolism in 198–200; structures 125–126, 394
chloropropham 329
chlorosis 88–91, 203, 262, 347

cholesterol 214–214, 251, 346
choline 212–213
Cholodny-Went theory 251, 279–283
chorismic acid 217
chromatin 2, 273
chromatography 2, 242
chromium 84
chromophore 398
chromoplasts 3–4, 214; carotenoids in 214, 234
chromosomal aberrations 397
chromosomes 2–3, 114, 273, 256, 276
chronobiology 314
Chrysanthemum 236, 261, 319, 332, 336, 345–346; maximum 336; morifolium 336, 346
Cichorium intybus 336
1:8 cineole 215–215
cinnamic acid 216–217–218, 220
circannual rhythm 308–309
circadian 307, 312, 330
circalunar 307
circulation 41
circumnutation 277
Cirsium arvense 228, 245, 323, 393
citric acid 87, 145, 181, 187, 211, 234; cycle (see Krebs cycle) 180
citrulline 196–197
Citrus 366; 395; reticulata 28
clay 15, 67–68, 380–382, 398
climacteric 191, 264, 270
climax community 351
clocks: resetting 342; two 310
clones 282
clover 297, 323; disease of sheep, 221
CMU 395
cobalt 83–84
cochromatography 136
cocklebur (Xanthium) 28, 46, 141, 233, 235, 245, 261, 265, 269, 280, 288, 290–291, 297, 305, 309, 311, 323, 330, 332–335–336–340–341–342–346; flowering 341; nitrogen metabolism in 197
cockroach 307
cocoa beans 222
coconut: milk 258; palms 314
codons 258
coenocyte 2
coenzyme 92, 114; A 92, 180–181, 205, 208–211, 249, 393
coffee 222, 357
cohesion 51, 58–59, 377–378; hypothesis 50–51–53, 58–60
colcemid 236
cochicine 282, 313, 222–222, 236, 239
Colchicum byzantinum 222
cold 75; hardiness 255, 302; requirement 318–319; resistance 347–348; storage lockers 367; treatment 321
coleoptiles 175, 239–240–247, 278–281, 297–298, 301, 394; curvature test 242–243–244, 281; phytochrome in 293
Coleus 241, 274; hybrida 337
collenchyma cells 280–281
Colletotrichum circinans 216, 401
colligative properties 26
colloids 29–30, 67, 70, 322, 378–380–383, 389
colors 384–384–385
color temperature 388
Commelina communis 39
community succession 351
companion cells 93, 101–102–102–103–104, 275
compensation point 163–164–165,

168 –169, 262
competition 351, 357; for nutrient 269
competitive inhibitors 120
Compositae (Asteraceae) 228
compost 68
concentration 12–14, 72–73, 75; gradient 76; of the test solution 24; series 24
condensing enzyme 181
conditioning effect 356
conduction 42
conductivity 388–389
coneflower 336
cone formation 345
conidiospores 307
conifer 39, 233, 256, 269, 277, 284, 301, 324, 345; lignin of 219; needles 45; roots 50; seed fats 207
coniferyl alcohol 219–219
conservation of energy 9, 404
constant volume method 24–25
continuum 351
contractile vacuoles 8
controls 308
convection 42–45, 369
Convolvulus arvensis 228, 245
cooling 42, 327
copalyl pyrophosphate 249, 251
copper 373–374; deficiency symptoms 89, 91; metabolic roles 92
cordycepin 246, 396
Coriaria 194
cork 68–69, 229; cambium 229, 258
corms 211, 228, 328–329
corn (Zea; maize) 32–33–34, 39, 55, 64, 79, 83, 88, 121, 140–142, 150, 158–159, 162, 164, 168–169–170, 172, 175, 197, 228, 233, 248, 254, 256, 258–259, 280, 282–283, 290, 292, 298, 299, 317, 336–337, 364–365–366, 394, 396, 400; nitrate accumulation in 198; nitrogen fixation in 193; seed composition 206, 208
cornflower 221
correlative effects 232
cortex 55–56–58, 65, 68–70, 150, 194, 202, 228–230
Corylus avellana 325
Corynebacterium fascians 261
cosmic rays 383–385
Cosmos 345; sulphureus 336
cotton (Gossypium) 89, 90, 141, 164, 168, 170, 237, 267, 274, 286, 323, 332, 366, 394; seed oils 206
cottonwood 28, 233
cotyledons 273, 261–262, 266, 282, 290, 296–301, 325, 329, 338; phototropism and 280
coumarins 216, 218–218, 267, 324, 401
coupling 183, 304, 311, 392
covalent bonds 376
cows 403
C-3 plants (species) 140–145, 164–167, 170–172
C-4 plants (species) 140–146, 163–169, 170–171
crabgrass 141, 143, 232–233, 280
cranberries 373
Crassula argentia 146
crassulacean acid metabolism (see also CAM) 145–147, 161, 362; sodium and 82
cream-type emulsion, 381
Creator 2
creosotebush 362, 389
cress 283
cristae 180–180, 182–183

critical: concentration 84; day 339, days 308; night 339; point for water 61
crocus 277
crop 399; ecology 350; physiology 350
crosscuts 60
cross vein 34
Crotallaria egyptica 323
Cruciferae 324
cryoscopic method 26–26–27
crystals 4, 13–14
cucumbers 36, 54, 86, 226, 244, 248, 261, 300–301, 317, 337
Cucumis sativas 337
Curcurbita 347–348; pepo 296
Cucurbitaceae 235
Cuppressaceae 277
Cupressus 66; arizonica 345
cushion plants 370
cuticle 34–35, 41, 45, 53–56, 98, 164, 170, 212–213, 245, 247, 252, 289, 393; waxes of 211, 213, 398
cuttings 244–245
cuvette 33–34, 109
cyanide 107, 182–184, 324
cyanidin 220–220; glycoside 325
cyclic: electron transport 135; growth 328; photophosphorylation 135; process 404; salts 192–193
cycloheximide 246, 262, 313, 395
cyclosis 100
cycocel see CCC
Cynodon dactylon 141
Cyperus: alternifolium 141; esculentus 141; inflexus and retroflexus 295; rotundus 141
cystathionine 204
cystein 112, 116, 119, 121, 187, 205; in papain 113; in ribonuclease 115; synthesis of 204; sulfur in 92
cytochrome 92, 182–183, 186–187, 198, 395, 397; b₃ 133; b₆ 133, 135; b₅₅₃, b₅₅₇, C₅₄₇, & C₅₄₉, 183; f 133–134; oxidase 110, 114, 170, 182–183, 267; phototropism and 278; system in mitochondria 182
cytokinesis 225, 228–229, 236–237, 258, 260–263, 275–276
cytokinins 108, 198, 229, 233–236, 240–243, 258–270, 275–276, 282, 297, 325, 328, 331, 347–348, 364, 400; in flowering 345
cytoplasm 2–4, 57, 69–70, 73–76, 102, 104; position of 391
cytoplasmic: pathway 70; streaming 57, 100–100, 103; viscosity 348
cytosine 91

2,4–D 99, 240–241–241–242, 245–246, 265, 324, 345, 394
Dactylis glomerata 141, 177, 337
Dahlia 217, 233
dalapon 324, 393
daltons 383
dandelion (Taraxicum) 28, 177; latex in 216, 245
Darnel ryegrass 336, 344
Datura stramonium 141, 336
Daucus carota (carrot) 141, 337
dawn 309–310
daylength 314, 318–319, 235–327, 332, 343; at various latitudes 333
day neutral 315, 333–334, 335–336–337
debranching enzyme 176
decarboxylations 111
deciduous perennials 321, 325

decimeters 3
dedifferentation 274–275–276
deep supercooling 366
deficiency symptoms 88–91
deficient zone 84
defoliation 343
degree symbol 10
dehydroquinic acid 217
dehydroshikimic acid 217
delay 310, 355
delphinidin 220–221
Delphinium: cultorum 336; barbeyi 222
delta-aminolevulinic acid 127–128, 187
denaturation 118–120, 317, 365–366
dendritic form 390
denitrification 192–194
density 382; of water 377
deoxyribonucleic acid see DNA 110, 114
deoxyribose 154–155
deplasmolysis 237–238
desert 8, 46, 52, 67, 324, 261; pavement 361; shrubs 65
desiccation 190
desmotubule 69–70
detached leaf method 33
detergents 213, 392
determinate growth 227
deuterium oxide 377
development 5, 224, 312, 356
devernalization 320, 327, 330
dew 42–43, 362
dextrins 176
dialysis 382–383
Dianthus 211; superbus 336
diatoms 92, 259
dicamba 245, 394
2,4-dichlorophenoxyacetic acid see 2,4-D
dichlorprop 393
dicot 37, 39, 55, 92, 393; leaf 53; stem 55; stomate 37
dictyosomes 3–4, 70, 72, 104, 158–159, 210, 228, 236, 237, 247, 280, 400–401
dicumarol 218
die back disease 91
dielectric constant 378
differentially permeable membrane 23, 57, 61–62
differentiation 5, 224, 273–276
diffraction 104, 383–384; grating 384
diffusion 13–14, 18–19, 23, 32, 34, 56–57, 72–73, 99, 190, 390–391; pressure deficit 16, 63; rate of 15; systems 15; technique 242, 251
digalactosyl diglyceride 213
Digitalis purpurea 214, 337
Digitaria: decumbens 256; sanguinalis 141, 143, 233
dihydrophaseic acid 267–268
dihydroxyacetonephosphate 139, 178–179, 208–209, 211
dihydrozeatin 259
diketovaleric acid 128
dill 336
dinitroaniline herbicides 394–395
dinitrophenol 107, 183, 393
dinoflagellate 307
Dionea muscipula 287
dioecious 233, 348
Dionea muscipula 287
dilution 22
1,3-diphosphoglyceric acid 138, 178–179
direct effects 354
discrete communities 351
disease 399–402; resistance 401

disorganization 11, 403
distal position 228, 232
distribution 351, 353, 361, 366
disulfide: bonds 116, 119–120, 145, 365; bridges 113, 116; in papain 113; in proteins 116; in ribonuclease 115
diterpenes 249
dithiocarbamate 394
dithiothreitol 145
diurnal 304–305
diuron 133
dividing cells 320
DNA 2, 4, 126, 127, 179, 204, 225, 228, 246, 255–258, 263, 272–273, 276, 320, 344, 366; polymerase 92; UV effects on 397
dollar plant 337
dormancy 226, 253, 268–269, 294, 302, 318, 321–322–328, 347–348, 370; of seeds 294–297; true 326
Douglas fir 49–53, 155, 215, 251, 302; seed composition 208
dream 312
driving force 77
droplets 58
Drosophila 307, 310–311, 313
drought 38, 146, 168, 270, 273, 283, 289, 361–366, 370, 374; resisters 362; spot of apples 91
dry: ice 59; matter 79; weights 97, 225
dual-daylength plants 337
ducks 310
duckweed 336, 340, 344
dusk 309–310
Dutchman's pipe 105
dye 59, 94

Eastern gamagrass 141
Echeveria corderoyi 146
Echinochloa: crus-galli 141; utilis 83
ecolosion rhythm 311
ecology 4, 332, 351; biophysical 47; research 348; UV effects on 397
ecophenes 356
ecosystems 46, 160, 162, 351, 357
ecotypes 332, 356–357, 359–360
ectendotrophic 66
ectomycorrhize 66–66–67
EDDHA 87–87
EDTA 87–87
effective concentration 13–14
einstein 11, 129, 384, 386
Einstein's law 129
elastic deformation 238
Eleagnus 194
electrical: double layer 380; neutrality 75–76; potential 396; resistance blocks 53
electrochemical potential gradient 76
electrolytes 378, 380, 382
electromagnetic 383; spectrum 384–385
electron(s) 383, 379, 386; microscope 2, 379; transport system 92, 182, 183, 198–199
electro-osmosis hypothesis 100, 108
electrophoresis 2, 116–117, 383
electropotential 76, 401; gradient 101, 108, 392
elicitor 401
elm 1, 253
elongation 365
Embden-Meyerhof-Parnas pathway (glycolysis), 178
embryo 5, 228, 325, 239; sac 234, 271
embryoids 276
Emerson enhancement effect 131
emission spectra 385, 387
emissivity 388

emotions of plants 308
emulsion 381
endodermis 34, 55–55–57, 61–62, 66, 68, 70, 98, 228–230, 280, 372
endogenous 305–306, 311–312, 315
endomycorrhizae 66
endoplasmic reticulum 2–3–4, 69–70, 104, 208, 210, 237, 247, 249, 280, 401; fat synthesis in 208
endopolygalacturonases 400
endoreduplication 276
endosperm 111, 175, 253–254, 258–259, 261, 267, 273, 290, 296–297, 323, 325
endotoxin 401
end product inhibition 122
endure drought 362
energy 1, 8, 9, 11, 21, 355, 384, 403–404; of activation 117; concept of soil water 388; exchange 42–48; flow 351; internal 11, 14, 403–404; kinetic 12, 381–382, 403–404; level 386; potential 12, 403–404; transformation 10
Englemann spruce 165
English: elm 323; ivy, 277
enolase 178
entelechy 1
entrained 309–311
entropy 11–14, 322, 403–404
environment 5–6, 356, 402, 404; factors of 359
environmental: effects on stomates 38; physiology 77, 350–360
enzyme 4, 71, 92, 322, 348, 350, 363, 382, 392–393, 395; activation energy 116; activity 368; activity and temperatures 317; chemical composition 111; combination with cations 114; compartmentalization 110; denaturation 352; induction 198, 273; inhibitors 120; mechanism of action 117; nomenclature 110; reaction 317, 401; repression 273; specificity 110; substrate complex 118–119; synthesis 396; systems 313
epicotyl 228
epidermal cell 37, 64–65, 69
epidermis 34–35, 55–56–57, 66, 70, 274–275; growth of 233
epinasty 246, 265–266, 345, 394
equilibrium 11–13, 404–405; constant 111, 354; dynamic 13; reaction 354
Equisetum arvense 83
equivalent: concentrations 380; solutions 67
Eragrostis: chloromelas 141; ferruginea 294; pilosa 141
ergosterol 214–214
ergs 385
Erwinia amylovora 401
erythrose-4-phosphate 139, 185–186, 216–217, 302
escape the drought 362
Escherichia coli 282
essential: amino acids 121; elements 81–83, 87–88, 92; oils 215, 324
estradiol 221, 346–347
estrogens 348
ethanol 157, 178–179, 187, 189
ethephon 265
ether 313, 383
ethrel 265
ethyl: alcohol 179, 313; carbamate 313
ethylene 189–191, 196, 233, 240, 243–246, 255, 258, 261–262, 265–267, 270, 288, 299, 328, 331,

347–384; 364; chlorohydrin 327; diaminetetraacetic acid (EDTA) 87; flowering and 345; seed dormancy and 297; synthesis 264
etiolation 290
etioplast 128–128, 262, 396–397; phytochrome in 293
Eucalyptus 277; regnans 49
eucaryotic 4, 304, 307, 390
Euchlaena mexicana 141
Euglena 307, 313
Euphorbiaceae 145, 216
Euphorbia: maculata 141; pulcherrima 334, 336
European: ash 325; beech 162
eutrophication 85
euxerophytes 362, 365–366
Evans solution 81
evaporation 10, 32, 34, 58, 63; cooling 42, 46, 376
evapotranspiration 33
evening primrose 233
evergreens 321, 326
excision repair 398
excitation stimulus 108
excited state 129, 386
excretory system 39
Exobasidium 261
exogenous 305, 311
explosions (in xylem) 60
extinction coefficient 386
extracts 344
eyes 310, 328, 385

Fabaceae (Leguminosae) 194
factor: intensity 352; interaction 354; X 306, 311–312
FAD 120, 134, 179, 181–183, 211; in nitrate reduction 198
FADH₂ 134–135, 179, 181–183, 209; in nitrate reduction 198
Fagus sylvatica 162, 268
false pulvinus 281
Faraday constant 132
farnesol 40, 213
farnesyl pyrophosphate 267
far-red 339, 342, 396–397; light 311, 326
fascicular cambium 274
fat(s) 187, 206–212, 234, 254, 264, 294, 299, 300, 302, 391–392; solvents 323
fatty acids 71, 92, 187, 206, 209, 211, 213, 359
feedback 5, 121–122; loop 40; mechanisms 356; respiration and 183; phosphofructokinase and 190; systems 313, 331
Fehling's solution 96, 147
femaleness 348
fermentation 178–178–179, 182
fern(s) 39, 194, 248, 259, 267, 282, 362; annuli 58; phenolics in 216; phytochrome in 293
ferredoxin 93, 111, 115, 133–134–135, 145, 204–205; glutamate formation and 220; nitrogen metabolism and 196, 198–199
ferredoxin-NADP⁺ reductase 134
ferryboat uncouplers 393
fertility 347
fertilized soils 66
ferulic acid 213, 216–220, 242
fescue, king's 170
Festuca arundinacea 141
fibers 55, 93, 101; protein 101
fibrous root systems 64 ⌐
fiddler crabs 307, 311
field: bindweed 228, 245; capacity 389, effects 275; studies 357–358

figs 235
filterpaper 379, 382
fire 324
fireblight 401
fireflies 314
first positive response 279
Fisher's analysis of variance 354
flag trees 369
flavin 114, 296; adenine dinucleotide see FAD
flavones and flavonols 220–221
flavonoids 92, 187, 205, 216, 220–221, 302–303, 397–398
flavoproteins 182–183, 301, 303, 395, 397; cytochrome b 310; phototropism and 278
flax 261, 263
flocculate 380–382
flooding injury 262
floral: characteristics 344; development 346, 348; stages 340–343
florigen 343–346
florigenic acid 344
flower 5, 329; development 346–347
flowering 313, 318, 329, 348, 370; hormone 108
flow velocity 59
fluids 13, 44; mosaic model 71
fluorescein 94, 106
fluorescence 129, 386
fluorescent lamps 44, 358, 384–385
fluoride 74, 84
fluoride 74, 84
food chain 351
foolish seedling disease 247
footnote law 13
forest 52, 366
forestry 5, 350
formic acid 168, 264, 377
foxglove 214, 319, 337
foxtail millet 141
Fragaria (strawberry) 336–337, 344
Fraxinus excelsior 325
fresh weight 224
free: energy 13–14, 20, 58, 388; space 75, 98
free-running period 306, 309
freeze drying 94
freeze fracture technique 71
freezing 10, 60; points 367; point method 26–26
freon 399
frequency 383–384
freshwater 52
freshweight 224
Froelichia gracilis 141
frost 42, 374; hardiness 325; 367, 370
fructans (fructosans) 147
fructokinase 178
fructosans (fructans) 124, 138, 147, 174, 177–178, 186, 237, 254, 256
fructose 96, 119, 124, 146, 151, 154, 176–178, 182, 234, 237–238, 256, 263
fructose-1,6-diphosphate 137, 139, 178, 145, 152, 189
fructose-6-phosphate 137–139, 178, 185, 147–150, 154, 179, 189
fruit 5, 324, 345; ripening 264; storage 267; trees 326, 328, 366, 394
Fuchsia hybrida 336
fucose 154–155
fumarase 181
fumaric acid 120, 181, 187, 211
function 4–5, 101–102
fungal hyphae 322
fungi 2, 31, 92, 194, 205, 214, 248, 259, 261, 268, 278, 307, 323, 344, 357, 360, 399
fungus root 65
Fusarium moniliforme 247

GA (see also gibberellins, gibberellic acid) 344–347–348; GA₃ 248, 251, 254, 256, 262, 277, 279
galactose 154–155, 213, 220
galactosyl glycerol 373
galacturonic acid 75, 154–155
gallic acid 216–217–218, 345
gallotannins 218
gametophyte 103, 111
gamma plantlets 225, 228, 275
gamma rays 225, 384–385
garlic 205
gas 9, 22, 377; chromatography 196, 205, 234, 264; constant 9, 12, 21; exchange 63
GDP (and GDPG) 15
Geiger-Müller tubes 136
gel 381
gelatin 30
gene(s) 2, 4, 115–116, 224, 272–276, 345, 395; activation 273; chloroplast 127; flow 356; host–parasite 402; repression 273
genetics 4, 313, 356–357
genotype 110
geomagnetic fields 311
geotropism 246, 255, 269, 277–286, 290
geranylgeranyl pyrophosphate 249
germanium 83, 274
germicidal lamps 397
germination 31, 308, 231–326, 347–348, 397; inhibitors 324, 362; in response to light 291, 297, 355
germ tube 401–402
Geum 337; rossii 370
gherkins 303, 397
Gibberella fujikuroi 247
gibberellic acid 154, 248, 354–355
gibberellin(s) 92, 187, 213, 230, 233, 235, 238, 240, 242–243, 247–248–257, 277, 288, 297, 301, 320–321, 325, 327–328–331, 396–397; geotropism and 280–281; flowering and 345; synthesis 249, 251, 253, 257–259, 261–262, 266–270
Gibbs: equation 13; free energy change 11–12, 20, 132, 182
Ginko biloba 293
girdling 93–94
glands 390
glass 367
global air movements 361
globeamaranth 141
globulins 273
glucose 96, 119, 124, 146–147, 150–151, 154–155, 174–178–179, 182, 185, 220, 222, 234, 237–238, 250, 259, 263, 391; -1-phosphate 147, 150–154, 176–178–179; -6-phosphate 137, 138, 147, 154, 178–179, 184–185; -6-phosphate dehydrogenase 185–186
glucosides: abscisic acid 268; auxin 242; cytokinin 259–260; gibberellin 251
glucuronic acid 154–155
glutamate synthase 199–200
glutamic acid 112–112–116, 186, 203–204, 364; dehydrogenase 92, 200; metabolism of 199–201; translocation of 196
glutamine 112–112–115, 187; metabolism of 199–201; synthetase 200; translocation of 196
glycerol 187, 206–213
glycine 112, 168, 259, 393
Glycine max (soybean) 105, 141, 162, 336, 339; seed composition 208
glycogen 178

glycolate 166–166–168; oxidase 168; pathway 168, 186
glycolipids 70–71, 206, 208, 212–213
glycolysis 92, 147, 182–184, 187–189, 267; reactions 177–178–179
glycophytes 361
glycoproteins 114, 401
glycosides 147, 150, 214, 220, 401
glyoxylate 168; cycle 209–211
glyoxysomes 3, 70, 207, 209–210, 210
glyphosate 393
God 1
golgi apparatus (see also dictyosomes) 2–3, 72, 159
Gomphrena globosa 141, 337
gonads 310
Gonyaulax polyedra 307, 309–311–314
goosefoot 336, 338
Gossypium hirsutum 141, 336
G₁ (G₂) phase 255, 275
gradient 13–15; activity 73; concentration 35; osmotic 106; pressure 106, 108; water potential 57
graft union 320, 343
grana 2, 125–126–126–127, 143, 163, 262; lamellae 126
grand phase of growth 327
grape 49, 60, 96, 191, 234–235, 255
grapefruit 191, 234
grass 37, 64, 332, 382, 392–393; guard cells 37
gravitational water 389
gravity 5, 12–13, 388
gray speck of oats 91
greasewood 363, 371
green: alga 8, 92, 366, 390; islands 400; light 383; sprangletop 141
Green and Stanton's method 29
Greenville farm 32–33
Gro-Lux lamp spectrum 385
ground state 386
growth 6, 42, 62, 224–239, 307, 326; cells and 75; chambers 358; correlative effect 232; curves 225–226; regulators 108, 330, 345–347–348; rings 94; temperature and 317; units 327; UV effects on 397
grunion 307
guanosine diphosphate 159
guard cells 34–35, 37–39–40, 238, 269, 275
guayule 218
guineagrass 196
guttation 7, 50, 77, 201
gymnosperm(s) 55, 60, 66, 102, 104, 259, 265, 267, 290; phytochrome in 293–294; sieve cells 98
gypsum 373

Haber-Bosch process 193
half-life 95
Halicystis 73–74
Halogeton glomeratus 82
halophytes 28–29, 51, 62, 145, 361, 364, 371–373
Hansenula wingei 402
hardiness 361–362, 366–368, 371
Hartig net 66, 66
haustoria 103
hazel tree 325
heart rot of beets 91
heat 11, 382, 386, 403–404; interaction 402; of fusion 26, 366–367; transfer 42–48, 315
Heavea 323
heavy water 313
Hedera helix 277
Helianthus (sunflower): annuus 28,

141, 208, 336–337; tuberosus (Jerusalem artichoke) 337
Helminthosporium toxins 400
helminthosporoside 400
hemicelluloses 154-155, 159, 400
hemoglobin 115
hemp 338
henbane 227, 320–321, 334, 338–340, 344; black 336; European 319
herbaceous plants 61
herbicides 41, 86, 94, 105, 245–246, 283, 393–395; as photosynthetic inhibitors 133, 394–395; nitrate accumulation and 198
heteroblasty 362
heterogenetic induction 274–275
heterophylly 255, 277
heteropolymers 116
Hevea brasilensis 216
hexadecanol 41
hexokinase 178
hexose monophosphate shunt 184
hibernation 308
Hibiscus syriacus 336
high energy reactions (HER) (see also HIR) 296
high irradiance reactions (see also HIR) 296–303
Hill reaction 124, 198, 394–395
Hipphophae 194
HIR 296–303, 340, 395, 396
histidine 112, 115
histones 114–115, 273
Hoagland's solution 81, 256
Höfler diagram 22, 27–29, 75
Hofmeister series 380
homeogenetic induction 274–275
homeostasis 5, 356
homocysteine 204
homopolymers 116
homoserine 204
honeybees 315
honeydew 95–96, 106, 260, 344
hooked bristlegrass 337
hops 338
Hordeum vulgare 141, 336, 339
hormone(s) 5, 97, 108, 240–271, 310, 343–348, 397, 400; accumulation 325; balance 394
horse latitudes 361
horseradish 242
horsetail 83, 267
horticulture 5, 350
hosts 399
Houblon japonais 338
hourglass 340–341
houseleek 338
humidity 45, 52, 358–360
humus 86
hydration 15, 29, 31, 50–51, 58, 365–367, 381–382
hydrocarbons 377, 381
hydrogen 9, 376; bonds 15, 51, 59, 70, 120, 157, 376, 376–378, 380, 395; ion (H⁺) 39–40, 76, 378–381, 392; peroxide 168, 220; sulfide 124, 205
hydrolytic activity 111, 176, 254, 257, 365
hydrophilic 28, 30, 71–72, 111
hydrophobic 29, 41 71–72, 112; membranes 392
hydrophytes 361
hydroponic cultures 80–81, 86
hydrostatic pressures 13, 18
hydrotropism 65
hydroxybenzoic acids 401
hydroxylamine 199
hydroxyl ion (OH⁻) 76, 378, 392
hydroxyproline 155, 238
Hyoscyamus 339; niger 227, 319, 334,

336
hypersensitive response 401
hypertension 20
hyphae 66, 70
hypnosis 315
hypocotyls 394, 397
hypodermis 34, 274
hyponasty 277
hypotheses 8, 99
hysteresis 381
Hyvar 395

IAA 91–92, 229, 240–241–242–248, 251, 259–260, 264, 266, 270, 275–283, 288, 301, 328, 345; oxidase 242
IBA 241–241, 244
ice 60, 368, 377, 381–382
ice-pick uncouplers 392
illite 67
illuminance 163, 385
imbibition 29, 322, 381
immiscible 381
immunological system 401
impaction 323
Impatiens 197
imperfect flowers 233
Impomea purpurea 334
incandescent light 43–44, 358, 384–385, 388
incipient plasmolysis 27, 27–28–29, 238
indeterminate growth 227
Indian lotus 323
indole 188, 215
indoleacetaldehyde 241–241
indoleacetic acid see auxins and IAA
indoleacetonitrile 241–241
indoleacetyl aspartic acid 242
indolebutyric acid (see also IBA) 241–241
indole ethanol, 241–241
indole pyruvic acid 241–241
induced-fit hypothesis 116, 118
induced state 321, 346
induction 274, 346
inductive 355; effects 318, 329; resonance 129
infection 399–400; thread 194
inflorescence 228; primordium 340
information 4, 6
infrared 43–44, 47, 378, 384–385, 388
inheritance of acquired characteristics 318
inhibitors 72, 107, 120, 313, 323, 325, 327, 330, 344–345, 352
inositol 213
insecticides 86
insects 307, 315, 346, 363, 399
intensity 339, 386
interact 402, 405
intercalary meristems 230, 256
intercellular spaces 35
interfacial movement 99
interference 383–384
intermediate daylength 334, 337
interfasicular cambium 274
International: Enzyme Commission 110; system 3
intervening light period 341–342
inulins 177–177
inversion layer 369
invertases 177–178, 250
iodine 74, 83, 150
iodinebush 371

ionic: bond 376; content of the atmosphere 306, 312
ipomeamorone 401
Ipomoea 348; purpurea 336; tricolor 233–234
Iris 177, 329; xiphium 329
iron 68, 99, 373; aconitase and 181; cytochromes and 133; deficiency symptoms 89–90; electron transport and 182; metabolic roles 92; nitrate reduction and 198–199; nitrogen fixation and 196; -sulfur proteins 182–183
irradiance 385
irreversible 11
irrigation 32, 67, 363, 371
irritability 6
isocil 133
isocitrate lyase 211
isocitric acid 145, 181, 187, 211, 234; dehydrogenase 92, 181
isocoumarin 218
isoelectric: pH 381–382; precipitation 382
isoenzymes see isozymes
isoflavones 221
isoleucine 112, 115, 121
isopentenyl adenine 258–259
isopentenyl pyrophosphate 249, 267
isoprene unit 214
isoprenoids 187, 213–216, 249, 267
isozymes 116–117, 220, 242, 254, 256
ivy 278

Japanese morning glory 248, 335–336, 338, 340, 343–344, 346
Jerusalem artichoke 177, 337
jasmine 215, 277
Jimson weed 141, 336
Johnsongrass 393
joule 385
Juglans 357
Juglone 357
Juncus effuses 239
Juniperus 66, 233, 277
juvenile phase 276–277, 338

K₍ₑq₎ 12, 150
K_m 119, 170, 200
kaempferol 242
Kalanchoe 38; blossfeldiana 310, 345, 336; laxiflora 337
kales 319
kaolinite 67
Karmex 395
kaurene 249, 251
Kentucky bluegrass 294, 323, 337
kidneybean 337
kilocalorie 377
kilometer 3
kinetic 11, 13, 72; theory 9, 379
kinetin 258, 262–263, 328
Kochia childsii 83, 141
kraft process 219
Kranz anatomy 140
Krebs cycle 140, 145–146, 174–180–181–181–184, 187, 189, 200, 209, 211
krummholz 369
kymograph 306

lactic acid 178–179, 187; dehydrogenase 92, 178
lactones 324
Lactuca sativa (lettuce) 141, 202, 337
LAI 172
Lambert's law 386
lambsquarters 36, 332, 335, 338; goosefoot 141, 337
lamella 125–126–128, 131

langley **385**
lanosterol 346
larch 326
larkspur 221–222; florists *336*
Larrea divaricata 362, 389
latent heat 42, **377**
lateral buds 242, 245, 260–261, 277
lateral: roots 228–229, 244, 258, 261; transport 106
laterite soils 67
latitude 332–*333*
lauric acid 206–*207*
law(s) 8; of the minimum **352**; of nature 402; of stable equilibrium **404**–405; of thermodynamics 405
layering 324
leaching 357, 372
lead 119, 374
leaf (leaves) **5**, 8, 53, 57, 93, 97, 232–233; *391*; area index **172**–*173*; cells *61*; detects photoperiod 343, 338, *347*–348; enlargement 365; fall *347*; growth 347–348, 370; mosaics 164, **278**, 280; movements **305**, 307, 311–315, 343; nodes, 103; primordia *231*, 277, 300; senescence 325; shapes 48; sheath 281; temperature 42–43, 46, 48
lecithin 212
leghemoglobin 195
legumes 68, 193–197, 221, 373
Leguminosae (*see also* Fabaceae) 194–197, 323
lemming burrows 322
Lemna: gibba 336, 340; paucicostata 336; perpusilla 336
lemons 191, 234
lenticels 68–69, 230
Lepidium: sativum 283; virginianum 294
Leptochloa dubia 141
lettuce (*Lactuca*) 86, *141*, *202*, 225, 228, 256, 261, 299–301, 303, 339, 360, *337*, 396; germination of 291–292, 294, 296–297; protein bodies in 202
leucine *112*, 115, 121
leucoanthocyanins 401
leucoplasts 3, 127, 150, 280
Leucopoa kingii 170–171
Leuresthes tenuis 307
levan *177*
lianas 60
lichens 170, 194, 367
Liebig's law 353–354
lie detector 308
life 11, 378, 382; cycle 317; definition of 5–6; span for seeds *323;* water and 378
ligand 87
light 39, 304, 359, **384**, *391*; action 395–397; compensation point 163, *164*, 168, 262; energy 339; intensity *38*, *329*, 370, **385**; interruption effects *339*; photosynthesis and 129–135, 145, 161–168; properties 129; sources **384**; waves 379
lignin 54, 67, 86, 92, 157, 186–187, 205, 216, 218–*219*–220, 270, 272, 284, 303, *391*
lilac *34*, 96, 327
Liliaceae 145
lily pads 37
lima bean 228
liming 373
limiting factor 352–*354*
Limnocharis flava 36
Limonium 371
linear phase 225
linoleic acid 206–*207*

linolenic acid 206–*207*
lipases 209
lipid: **206**, bodies 207; matrix with watery holes model **72**
lipids 70–*71*, 206–216, 272, 368, *393*
lipoamide 181
Liquidambar styraciflua 65, 288
liquid 9, 59, 377; crystalline 368
Liriodendron tulipifera 229
litter 324
little leaf of apples, peaches, pecans 91
liverworts 194, 216, 220, 293
loams *389*
loblolly pine 227, 229
Lockhart's method 27
locoweed *323*
lodgepole pine 302
logarithmic phase **225**
Lolium: multiflorum 141; perenne 177; temulentum 336, 340
Lombardy poplar 244
long day(s) 319, 325–326, 328–329; leaf 344; plants **315**, 321, *334*, *336*, 338
long-short-day plants **334**
Lophorphora williamsii 222
Lotus uliginosus 323
lovegrass *141*
low salt roots 74–75
low temperature **336**–*370;* response 318–322, 325–331
lumen **104**
Lunaria annua 337
lunar rhythms 304, 307
lunularic acid 268–*268*
lungs 390
lupine 197, 201, *317*, 322
Lupinus arcticus 322–323
lutein 126, **130**, 214
lux 385
luxury consumption **84**–85
lycoctonine 222
lycopene 126, **214**
lyophilizing **94**
lyotropic series 382
lysimeter 33–*33*
lysine *112*, 114–116, 121, 222
lysosomes 3

macromolecules 4, 6
macro nutrients **82**
macropores 60
macroscopic scale 404
Madia elegans 337
Magnesium: deficiency symptoms 89–90; metabolic roles 92, 178, 181
maize *see* corn
major elements **82**
maleic hydrazide 99, 329
maleness 348
malic acid 40, 76, 111, 117, 142–147, 178, *181*, 178, 198, 211; dehydrogenase 110–111, 117, **140**–145, 181, 199; enzyme 142; in fruits 234
malonic acid 120
malonyl maltase **176**
Malvaceae 221
malvidin *220*–**221**
mammals 346
man 307–308, 315–316
mandarin orange *28*
manganese 373; deficiency symptoms 89, 91; metabolic roles 92, 131, 181, 208
mangel *323*
mangos 264–266
mangrove *19*, 61–62, 371
mannans **297**
mannitol 21, 27, 96, 263, 296

mannose **154**–*155*, 297
manometer *29*, 49
manure 68
maple 169, 177, 233, 253, 325–326; sugar 150, *164*
marsh rosemary 371
marsh spot of peas 91
Maryland Mammoth tobacco 314, 321, *336*, 339
masking 313
mass: action 392; flow 77, 99–100, 107; law relationship **378**; spectroscopy 242; transfer rate **97**
mating 307
matric: component 31; effects 15; potential 29–*30*–31, 382, 388–*389;* surfaces 30
matrix 15, *72*, 107, 388
matter 9
mature phase 276
Maximowiczia sonorae 362
maximum *352*–352; temperature **317**
Maxwell-Boltzmann distribution 10
MCPA 241, 245, 265
mean 309–*309*
mean-free-path 9
mechanism 1
Medicago: orbicularis 323; sativa 55, 323
medicarpin **401**
medicine 5
Mediterranean climates 361
Melandrium 345; *rubrum* 348
Melilotus alba 323
melon 226, 235
melting 9
membrane(s) **2**–*3*–4, 8, 56–57, 64, 74, 92, 108, 313, 331, 381, 388, 392, 395–397; filter 30; model *22*–*23*, 70–72; osmosis and 18; permeability of 401
meniscus 29, **50**, 58
menstrual cycle 308
menthol 215–*215*
menthone 215–*215*
Menyanthes 198
mercaptans 205
mercaptides 120
mercury 119–120; lamps 44, *385*
meristems 227–229, 319–320; cells in 53, 56, 68, 398; phytochrome in 293
meromyosin 104
mescaline *222*–222
mesocotyl 297–298
mesophyll cells 35, 37, 53, 58, 98, 141–*143*–145, 233, 239, *391*
mesophytes **361**, 365
mesquite 245, 362
messenger RNA 258, 272, 395
metabolic: activity 307, 323; energy 99; pathway 110; roles of elements 92
metabolism 6, 72, 75, 355; and transport 107
metal activators **114**, 382
metastable *61*
metaxylem 180
meter 3
methane *377*–378
methionine *112*, 115, 121, **205**, 264–265; sulfur in 92; syntheses 204–205
methyl 92, *377*
2-methyl-4-chlorophenoxyacetic acid 241
methylene blue 25
3-methyleneoxindole 242
methyl mercaptan 205
6-methylpurine 246

methylthioadenosine 264
2-methylthioisopentenyl adenine 258
2-methylthiozeatin 258
metric units 3
mevalonic acid 249
Michaelis-Menten constant 119, 170, 200
microbial action 323
microbodies 3–4, 70, **209**
microcapillary 107
microdimensions 61
microelements 82
microfibrils *see* cellulose microfibrils
micrometer (micron) *3*
micronutrients 82, 87
microorganisms 67
microphones 60
microscope 379
microscopic scale 403–405
microtubules 3, 111, **236**–*237*, 239, 273
middle lamella *2*, 4, 70, 90, 92, 103, **236**–*237*, 270, 400; lignin in 218
millimeters *3*
millimicron 3
Mimosa 108, 305, 376; *glomerata 322*–*323; pudica* 277, 284–*287*, 396
Mimulus cardinalis 44
mineral(s) 8, 77, 99; absorption 64, 67–*77;* ions 97, 106, 390; nutrients 65, 326, 338, *352*, 373; transport 41
mine tailings 374
minimum *352*–352; leaf number **338**; temperature **317**
minor elements 82
mint 395
Mitchell chemiosmotic theory 183, 392–393
mitochondria *2*–*2*, 4, 70, 104, *128*, 141, 143, 166–*167*–168, 177–**180**–*180*–182, 188–189, *195*, 200, 205, 209–*210*–211, 280, 300, 346, 368, 392–393; in root nodules 195; phytochrome in 293
mitosis 3, 110, **225**, 229, 255, 275–276, 346, 395
mobility and food reserves 391
model 8–9, 13, 99, 108
modulated responses 355
Mo-Fe protein 196
moisture release curve 30–*30*, *389*
molality 21
molal water potential **21**
molar extinction coefficient 386
molarity 21
molds 367
molecular: sieves 382; sieve hypothesis **72**; structures 1; weight 9
molecule 72, 384, 404
mole fraction 14, 16–17
molybdenum: deficiency symptoms 91; metabolic roles 92, 196, 198
monocarpic 227–228, 233, 338
monocot *36*, 39, *55*, 92, 104, *393;* stomate 37
monoecious 348
monogalactosyl diglyceride 213
montmorillonite 67
monuron 133, 395
moon 308, 316
moonlight 312
morning glories 233, 245, 265, 334, *336*
morphine *222*–222
morphogenesis 5, **224**, 290, 318
morphology 4
mosses 39, 194, 259, 267, 303, 362, 367; phenolic compounds in 216, 220; phytochrome in 293
Mougeotia 303, 396

mountain sorrel 370
mouse ear cress *336*
mucigel **228**
Muhlenbergia schreberi 141
mullein 177
multiphasic absorption 72
multiple epidermis *34*
multiplicative factors *354–355*
Münch hypothesis 106, 108
mung beans 396
mushrooms 382
music 308
muskmelon 261, *335*
mustard 394–395; oils 324
mutation 397, 399
mycoplasms 4
mycorrhizae 65–67, 77, 85, 197, 228, 259
Myrica gale 194
myristic acid *207*

NAA 241–*241*, 244–247, 265, 345
NAD⁺ 111, **134**–*134*, 142, 147, 178–179, 181–184, 209, 211, 222; in nitrate reduction 198–199
NADH 128, **134**–*134*, 142, 147, 174, 178–179, 181–183, 190, 200, 204, 209, 211, 256; in nitrogen metabolism 196, 198–199
NADP⁺ **124**, 131–134–*134*–135, 139–147, 179, 181, 184–185, 199, 208, 217, 222, 249
NADPH 124–125, 131–*134*–135, 138–147, 165, 167, 174, 179, 185–186, 190, 198, 204–205, 208, 216–217, 220, 224, 249, 256; in nitrogen metabolism 196, 200
nanometers *3*
naphthaleneacetic acid *see* NAA
naphthoquinones 132
napiergrass *141*
narcotics **357**
narringenin *347*
nastic responses **277**, 285–287
natural selection 40
N-dimethylaminosuccinamic acid 236
nectaries 103
negative: feedback loops 5; phototropism **278**
Nelumbo nucifera 323
nematodes 399
Nemophila insignis 94
Nerium 34
nervous pudding 381
net radiation 43, **387**
Neurospora crassa 307, 310–*311*, 313
niacin 222
niche 315
nickel 83–84, 374
Nicotiana (tobacco) 339; *sylvestris 336*; *tabacum* 222, 323, *336–337*
nicotinamide 134; dinucleotide *see* NAD⁺
nicotine 222–*222*
night-blooming jasmine 337
night interruption **339**, 341–344, *347–348*
nightlengths *341*
nimblewill *141*
Nitella 29, 73–74, 104; *axillaris* 29
nitrate (NO₃⁻) 65, 68, 73–74, 192, 324; metabolism of 197–199; reductase 92, **198**, *364*, 396
nitrification 192, **194**
nitrite: metabolism 198–199; reductase 92, 198–199, 273; toxicity 198

factor *352–353*; metabolism 42, 92, 192–204
nitrogenase 195–197
nitrous oxides 399
nodes 106, 227
nodules 194
Nomarski optics 104
noncompetitive inhibitors **120**
noncyclic: electron transport 135; photophosphorylation 135
noninductive effects **354**
nonpolar molecules 378
nonreducing sugars 147
nonsucculent perennials **362**
Norway spruce 325
n-propyl mercaptan 205
nuclear: energy 11–12; magnetic resonance 242; size 346
nuclei 4, 70, 93, 102, 110, 258, 293
nucleic acids 4, 173, 92, 187, 190, 219, 234; synthesis 345; UV effects on 397–398
nucleolus 2–*2*, 246, 263, 346
nucleotides 4, 192, 203; as cytokinins 258–259
nucleus 2–*2*, *103*–104, *202*, 247, 280, 307
null response technique 340
nutrient solutions 81
nyctinasty **277**
Nymphaea odorata 28

oak 208, 227, 245
oats (*Avena*) 91, 121, *125*, 140–*142*, 170, 197–198, 203, *240*, 243, 247, *256*, 278–281, 283, 292–293, 297–298, 301, 330, *336*, 396
obligate: halophytes 371; parasites **400**
Oenothera 233
Ohm's law 44, 77
oil(s) 187, 206, 391; wells 366
oleander *34*, 37
oleic acid 174, **206**, *207*
oleosomes *202*, **207**–208, *210*, *264*
onion (*Allium*) 205, *337*, 347–348, *336*, 401; smudge 216–217
o-phosphorylhomoserine 204
opium poppy 222
optical density 386
optically active **384**
optimum **352**–*352*; temperature **317**; turgidity 42
Opuntia 146
oranges (*Citrus*) 191, 215, 226, 234, 255, 264
orchardgrass *141*, 164, 177, *337*
orchard trees 327
Orchidaceae 145, 322
orchids 215, 218, 277, *336*, 392
orchinol 218
orchirol 401
organelles 2, 104, 391
organic: acids 38, 40, 76, 92, 136, 174, 348, *377*; gardening 86; matter 67–68; solvents 382
organization **403**
organs 5
ornamentals 345
Orobancheae 322
orotidine-5′-phosphate decarboxylase 92
orthogeotropism **280**
Oryza sativa 141, *336–337*
oscillating timer 340–341, 384
osmolality 21
osmometer 18, *18–19*, *99*–100
osmosis 8, **14**, 37, 42, 62, 72, 77, 95, 99, 108
osmotic: filtration 27; forces 388; po-

tential (*see next entry*); pressure 22; system 18, 57
osmotic potential(s) 20–22, 30, 39–40, 62, 98, 100, 243, 252, 256, 297, 324, 364, 370–371, 373, 388–389; empirically determined 28; fruits 234; growth and 237–239; guard cells 38; leaves 51; measurement 26; nutrient solutions 81
ovule 234
oxaloacetic acid **140**, 144–147, *181*, 186–187, 199–200, 211
oxidation 68
oxidative phosphorylation 124, **182**–186, 190, 211
oxides 67
oxilates, calcium 92
oxygen 10, 40–41, 67–68, 76, 323–324, 376, 390, 404; and seed viability 322
Oxyria digyna 332, 370
ozone 160, 162, 193, 397–399

P₆₈₀ and P₇₀₀ 131–135
Pᵣ and Pfᵣ 292
palisade *34*; parenchyma *35*–*35*
palm 42, 45–46, 104, 362; Palmyra 106; sugar 106
palmitic acid **206**–*207*–208
Pangola digitgrass 256
Panicum, several species *141*; maximum 196
pantoic acid 393
papain *113*
Papaver somniferum 222
papaya 77
paper chromatography 136, 149, 153
Paramecium 307, 310–*311*
parasites 399
parasorbic acid 324
parenchyma *34*, *55*
Parthenium argentatum 218
parthenocarpic fruits **235**, 254, 258
particles 380; of radiant energy 383–384
Paspalum notatum 141
passive: theories 99–100, 107; uptake 57, **76**–77
Pasteur effect **189**–190, 267
pathogen 399–401
pathotoxins **400**–401
pathway 53–58, 61, 70
pattern **309**
p-coumaryl alcohol **216**, *219*–220
p-coumaric acid 216–217–220, 242
peach 91, 226, 235, 324, 327, 329, 256
peanut (*arachis*) *141*, 164, 209, 297; nitrogen fixation in 196–197; seed oils 206
pears 226, 234–235, 244, 264
pearl millet 196
peas (*Pisum*) 70–71, 91, 115, 121, 150, 170, 174, 196–197, 203–204, *225*, *243*, 256, *266*, 277, 279, 283, 286, 291, 296, 299, 225–226, 229, 243, 248, 261–262, 265, 273, 330, *337*, 356, 397; pisatin in 218; seed composition of 208
pecan 91
pectin 4, 92, 139, **154**–155, 159, 205, 236–237
pectinase 270, 400
Pectinophora 310
pelargonidin 220–221
Pelargonium 274
pendulum 21
Pennisetum: americanum 196; *purpureum 141*
pentose phosphate pathway **184**–*185*–186, 208, 216, 400

peonidin 220–**221**
PEP **140**–147, 211, 216–217; carboxykinase **146**, 211; carboxylase **140**, 142, 145, 165–166, 170, 186, 200
peppergrass 297
peptidases 203
peptides 112–113, 401
percent transmission **386**
perennating buds **227**
perennial(s) **227**, 228, 276–277; grasses 319; ryegrass 340
perfect: flowers **233**; gas law 21–22; osmometer **18**; solutions 16
perforation plates 55
perfume discharge 307
periclinal divisions **232**
pericycle *55*–*55*, 57, 194, 228–*230*, 244
Perilla **336**, 339, 346; *crispa 336*; *fruticosa* 197
period **309**–*309*, 313
perisperm 111
peristalsis 101
perlite 81, 169
permeability 15, 368, 396
permeases 72
permanent wilting percentage 62–63, *389*–*389*
peroxidase 92, 220, 242
peroxisomes 3, 70, 166–**167**–*167*, 207, 209, 210
perpetual motion machine 11
petal movements 307
petkus rye 318, *320*
Petunia 276, *335*, 403, 405; *hybrida* 377
petunidin 220–**221**
3-PGA **136**–*139*–143, 146, 152–153, 166–168, 199, 211
p*H* 39, 81, 85, 87, 90, 115–116, 119–120, 132, 135, 145, 183, 373, 382–383; anthocyanin colors and 221; changes 75–76; gradient 392; nitrification and 193; of fruits 234; PEP carboxylase and 198; scale *378*, 381
Phacelia tanacetifolia 294, 296–297
Phalaris canariensis 141
Pharbitis nil (Japanese morning glory) 233, 238, *336*, 338
phase **309**; diagram *61*; shifting **309**–*310*, 343
phaseic acid **267**–*268*
phaseolin 218, 401
Phaseolus: coccineus 397; *vulgaris* (common bean) *141*, 248, 305, 312, 337
phasing dark period 341–*342*
phellem **229**
phelloderm **229**
phenolic compounds 67, 92, 216–221, 324, 401
phenological data 327
phenotypes 115
phenylacetic acid *241*
phenoxy herbicides 394
phenylacetic acid *241*
phenylalanine 303, *112*, 115, 121, 187, 216–218, 220, 222; ammonia lyase **216**, 267, 303, 396
phenylpyruvic acid 217
phleins 177
Phleum pratense 177
phloem 5, *34*, **55**, 57, 93, 97, 99–100, 154, 177, 196, 203, 228–229, 242, 245, 253–254, 258, 260, 267–268, 274–276, 285, 343, 400; anatomy 101–105; development 103–104; exudate 95, 97; fibers 55, **102**; loading 88–*88*, *100*, 102–103, 107–108;

parenchyma 93, 102, 104; sap 96–97; transport in 92–93
Phoenix dactylifera 208
phosphatases 111, 149, 167
phosphate ($H_2PO_4^-$, $H_2PO_4^=$ $H_2PO_4^\equiv$) 65–66, 68, 73–74, 373, 391
phosphatidyl compounds 71, 212–212–213
phosphoenolpyruvate 140, 178, 187, 302; carboxylase (PEP carboxylase) 92, 140
phosphofructakinase 178, 189
phosphoglucomutase 178, 184
6-phosphogluconic acid 185–185; pathway 184
3-phosphoglyceraldehyde 138–139, 147, 178–179, 185, 199, 211, 222; dehydrogenase 139, 145, 178, 199
3-phosphoglyceric acid 136, 138, 178–179; kinase 139
phosphoglycolic acid 166–167
phosphohexoseisomerase 178, 185
phospholipids 70–77, 92, 187, 206, 209, 212–213
phosphon D 236, 249, 251
phosphorescence 386
phosphoric acid 393
phosphorolytic enzymes 176
phosphorolysis 150
phosphorus 77, 99, 352–353; deficiency symptoms 89; metabolic roles of 92; limiting factor 352–353, radioactive 95
5-phosphoshikimic acid 217
photochemical: reactions 397; active irradiance 386
photodormancy 294–297, 397
photoelectric effect 384
photomorphogenesis 290, 395–397
photons 129–130, 135, 384, 386, 397
photooxidation 214
photoperiodism 302, 312, 315–316, 326, 328, 330, 332–348, 356–358, history of 338
photophil 309
photophosphorylation 124
photoreactivation 398
photoreceptor 310
photorespiration 164–166–170, 172–173, 283
photosynthates 93, 97
photosynthesis 3–4, 8, 11–12, 32, 37, 40, 42, 47, 63, 92, 100, 123–173, 214, 287–288, 307, 344, 350–351, 354, 357–362, 364, 370, 384, 395; in infected leaves 400; nitrogen metabolism and 196, 199–200; summary equation 135; UV effects on 399
photosynthetically active irradiance 386
photosynthetic phosphorylation 124, 132–135, 149–150, 183
photosystem I 131–135, 145
photosystem II 131–135, 143, 165, 198
phototaxis 164, 303, 307
phototropism 164, 240, 243, 246, 250–251, 277–280, 286, 290, 296, 299, 301, 303, 355
phycobilins 131, 292
phyllotaxis 232, 275
phyllotaxy 77
physics 1, 47, 308
physiological: ecology 350–351, 353; experiment 320; plant pathology 399–401
physiology 4
phytins 202, 254
phytoalexins 216, 218, 401

phytochrome 187, 198, 261, 272, 284, 286, 290–303, 311, 326, 330, 339–341, 347–348, 352, 354, 395–397; destruction 294; discovery of 290–292; distribution of 292; properties of 292; transformations of 293–294
phytoferritin 90
phytol alcohol 126–127, 187, 213
Phytophthora parasitica 401–402
phytoplankton 314
phototoxins 215, 400
phytotrons 251, 358, 384
Picea: abies 325; *engelmannii* 165; *pungens* 28
pickleweed 28, 371
picloram 245
pigment 5, 307, 310, 384, 386; shift 341
Pilobolus 307
Pinaceae 248, 345
pine (*Pinus*) 28, 34, 37, 54, 90, 210, 226–228, 284, 286, 293; strawberry 336; sugar 228
pineal gland 310
pineapple 191, 226, 234–235, 264–265
Ping-Pong balls 404
Pinus: cembra 370; *contorta* 66–67; *lambertiana* 228; *monticola* 28; *strobus* 34; *sylvestris* 162; *taeda* 66, 229; *thunbergii* 396
pisatin 218, 401
Pisum (pea): *arvense* 203; *sativum* 337
pith 55, 150
pits 54, 57–58
plagiotropism 280, 284
Planck's constant 384
plant: -atmosphere interface 63; cells 4; distribution 338; pathology 360; physiology 1, 4, 8, 16, 39, 77; -water relations 9, 358
Plantago maritima 373
plantlets 348
plasmalemma 4, 70, 74, 102–105
plasma membrane 70–71–72, 237, 247, 252, 269, 278, 401; origin 158–159, 236–237; phytochrome in 293
plasmodesmata 4, 37, 57, 69–70, 98, 102–103, 142–143, 272
plasmolysis 20, 27, 29, 237–238
plastic deformation 238–239
plastids 2–3–4, 263, 267, 274, 300; geotropism and 280
plastocyanin 92, 133–134
plastoquinone A (PQA, $PQAH_2$) 132–135
Plum 226, 324
Poa pratensis 28, 141, 289, 294, 323, 337
podophyllotoxin 236
podzolic soils 67
poinsettia 334, 336
poise 378
poisons 72, 74–75, 357
polar 376–377
polarization 274, 383–384
political science 4
pollen 235, 266, 276, 391, 402; fertility 348; tube 91, 271
polyacrylamide gels 116, 382
polycarpic 227–288, 233, 338
polyethylene glycol (PEG) 23, 27, 296–297
polygalacturonic acid 400
Polygonum 324
polygraph 308
polymers 368

polypeptides 111, 113, 115–118, 272
polyphenol 401; oxidases 217
polyribosomes 270, 396
polysaccharides 92, 151, 154, 156, 159, 172, 176, 254, 368, 391
pomologists 322
ponderosa pine 210
poplar 226, 259
population 356
Populus: deltoides 28; *tremuloides* 64
pores 34–35, 72, 104
Portulaca: gradifolia 233; *oleracea* 141
Potamogeton matans 274
potassium 40, 67, 73–74, 99, 243, 269, 373; concentrations 39, 307; deficiency symptoms 89, 90; ions (K^+) 39–40, 100–101, 313, 396; metabolic roles 92; nyctinasty and 284; pyruvate kinase and 178; starch synthesis and 150; thigmonasty and 285
potato 30, 70, 73, 88, 150, 164, 170, 172, 175–176, 189, 200, 213, 236, 258, 263, 267, 328–330, 348; cellar 312; cholesterol in 214; chlorogenic acid in 217
Potentilla glandulosa 356–357
potometer 33
P-protein 103–104, 108; bodies 104; fibers 105
prairie: cordgrass 170; grasses 66
precipitation 359, 380, 382
precursor 354
preocenes 218
prephenic acid 217
presewing drought hardening 365
pressure 13–14, 21, 50, 52, 58, 348, 382; bomb method 26–28, 51–53, 59, 61–62; chamber 30; cuff 107; flow 99–99–100, 108; gauge 107; membrane 389; negative 51; osmosis and 19–20; phloem 106–107; potential 20, 22, 30, 62, 100, 243, 297; potential measurement 29; potential role in growth 237–239; -volume product 12
prestressed structure 391
primary phloem 55, 229
primary wall 54; growth of 237–239; structure of 154–159
primary xylem 55–55, 229
Primula obconica 101
prism 384
probability 403–405
procambium 244, 274
procaryotes 4, 214, 392
productivity 42, 361, 366
progesterone 214
prolamellar body 128–128, 262
proline 42, 112, 364, 373
promotors 344–345
propionic acid 209
proplastids 127–128, 198, 200, 205, 267
Prosopis tamarugo 20, 362
prosthetic groups 114, 134, 180, 292
proteases 113–114, 203, 233, 254, 263, 270, 294
protein(s) 2, 4, 15, 30–31, 70–71, 73, 92, 322, 325, 377, 381–383; amino acid sequences 114–115, 117; bodies 201–202–202–203; degradation 200–203; derivatives 117; host-pathogen 401; iron 196; molecular weights 111; molecules 380; purification 382–384; structures 115; synthesis 42, 92, 313, 345, 363–364–365, 395–396; terminal ends 113; UV effects on 397

protoanemonin 324
protocatechuic acid 216–217
protochlorophyll 300; formation 363–364
protochlorophyllide a 127–128
protons 376
protoplasm 2, 376, 378–381, 390–391
protoplasmic 365, 368; viscosity 366
protoplast 27, 55–56, 62, 282
protoxylem 229
proximal position 228
pseudocyclic photophosphorylation 135
pseudoscience 308
Pseudotsuga menziesii 208, 302
pteridophytes 245
pterocarpin 216
pulse 307; of heat 59
pulvinus 396; false 381; nyctinasty and 284; thigmonasty and 285
pump 73–74, 76, 100; cells 50
pumpkin 95, 97, 233, 261
puromycin 246, 313, 395
purslane 141
pyridoxal phosphate 201
pyrophosphate 143, 147, 154, 204, 209, 249, 259
pyrrole rings 127
pyruvate kinase 92, 178–179
pyruvate, phosphate dikinase 143–145
pyruvic acid 142–147, 177–181–182, 187, 189, 204, 209, 211; dehydrogenase 180–181

Q 132–133
Q_{10} 10, 252, 294, 311, 313, 330–331; for diffusion 15; of respiration 190
quackgrass 141, 393
qualitative response 334, 336
quantitative response 334, 336
quantum 129, 384; mechanics 9; requirement 135
quercetin 242
Quercus: alba 28; *robur* 208
quiescent 321, 326
quiescent center 228–229, 280
quinic acid 216
quinine 222
quinones 132, 217, 397, 401

radial micellation 37–37
radiant: energy 42–44, 47, 359, 378, 383–388; flux 385
radicle 175, 228, 253, 261, 297, 323, 325
radioactive tracers 94, 97, 106
radioautography *see* autoradiography
radio waves 383–385
radish 64, 128, 197, 200–201, 259, 261–262–263, 279, 299, 336
raffinose 96, 237
rainfall 41, 67, 373, 380
rain shadow effect 361
randomness 11, 13, 403
range 309–309; management 5, 350
Ranunculus 55, 322
Raoult's law 16–17, 25
raphanatin 259–260
Raphanus sativus 259, 336
rattan vines, 60
rays 55
reactant 354
reaction center 129, 131, 133
reaction: rates 330; wood 284–284
red: algae 214, 259, 292–293; clover 169, 218, 323; goosefoot 336; light 291–303, 311–312, 326, 339–342, 354–355, 396–397; maple 28; oak 227; pine 227

redifferentiation 274–275
redroot pigweed 141, 394
redtop 141
reducing sugars 96
reduction potential 132
redwood tree 5
reflectance 44
reflection 386
refraction 383–384
regression analysis 354
relative humidity 17, 25, 41–42, 45–46
R-enzyme 176
repair mechanisms for UV 398–399
rephased 309
replication 276
reproduction 6
reptiles 346
reset 204, 306
resin(s) 34, 215–215
resistance 44–45, 49, 52, 55, 61, 77, 361–362, 366, 402; leaf 46
respiration 42, 68, 75–76, 107, 150, 161–164, 166, 169–191, 256, 270, 307, 345, 350–351, 364; disease and 400; factors affecting 186–191; inhibitor 313; minerals and 92; nitrogen metabolism and 196, 199
respiratory quotient (RQ) 174
response types 318, 333–334–335–337, 343
rest 322
reverse flap technique 94–94
reversible 111, 381
reversion 294
rhamnose 154–155
Rhizobia 157, 194, 402
rhizocalines 245, 271
rhizomes 228, 280, 329
rhizosphere 196
rhodesgrass 141–142
Rhododendron 370
rhubarb 295
Ribes nigrum 268
riboflavin 98, 168, 182, 278, 301
ribonucleases 115, 150, 203, 233, 254, 263, 270
ribonucleic acid see RNA
ribose 154, 203, 259, 264; in NADP+ and NAD+ 134
ribose-5-phosphate 137, 139, 185–186
ribose-phosphate isomerase 139
ribosomal RNA 263
ribosome(s) 2–2–4, 69, 237, 258, 264, 272, 280, 346, 396; subunits 92
ribulose 153–154
ribulose-1, 5-diphosphate 136, 138–140, 147, 152–153, 166–168; carboxylase 92, 111–111, 116, 137, 139, 141–142, 145–146, 164–166, 168, 170; oxygenase 166
ribulose-5-phosphate 138–139–140, 185; epimerase 139, 185; kinase 139, 145
Ricca's factor 287
rice 76, 121, 140–141, 150, 189, 196, 248, 293, 332, 336–337, 345, 366
ricinoleic acid 207, 213
Ricinus communis 207–208
ripeness: to flower 338; to respond 334, 338, 345
RNA 110, 113, 127, 203–204, 234, 246, 254, 256–260, 263, 269–270, 273, 300, 346; polymerases 273
Robinia pseudoacacia 98
Rocky Mountain National Park 304
rodents 307
root(s): cap 56, 228–229; crops 345; hairs 55–56, 64–68; 228–229,

274, 390; formation on cuttings 347–348; geotropism in 280, 283; growth of 228–229; low-salt 74; nodules, 83, 193–194–195, 197, 201; pressures 13, 49–50, 56–57, 77; phytochrome in 293; system 362; 390; temperature 370; tip 55, 68, 70; tissue 388
roots 5, 55, 64, 68, 70, 74, 77, 97, 326, 382
rose 233, 324
rosette of apples, peaches, pecans 91
roundup 393
rubber 206, 213, 216
rubidium ion (Rb+) 74
Rudbeckia: bicolor 336; hirta 336
Rumex 348
running cycles 307
Russian thistle 141
rye (Secale) 64, 121, 227, 293, 319, 371
ryegrass 177, 280, 344

Saccharum: officinarum (sugarcane) 141; spontaneum 336–337
S-adenosylmethionine 92, 205, 216, 264
sagebrush 28, 245, 363
Salicornia 28, 371; rubra 28
salicylic acid 344
salinity 371–373
Salix: babylonica 28; japonica 322; pierotti 322
Salsola kali 141
salt 61, 67–68, 77, 372, 374, 380; accumulators 371; brush 141; bush 372; concentrations in soil 363; glands 103, 372; incrustations 371; marshes 8; regulators 372; respiration 183; water (see also seawater) 62, 73
salting out 382
Salvia leucophylla 215
Samanea saman 396
samphire 371
sand 81, 398; dropseed 141
sap 26, 366; flow 50; movement 61; pressure 52
Sarcobatus 371; vermiculatus 363
saturation 72, 351–352; kinetics 73
Saugkraft 16
Sauromatum guttatum 184, 188–189
scarification 323–324
scarlet pimpernel 336, 338
science 1, 8, 99
scintillation counters 136
scopoletin 218–218
scouring rush 83
scutellum 175
screening UV 399
Scrofularia hyecium 345
scutellum 254, 297–298
SD & SDP 335–336
sea lavender 371
seashore 52
seasonal responses 332, 340
seawater (see also salt water) 19, 61, 71
Secale cereale 64, 318, 336–337
secondary: wall 53; xylem tissue 55
second positive response 279
sedges 141
sedoheptulose-1, 7-diphosphate 137, 139, 145
sedoheptulose-7-phosphate 137, 139, 185
Sedum spectabile 336
seed(s) 5, 31, 308, 324–325, 382, 390–392; coat 323, 325–326; dormancy 253, 269, 273, 291; germina-

tion 269, 274, 294–297; pieces 328
seedlings 338
Selaginella 282
selection value 312
selenium 74, 83–84
self-assembly 272–273
semipermeable 18, 26
Sempervivum funkii 338
senescence 187, 191, 225–227, 235, 255, 260–261, 263–266, 269–270, 273, 303; factor 270–271; phase 225
sensitive plant (Mimosa pudica) 108, 287, 305, 323, 376
sensitivity (emotional) cycle 308
sephadex 382
Sequoia 44–50, 66
serine 112, 115, 168, 185
serpentine 360
sesquoxides 67
Setaria: italica 141; lutescens 141, 323; verticillata 337
sex: expression 347–348; hormones 348
shade: leaves 163–164; plants 163, 165
shadscale 28, 323–324, 371
sheath 55
Shepherdia 194
shikimic acid 216–217; pathway 216–217–218, 220, 303
shock wave 60
shoot apex 175, 231–233, 255–256, 264, 274; phytochrome in 293, 298
shoot-to-root ratios 89
short-day 319, 321, 325, 328, 367; induction 326; plants 315; 334–336–339
short-long-day plants 334
sideoats grama 141, 332
sieve: cells 102, 104; elements 93, 98, 102–106; plates 93, 102–105; pores 103; tubes 93, 97, 100–102, 104, 107–108, 147, 228; -tube element 96, 101, 103, 275; -tube member 102
Silene 336, 345; acaulis 370; armeria 334, 336, 343
silicon 67, 79, 80–84
silk tree 284
silver 119–120
simazine 133
simple pits 54
Sinapis alba (white mustard) 300, 336, 355
sinapyl alcohol 219–219–220
Sinbar 395
sink 94, 97, 100, 106–107
Sirius 388
sitosterol 214, 346
skatole 188
slash pine 227
sleep 307
slime: bodies 103–103–104; molds 2
smooth bromegrass 336
snorkel system 76
snow accumulation 370
sociology 4
sodium 73–74, 363, 373; chloride 324, 376; ion (Na+) 380; pump 73, 372; vapor lamps 44
softwoods 367
soil 15, 57, 65–65, 67–68, 350–359; moisture 32, 61; pH 373–374; solution 57, 67, 73, 388; sterilant 395; structure 373; temperatures 326; texture 389; water 14, 20, 30, 388–389
Sol 381–382
Solanaceae 235

solar 384–385
solar: constant 162; radiation 162, 168, 398; spectrum 388, 397
solarization 165
sol-gel system 381
solids 10, 377
solstices 332
solute 14, 17, 23, 30, 56–57, 64, 70, 74, 365, 380–381, 388; absorption 72; in osmosis 20; particles 379, 382
soliths 8; mining process 64
solvent 14, 17, 23, 378
sorbitol 96, 234
Sorghum 40, 121, 141, 143, 164, 170, 303, 339, 366; bicolor 141; halepense 393
sound waves 308
source 94–97, 100, 106–108
Southern corn blight 400
soybeans (Glycine) 83–84, 95, 105, 121, 141, 162, 164, 166, 170, 194–195–197, 235–236, 246, 260–263, 269, 282, 312, 314–315, 332, 336, 339, 341–342, 365–366, 394, 402; seed composition 206, 208
sparrows 310
Spartina pectinata 170–171
specific: extinction coefficient 386; heat 377–377; water potentials 21
speckled yellows of sugar beets 91
spectrophotometer 47, 130
spectrum 384–385
S phase 255, 275
spherosomes 207
spinach (Spinacia) 127, 141, 233, 269, 315, 335–336, 339
Spinacia (spinach) 339; oleracea 141, 336
spindle apparatus 92, 236
Spirillum lipoferum 196
spongy parenchyma 34–35–35–36
spontaneous process 11–12
spore: discharge 307; tissue 5
Sporobolus cryptandrus 141
sporophyte 103
spring and summer wood 55
sprinkling 327
sprouts 65
spruce 165, 245
squash 233, 248, 261–262
squirrels 308
stable: isotopes 94; equilibrium 404; states 405
stachyose 234, 237
standing waves 104
starch 4, 15, 30–31, 40, 124, 138–147, 150–151, 154, 168, 176–178, 184, 186, 188, 211, 234, 237, 254, 256, 261, 280, 298–299, 302, 325, 377, 382, 391–392; grains 97, 125; phosphorylase 150–151, 175–176, 178, 184; sheath 280; synthetase 92; to sugar 39
statistical methods 358
statoliths 280–281
stearic acid 206–207
Stefan-Boltzmann law 44, 387–387
stele 55–57
Stellaria media 28, 197
stem(s) 5, 55, 97; elongation 345, 347–348, 397; growth of 320–232, 354; underground 328
steric acid 260
steroisomers 4
steroids 347
sterols 70, 92, 187, 213, 214, 251, 267; in flowering 346
stigmasterol 214–214, 251
stimulus 347–348, 367

stolons 280

stomatal: apparatus **35**; **53**; closure 62; crypt *34*; mechanics 35–40; movement 92; opening 307, *364*

stomates 5, 32, *34*–**35**–*36*–37, 41–42, 45, 63, 144, 147, 162, 170–171, 189, 275, 300, 364

stonecrop *336*

storage: conditions 322; organ 94, *347–348*

straight growth test **243**–244

strands 105

stratification 318, **324**–325, 329

strawberry *(Frageria)* 191, 226, 234– 235, 328, *336–337*, 344, 348

stress 62–63, 77; physiology **361**– 374

strobili 345

stroma 2, *125*–**126**–*126*, 128, 145, *164*, *392*; lamellae 126, *143*, 262

strontium ion (Sr⁺²) 74

strophiolar plug 323

structure 4–5, 101–102

strychnine 222

Strychnos nuxvomica 222

stylets 95–96, 106

suberin **55**, 65, 69–70, 77, 170, 206, **213**, 217, 229, 270

subjective day and night **307**; 310, *342*

subsidiary cells 35, *39*, 275

substomatal chamber *34*

substrates of enzymes **110**

subterranean clover 221

successional sequence 351

succinic acid 120, *181*, 187, 211

succinyl coenzyme A **181**, 187

succulence 372

succulents 38, **145**, 161, 348, **362**, 391

sucrose 17, 21, 27, 96, 98, 106, 138– 139, 141–*142*, 144, 146–*147*, 149–**151**, 154, 171, 174–175, 177–178, 186, 196, 208–209, 211, 234, 237–238, 240, 243–244, 254, 256, 275, 282, 296, 301, 354, 391; -6-phosphate 147, 149; phosphate synthetase **147**, 150; synthetase 149

suctions 388

sudangrass *141*

sugars 40, 66, 72, 92, 98, *364*; con- centrations 370; maple 164, *323*; nonreducing *96*; solution 21; beets *(Beta)* 91, 96, 98, 107, 147, 164, 169, 236, 295, *319*, 330, *336*, 345; cane 140, 143, 147, *141*, 164, 168, 196, 256, 280, *323*, *336–337*, 366

sulfate ion (SO₄⁼) 68, 73–74; metabolism of 204–205

sulfhydryl groups 116, 119–120, 145, 186, 401

sulfide 204–205

sulfite process **219**

sulfolipid 92, 213

sulfoquinovosyl diglyceride **213**

sulfoxides 205

sulfur 373; deficiency symptoms 89, 90; metabolic roles 92, 204–205; presence in coenzyme A 180

sulfuric acid 323

sulfur dioxide 90, 204

sun 11, 43, *387*–388; burn 397–399; spot cycle 304; leaves *163*, 165; light from 32, 44, 46, *52*, *387*, 390; plants 165; spot cycle 304

sunflower *(Helianthus)* *28*, 77, 85, *141*, 164, 169, 172, 190, 197, 238, 260, 262, 281, *336–337*, 344, *365*; seed composition of 208

supercooling *26*, 367

superorganism 351

supersaturated solution 61

surface(s) 379–380; charges 380, 382; tension 30, 58, **378**

surfactants 41

surgical experiments 93–94

surroundings 402–405

susceptibility 399, 402

suspension particles 379

supensor 274–275

sweet: clover 218, 323; gum *65*, 288, 326; peas 261; potatoes 177, 366

swelling 382

sycamore-maple 155, 268

symplast 57, 69, 98, *100*, 102, 108, 394

synchronizer 309, 311

synecology 351

syneresis 381

Syringa 34; vulgaris 325

system 402–404

2,3,4-T 99

2,4,5-T *241*, 245, 265

talking to your plants 308

Tamarix pentandra 371

Tanada effect **396**

tannins 218

Taraxacum officinale 28, 216

Tarweed *337*

Taxodiaceae 277

Taxodium 66

taxonomy 4

TCA 393

TDEAP 346

tea 322, 357

teleological 40

temperate zone 361

temperature 5, 12, 15, 25, 72, 273, 304, 307, 348, 370, 377; absolute 387–388; alternating 330; coeffi- cient 331; compensation system **313**, 315; control 358–359; ele- vated 366; gradients 14; hydration and 382; independence 311; leaf 32; limits for survival 366; night *341*, 328; phloem transport and 107; photosynthesis and 170–171; respi- ration and 190; starch to sugar and 176; storage 328–329; stomates and 38; storage 328–329; transpira- tion and 45; vernalization and *320*

tensile strength 51, 58–59, 377, 391

tension 14, **20**, 23, 50–*52*, 57, 59– *61*–62, 77, 381, 388–**389**

tent and cuvette methods 32–33

teosinte, Mexican *141*

terbacil **395**

terminal bud 325, 327

terminology for water potential 63

terpenes 187, *215*

test dark period *342*–*342*

testes 310

testosterone *347*–348

tetrahedrons 377

tetrazolium hydrochloride 367

Thea 323

Thelephora terrestris 66

theobromine **222**–*222*

theories 8

thermal: analysis *367*; radiation 388

thermocouple 26; psychrometer 53, 77

thermodynamics 402–405; first law of **10**, 404; laws of 1, 8–**9**–**10**, 16, 72–73; probability and **403**; second law of 11, 250, 404–405; third law of 11

thermoperiodism **330**

thiamine 92, **180**; pyrophosphate 139,

181

thigmomorphogenesis 264, 286–**288**

thigmonasty **277**, 285

thigmotropism **277**

thiocarbamate 394–395

thiourea 324, 329

thixotropic gels **381**

threonine *112*, 121

threshold 351; turgor 238

Thuja 66; *occidentalis* 368

thylakoids **126**–*126*–129, 131–133, 213, *392*

thymidine 229

thymine 91; dimers **398**

TIBA 242

tidal cycles 304

Tidestromia oblongifolia 141, 163– 164, 171, 366

timberline 370

time 5; during day 315; lapse photog- raphy *306*; maximum sensitivity and 343; measurement 304, 313, 315– 316, 338, 340–343, 346; memory **315**

timothy 177

tin 84

tissue 5; analysis 84; culture 258, 282–283; volume method **23**

tobacco *(Nicotiana)* 222, 235, 246, 251, 258, 260–261, 263, 282–283, 314–315, 323, 336–337, 344, 357, 402

tocopherylguinones 132

tomato 41, 77, 81, 86, 88, 91, 97, 126, 164, 214, 220, 226, *235*–236, 262, 266, *276*, *317*, 330, *335*, 338, *347*–348, *352*–*353*, 360, 371; juice 324

tonoplast 4, *70*, 74, *103*–104, 203, ·237, 265

Tordon 245

torus 54–*54*, 58

total: cycle lengths *342*; soil moisture stress 63

totipotent **276**

toxic 352; chemicals 307; effects 373–374; zone 84

toxin 400

trace elements **82**

tracers 11, 94, 106–107

trachea 390

tracheids 53–54, 56–57, 60–61

Trail Ridge Road 350

transcription 296

transfer cells 390

transamination **140**, 142, 217

transcellular strand theory *101*–*101*, 104, 108

transcription 246–247, 272, 274

transducers 272, 318, 365

transfer: cells 98, **102**–*103*; of genetic information 344; RNA 258–259

transfusion tissue *34*

transients **310**

transketolase 139, 185

translation 396

translocases 72

translocation 76, **93**; 150, 171–172, 175, 246, 272, 274, 344, 350, 360, 364, 400; of florigen *343*, 345; time of 97

transmission **386**

transmittance *44*, **386**

transpiration 8, **32–33**, 37–38, 40–41, 43, 50–51, 57, 62–63, 77, 93, 165, 170, 350, 355, 360, 362, 370; as a cooling process 42, 45–47; mea- surement of 32–34; stream 41, *364*, 366

transport 39, 72, 92, 99–108, 320; ve-

locity of 97; 107

tree 60; limit 370; rings 42; tall 49, 61

triazine 394

tricarboxylic acid cycle (*see also* Krebs cycle) 180

2,4,5-trichlorophenoxyacetic acid *see* 2,4,5-T

trichoblast 275

trichome *34*

trifluralin 236

trifolirhizin 218

Trifolium: pratense 323; repens 337; striatum 323

triggered responses **355**

Triglochin maritima 373

triglycerides (*see also* fats) **206**

Trigonella arabica 323

2,3,5-triodobenzoic acid 242

triosephosphate isomerase 139, 178

triple respons of peas 264–265

Tripsacum dactyloides 141

triticale 236

Triticum (wheat) *36; aestivum 141*, *336; vulgare* 208

tritium 93

tropical plants 332

tryptamine 92, 241

tryptophan 92, *112*, 121, 187, 215– 217, 222, 241

tubers 150, 170–172, 175–177, 189, 200, 211, 213, 217–218, 228, 236, 258, 263, 267, 271, 328–329, *347*– 348

tubulin proteins 236, 272

Tulipa 177, 233, 277; *gesneriana* 329

tundras 8, 14, 304, **368**–371

tungsten 382, *385*

turgor 27, 61–62, 70, 75; maintained by potassium 92; movements **277**; pressure **29**, 42, 243–297, 365, 389; threshold 238

turnip 91, 200, 291, 303

turpentine 213–214

Tyndall effect 379

tyrosine *112*, 115, 121, 187, 216–217, 222; ammonia lyase 216

ubiquinone 182–183

UDP 149, 151, 154, 159, 177; various derivatives 159

UDPG 147, 149, **151**, 154, 159, 176, 211

Ulmus: americana 323; campestris 323

ultracentrifuge 2, **383**

ultradian **307**

ultramicroscopic world 18, 404–405

ultraviolet radiation 43, 163, 378, 384–*385*, 397–399; flavonoid ab- sorption of 221; ozone formation and 160, 397–398; vitamin D₂ 214

uncouplers **392**

underground storage organs 328

uniform garden **356**

unit membrane 4, **70**

unsaturated **206**; fatty acids 368

unstirred layer 44

upland cotton *336–337*

uracil 258; herbicides 395

ureides 197

urethane 313

uridine diphosphoglucose *see* UDPG

urine excretions 307

UTP 147, 149

vacuole 2, 4, 8, *39*, 69–70, 73–76, 92, 104, 118, 146–147, *202*, 218, 265, 275, 293, 390–391; formation during seed germination 203; sap in 26

vacuum: boil 49, 61, 381; infiltrated 41, 98
valine *112*, 115, 121
valinomycin 313
Valonia 365
valve 54
vanadium 83–84
Van der Waals forces 116, 118, **376**–377, 380–381
van't Hoff equation 21–22, 26
vapor 14, 57–58, *61*, 63, 377, 394; gap 57; lock 57, *60;* pressure 15–**16**–*16*–17, *24*, 52, 58, 382; pressure difference 45–46; pressure method **25**–**26**, 30; water *46*, 51
variability 370
vascular: bundles **34**, **53**; cambium 226–230, 242, 244, 255, 266, *274*–275, 303; elements 53; plants 102, 390; tissue pattern 97
vegetables 319
vegetative reproduction *347*–348
velocity 9, 100, **383**–384
velvet *36*
Venus'-flytrap 277, 286–287
Verbascum thapsus 177
verification 308
vermiculite 169
vernalin **320**–321
vernalization 253, **318**–321, 330–331, 334–345, 355
vesicles 390, 401–402
vessel(s) **55**–*55*–56, 60–61; elements **55**–*55*, 57, *228*
viable **322**, 367
Vicia faba (broad bean) 38–39, *106*, 203
vicinal hydroxyl groups 92
Victoria blight 400
victorin **400**–401
villae 390
vinblastine **236**

violoxanthin **397**
viruses 6, 94, 105, 272, 360, 299–400
viscosity 377–*378*
visible light 384–*385*
vitalism 1, 50
vitamin(s) 86, 114, 258, 282; A 215; B 124; B$_1$ **180**; B$_2$ 168; B$_{12}$ 83; D$_2$ 214; E 132
Vitis 347
vitrifies 367
volume *22*, 24, *379*
volumetric water potential **21**
voodoo lily 188

wall: loosening 247, 252–253, 257, 263, 297–298; plasticity 256; synthesis *364*–365
walnut 325, 357
wandering Jew *28*
Warburg-Dickens pathway (*see also* pentose phosphate pathway) 184
Warburg effect **164**, 166
Washingtonia fillifera 42, 45
water 8, 14, *61*, 72–73, 323, 359, 374, 376–378, 380, 390; columns 59; concentration 14; content *30*; core of turnip 91; layers 381–382; molecule *376*; potential *(see below)*; relations of plants 8, 10–12, 16; relations of growth 237–238; stress *38*, 40, *42*, 362–365; uptake 75; vapor *46*, 51
waterlily *28*
waterlogged: plants 262, 266; soils 78, 189, 194, 264, 266
watermelons 248
water potential **13**–22, 30, 57–58, 61, 63, 77, 95, 162, 237–239, 297, 350, 362–365, 370, 374, 382, 388–389; atmospheric *51*; gradient 19–20; measurement 23–*24*–*25*–26: sto-

mates and 38
watt 385
wave: motions of electrons 386; nature of radiation **383**; number **383**; 385; length of light 40, *44*, 339–340, **383**–385, 387
waxes 206, 209, 211–*212*–213, 393
weather cycles 304
weed control 393
wheat *(Triticum)* 36, 85, 97, 121, 140–141, 150, 168–*169*–170, 203, 225, 227, *231*, 249, 254, 267, 283, 297–298, 330, 332, *336*, 396; lancer 318; seed composition of 208
whiptail disease of cauliflower and broccoli 91
white: ash 96, 227; clover 197, *337*; lupine 197; mustard 261, 299–300, 303, *336*, 354–355, 396; oak *28*, 277; pigweed *141*; pine *28*, *34*, 227
Wien's law **387**–*387*
wild: carrot *337*; oats 254, 323
willow *28*, 106, 233, 244
wilting *95*, 376, 389, 391
wind 32, 48, 60, 304, 359, 361–362, 369; tunnel 46–47; velocities *46*
winter: annuals 318–319; barley *336*; burn 368; dormancy 226; heat 320; rye 320, *336*, 367, 377; wheat *336*, 367
wintex barley 315
wisteria vine 50
witches'-broom 261
witchgrass *141*
Wolffia microscopia 336
women in science **314**
wood **53**, 93, 391
work **11**–13, 21, 203–404; interaction 402
wound healing 267, 270
wristwatch 307

Xanthium (cocklebur) 28, 339; *strumarium* 46, *141*, 235, 309, *323*, 332, *336*, 343
xanthophylls **126**, 214, 397
xanthoxin *396*–397
xenon 384–*385*
xeroderma pigmentosum 399
xerophytes 361, 363
X rays 94, 307, 384–*385*
xylem 5, *34*, *55*–59, 61–62, 68, 70, 93, *100*, 150, 157, 196–197, 201, 228–229, 275–276, 284–285, 287, 303, 381; exudates 260; ion transport in 92; lignin in 218; parenchyma **55**; sap 62, 77; tissue **53**; tubes 50
xyloglucan **155**
xylose 154–*155*, 368
xylulose-5-phosphate 137, *139*, *185*

yeasts 258, 357, 402
yellow: bristlegrass *141;* foxtail *323;* popular 227, 229, 326
yields 351, 363
young root 55

Zea mays (corn, maize) *34*, *55*, *141*, 162, 258, *336*–337
zeatin 258–259, 261–263; riboside **258**–260
Zebrina pendula 28
Zeitgeber 309, 311
zeolite 380
zinc 178, 200, 373–374; deficiency symptoms 89, 91; metabolic roles 92
Zizana aquatica 322
zone **270**; of elongation 228–**229**
zoology 4
Zoysia japonica *141*
Z-tube *59*–60
zygote 5, 11, 274